ROCKSLIDES AND ROCK AVALANCHES OF CENTRAL ASIA

ROCKSLIDES AND ROCK AVALANCHES OF CENTRAL ASIA

Distribution, Morphology, and Internal Structure

Alexander Strom

Chief expert, Geodynamics Research Centre - branch of JSC "Hydroproject Institute", Moscow, Russia

Kanatbek Abdrakhmatov

Director, Institute of Seismology, National Academy of Sciences of Kyrgyz Republic, Bishkek, Kyrgyzstan

ELSEVIER

Elsevier
Radarweg 29, PO Box 211, 1000 AE Amsterdam, Netherlands
The Boulevard, Langford Lane, Kidlington, Oxford OX5 1GB, United Kingdom
50 Hampshire Street, 5th Floor, Cambridge, MA 02139, United States

British Library Cataloguing-in-Publication Data
A catalogue record for this book is available from the British Library

Library of Congress Cataloging-in-Publication Data
A catalog record for this book is available from the Library of Congress

ISBN: 978-0-12-803204-6

Publisher: Candice Janco
Acquisition Editor: Marisa LaFleur
Editorial Project Manager: Brianna Garcia
Production Project Manager: Anitha Sivaraj
Cover Designer: Miles Hitchen

Typeset by MPS Limited, Chennai, India

Dedication

We dedicate this book to our wives—Mrs. Tatiana Strom and Mrs. Gulnara Djamankulova. Without their understanding of our aspirations, their permanent support and readiness to accept husbands' every year long-term traveling during field seasons and everyday sitting in front of the computer screen when we are at home, this book could not be written.

Contents

Appendix section for this book will be available in entirety only in the online version of the book.

CHAPTER

1

Introduction

Large-scale catastrophic bedrock landslides (rockslides) belong to the most hazardous natural phenomena endangering people in mountainous regions. Unlike "common" landslides and rock falls affecting just collapsing slopes themselves and areas directly at their feet, the simultaneous failure of millions and, sometimes, billions of cubic meters of rocks can devastate vast areas, extending up to 5–10 km and, in the exceptional cases, up to 30–40 km from the source zone (Kurdiukov, 1950, 1964; Crosta et al., 2012, 2015; Strom, 2014a; Robinson et al., 2015; Reznichenko and Davies, 2015; Reznichenko et al., 2017). Some of these abnormally mobile rockslides claimed hundreds and even thousands of lives (Evans, 2006; Evans et al., 2006).

Numerous disastrous rock slope failures occurred in rather densely populated Alps (Heim, 1932; Abele, 1974; Bonnard, 2011). Among them we can mention the 1881 Elm rock avalanche in the Swiss part of this mountain system, the first long runout rock avalanche that have been described scientifically (Heim, 1882, 1932; Hsü, 1975, 1978), and the 1963 Vajont rockslide in Italy, the "nickpoint" disaster that changed the approach to the slope stability assessment for hydraulic engineering (Muller, 1964; Semenza, 2010; Wolter et al., 2016).

The worldwide known historical bedrock landslides in Northern and Southern America can be exemplified by the 1903 Frank rockslide in Canada (McConnell and Brock, 1904; Krahn and Morgenstern, 1976), the 1959 Madison rockslide triggered by the M7.5 Hebgen Lake earthquake in the United States (Witkind et al.,1962; Meyers and Hamilton, 1964; Hadley, 1964), the 1962 and 1970 Huascaran ice-rock avalanches (Plafker and Eriksen, 1978; Evans et al., 2009a) and the 1974 Mayunmarca rockslide in Peru (Hutchinson and Kojan, 1975; Kojan and Hutchinson, 1978), by rock avalanches triggered by the 1964 Alaska (McSaveney, 1978) and the 2002 Denali fault earthquakes (Jibson et al., 2006).

Many of the similar events have occurred in Asian mountains outside of the Central Asian region, which is the focus of our studies. Among the most recent events, we can mention the 2005 Hattian Bala rockslide triggered by the M7.6 Kashmir earthquake (Dunning et al., 2007; Owen et al., 2008; Delaney and Evans, 2011; Schneider et al., 2013), the 2010 Attabad rockslide (Delaney and Evans, 2011; Schneider et al., 2013) in Pakistan, and the 2006 Leyte rock avalanche in the Philippines (Catane et al., 2007, 2008; Evans et al., 2007). Numerous large-scale rockslides that buried thousands of people were triggered by the 2008 M8 Wenchuan earthquake in Sichuan (China). About 30 of them formed large landslide dams, and only enormous efforts undertaken by Chinese authorities prevented the uncontrolled catastrophic outburst floods (Yin et al., 2008; Wang et al., 2009; Cui et al., 2009, Huang and Li, 2009; Chigira et al., 2010; Dai et al., 2011).

Large number of rock slope failures has been reported from the Southern Alps in the New Zealand (McSaveney, 2002; Korup, 2002).

The most hazardous consequences of large-scale bedrock slope failures, sometimes much more disastrous than a collapse and rapid motion of a gigantic amount of crushed rock debris itself, are river damming resulting in valley inundation and subsequent outburst floods (Costa and Schuster, 1988, 1991; Evans et al., 2011). It can be exemplified by the 1786 earthquake-triggered Dadu rockslide in China that formed a dam, which subsequent failure caused the lake outburst and flooding that killed about 100,000 people downstream, thus being the most disastrous rockslide-related catastrophe ever reported (Dai et al., 2005; Lee and Dai, 2011). Two catastrophic floods occurred in Punjab in 1841 and 1858, being caused by failure of rockslide dams that had blocked the Indus River and its large tributary, the Hunza River, correspondingly (Drew, 1875; Mason, 1929; Burbank, 1983; Delaney and Evans, 2011; Ahmed et al., 2014). The 1914 breach of the Rio Barrancos, a natural dam in

Rockslides and Rock Avalanches of Central Asia.
DOI: https://doi.org/10.1016/B978-0-12-803204-6.00001-6

Argentina that had existed for millennia, resulted in the outburst flood that devastated the entire 900 km-long Rio Colorado valley up to its mouth (Groeber, 1916; González Diaz et al., 2001; Hermanns et al., 2004). One of the most recent events was the 2000 Yigong disaster in China (Shang et al., 2003; Evans and Delaney, 2011). These tragic lists can be extended easily (see, e.g., Evans et al., 2006, 2011).

All the above mentioned, as well as many other case studies from these regions, are world-famous and have been described in numerous papers and referred in the students' textbooks. However, the Central Asian rockslides are much less known by the international community of landslide researchers, despite the fact that this region, embracing the Pamir, the Tien Shan, and the Dzungarian mountain systems (Fig. 1.1), is considered as one of the landslide "hotspots" (Nadim et al., 2006). Because most of the publications describing these features are in Russian (for the former Soviet part of the study region) (Landslides ..., 2010) and in Chinese (for its eastern part that belongs to the People's Republic of China), it makes their usage difficult for researchers from other counties. Analysis of the worldwide landslide databases compiled to study general regularities of landslide damming (Costa and Schuster, 1988, 1991), and quantitative relationships between landslide size and mobility (Sheidegger, 1973; Hsü, 1975; Li, 1983; Shaller, 1991; Kilburn and Søremsen, 1998; Legros, 2002) also shows that just very few of the Central Asian case studies, such as the 1911 Usoi rockslide, have been analyzed. Our book aims to fill this gap to some extent and to provide brief information about numerous interesting case studies that could attract researchers' attention. Following the proverb existing in many languages that "a picture is worth a thousand words," we tried to illustrate most of our descriptions and statements by handmade photographs, schemes, space and aerial images, and 3D Google Earth views. We believe that it will help the reader to select the most interesting case studies requiring more detailed study.

As we mentioned above, most of the publications describing Central Asian case studies and sited hereafter are in Russian. Considering the poor availability of many of them outside former Soviet Union, we tended to refer to summarizing Russian publications, as well as to primary sources of information in which the phenomena in question were described for the first time. We must also mention that a lot of data have been presented in the unpublished technical reports that are hardly accessible now.

Rockslides in Central Asia were investigated by the Chinese researchers as well (Fen, 1987; Yi et al., 2006; Yuan et al., 2013). However, as far as results of their studies were often published in Chinese, these publications are inaccessible for us and, thus, some data available on rockslides in the Eastern (Chinese) part of the study region could be missed. Most of the site names mentioned in the book have been given according to Russian topographic maps. It could be, however, that

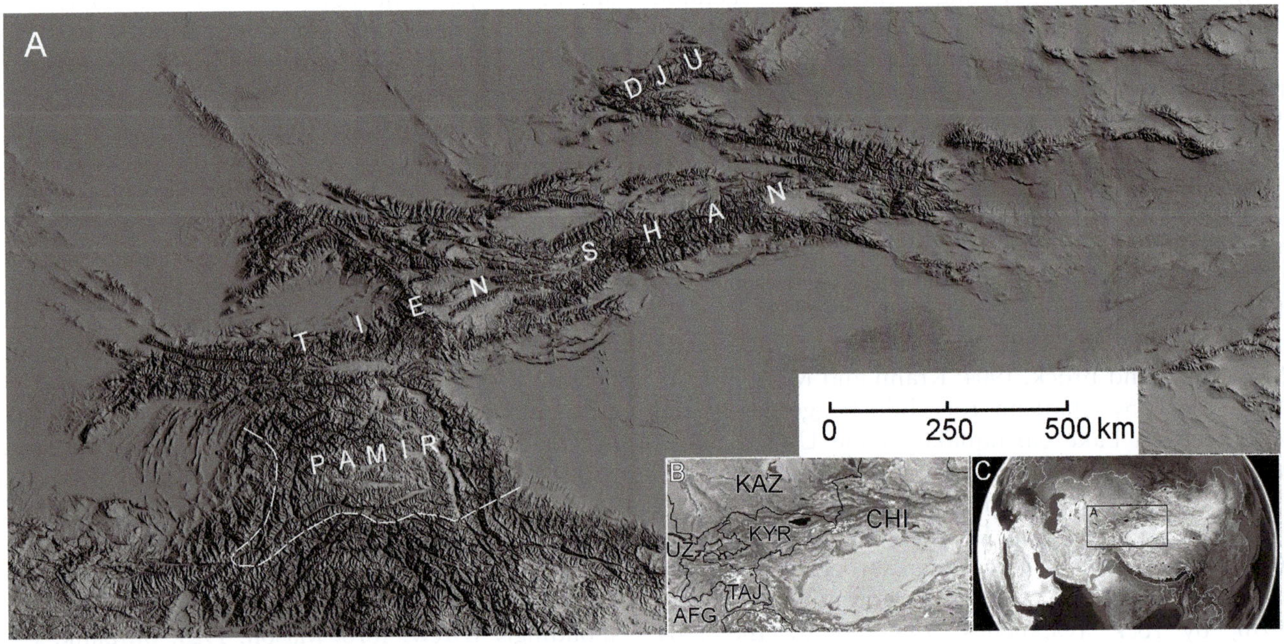

FIGURE 1.1 (A) The 30″ SRTM DEM of the study region; (B) political boundaries in the Central Asian region (from Google Earth); (C) location of the study region (outlined) in Asia. DJU, the Dzungarian Range; AFG, Afghanistan; CHI, China; KAZ, Kazakhstan; KYR, Kyrgyzstan; TAJ, Tajikistan; UZ, Uzbekistan. Dashed line indicates the conventional western and southern boundaries of the Pamir.

local toponymies, especially for the Chinese part of the region, differ from those on Russian maps. Authors will be thankful for colleagues from the Central Asian countries for corrections.

The Central Asian Mountains are located in six states—Afghanistan, China, Kazakhstan, Kyrgyzstan, Tajikistan, and Uzbekistan (Fig. 1.1B) (we do not include the area south of the so-called Wakhan Corridor). Geologists and geomorphologists in all these countries, including those that previously were parts of the Soviet Union, have their own traditions of landslide studies with somehow different approaches to landslide identification, classification, and mapping. One of the aims of this book is to summarize data on the Dzungaria, Tien Shan, and Pamir large-scale bedrock landslides to analyze these features throughout the entire region from the joint stance and to characterize bedrock landslides distribution and regularities regardless of the political boundaries that have no relation to the geomorphic, geological, and tectonic constraints. Such work had been done twice for the Alps that also belongs to several countries—France, Switzerland, Germany, Austria, and Italy, first by Albert Heim in his famous "Bergsturz und Menschenleben" (1932) and, about 40 years later, by Gerhard Abele in "Bergstürze in den Alpen, ihre Verbreitung, Morphologie und Folgeerscheinungen" (1974). These outstanding monographs still serve as the main data sources for researchers studying rockslides in the Alpine region, and we considered them as an example to follow.

It is undeniable that many of the case studies deserve much more detailed and thorough description than it is allowed by the size of this book. That is why, besides presenting the most interesting case studies on the regional basis, which will be inevitably incomplete, we will focus on those peculiarities of bedrock landslides morphology and internal structure that allow better understanding of their motion mechanisms and more grounded hazard assessment. Regular detailed description of more than 900 large-scale rockslides and rock avalanches that have been identified could be done based on field investigations of each of them, which could be performed in far future only and would require multivolume proceedings to be presented.

Special emphasis will be paid on controversial issues such as distinguishing between the rockslide and the moraine blockages and selection of seismically triggered rockslides that predetermine at a large extent both landslide and seismic hazard assessment, and elaboration of technical and social measures aimed to predict and to minimize or, if possible, to prevent related disasters (Schuster and Evans, 2011; Strom, 2013a). Regularities that can be derived by the analysis of numerous features from this region could be applied for case studies in other mountainous systems of the World.

We did not intend to carry out regular analysis of the landslide susceptibility for the study region considering enormous variability of its geological and geomorphic conditions that predetermine formation of large-scale bedrock landslides, as well as of the triggering factors. Most of the features, which more or less detailed description is presented hereafter, have been formed already and, except those still blocking river valleys or reflecting ongoing slope deformations, do not pose direct threat to local communities and infrastructure. But their study forms the ground for better understanding of the regularities of such phenomena occurrence and evolution, their spatial and temporal distribution.

We believe that our book will attract attention of landslide researchers interested in new impressive examples of large-scale catastrophic bedrock landslides and in new data supplementing case studies from other regions, of the emergency experts working on landslide and outburst flood hazard assessment, and of researchers and practitioners in the fields of paleoseismology and seismic hazard assessment, because large-scale bedrock landslides are often considered as being triggered by large past earthquakes.

We also believe that it could be interesting for hydraulic engineers working on dam projects in the Tien Shan and Pamir mountainous rivers, considering the high potential of the Central Asian Rivers for the hydraulic energy production and the significant threat posed by large landslides for the hydraulic schemes safety.

Some of the data presented hereafter were collected and analyzed during site investigations for hydraulic schemes in the Central Asia region—the Rogun and Pamir-1 hydraulic power plants (HPP) in Tajikistan, the Kambarata 1 and 2, Tashkumyr and Upper Naryn HPP, and the Sarydjaz diversion project in Kyrgyzstan, the Mainak HPP in Southern Kazakhstan performed by Hydroproject Institute in 70th—90th of the twentieth Century.

Since the end of 90th, our study of these phenomena was supported by the International research projects: INCO-COPERNICUS CT97 0202 (Landslides triggered by earthquakes in Kyrgyzstan, Tien Shan), NATO SFP Project 983289 LADATSHA (Prevention of Landslide Dam Disasters in the Tien Shan, Kyrgyz Republic), the GEF/UNEP/UNU PALM Project (Sustainable Land Management in the High Pamir and Pamir-Alai Mountains), the ISTC KR 1281 Project (Geodynamics of the Talas-Ferghana fault of Tien Shan and natural disasters on territory of Central Asia). Detailed data on numerous rockslides in the Central Tien Shan were collected during preparation of the NATO Advanced Research Workshop "Natural and Artificial Rockslide Dams" (Bishkek, 2004) field trips and in the course of the International Summer School on Rockslides and Related Phenomena supported by the International Consortium on Landslides (ICL) and International Program on Landslides (IPL) Projects M111, M126, and C-106-2. This Summer School has been organized annually since 2006 in the Kokomeren River basin, Kyrgyzstan (Strom, Abdrakhmatov, 2009). Actually, this book "grew" from the Kokomeren Summer School full-color Guidebook, and many of the case studies described hereafter have been investigated during these field training courses.

Compilation of the Central Asian rockslide database was based on the analysis of space images provided by such services as the Google Earth, SAS Planet, and ArcGIS Earth. We also used the 3″ SRTM DEM for quantitative terrain representation and the Global Mapper software for compilation of all data collected in the GIS.

Such work could not be compiled without close cooperation with our colleagues. We want to express the warmest gratitude to Stephen Evans and Gabriele Scarascia Mugnozza, who organized the NATO Advanced Research Workshops in Celano, Italy, in 2002, and in Bishkek, Kyrgyzstan, in 2004, which intensified our involvement in the international community of landslide researchers; to Hans-Balder Havenith and Isakbek Torgoev with whom we carried out field study of many rockslides in Kyrgyzstan and to Oleg Chedia, Alexey Trofimov and Vladimir Makarov with whom we started our studies in Tien Shan. We want to thank Kyoji Sassa for permanent support of the International Summer School on Rockslides and Related Phenomena. We had useful discussions with Vitaliy Adushkin, Masahiro Chigira, David Cruden, Tim Davies, Damien Delvaux, Atyrgul Djumabaeva, Zhildyz Dyikanalieva, Stuart Dunning, Xuanmei Fan, Eugen Gaspirovich, Tatiana Grigorieva, Fausto Guzzetti, Filip Hartvich, Reginald Hermanns, Kenneth Hewitt, Runqiu Huang, Oldrich Hungr, Anatoly and Nikolai Ischuk, David Jarman, Alexander Kolichko, Vilen Korchevskiy, Andrey Korjenkov, Oliver Korup, Nikolai Leonov, Olga Maksimova, James McCalpin, Mauri McSaveney, Natalia Mikhailova, Konstantin Nagorniuk, Hushnud Nazrishoev, Roman Nepop, Andrey Nikonov, Igor Nosenko, Cholponbek Ormukov, Galina Osipova, Marc-Andre Ostermann, Tomas Panek, Leonid Pernik, David Pettley, Andreas von Poschinger, Natalia Reznichenko, Vladimir Rodionov, Ekaterina Romanenko, Swenja Rosenwinkel, Anatoly Savich, Robert Schuster, Eugen Shemiakin, Veronica Smolkova, Olga Solomina, Petra Stepanchikova, Yuri Svoiskiy, Vladimir Trifonov, Svetlana Vinnichenko, Fawu Wang, Gonghui Wang, Patrick Wassmer, Goldie Walker, Andrea Wolter, Yin Yueping, Oleg Zerkal, Anatoliy Zhirkevich, and many other researchers to whom we want to express our sincere thanks. Finally, we want to thank the Elsevier Publishers for their interest to our work and for permanent support during compilation of this book.

PART I

GENERAL INFORMATION

CHAPTER

2

Brief History of the Bedrock Landslide Studies in the Central Asia

We mentioned already in Chapter 1, Introduction, that the Central Asian region is considered as one of the global landslides hotspots (Nadim et al., 2006). Indeed, numerous slope failures occur here in soft sediments every year often causing significant losses and even casualties (Niyazov, 1974, 1982, 2009, 2015; Schlögel et al., 2011; Torgoev et al., 2013; Havenith et al., 2015; Pettley, http://blogs.agu.org/landslideblog/). These phenomena have been studied in details for many decades, but most of the investigations were focused on landslides in loess and other soft sediments that provide major hazard in the densely populated foothill areas of the Ferghana Valley rim, and in the Central and Southern Tajikistan due to their almost annual occurrence (see Mavlianov, 1968; Presnuhin, 1971; Niyazov, 1971, 1974, 1982, 2015; Landslides and Mudflows, 1984; Torgoev and Alioshin, 2005; Schlögel et al., 2011; Pettley, http://blogs.agu.org/landslideblog/).

First studies focused on large-scale bedrock landslides in the Central Asia were performed by Russian researchers at the end of the 19th—beginning of 20th centuries—when several large earthquakes occurred in this region being accompanied by numerous, sometimes extraordinary rockslides. In 1887 the M ~ 7.3 Vernyi earthquake near the present-day Almaty City (the former name—Vernyi town) caused significant large-scale slope failures that were mapped and studied in details during special field mission organized by the Geological Committee of Russia (Mushketov, 1890). It was the first professional description of such phenomena in Central Asia. We should mention that 2 years before, in 1885, the M ~ 7.0 Belovodsk earthquake affected the area just west of the present-day Bishkek City, the capital of Kyrgyzstan, about 200 km west of the region affected by the Vernyi earthquake. This event was also accompanied by slope failures but the reconnaissance team that investigated its effects paid attention on the structural damage mainly (Ignatiev, 1986), and very limited information about landslides could be found in this publication. The same was with the gigantic (Ms ~ 8.3) Chilik earthquake that occurred in the northeast of the Issyk-Kul Lake in 1889, 2 years after the Vernyi earthquake. Unlike the two preceding earthquakes, this extraordinary event was not studied by experts in the field. Only questionnaires filled by local administration officers were collected and analyzed (Mushketov, 1891). Unfortunately, quite limited information about slope failures in the mountains during this earthquake is available from these descriptions. Only recently its geological effects were reinvestigated but with main emphasis on surface faulting (Tibaldi et al., 1997; Abdrakhmatov et al., 2016).

In 1907 two large ($M = 7.4 \pm 0.1$ and 7.3 ± 0.5; Kondorskaya and Shebalin (1982)) Karatag earthquakes affected the foothills of the Gissar Range, west of the present-day capital of Tajikistan—the Dushanbe City (those times it was a small town). Earthquakes resulted in significant damage, and more than 1500 casualties but only general description of their geological effects could be found in published reports (Bronnikov, 1908; Korolkov, 1912; Niyazov, 2009).

In February 1911 the largest historical nonvolcanic rockslide triggered by the large earthquake originated in the very remote and hardly attainable area of the Central Pamir. Its magnitude estimates vary from $Ms = 7.4 \pm 0.1$ (Kondorskaya and Shebalin, 1982) to $Mw = 7.7 \pm 0.2$ (Ambraseis and Bilham, 2012) and to $Ms = 7.7 \pm 0.1$ (Kulikova et al., 2016). Rockslide buried the small Usoi village with all its inhabitants except three men who went that day to the Sarez village, about 20 km upstream (Luknitskiy, 1955; Agakhanjanz, 1989). The Usoi Landslide (it inherited the name of the village) was so large that it was discussed initially if the slope failure was triggered by earthquake or the impact of sliding mass on the valley floor and opposite slope was recorded worldwide as an earthquake

Rockslides and Rock Avalanches of Central Asia.
DOI: https://doi.org/10.1016/B978-0-12-803204-6.00002-8

(Galitzin, 1915). The Usoi rockslide (Tajik 090) dammed the Murgab River at 38.274° N, 72.6° E and formed the 500-m deep Sarez Lake that still poses a threat for the large part of the Amu Darya River basin (Tajik XXX—hereafter indicates the ID of the cited rockslides. ID includes the Country indication (Afghan, China, Kazakh, Kyrgyz, Tajik, Uzbek) and the three-digit number. They correspond to the IDs of the database provided in the digital form as Appendix).

It was studied soon after the disaster (Bukinich, 1913; Shpilko, 1914; Preobrajensky, 1920) and attracts researchers' attention till now (Sheko and Lekhatinov, 1970; Gaziev 1984; Agakhanjanz, 1989; Alford and Schuster, 2000; A. Ischuk, 2011; Strom, 2014b). This extraordinary rockslide is one of the few worldwide known Central Asian case studies. More detailed description of the Usoi rockslide will be presented in Chapter 9, Pamir. On December 7, 2015, another large (M7.2) earthquake occurred about 20 km northeast of the Usoi blockage. Luckily, according to the data available, it did not affect the dam and the Sarez Lake banks significantly (Ischuk and Strom, 2016).

Impressive surface ruptures and seismically triggered landslides originated when the catastrophic M8.2 Kemin earthquake occurred in December 1911 in the northern part of the Tien Shan, north of the Issyk-Kul Lake. They were studied by the special reconnaissance team soon after the event, and this report (Bogdanovich et al., 1914) still provides one of the most detailed descriptions of the natural catastrophe in the entire Central Asian region. In 60th its effects were studied by V.K. Kuchai (1971). Recently surface ruptures and rockslides triggered by the Kemin earthquake were reinvestigated (Delvaux et al., 2001, Havenith et al. 2003). Some of these slope failures will be analyzed in Section 8.3 of Chapter 8, Tien Shan.

Brief description of several earthquake-induced rockslides in the western part of the Tien Shan, in Uzbekistan, that occurred in the 19th century and survived in memory of local communities is presented in (Niyazov, 2009). Landslide catalogue included in this book was used for compilation of the rockslides database (see Appendix). Unfortunately location of landslides included in this catalogue is described by the locality name and the name of river basin where it occurred, not by coordinates.

Several large (M ~ 8.0) earthquakes, likely being accompanied by rockslides, have occurred in the eastern (Chinese) part of the study region as well—the Nilki earthquake (Yan et al., 1985; Guang-hua et al., 2001; Wu, 2006) in 1812, the Kashgar earthquake in 1902 (Omori, 1907) (the other name of this event is the Artush earthquake; Huan and Shi, 1987), and the Manas earthquake in 1906 (Huan and Shi, 1987). However, because most of the papers describing these events are in Chinese (Chongshao and Hongwei, 1997; Yang et al., 2005) very limited data on their geological and geomorphic effects, including slope failures, are available for the authors.

Series of the earthquake-induced landslide disasters at the end of the 19th—beginning of the 20th centuries accompanied by numerous large-scale bedrock landslides was followed by about 30 year-long period when such strong earthquakes did not affect the Central Asian region. Absence of such hazardous phenomena slowed down research activity focused on the large-scale bedrock landslides. Next activation of seismicity took place in 40th when several devastating earthquakes, also accompanied by large-scale slope failures have occurred, which attracted researchers' attention to these phenomena again.

It started with the M6.4 1941 Garm (Gubin, 1943) and the M6.0 1943 Faizabad (Gubin, 1960; Nikonov, 1974) earthquakes that caused several landslides, though none of them could be classified as the large-scale bedrock slope failure. In 1946 the M7.5 Chatkal earthquake hit the northwestern Kyrgyzstan resulting in numerous landslides and rockslides (Leonov, 1965, 1970; Kuchai, 1971), one of which dammed the Naryn River upstream of the Toktogul dam site at 41.73° N, 72.71° E. It caused valley inundation, and special measures (blasting) were undertaken to prevent most catastrophic consequences (now remnants of its body are covered by the about 150 m deep water of the Toktogul reservoir formed by the dam that was built in 70th). Unfortunately no regular mapping of the environmental effects of this earthquake was performed immediately after this earthquake. Most of the data were obtained during field reconnaissance carried out in late 60th, 15 years or even more after the event (Leonov, 1970; Kuchai, 1971).

The most catastrophic event of this period was the 1949 M7.4 Khait earthquake in Tajikistan that triggered the 7-km long rock avalanche. Its 12- to 75-m thick deposits buried the small Khait town with several thousands of its inhabitants (Panasenko, 1952; Gubin, 1960; Leonov, 1960; Evans et al., 2009b). Besides, numerous shallow landslides in loess-like loam occurred being transformed into gigantic earth flow that also buried several villages in the adjacent Yasman River valley. It was the most disastrous landslide event in Central Asian region that claimed thousands of lives (likely up to 19,000, though exact number of casualties caused by this earthquake is debatable; Evans et al., 2009b). The Khait rock avalanche (39.19° N, 70.9° E, Tajik 038) attracts researchers' attention for many years till now.

Next impulse to study large-scale bedrock landslides in the Central Asian region was provided by several landslide-related disasters associated with catastrophic breaches of landslide-dammed lakes that have occurred in the Soviet part of Central Asia in 60s of the 20th century. On July 7, 1963, the catastrophic breach of the rockslide-dammed Issyk Lake (43.257° N, 77.487° E, Kazakh 021) occurred east of the Almaty City that was the capital of Kazakhstan at that time (Gerasimov, 1965; Litovchenko, 1964; Strom, 2010b, 2013a). It caused the devastating debris flow that passed through the Issyk River valley where it traces could be visible on the valley slopes till now, more than 50 years after the event.

Breach of this almost 200 m high prehistoric rockslide dam resulted from its overtopping, which, in turn, was caused by debris flow that originated in the upper reaches of the catchment due to the Jarsay glacial lake breach. This powerful debris flow lasted for about 3–4 hours and brought up to 6–7 million m^3 of debris into the Issyk Lake resulting in a rapid increase of the lake level and creation of waves that destroyed the upper part of the rockslide blockage. Water stored in the lake rushed downstream eroding the dam body as well as river banks and valley bottom and forming the "secondary" debris flow. Peak discharge recorded about 10 km downstream of the breached dam, close to the Issyk town, where the Issyk River leaves the Zailiyskiy Range, was 745 m^3/s, 150 times more than mean annual discharge of the river at this site (4.96 m^3/s according to the State Water Inventory, vol. V, 1988). Only small remnant lake remained after the disaster. Later on, an artificial dyke was built in the eroded channel to protect downstream valley and the Issyk town from new debris flows and to renew this beautiful touristic site (Fig. 2.1).

FIGURE 2.1 The Issyk rockslide dam and the Issyk Lake after dam restoration. The rockslide head scarp is on the right-bank slope of the valley. Large alluvial fan entering the lake was formed by the 1963 debris flow that caused the dam's breach. Traces of the outburst are visible downstream from the blockage. The artificial dyke was built afterward in the eroded channel. 3D Google Earth view.

Next year, on April 24, 1964, 22 Mm^3 of rock debris blocked the Zeravshan River valley in Tajikistan downstream of the Zeravshan and Fan Darya Rivers junction and just upstream of the Aini town at 39.385° N, 68.545° E (Tajik 022) (Fedorenko, 1988). It formed the dam with an effective height (the minimal height of the blockage crest where overtopping would start first) about 50–60 m. The 1.3 km wide (along the stream) rockslide dam had blocked the valley and the dammed lake fed by both rivers with cumulative discharge of about 90 m^3/s (in April–May, 1964; mean annual discharge at this gauge is 147 m^3/s, according to the State Water Inventory, 1990) could store up to 126×10^6 m^3 of water, which uncontrolled release could devastate lower part of the Zeravshan River valley with the historical Samarkand City. To take the situation under control, the 865-m long and up to 23 m deep trench across the dams' crest was excavated by directed blasts. On May 6, 1964, water started passing through this artificial canal and 2 days later the discharge exceeded the inflow. The peak discharge of 1200 m^3/s was recorded on May 31. Lake was almost emptied at June 20, when its level dropped for 32.5 m. Three million cubic meters of debris was eroded from the dam resulting in significant aggradation of the Zeravshan River channel downstream (Neshikhovskiy, 1988; Strom, 2010b, 2013a) (Fig. 2.2). It was one of the first cases of the successful prevention of large rockslide dam breach disaster.

One of the most catastrophic outburst floods caused by the rockslide dam breach occurred on June 18, 1966, in the Isfairamsay River basin in Kyrgyzstan at 39.927° N, 72.297° E (Kyrgyz 129). This rockslide was, likely, triggered by an earthquake, though exact time of this event mentioned in different publications varies from AD 1833 (Pushkarenko and Nikitin, 1988) to about 300 years before its breach, i.e., ca. AD 1666 (Rezvoi and Rezvoi, 1969) (we must mention, however, that no large earthquakes at these times in the region in question have been listed in the earthquake catalogues available: Mushketov and Orlov, 1893; Semionov, 1958; Kondorskaya and Shebalin, 1982). The nearly 100 m high dam had blocked the Tegermach River valley (right tributary of the Isfairamsay River) and formed the Yashinkul Lake. The dam was overtopped and breached due to continuous raining. The resultant outburst flood lasted for about 7 hours with peak discharge of 5000 m^3/s and caused significant damage not only in Kyrgyzstan but also in Uzbekistan farther downstream (Krukovskiy, 1969, Rezvoi and Rezvoi, 1969; Rezvoi et al., 1971; Reizvikh et al., 1971; Pushkarenko and Nikitin, 1984, Strom, 2010b, 2013a). Large portion of the rockslide body was eroded (Fig. 2.3).

Later on similar disastrous phenomena occurred twice, in 1977 and in 1998, in the Shahimardan River

FIGURE 2.2 Remnant of the 1964 Aini rockslide dam just upstream of the Aini town (Tajikistan) and downstream of the junction of the Zeravshan River (coming from the left with clear blue water) and the Fan Darya River (with muddy water, coming from the gorge at the background). Rockslide occurred in the head scarp of the much larger and older past rockslide (Tajik 022a). Red arrow marks small scarp indicating possibility of new slope failure at this site in future. 3D Google Earth view. *Reprinted from Strom, A., 2013a. Geological prerequisites for landslide dams' disaster assessment and mitigation in Central Asia. In: Wang F., Miyajima M., Li T., Fathani T.F. (Eds.), Progress of Geo-Disaster Mitigation Technology in Asia, Springer-Verlag Berlin, Heidelberg, pp. 17–53 (ISBN 978-3-642-29106-7) with the permission of Springer.*

FIGURE 2.3 The breached Yashinkul dam. Note the large difference of the benchmark elevations upstream and downstream of the dam resulting from the significant sedimentation of the presumably 300-year old lake. Just downstream of the dam evidence of valley aggradation is still visible. 3D Google Earth view. *Modified from Strom, A., 2013a. Geological prerequisites for landslide dams' disaster assessment and mitigation in Central Asia. In: Wang F., Miyajima M., Li T., Fathani T.F. (Eds.), Progress of Geo-Disaster Mitigation Technology in Asia, Springer-Verlag Berlin, Heidelberg, pp. 17–53 (ISBN 978-3-642-29106-7) with permission of Springer.*

basin that, similarly to the Isfairamsay River, originates in Kyrgyzstan and flows to Uzbekistan that complicated arrangement of the emergency measures. In some publications it was mentioned that these events resulting in significant damage and more than 100 fatalities were caused by partial breach of the Kurban-Kul Lake (39.936° N, 71.845° E, Kyrgyz 132) (Ecological State, 2006). However, more likely they occurred due to glacial lakes breach.

Several other catastrophes caused by breach of the existing rockslide dams have been reported from this region since the end of the 19th century (Mallitskiy, 1929; Pushkarenko and Nikitin, 1984). Obviously, similar slope failures and subsequent secondary effects have occurred in the Eastern, Chinese part of the study region in the historical times too. However, authors could not find descriptions of their effects in the available publications.

The most recent well-known disaster in the study region caused by large-scale bedrock landslide and river damming occurred in 1992–93 in the Belaldy River valley—the small right-bank tributary of the Naryn River that falls now in the Toktogul reservoir. Large-scale slope failure occurred on August 19, 1992, in the upper reaches of this valley at 42.06° N, 73.279° E (Kyrgyz 254) being triggered by the M7.3 Suusamyr earthquake (Bogachkin et al., 1997; Ghose et al., 1997). It blocked rather steep but wide glacial valley (Fig. 2.4). Rockslide buried the shepherd's family who tended a herd and dammed two minor streams forming lakes up to 200–300 m in size only. Nevertheless, 10 months later, in June 1993, likely after intensive snowmelt at elevations exceeding 3000 m above sea level (a.s.l.), these lakes burst out producing powerful debris flow that partially devastated the Belaldy village 17 km and the Torkent village 30 km downstream of the damming site (Korjenkov et al., 2004).

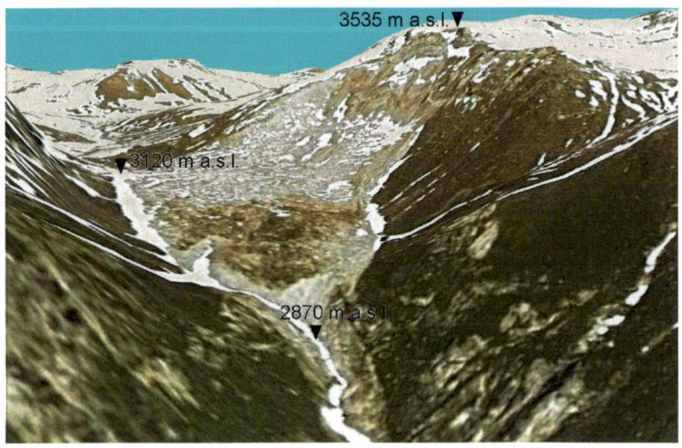

FIGURE 2.4 1992 rockslide at the upper reaches of the Belaldy River—source of the powerful debris flow of June 1993. 3D Google Earth view. Furthermore, we will skip "m a.s.l." at the elevation marks and will indicate altitude values only.

Besides studying the historical rockslides and rockslide-related disasters, regular study of the prehistoric large-scale bedrock landslides was performed in some parts of the former Soviet Central Asian region as well (Sapov, 1970; Pushkarenko and Nikitin, 1984; Fedorenko, 1988; Bochkariov and Medeuov, 1990; Vinnichenko et al., 1990). These studies were stipulated by the above-mentioned disasters in 60th, by the anxiety on the Sarez Lake safety and by construction of several major hydraulic projects that had started in the Soviet part of Central Asia in the same period. More than 50 rockslide-dammed lakes have been identified here, but most of these data were presented in unpublished reports written in Russian, some of which are hardly to get now even for Russian-speaking researchers. We must mention that during compilation of the 1:200,000 State geological map of the USSR that covers the entire country, in many cases coarse debris accumulations at the feet of mountain ranges were mapped as moraines while in fact they are bodies of long runout prehistoric rock avalanches (see, e.g., Strom, 1998).

The pioneering work in this field was performed by K.V. Kurdiukov—the first researcher who identified deposits of the extraordinary mobile prehistoric rock avalanches in the Alai valley (Kurdiukov, 1950, 1964). One of them—the Komansu (Koman) rock avalanche (39.56° N, 72.69° E, Kyrgyz 345) had the runout of about 34 km (Strom et al., 2014a; Robinson et al., 2015; Reznichenko and Davies, 2015; Reznichenko et al., 2017), thus being the world-second event, according to its travel distance, after the 43-km long Cerrillos Negros Tomasiri rock avalanche in Peru (Crosta et al., 2012, 2015), which, however, was much larger in volume. Landslide origin of these deposits was contested by A. Nikonov and his co-authors (Nikonov et al., 1983), who treated them as moraines, but the correctness of the Kurdiukov's assumption was proved recently based on the detail study of the Komansu and the Achiktash (Kyrgyz 346) deposits morphology and internal structure (Strom et al., 2014a; Robinson et al., 2015; Reznichenko and Davies, 2015; Reznichenko et al., 2017).

Detail studies of the bedrock landslides in the Zeravshan River basin and in the Naryn River valley near the Toktogul dam site and in its vicinity were carried out by researchers from Moscow State University (Fedorenko, 1968, 1976; Matveev, 1968; Zolotariov, 1974; Zolotariev et al., 1969). Results of these investigations were summarized by Fedorenko (1988) whose book includes the comprehensive reference list. However, interpretation of the Sarychelek (41.85° N, 71.98° E, Kyrgyz 066) and the Karakul (41.633° N, 72.645° E, Kyrgyz 061) rockslides' structure and origin provided in some of these publications (see, e.g., Matveev, 1968) is erroneous from our point of view. It will be described later, in Part 2 in Sections 8.4 and 8.5, during regional description of the selected rockslides.

Numerous investigations were performed in Soviet times at the Sarez Lake aimed to specify the Usoi blockage and the Sarez Lake banks stability and, finally, to access risk of the Lake outburst (Denikaev, 1970; Sheko and Lehatinov, 1970; Fedorenko et al., 1981; Fedorenko, 1988). More recently this area was reinvestigated within the frames of the UN ISDR Project (Alford and Schuster, 2000; A. Ischuk, 2011). Some controversial aspects of the Sarez Lake problem were highlighted by N. Ischuk (2013b, 2014).

In Tajik part of the region that includes most of the Pamir and Southern Tien Shan, the bedrock landslides were studied by S. Vinnichenko (1977, 1989, 2006). She identified numerous types and subtypes of large-scale bedrock slope failures, so that almost each feature can be ascribed to a special subtype. Our approach to landslides classification is different: we try to highlight some general characteristics that can reflect rockslide origin or motion mechanism irrespective of the geological and geomorphic conditions of the site (see Section 5.3 in Chapter 5: Bedrock Landslide Types and Classification Systems).

Recently N. Ischuk published series of papers (2008, 2011, 2013a,b) in which he argues that most of the natural blockages in Pamir and Southern Tien Shan including well-known features such as the existing dams of the Yashinkul, Shiva, Iskanderkul Lakes and the breached Kudara-Pasor blockage were formed by end moraines. We disagree with such interpretation and this problem will be discussed in details in Chapter 4, Principles and Criteria of Landslides Identification and Discrimination From the Features Formed by Other Mass Wasting Phenomena.

Many large-scale bedrock landslides were identified in the course of paleoseismological studies, being considered as effects of large prehistoric earthquakes (Kuchai, 1971; Guschenko and Leonov, 1972; Utirov, 1975, 1991; Vinnichenko, 1977, 1989; Nikonov, 1977, 1981; Leonov and Strom, 1979; Strom, 1982; Nikonov et al., 1983; Nikonov and Shebalina, 1979, 1986; Hodjaev, 1974, 1985; Korzhenkov, 2006; Tibaldi et al., 2015). However, the problem of correct and well-grounded discrimination of seismically triggered and nonseismic rockslides is far from final solution (Strom, 2013a,b,c, 2015a) and will be discussed in Chapter 13, Seismically Induced Rockslides: Their Discrimination.

Special investigations of highly mobile rock avalanches and their motion mechanism were performed in the Soviet Union in 70th–90th. Besides providing more grounded and justified landslide hazard assessments, an important goal of these investigations was to predict shape of the artificial rock avalanches that were proposed for the blast-fill dams' construction. This technology was considered as an alternative to the conventional technologies of high dam construction in narrow river valleys (Korchevskiy and Petrov, 1989; Korchevskiy and Muratova, 1991; Adushkin, 2000, 2006, 2011). These studies included not only field geological reconnaissance but also extensive laboratory numerical and physical modeling (Grigorian, 1979; Fedorenko et al., 1979) and even observations during and after large-scale underground nuclear (Adushkin, 2006) and chemical (Korchevskiy and Muratova, 1991; Korchevskiy and Petrov, 1989; Korchevskiy et al., 2011) explosions. Several dams were created in the Soviet part of the Central Asia by using this technology (Adushkin, 2011), most of them by the directed blasting, while the proposed 270-m high Kambarata-1 dam had to be created by artificial rock avalanche (after collapse of the USSR, this project was stopped).

In the last years several rockslides and rock avalanches in the Chon-Kemin River valley and in the valleys dissecting the northern slope of the Kyrgyz Range were studied during preparation of the field trip of the NATO Advanced Research Workshop that was held in 2004 in Bishkek (Strom and Abdrakhmatov, 2004). Numerous prehistoric bedrock landslides of various types were identified and described in the Kokomeren River basin and in some other parts of the Kyrgyz Tien Shan (Abdrakhmatov and Strom, 2006; Hartvich et al., 2008; Korup et al., 2006; Strom, 1994a,b, 1996, 1998, 2006, 2012; Strom and Abdrakhmatov, 2009; Strom and Korup, 2006). These data allowed elaboration of rock avalanches classification (Strom, 1996, 2006, 2010a) (see Section 5.3 in Chapter 5) that will be widely used hereafter, when describing numerous case studies.

Regional analysis of landslide-related hazards of the former Soviet part of the Tien Shan and of some of local regions, in Kyrgyzstan mainly, was performed recently (Roessner et al., 2005; Havenith et al., 2006a, b, 2015a, b; Schlögel, 2009; Schlögel et al., 2011; Golovko et al., 2014, 2017).

In 2014 the Landslide Hazard Map of the Territory of Tajikistan was conducted with the support of the UNDP Disaster Risk Management Program (A. Ischuk, personal communication).

Similar studies were carried out in the Chinese part of the study region as well (Fen, 1987; Yi et al., 2006; Yuan et al., 2013). Several papers present interesting data on the ages of prehistoric rockslides in the study region (Sanhueza-Pino et al., 2011; Yuan et al., 2013).

We have to note that some recent papers provide quite unrealistic assumptions of the rockslide-related hazards. As an example we should mention the possibility of $\sim 1\ km^3$ failure in the Toktogul reservoir (Tibaldi et al., 2015). No evidence of so voluminous present or past slope deformations on the reservoir bank is available to propose such a catastrophic scenario.

No doubts that we are just at the starting point of the comprehensive study of Central Asian bedrock rockslides and related phenomena. Much more efforts are needed to solve numerous controversial problems of the nature of these features and triggering phenomena such as large earthquakes or climate change that can destabilize high rocky slopes. Sound conclusions on the origin of the prehistoric rockslides in the study region are prevented at a large extent by lack of their ages. It seems to be the most critical gap in our knowledge and should be considered as one of the main targets of future research projects. Limited data on the ages of several features that are available at present will be presented hereafter along with the morphological and geological description of the selected case studies.

As we mentioned already in Chapter 1, authors hope that the inevitably brief description of the impressive bedrock landslides from the study region presented hereafter will attract attention of the next generation of landslide researchers whose studies will shed new light on the problems that we could just mark out.

CHAPTER

3

Overview of the Study Region

OUTLINE

3.1 THE STUDY REGION: ITS TOPOGRAPHY AND BOUNDARIES

The Central Asian region includes three main mountain systems—the Dzungaria in the north, the Tien Shan in the center, and the Pamir in the south (see Fig. 1.1).

The Dzungarian Mountain system stretches for about 350 km from the west to east in Kazakhstan mainly. Its southern part is in Xinjiang Province of China. Maximal width of this system of mountain ranges is about 110 km. Its highest summit—the 4622 m a.s.l. Semionov-Tian-Shanskiy peak. While from the westward and northward the Dzungarian mountains height declines gradually towards the Balhash Plain, from the east it is cut sharply by the northwest trending escarpment associated with the Dzungarian fault dividing it from the depression with several lakes, the largest of which is the Alakol Lake. From the south the Dzungarian Range is bounded by the Ili intermontane depression. In Xinjiang, east of the meridionally trending part of the China–Kazakh border, the southern chain of the Dzungarian Mountains named the Boro-Khoro Range stretches in the east–south–eastern direction merging at its eastern terminus with the Iren-Khabyrga Range, the northern mountain chain of the Tien Shan.

The Tien Shan is the largest mountain system of the study region. It stretches in the east-western direction for about 2500 km, being up to 300 km wide and has very well defined natural boundaries. From the west and northwest the Tien Shan is bounded by the plains occupied by the Karakum, Kyzylkum, and Muyunkum deserts; further to the east by the Ili intermontane depression, dividing the Tien Shan from the Dzungarian Mountain system. Eastern (Chinese) Tien Shan is bounded by the plains of Northern Xinjiang extending towards the Altai Mountains and occupied by the Gurbantunggut desert. From the south the Tien Shan is bounded in the west by the so-called Tajik depression—the virgation of mountain chains up to 2300 m high between the south-western Tien Shan and Pamir. In the central part its southern boundary passes along the Kyzylsuu–Surkhob–Upper Vakhsh Rivers dividing Southern Tien Shan (the Karategin and the Alai Ranges) from Northern Pamir (the Trans Alai Range and the Peter the First Range). The most prominent geomorphic feature of this boundary zone is the Alai valley—the 130 km long and up to 25 km wide intermontane neotectonic depression. Further to the East, in China, the Tien Shan is bounded from the south by the Tarim depression occupied by the Takla-Makan desert. Most of the Tien Shan ridges are up to 4.0–4.5 km high, and they rise up to more than 6000–7000 m a.s.l. in the central part of this mountainous system close to the China–Kyrgyz border (the highest summits are the Victory Peak—7439 m a.s.l. and the Khan Tengri Peak—6995 m a.s.l.).

Rockslides and Rock Avalanches of Central Asia.
DOI: https://doi.org/10.1016/B978-0-12-803204-6.00003-X

The Tien Shan can be divided, conventionally, into Eastern, Highest, Northern, Central, North-Western, and Southern parts. This division was used to simplify regional description of rockslides widely developed in this mountain system and will be followed in Chapter 8, Tien Shan.

The Pamir is the plexus of the Karakoram (northwestern part of the Himalayas), Kunlun, and Hindu Kush mountain systems. Pamir is the most elevated mountain system of the entire Central Asia. The highest summits—the Kongur Peak (7719 m a.s.l.) and the Mustagata Peak (7546 m a.s.l.) are in its eastern (Chinese) part. The Ismail Somoni Peak (formerly the Kommunism Peak) (7495 m a.s.l.) and the Korjenevskaya Peak (7105 m a.s.l.) are in the Central Pamir in Tajikistan, and the Lenin Peak (named in Tajikistan the Ibn Sina Peak) (7134 m a.s.l.) crowns the Trans Alai Range at the Kyrgyz–Tajik border. Boundaries of the Pamir are well defined from the north (by the above mentioned Kyzylsuu–Surkhob–Upper Vakhsh river system), from the east, where it is bounded by the Tarim depression and from the west, where the Pamir composed of the predominantly Paleozoic and Precambrian rocks is bounded by the Tajik depression composed of folded Mezo-Cenozoic sediments. However, it cannot be delimited geomorphologically with the same confidence from the south, southwest, and southeast. Here we consider the conditional boundary that partially follows the water divide between the Pianj River basin and the Kokcha River basin in the west, the Kabul River basin in the southwest, and the Hunza River basin in the south. We also included area in the upper reaches of the Kokcha River basin in Afghanistan where large-scale rockslides are widespread. In the Eastern (Chinese) Pamir this conditional boundary follows the 37° N latitude approximately, that corresponds to the transition from the South-Eastern Pamir to the east-west trending Western Kunlun mountains (see Fig. 1.1).

Geomorphologically Pamir can be divided into three main parts—Western, Central, and Eastern. Western Pamir embraces high ranges divided by deep valleys of the Pianj River and its main tributaries—Vanch, Yazgulem, Bartang, Gunt, Shakhdara, and several smaller rivers. These valleys up to 2.0–2.5 km deep, often with extremely steep walls were incised by glaciers and river erosion. This part of the Pamir, often named as Badakhshan, belongs to Afghanistan and Tajikistan.

Central Pamir, between 73° and 74.5° E approximately, that belongs to Tajikistan, can be classified as high-altitude Tibet-like plateau with valleys at 3500–4500 m a.s.l. and ranges raising up to ca. 5000 m a.s.l. Most of river valleys here are much wider and much gentler than in the west. Some of them inherit shallow neotectonic depressions and their deepest parts are filled by the Karakul and Rangkul drainless lakes.

Eastern (Chinese) Pamir differs from both its western and central parts. Its ranges raising for more than 5 km on average above the adjacent Tarim block and crowned by the summits exceeding 7500 m a.s.l., are divided from the plateau-like Central Pamir by deep Sarykol and Tashkurgan intermontane depressions up to 40–80 km long and 7–15 km wide bounded by active faults. The latter are especially pronounced along eastern boundaries of these depressions. Such topographic peculiarities make Eastern Pamir similar to some parts of the Tien Shan with its well pronounced intermontane neotectonic depressions.

Rugged topography with high ranges bounded by kilometers-deep tectonic depressions and erosional river valleys with steep slopes typical of most of the study region preconditions the formation of large-scale landslides and of other types of slope instabilities.

3.2 GEOLOGY AND TECTONICS

Geology of the Central Asia has been studied for more than a century. Its western part—in Uzbekistan, Tajikistan, Kyrgyzstan, and Kazakhstan—was mapped regularly at 1:1000,000–1:200,000 scale. More general geological maps are available for all this territory (Geological map, 1978, 1978, 1979, 1984). The 1:250,000 scale geological maps are available for large part of the Chinese Pamir and Tien Shan at the website of the China Geological Survey [http://www.ngac.org.cn]. The Dzungaria, Tien Shan, and Pamir are composed of rocks of different types—igneous, sedimentary, metamorphic, whose ages vary from the Precambrian to Late Cenozoic. Most of mountain ranges are composed of Paleozoic sedimentary and volcanic rocks partially metamorphosed and intruded by Caledonian and Variscan granites and igneous rocks of other types. Mesozoic sediments are widely developed in the Tajik depression, in the Ferghana Range east of the same name intermountain depression, in the Central and South-Eastern and Eastern (Chinese) Pamir, and in the Northern and Southern foothills of the Eastern (Chinese) Tien Shan. Cenozoic sediments form inner parts of intermountain neotectonic depressions and the foothills areas (Fig. 3.1).

Lithology of the region varies in a very wide range. Slopes affected by gravitational processes are composed of almost all types of rocks—sedimentary (terrigenous and carbonate), igneous (granites mainly), metamorphic (gneiss, phyllites, etc.) of different ages. Information about rock types and ages affected by rockslides will be provided in Appendix and indicated when describing particular case studies.

Northern parts of the study region (Dzungaria and Northern chains of the Tien Shan) were formed during

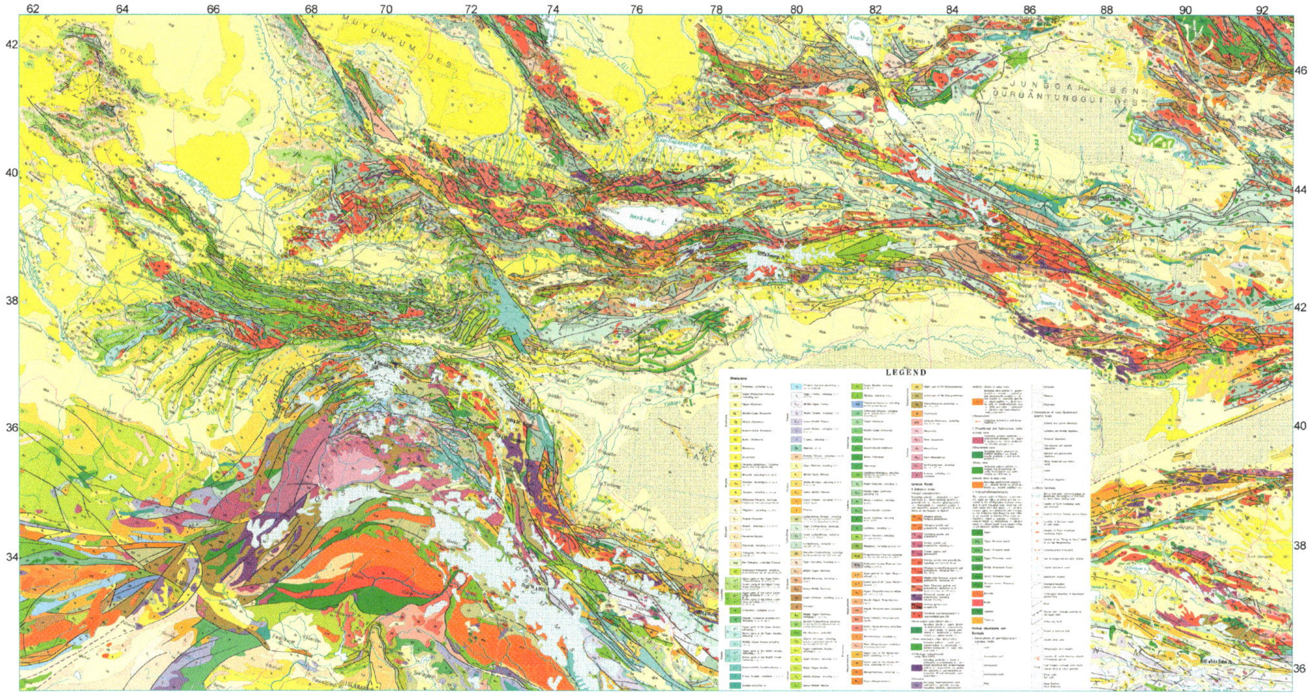

FIGURE 3.1 Fragment of the 1:2,500,000 Geological map of Northern-Central-Eastern Asia and Adjacent Areas (Atlas, 2012).

Caledonian stage, while most of the Tien Shan south of the so-called "Nikolaev Line" was formed during the Variscan stage (Khain, 1979). Pamir is often classified as a part of the Alpine belt (Khain, 1984) and forms the horseshoe-shape system of the northward convex belts (see Fig. 3.1). From the north Central Asian mountain system is bounded by the Epi-Paleozoic Kazakh platform and from the southeast by the Tarim platform block located south of eastern Tien Shan and east of Pamir and occupied by the Takla-Makan desert.

Despite significance of the Caledonian, Variscan, and Alpine stages, age of the main tectonic deformations itself does not play an important role in slopes stability assessment as compared to the lithology and local geological structure. Premesozoic basement in most parts of the study region have been subjected to intensive folding, thrusting and reworking by intrusions and regional, contact and dynamic metamorphism. Paleozoic thrusting and formation of nappes was especially intensive in the Southern and Central Tien Shan and in central part of Pamir (Ruzhentsev, 1968; Burtman, 1975, 2012). Structural pattern of the areas composed of Mesozoic and Cenozoic sedimentary rocks is characterized by intensive folding and thrusting too. Strike-slip faults are widely developed in the study region as well. Extraordinary variability of the geological structures along with good state of their knowledge all over the Central Asian region makes it impossible to describe structural pattern of the study area here at a regional basis even in brief. Interested reader could find this information in special literature. In those cases when bedrock structure plays an important role in the formation of a particular rockslide and such data are available, we will present them describing particular case studies.

3.3 NEOTECTONICS AND GEOMORPHOLOGY

The latest neotectonic stage of the study region evolution plays the most important role in slope processes predetermining their distribution and magnitude. It is just this stage when modern relief was formed. Neotectonic processes formed the modern tectonic structure and govern the seismicity, which is one of the main, probably the most important trigger of large-scale slope failure in the Central Asian region with arid climate. This stage is extremely well pronounced in the Tien Shan where the "Neotectonics concept" itself was introduced first (Shultz, 1948). Excellent descriptions of the neotectonic structure of the former Soviet part of the Tien Shan were presented by our teachers (Makarov, 1977; Chedia, 1984). More recently various peculiarities of the Tien Shan neotectonics and modern geodynamics were analyzed in (Avouac et al., 1993; Abdrakhmatov, 1995; Omuraliev and Omuralieva, 2004; Makarov, et al., 2005; Abdrakhmatov, et al., 1996, 2007; Zubovich et al., 2010; Burtman, 2012).

FIGURE 3.2 Complex boundary between the Djumgal Range and the Djumgal intermontane depression. Main boundary fault is marked by red arrows. Each basement massif marked as PZ is an individual basement anticline that was overlaid by Neogene sediments (N) folded conformably with the planation surface formed on the Paleozoic basement. Quaternary sediments (Q) overlay Neogene with angular discontinuity.

The main peculiarity of the Dzungaria and the Tien Shan neotectonic and modern structure is the presence of large-scale basement folds. Practically each range is a neotectonic anticline from few up to 15–20 km wide and up to several dozen kilometers long, while most of the east-west trending linear and isometric intermontane depressions occupied by lakes and river valleys are synforms. Usually these folds are asymmetric with one limb connecting the adjacent anticline and syncline, while on the other side anticlines are thrusted over the synclines, sometimes with very complex system of the reverse faults and minor relatively sharp basement folds (Fig. 3.2). Vergency of these structures is variable either northward or southward ($\pm 20^\circ - 30^\circ$) (Sadybakasov, 1972, 1990). Several neotectonic depressions in the Central Tien Shan could be classified as bilateral ramp grabens (Sadybakasov, 1972; Chedia, 1984).

Similar combination of neotectonically uplifted ranges and intermontane neotectonic depressions divided by fault zones is typical of Eastern (Chinese) Pamir, while the Central and Western parts of this mountainous system reflects more uniform plateau-like block uplifted during neotectonic period and dissected at its western part by kilometer-deep river valleys formed by glaciation and water erosion.

The entire region is crossed by numerous tectonic faults many of which are long-living, inherited from the Paleozoic stages of the tectonic evolution. However, faults originated during neotectonic stage are known in the Central Asia as well. Large proportion of all these faults can be considered as being active (Allen, 1975; Trifonov and Machette, 1993; Nikonov, 1995; Strom, 2017). We want to point out that active faults play dual role in slopes destabilizing. First, being causative structures of large earthquakes, active faults delineate zones of most intensive seismic shaking that trigger large-scale rockslides. Besides, any large fault zone is characterized by intensive fracturing, hydrological anomalies, etc., conditions favorable for slope failure.

Regional-scale map of active faults of the study region and their inventory have been published in (Trifonov et al., 2002). Fragment of this map embracing the former Soviet part of the region in question is shown on Fig. 3.3. Some of active faults were ruptured by the historical earthquakes, such as the 1812 Nilki earthquake in Xinjiang (Yan et al., 1985; Feng, 1990; Guang-hua et al., 2001; Yin et al., 2001, 2002; Wu, 2006), 1911 M8.2 Kemin earthquake in Northern Tien Shan (Fig. 3.4) (Bogdanovich et al., 1914; Delvaux et al., 2001) and 1992 M7.3 Suusamyr earthquake in Central Tien Shan (Fig. 3.5) (Bogachkin et al., 1997; Ghose et al., 1997).

Much more active faults ruptured in the prehistoric times have been identified and studied by detailed mapping, profiling, and trenching (Abdrakhmatov et al., 1994, 2007, 2016; Arrowsmith and Strecker, 1999; Korjenkov, 2006; Korjenkov and Chedia, 1986; Korjenkov et al., 2012; Trifonov et al., 1992; Tibaldi et al., 1997, 2015; Strecker et al., 2003; Makarov et al., 2005; Djumabaeva, 2012; Campbell et al., 2013, 2015; Landgraf et al., 2016; Grützner et al., 2017). Most of them are reverse (thrust) faults (Figs. 3.6 and 3.7A). Several north-west trending faults, such as the

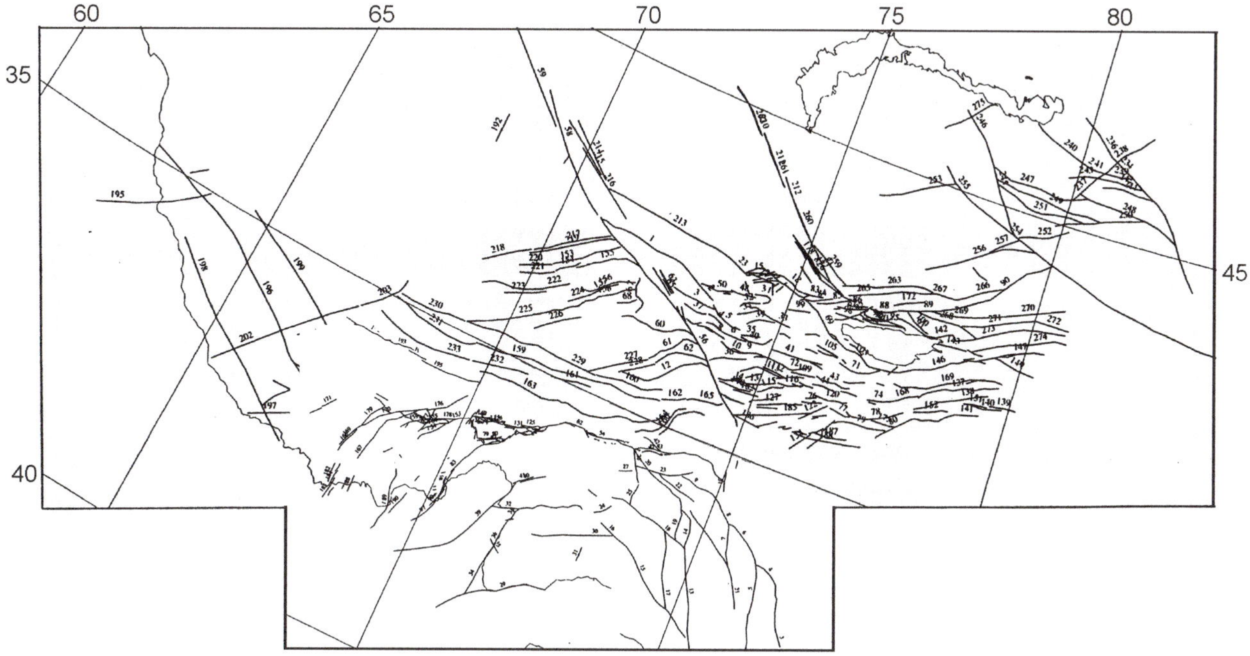

FIGURE 3.3 Regional map of active faults of the former Soviet part of the Central Asian region. *Reprinted with some modifications from Trifonov, V.G., Soboleva, R.V., Trifonov, R.V., and Vostrikov, G.A., 2002. Recent Geodynamics of the Alpine-Himalayan Collision Belt. Transactions of the Geological Institute of RAS, Vol. 541, Moscow, GEOS, 224 p. (in Russian), with kind permission of V.G. Trifonov.*

FIGURE 3.4 Small ridge passing along the right slope of the Chong-Aksu River valley (Northern Tien Shan, 42.837° N, 77.36° E) formed by recurrent surface faulting. The latest episode was associated with the 1911 M8.2 Kemin earthquake. The lake with submerged trees was dammed by this reverse fault. Giant erosional canyon at the background was formed after the earthquake (the exact date is unknown) by the breach of the glacial lake associated with the same surface rupture.

FIGURE 3.5 Surface rupture (thrust fault with vertical offset of ca. 2.5 m) of the 1992 M7.3 Suusamyr earthquake (Central Tien Shan, 42.2055° N, 73.66° E). Above: 5 days after the event; below: 14 years later.

Dzungarian fault forming eastern boundary of the same-name mountain system and the Talas-Ferghana fault dissecting the entire Tien Shan demonstrate distinct strike-slip motions (Burtman, 1964; Burtman et al., 1996; Tibaldi et al., 2015) (Fig. 3.7B, 3.8A and B, and 3.9D).

Numerous case studies of spatial and, likely, temporal coincidence of active faults and large-scale slope failures (the latter's proof requires extensive dating) will be described in the following chapters.

3.4 SEISMICITY

3.4.1 Instrumental and Historical Records

Central Asia is one of the most seismically active parts of the World. Despite short period of observations—generally less than 150–200 years—hundreds of strong earthquakes have been recorded here, several of which—in 1812, 1889, 1902, 1906, 1911—had magnitudes (Ms)

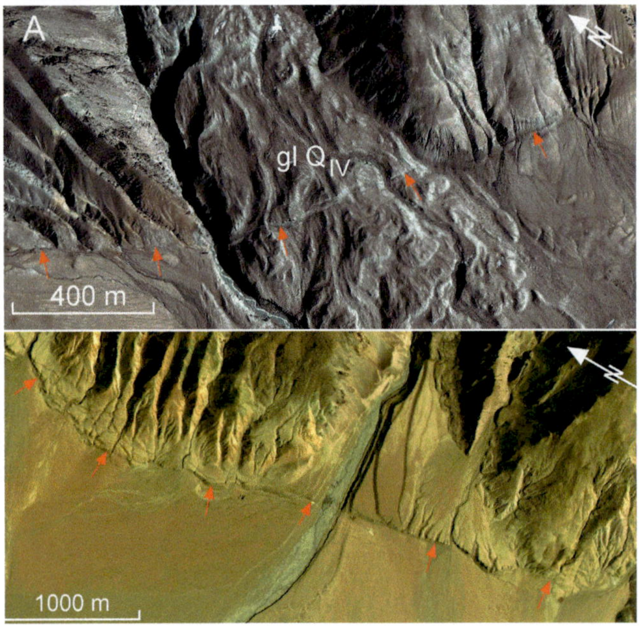

FIGURE 3.6 Surface ruptures along eastern boundary of the Sarykol intermontane depression (Eastern Pamir) displacing recent landforms. (A) Rupture displaces fresh, likely Holocene moraine at 38.867° N, 74.97° E; (B) Rupture displaces series of low terraces at the alluvial fan at 39.08° N, 74.59° E. Only modern floodplain has not been affected. Google Earth images.

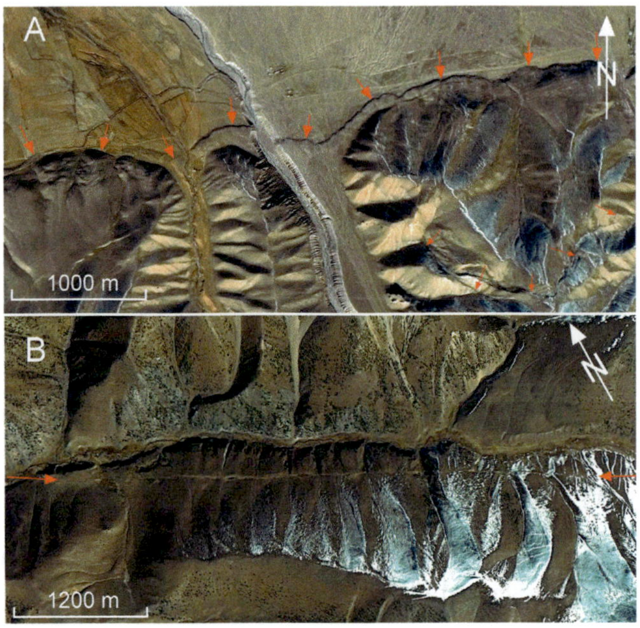

FIGURE 3.7 Surface ruptures of different kinematic types. (A) Evidence of recent thrusting along the Trans Alai fault west of the 34-km long Komansu rock avalanche (39.48° N, 72.54° E). Main rupture is marked by large red arrows, the branch fault in its hanging wall by small red arrows. (B) The strike-slip surface rupture marked by red arrows along the right-lateral Talas-Ferghana fault at 42.1° N, 71.88° E. Google Earth images.

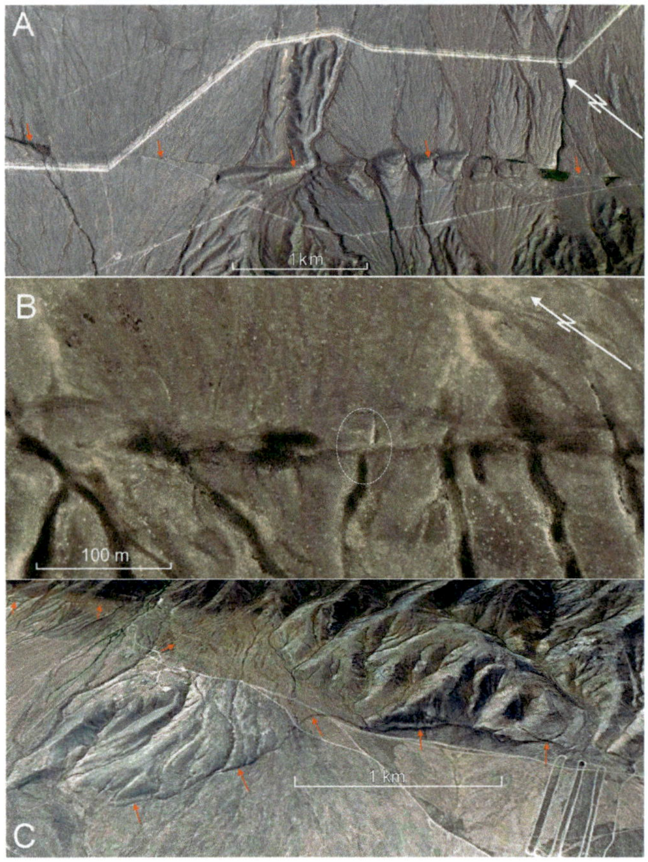

FIGURE 3.8 Surface ruptures along the Dzungarian fault (see locations of A–C on Fig. 7.1). (A and B) With right-lateral strike-slip kinematics; (C) With reverse (thrust) kinematics. (B) Dashed oval marks the most prominent ~10 m right-lateral offset. Google Earth images.

FIGURE 3.9 Fragments of the Lepse fault (see locations of D and E on Fig. 7.1). (D) At the western section with right-lateral strike-slip kinematics highlighted by the pull-apart basin marked by red arrows and well visible Riedel shears; (E) At the eastern section with prevailing vertical offsets (marked by + and −). Small lake was formed in the gulley at the downthrown side; larger red arrows mark the main rupture, smaller arrows marks the auxiliary rupture north of it. Google Earth images.

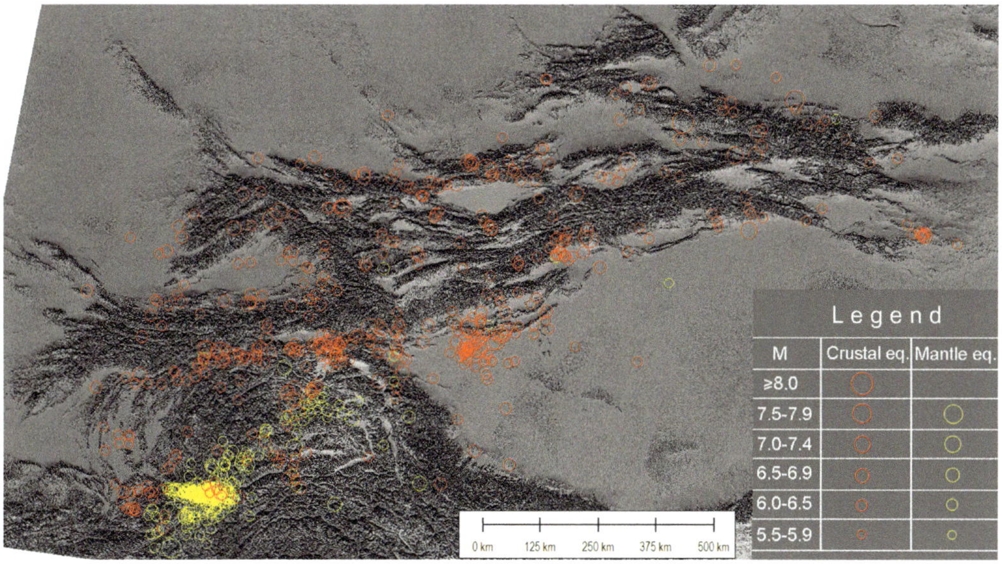

FIGURE 3.10 Spatial distribution of earthquake epicenters in the Central Asian region.

about 8.0 and even more, and many other exceeded M7.0. Data on the historical and instrumentally recorded earthquakes were summarized in several catalogues (Mushketov and Orlov, 1893; Kondorskaya and Shebalin, 1982; Pasheko and Sykes, 1992; Min et al., 1995; Engdahl and Villaseñor, 2002). Largest known historical earthquakes of the Central Asia region accompanied by large-scale rockslides were listed above in Chapter 2, Brief History of the Bedrock Landslide Studies in the Central Asia.

In the Tien Shan and Dzungaria practically all earthquakes occur within the earth crust, at a depth from 5 to 25 km mainly. In contrast, in the Pamir most of the historical earthquakes occur at a depth of 70–250 km. This belt of deep-focus earthquakes passes in the north-eastern direction from Hindu Kush Range in Afghanistan towards the Karakul Lake in northern part of Pamir – the so called Hindu Kush focal zone (Fig. 3.10). Largest of these earthquakes are felt over vast area, usually much larger than that affected by crustal earthquakes of the comparable magnitudes.

Spatial distribution of crustal seismicity is quite uneven (Fig. 3.10). No large earthquakes ($M > 5.5$) have been recorded in the historical times in Dzungaria (evidence of the numerous prehistoric earthquakes in this region will be mentioned in Section 3.4.2).

In the Tien Shan several distinct seismic belts can be delineated, especially in its western and central parts west of the 80° E, where width of this mountainous system is maximal. Most of the earthquakes, the largest of which are the 1907 M ~ 7.5 Karatag earthquake, the 1941 M6.4 Garm earthquake, the 1943 M6.0 Faizabad earthquake, the 1949 M7.4 Khait earthquake, the 1974 M7.3 Markansu earthquake, and the 1902 M ~ 8.0 Kashgar earthquake, occur along the southern feet of the Gissar, Karategin, Alai and Kokshaal ranges. In 1716 (±1 year) large earthquake destroyed the Aksu town at 41.2° N, 80.3° E (±0.5°) (Kondorskaya and Shebalin, 1982).

Another seismic belt coincides with the northern chains of the Tien Shan. While total number of earthquakes here is much less than in the southern boundary belt, it has much higher proportion of large and giant events. The largest of them are the 1885 M ~ 7.0 Belovodsk earthquake, the 1887 M ~ 7.3 Vernyi earthquake, the gigantic 1889 M ~ 8.3 Chilik, and the 1911 M8.2 Kemin earthquakes. They were followed by the 1938 M6.9 Kemin-Chu earthquake, the 1970 M6.8 Sarykamysh earthquake, and the 1978 M6.6 Jalanash-Tiup earthquake. One more large seismic event that can be ascribed to the same seismic belt is the 1812 M ~ 8.0 Nilki earthquake at the southern boundary of the Boro-Khoro Range (Yan et al, 1985; Feng, 1990; Guang-hua et al., 2001; Yin et al., 2002; Wu, 2006). The preceding large earthquakes, whose descriptions are quite scarce, occurred in this zone east of the modern Bishkek in ca. 1475, in 1770 close to the 1885 Belovodsk earthquake epicentral area, and in 1807 close to the modern Almaty (Mushketov and Orlov, 1893; Kondorskaya and Shebalin, 1982).

The only one large earthquake recorded in the inner part of the Central Tien Shan was the 1992 M7.3 Suusamyr earthquake (Bogachkin et al., 1997; Ghose et al., 1997).

Largest earthquake with magnitude ~7.5 occurred in 1946 in the western part of the Tien Shan, west of the Talas-Ferghana fault, at the junction of the Chatkal and the Atainok ranges (Leonov, 1970; Kuchai, 1971).

Most of other earthquakes in this part of the Tien Shan occurred at the Ferghana valley rim. The most destructive one was the 1902 M6.4 Andizhan earthquake. Several large earthquakes also with magnitudes more than ca. 6.5 occurred not far from Tashkent in 1868, 1886 and 1896 (Mushketov and Orlov, 1893; Kondorskaya and Shebalin, 1982).

Largest seismic event in the Eastern part of the Tien Shan—the Ms ~ 8.0 (Mw ~ 7.7) Manas earthquake occurred in 1906 at the southern feet of the Iren-Khabyrga Range. Several other large earthquakes with magnitudes up to 7.0 have been recorded at the southern and northern boundaries of the eastern (Chinese) Tien Shan (Fig. 3.10).

In the southern part of the study region, in Pamir, only few significant crustal earthquakes have been recorded. Largest of them are the 1911 M7.4–7.7 and the 2015 M7.2 Sarez earthquakes (Ambraseys and Bilham, 2012; Kulikova et al., 2016). The 1911 seismic catastrophe triggered the Usoi rockslide and formed the Sarez Lake. The second, 2015 Sarez earthquake, caused some temporary variations of water level at the blockage and head part of the lake, likely due to coseismic crustal deformation (Ischuk and Strom, 2016). Among earthquakes in the Eastern (Chinese) Pamir, we can mention the most recent one with M6.6 that occurred on November 25, 2016, at the northern end of the Sarykol intermontane depression, about 50 km east of the Karakul Lake.

It should be pointed out that analysis of the spatial distribution of large-scale bedrock landslides does not reveal close correlation between these phenomena and spatial distribution of large historical and instrumentally recorded earthquakes at a regional level. Rockslides, despite their uneven distribution over the study region and pronounced spatial clustering that will be described in Chapters 7–9, Dzungaria, Tien Shan and Pamir, respectively, often concentrate in the areas that have not been subjected to any strong seismic shaking in the historical times.

The Central Asian historical and instrumentally recorded seismicity is characterized not only by quite uneven spatial distribution but also by temporal clustering. Most of largest earthquakes mentioned above occurred in few decades at the end of 19th and the beginning of the 20th centuries—in 1885, 1868, 1886, 1887, 1889, and 1896, 2 events in 1902, in 1906, 1907, 2 events in 1911. Another activation of seismicity in the form of the M > 6.0 earthquakes occurred in 40th of the 20th Century—in 1938, 1941, 1943, 1946, 1949. Though most of them triggered significant bedrock landslides (see Chapter 2), the recording period is too short to make sound conclusions about regular temporal coincidence between rock slope failures and seismicity.

3.4.2 Paleoseismological Data

Due to quite short period of the historical recording of Central Asian earthquakes, it is obvious that paleoseismological data should have the first priority for seismic hazard assessment in this region. Here we will present in brief several examples of the "primary" paleoseismic evidence (McCalpin, 2009), i.e., surface faulting. Paleoseismological interpretation of rockslides is much more ambiguous. It will be described case-by-case in Chapters 7, 8 and 9, and will be discussed further in Chapter 13, Seismically Induced Rockslides: Their Discrimination.

Paleoseismological investigations in the Central Asia have been performed since 1970s in Tajikistan (Nikonov, 1974; Vinnichenko, 1977; Leonov and Strom 1979; Nikonov and Shebalina, 1979, 1986), Kyrgyzstan (Kuchai, 1971; Utirov, 1975, 1991; Strom, 1982; Abdrakhmatov, 1995; Korjenkov and Chedia, 1986; Arrowsmith and Strecker, 1999; Delvaux et al., 2001), Uzbekistan (Hodjaev, 1985), and Kazakhstan (Tibaldi et al., 1997). Results of such investigations in Kyrgyzstan were summarized by Korjenkov (2006). During the last decade, these studies continued in Kyrgyzstan and Kazakhstan mainly (Abdrakhmatov et al., 2007, 2016; Djumabaeva, 2012; Campbell et al., 2013, 2015; Landgraf et al., 2016; Grützner et al., 2017).

In the Dzungaria the most prominent evidence of the Late Quaternary surface faulting were found along the Dzungarian fault (Suvorov, 1963; Voitovich, 1969; Campbell et al., 2013). The extremely impressive surface ruptures featuring dextral strike-slip (Fig. 3.8A and B) and, surprisingly, reverse (Fig. 3.8C) kinematics were found here. Not less pronounced is the surface rupture along the Lepse active fault (Kurdiukov, 1956; Campbell et al., 2016) that passes for about 100 km from east to west crossing both mountains up to ~1.5 km high of the northern part of this system and flat plains west of them. While in the west, where crossing the lowland, this fault demonstrates purely dextral strike-slip kinematics (Fig. 3.9D), in the east it has the predominantly vertical component and forms and expressive scarp several meters high that dams small valleys forming a permanent lake at one site (45.922° N, 81.012° E, Fig. 3.9E).

In the Tien Shan numerous evidence of the prehistoric surface faulting were found both at the faults that have been ruptured by large historical events, such as the 1885 Belovodsk earthquake (Abdrakhmatov et al., 2007; Landgraf et al., 2016), the 1889 Chilik (Tibaldi et al., 1997; Crosby et al., 2007; Abdrakhmatov et al., 2016), the 1911 Kemin (Delvaux et al., 2001; Arrowsmith et al., 2015; Abdrakhmatov et al., 2016), and the 1812 Nilke (Yan et al, 1985; Feng, 1990;

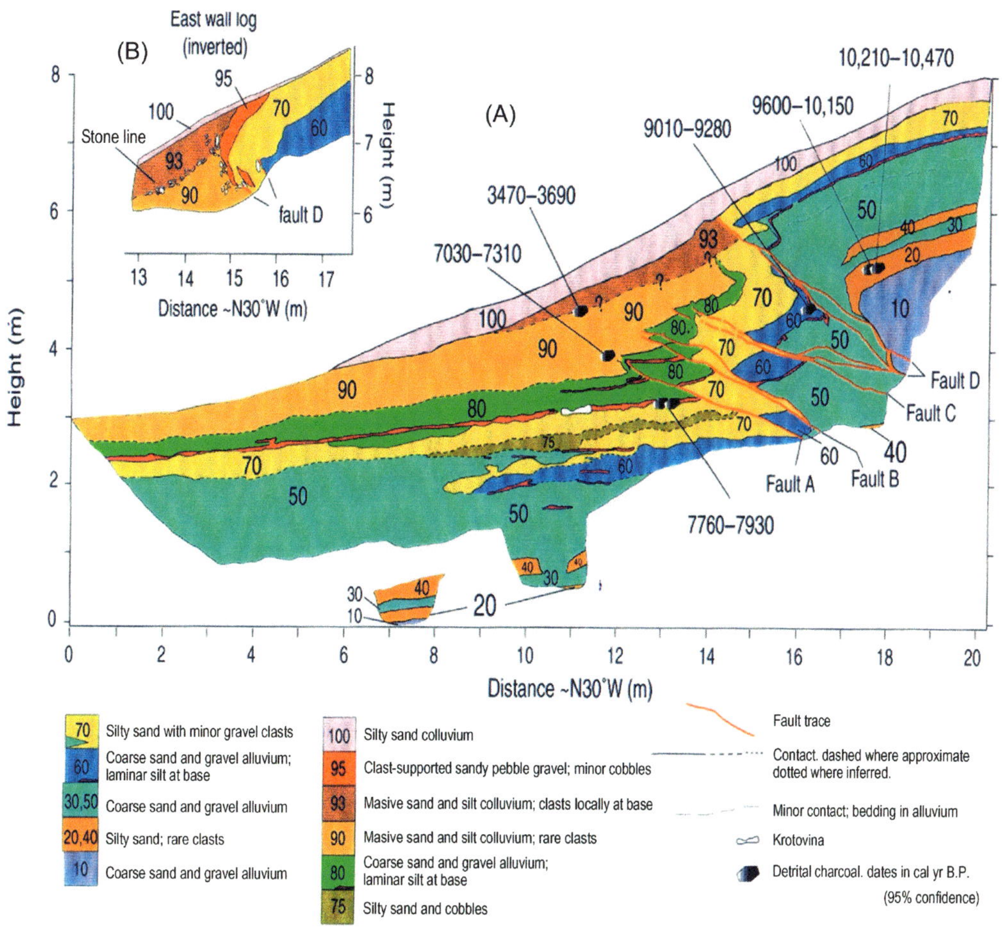

FIGURE 3.11 Logs of the Oinakdjar trench across the Atbashi fault excavated at 41.06207°N, 75.41781° E and featuring recurrent offsets. (A) General log of the western wall of the trench with multiple ruptures (Faults A–D). (B) Log of the part of eastern wall illustrating relationships of the upper part of Fault D with lithological units. *Reprinted from Abdrakhmatov, K.E., Thompson, S., Weldon, R., 2007. Active tectonics of the Tien Shan. Bishkek, Ilim Publishers, 71 (in Russian).*

Guang-hua et al., 2001; Yin et al., 2002; Wu, 2006) earthquakes and at the faults that did not experience historical faulting (Strom, 1982; Belousov et al., 1994; Abdrakhmatov, 1995; Korjenkov, 2006; Djumabaeva, 2012; Tibaldi et al., 2015). Numerous trenches excavated across active faults revealed evidence of the recurrent faulting associated with large prehistoric earthquakes (an example is shown on Fig. 3.11) that confirms geomorphic evidence of multiple offsets visible along most of active faults (Figs. 3.4, 3.6, and 3.12).

Besides surface faulting, indications of seismically induced liquefaction were revealed and described in some parts of the Central Asian region, at the banks of the Issyk-Kul Lake mainly (Korjenkov, 2000, 2006). This promising method of paleoseismological studies can be applied to lake sediments accumulated in the landslide-dammed lakes as well (Strom, 2015a), and results of these studies will be presented in Section 8.4.

Unfortunately we have no data about similar detailed paleoseismological studies with trenching carried out in Pamir, both in its Tajik, Chinese or Afghan parts, though there are distinct geomorphic evidence of recent recurrent surface faulting in this mountainous system too (Figs. 3.6 and 3.13).

Magnitudes of the prehistoric earthquakes identified by studying of surface ruptures were estimated based on relationships between surface rupture length and displacements on one hand and earthquake magnitude on the other hand (Wells and Coppersmith, 1994; Strom and Nikonov, 1997, 1999; Chipuzubbov, 1998). It was found that numerous earthquakes with magnitudes 7+ had affected almost the entire Central and Northern Tien Shan in Kyrgyzstan and Kazakhstan in the past, being associated with active faults dividing neotectonic anti- and synforms. Similar seismic activity could be assumed for much wider areas than in the historical times in the Eastern (Chinese), North-Western (in Kyrgyzstan and Uzbekistan), and

FIGURE 3.12 The Kalmak-Ashu surface rupture at 41.579° N, 76.395° E with evidence of two recurrent offsets. Above: KFA-3000 space image superimposed on the 3″ SRTM DEM. LS: landslides in Neogene-Quaternary deposits, likely triggered by earthquakes associated with surface faulting; 8 and 4: height of the upslope-facing scarp in meters; white angle indicates site from where the photograph was made. Below: view along this surface rupture that is about 8 m high at the background and about 4 m high at the foreground where it displaces deposits of the small rock fall of granites dated by the ^{10}Be method. Age of the rock fall boulder from the uplifted downslope side of the fault was determined as 3454 ± 986 BP, which corresponds to the maximal age of the last faulting event; age of the boulder with similar position (on the uplifted fault side) where scarp height is ~8 m is 18680 ± 5711 BP, indicates maximal age of the penultimate displacement with similar ~4 m vertical offset.

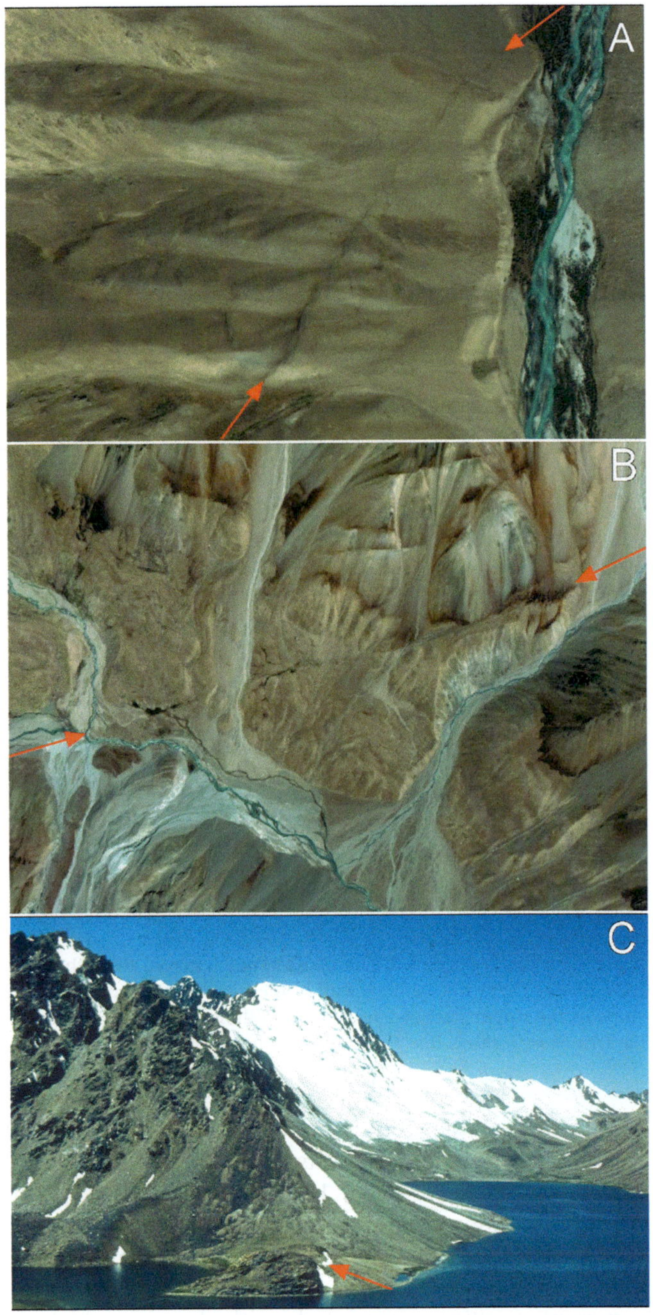

FIGURE 3.13 Surface ruptures marked by red arrows along the E-N-E trending Pathur-Nemos active fault that crosses the Rushan Range in Western Pamir in Tajikistan. (A) At the right bank of the Pathur River valley (37.741° N, 72.203° E); (B) across moraine deposits in the upper reaches of the Shazud River valley (37.815° N, 72.355° E); (C) small scarp at the southern bank of the Chapdara Lake (37.948° N, 72.663° E) that reflects latest offset, while high escarpment covered by snow (at the background)—cumulative effect of multiple displacements. (A and B) Google Earth images. *Reprinted with some modifications after Strom, A., 2014b. Sarez Lake problem: ensuring long-term safety. In: Sassa, K., Canuti, P., Yin, Y. (Eds.). Proceedings of World Landslide Forum 3, June 2–6, 2014, Beijing, v. 3, 633–639, with permission of Springer.*

Southern (Tajik) Tien Shan, in Dzungaria and Pamir. These findings strengthen the assumption of seismic triggering of bedrock landslides, which will be discussed further in Chapter 13.

3.5 CLIMATE, GLACIATION, AND HYDROLOGY

Climate of the Central Asian region can be classified as extremely continental, which is caused by its location at rather low latitude in the inner part of the huge Asian continent, far away from the oceans. Elevated mountainous region is surrounded by plains occupied by deserts that affect the climate not only of the foothill areas but of the entire mountain system. However, large topographic diversity determines sharp moisture and temperature contrasts (Figs. 3.14 and 3.15). Mean

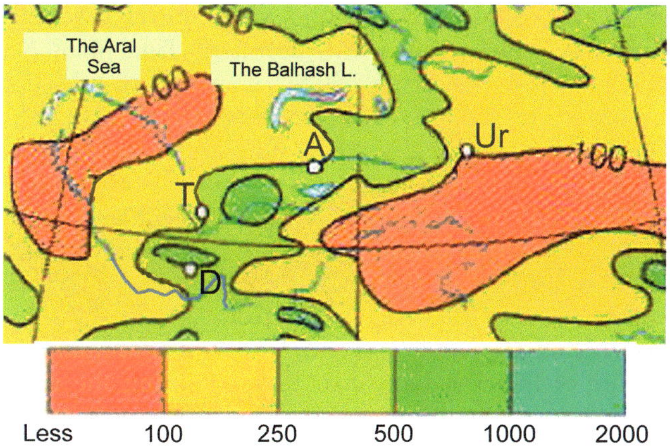

FIGURE 3.14 Annual mean precipitation (mm/year) in the Central Asian region. A, Almaty; D, Dushanbe; T, Tashkent; Ur, Urumchi. *Simplified from http://www.200stran.ru/maps_group28.html.*

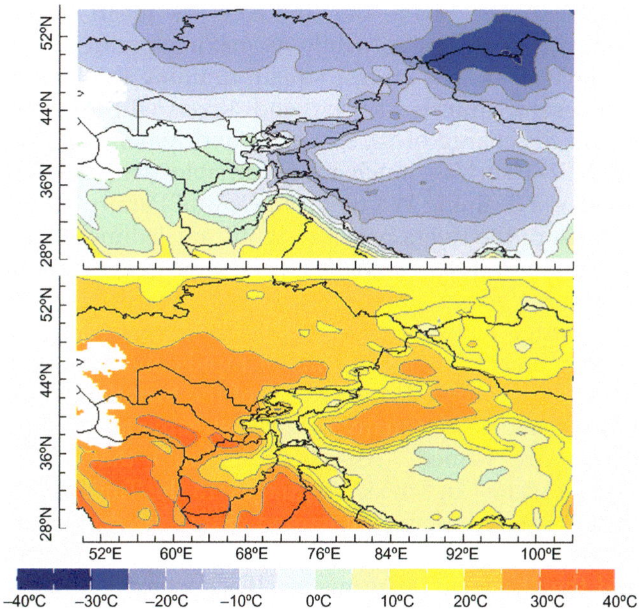

FIGURE 3.15 Long-time annual average temperatures in January (above) and July (below) of the study region. State boundaries are shown. *Simplified from http://www.200stran.ru/maps_group28.html.*

annual precipitation vary from less than 100 mm/year (90 mm/year in some parts of the central Pamir and Eastern Tien Shan) to more than 1000 mm/year (up to 1600 mm/year in some parts of Western Pamir and Tien Shan (Glaciation, 2006).

Annual temperature variation in the study region can reach 40°C–50°C and even more (locally up to 70°C–80°C), resulting in intensive physical weathering. Saturated Atlantic air masses carried by the eastward air transport high above the Central Asian deserts reach high mountain ranges providing rather significant precipitation on their western slopes. In contrast, eastern part of this mountain system and inner depressions of the Tien Shan and Pamir are characterized by much higher aridity (Fig. 3.14).

Most of the precipitation occurs in summer period. In winter thickness of snow in the valleys looking westward reaches 2–3 m, while some eastward looking valleys are almost snow-free, being used for winter pasture.

Intensive summer air heating above the Central Asian deserts facilitates high level of the snow line in this region that varies from 3600–3800 m a.s.l. in the North-Western Tien Shan and Dzungaria, to 4200–4450 m a.s.l. in the highest parts of the Tien Shan and up to 5000 m a.s.l. in Pamir. Many passes at more than 4000 m a.s.l. are free of snow in summer time.

Climatic conditions at each particular site strongly depend on its altitude. While mean summer (July) temperature in low mountain valleys is +20–25°C, it decreases below zero at the watersheds of high mountain ranges. Mean winter (January) temperature is negative throughout most of the mountainous part of the study region (Fig. 3.15). It results in the formation of the high-altitude permafrost that is widely developed in the most elevated parts of the study region. Considering the present-day climate change (global warming), permafrost degradation could facilitate destabilizing of high slopes (Huggel, 2008; Huggel et al., 2010; McColl, 2012). Several examples of sudden formation of large-scale bedrock landslides in the high-altitude zones of the study region that occurred without any identifiable trigger will be described hereafter in Sections 4.5 and 8.4 in Chapters 4 and 8, Principles and Criteria of Landslides Identification and Discrimination From the Features Formed by Other Mass-Wasting Phenomena, and Tien Shan, respectively.

The topmost parts of most of mountain ranges are covered by snow throughout the year and many of them are glaciated. Most intensive glaciation characterizes Central Pamir with the largest ~80 km long Fedchenko glacier (the World longest glacier outside the Polar zones); the Trans Alai Range at the northern boundary of Pamir; the Highest Tien Shan at the Kyrgyzstan–China border with the 60.5 km long Southern Inylchek Glacier at its western side and several large glaciers at its eastern side; and the Akshiyriak Plateau in Central Tien Shan at the water divide of the Naryn and Sarydjaz-Aksu River basins. The specific feature of the latter, as well of some other glaciated regions, is the presence of the so-called "flat summit glaciers" developed not in the valleys but on the elevated plateau-like remnants of the Preneogene planation surface. Some of the Pamir and Tien Shan glaciers are surging, that sometimes result in rivers' damming and outburst floods.

Total area occupied by glaciers in the Pamir is estimated to reach 12000 km^2 (Aizen, 2011). Values provided for the Tien Shan and Dzungaria vary from 6000 to 8000 km^2 (Aizen et al., 1997; Cao, 1998; Solomina, 1999; Solomina et al., 2004). At present most of the Central Asian glaciers degrade (Cao, 1998; Glaciation, 2006), while in the past it was much more widespread. Practically all Pamir valleys both in its central plateau-like section and in the deeply incised western and eastern parts were occupied by glaciers (Chedia, 1971, 1972). In the Tien Shan and Dzungaria the Pleistocene glaciation was also much larger than at present; however, many of their intermontane depressions, except the most elevated, have not been covered by ice sheets totally.

It is assumed that Quaternary glaciation in this mountainous region occurred repeatedly being divided by warmer periods. Such processes when glaciers prograde first and then melt, providing powerful streams, whose discharge could exceed the present-day values significantly, should be taken into consideration for studying past landslides, because they must rework, if not destroy completely evidence of slope failures that had occurred before or during past stages of glaciation. It can explain deficit of landslide-related landforms at the upper parts of mountain ranges with extremely steep slopes and scarcity of the evidence of older (pre-Late-Pleistocene) landslide-related landforms despite the longevity of the preceding evolution of this mountainous region.

All rivers draining the study area belong to the closed drainage basins the largest of which are the Aral Sea basin in the west (Amu Darya and Syr Darya Rivers), the Balhash and Alakol basins in the North (rivers of the Dzungaria), the Tarim Basin (the Sarydjaz-Aksu River, the Jarkend River, other Rivers of Eastern Tien Shan and Eastern Pamir). Some rivers just do not reach any water body but disappears in the deserts and semideserts surrounding the Central Asian Mountains such as the Chu River draining parts of the Central and Northern Tien Shan that flows towards the Syr Darya River but does not reach it at present. All largest rivers that start in high-altitude glaciated ridges are fed by glaciers and snow melt water mainly and are characterized by pronounced summer peak (State Water Inventory, 1987a,b,c, 1988, 1990; Rivers and Lakes, 2012). For example, the Vakhsh River that drains Northern Pamir and Southern Tien Shan has mean annual discharge 580 m^3/s, maximal July discharge up to 3120 m^3/s, and minimal value recoded in February 130 m^3/s only. Mean annual discharge of the Pyandj River that accumulates drainage from most of Pamir is 845 m^3/s. Maximal recorded discharge of the Naryn River that drains large part of the Central Tien Shan is 2660 m^3/s (May, 24, 1969), while its annual values vary in different years from 270 to 709 m^3/s (mean annual value is 418 m^3/s). Mean annual discharge of the Sarydjaz-Aksu River draining the western part of the highest Tien Shan is much less—208 m^3/s only, but it is characterized by more or less regular short-term powerful floods with peak discharges more than 1000 m^3/s caused by repeated outburst of the ice-dammed Mertsacher Lake (Glazirin, 2010; Jansky et al., 2010). The largest river of the Northern Tien Shan—the Ili River has mean annual discharge of 479 m^3/s, while its maximal values reach ca. 1500 m^3/s. The largest stream draining the Eastern (Chinese) Pamir is the Jarkend River with a mean annual discharge of ca. 170 m^3/s and maximal values up to 2710 m^3/s—much more variable than those of the rivers draining Western and Central Pamir.

Most of the rivers provide significant sediment yield (ten million tons annually for the largest streams). Both large rivers and, especially, smaller streams that drain ridge watersheds and slopes are quite energetic, often representing just chains of rapids and waterfalls with white color of surfs typical of the turbulent water flows. That is why many of these streams all over the region are named "Aksu" ("Ak Suu") that means "White water" in Turkic languages. In the Tajik part of the area such streams are named "Safedob" that means the same. Another common river names are "Karasu" and the "Kyzylsu" that means "Black water" and "Red water" in Turkic, due to dark gray or red or brown color of their waters enriched by silt. Such streams, even if they have rather small discharge, have large erosion power and can destroy any obstacle interrupting their flow in a short time. These characteristics should be considered when analyzing siltation and longevity of the landslide-dammed lakes.

CHAPTER

4

Principles and Criteria of Landslides Identification and Discrimination From the Features Formed by Other Mass-Wasting Phenomena

OUTLINE

Large-scale bedrock landslides could be identified based on the combination of morphological and sedimentological characteristics allowing their recognition both on the remote sensing imagery and in the field, and distinguishing their bodies from the deposits left by other geological mass-wasting phenomena—glaciers, rock glaciers, and extreme debris flows. While rockslides and rock avalanches can be characterized, at least in many cases, as dry granular flows, the alternative phenomena could not take place without the "intermediate" agent—water in liquid or frozen state.

Sometimes, however, bedrock landslide identification is not so obvious. First, the gigantic size of these features complicates their recognition both on aerial images, each of which covers just a small part of the affected area and in the field if they could not be visible as a whole. Some of the largest features were identified only when space images with relatively high resolution (5–8 m and better) became available, allowing one eye to embrace the whole feature and, at the same time, to see the details of their morphology. It can be exemplified by the Beshkiol landslide 8–12 km^3 in volume (41.41° N, 74.48° E, Kyrgyz 253)—the largest one in the entire Central Asian region, whose body exceeds 40 km^2 is size and the total affected area (headscarp + deposits) is 55.6 km^2 (Fig. 4.1) (Strom, 1998) or by the Ters rockslide (41.5° N, 72.53° E, Kyrgyz 313) with the affected area of about 25 km^2 (Fig. 4.2, see also Fig. 5.16).

Morphological identification becomes more complex and questionable for older—Pleistocene—features that had been significantly reworked by erosion. The additional complexity is provided by morphological similarity of rockslide and rock avalanche deposits with the deposits left by other natural phenomena widely developed in mountainous regions that bring large amount

Rockslides and Rock Avalanches of Central Asia.
DOI: https://doi.org/10.1016/B978-0-12-803204-6.00004-1

of debris down slopes and/or down valleys, sometimes resulting in the complete or partial blocking of river valleys (see, e.g., Hewitt, 2001 and Iturrizaga, 2012).

Hence we found it necessary to describe first the basic criteria allowing identification of the prehistoric rockslides and distinguishing their deposits from those created by glaciers, rock glaciers and debris flows. These criteria were used when compiling the large-scale bedrock landslides database of the study region (Appendix).

FIGURE 4.1 KFA-1000 space image (5–8 m resolution) of the Beshkiol landslide. (A-B) Cross-section shown on Fig. 5.13.

The correct interpretation of such deposits, especially of those forming natural barriers in the formerly and/or presently glaciated river valleys is not just a scientific problem but is critically important for landslide hazard assessment. If past natural blockages, both intact and breached, were formed by glacial processes, they, first, indicate recurrent late Quaternary glacier advances and retreats reflecting the differently directed climate change during this period. Second, considering present-day deglaciation typical of the study region (Solomina, 1999; Kotlyakov, 2006), we can expect quite low probability of the occurrence of such events in the near future, and, correspondingly, low hazard and related risks of river damming and subsequent outburst floods unless significant inverse climate change. The only exception is the possibility of floods and debris flows caused by breach of glacial lakes forming behind end moraines left by the degrading glaciers or being dammed by glacial surges and gradually moving rock glaciers. If, however, the opposite—rockslide—origin of these features is true, then, in contrast, no general restrictions of large-scale slope failures and river damming at any time in future exist and related risks should be much higher. Besides, it indicates absence of the significant glacial advances younger than these accumulations; otherwise, the latter would not survive retaining undisturbed shape and morphological peculiarities.

Moreover, the correct interpretation of such prehistoric features in mountainous regions plays an important role in seismic hazard assessment (see Chapter 13:

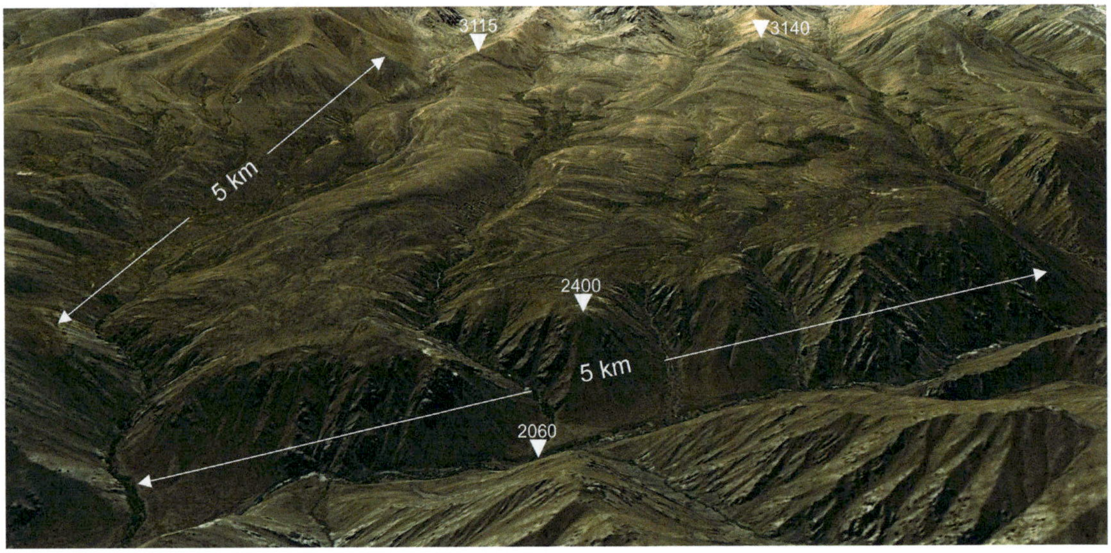

FIGURE 4.2 3D Google Earth view of the Ters rockslide. Headscarp crown is at 3115–3140 m a.s.l.

Seismically Induced Rockslides: Their Discrimination). While formation of moraines, rock glaciers or debris flows are climate-driven processes, they have no or quite limited and indirect relation to seismicity (Larsen et al., 2005). On the contrary, landslides, those in bedrock especially, are often triggered by large earthquakes. Thus, if the seismic origin of past rockslides is proved, and their ages and recurrence intervals are determined, it allow estimating the probability (or, at least, the possibility) of the occurrence of similar earthquakes in future and of all the consequent hazard chains such as new large-scale slope failures, formation of river-damming blockages, and subsequent outburst floods (Strom, 2015a).

FIGURE 4.3 The Aik-Kiol rockslide with the distinct headscarp and body, well-balanced in volume (~20 Mm^3). 3D Google Earth view.

4.1 GENERAL CRITERIA OF ROCKSLIDE/ROCK AVALANCHE IDENTIFICATION

In many cases, identification of the prehistoric rockslide/rock avalanche is a simple task and is based just on the visual morphological analysis. Combination of debris accumulation at the slope base and of the concave headscarp above it, whose volume corresponds well to the deposits' volume, allows unequivocal interpretation of the feature in question as a rockslide as, for example, in the Aik-Kiol rockslide case in Kyrgyzstan (41.726° N, 73.96° E, Kyrgyz 076) (Fig. 4.3).

Sometimes, however, it is not so obvious and additional evidence must be taken into consideration to identify such features and to interpret their origin correctly. Quite often, we can hardly reconstruct the preslide relief and it looks as if the deposits' volume is much larger than that of the source zone. It can be exemplified by the gigantic, about 400 Mm^3, Aidyn-Ula rockslide in the Chinese Tien Shan at 42.8° N, 87.537° E (China 131), near the mountain with the same name (Fig. 4.4). Its ~1.5 km wide and more than 600 m high rectangular headscarp on the slope composed of the Ordovician granites looks much smaller than more than 3.5 km long and up to 2.1 km wide rockslide body, whose frontal part marked by glQ (?) on Fig. 4.4 is composed of the Quaternary, likely glacial deposits bulldozed by the moving rockslide debris during its emplacement. This size discrepancy, however, is erroneous. In this case, slope failure had affected the convex rocky spur that had extended for more than 250 m off the main ridge, though its original shape cannot be reconstructed precisely. Additional criteria, such as the presence of large angular boulders on top of the main part of the deposits and the sharp boundary between the main (blocky) and the frontal (with much more smooth surface) parts of the deposits also indicate the rockslide origin of this feature. Absence of the simultaneous fresh cirque-like scarps in

FIGURE 4.4 The Aidyn-Ula rockslide whose deposits' size looks as if it exceeds the size of the source zone significantly. glQ (?), the assumed Quaternary, likely glacial deposits bulldozed by the moving rockslide debris. White line marks the cross-section shown on Fig. 6.3. 3D Google Earth view.

the vicinity at the same altitude level excludes its alternative—glacial—origin.

As we mentioned above, some older (Pleistocene) rockslides could not be identified just morphologically since both their headscarps and the deposits have been so strongly reworked by erosion that none of them could be identified univocally on the remote sensing data and even in the field just visually. However, careful analysis of the deposits composition and internal structure, as well as of their relationships with other geomorphic features helps to make sound conclusions on their origin. Such analysis allowed us to identify several large-scale Pleistocene rock slope failures, which are not evident from the purely geomorphic analysis.

One of them is located in the Kokomeren River valley at 41.9° N, 74.25° E, about 6 km upstream of the Djumgal intermontane depression and was named the Displaced Peneplain rockslide (Kyrgyz 087) since its headscarp had affected the exhumed pre-Neogene plantation surface. This rockslide about 400 Mm^3 in volume originated on the 500 m high slope composed of the Precambrian gneiss with schist interbeds and some mafic dykes that differ from the bedrock lithology of the depositional area. It had blocked the Kokomeren River valley so that the river had to erode a new bypass valley through the left-bank granite. Afterward, the entire area was reworked by the erosion so significantly that now both the source zone and deposits can be hardly identified on space and aerial images (Fig. 4.5) and even in the field (Fig. 4.6).

However, a closer look on the deposits composed of crushed gneiss with stripes of crushed schist and mafic rocks and with some evidence of the inverse grading typical of rockslide and rock avalanche deposits that will be described in details in Chapter 11, Internal Structure and Grain-Size Composition of Large-Scale Bedrock Landslides, allows sound conclusion that the deposits that form hilly area at the foot of high rocky slopes are allogenic. It is supported by the presence of the abandoned river valley on the present-day right bank of the Kokomeren River valley, whose "entrance" filled by coarse well-rounded granite boulders (stream alluvium) can be seen above the road just upstream of this Pleistocene blockage (see Fig. 4.6).

Another site where identification of a large Pleistocene rockslide was based on the detailed sedimentological analysis of the deposits is located in Southern Kyrgyzstan, in the Tar River valley close to the Terek-Suu village that gave the name to the Terek-Suu rockslide (40.53° N, 73.825° E, Kyrgyz 235). Here smooth hills covered by the well-cemented coarse limestone breccia affected by karst processes could be traced at a distance of about 3 km on the river's right bank locally outcropping in the road cut (Fig. 4.7). They rest on the ~50–70 m high terrace. Such lithification of the carbonate-rich rockslide deposits was described also for the Alpine rockslides in Sanders et al. (2010). The ridge at the opposite side of the valley rising above the river bed for about 2 km (Fig. 4.8) is composed of Carboniferous sedimentary rocks, mainly limestone, thrusted over the Cretaceous red beds. The

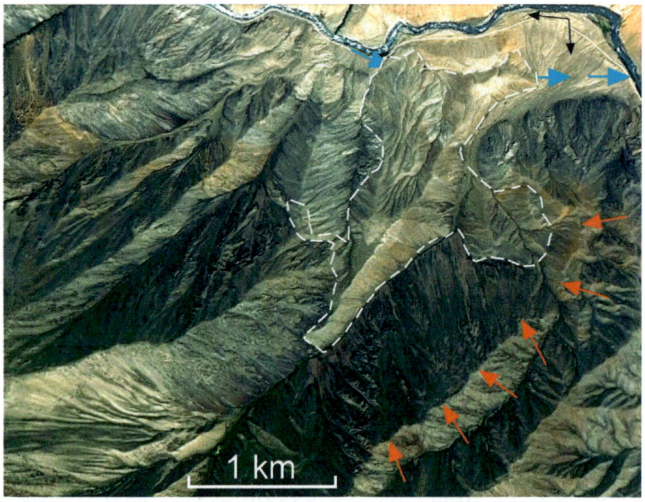

FIGURE 4.5 High-resolution Google Earth image of the Displaced Peneplain rockslide. Red arrows mark the headscarp; white dashed line, the remaining part of the deposits; blue arrows, the assumed position of the pre-rockslide Kokomeren River (the truncated arrow is at the abandoned valley "entrance" shown in the inset of Fig. 4.6); small black arrows indicate the point from which the panorama shown in Fig. 4.6 was made.

FIGURE 4.6 Panoramic view on the Displaced Peneplain rockslide. The inset shows the abandoned river valley with stream alluvium (aQ) blocked by the rockslide. The small dark blue arrow on the inset indicates the position of the present-day Kokomeren River valley.

FIGURE 4.7 The Terek-Suu Pleistocene rockslide. (A) An overview; (B) an outcrop with the pseudo-stratified rockslide (RS) deposits with heavily crushed red sandstone and siltstone and whitish marl overlaid by coarse well-cemented limestone breccia. This succession rests on the alluvial gravels (alQ). Position of the outcrop shown on (B) is marked on (A) by the white arrow.

FIGURE 4.8 The Terek-Suu rockslide headscarp area. Whitish outcrops above ~1570 m a.s.l. with the pyramid-like badlands and fan-shape surface up to 1620 m a.s.l. on the left and 1710 m a.s.l. on the right are the remnants of the reshaped rockslide deposits.

same succession of the crushed rocks, sometimes pulverized up to powder, forms the deposits in question. Their characteristic peculiarities are the preservation of the lithological succession typical of the source zone in the comminuted debris and inverse grading. Both are typical of rockslides and rock avalanches (Heim, 1932; Hsü, 1975, 1978; Erismann, 1979, 1986; Davies, 1982; Strom, 1994, 2006). Since no evidence of the stream damming was found out, it is most likely that slope failure had occurred long ago, before the incision of the canyon-like lowermost part of the river valley. The source area also underwent strong reworking, so that the headscarp could be hardly identified (see Fig. 4.8). Nevertheless, the presence of a large ancient rockslide at this part of the Tar River valley has been proved with confidence.

These case studies demonstrate the possibility to identify large-scale catastrophic bedrock landslides with poor geomorphic expressiveness. We should point out that both of them were identified first during a field study of the deposits and only later the entire features (headscarps and rockslide bodies) have been delineated and mapped. Moreover, the first description of the feature that was identified later as the Displaced Peneplain rockslide was incomplete and its headscarp was placed erroneously in the upper reaches of deep dry gulley that passes from southwest along the foot of its real headscarp (Strom, 1994). Only further field studies allowed correct identification and delineation of the entire feature visible on Fig. 4.5.

Revealing such features raises a question on the completeness of the Central Asian rockslide inventory compiled from the remote sensing data mainly. Most of the features described in our book were identified from the satellite imagery analysis and only for some regions they have been checked by our own field studies or by comparison with published and unpublished data collected by other researchers. Besides erroneous interpretation of debris accumulations, which will be discussed in the following sections of this chapter, it is important just not to miss the features requiring a more thorough analysis. It is obvious that more recent (Holocene and Latest Pleistocene) rockslides with better preserved deposits are more impressive and could be identified and interpreted with more confidence. Older (Pleistocene) features that, theoretically, must be more multitudinary, considering the longevity of the Pleistocene in comparison with the Holocene epoch, are, nevertheless, much more rare. Presumably, most of them were reworked through time so deeply that only careful field studies, both morphological and sedimentological, could help their identification.

Several case studies of such older feature identified on space images are presented hereafter along with the reasons that allowed us to consider them as old rockslides. One is located at the southern slope of the Sarmin-Ula Ridge (Xinjiang) composed of Paleozoic granites. Here, about 15 recent and at least 4 much older rockslides were identified within a small area extending for about 40 km from west to east (85.35° E–85.85° E) and 20 km from north to south (42.4° N–42.6° N). More detailed description of

FIGURE 4.9 Old deeply eroded rockslide with the hummocky relief of the deposits' surface. 3D Google Earth view.

rockslides forming this cluster will be presented in Section 8.1 of Chapter 8, Tien Shan, and possible causes of clustering (Strom, 2015c) will be discussed in Part III, Analysis, of this book. Here, we focus on the criteria allowing identification of the suspicious geomorphic features at 42.463° N, 85.776° E (China 046, Fig. 4.9) and at 42.5° N, 85.683° E (China 047, Fig. 4.10), and their interpretation as rockslides.

For the rockslide shown on Fig. 4.9, besides the distinct concave cirque-shape slope bounding the deeply dissected flat surface about 400 m above the valley bottom, considered as the rockslide headscarp and its body surface correspondingly, the additional important argument in favor of such interpretation is the hummocky microrelief well preserved on this surface and typical of rockslide deposits. Such combination allows assumption that this feature represents strongly reworked old rockslide. Additional reason for such decision is the concentration of the few similar features within the same area where much more recent rockslides form a cluster. It could indicate recurrent occurrence of some triggering phenomena, most likely large earthquakes in this particular area.

Similar peculiarities are typical of the China 047 case study. Its body is not as thick as in the previous case (about 200–330 m) but it is wider, with large smooth hummocks on it. The presence of rather large ($\sim$3.5 Mm^3) recent landslide (Fig. 4.10B) and the material visible in its headscarp show that this body is composed of strongly comminuted debris. This old (Pleistocene) rockslide had dammed the river, and remnants of its body can be recognized on the left bank just downstream from the recent landslide.

FIGURE 4.10 Old rockslide with deeply eroded headscarp and body. (A) recent rock avalanche (China 041); (B) recent landslide (China 147) in the rockslide deposits. 3D Google Earth view.

Another set of evidence in favor of the landslide origin of the cirque-shape lowering on the slope was used when analyzing the giant deeply eroded conic-shape body about 200 Mm^3 in volume at the foot of the 1.6 km long cirque-like "bite" at the 1.2 km high left slope of the Ashu River in the northern-most chain of the Chinese Tien Shan (44.2° N, 83.644° E, China 095) (Fig. 4.11).

Its scar beheads small gullies at the opposite—northern slope of this ridge composed of the Paleozoic rocks indicating the superimposed character of this morphological feature. Several distinct "layers" can be seen in the deeply eroded, almost 400 m thick body resting at the foot of this scar. It is very similar to the internal structure of the Kokomeren rockslide in Kyrgyzstan (41.93° N, 74.22° E, Kyrgyz 086) (Strom, 1994, 2006; Abdrakhmatov and Strom, 2006; Hartvich et al., 2008) representing one of world-best examples of the pseudo-stratified rockslide body (see Section 8.4). All these gave us the reason to interpret this case study as an old bedrock landslide.

FIGURE 4.11 Assumed rockslide about 200 Mm^3 in volume on the left bank of the Ashu River (Xinjiang). Several "layers" of debris can be identified in its nearly 400 m thick body. 3D Google Earth view.

Finishing the description of the general criteria of past rockslides' identification, we want to describe the case study where the conclusion on the origin of the suspicious feature was made by contradiction, based on the impossibility to explain its origin by the alternative natural phenomena. Strange ~50 m high debris accumulation composed of angular granite boulders was found at the bottom a deep dry wedge-like gulley (Fig. 4.12) in the upper reaches of the Kokomeren River valley in Kyrgyzstan (42.1° N, 74.137° E, Kyrgyz 241) adjacent to the extremely impressive Late Holocene Seit rock avalanche (Kyrgyz 240) that will be described in Section 5.3.1 of Chapter 5, Bedrock Landslide Types and Classification Systems. The entire area is composed of Paleozoic granite and granodiorite differing in strength and erosion resistance. Same angular bouldery material composes the fan-shape surface below and the deposits at 2140 m a.s.l. overlying the granite spur marked by γPZ on Fig. 4.12. These deposits are located too low to be classified as moraine. At the same time the gulley catchment below the summit at 2890 m a.s.l. is too small to produce debris flows powerful enough to displace blocks of

FIGURE 4.12 The Late Holocene Seit rock avalanche (arrows indicate its headscarp—H/S and the deposits—BODY) and the assumed headscarp of the much older Pre-Seit rockslide.

FIGURE 4.13 Closer view on the remnants of the assumed Pre-Seit rockslide deposits.

granite some of which are more than 1 m in size (Fig. 4.13). Moreover, this bouldery hill does not look as the fan-shape debris flow or snow avalanche deposits. Impossibility to explain its origin by any ice- or water-driven process forced us to consider rockslide origin as the most logical and consistent explanation. Such explanation contradicts neither with the wedge-like shape of the assumed source zone, nor with the size and shape of large angular boulders forming the deposit. Its closeness to the incontestably identified Seit rock avalanche is in line with the commonly observed phenomenon of their spatial clustering (Strom and Abdrakhmatov, 2004; Strom, 2015) mentioned above for rockslides at the southern slope of the Sarmin-Ula Ridge. Thus the conclusion that the feature in question is the remnant of a very old Pre-Seit rockslide that has been reworked for millennia sounds reasonable and well grounded.

We did not follow the approach used by researchers from the United States and Pakistan who compiled the landslide inventory for the area just south of the region in question—in Northern Pakistan, drained by the Indus River and its large tributary the Hunza River (Ahmed et al., 2014, 2016), based just on slope topography. Each feature included in our database was identified on space or aerial images based on the combination of the geomorphic and the geological evidence and, in some cases, proved by field investigations.

4.2 ROCK AVALANCHES VERSUS MORAINES

As we mentioned above, rockslide and rock avalanche deposits are often misinterpreted as moraines and opposite mistakes are possible as well. Numerous debris accumulations left by rockslides that are considered now as classical case studies had been interpreted initially as being left by glaciers. It was for the Alps prior to Albert Heim investigations (1878, 1883), in the Karakoram, before reinterpretation of numerous extraordinary river valley blockages made by Kenneth Hewitt (1988, 1998, 1999, 2001, 2002, 2006a, b, 2011), in the Southern Alps in the New Zealand (Barth, 2013). Rock avalanche deposits were misinterpreted as moraines in the study region as well (see examples in Section 8.4). There was a long story of back and forth reinterpretation of the Koman (39.54° N, 72.69° E, Kyrgyz 345) and the Achiktash (39.54° N, 72.9° E, Kyrgyz 346) hummocky fields in the Alai valley (they will be described in Chapter 9: Pamir), that were interpreted first as glacial deposits (Korjenevskiy, 1930, 1979 (reprinted)), then as rock avalanches (Kurdiukov, 1950, 1964), again as moraines (Nikonov et al., 1984) and, recently, their rock avalanche origin was confirmed based on complex study of both the deposits microrelief and their composition and internal structure (Strom et al., 2014a; Robinson et al., 2015; Reznichenko and Davies, 2015; Reznichenko et al., 2017). Same erroneous interpretation was for the long runout Chaartash rock avalanches (41.21° N, 73.95–74.0° E, Kyrgyz 263–265) and for rock avalanche that collapsed from the back wall of the Gigantic Beshkiol rockslide (see Fig. 4.1) at the southern feet of the Western Akshiyriak Range that were mapped as glacial deposits on the 1:200,000 State Geological maps, sheets K-43-XXI, K-43-XXVI, compiled by the Geological Survey of Soviet Union in 1960s–70s (Strom, 1998).

Recently the problem was raised again for the Central Asian region in publications of Nikolai Ischuk (2008, 2011, 2013a,b, 2016), who considers majority of natural barriers in river valleys, including well-known features such as the existing dams of the Yashilkul (37.78° N, 72.735° E, Tajik 091), Shiva (37.39° N, 71.4° E, Afghan 007), Iskanderkul (39.09° N, 68.395° E, Tajik 014) lakes, and the breached Kudara-Pasor blockage (38.39° N, 72.58° E, Tajik 092) as being formed, partially or completely, by end moraines.

However, several reasons do not allow agreeing with his interpreting of most of these valley barriers either as end moraines or as a combination of end moraines as main river-damming features accompanied by subordinate rock slope failures. One of the main arguments in favor of the moraine origin of natural barriers is the presence of large amount of moraine material (rounded and semirounded boulders in sandy-silty matrix) in their bodies (N. Ischuk, 2011). However, these observations can be explained in different way, consistent with the rockslide origin of the features in question. Most of the Pamir river valleys and some river valleys in the Tien Shan and Dzungaria had been intensively glaciated in the past. These glaciations left the material from the lateral and/or bottom moraines that cover lower parts of their slopes. Besides, thick debris flow fans come out from the tributary valleys with modern glaciers in their upper reaches. When large-scale slope failure occurs, moving bedrock material bulldozes these boulders in sandy matrix and often overrun and overlay it (Strom, 1994, 2006).

FIGURE 4.14 The Imom rockslide at the left bank of the Gunt River valley. Frontal and flank parts of its body are composed of moraine material resting at the slope base that had been bulldozed by moving rockslide debris. 3D Google Earth view. *Reprinted from Strom, A., 2013a. Geological prerequisites for landslide dams' disaster assessment and mitigation in Central Asia. In: Wang F., Miyajima M., Li T., Fathani T.F. (Eds.). Progress of Geo-Disaster Mitigation Technology in Asia, Springer-Verlag Berlin: Heidelberg, pp. 17–53 (ISBN 978-3-642-29106-7), with permission of Springer.*

It can be well exemplified by the Imom rockslide in the Gunt River valley about 15 Mm^3 in volume resting on the 10–20 m high left-bank river terrace at 37.692° N, 72.327° E (Fig. 4.14, Tajik 093). It also overlays large Late Holocene alluvial fan fed by debris flows from the steep gulley with a small glacier in its upper reaches. Such relationships indicate very young age of the rockslide. Typical moraine material—rounded and semirounded boulders with sandy matrix composes the ~300 m wide belt at the front of this 1.4 km long body and 50–100 m wide stripes bounding it from both sides. This zone has different color and smoother surface than the inner part of the deposits composed

FIGURE 4.15 Front view on the Imom rockslide. Notice different microrelief of the outer and inner parts of the body.

FIGURE 4.16 The Liangar rockslide, which frontal part is composed of glacial deposits bulldozed and scraped down from the headscarp base. The upper part of the Yashilkul rockslide headscarp (Y RS) is visible on the background. 3D Google Earth view.

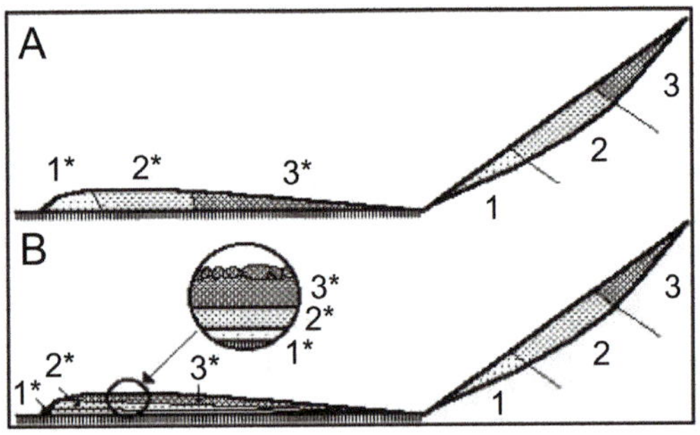

FIGURE 4.17 Typical deposits' structures of rockslides that occurred on slopes composed of different rock types or with loose debris at the headscarp foot. 1, 2, 3: initial position of the affected rock/soil types; 1*, 2*, 3*: same lithologies in the resultant deposits. (A) Retention of the initial sequential order (the "train-like" structure); (B) Conversion into a succession of "layers" ("pseudo-stratified" structure). *Modified from Strom, A.L., 2006. Morphology and internal structure of rockslides and rock avalanches: grounds and constraints for their modelling. In: Evans, S.G., Scarascia Mugnozza, G., Strom, A., Hermanns, R.L. (Eds.). Landslides from Massive Rock Slope Failure. NATO Science Series: IV: Earth and Environmental Sciences, Vol. 49, 305–328, with permission of Springer.*

FIGURE 4.18 3D Google Earth view of the Yashilkul rockslide. Dam surface is covered by hardly erodible blocks of gneiss more than 1 m in size. Small back arrows mark the frontal boundary of the rockslide (rock avalanche) outer zone composed of the moraine material displaced and overlaid by rapidly moving blocky material. Small white arrows mark the site from where photograph shown on Fig. 4.19 was made in 1977, before the construction of the artificial spillway.

of crushed crystalline rocks exclusively that form the 700 m high slope from which it caved and have significantly different hummocky microrelief (Fig. 4.15). Quite fresh headscarp with a crown at ~3800 m a.s.l., which is much lower than the present-day glacial cirques (they are at the level of about 4500 m a.s.l. at this part of the Pamir) is unique for this section of the Gunt River valley that also contradicts the glacial origin of the study feature.

Similar phenomena was observed at the Liangar rockslide that had blocked the same-name river valley, 3 km upstream of its junction with the Gunt River at 37.814° N, 72.71° E. Its frontal 100–150 m wide zone is composed of the glacial deposits bulldozed from the slopes' foot (Fig. 4.16, Tajik 094).

In both cases described above the succession of the lithologies forming rockslide bodies retains their initial order—rock mass that originated from the upper part of the headscarp pushes the material resting at its foot, so that all of them move like a series of train cars (Fig. 4.17A). However, quite often the material from the headscarp crown with higher unit potential energy overruns and overlays the material from the headscarp lower part forming the pseudo-stratified body (Fig. 4.17B) (Strom, 1994, 2006). When such transformation occurs within moving rockslide in which the moraine material from the slope base is involved, it looks as if coarse apron of angular bedrock boulders and debris overlays glacial deposits. Just such interpretation—valley blocking by end moraine and its subsequent overlapping by rockslide—was proposed by N. Ischuk (2011) for the Yashilkul blockage in the upper reaches of the Gunt River valley in the Central Pamir (37.78° N, 72.735° E, Tajik 091) (Fig. 4.18).

FIGURE 4.19 Displaced moraine material bounding the Yashilkul rockslide debris at its right side (see Fig. 4.18). Note the distinct difference between the slope base formed by the reworked old moraine, and its part covered by the moraine material mobilized during rockslide motion.

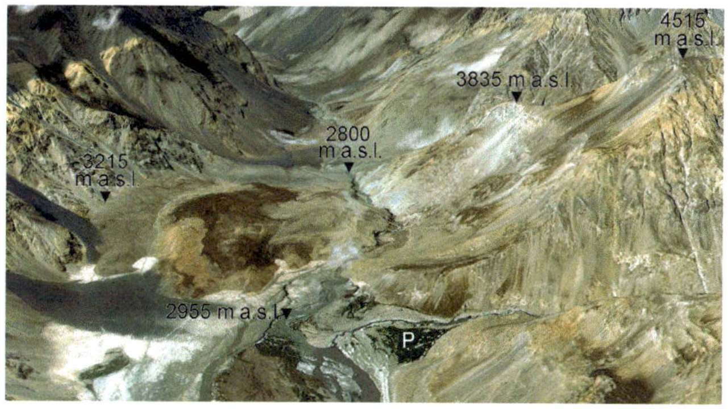

FIGURE 4.20 3D Google Earth view of the breached dam in the Kudara River valley downstream of the Pasor village (P). Varicolored lithologies form concentric bands on the dam's surface, whose position clearly indicates that material moved across the valley from the headscarp with top mark at about 3835 m a.s.l. The concave slope above the headscarp up to 4515 m a.s.l. is the remnant of an old glacial trough valley.

From our point of view, there is no need for such complicated interpretation. All observable phenomena can be explained within the frames of the single-event rockslide model (Strom, 2004). Catastrophic failure of several tens of millions cubic meters of gneiss from more than 1 km high slope, which base was covered by the moraine of the Pleistocene Alichur valley glacier and, likely, by moraine or fluvio-glacial deposits from the adjacent valley visible true left from the headscarp on Fig. 4.18, bulldozed this material and threw it into the stream blocking the river, while coarse gneiss fragments spread over, forming relatively thin blocky apron. Some moraine material spread in front and aside the blocky part and formed this rock avalanche outer zone (Fig. 4.19).

One of the most didactic case studies is the giant breached Kudara-Pasor rockslide dam at 38.39° N, 72.58° E (Tajik 092), also considered by Ischuk (2011) as a moraine dam. However, mutual position of variegated concentric lithological bands expressively visible within this 3×3.7 km and up to 250 m thick body about 1.5 km^3 in volume (Fig. 4.20) shows unambiguously that this material had caved from the distinct headscarp on the right slope of the Kudara River valley and could not be deposited by a glacier moving either along this valley or entering it from right tributary just upstream of the blockage. We should also mention that the frontal part of this blockage is characterized with more than 100 m run-up that decreases gradually up- and downstream. It also shows that this blockage was formed by the avalanche-like motion of the gigantic rockslide that had rushed across the wide trough-like glacial valley and splashed over its opposite slope base.

FIGURE 4.21 The Rivakkul Lake rockslide dam and much smaller rockslides downstream of it. Fronts of rockslide bodies are marked by small black arrows. Flat green meadow at 3700 m a.s.l., is the bottom of the small emptied lake. gl, modern glacier deposits contemporary with the blockages originated much higher than headscarp crowns. 3D Google Earth view.

Another important consideration opposing glacial origin of such natural blockages is the common coincidence of their bodies and of the concave scars on the valley slopes just above them. N. Ischuk (2011, 2013a) interpreted these features as being formed by small cirque (niche) glaciers or "preglacial" nivation hollows. However, practically in all cases, these scars are distinctly younger than the adjacent slope surfaces (more "fresh") and their crowns are lower, often significantly, than the contemporary niche glaciers. It can be exemplified by the Rivakkul dam (37.615° N, 72.06° E, Tajik 095) at the upper reaches of the Rivak River—the left tributary of the Gunt River in Pamir (Fig. 4.21). It should be pointed out that glaciation is a climate-driven process depending (at a valley scale) on the elevation, precipitation distribution, and slope aspect in

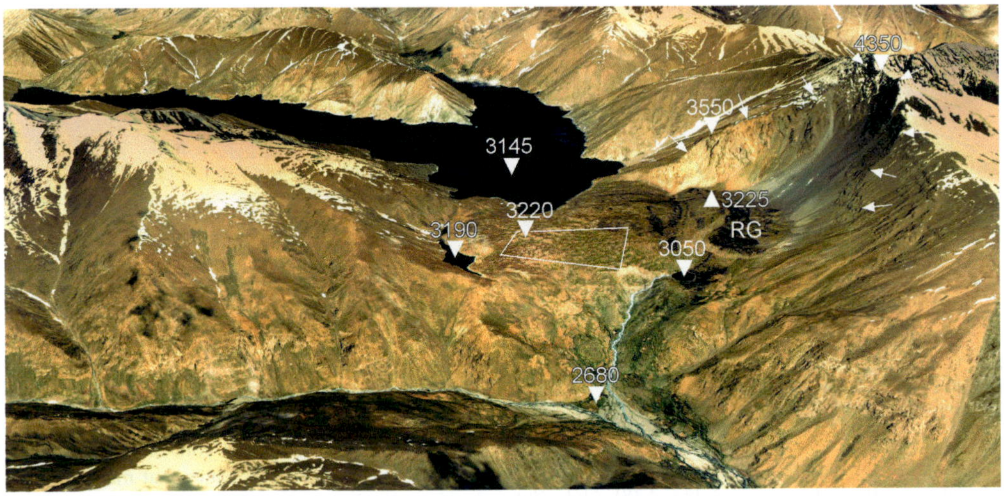

FIGURE 4.22 The Shiva Lake rockslide dam. It is clearly visible that the headscarp marked by small white arrows is a unique feature much younger than other parts of the valley slope. RG, rock glacier in the rockslide headscarp. Outlined area is shown in Fig. 9.110. 3D Google Earth view.

FIGURE 4.23 Impressive conical molards on top of the Spur rock avalanche body, Central Tien Shan (41.975° N, 75.902° E, Kyrgyz 107); see general view of this feature on Fig. 5.4.

such way that more or less similar local glacial features develop on the entire uniformly oriented slope at the same elevation, while the study features are rare or even unique for the large areas as we can see in the Rivak River valley on Fig. 4.21 or, even more expressively, in the vicinity of the Shiva Lake in Afghan Badakhshan at 37.39° N, 71.4° E (Fig. 4.22, Afghan 007). Specific hummocky topography observed at some of blockages and interpreted as indicator of their glacial origin is also typical of many rockslide bodies (Fig. 4.23) (Cassie et al., 1988; Reznichenko and Davies, 2015; Robinson et al., 2015; Reznichenko et al., 2017).

What about the hypothesis that the main damming features are end moraines, blanked by the material of bedrock slope failures, we cannot accept very high probability of association of end moraines and subsequent rockslides—there is no reason to interconnect these totally different phenomena. Of course, sometimes such coincidence could occur, especially at the upper pasts of river basins that had been glaciated rather recently. In such cases, it is difficult to distinguish the exact boundary between the underlying glacial deposits and overlying rockslide debris on the space images with confidence. It can be exemplified by the Sarybiriuk rockslide dam in the upper reaches of the Kyzyl-Tentek River at 45.225° N, 81.00° E (Fig. 4.24, Kazakh 077). Here about 20 Mm^3 of rockslide debris that caved from the ~800 high right slope

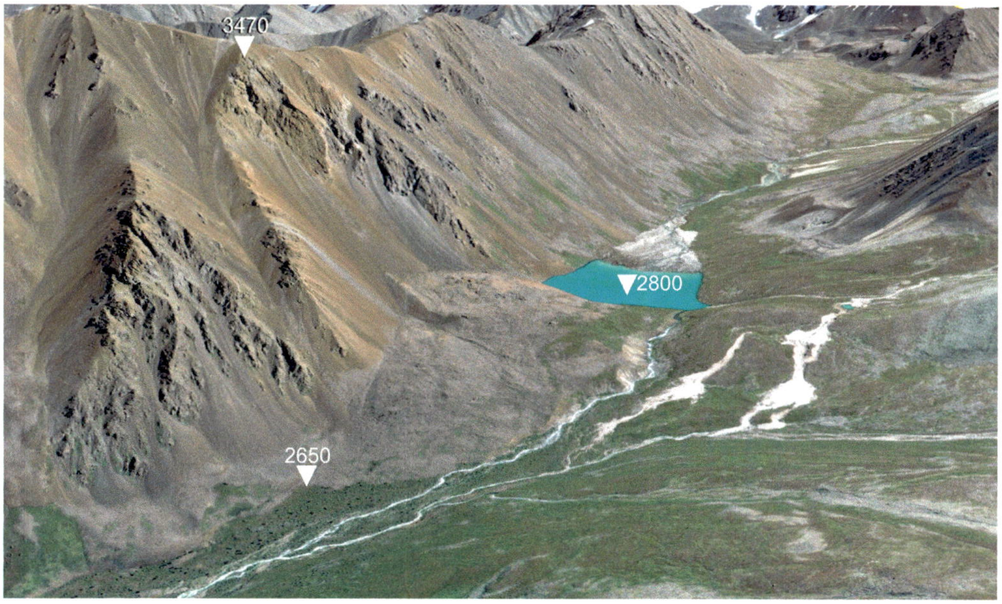

FIGURE 4.24 The Sarybiriuk rockslide in the upper reaches of the Kyzyl-Tentek River overlying glacial deposits. 3D Google Earth view.

of this formerly glaciated valley blocked its bottom just opposite two small left tributaries filled by moraines. Despite the presence of the distinct headscarp and the specific shape of debris accumulation at its base, indicating its slope failure origin, it is difficult now to say with confidence if the lake has been dammed by the slope failure or had existed already being dammed completely or partially by end moraines from the left-side tributary valleys. Such conclusion requires field check.

The sound argument opposing rockslide origin of several natural blockages is the smaller depth of some of the dammed lakes than it could be expected from the longitudinal river profile (see Fig. 36 from N. Ischuk, 2011). Dammed lakes siltation starts from their upstream parts, so it can be hypothesized that the lake bottom level just at the dam should correspond to the original (pre-damming) thalweg level, while the lakes dammed by end moraines would be shallower initially being underlain by bottom moraine (Fig. 4.25).

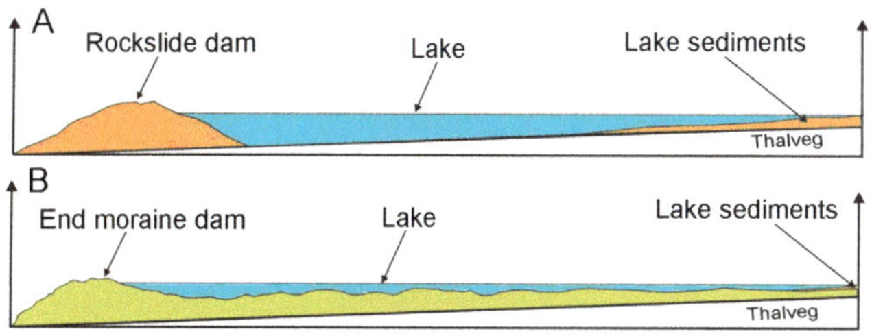

FIGURE 4.25 The schematic longitudinal profiles of dammed lakes in mountainous regions demonstrating difference between lakes dammed by rockslides (A) and by end moraines (B). *Modified after Ischuk, N.R., 2016. Origin of lake dams at the northern slope of the Zeravshan Range GeoRisk, 2, 44–55 (in Russian).*

Despite general correctness of this statement we must take into consideration that numerous rockslide-dammed lakes have been silted completely (some of them will be described in the following chapters). Water bodies in deep narrow valleys with intensive talus formation, often fed by debris flow prone creeks, would be filled by sediments over their entire length. Moreover, if river-damming rockslide has been triggered by strong earthquake (which is quite a common phenomenon), it could be accompanied by numerous minor rock falls and rockslides that could bring significant amount of debris into this lake, while their headscarps could be hardly recognizable centuries and millennia after the event.

Lakes' siltation along their entire length can be exemplified by two rock avalanche dams in the central part of the Gunt River valley in the Pamir (Fig. 4.26). The most recent Chartym blockage named after the Chartym (Charcem) village (37.703° N, 72.174° E, Tajik 096) formed the lake that is now silted completely (Fig. 4.27) both by alluvium transported by Gunt River itself and by debris flows from its tributaries. About 9 km downstream, there is another distinct break in the longitudinal profile associated with old, likely Pleistocene blocking of the Gunt River by the enormous, about 6.5 km long, Vir-Tangiv rock avalanche (Fig. 4.28). Rock avalanche came out from the tributary valley and had formed a dam up to 300 m high, which later on was breached and deeply eroded so that its

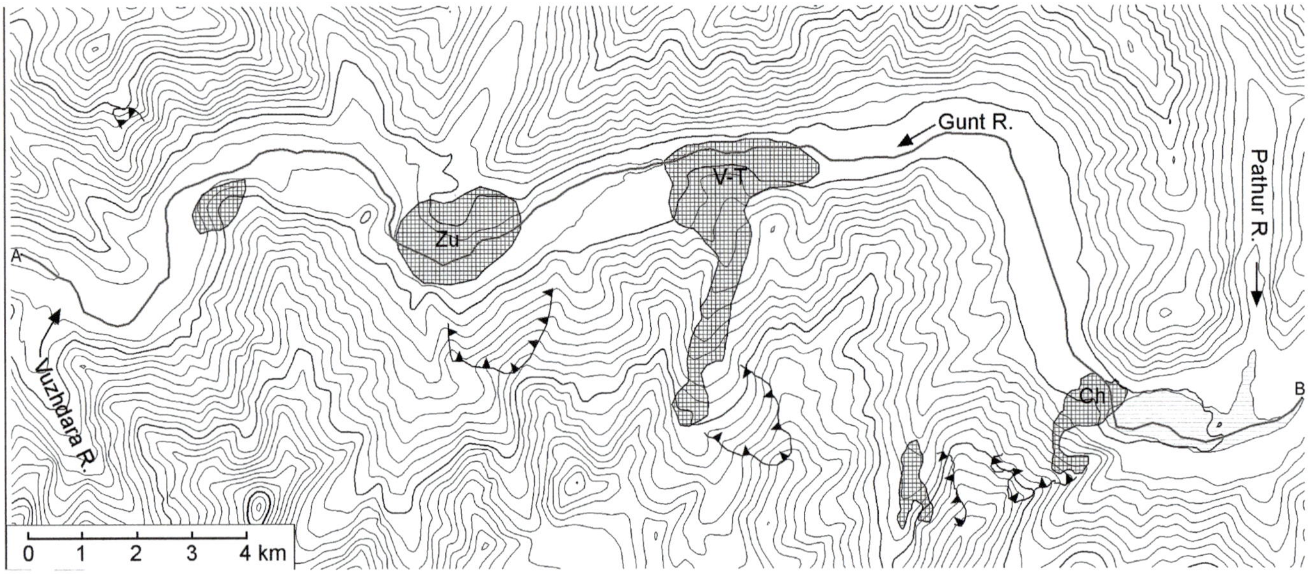

FIGURE 4.26 Schematic map of the Gunt River valley section between the Pathur River and the Vuzhdara River mouths. Contour lines with 100 m interval. Cross-hatched areas—rockslide deposits: Ch, the Chartym, V-T, the Vir-Tangiv, Zu, the Zuvor (Tajik 098); hatched area upstream of the Chartym dam, the silted lake; bold lines with triangles, rockslide headscarps; (A and B) the longitudinal profile along Gunt River shown on Fig. 4.27.

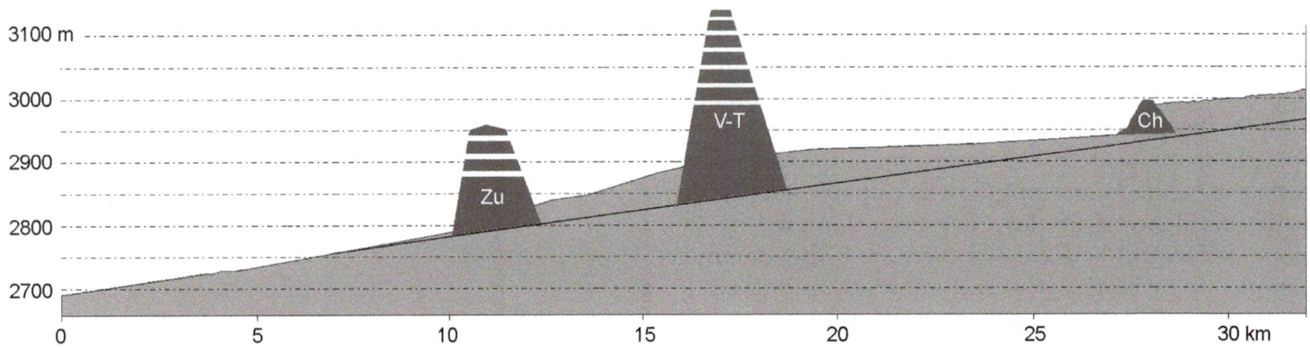

FIGURE 4.27 The longitudinal profile along Gunt River between the Pathur River and Vuzhdara River mouths. The existing Chartym and past Vir-Tangiv and Zuvor blockages are indicated. Black line—the assumed original thalweg.

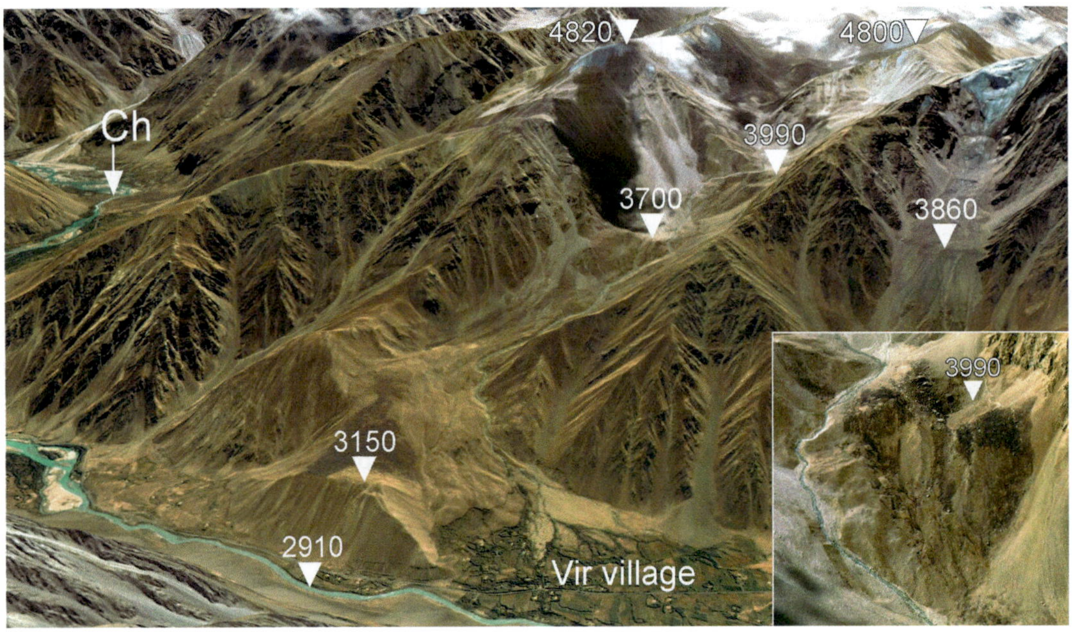

FIGURE 4.28 Breached blockage in the Gunt River valley formed by the Vir-Tangiv rock avalanche. Its 1.12 km high headscarp crown is at 4820 m a.s.l. Ch, the Chartym rockslide dam. Giant blocks up to 30–50 m in size at the proximal part of the deposit can be seen in the inset. 3D Google Earth view.

260 m high remnant can be seen on the left bank of the river. Upstream of this site up to the Chartym rockslide dam, the valley bottom is nearly horizontal indicating significant siltation of the past dammed lake. Assuming smooth profile of the prefailure thalweg, thickness of lake sediments just upstream of the Vir-Tangiv blockage could exceed 50–60 m (see Fig. 4.27), thus demonstrating situation shown on Fig. 4.25B. In this particular case, Gunt River valley blocking was caused by the body that came out from the tributary valley.

Comparable thickness of the deposits downstream of this breached dam (see Fig. 4.27) could be explained as follows. About 3 km downstream of the Vir-Tangiv rock avalanche the Gunt River was blocked by the Zuvor rockslide (37.726° N, 72.045° E, Tajik 098). It originated on much lower slope composed of the same Mesozoic granite as the source zone of the Vir-Tangiv rock avalanche. We assume that both failures had occurred simultaneously and the larger upstream dam had protected the downstream one from breaching and the valley section between these dams from inundation and siltation, but, when, finally, the Vir-Tangiv rock avalanche dam had breached, material eroded from its body had to accumulate in the "shelter" between two dams, at least before the second dam was breached too. More detailed analysis of the interaction between several rockslide blockages in the same river valley will be presented in Chapters 9, Pamir, and 14, Rockslide Dams—Origin, Evolution, and Failure.

Case studies described above demonstrate that neither the presence of moraine material in the river-damming barriers nor the debris accumulation along the entire lake length can exclude their rockslide origin. Combination of blockages and headscarps just above them along with the specific microrelief of their bodies and mutual position of bedrock debris and Quaternary material involved proves that most of large natural barriers in the Central Asian river valleys, both existing and breached have been formed by rockslides, not by end moraines. Some controversial case studies will be discussed in the next chapters.

Addition argumentation either pro or contra of rockslide origin could be provided by the analysis of the presence or absence of agglomerate signature typical of rock avalanche materials despite their original lithologies and identifiable under a scanning electron microscope (Reznichenko et al., 2012a). This technique was successfully applied for the analysis of the Koman and Achiktash rock avalanches and proved its efficiency (Reznichenko et al., 2017). One more method that could shed light on this still controversial problem is age determination, especially for those cases where several rockslides block the same river valley, like the Marguzor Lakes in the Shing River valley in western Tajikistan, which have been treated either as rockslide-dammed (Fedorenko and Kozhevnikov, 1974) or as moraine-dammed (N. Ischuk, 2016) water bodies. If the latter assumption is correct, their ages must decrease from downstream to upstream, otherwise older dams would be destroyed by the advancing glaciers. Results of ^{10}Be dating of the Aksu rockslide (42.54° N, 74.0° E, Kyrgyz) in the Kyrgyz Range (Northern Tien Shan) show that glaciers did not extend towards the northern front of this range after ~63 ka (Sanhueza-Pino et al., 2011), which contradicts glacial origin of relatively young natural blockages. Otherwise glacial advances during Last Glacial Maximum that was dated in the northern Tien Shan as ~49 ka (Koppes et al., 2008) would destroy its body inevitably. These data, though quite limited, provide some constraints on the extent of glaciation in the Tien Shan. More sedimentological and geochronological studies are needed for final solution of the controversial issue of the origin of some river-damming features in this region.

4.3 ROCK AVALANCHES VERSUS ROCK GLACIERS

Another natural phenomenon typical of high mountainous regions that can be misinterpreted as rock avalanche is a rock glacier—gradual and relative slow movement of debris enriched by the interstitial ice (Whalley, 1974; Barsch, 1977, 1988; Haeberli, 1985; Owen and England 1998). In general, debris transported by rock glaciers moves like a crawler belt so that boulders laying on top of the body move faster and gradually fell from its front being overlaid by the inner part of the rock glacier [ibid], which is totally different from rock avalanches that, as can be derived from the internal structure of the deposits (Heim, 1932; Strom, 1994, 2006), move entirely as a laminar flow. Debris is supplied to rock glaciers either by real glaciers upstream or (and) by the numerous small rock falls and snow avalanches from the steep valley walls. The frontal and lateral slopes of these accumulations are inclined at an angle of repose typical of the clast-supported mixture of angular boulders and fines.

Rock glaciers are widespread in the Central Asian region, usually above 3000 m a.s.l. (Gorbunov and Titkov, 1989; Titkov, 1979; Gorbunov et al., 1992; Schröder, 1992; Schröder et al., 2005) and sometimes really look very similar to rock avalanches (Fig. 4.29). They often dam small river valleys.

Sound discrimination between rock glaciers and rock avalanches is based on the absence of the headscarp somewhere above the rock glacier comparable in volume with the deposits, which is essential for rock avalanches. Another discriminating feature is the run-up

FIGURE 4.29 Rock glaciers in the Chong-Aksu River valley (Northern Tien Shan, Kyrgyzstan) in the epicentral zone of the 1911 M8.2 Kemin earthquake. (A) (42.85° N, 77.263° E), fed by the present-day small glaciers in the upper reaches of the catchment mainly; (B) (42.853° N, 77.283° E), fed by rock falls and snow avalanched from the valley walls. Both rock glaciers dammed the river partially. It can be hypothesized that some acceleration of their motion could be caused by the 1911 earthquake, which surface rupture passed 1.0–1.5 km south of the river.

over the opposite valley slope or any other significant obstacle, which is impossible for rather slowly moving rock glaciers but is common for rock avalanches that move by inertia releasing enormous momentum gained during initial downslope acceleration.

4.4 THE ABNORMALLY LARGE FAN-SHAPE DEPOSITS AND THEIR ORIGIN

Sometimes one can see the abnormally large fan-shape debris accumulations at the mouths of the gullies with relatively small catchments. Such features morphologically similar to the alluvial fans were described in the Alps by Jarman et al. (2011) who hypothesized that they have been formed by catastrophic collapses in the catchment areas.

The same or very similar features were identified in the Tien Shan. They can be exemplified by several case studies. One is from the Muzart River valley in Xinjiang, southwest of the Victory peak. Here, between the Atanbulak and Yangimalia villages at 41.95° N, 80.87° E (China 009), the Muzart River valley had been blocked at a distance of about 2 km by the gigantic fan-shape debris accumulation up to 150 m thick that, according to classification system described hereafter in Section 5.3.3 in Chapter 5, Bedrock Landslide Types and Classification Systems, represents the deflected secondary rock avalanche about 250 Mm^3 in volume (Fig. 4.30). It traveled for about 4.3 km from the downstream side of the rockslide dam formed by failure of about 400 Mm^3 of the Proterozoic metasediments and igneous rocks. More than half of it was involved in the Secondary avalanche-like motion (see Section 5.3.3.1).

The fan-shape body could not be formed either by the recurrent or single catastrophic debris flows, since, first, the dam in the tributary valley formed by the initial slope failure has never been breached and, second,

FIGURE 4.30 The gigantic fan-shape blockage in the Muzart River valley formed by the secondary rock avalanche (see Section 5.3.3 in Chapter 5: Bedrock Landslide Types and Classification Systems). 3D Google Earth view.

FIGURE 4.31 The abnormal ~80 m thick fan originated from the small dry catchment. Assumed rockslide headscarp crown is to the left from the point marked as 2490 m a.s.l. 3D Google Earth view.

frontal part of this body resting on the opposite left bank of the Muzart River, rose for about 100 m above the blockage crest, which is typical of rapidly moving dry rock avalanches but not of much less viscous water-rich debris flows that would spread up- and downstream the valley.

One more similar, but, likely, older fan-shape feature was identified in the western part of the Tsagan Giunge Mountains at 42.31 N, 87.324 E (Fig. 4.31, China 052). Unlike the Muzart case where rock avalanche origin of the fan-shape river-damming body is much more obvious, here it is not so unequivocal. The fan-shape deposits differing significantly from the rocks outcropping at the valley bottom had blocked it at a distance of about 1.1 km, so that the river had to turn left and to erode a new bypass gorge. Size of this body in plan view and its thickness are several orders of magnitude larger than those of debris flow deposits upstream that originated in the gullies cross-cutting the ridge composed of the same Devonian metasediments, while these gullies are only 2–4 times smaller than the source zone of the feature in question. Besides it is unlikely that gulley with such a small catchment could produce debris flows powerful enough to dam the river forcing it to erode the bypass channel that exclude formation of the deposits in question by recurrent debris flows. Most likely it is the body of a single-event rock avalanche of the Primary type (see Section 5.3.1) about 25 Mm^3 in volume that originated in the upper reaches of the gulley and traveled 2 km being stopped by the opposite valley wall. Later on, both the headscarp and the deposits were smoothed significantly with time.

4.5 BEDROCK LANDSLIDES IN THE CONTEMPORARY GLACIATED REGIONS

Identification of rockslides in high-altitude zones affected by glaciation at present or just recently is a special and, sometimes, complicated task. Though high and often over-steepened walls of glacial troughs seem to be favorable for large-scale slope failures, rockslide bodies that caved on glaciers are destroyed so rapidly, within decades, by moving ice that only most recent events could be identified on the contemporaneous remote sensing images. What about headscarps, they are often masked by snow and can be hardly distinguished from glacial landforms. Because the topmost parts of the ranges (above 3500–4000 m a.s.l. on average) are scarcely populated, very few eyewitness reports are available even if such events occurred recently. That is why such phenomena are reported so rarely for the large part of the study region subjected to modern glaciation, and the vast majority of the identified bedrock landslides included in the database originated below the presently or recently glaciated topmost parts of mountain ranges.

Distinguishing of the glacial and rockslide-generated landforms is also complicated by their significant similarity predetermined by the flow-like motion of both glaciers and rock avalanches. Both processes create hummocky relief, longitudinal and transverse levees, etc. observed in the resultant deposits.

The main and the most widely used indicator of recent large-scale single-event accumulations of rock debris on the glacier surface is high-color contrast between ice and rock debris. It allowed identification of rockslides and rock avalanches that collapsed on the Bualtar glacier in Karakoram in 1986, likely triggered by snowmelt and rainstorm (Hewitt, 1988), on the Sherman glacier in 1964 (McSaveney, 1975, 1978), and Black Rapids glacier in 2002 in Alaska (Jibson et al., 2006; Shugar and Clague, 2011) being triggered by large earthquakes, on Ice Valley Glacier in 1997 in British Columbia (Evans and Clague 1999; Evans and Delaney, 2014) and in the Glacier Bay National Park and Preserve in Alaska in 2012 (Geertsema, 2012). High-color contrast allowed identifying rock avalanche that had collapsed, likely, few years before 1959 (when aerial photography was performed) on the glacier in the upper reaches of the Ukukh River valley in Greater Caucasus (Strom, 2014a). We should point out that remnants of this rock avalanche became quite hardly identifiable on space images obtained in 2004 due to their reworking by moving glacier. The detailed analysis of rock avalanches emplaced over glaciers reworking was performed for the New Zealand Mueller glacier where their deposits were identified within the terminal moraine (Reznichenko et al., 2015) and for $\sim 3\,Mm^3$ 2004 Thurwieser rock avalanche that collapsed on the Alpine glacier (Frattini et al., 2016).

Some general aspects of rockslides' formation in the presently glaciated regions and of their motion over glacier ice were described in (Evans and Clague, 1988; Huggel et al., Hewitt, 2009; Shugar and Clague, 2011).

Hereafter, we describe several rock (ice-rock?) avalanches identified in high-altitude zone of the study region focusing on the criteria indicating their origin and relationships with glaciers' motion. Though some of these rock avalanches do not reach the threshold value mentioned above ($\sim 1\,Mm^3$ in volume), we describe them considering uniqueness of such phenomena.

Evidence of at least two successive rock avalanches were found in the upper part of the Koprsay glacier in Xinjiang, in the central most elevated part of the Tien Shan, at 42.309° N, 81.0705° E (China 132) and at 42.35° N, 81.082° E (China 132a). Despite impossibility to identify the source zone of these bedrock slope failures masked by fresh snow, their deposits can be identified and mapped on the glacier surface (Fig. 4.32). Just at the foot of the tremendous ~ 2600 m high back

FIGURE 4.32 Fresh rock avalanche about $0.3\,Mm^3$ in volume at the foot of ~ 2600 m high back wall of the Koprsay glacier (Xinjiang). 3D Google Earth view. Image was obtained on September 11, 2007. Its plan view is shown in Inset 1. Shape and position of these deposits 3 years later is shown in Inset 2. Point A is marked to demonstrate its front displacement. Dashed white line outlines assumed remnant of older rock avalanche debris shown on Fig. 4.33.

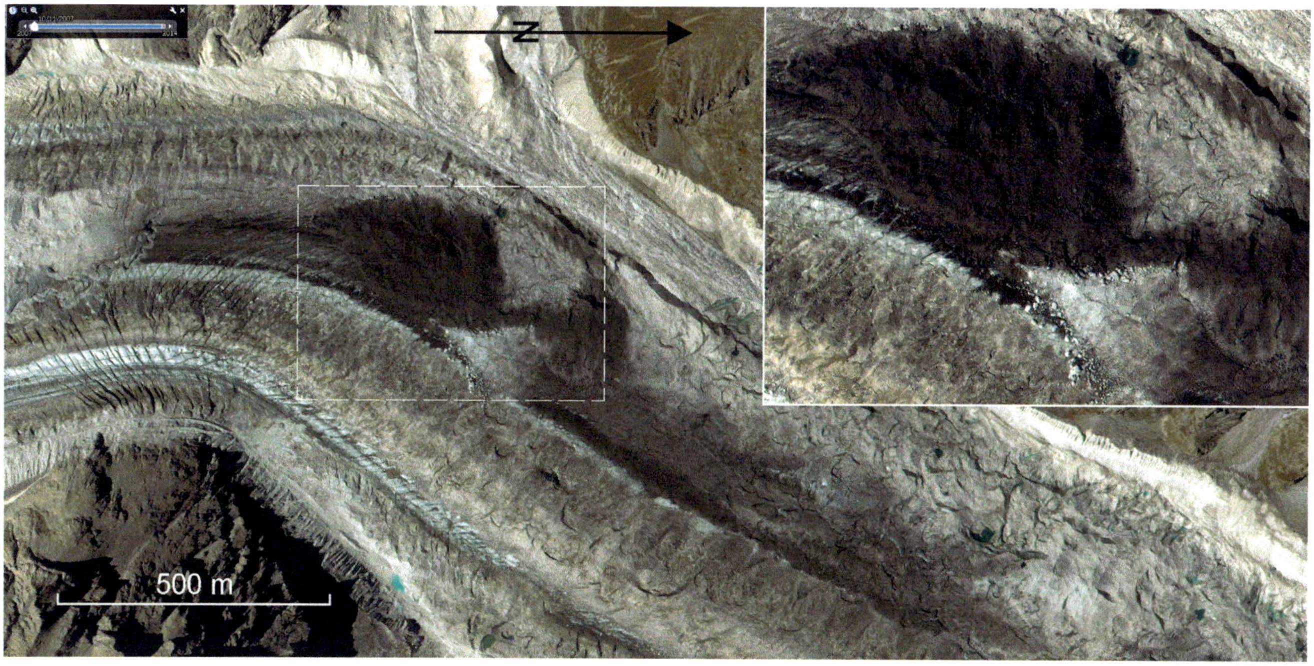

FIGURE 4.33 Assumed reworked deposits of older rock avalanche on the same glacier 3 km down valley from fresh rock avalanche shown on Fig. 4.32. Large angular boulders can be seen along its frontal boundary on the zoomed fragment shown in the inset. Google Earth image.

wall of the glacier valley, one can see ~800 m long stripe of dark-gray debris practically not affected by fractures visible in the ice (China 132, see Inset 1 on Fig. 4.32) that indicates that this failure occurred very recently. Assuming ~3 m average thickness of the deposits, we get 0.3 Mm^3 volume of this rock avalanche. Three years later, as it can be seen on the WorldView 2 image made on August 5, 2010, available on Arc GIS Earth, its tailing part almost disappeared, likely being covered by snow avalanches, while the remaining frontal one about 430 m long was transported 400 m down valley (see Inset 2 on Fig. 4.32). It allows estimating glacier velocity as about 130 m/year.

About 3 km down valley, another rock debris accumulation (China 132a), composed locally of large angular boulders can be identified along the central part of the glacier not being connected with the medial moraine levees (see Figs. 4.32 and 4.33). We hypothesize that it is the reworked deposit of similar, but older and larger (likely up to 1 Mm^3 in volume) rock slope failure, transported and reshaped by moving glacier. Study of such features, along with the data on the rate of the ice movement, could help to assess the recurrence of rock slope failures in the upper reaches of the Koprsay glacier. This body moved ~100 m during the same period of time, from September 2007 till August 2010, providing the velocity of about 33 m/year only. Averaging above estimates of the glacier velocity gave us ~80 m/year that allows assumption that this penultimate rock slope failure occurred 37–40 years earlier, around 1970.

In the Central Pamir, very large rock slope failure was reported in the upper reaches of the Vanch River valley at 38.761° N, 72.143° E. It came out from the right tributary glacier valley and covered ~3 km long section of the RGO glacier (Tajik 100, Fig. 4.34) (Kotlyakov et al., 2003). The exact date of this event is unknown, and time constraints are based on dates when the last space photograph without evidence of this feature existence (June 25, 2001) and the first one where it is visible (September 30, 2002) were obtained. More recently Stark et al. (2012), based on the successive satellite imagery analysis and reexamination of teleseismic data assumed that large-scale rock slope failure had occurred in August–September 2001. The assumed source of this failure was at about 6000 m a.s.l., 3–4 km southwest of the Garmo Peak (6595 m a.s.l.) (G. Osipova, personal communication). Thus, it traveled about 11 km along the right source glacier of the dendritic RGO glacier (No 96; Kotlyakov et al., 2003) before coming to a halt at the junction of the RGO and the Krasnoarmeiskiy (Red Army) glaciers (see Fig. 4.34). Style of its reworking with almost complete

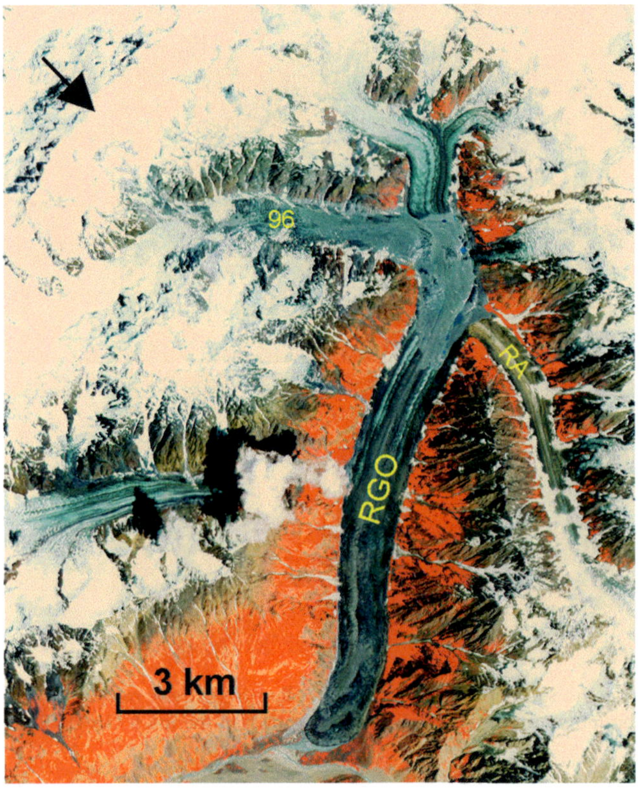

FIGURE 4.34 Rock avalanche on top of the RGO glacier. Upper reaches of the Vanch River, Central Pamir. The false color ASTER space image of August 8, 2003. No 96, RGO and Red Army (RA) glaciers are marked. Black arrow indicates assumed source zone. *Courtesy G. Osipova.*

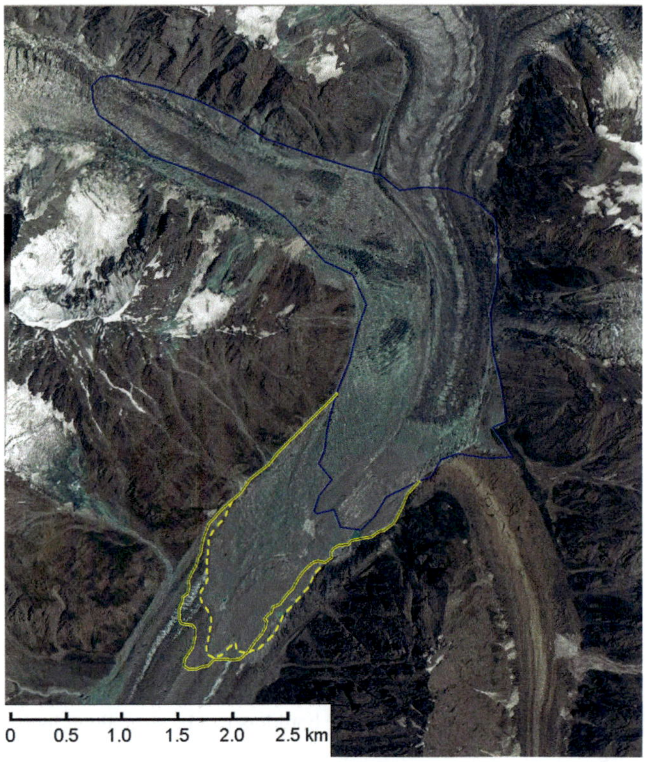

FIGURE 4.35 Reworking of the 2001 rock avalanche by the RGO Glacier. Blue line marks rock avalanche deposits boundary at 08.08.2003 (see Fig. 4.34). Solid yellow line marks its front in 2014 and the dashed yellow line shortly earlier, when the SAS Planet space image used as the background was obtained (the exact date is unknown). The upstream part of the rock avalanche apron has been reworked almost completely by the advancing glaciers, while its front just moved ~2 km downstream remaining almost intact and well defined.

reshaping of the upstream boundary and much better preservation of the downstream front that can be seen on the space image obtained several years later (Fig. 4.35) (the exact date is unknown) is in line with the model proposed in (Reznichenko et al., 2015) and based on thorough study of the Mueller glacier terminal moraine.

Stark et al. (2012) assumed that weight of the collapsed rock and ice that formed the RGO Glacier rock avalanche was 2.8×10^{11} kg that corresponds to the volume of about 100 Mm^3. From our point of view this estimate is exaggerated significantly. Considering the approximately 5 km^2 area of the deposits and its mean thickness of about 3–5 m (according to photographs in Kotlyakov et al., 2003 and G. Osipova personal communication), we estimate its volume as about 20 ± 5 Mm^3. Because slopes at this elevation are covered by snow and the SRTM DEM is quite rough, we could not delineate the headscarp with confidence and compare volume of the deposits with that of the source zone.

Another interesting example of recent rock slope failure on the glacier (Tajik 101) was found also in Pamir, at the junction of the Valter and Soldatov glaciers (39.03° N, 71.963° E, Fig. 4.36). Failure occurred somewhere on the slope north of the Eugenia Korjenevskaya peak (7105 m a.s.l.), though the exact source cannot be identified with confidence on space images available. Debris passed over the Soldatov glacier and formed a fan-shape body that crossed the Valter glacier and covered 1.564 km^2 of ice and of the surrounding ice-free valley bottom by thin, definitely less than 5 m, blanket of fine gray-color debris. Rather fussy frontal contour of this splash allows assumption that it could be formed by the dust storm often accompanying rock avalanches (Bhandari and Kumar, 2000; Geertsema, 2012) rather by the direct run-up of debris. Such dust blows could leave their traces on the surrounding slopes far beyond the debris limits that could be recognizable for some time after the event.

Analysis of the successive space images kindly provided by Dr. Galina Osipova from the Institute of Geography of Russian Academy of Sciences shows that this slope failure occurred between 1990 and September 2000 (Fig. 4.37). After its emplacement, the

FIGURE 4.36 Ice-rock avalanche deposits on the Valter glacier, Central Pamir. Space image obtained on July 27, 2014. SG, the Soldatov glacier; FG, the Fortambek glacier. More than 200 m high splash of debris could be traced on the opposite side of the Valter glacier valley up to ~4270 m a.s.l. (see also Fig. 4.38). Google Earth image.

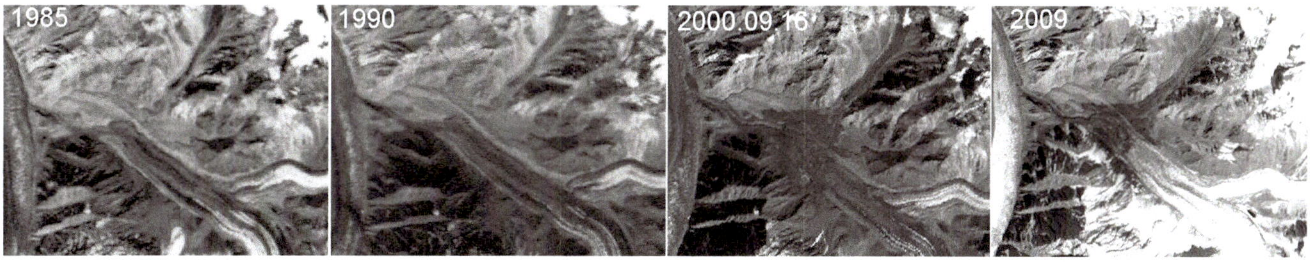

FIGURE 4.37 Series of the successive space images of the Valter glacier lower part. No evidence of the material surge from the Soldatov glacier valley can be seen in 1985 and 1990, but it is clearly visible on space image obtained on September 16, 2000, and later. *Space images courtesy G. Osipova.*

advancing Valter glacier moved 55–57 m/year destroying rock avalanche apron from the upstream side. As can be seen on Fig. 4.37, in 2009, it had advanced for about 500 m, while in July 27, 2014 this advance reached 800 m (from D to C on Fig. 4.38) that can be derived from the color of the advancing glacier similar to those part of it that had never been covered by rock avalanche debris. About 900 m of the glacier frontal part is still covered by gray-color debris (between C and B on Fig. 4.38). Narrow "tongue" of debris that spreads downstream to 3550 m a.s.l. (from B to A on Fig. 4.38) and reached the side of the Fortambek glacier could be formed by the seasonal melt water of the Valter glacier enriched by rock avalanche material. Since on the September 2000 space image the upstream boundary of the rock avalanche apron is, practically, undisturbed (see Fig. 4.37), it can

FIGURE 4.38 The Valter glacier destroying rock avalanche debris apron. Part of the glacier surface from (D) to (C) is not covered by gray debris distinctly visible outside. Narrow stripe of debris from (B) to (A) is, likely, formed by melt water of the Valter glacier. Google Earth image.

be assumed that this image was obtained shortly after slope failure.

Unlike the RGO rock avalanche that was emplaced on the central part of the glacier, far from its front, the Soldatov rock avalanche covered the Valter glacier almost at its front. In both cases, we see significant reshaping of the upstream limit of rock avalanche deposits due to glaciers' advance. But behavior of the downstream boundary is quite different. In the RGO case it is "riding" on the glacier with rather limited reshaping. More or less the same can be observed at the older rock avalanche apron on the Koprsay glacier which frontal boundary is more pronounced than the tailing one (see Fig. 4.33). In contrast, the Soldatov rock avalanche apron where it covers the Valter glacier is reworked by melting ice significantly. We can still trace debris limit because it spread over the presently ice-free valley bottom outside the glacier. Such complex multifactor interaction of glacier ice and rock avalanches blanketing it makes the analysis described in (Cook et al., 2013; Reznichenko et al., 2015) possible only for end moraines studied in the field. At the same time, results of their investigations allow assumption that large rock slope failures in the presently glaciated regions occur much more frequently than it can be derived from the analysis of the remote sensing data available for several last decades only.

Distinguishing of landslides that occur at high-altitude zone covered by glaciers just recently but ice-free at present is even more complex, though detailed field study and landforms dating allow such discrimination. Such studies were carried out, for example, in the uppermost part of the Northern Japanese Alps (Kariya et al., 2011).

Sometimes differentiation between glacial or rockslide origin of the landforms in question can be done based on their relative age and morphological characteristics as in the case of the rockslide feature near the watershed of the Djumgal Range at 41.986° N, 74.315°° E (Kyrgyz 347). Here the horseshoe-shape lowering is superimposed on the older glacial cirque with smooth slopes (Fig. 4.39). This feature is, definitely, younger than the slopes surrounding this glacial landform. Just below it one can see the accumulation composed of angular blocks that cannot be the moraine of the niche glacier because it is almost noneroded, while glacial deposits at the same elevation all around have been reworked almost completely. Uniqueness of this feature within large area with evidence of past glaciation allows its sound interpretation as a postglacial rockslide. This rockslide about 2.5 Mm^3 in volume only is located 2.7 km southwest of the headscarp of the much larger (~80 Mm^3) Mingteke rock avalanche (42.00° N, 74.355° E, Kyrgyz 092) that will be described in Section 8.4 in Chapter 8, Tien Shan. We should mention that both of them occurred on the slope with almost the same aspect.

FIGURE 4.39 The postglacial rockslide in the glacial cirque at the watershed of the Djumgal Range. GL, glacial circus; RS—rockslide body.

One more example of the relatively young (likely Late Holocene) rock avalanche in the recently glaciated watershed of the Shugnan Range (South-western Pamir) was found in the upper reaches of the Urmochdara River that is the left source of the Bogevdara River—left tributary of the Gunt River. The 5.5 km long Urmochdara rock avalanche (37.423° N, 71.782° E, Tajik 103) is the only one rockslide-related feature surrounded by numerous Late Holocene glacial landforms and rock glaciers (Fig. 4.40). Its headscarp itself is hardly distinguishable from the nearby glacial niches reworked by recent erosion. However, the striated microrelief of its deposits (Fig. 4.41) with numerous conical-shape molards differs significantly from that of both moraines and rock glaciers widely distributed at this altitude, allowing univocal determination of this feature as a rock avalanche.

The above-mentioned Vir-Tangiv rock avalanche (see Fig. 4.28) also originated at the upper part of the Shugnan Range with widespread glacial landforms. Its rockslide nature can be derived from the composition of its gigantic (about 300–400 Mm^3 in volume) body composed of the monotonous intensively crashed whitish gneiss debris, totally different from the moraines. It can be also seen at the nearby Zuvor rockslide (37.726° N, 72.045° E, Tajik 098), whose body was overlaid by moraine from the Zuvordara River valley—the right tributary of the Gunt River (Fig. 4.42). Besides, the Vir-Tangiv rock avalanche body have coarse blocky carapace with some blocks on top of its proximal part up to 30–50 m in size (about 50,000–60,000 tons) (see Inset on Fig. 4.28), which is also not typical of glacial deposits.

FIGURE 4.40 Central part of the Shugnan Ridge above the 3500 m a.s.l. Area shown on Fig. 4.41 is outlined by green rectangle. Red arrows mark surface rupture damming a small lake at the terminus of large moraine field in the upper reaches of the Bogevdara River. Google Earth image.

FIGURE 4.41 Striated microrelief with numerous hummocks (molards) of the Urmochdara rock avalanche differing from that of moraines (see Fig. 4.40) and rock glaciers (RG). Google Earth image. *From Strom, A.L., Abdrakhmatov, K.E., 2017. Large-scale rockslide inventories—from Kokomeren River basin to the entire Central Asia region—IPL106-2 and WCoE. Proc. 4WLF, with permission of Springer.*

One more rock avalanche in the recently glaciated zone was identified on space images in the Suusamyr Range at the upper reaches of the Kashkasu River (41.867° N, 74.15° E, Kyrgyz 314). The upper part of its valley bottom is filled by bouldery deposits with distinct striation (Fig. 4.43). Both these deposits and the assumed headscarp differ significantly from what can be seen in the adjacent valleys at the same altitude (Fig. 4.44). Microrelief of the deposits filling other valleys is totally different, and their watershed slopes are much more curved by nival processes than the smooth headscarp of this rock avalanche covered by talus. Combination of these morphological peculiarities indicates that the Upper Kashkasu valley fill differs from that of all the surrounding valleys, which, in turn, reflects difference of the processes that formed this

FIGURE 4.42 Zuvor rockslide body (RA at the foreground) overlaid by glacial deposits of the past Zuvordara River glacier (glQ). Its head-scarp is on the opposite—left side of the Gunt River valley with its crown at 3900 m a.s.l. "RA" at the background, behind the Vir village is the distal part of the Vir-Tangiv rock avalanche (Tangiv is a small village just upstream of this past blockage). 3D Google Earth view.

FIGURE 4.43 5.7 km long Upper Kashkasu rock avalanche that originated in the recently glaciated part at the eastern termination of the Suusamyr Range. Its debris rose on the opposite slope up to 3510 m a.s.l. Freshness of the assumed headscarp, whose relief differs from that of the adjacent slopes is visible on the inset. The elevation mark 3125 m a.s.l. indicated the headscarp crown of a smaller rockslide whose body merges the main rock avalanche. 3D Google Earth view. *From Strom, A.L., Abdrakhmatov, K.E., 2017. Large-scale rockslide inventories—from Kokomeren River basin to the entire Central Asia region—IPL106-2 and WCoE. Proc. 4WLF, with permission of Springer.*

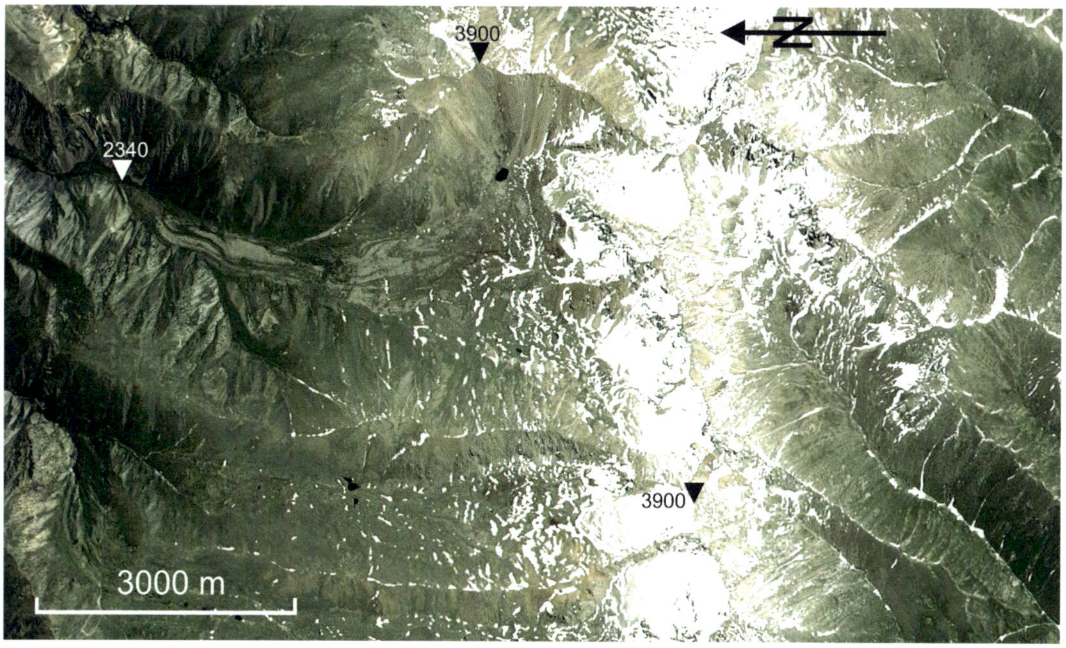

FIGURE 4.44 Google Earth space image of the part of the Suusamyr Range watershed.

relief. Its similarity with the Urmochdara rock avalanche (see Fig. 4.41) allows sound assumption that it is the rock avalanche, the longest one in the entire Kokomeren River basin.

Concluding description of the criteria allowing distinguish bedrock landslides in the presently and recently glaciated regions, we must point out that number of such rockslide-related landforms identified within the topmost parts of high mountain ranges of the Central Asian region is surprisingly small, considering their extremely rugged relief. The most logical explanation of this peculiarity, from our point of view, is that, due to cyclic climate changes resulting in alternating glaciation and deglaciation, spoon-shape landslide headscarps formed on high-altitude slopes during warmer periods, irrespective of their triggering factors, later, when the climatic conditions became favorable for glaciation, are occupied by niche glaciers. With time they are reshaped and become indistinguishable from the "purely glacial" landforms. Similarly, while in the lower parts of the region large-scale rockslide bodies are preserved for a long time, so that we can identify deposits of the Pleistocene features, similar accumulations affected by glacial processes are reworked and transported far from their initial position, so that they could be distinguished in rare cases by the thorough field studies only.

CHAPTER

5

Bedrock Landslide Types and Classification Systems

OUTLINE

Most of the landslide classifications (Varnes, 1954, 1978; Emelianova, 1972; Zolotariov, 1964, 1983; Zolotariev et al., 1969; WP/WLI, 1993a,b, 1994; Cruden and Varnes, 1996; Cruden and VanDine, 2013; Hungr et al., 2012, 2014) describe conditions of the initial stages of slope failure foremost. In the latest and most comprehensive classification (Hungr et al., 2014), the following movement types are proposed for bedrock landslides, which are the target of our research: fall, topple (block and flexural), slide (rotational, planar, wedge, compound, and irregular), spread, flow, and slope deformation. We must mention that in case of the bedrock landslides, flow (rock avalanche—see table 5 in Hungr et al., 2014) characterizes style of debris motion rather than that of the initial slope failure.

Practically all types of such landslides and slope deformations could be found in the Central Asian region. Some examples are presented hereafter, most of which are from the Kyrgyzian Central Tien Shan where we performed our field investigations mainly. Considering size of the studied features, many of them demonstrate peculiarities typical of different landslide types. What hits the eye first is the drastic transformation of the motion style at different stages of bedrock landslide evolution. One of the most hazardous type of rockslides—highly mobile and long runout rock avalanches characterized by the flow type of debris motion—could originate as a planar slide, which can be exemplified by the Flims (Pollet et al., 2005; Poschinger, 2011), the Seimareh (Harrison and Falcon, 1937; Watson and Wright Jr., 1969; Shoaei and Ghayoumian, 2000; Roberts and Evans, 2013), or the Avalanche Lake rock avalanches (Evans et al., 1994), as a rotational slide exemplified by the gigantic Karivhoh rock avalanche in Northern Caucasus about 2 km^3 in volume that climbed almost 500 m on the opposite slope of the Ardon River valley (Strom, 2004), as a wedge slide—the disastrous 2006 Leyte rock avalanche in the Philippines (Evans et al., 2007), as compound or irregular slides like the 7.5 km long 1949 Khait rock avalanche (Gubin, 1960; Evans et al., 2009b), and many other large-scale rock slope failures. Their conversion into flow-like motion of dry crushed debris had occurred "on the way" being governed by the morphology of the transition and deposition zones and by internal processes acting in the rapidly moving rock/debris mass. The "incomplete" transformation can be exemplified by the Waikaremoana rockslide in the New Zealand (Beetham et al., 2002; Davies and McSaveney, 2011) that had moved as a gigantic planar block slide, while just its frontal part had converted into a flow-like rock avalanche.

Rockslides and Rock Avalanches of Central Asia.
DOI: https://doi.org/10.1016/B978-0-12-803204-6.00005-3

Considering such transformations, we used the classification system proposed in Hungr et al. (2014) to characterize styles of the initial stages of slope failures, where it is possible. However, hazard assessment for large-scale bedrock landslides in high mountains depends at first not on the conditions of the initial slope destabilization but on correct estimate of the area that could be affected by moving debris, of its runout, and of the ability or disability of the particular slope failure to block river valley. Such hazard assessment requires classification of rockslides according to their mobility and shape of the resultant deposits, rather than of the initial slope failure style. To describe these most hazardous phenomena we followed rockslide dams' classification proposed in Hermanns et al. (2011b) and rock avalanches' classification proposed in Strom (1996, 2006, 2010a). These classifications are based on the analysis of world-wide case studies of rockslide dams and mobile rock avalanches regardless of the style of the initial slope failure.

Before describing different classification systems and types of bedrock landslides, we want to dwell on the terms "rockslide" and "rock avalanche" widely used in this book and on differentiation between them. Similarly to the most general term "landslide," we use the term "rockslide" to describe any type of the catastrophic slope failures in bedrock (except rock falls and deep-seated gravitational slope deformations—the DSGSDs (Agliardi et al., 2001, 2013; Crosta et al., 2013; Pánek and Klimeš, 2016)) as "rockslide." Indeed, all of them originate as a "slide" (planar, rotational, wedge, and compound). Topples also convert into slides, being detached from the host massif. Further evolution of such "rockslides" depends on various factors. In most of the cases, their bodies disintegrate and move in a flow-like manner regardless of the morphology of the transitional and depositional zone. We always use the term "rockslide" describing those cases where bedrock landslide formed rather compact body, e.g., natural dam, whose height is more or less comparable with its width and length (see, e.g., Figs. 2.1 or 4.3). Those cases, where deposits' thickness is much less than its linear dimensions, they are described as rock avalanches (see Figs. 4.32, 4.41, or 4.43). The compound terms "Primary rock avalanche," "Jumping rock avalanche," and "Secondary rock avalanche" proposed in Strom (1996, 2006, 2010) will be widely used hereafter to define rockslides with the particular styles of debris motion regardless of their compactness/mobility. Their meaning is specified in Sections 5.3.1, 5.3.2, and 5.3.3.

5.1 BEDROCK SLOPE FAILURE TYPES

Impressive example of the typical rock fall on the slope composed of Paleozoic granites can be observed in the Kokomeren River valley, downstream of its left tributary—the Oikaing River mouth (42.043° N, 74.171° E, not included in the database). Large angular blocks of granite rest at the foot of the 500 m high slope with rocky cliffs on its top (Fig. 5.1). It was a real fall, where individual blocks collapsed, jumped, and rolled. Such hazardous natural phenomena are very common in mountainous regions, but, fortunately, usually have limited size, rarely exceeding 100,000 m^3 in volume. Thus, most of them are much less than the threshold value of slope failures analyzed in this

FIGURE 5.1 Typical rock fall deposits at the foot of a slope composed of fractured granite.

study—one million cubic meters approximately—and have not been included in the database (see Chapter 6: The Bedrock Landslides Database).

One of the largest features that can be classified as a rock fall and has the same conical-shape body resembling the famous 1991 Randa event in the Swiss Alps (Gotz and Zimmermann, 1993; Sartori et al., 2003) was found in Xinjiang, north of the Bagrashkol Lake at 42.32° N, 87.288° E. Here about 0.5 Mm^3 of the Devonian metasediments collapsed from the ~300 m high rocky slope (Fig. 5.2, China 103).

Toppling, the flexural one at first, is one of the most common mechanisms of slopes destabilization. Good example of high slope undergoing flexural toppling could be found on the right bank of the unnamed gulley at the south-western flank of the small Kyzyl-Oi intermontane depression at 41.94° N, 74.095° E (Kyrgyz 080). Here the northern slope of a 500–600 m high ridge composed of Paleozoic metasediments with distinct bedding planes steeply dipping south, "in the slope," is crossed by numerous upslope-facing scarps up to several meters high at a distance more than 1 km (Fig. 5.3). They reflect downslope bending of bed heads due to gradual creep deformation.

FIGURE 5.2 Recent (though its exact age is unknown) rock fall from the ~300 m rocky slope in Xinjiang. 3D Google Earth view.

According to Nichol et al. (2002) "*... toppling can initiate a brittle catastrophic rockslide.*" If it will occur at this site, about 40 Mm^3 of metasediments could fail catastrophically. Fortunately, no settlements or important infrastructure facilities are located within the endangered area.

Another mechanism of the rock slope deformation preceding its catastrophic failure is the so-called buckling characterized by gradual sagging of the upper part of the slope and bulging of its lower part. Though identification of the sites with ongoing long-term buckling requires regular monitoring, which is carried out at a very few sites in the study region, evidence of similar mechanism could be derived from debris distribution of some past rockslides, at the Spur rock avalanche (41.975° N, 75.902° E, Kyrgyz 107) in particular. This rather young (the exact age is unknown) rock avalanche about 10 Mm^3 in volume caved from the 700 m high rocky spur dividing headscarps of the gigantic Pleistocene Karakudjur rockslides (Kyrgyz 106, see also Section 8.4 in Chapter 8: Tien Shan) and composed of the Proterozoic metamorphosed volcanic rocks. At the middle part of its headscarp, rock avalanche debris starts expanding outside from the headscarp limits and covers the slope by a wide triangular-shape apron up to 5–6 m thick (Fig. 5.4). It is hypothesized that slope failure that initiated from top of the headscarp area, stressed the lower part of the affected rock mass, which bulged and, finally, when this overstressed bulge failed, debris moved not only forward but sidewise as well.

FIGURE 5.3 Upslope-facing scarps along the slope undergoing flexural toppling.

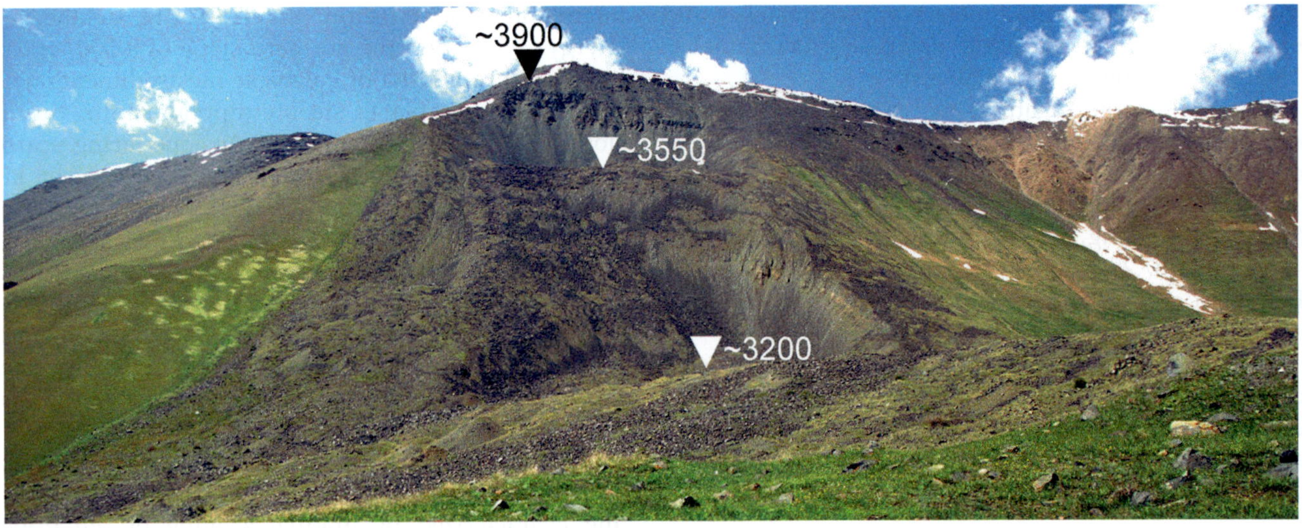

FIGURE 5.4 Spur rock avalanche with debris apron expanding over the slope outside from the headscarp limits.

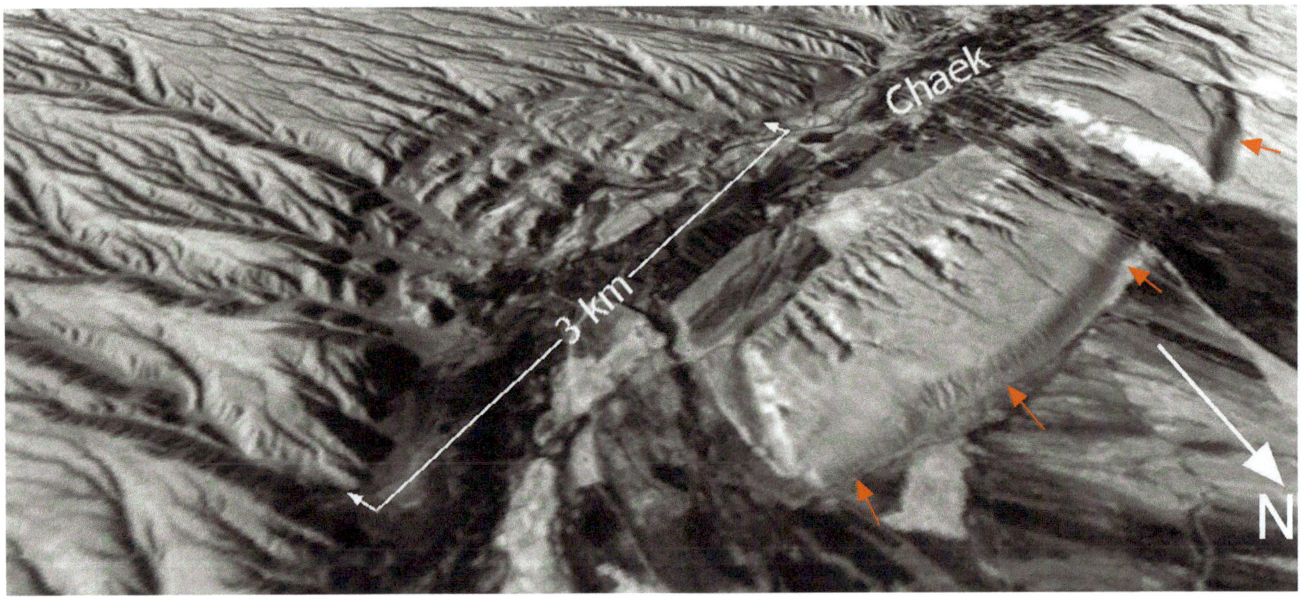

FIGURE 5.5 Gigantic landslide with rotated blocks on the left bank of the Djumgal River near the Chaek town. Quaternary tectonic uplift bounded by active fault (red arrows) is well expressed at the opposite side of the valley. 3D view of the 3″ SRTM DEM visualized by the Global Mapper software.

Large-scale rotational landslides characterized by block's rotation along the cylindrical (in 2D) or spherical (in 3D) sliding surface are rare in the study region, because they occur mainly in poorly lithified soils whose mechanical properties are more or less uniform in all directions. In sedimentary rocks with well-pronounced bedding planes, planar or wedge types of failures occur more often, whereas in igneous or deeply metamorphosed rocks, wedge or irregular slides are common. However, the good example of the rotational landslide that resulted in back-tilting of the preslide daylight surface of the rotated blocks could be found in the central part of the Djumgal intermontane depression at 41.93° N, 74.57° E. Here, opposite the Chaek town, the tremendous amphitheater about 3.0×1.8 km in plan view with stepped relief disturbs the gently inclined surface formed by merged alluvial fans (Kyrgyz 102, Fig. 5.5). Each step within the amphitheater is formed by the rotated and tilted block 100–200 m wide. It can be hypothesized that this feature could be triggered by earthquake associated with the evolution of an impressive tectonic uplift on the

FIGURE 5.6 Sliding surface of the Lower Ak-Kiol rock avalanche that originated as a block slide in the inclined coarse-bedded Neogene conglomerates well visible in the sidewall cliff with a summit at 2040 m a.s.l. Large blocks of conglomerates on top of the rock avalanche deposits can be seen on the foreground. Behind, at 1760 m a.s.l, there is a proximal depression more than 20 m deep.

opposite side of the river valley bounded from its northern side by active fault (see Fig. 5.5).

The planar slides can be exemplified by the Lower Ak-Kiol rockslide (41.705° N, 74.284° E, Kyrgyz 009) about 30 Mm^3 in volume that had blocked the small Unkursai River valley—left tributary of the Minkush River, which, in turn, falls into the Kokomeren River. Huge block of the coarse-bedded Pliocene conglomerates about 1000 × 500 m in size and more than 60 m thick slid down along the bedding plane dipping ~25° towards the river (Fig. 5.6). It converted into rock avalanche with 1.9 km runout that formed the dam that was breached subsequently and, likely, catastrophically.

The wedge slides are rather common type of bedrock slope failures in the study region with widely developed intensively folded and faulted sedimentary and metamorphosed rocks. Largest of such rockslides usually had left very deep (up to 1 km) scars on slopes. Just such origin has the famous 1911 Usoi rockslide in Pamir (Fig. 5.7, Tajik 090). According to A. Ischuk (2011), its headscarp more than 1.2 km high (if being measured from the top of the deposits, while its total height, measured from the sliding surface, could be about 1.7 km), more than 2.2 km wide and about 1 km deep (in horizontal direction) is composed of the Permian–Triassic dolomite and limestone with gypsum and anhydrite in its upper part (whitish unit on this photo) underlain by the gray sandstone, schist, and quartzite of the Carboniferous Sarez Formation. These two units are divided by the Usoi Thrust dipping 60–80 degrees in the south-eastern direction. One more shear zone at the eastern part of the headscarp dips 50 degrees towards north-west. It is assumed that both these zones of weakening bounded the gigantic wedge along with the anticipated sliding surface dipping towards the river and forming its base.

We want to mention that this seismically triggered failure had affected not the topmost part of the slope close to the water divide where maximal topographic amplification could be anticipated, but its middle and lower parts, downslope of the well-defined topographic bench, where frequency of shaking could be closer to the natural frequency of the rock massif (H.-B. Havenith, personal communication).

Another impressive example of the wedge-like slope failure is the Ak-Kiol rockslide in the Unkursai River valley (41.68° N, 74.28° E, Kyrgyz 009), few kilometers upstream of the above-mentioned Lower Ak-Kiol rockslide. The Ak-Kiol rockslide forms the dam with a beautiful lake behind (Fig. 5.8).

This rockslide originated on the left-bank slope of the valley composed of the Carboniferous brown–red conglomerate and sandstone with thin gypsum interbeds of the Dungurminskaya Series and has about 0.6 km deep headscarp that affected the ridge far beyond its watershed (Fig. 5.9).

Carboniferous red beds are folded here in the syncline with hinge line gently dipping east towards the river valley and limbs inclined up to 30–40 degrees. In fact, the entire syncline core about 20 Mm^3 in volume

FIGURE 5.7 The gigantic wedge-like headscarp of the 1911 Usoi rockslide.

FIGURE 5.8 The Ak-Kiol rockslide-dammed lake and the northern wall of the very deep headscarp (on the left).

caved out from the valley slope leaving about 600 m deep scar. The southern side of the headscarp strictly corresponds to the bedding and additional subsequent failures likely had occurred here later on. The opposite—northern side—of the headscarp is slightly steeper than the bedding. The presence of the gypsum with low strength in these strata should facilitate formation of a very deep rockslide scar.

Rockslide debris patches on both sides of the headscarp that remain 130–140 m higher than the surface of the rockslide body (Fig. 5.10) allow reconstruction of rockslide motion at the initial stage, before it had left the headscarp and entered the Unkursai River valley. It is especially impressive on the true right side of the headscarp where rockslide debris fills the dry gully mouth (Fig. 5.11).

Though the rock massif that formed the rockslide had a syncline core structure, its deposits' structure well visible in the canyon eroded through the dam body, looks as if it is composed of nearly horizontal layers of sandstones and conglomerates (Fig. 5.12). It can be hypothesized that the gigantic synclinal wedge slid down initially as a block slide following hinge line and leaving patches of debris on the headscarp walls.

As soon as it left the deep headscarp and lateral confinement provided by headscarp walls disappeared, this synclinal structure was unfolded and, finally, the valley was blocked by the dam retaining the original succession of strata but resting nearly horizontal.

The compound slides where proximal block with high potential energy "pushed" low-energy "passive" frontal block (Hutchinson, 2006; Hungr et al., 2014) can be exemplified by the Beshkiol landslide 8–12 km^3 in volume (see Fig. 4.1, Kyrgyz 253) (Strom, 1998; Strom and Korup, 2006; Korup et al., 2006). Its proximal part had affected steep slope of the Western Akshyiriak Range rising for about 2350 m above the Naryn River valley and composed of Paleozoic rocks thrusted over much weaker Neogene sandstone, siltstone, and conglomerate with gypsum interbeds forming wide and rather gentle syncline (Fig. 5.13). Its proximal part subsiding from the 1.5 km high steep slope likely had pushed the gigantic block of Neogene sediments resting at the foot of the ridge along very gentle sliding zone coinciding with some weak bedding plane(s).

Landslide had moved several kilometers northward and filled the Naryn River valley at a distance of about 7 km forming a giant long-living dam. Collapse of the steep high slope composed of Paleozoic rocks produced rock avalanche, whose deposits cover the proximal part of the landslide body. This extraordinary case study will be described in more details in Section 8.4.

Many if not most of large-scale bedrock landslides in the study region can be classified as irregular slides. According to Hungr et al. (2014), such rock slope failures occur in strong rocks with nonsystematic structure and their sliding surfaces "consist of a number of randomly oriented joints, separated by segments of intact rock ('rock bridges')." Such rockslides are

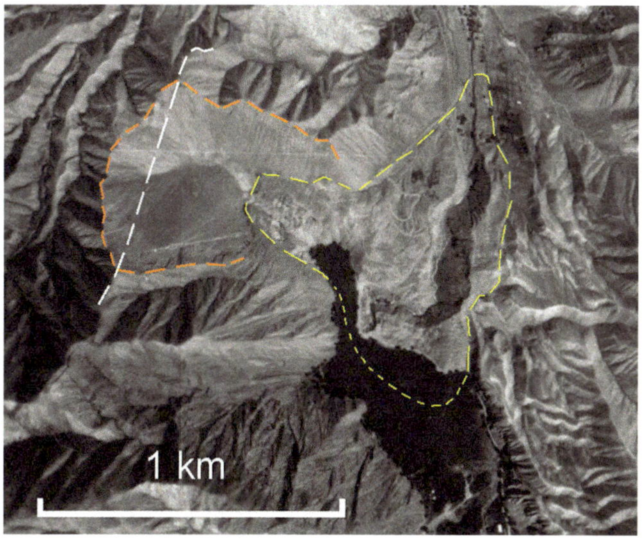

FIGURE 5.9 Air photograph of the Ak-Kiol rockslide. (A) Rockslide-dammed lake; headscarp is marked by light-brown dashed line; dam's body is marked by thin dashed yellow line; the position of the prefailure local watershed is marked by dashed bold white line.

FIGURE 5.11 Mouth of the gully just near the headscarp that had been blocked by rockslide debris. Nearly horizontal top of debris is about 130–140 m above the surface of the main body.

FIGURE 5.10 Panorama of the Ak-Kiol rockslide headscarp. The outlined areas show the position of debris patches at the headscarp outlet (that on the left is shown on Fig. 5.11).

FIGURE 5.12 Ak-Kiol rockslide body composed of nearly horizontal strata despite the syncline structure of the host rock mass.

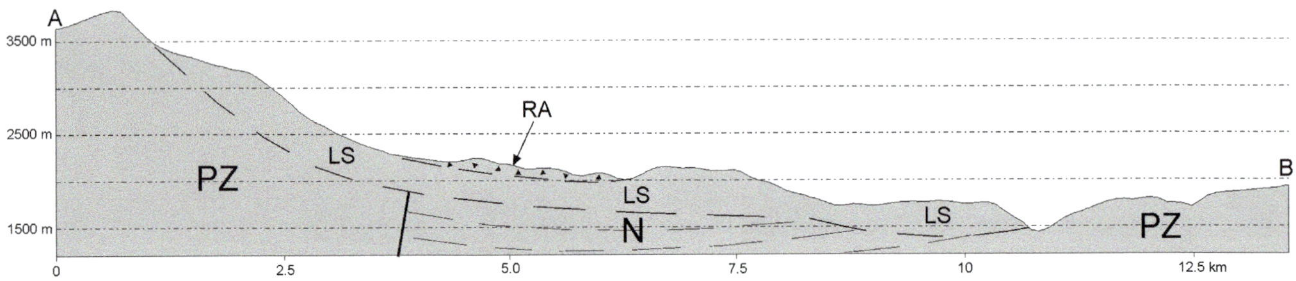

FIGURE 5.13 Schematic cross-section of the Beshkiol landslide along the line shown on Fig. 4.1.

common on slopes composed of igneous and deeply metamorphosed rocks widely developed in the Tien Shan and Pamir mountains (Fig. 5.14).

Hungr et al. (2014) mentioned that such slope failures are *"often very sudden and extremely rapid,"* which can be proved by the most catastrophic bedrock slope failure of the entire Central Asia—the 1949 Khait rock avalanche (Tajik 038) triggered by the M7.4 Khait earthquake (Fig. 5.15).

One of the most hardly identified types of the bedrock landslides in the study region is the spread slides. As indicated in Hungr et al. (2014) *"Rock slope spreading, involving the displacement and rotation of rigid blocks of stronger rock, because of severe plastic deformation of an underlying layer of weak rock is very common in horizontally bedded, weak sedimentary sequences."*

Such situation, quite common in some Alpine mountain systems—in the Apennines (Canuti et al., 1990; Picarelli and Russo, 2004) or Carpathians (Nemčok, 1982)—is rare in the Tien Shan and Pamir where metamorphosed sedimentary rocks with higher strength are developed more widely. Nevertheless, some features, identified in this region, allow assumption of this mechanism of motion. Large-scale deformations that, likely, can be ascribed to this type can be seen on some gentle slopes composed of hard rocks (granites) and just blanket by rather thin apron of Quaternary deposits. One of the most impressive examples is the above-mentioned Ters rockslide on the northern slope of the Chatkal Range in Western Kyrgyzstan (41.485° N, 70.55° E, Kyrgyz 313). Here vast area—about 5 × 6 km in size with mean slope

FIGURE 5.14 The irregular headscarp and the debris-free transition zone of the 3.2 km long Holocene Seit rock avalanche in the upper reaches of the Kokomeren River valley (42.11° N, 74.14° E, Kyrgyz 240). Failure occurred on the slope composed of Paleozoic granite and diorite. Small arrows mark the right-bank trimline about 100 m above the debris-free gully bottom.

FIGURE 5.15 Irregular headscarp and upper part of the 1949 seismically triggered Khait rock avalanche. Failure occurred on the slope composed of metasediments. Sagging is well expressed on the adjacent slope. *Photo courtesy: Steve Evans.*

inclination of 6–7 degrees only bounded from top by the horseshoe-shape scarp up to 50–70 m high—is crossed by numerous 2–3 km long upslope- and downslope-facing scarps (see Fig. 4.2 and Fig. 5.16). Large length of local scarps indicates that they should extend rather deep into bedrock. At the same time, no evidence of deformations of the 450 m high right bank of the Ters River gorge is visible on space images, preventing to classify this slope deformation as the planar or compound slide. Such combination of geomorphic features allows classifying this rockslide as a spread type, though the geological situation is not favorable for such type of slope deformations.

Rockslides of the spread type can be linked with DSGSDs, which reflect gradual long-term process and, at the same time, can precede catastrophic collapses (see the review in Pánek and Klimeš, 2016). The DSGSDs are very common in the study region and could be found on ridges composed of various types of rocks.

Most of the DSGSD represent the "single-side ridge-top spreading" (Hungr et al., 2014) or "asymmetric sackung" (Pánek and Klimeš, 2016)—unidirectional motion of large part of the ridge towards the bounding valley. It took place, in particular, at the flexural toppling site described above (see Fig. 5.3). Another subtype of the single-side slope deformations that can be classified as large-scale bedrock slope sagging is exemplified by the site at the Southern slope of the Kokshaal Range in the Chon-Uzengiogush River basin, on the left bank of the Betdjin Creek (41.146° N, 78.354° E, China 081; Fig. 5.17). Here the ridge slope about 500 m high composed of the Devonian metasediments is deformed by the series of steps up to 3 km long in toto, whose strike, generally, coincides with the foliation. Though the stream at the foot of this ridge is narrower than upstream, where it passes through the same geological unit, the absence of damming indicates slow gradual slope sagging compensated by river erosion rather than single-event catastrophic subsidence.

Another impressive example of the DSGSG can be found in Tajikistan, on the northern slope of the Zeravshan Range above the Rogiv village (39.374° N, 69.335° E, Tajik 078) where the entire watershed composed of the intensively folded Silurian terrigenous rocks is displaced by multiple bow-shape scarps at a distance of about 2.5 km (Fig. 5.18). Unlike the previous case where deformations had affected the ridge slope mainly, here the scarps expand far beyond the crest of the ridge and are accompanied by the upslope-facing scarps on the other side, so that the entire feature has some elements of the "double-sided ridgetop spreading" (Hungr et al., 2014), though unidirectional deformation with main vector of rock mass motion towards the gulley with elevation mark 1740 m a.s.l. prevails.

Sometimes one can find even more pronounced "double-sided ridge-top spreading" that can be classified as the nearly symmetric "ridge-top graben" (or "summit graben"). In the study region it can be exemplified by the deformational structure formed on top of the small Honaihol Ridge (38.653° N, 69.82° E) composed of nearly vertically dipping coarse Lower Cretaceous sandstone striking along the ridge. This site is located close to the

FIGURE 5.16 Upper part of the gigantic Ters rockslide classified as a spread type feature. 3D Google Earth view.

FIGURE 5.17 Stepped slope sagging in the Betdjin Creek valley (Chinese Tien Shan). 3D Google Earth view. *From Strom, A.L., Abdrakhmatov, K.E., 2017. Large-scale rockslide inventories—from Kokomeren River basin to the entire Central Asia region—IPL106-2 and WCoE. Proc. 4WLF, with permission of Springer.*

Rogun Dam site in Tajikistan. The nearby Djuruj Ridge composed of the Upper Cretaceous and Paleogene limestone and marl was also affected by the DSGSD, but this one is definitely the unidirectional feature (Fig. 5.19).

It was assumed (Vinnichenko, 1977; Leonov and Strom, 1979) that such structures could be triggered by strong earthquakes with high accelerations, because much more loading is needed to bang down the ridge summit into hard rock massif than it can be provided just by its weight. This assumption was supported by the results of the finite-element numerical modeling that showed that 10^5–10^6 m^3 volume deficit of such

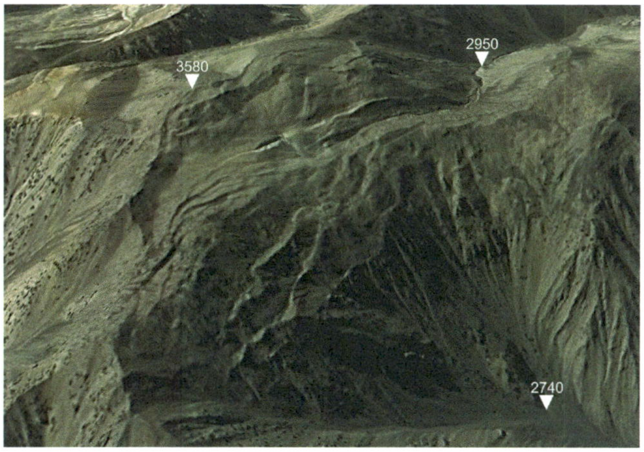

FIGURE 5.18 The DSCSD in the Zeravshan River basin above the Rogiv village in Tajikistan. 3D Google Earth view.

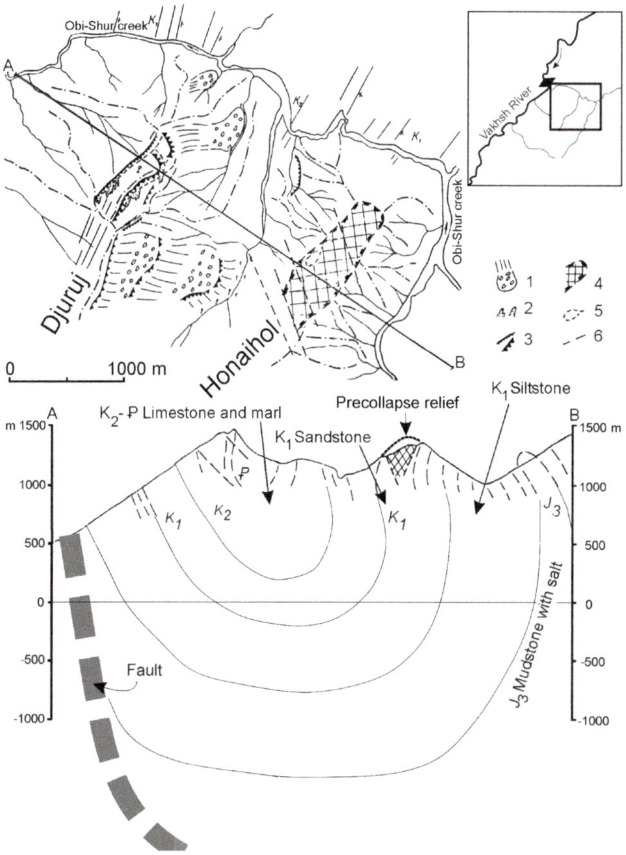

FIGURE 5.19 The schematic geological map and cross-section of the Honaihol and Djuruj ridges close to the Rogun dam site (rhombus on the inset). 1: rockslides; 2: talus accumulated above the upslope facing scarps (3); 4: boundaries of the summit graben; 5: small dried pond dammed by the upslope-facing scarp and filled by sediments; 6: assumed rupture surfaces. *Reprinted from Strom, A.L., Groshev, M.E., 2009. Mysteries of rock massifs destruction. In: M. Abbie, J. S. Bedford (Eds.) Rock Mechanics: New Research. Nova Science Publishers, New York, 211–231, with permission from Nova Science Publishers, Inc.*

summit graben can be balanced by ridge sidewise expansion in the free space of the adjacent valleys caused by strong shaking (Strom and Groshev, 2009).

We should mention, however, that morphologically very similar structures in some parts of the World, in the Alps in particular, are often interpreted as the effect of gradual processes, such as deglaciation and long-term exhumation of mountain ranges (Agliardi et al., 2013), whereas in other regions—as evidence of large earthquakes (see McCalpin, 2009 and Chapter 8). This controversial issue will further be discussed in Chapter 13, Seismically Induced Rockslides: Their Discrimination.

It is well known that some of the DSGSD are closely related to the neotectonic faulting (see, e.g., Gori et al., 2014). Same association in the study region can be found as well and is most pronounced at the northern slope of the Peter the First Range totally affected by large-scale DSGSD. This range stretching for about 200 km from the Alai Valley in the east to the Obikhingou River valley in the west is the northernmost unit of the Pamir mountain system and the closest one to the Southern Tien Shan. The entire northern slope of the Peter the First Range up to 10 km wide is crossed by numerous bow-shaped scarps with northern sides down indicating sagging of the entire slope towards the Surkhob River valley (Fig. 5.20). Similar, though much smaller structure characterizes the northern slope of the 50-km long Surhku Range—the northernmost range of the so called Tajik depression—the trust and fold belt west of the Pamir composed of the intensively folded Cretaceous and Cenozoic sedimentary rocks detached over the salty Jurassic sediments. These slope deformations are so large and complex that it is difficult to delineate separate "individual" landslides. That is why we did not include them in the database.

These slow moving gravitational slope deformations were identified first at the Surkhu Range in the course of the engineering–geological investigations for the Rogun Dam project (Kolichko, 1975; Kolichko and Fil, 1980). These researchers considered the phenomena in question as purely gravitational. However, morphological analysis of these ranges, the Peter the First range at first, allows assumption that the gigantic DSGSD just reflect the continuous destruction of the hanging wall of the active frontal Pamir thrust (Fig. 5.21). It should be considered when analyzing the geodetically recorded recent movements within this zone (see, e.g., Rastvorova and Enman, 1997), which may reflect more "secondary" gravitational rather than "primary" tectonic deformations. The latter, resulting in distinct evidence of surface faulting, are well expressed east of the Peter the First Range, along the northern feet of the Trans Alai Range (Arrowsmith and Strecker, 1999) and

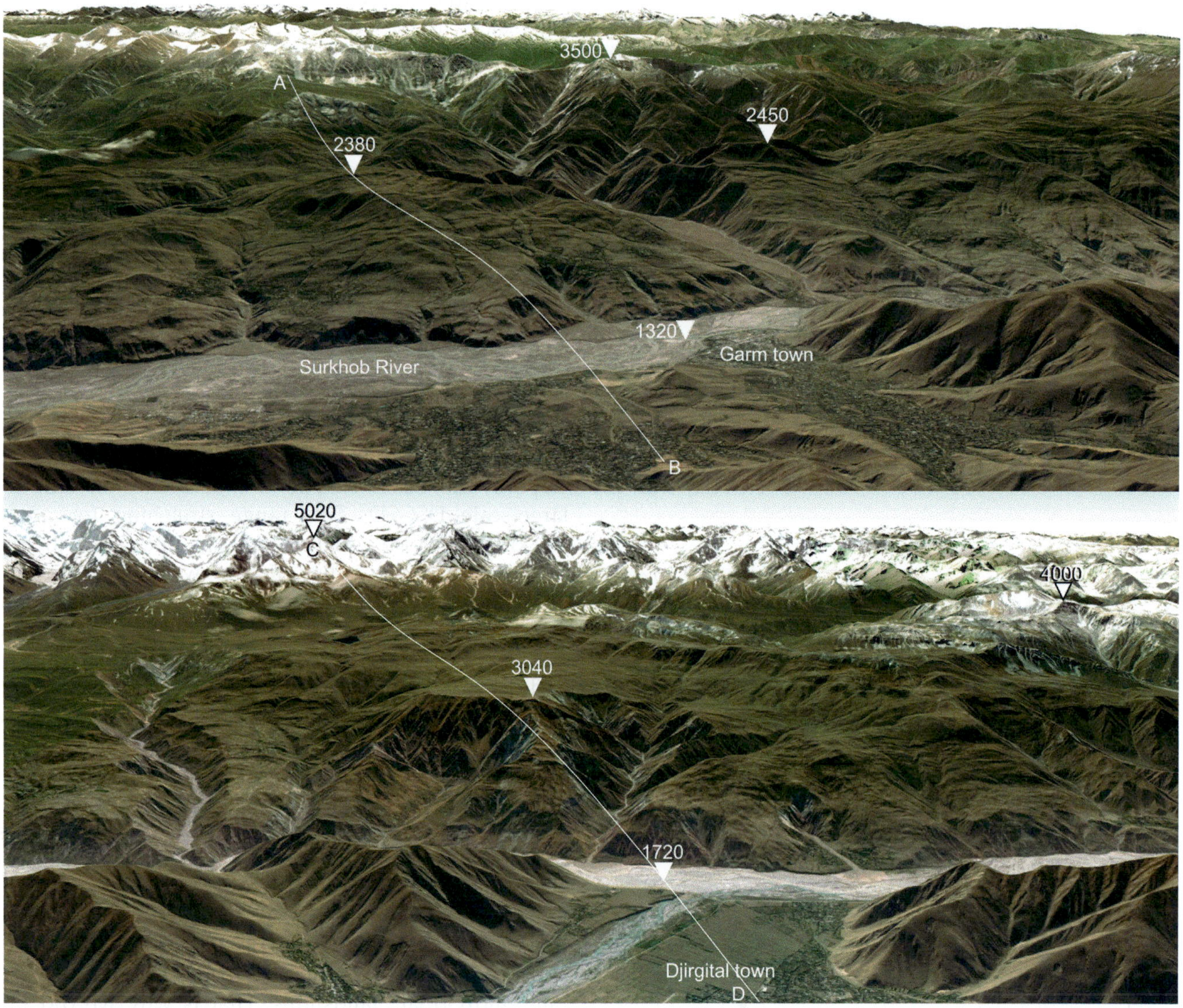

FIGURE 5.20 Southward views of the about 20-km long sections of the Peter the First Range opposite the Garm town, about 70.4° E (above) and the Djirgital town, about 71.2° E (below). Numerous bow-shaped scarps indicate sliding of the entire northern slope of the Range towards the Surkhob River valley. 3D Google Earth views.

west of it, along the feet of the Vakhsh Range. These sections of the Pamir northern front lack large-scale northward overthrusting and, thus, gravitational destruction of the thrust hanging wall.

All types of the DSGSD briefly described above, as well as any landslide, regardless of the combination of their causative/triggering factors, represent downward gravity-driven motion of rock masses from the ridge towards the adjacent free space of the erosional or tectonic depression. Volume that disappeared from the headscarp area is balanced by volume accumulated downslope or removed by erosion. However, at least two sites where large, up to several cubic kilometers rock massifs had really "disappeared" in the ranges' interiors have been identified on the neotectonic anticlines north of the east-west trending Naryn River valley in the Central Tien Shan. The most impressive is the Kyzylkiol cavity on the Santash Ridge (41.8° N, 73.75° E). This ridge is formed by the asymmetric neotectonic anticline 30 km long and about 10 km wide bounded from the south by active reverse fault (Orlov, 1980; Strom, 2000; Strom and Groshev, 2009).

Morphologically the Kyzylkiol cavity is the rhomboid-shape depression 3 × 2 km in plan view and up to 750–800 m deep (mean depth is about 500 m) with steep headscarp-like walls (Fig. 5.22). According to geomorphic expressiveness, it can be dated as Late Pleistocene—Early Holocene. Well-developed relief on top of the Santash Range disrupted by the cavity walls indicates that the collapse had occurred

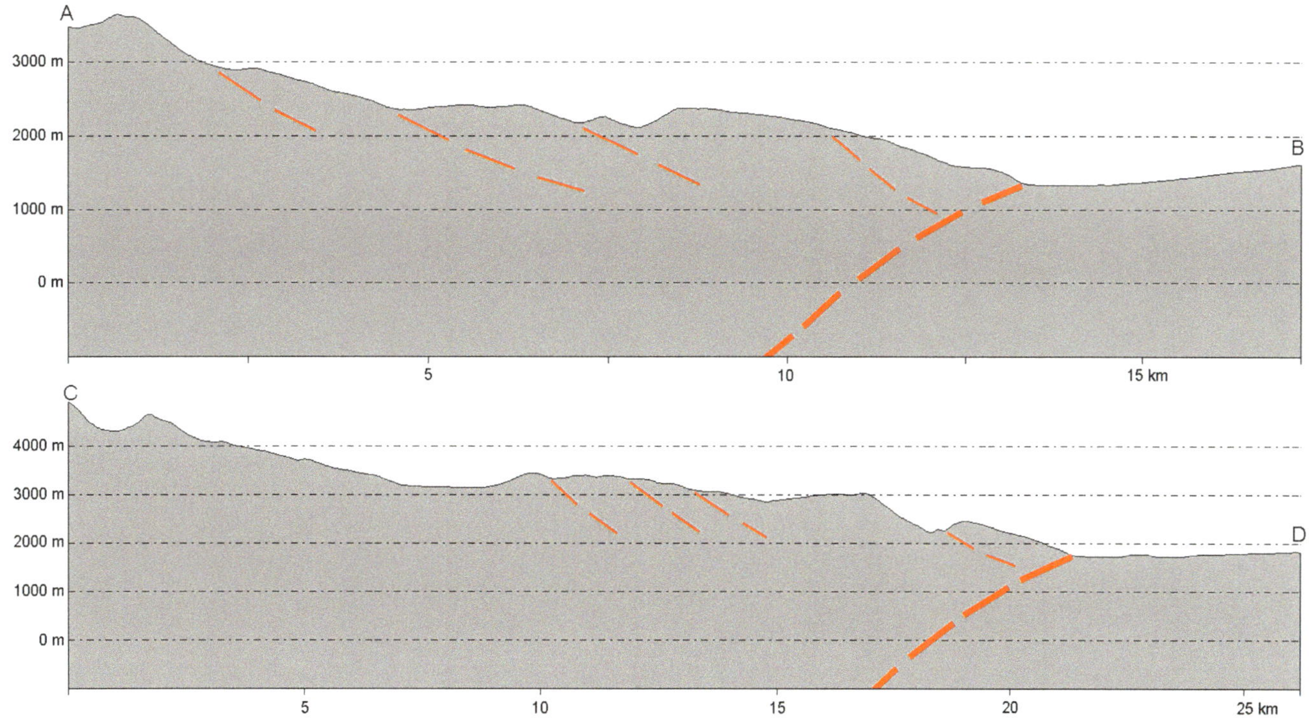

FIGURE 5.21 Schematic cross-sections of the DSGSD at the northern slope of the Peter the First Range affecting hanging wall of the Pamir frontal thrust. Their position is shown on Fig. 5.20.

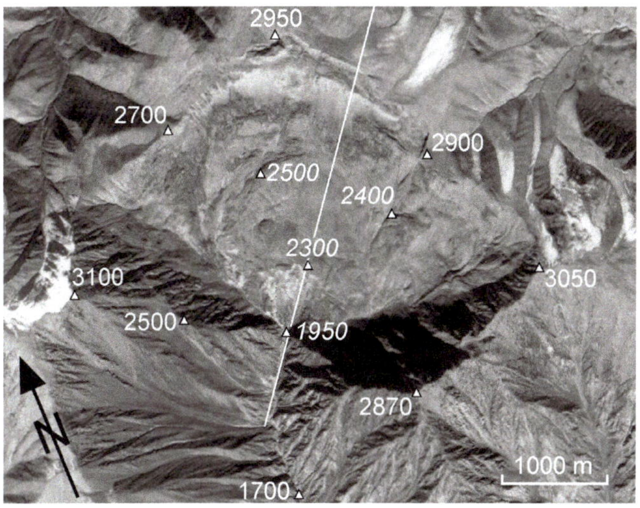

FIGURE 5.22 Aerial photograph of the Kyzylkiol cavity. Figured triangles: point elevations (those inside the cavity is given in italics). Thin white line: geological cross-section shown on Fig. 5.26. *Reprinted with some modification from Strom, A.L., Groshev, M.E., 2009. Mysteries of rock massifs destruction. In: M. Abbie, J.S. Bedford (Eds.) Rock Mechanics: New Research. Nova Science Publishers, New York, 211–231, with permission from Nova Science Publishers, Inc.*

catastrophically when the anticline formation had been completed (Fig. 5.23) and minor geomorphic features on top of it had been formed by long-term erosion (Fig. 5.24).

In contrast, smooth cavity bottom, gently inclined towards the south, is not an erosional feature but represents the top of the subsided block. Its only eroded part is located just where the cavity southern wall had been crossed by the V-shape Kyzylkiol Creek gorge (Fig. 5.25) and only about 50 Mm^3 of rocks have been removed by erosion through this valley. Therefore, the entire volume of the cavity—about 3 km^3—had subsided and "disappeared" in the ridge interior (Strom, 2000; Strom and Groshev, 2009).

The Santash Ridge is composed of igneous and metamorphic Proterozoic (?) and Paleozoic rocks overline by Carboniferous red beds (Fig. 5.26). The latter are developed at the northern slope of the ridge only, while its watershed and southern slope crossed by the Kyzylkiol Creek gorge are composed of gray-greenish igneous and metamorphic rocks. Thus, presence of red-color debris eroded from the cavity bottom just at the lowermost several dozen meters of the creek gorge slopes indicates that originally the cavity was enclosed and that its southern wall was "unlocked" by the stream due to backward erosion quite recently. Only after that, the red-colored debris eroded from the cavity bottom could start "painting" canyon walls. We should note that in Kyrgyz language, the "Kyzylkiol" means "Red lake." May be there was really a lake here in the antiquity?

FIGURE 5.23 View from the north-east on the Santash Range with the distinct lowering of the watershed line corresponding to the Kyzylkiol cavity. *Reprinted with some modification from Strom, A.L., Groshev, M.E., 2009. Mysteries of rock massifs destruction. In: M. Abbie, J.S. Bedford (Eds.) Rock Mechanics: New Research. Nova Science Publishers, New York, 211–231, with permission from Nova Science Publishers, Inc.*

FIGURE 5.24 The Kyzylkiol cavity northern wall. Note disrupted erosion relief at the foreground and rock fall deposits at the foot of the about 400 m high wall. *Reprinted with some modification from Strom, A.L., Groshev, M.E., 2009. Mysteries of rock massifs destruction. In: M. Abbie, J. S. Bedford (Eds.) Rock Mechanics: New Research. Nova Science Publishers, New York, 211–231, with permission from Nova Science Publishers, Inc.*

One more similar feature, though much smaller in size and, likely, much older, was found about 30 km west of the Kyzylkiol one, at the upper reaches of the Djuzumdybulak Creek (41.814° N, 73.385° E), also on top of the neotectonic anticline. Unlike the much larger and much more expressive Kyzylkiol cavity, the Djuzumdy one originated on the ridge composed of Paleozoic granite only, and not on the more or less flat watershed but on the slope of the small and rather deep valley. Further incision taped the boundary between the bedrock granite and more fractured and "disturbed" granite of the block that forms the cavity bottom (Fig. 5.27). Volume of rocks that had "disappeared" in this case is about 120 Mm^3.

These features are really unique. The only one historical analogue comparable in size with the Djuzumdy cavity is the so called "Bitut structure" that originated in 1957 in the central part of the Ihe-Bogdo Range in the epicentral zone of the M8.1 Gobi Altai earthquake (The Gobi Altai earthquake, 1963). Volume of rocks that "disappeared" in the Kyzylkiol cavity is about 30 times larger.

To explain their origin, we must find "free space" in the ranges' interiors large enough to accommodate

FIGURE 5.25 Eroded part of the Kyzylkiol cavity bottom—the source of the Kyzylkiol Creek. Note sharp difference between Carboniferous red beds outcropping in the cavity bottom and gray-color rocks, forming the cavity southern wall. *Reprinted with some modification from Strom, A.L., Groshev, M.E., 2009. Mysteries of rock massifs destruction. In: M. Abbie, J.S. Bedford (Eds.) Rock Mechanics: New Research. Nova Science Publishers, New York, 211–231, with permission from Nova Science Publishers, Inc.*

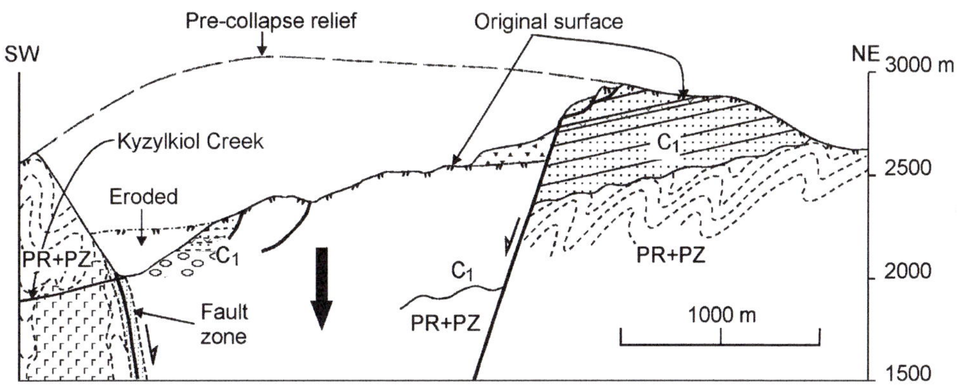

FIGURE 5.26 The Schematic geological cross-section of the Kyzylkiol cavity. C_1: Carboniferous red beds (sandstone and conglomerate with gypsum interbeds; PR + PZ: Proterozoic and Paleozoic metamorphic and igneous rocks. *Reprinted with some modification from Strom, A.L., Groshev, M.E., 2009. Mysteries of rock massifs destruction. In: M. Abbie, J.S. Bedford (Eds.) Rock Mechanics: New Research. Nova Science Publishers, New York, 211–231, with permission from Nova Science Publishers, Inc.*

the amount of rocks that had "disappeared" from above. Because evidence of neither volcanism nor karst processes, nor pull-apart structure formation that could explain somehow formation of the gigantic underground cavern inside neotectonic anticlines were found, it was hypothesized that the required "free space" could appear in the core zones of neotectonic anticlines during their formation in the course of the detachment folding of the tectonically stratified upper crust (Fig. 5.28) (Strom and Groshev, 2009). We must point out that the assumed mechanism of these unique features' formation is completely different from that of any type of landslide.

5.2 ROCKSLIDE DAMS CLASSIFICATION

As we mentioned above, the most destructive and hazardous consequence of bedrock landslides is formation of rockslide dams and their failure. Special classifications of such features were proposed in Costa and Shuster (1988) and in Hermanns et al. (2011b). Types and subtypes of the latter classification system shown on Figs. 5.29 and 5.30 were selected based on the relationship between the shape and size of the rockslide deposits and those of the river valley and consider neither type of slope failure described above nor debris motion mechanism(s). Though it allows predicting evolution of natural blockage to some extent, more sound prediction requires information about grain-size composition and internal structure of these deposits (Dunning et al., 2005, Dunning and Armitage, 2011). The complexity of such prediction can be exemplified by the 2010 Attabad blockage on Hunza River in Pakistan (Delaney and Evans, 2011; Chen et al., 2017) that was expected to fail catastrophically soon after its formation, but still exist more than 6 years after the event, despite the fact that, according to quite large dimensionless blockage index (4.62~4.85) (Ermini and Casagli, 2002, 2003), it should be classified as an unstable feature.

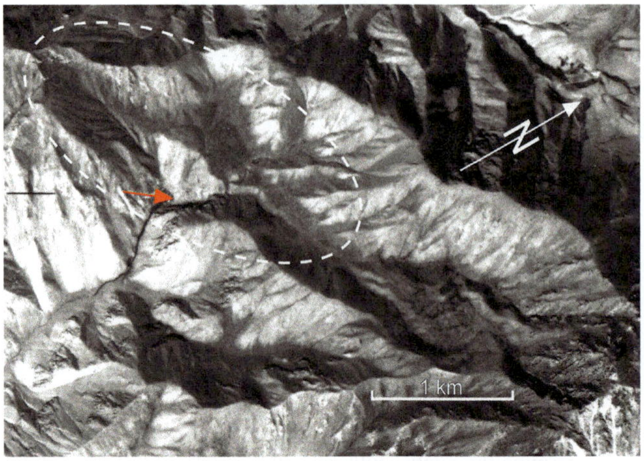

FIGURE 5.27 Aerial photograph of the Djuzumdy cavity on the ridge composed of PZ granite. The site is outlined by dashed white line. Northeast of it, just above the scale bar, there is ~1-km-long rock avalanche. Red arrow indicates contact between the block of the cavity bottom and the bedrock. *Reprinted with some modification from Strom, A.L., Groshev, M.E., 2009. Mysteries of rock massifs destruction. In: M. Abbie, J.S. Bedford (Eds.) Rock Mechanics: New Research. Nova Science Publishers, New York, 211–231, with permission from Nova Science Publishers, Inc.*

While compiling the Central Asian landslides database, whose principles will be discussed in Chapter 6, The Bedrock Landslides Database, we modified some of landslide dams' classification criteria, especially of those based on the across-valley profiles shown in the first row of Fig. 5.30. We focused more on the dam crest profile rather than on the relationship between river damming body and its basement. Such morphological peculiarities, along with some additional factors influencing rockslide dammed lakes evolution will be discussed in Chapter 14, Rockslide Dams—Origin, Evolution, and Failure.

5.3 ROCK AVALANCHES CLASSIFICATION

Likely the most hazardous type of bedrock landslides in rugged terrain typical of the Central Asian mountains is rock avalanches characterized by the

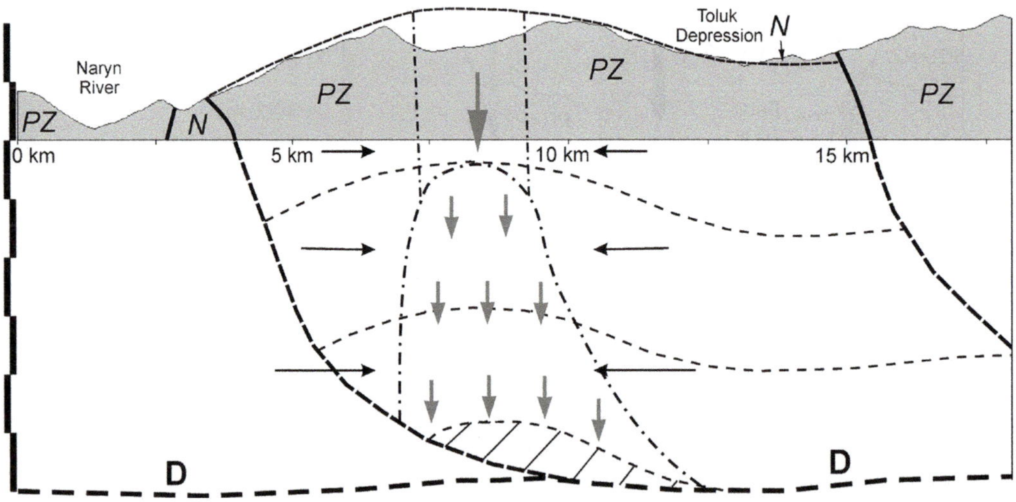

FIGURE 5.28 Schematic model of the Kyzylkiol cavity formation. The upper part of the Santash ridge cross-section represents the reconstructed pre-orogenic planation surface, whose deformation reflects well-pronounced basement folding. Hatched area—the assumed 'initial free space' in the core of an anticline above the detachment (D); black arrows show qualitative distribution of the horizontal tectonic stress; small gray arrows below the dashed-dotted line—gradual failure and formation of the cavern roof arch; large gray arrow—final catastrophic collapse that formed morphologically pronounced cavity. *Reprinted with some modification from Strom, A.L., Groshev, M.E., 2009. Mysteries of rock massifs destruction. In: M. Abbie, J.S. Bedford (Eds.) Rock Mechanics: New Research. Nova Science Publishers, New York, 211–231, with permission from Nova Science Publishers, Inc.*

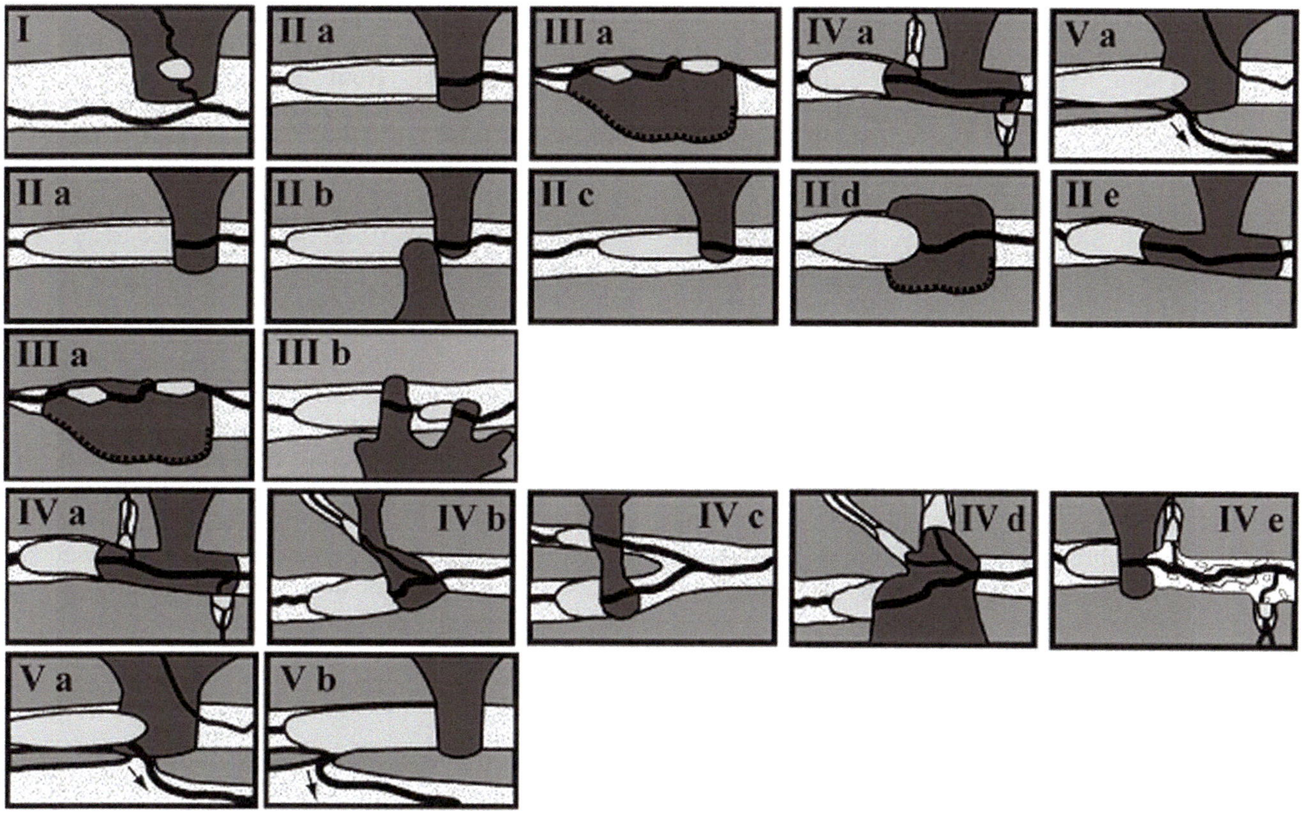

FIGURE 5.29 Types and subtypes of rockslide dams based on plan view debris distribution. (I–Va) Main types; (IIa–IIe, IIIa,b, IVa–IVe, Va,b) subtypes. *Reprinted from Hermanns, R.L., Hewitt, K., Strom, A., Evans, S.G., Dunning, S.A., Scarascia-Mugnozza, G. 2011b. The classification of rockslide dams. In: Evans, S.G., Hermanns, R., Strom, A.L., Scarascia-Mugnozza, G. (Eds.) Natural and artificial rockslide dams. Lecture Notes in Earth Sciences, vol. 133. Springer, Heidelberg, 581–593, with permission of Springer.*

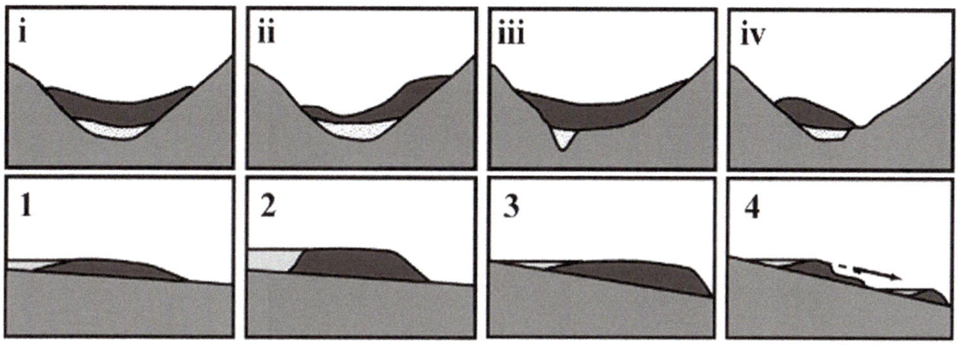

FIGURE 5.30 Types of across-valley (above) and along-valley (below) profiles of rockslide dams. *Reprinted from Hermanns, R.L., Hewitt, K., Strom, A., Evans, S.G., Dunning, S.A., Scarascia-Mugnozza, G. 2011b. The classification of rockslide dams. In: Evans, S.G., Hermanns, R., Strom, A.L., Scarascia-Mugnozza, G. (Eds.) Natural and artificial rockslide dams. Lecture Notes in Earth Sciences, vol. 133. Springer, Heidelberg, 581–593, with permission of Springer.*

"extremely rapid velocities" (see footnote of table 5 in Hungr et al., 2014). They can affect vast areas at the feet of mountain ranges, sometimes endangering settlements kilometers far from the potentially unstable slopes that have been considered erroneously as being safe. The mobility and final shape of the rock avalanche deposits (or, in other words, of the affected area) strongly depend on debris motion mechanism(s), which peculiarities can be derived from the deposits' morphology and internal structure (Strom, 2006).

The rock avalanches' classification based on the deposits' shape, on debris distribution along the runout path at first, was developed based on the analysis of numerous bedrock landslides from the Tien Shan and Pamir mainly (Strom, 1996, 2006, 2010a). Similar to the rockslide dams' classification presented above, it ignores type of the initial slope failure. However, in contrast with rockslide dams' classification, rock avalanches have been classified regardless of the depositional area morphology, but, at the same time, with due regard to such morphological peculiarities of the resultant deposits that allow some conclusions/assumptions about debris motion mechanism for each of the selected rock avalanche type and subtype.

Three main types of rock avalanches are identified (Fig. 5.31): 1: the Primary type (1a: in unconfined or in laterally confined conditions when debris moves down-valley without significant obstacles on its path, and 1b: when debris collides with and/or rises on an opposite slope of the valley); 2: the Jumping type characterized by sliding surface that daylights much above the foot of a steep slope, so that rockslide body really jumps like from the springboard; 3: the Secondary type characterized by bipartite morphology with a compact body at its proximal part, while other portion forms a long-runout "secondary" avalanche. The Secondary rock avalanches can be ascribed to the "Classical" (Fig. 5.31-3) and the "Bottleneck" (Fig. 5.31-4) subtypes. The "Classical" subtype characteristic feature is a concave shape of the compact proximal part frontal side—the so-called "secondary scar" that marks the boundary between the compact and highly mobile avalanche-like portions of debris (Strom 1996, 2006). The "Bottleneck"" subtype is also characterized by bipartite mass distribution, but often without pronounced secondary scar. Here the boundary between the proximal compact and highly mobile distal parts is linked with a sharp valley constriction in plan view (see Fig. 5.31-4). Portions of debris that pass through such

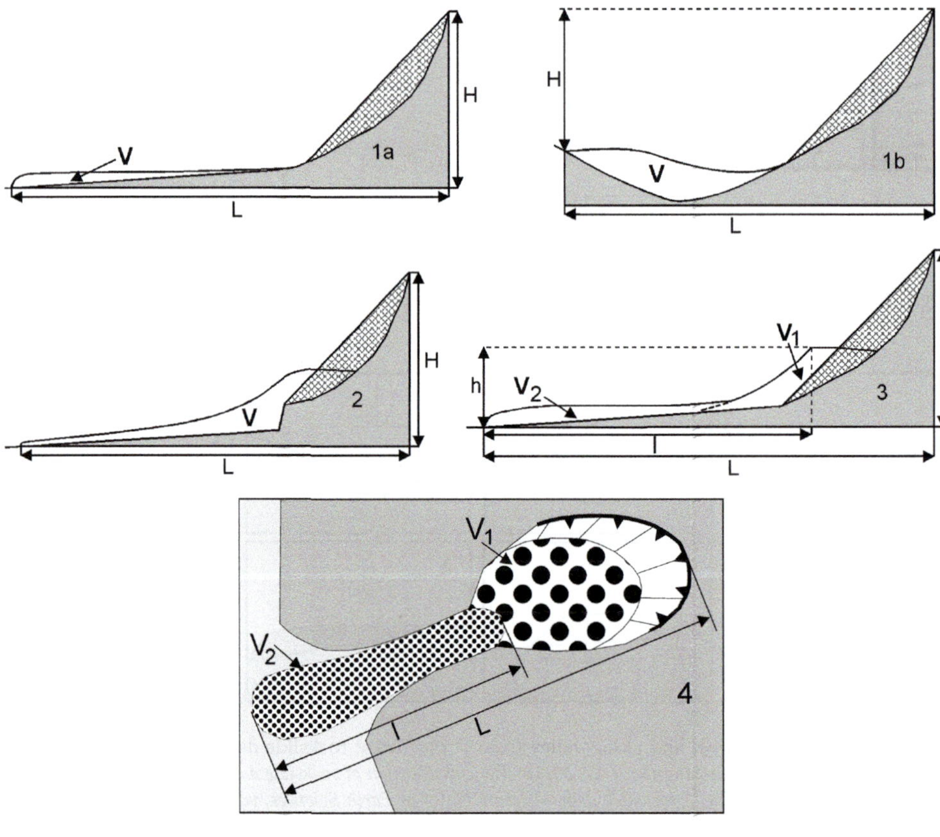

FIGURE 5.31 Rock avalanche types. 1: Primary rock avalanche. The entire rock mass is involved in the avalanche-like motion. 1a: in unconfined or laterally confined conditions, ignoring debris spreading in plan view, which can be significant for unconfined events; 1b: in frontally confined conditions. 2: Jumping rock avalanche with compact proximal and gradually thinning avalanche-like parts. 3 and 4: Secondary rock avalanches. H: height drop (vertical distance between the headscarp crown and the deposits tip); h: that of the secondary rock avalanche; L: runout; l: secondary rock avalanche runout; V: entire volume; V_1: volume of the compact part of the secondary rock avalanche; V_2: that of its avalanche-like part. 3: Secondary rock avalanche of the "classical" subtype characterized by compact proximal accumulation with the prominent secondary scar above the avalanche-like part. 4: Secondary rock avalanche of the "bottleneck" subtype, which bipartition occurs when the rapidly moving rock mass meets sharp valley constriction. *Modified after Strom, A.L., 2010a. Evidence of momentum transfer during large-scale rockslides' motion. In: Williams A.L., Pinches G.M., Chin C.Y., McMorran T.G., Massei C.I. (Eds.) Geologically Active. Proceedings of the 11th IAEG Congress, Auckland, New Zealand, September 5–10, 2010, Tailor & Francis Group, London, 73–86, with permission of Tailor and Francis Group.*

"bottleneck" often demonstrate an abnormally long runout sometimes exceeding that of similar-sized rock avalanches moving over unconfined areas without visual barriers restricting their motion. Besides, Secondary and Jumping rock avalanches with distinct bipartition of their bodies could be classified as unidirectional or deflected. Unidirectional are characterized by the same motion direction of the mobile (avalanche-like) part as that of the initial slope failure; deflected—by the pronounced, up to right angle and even more, change of motion direction of the mobile part after collision with valley bottom or its opposite slope (see Sections 5.3.2 and 5.3.3).

We should point out that the proposed classification can be applied not only to rock avalanches with kilometers-long runout, but to any bedrock landslide that converts from the initial block slide to flow-like motion of debris during the emplacement. Indeed, it is almost impossible to discriminate rockslides from rock avalanches strictly, considering that their runout depends not only on motion mechanism but at a very large extent on the confinement (see, e.g., Shaller, 1991; Strom, 2006, 2010a).

5.3.1 Primary Rock Avalanches

Those catastrophic rockslides and rock avalanches which debris accumulates at the distal part of the runout mainly are ascribed to the Primary type (see Fig. 5.31-1a and 1b). Catastrophic bedrock landslides of this type occur both in unconfined conditions and when debris motion is confined either laterally or/and frontally.

A good example of the unconfined Primary rock avalanche that formed an elongated thin body whose maximal width just slightly exceeds the width of the headscarp base can be found in Western Xinjiang at 42.699° N, 87.955° E (Fig. 5.32, China 028). Besides rock avalanches with such shape of the unconfined bodies, the deposits of some other Primary type events are fan-shape or even pancake-shape indicating not only forward but also sidewise debris spreading that must be considered for hazard assessment. It can be exemplified by rock avalanche about 20 Mm^3 in volume in the Kash River valley (Xinjiang, 43.622° N, 84.533° E, China 124) shown on Fig. 5.33 that caved from the headscarp at 3140 m a.s.l. and spread up- and downstream. Rock avalanche debris extends from the tributary valley No. 1 to tributary valley No. 2, at a distance four times larger than the headscarp outlet width. It covers 3.7 km^2 of the valley bottom with thin apron. Some limited confinement took place when debris reached the foot of the opposite valley bank only, but it could not result in such a sidewise spreading that started immediately when sliding rock mass had left the headscarp.

FIGURE 5.32 A 2.5-km long elongated unconfined Primary rock avalanche about 55 km west of the Toksun town, Xinjiang. 3D Google Earth view.

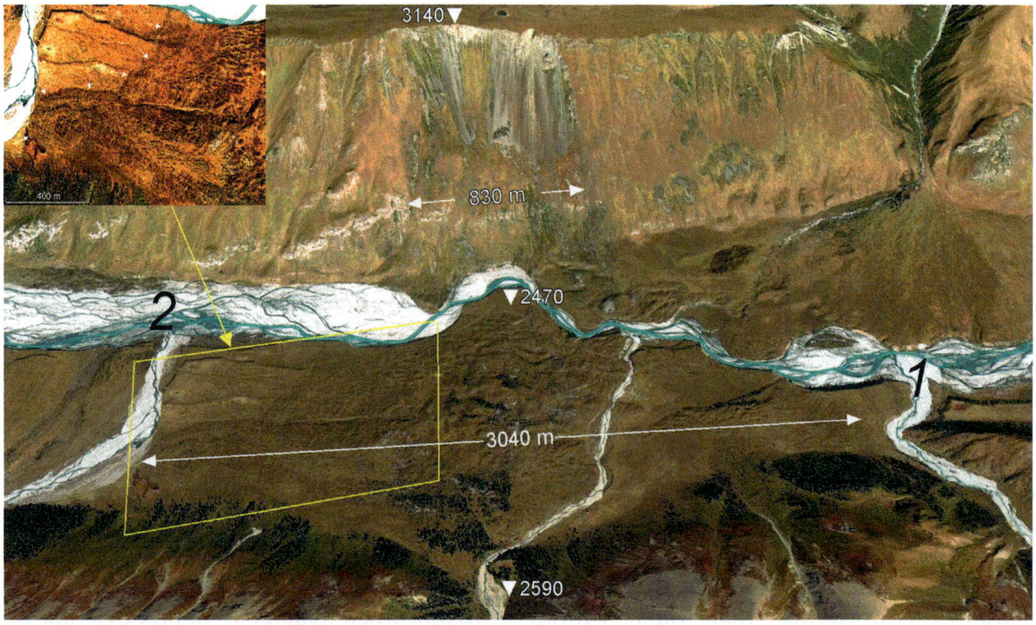

FIGURE 5.33 The Primary rock avalanche in the Kash River valley. Arrows show width of the headscarp outlet and of the deposits. The inset at the top left shows the distinct boundary of rock avalanche debris resting on the river terrace (in false colors). 1 and 2 mark the tributary valleys mentioned in the text. 3D Google Earth view.

We should mention that significant sidewise debris spreading was observed at several fan-shaped rock avalanches that fell onto glaciers during the 1964 Alaska and 2002 Denali fault earthquakes (McSaveney, 1975, 1978; Jibson et al., 2006).

From our point of view, just such case studies allow the most correct analysis of the relationships between rockslide volume and H/L ratio ("fahrborshung"—the term introduced by A. Heim, 1932) or runout (L) (Sheidegger, 1973; Hsü, 1975, 1978; Shaller, 1991; Davies, 1982; Nicoletti and Sorriso-Valvo, 1991; Corominas, 1996; Kilburn and Sørensen, 1998; Legros, 2002; Federico and Cesali, 2015), or volume and affected area (Li, 1983), because the basal friction and processes in moving debris (comminution, internal friction, etc.) are the only mechanisms governing such rock avalanche slowing-down and halt. Rockslides of any other type came to a halt due to combined effect of the same processes plus work against gravity force when rising on the opposite slope and energy release during the impact(s) with steep, sometimes up to vertical obstacles. Each of these processes can be quantified following different equations that complicate revealing of strict quantitative correlations.

Indeed, in the situation shown on Fig. 5.34, both unconfined and frontally confined rock avalanches, have the same fahrborshung (H/L ratio). The kinetic energy of moving debris when it reached Point A should be the same too, but physical regulations governing its motion beyond this point should differ significantly.

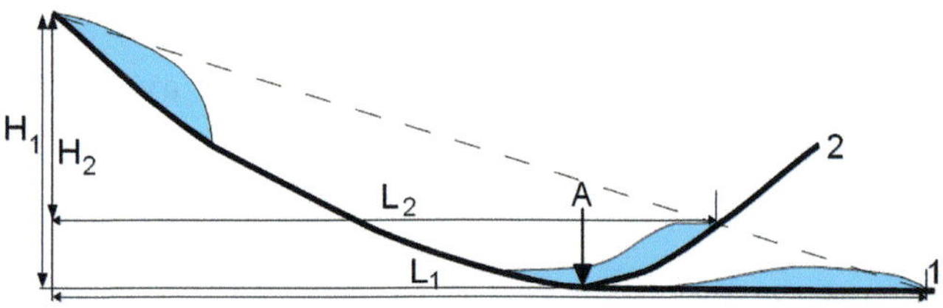

FIGURE 5.34 Scheme of the Primary rock avalanche without (1) and with (2) frontal confinement. H/L ratio is the same for both cases.

Considering morphological variability of real rock avalanches we cannot substantiate if H/L ratio of the rockslide of the same volume that abutted on the opposite slope of the valley will be the same or not without such confinement. Moreover, compiling such correlations, one should consider not only runout distance but also the size of the affected area (Li, 1983). Mixing of case studies with and without confinement in one database used for the statistical analysis can lead to erroneous quantitative estimates. Such relationships based on the Central Asian case studies will be presented and discussed in Chapter 10, Quantitative Relationships of Central Asian Bedrock Landslide Parameters.

Primary style of debris motion is quite well expressed at some long runout rock avalanches that moved along narrow gullies, whose steep slopes provide distinct lateral confinement. One of the best examples is the Seit rock avalanche in the upper reaches of the Kokomeren River valley in Central Kyrgyzstan (42.112° N, 74.15° E, Kyrgyz 240) (Fig. 5.35). This feature was the first one for which the Primary type of rock avalanche was proposed (Strom, 1996). About 17 Mm^3 of granodiorite caved from the steep 400 m high slope in the upper reached of the dry gulley and traveled 3.2 km along it up to the sharp turn of the valley where it halted abruptly. The thickest Archa (Central Asian juniper) trees growing on its surface are up to 320 years old according to tree ring dating (O. Maksimova, personal communication), which gives the upper limit of this slope failure age.

The upper half of the Seit rock avalanche travel path is almost debris-free (Fig. 5.36), so that all collapsed material accumulated at the lower half forming the drop-shaped body up to 100 m thick with wavy surface and interchanging stripes of large angular blocks and more crushed debris of the same igneous rocks (Fig. 5.37). Thickness of debris that traveled through the presently debris-free zone can be reconstructed from the trimlines that remain on the slopes more than

FIGURE 5.35 The 3.2 km long Primary Seit rock avalanche (Kyrgyzstan) with all the way lateral and partial frontal confinement.

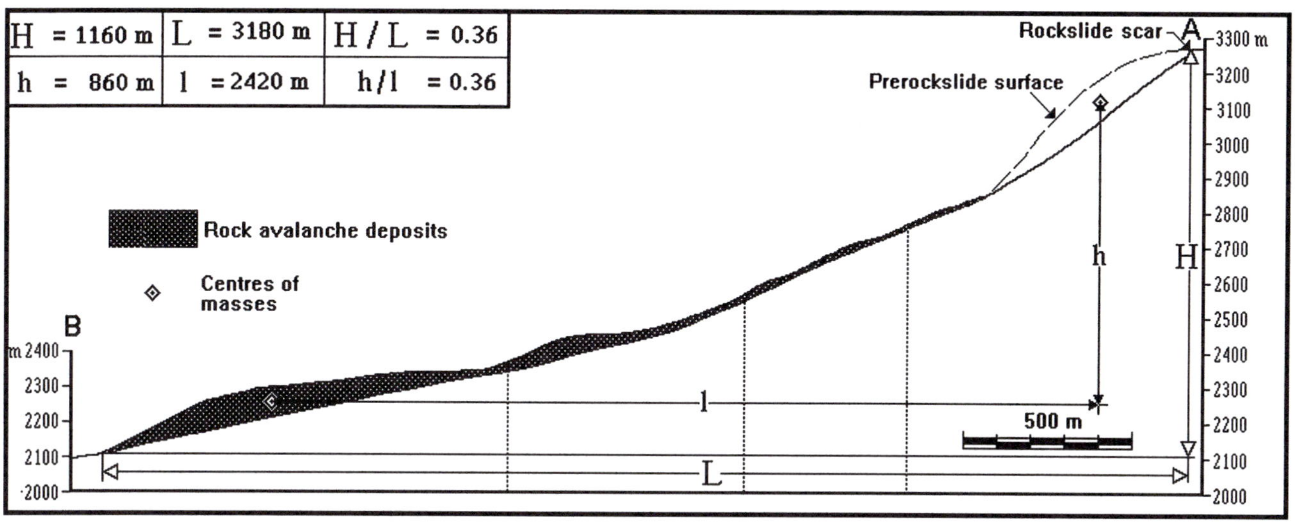

FIGURE 5.36 Longitudinal cross-section along the Seit rock avalanche.

FIGURE 5.37 Wavy surface of the Seit rock avalanche body with stripes of coarser and finer debris.

FIGURE 5.38 3D Google Earth view of the Seit rock avalanche with marked trimlines 100–140 m above the gully bottom in the upper part of the transition zone.

100 m above the resultant deposit's surface (Fig. 5.38). The distinct trimline on the right side of the gully observed above the middle part of the transition zone (Fig. 5.39), likely, reflects superelevation of the rapidly moving debris due to some bending of the path, and, thus differs from the trimlines above its upper part that are identified on both slopes. Such superelevation was described at the 1959 Pandemonium Creek rock avalanche in Canada (Evans et al., 1989) and at the 2002 Karmadon rock-ice avalanche in North Ossetia (Nikitin, et al., 2006). Both grain size composition and morphology of the deposits indicate that the Seit rock avalanche moved as a dry granular flow.

The Primary style of motion can also take place if rockslide moves into relatively narrow river valley with distinct frontal confinement, thus forming a dam. In such cases, rockslide debris accumulates at the opposite side of the valley mainly with a pronounced lowering at the headscarp foot. Such debris distribution is typical of the above-mentioned 1911 Usoi blockage in Pamir, whose crest at its distal part is ~180 m higher than at the lowermost proximal part and, though less pronounced, of the Ak-Kiol rockslide dam described in Section 5.1. Numerous similar case studies will be described hereafter in Chapters 7–9, Dzungaria, Tien Shan, and Pamir, respectively.

We must point out that natural dams formed by the Primary rock avalanched with frontal confinement are more susceptible to catastrophic breach caused by overtopping, because the lowermost part of their crests are at the dams' sections composed of rock avalanche debris, which is much more erodible than the bedrock, through which incision starts if the lowering coincides with blockage's front at the contact with the bedrock of the opposite slope (Korchevskiy et al., 2011). Unfortunately, this variability was not considered completely in the rockslide dams' classification proposed in (Hermanns et al., 2011b) (see Fig. 5.30 i–iv).

One of the most impressive examples of the Primary rock avalanche with frontal confinement could be found in western Xinjiang in the Muzart River basin (42.125° N, 80.765° E, China 002) (Fig. 5.40). About 90 Mm^3 of the metamorphosed rocks had collapsed from the ~1 km high slope, and after blocking the river by almost 250 m high dam rose 170 m on the opposite slope of the valley. The lake dammed by this partially breached blockage was almost filled by sediments, likely due to small (relative to the dams' size) catchment area upstream of the dam and closeness to the large glacier in its upper reaches providing large sediment yield of the river. When rockslide with the similar debris distribution blocks larger rivers, they are usually breached rather fast, which can be derived from a limited amount of lake sediments or, even, of their complete absence as in the case of the Aksu rockslide dam (42.54° N, 74.0° E, Kyrgyz 203) that can be classified as a Primary rock avalanche and will be described in Section 8.3.

FIGURE 5.39 Lower part of the Seit rock avalanche transition zone. The trimline marked by arrows rises for about 100 m in the central part of the panorama blocking the small gully and then decreases gradually towards the distal accumulation.

FIGURE 5.40 Primary rock avalanche with frontal confinement in the upper reaches of the Muzart River basin (Chinese Tien Shan). The headscarp crown is at 3900 m a.s.l. Frontal part of the deposits is 170 m higher than the lowermost part of the dams' crest. 3D Google Earth view.

The world-wide known breached rockslide dam with the predominantly distal accumulation of debris within the blockage body and with distinct run-up on the opposite slope was formed by the 2–3 km^3 Köfels rockslide that had blocked Ötz River valley in the Tyrolean Alps (Erismann, 1979; Sørensen and Bauer, 2003). In the study region the completely breached rockslide dam formed by the Primary rock avalanche can be exemplified by the Mini-Köfels rockslide in the Kokomeren River valley few kilometers upstream of the Djumgal intermontane depression (Central Tien Shan, Kyrgyzstan, 41.906° N, 74.28° E, Kyrgyz 085). The original shape of its body (Fig. 5.41) resembles that of the Köfels, though volume of the deposits is about two orders of magnitude less than that of the Alpine "big brother." The reddish and gray-brownish gneiss debris that descended from the right slope of the valley rests on the left bank of the Kokomeren River composed of gray-colored Paleozoic terrigenous deposits. The rockslide had formed a dam several dozen meters high only, thought its distal part rose about 170 m above the assumed dam crest and filled the prominent lowering on top of the opposite bank. The dam had been overtopped shortly after the event and breached soon at its proximal part by the powerful Kokomeren River. No lacustrine sediments related to this blockage have been identified upstream with confidence.

5.3.2 Jumping Rock Avalanches

Rock avalanches ascribed to the Jumping type are relatively rare, because their formation requires specific combination of the slope topography and rock massif structure due to which the base of the sliding surface daylights on the slope well above its foot. The sketch model of the Jumping rock avalanche formation is shown on Fig. 5.42. The simultaneous detachment of the entire body and its motion along the concave spoon-shape sliding surface results in the springboard-like jump of its frontal part and its collision with the slope base almost at a right angle. Due to large linear dimensions of the sliding block, its proximal part that originates from top of the ridge and, thus, have larger unit potential energy, is still sliding down and accumulating on top of the debris portion that is already at place, compressing it. Moreover, it is not just a static loading but highly dynamic process resulting in the rockslide frontal part compression, and in its extrusion

FIGURE 5.41 The Mini-Köfels breached dam in the Kokomeren River valley. Thin dashed line: reconstructed profile of the dam; bold dashed arrow: direction of debris motion.

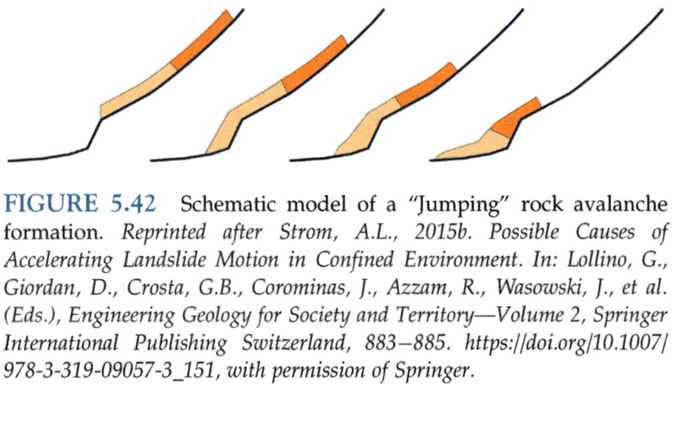

FIGURE 5.42 Schematic model of a "Jumping" rock avalanche formation. *Reprinted after Strom, A.L., 2015b. Possible Causes of Accelerating Landslide Motion in Confined Environment. In: Lollino, G., Giordan, D., Crosta, G.B., Corominas, J., Azzam, R., Wasowski, J., et al. (Eds.), Engineering Geology for Society and Territory—Volume 2, Springer International Publishing Switzerland, 883–885. https://doi.org/10.1007/978-3-319-09057-3_151, with permission of Springer.*

from under gradually accumulating tail part. The extruded material fluidizes and spreads down valley as a dry viscous granular flow. Such rockslides have thick proximal accumulation with convex slopes and gradually thinning avalanche-like part.

We should mention that just such type of motion characterized the famous 1882 Elm sturzstrom (rock avalanche) that really jumped from the quarry bottom and had the debris distribution same as described above (Heim, 1882; Hsü, 1975, 1978).

In the study region the Jumping type can be exemplified by the Northern Karakungey and the Kashkasu rock avalanches (Strom, 1996, 2006, 2013d). The 10 Mm^3 Northern Karakungey rock avalanche (Kyrgyz 321) caved from the nearly 500 m high slope composed of granites and dammed the valley at the western termination of the Djumgal intermontane depression at 41.87° N, 74.24° E. It jumped from the bench that had existed about 200 m above the valley bottom and formed a compact body about 70–80 m thick with convex slopes, rising above its avalanche-like part (Fig. 5.43). While jumping from this bench, the descending rock mass divided into two parts—the larger one converted into a Jumping rock avalanche and the smaller one formed a separate deposit upstream of the main dam (see map on Fig. 5.44).

As can be seen on Fig. 5.44, the rock avalanche followed the creek's channel in a direction almost transverse to the direction of the initial slope failure. Unlike Primary rock avalanches described above and Secondary rock avalanches that will be described hereafter in Section 5.3.3, the mobile portion of the Northern Karakungey debris did not spread over the entire valley bottom but hugged the true left bank of the valley following its channel (see Fig. 5.43). We hypothesize that such a difference reflects relatively high viscosity of this granular flow.

One more example of the Jumping rock avalanche have been found at 41.91° N, 74.17° E in the Kashkasu River valley—right tributary of the Kokomeren River (Kyrgyz 245). Here 5–10 Mm^3 of alternating gneiss-like granites, prevailing within the headscarp area, and dark gray to black mafic igneous rocks collapsed from the ~700 m high steep slope and formed the ~150 m high dam just upstream of the junction of two merging valleys (Fig. 5.45). The small dammed lake had been

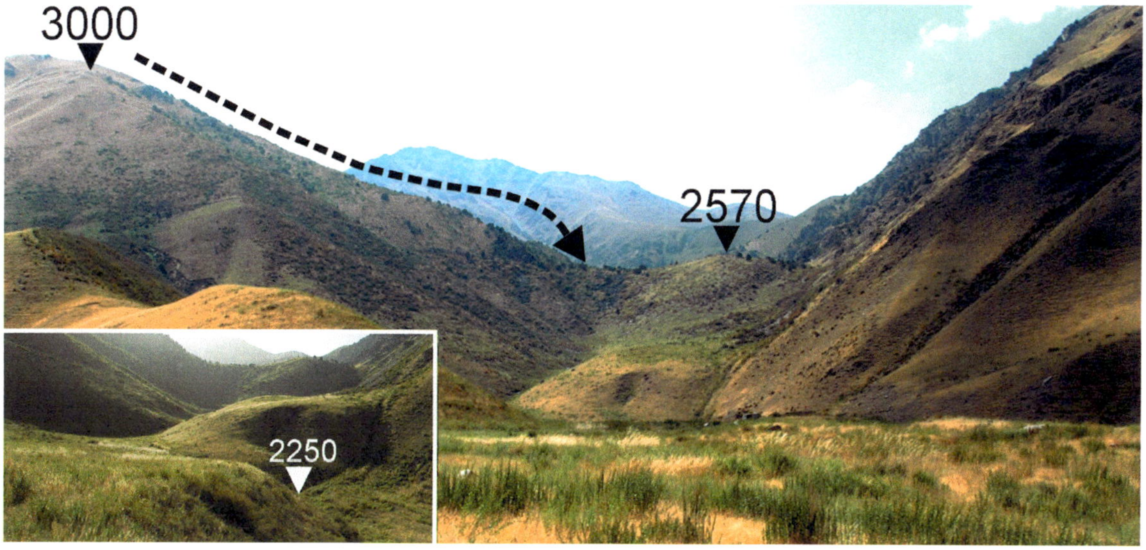

FIGURE 5.43 Overview of the Northern Karakungey rockslide of the Jumping type. It caved from the true right slope of the valley composed of granites (direction of debris motion is shown by the dashed arrow). Filling of the stream by avalanche-like portion of debris is shown in the inset.

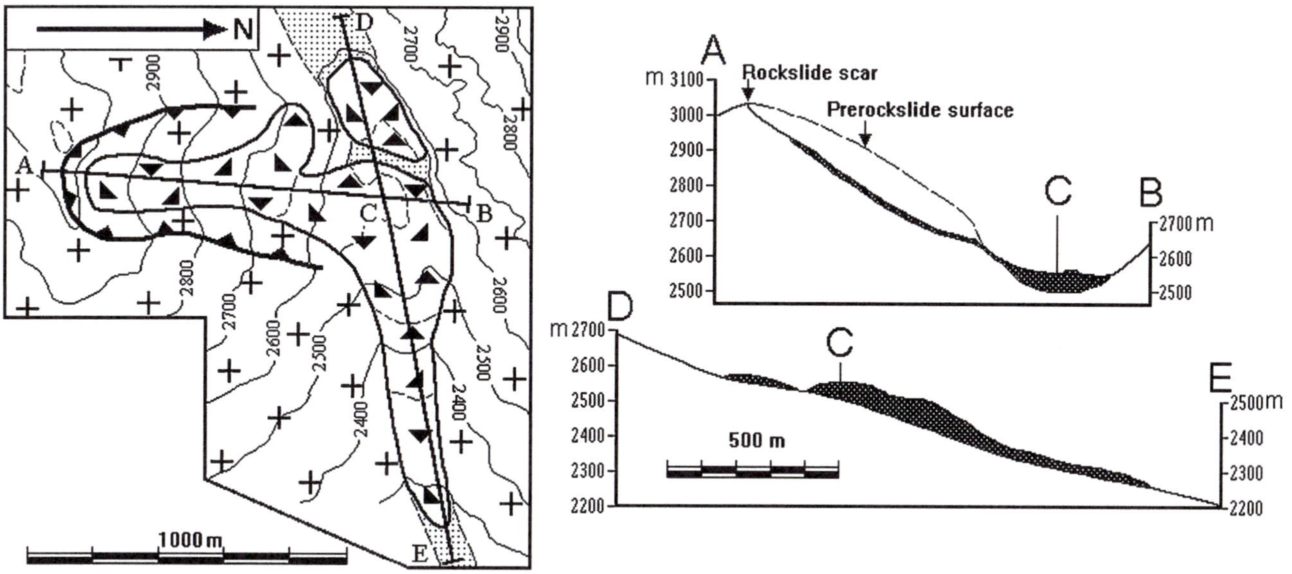

FIGURE 5.44 Schematic map and cross-sections of the Northern Karakungey rockslide. *Modified from Strom, A.L., 2006. Morphology and internal structure of rockslides and rock avalanches: grounds and constraints for their modelling. In: Evans, S.G.; Scarascia Mugnozza, G.; Strom, A.; Hermanns, R.L. (Eds.) Landslides from Massive Rock Slope Failure. NATO Science Series: IV: Earth and Environmental Sciences, Vol. 49, 305–328. Reprinted with permission of Springer.*

filled by sediments completely. The base of the spoon-shape sliding surface daylights on the slope about 200 m above the preslide valley bottom.

The topmost part of this dam has a triangular along-valley profile and rests on much wider basement up to 100 m thick well visible downstream (flat surface at 2000 m a.s.l. on Fig. 5.45). Some poorly defined evidence of such widening could be found upstream too, but this part is masked by the lake fill. This distinctly widened part of rockslide body represents the mobilized basal portion of debris that collided with valley bottom first and was extruded from under compact accumulation. Steep high slopes of both upper and lower parts of this body with inclination at an angle of repose indicate large viscosity of the material.

The additional peculiarity of the Jumping rock avalanches that occurred in natural conditions (contrary to the artificially triggered Elm rock avalanche (Heim,

FIGURE 5.45 Blockage formed by the Jumping Kashkasu rock avalanche. Flat surface at 2000 m a.s.l. is the top of the ~100 m thick avalanche-like part of this natural blockage. It can be seen that the base of the sliding surface daylights on the slope at about 2100 m a.s.l.

1882; Hsü, 1975, 1978)) is that they always caved from the upper parts of the slopes. Most likely, they have been triggered by strong earthquakes, rather than by slope undercutting.

5.3.3 Secondary Rock Avalanches

Rockslide body bipartition in the compact proximal and much more mobile distal (avalanche-like) parts typical of the Jumping rock avalanches is the main characteristic feature of the third, most complex type of highly mobile and, thus, hazardous large-scale bedrock slope failures named in (Strom, 1996, 2006, 2010a)—the Secondary rock avalanches. They originate when rapidly moving rockslide that gained its momentum during initial descend, strikes either the slope base (not at an almost right angle as in the Jumping case, and, at the same time, not converting into inertial motion gradually as in the Primary case) or the opposite slope of the valley. In such cases, part of the sliding mass accumulates at the foot of a headscarp forming a compact body, while portion of debris retaining possibility of further motion converts into long runout secondary rock avalanche. A characteristic feature of such rockslides ascribed to the "Classical" subtype is a concave slope of the compact part above the avalanche-like part—the so-called "secondary scar" (see Fig. 5.31-3). Similar phenomenon takes place if rapidly moving debris passes through a sharp valley constriction, resulting in formation of the secondary rock avalanches of the "Bottleneck" subtype (see Fig. 5.31-4).

It was hypothesized (Strom, 2010a) that secondary rock avalanches of both subtypes originate due to momentum transfer from the rapidly decelerating portion of debris that accumulates at the proximal part of the deposition zone to the portion retaining possibility of further motion. The latter accelerates that cause its extra-mobility.

5.3.3.1 *Secondary Rock Avalanches of the Classical Subtype*

The unidirectional case study of this subtype can be exemplified by the ~7 Mm^3 Chongsu rockslide (Fig. 5.46, Kyrgyz 094) that descended from the 400 m high slope composed of granites at the northern boundary of the Kyzyl-Oi intermontane depression in the upper reaches of a short but deep gulley (41.99° N, 74.02° E).

It demonstrates all elements of such type of rock avalanches—the bench-like accumulation at the headscarp foot, concave secondary scar, and more than 2 km long avalanche-like part (Fig. 5.47), whose volume (~5 Mm^3) exceed that of the compact body (~2 Mm^3). It seems also that the secondary rock avalanche volume exceeds the amount of debris that could have been stored in the secondary scar. This excludes the possibility that the secondary rock avalanche could collapse later, with some time interval after the initial slope failure. From our point of view, the secondary scar just marks the abrupt change in debris motion mode.

Further motion of rock avalanche followed the same regulations as those of the laterally confined Primary

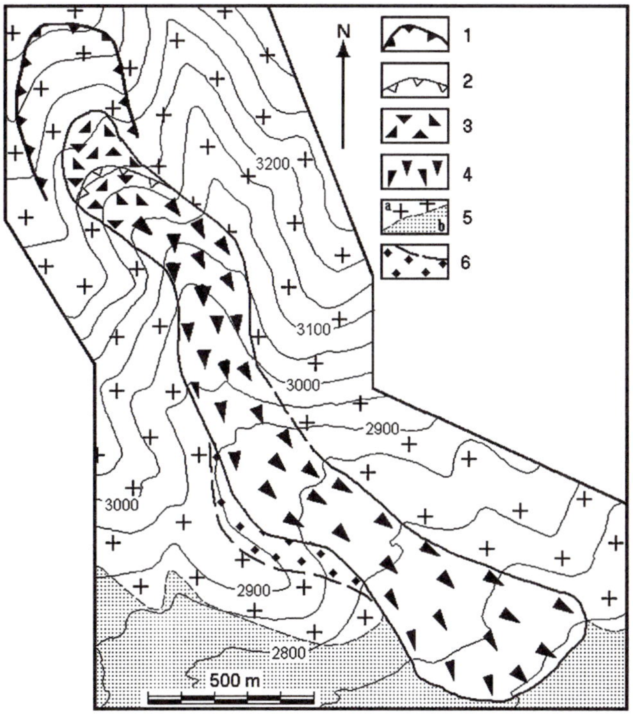

FIGURE 5.46 The schematic map of the Chongsu rock avalanche. 1: main headscarp; 2: secondary scar; 3: compact part deposits; 4: rock avalanche deposits; 5: bedrock granite (a) and Quaternary deposits (b); 6: remnants of debris splash. *Reprinted from Strom, A.L., 2006. Morphology and internal structure of rockslides and rock avalanches: grounds and constraints for their modelling. In: Evans, S.G.; Scarascia Mugnozza, G.; Strom, A.; Hermanns, R.L. (Eds.) Landslides from Massive Rock Slope Failure. NATO Science Series: IV: Earth and Environmental Sciences, Vol. 49, 305–328, with the permission of Springer.*

rock avalanches (see the description of the Seit rock avalanche in Section 5.3.1). Expressive trimlines on the valley slopes that confined the rock avalanche transition zone laterally, downstream of the secondary scar, are visible much above the final rock avalanche surface (see Fig. 5.47). They stretch parallel to it for a long distance—about 700 m on the left and up to 1 km on the right slope. Because they are visible on both sides of the transitional zone, it is unlikely that these trimlines mark just the inertial run-ups, when rockslide debris splashes either on one or another slope like the Karmadon rock-ice avalanche (Nikitin et al., 2006) or the Pandemonium Creek rock avalanche (Evans et al., 1989). We think that they were left by thick body of crushed granite debris moving rapidly along the valley after it was ejected due to momentum transfer.

In other words, we assume that the trimlines indicate constancy of moving debris thickness up to the middle part of the transition zone where they sharply come down to the resultant rock avalanche surface level. This means that at some stage of motion the thickness of moving debris decreased rapidly.

About 1 km downstream of the secondary scar, where the rock avalanche turns following the valley direction, the right-side trimline rises above its general level. Besides, there is a thin, "one boulder thick" apron of angular granite boulders on the right slope of the valley that rises up to 15 m above the bench-like trimline (Fig. 5.48).

Moreover, several boulders rest on the slope outside the main apron (Fig. 5.49), allowing the assumption

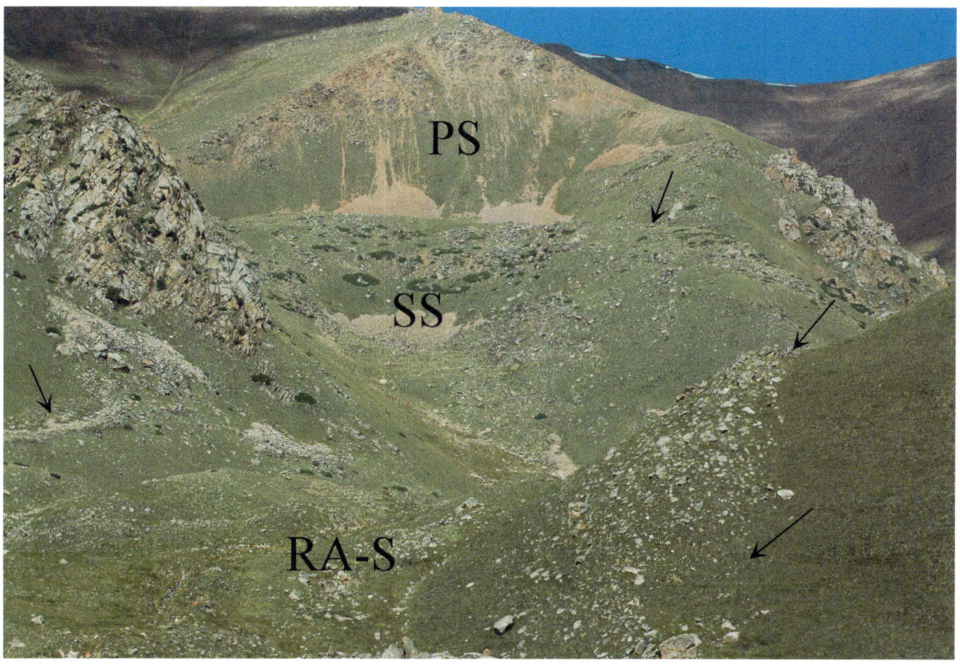

FIGURE 5.47. The upper part of the Chongsu secondary rock avalanche with cirque-like primary (PS) and secondary (SS) headscarps. Arrows mark the trimlines on the valley slopes much above the rock avalanche surface (RA-S).

FIGURE 5.48 The trimline on the right slope of the Chongsu secondary rock avalanche transit zone. Arrows mark a trimline corresponding to the thickness of moving debris. Splash of debris above the trimline is visible at the left part of the photograph.

FIGURE 5.49 Angular boulder (marked by black arrow) resting above the nearly horizontal trimline bench (HB), outside the splash indicating the ballistic trajectory of debris motion.

that boulders forming both thin apron and resting separately were displaced here by ballistic trajectory. Similar ballistic motion was observed at the Avalanche Lake rock avalanche in Canada (Evans, et al. 1994). The position of both the thin continuous apron and separate angular boulders is not random—this splash of debris outside the main mass occurred just at that section where debris motion direction changed sharply.

The ~60 m amplitude of this superelevation corresponds to the ~125 km/hour velocity of the moving debris at this site. We should note that the velocity estimates based on simple equating of the potential and the kinetic energy are quite rough. More comprehensive analysis that takes into consideration the real geometry of the flow-path was proposed in Rahman et al. (2014). However, considering that, as they pointed out, velocity estimated based on mud-marks far from the source area were close to those obtained by the numerical modeling, our velocity estimate seems to be realistic.

Before leaving the valley, rock avalanche forms a flat levee with sharp lateral boundaries composed of large angular blocks (Fig. 5.50) indicating that debris moved at this section ahead not spreading sidewise. The distal part forms a distinct fan-shape body from 3–5 to 7–10 m thick composed of the same large angular boulders with sharp lateral and frontal edges (Fig. 5.51), indicating the abrupt halt of motion.

The Southern Karakungey rock avalanche (41.86° N, 74.25° E, Kyrgyz 323), about 20×10^6 m^3 in volume (KK-S on Fig. 5.52), is a classic example of the ***deflected***

FIGURE 5.50 The Chongsu rock avalanche left boundary downstream of the site shown on Fig. 5.48. Person marked by a circle is about 1.6 m high.

FIGURE 5.51 Lateral (behind sitting geologists) and frontal (right of them) edges of the Chongsu rock avalanche indicating abrupt halt of debris motion.

FIGURE 5.52 The Karakungey rock avalanches at the western frame of the Djumgal intermontane depression. KK-S, the Southern Karakungey rock avalanche; KK-N, the Northern Karakungey rock avalanche; KK-O, assumed headscarp of much older (Pleistocene?) rock avalanche.

FIGURE 5.53 Compact body and secondary scar of the Southern Karakungey rock avalanche. Headscarp is visible in the background, while compact blockage with raised frontal part and the concave downstream slope—the secondary scar—is visible in the foreground.

secondary rock avalanche. It is located close to the Northern Karakungey rock avalanche of the Jumping type (KK-N on Fig. 5.52). Both bedrock landslides originated on slopes of the nearly same height, steepness, and aspect composed of Paleozoic granite and, likely, simultaneously, but shape of their bodies and emplacement mechanisms differ significantly.

It differs also from the above Chongsu case, though, similar to it, have the same features typical of the secondary rock avalanches of the Classical subtype—compact accumulation at the foot of the headscarp, impressive concave secondary scar (Fig. 5.53), and long runout avalanche-like part (Fig. 5.54). However, unlike the Chongsu case, the direction of the

FIGURE 5.54 Mobile part of the Southern Karakungey rock avalanche. Thin arrows mark the distinct left-side trimline. Bold arrow indicates remnant of the assumed older rock avalanche.

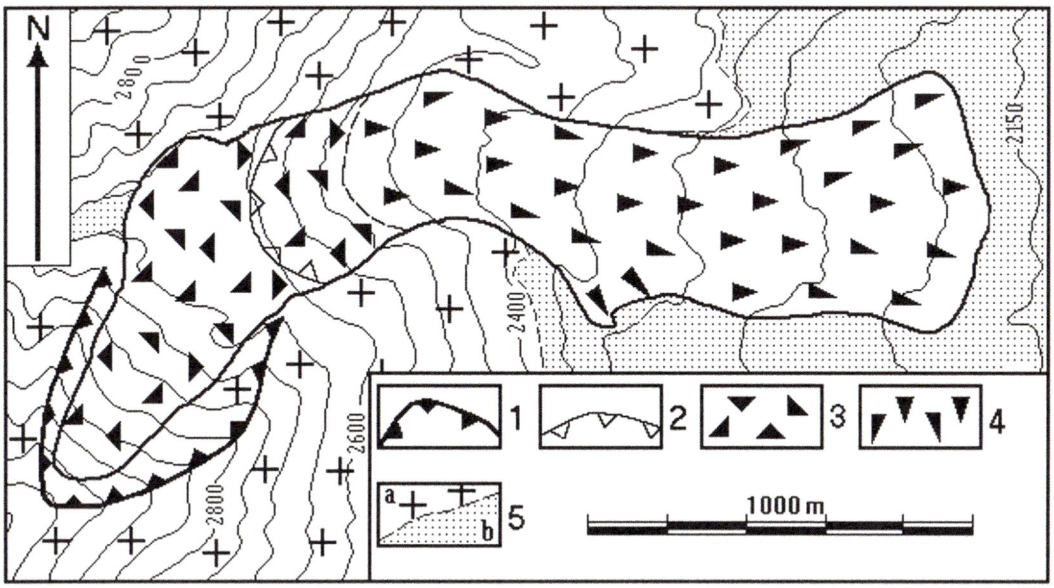

FIGURE 5.55 Schematic map of the Southern Karakungey rockslide. 1: Main headscarp; 2: Secondary scar; 3: Compact part deposits; 4: Deposits of the secondary rock avalanche; 5: Bedrock granite (a) and Quaternary deposits (b). *Reprinted from Strom, A.L., 2006. Morphology and internal structure of rockslides and rock avalanches: grounds and constraints for their modelling. In: Evans, S.G.; Scarascia Mugnozza, G.; Strom, A.; Hermanns, R.L. (Eds.) Landslides from Massive Rock Slope Failure. NATO Science Series: IV: Earth and Environmental Sciences, Vol. 49, 305–328, with the permission of Springer.*

avalanche-like debris motion in this case is not in line with the direction of the initial slope failure (Fig. 5.55). Almost 20 Mm^3 of the collapsing granites hit the opposite slope at a rather sharp angle (~40 degrees) and about half of it stopped forming a dam up to 150 m high. But nearly half of debris continued moving down valley and formed a tongue-like body 2000 m long, 300–530 m wide, and 15–20 m thick as shown on Fig. 5.54. A distinct secondary scar highlights the boundary between the compact and long runout parts of the rockslide deposits (see Fig. 5.53). It is quite unlikely that this secondary failure could occur with a significant delay, thus the secondary scar just marks the transfer from the first stage of emplacement to the second one.

The outer zone of the long runout rock avalanche several dozen meters wide is composed of huge angular boulders, whereas its axial part and boundary slopes is composed of finely crushed debris. Besides, the central part of this "tongue" is lower than the outer bouldery zone (see Fig. 5.54). Numerous large boulders were found also just at the base of the secondary scar. Such grain-size distribution can be explained as follows. Most of catastrophic rockslides are characterized by coarse blocky carapace and intensively crushed internal part (see description of deeply incised rockslide bodies where their internal structure is visible in Chapters 7–9, and the analysis of this phenomenon in Chapter 11: Internal Structure and Grain-Size Composition of Large-Scale Bedrock Landslides). We assume the same for this case study. During secondary rock avalanche formation, its sliding surface cut both blocky carapace and comminuted internal zone. The internal material within which the hypothesized momentum transfer occurred pushed carapace ahead and aside, and it formed bouldery boundary levee, while part of the carapace from the secondary scar crown just subsided and formed the proximal bouldery accumulation. In contrast, the central part of the secondary rock avalanche is formed by the material from the internal crushed zone.

On the true right side of the avalanche, where it leaves the rather narrow deep valley, one can see the remnant of older bouldery deposits overlain by the deposits of the Holocene (?) Southern Karakungey rock avalanche. Though the origin of these deposits is questionable, it could be hypothesized that it is a remnant of a much older rock avalanche whose source zone cannot be recognized with confidence.

On its path, the secondary rock avalanche left trimlines that are most obvious on the true left slope (see Fig. 5.54). Moreover, there is a patch of coarse angular boulders above the distinct trimline resting on the slope composed of much finer material. Because all Quaternary deposits here are granite debris from the adjacent bedrock slopes, it is difficult to make a distinction between angular and semirounded boulders of different age and origin. However, in this case it can be assumed that this patch was formed by boulders ejected from a rapidly moving rock avalanche that had turned here about 30 degrees from E-N-E to E-S-E (see Fig. 5.55). Similar phenomenon was described at the Chongsu rock avalanche above.

An impressive example of the much larger deflected secondary rock avalanche in which about 250 Mm^3 of the entire 400 Mm^3 that failed initially was involved was found in the Muzart River valley at 41.95° N, 80.87° E and can be seen on Fig. 4.30 (China 009).

5.3.3.2 *Secondary Rock Avalanches of the Bottleneck Subtype*

As mentioned above, the Secondary rock avalanches of the "Bottleneck" subtype originate when rapidly moving debris passes through a sharp valley narrowing. Such subtype can be exemplified by the Snake-Head rock avalanche—an impressive and in many respects still mysterious feature that occurred in prehistoric times (most likely in Holocene) in the eastern part of the Suusamyr intermontane depression, in the upper reaches of a small unnamed gully on the right bank of the Karakol River valley (42.35° N, 74.34° E, Kyrgyz 099). Failure of 5–7 Mm^3 of the Devonian sandstone at the southern slope of a small hill (A on Fig. 5.56) converted into a rock avalanche that traveled 2815 m (runout measured from the head-scarp crown towards rock avalanche tip) along a narrow and gently inclined valley. After reaching an alluvial fan, the rock avalanche debris formed an elongated blade bounded by longitudinal levees with transverse levees and furrows between them (B on Fig. 5.56).

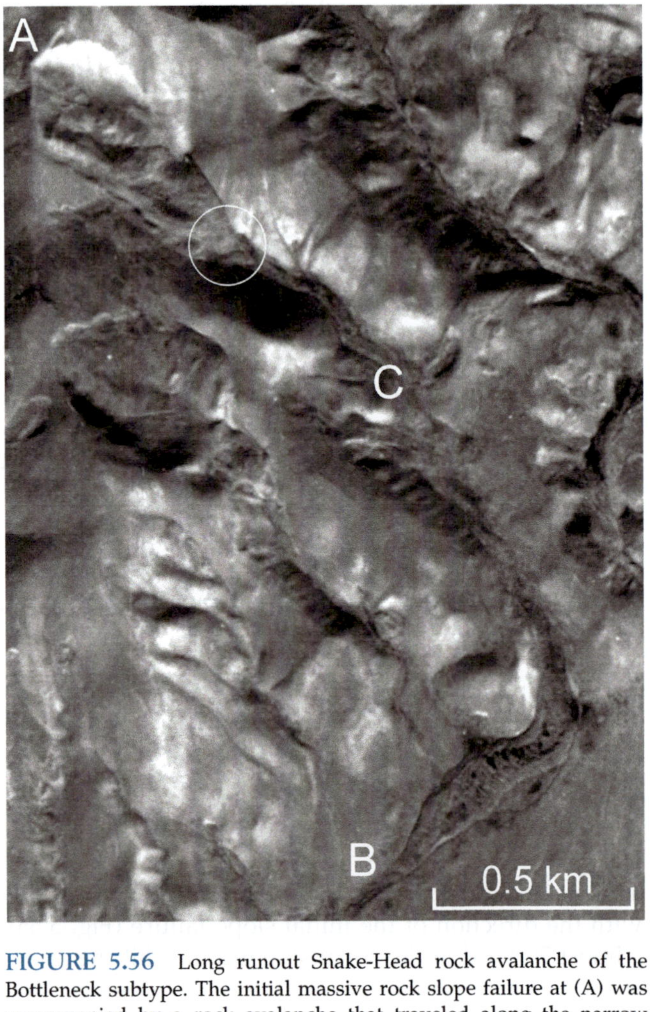

FIGURE 5.56 Long runout Snake-Head rock avalanche of the Bottleneck subtype. The initial massive rock slope failure at (A) was accompanied by a rock avalanche that traveled along the narrow gorge about 3 km up to (B); (C) central part of the transitional zone where debris of landslides from the valley slopes was added to the main rock avalanche. The "bottleneck" constriction is marked by a circle. Fragment of KFA-3000 space image.

FIGURE 5.57 The direction of the Snake-Head rockslide body motion (solid arrow) differing 50 degrees from the headscarp main wall dip direction (dashed arrow).

The initial rockslide at the site in question has an unusual asymmetric headscarp. Destruction affected almost the entire hillside, but the rockslide mass moved more along the slope rather than towards its dip direction (Fig. 5.57). We hypothesize that this could be due to high southward-directed seismic acceleration that had caused this bedrock slope failure.

The proximal part of the rockslide body is several dozen meters thick and about 800 m long; it rests at the scar's base filling the rather wide upper reaches of the unnamed creek (see Figs. 5.56 and 5.57). This part has an expressive wavy surface with several small enclosed depressions. At the western, lower, part of the headscarp, a low pressure ridge can be identified (Fig. 5.58). This outer zone composed of heavily weathered surficial material, likely bulldozed by the descending rockslide mass, has a rather sharp boundary with the inner zone composed of crushed bedrock. Part of the bulldozed material formed the above-mentioned pressure ridge.

The proximal body extends up to the rather sharp narrowing of the valley (Fig. 5.59), which formed an obstacle on the way of the rapidly moving rockslide debris. But motion did not halt completely. Rather small part of debris continued moving and formed a long runout rock avalanche that traveled nearly 2000 m along the creek's valley with a mean angle of about 5 degrees only. Rock avalanche debris fills the creek's bottom forming more or less planar surface. Though there is a minor stream at present, it has not produced any prominent erosional features. The rock avalanche surface is turf-covered, boggy in some parts. Locally, angular debris and boulders can be seen through the grass.

Several hundred meters downstream, at a local widening of the creek's valley, small landslides on valley slopes added some material to this rock avalanche (at C on Fig. 5.56). At the foot of the hilly area, the rock avalanche sharply turned to the right along the lowermost part of the preexisting alluvial fan and finally formed an elongated palm leaf-like body 650 m long, up to 150 m wide and, as it can be estimated, 4–6 m thick. It looks from above like the head of a snake; hence we called the entire feature the Snake-Head rock avalanche. The total volume of this distal part of the rock avalanche can be roughly estimated as $\sim 3 \times 10^5$ m^3. The volume of deposits that filled the narrow bottom of the creek is of the same order. Thus, only $\sim 10\%$ of the collapsed rock mass was involved in the avalanche-like motion.

The distal blade is bounded by levees that rise up to 4–5 m above the surrounding alluvial fan surface and 1–3 m above the internal part of the rock avalanche deposits. The latter represents a system of levees and furrows that strike transversely to the direction of rock

FIGURE 5.58 Wavy surface of the Snake-Head rockslide proximal part and the pressure ridge formed by the initial slope failure.

FIGURE 5.59 Sharp transition from the ~240 m wide compact part of the Snake-Head rockslide (marked by white arrows) to the narrow (≤100 m) avalanche-like part (marked by yellow arrows). View down path.

avalanche motion (Fig. 5.60). Such relief is completely different from that typical of debris flow deposits and, at the same time, is not typical of most of rock avalanche deposits. However, similar morphology, though at a much larger scale, was described at the distal part of the gigantic, 43 km long and up to 15 km^3 in volume, Pleistocene Cerro Caquilluco–Cerrillos Negros Rock Avalanche in Peru (Crosta et al., 2012, 2015). In the outcrops along the small stream that has eroded the right edge of the rock avalanche distal part, one can see that it is composed of intensively shattered debris of the Devonian sandstone (3–5-cm fragments in fine matrix), whereas the surface of the blade is covered by angular boulders of the same lithology 20–30 cm and, rarely, 40–50 cm in size. The rather steep edges of the distal blade (see general view on Fig. 5.60) indicate that it was formed, most likely, by dry granular flow. If it would move as a debris flow with high water content, it would spread much wider forming a fan-shape body.

Considering quite small amount of the material of both the entire failure and, especially of its part involved in the long runout motion, its mobility is extraordinary with H/L ratios of 0.14 (for the entire failure from the topmost part of the headscarp crown till the deposits tip) and 0.11 (for the avalanche-like part), which are much lower than that of most of rock

FIGURE 5.60 Transverse levees and furrows at the central part of the Snake-Head rock avalanche distal blade. General view above and zoomed outlined section below.

avalanches of the comparable volume. According to Hsü (1975), such values are typical of rock avalanches about 500 Mm^3 in volume, almost two orders of magnitude larger than the entire Snake-Head rock avalanche. We must note that further investigations should be carried out to explain the mechanism of its formation from the very beginning of the rock slope failure to the final halt of debris motion.

The extreme mobility of the Bottleneck secondary rock avalanches sometimes exceeding that of the same-size rock avalanches that moved in unconfined conditions can be exemplified well by the long runout Chaartash rock avalanches (Fig. 5.61) that collapsed at 41.22° N, 74.00° E on the northern slope of the Western Akshiyriak Range composed of Paleozoic granite, about 50 km W-S-W of the gigantic Beshkiol landslide (Strom, 1998). Two of them (Fig. 5.61A and C) can be classified as the Secondary Bottleneck rock avalanches, while the third one (B, ibid) moved without any visible confinement, though with some indications of the unidirectional Classical Secondary subtype (much thicker proximal accumulation with several concave secondary scars on its frontal slope). As far as the Chaartash-1 rock avalanche split into several branches divided by small ridges, we compare parameters of the Chaartash-2 and 3 rock avalanches only because both moved over the gently inclined surface of the intermountain depression (Table 5.1).

The longest of them—the Chaartash-2 rock avalanche (Fig. 5.61A, Kyrgyz 264), about 150 Mm^3 in volume—initiated on 600 m high slopes of the deep gully so that collapsing masses collided with each other and with gulley bottom. After that, a large portion of debris (50–70 Mm^3) passed through the gully mouth as through a funnel (marked by a circle on Fig. 5.61) and moved 7 km (!) along the surface with a mean slope of 6–7 degrees only. The total runout (L_{1-3}) of this rock avalanche is about 9 km and its H/L ratio is 0.18. At the end of its way, it crossed the ~50 m deep Kiok-Art River valley and stopped on its left bank.

In contrast, the Chaartash-3 rock avalanche (Fig. 5.61B, Kyrgyz 265), of the same order of size (slightly larger—about 170 Mm^3) that fell from a 800 m high slope composed of the same granite moved along the unconfined depression bottom. However, mobile tongue-like part of its body traveled about 3.3 km only—more than two times less than that of the Chaartash-2 case. The total runout of the Chaartash-3 is about 5 km and H/L ratio is 0.28, though nothing had hampered its motion. This comparison (see Table 5.1) illustrates an exceptional hazard that can be provided by the Secondary rock avalanches of the Bottleneck subtype.

5.3.3.3 *Combined Cases*

The proportion of debris involved in the secondary avalanche-like motion in almost all case studies

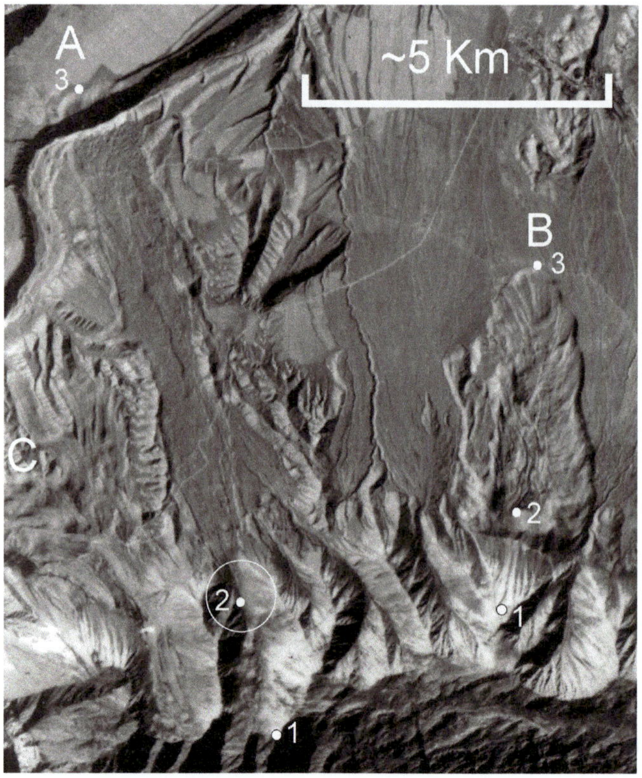

FIGURE 5.61 KFA-1000 space photograph of the long runout Chaartash rock avalanches at the western termination of the Western Akshiyriak Range. (A) The Bottleneck Chaartash-2, (B) the unconfined Chaartash-3, (C) the Bottleneck Chaartash-1 (Kyrgyz 263, partially visible). White dots marked 1, 2, 3—key points. *Modified from Strom, A.L., 2010a. Evidence of momentum transfer during large-scale rockslides' motion. In: Williams A.L., Pinches G.M., Chin C.Y., McMorran T.G., Massei C.I. (Eds.) Geologically Active. Proceedings of the 11th IAEG Congress, Auckland, New Zealand, September 5–10, 2010, Tailor & Francis Group, London, 73–86, with permission of Taylor & Francis Group.*

TABLE 5.1 Comparison of the Chaartash-2 and 3 Rock Avalanches Parameters

Chaartash-2 rock avalanche (A)	Chaartash-3 rock avalanche (B)
$V \sim 120 \times 10^6 m^3$	$V \sim 170 \times 10^6 m^3$
$H_{1-3} = 1600$ m	$H_{1-3} = 1140$ m
$H_{1-2} = 600$ m	$H_{1-2} = 800$ m
$L_{1-3} = 9$ km	$L_{1-3} = 4.96$ km
$L_{2-3} = 7$ km	$L_{2-3} = 3.3$ km
$Slope_{2-3} = 6-7°$	$Slope_{2-3} = 8-9°$
$H/L = 0.18$	$H/L = 0.23$

L and H, distance and elevation difference between key points 1, 2, 3 marked on Fig. 5.61.

described above was large enough—up to one half of its total amount and even more (except the Snake-Head rock avalanche, which is an abnormal feature in many of its aspects). It excludes, in particular, the possibility that secondary failure could occur later as an independent phenomenon. We consider their formation as an element of the continuous multistage process: the initial slope failure → accelerating downslope motion → collision with any obstacle → momentum transfer from the rapidly decelerating and, likely, elastically compressing part of debris to its portion retaining possibility to move further → the latter's ejection and secondary rock avalanche formation.

However, numerous rockslides all over the study region demonstrate a combination of features typical of both the Primary and the Secondary rock avalanches with avalanche-like motion affecting less than, say, 10% of the entire rockslide volume. Such "combined" case studies include both the unidirectional and the deflected Secondary rock avalanches. Due to relatively small volume of the portion of debris involved in more mobile secondary motion (in comparison with the entire collapsed rock mass), and, because all these features are prehistoric, we cannot prove that the secondary failure did not occur independently but reflect the continuous multistage process described above with confidence. Nevertheless, their morphological and structural similarity to the case studies described above allows grounded assumption that they reflect same mechanism—the momentum transfer from the rapidly deceleration debris to its "free" portion (Strom, 2010a). Several case studies of such combined type are presented hereafter.

The small unidirectional secondary rock avalanche with distinct headscarp can be seen at the frontal part of the Lower Ak-Kiol rockslide (Fig. 5.62) described above (Section 5.1) as an example of the planar rockslide. It can be hypothesized that such secondary failure could be provoked by some constriction of the main body travel path in plan view, thus representing the Bottleneck effect. If such type of motion could be anticipated, hazard assessment for the sites located "on the way" of the potential rock avalanches requires their correct runout prediction.

Formation of the deflected rock avalanches that can affect sites located not only far from the source zone but also in the direction totally different from that of the initial rock slope failure attracts special emphasis. Even if the slope failure is anticipated and parameters of its initial motion (volume and direction) are predicted correctly, ignorance of the phenomenon in question could lead to the significant underestimation of hazard as exemplified by the following events.

Small-size (3% of the entire volume only) secondary rock avalanche accompanied the ~ 10 Mm^3 Kiol-say rockslide in the Kokomeren River basin (Central Tien Shan, Kyrgyzstan) (41.795° N, 73.955° E, Kyrgyz 078) that occurred on the ~ 500 m high valley slope composed of granites and formed a 150 m high dam with a

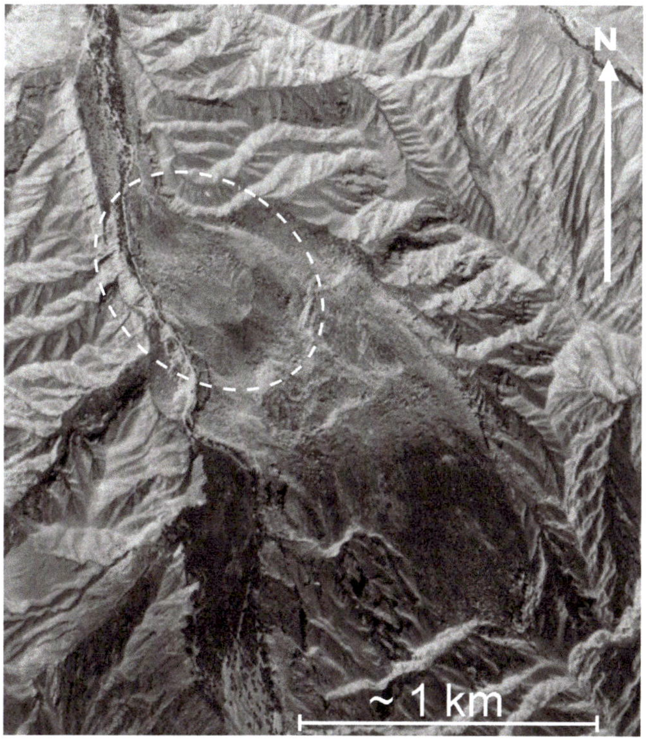

FIGURE 5.62 Small unidirectional secondary rock avalanche (marked by white oval) at the frontal part of the Lower Ak-Kiol rockslide. Fragment of the aerial photograph.

FIGURE 5.63 The Kiol-say rockslide ~10 Mm^3 in volume accompanied by a small-size deflected secondary rock avalanche that traveled ~0.75 km up to 1810 m a.s.l. Inset shows part of the valley bottom with large angular boulders. The largest one, about 10 m in size, is marked by white circle in the inset. 3D Google Earth view.

small lake behind it with evidence of insignificant, noneroding seasonal overflow (Fig. 5.63). The dam looks as an intact drop-shaped body with the only exception—the down-valley slope with distinct concave headscarp-like shape. The valley bottom downstream is covered by large angular boulders up to 5–10 m in size (about 1000 tons) at a distance of about 0.75 km. Because there is no evidence of a catastrophic dam breach, their displacement by a debris flow should be excluded. The only explanation is that these blocks are the remnants of the secondary rock avalanche partially reworked by erosion that removed fines during flood periods. It can be assumed that secondary rock avalanche formation resulted from the rockslide collision with the opposite valley wall, its compression, and the elastic springback of a small portion of debris towards the free space.

Much larger secondary deflected rock avalanche accompanied the 200–250 Mm^3 Karasu Lake rockslide dam (Kyrgyz 238) resting just over the Talas-Ferghana active fault at 41.57° N, 73.22° E (Belousov et al., 1994; Tibaldi et al., 2015), whose activity in the form of large earthquake likely had triggered this prehistoric rock slope failure (Korjenkov, 2006). Rockslide collided with an opposite slope of the Eastern Karasu River valley and formed a dam that has never been overtopped. The water balance in the Karasu Lake is maintained by filtration through its proximal part mainly, according to the position of the springs (Fig. 5.64). Minimal age of this rockslide composed of carbonate rocks was estimated as ~1500 year based on the lichenometry (Belousov et al., 1994).

We must note that the assumption of the two-stage formation of this natural dam due to recurrent slope failure proposed in Belousov et al. (1994) seems to be erroneous according to the morphology of the dams' surface. What could be treated as the second stage of its evolution is the formation of the 1.4 km long flow-like failure on the dams' downstream face. We cannot exclude soundly the possibility that flow-like sliding occurred later on, being supported by filtration. There is, however, some evidence in favor of its formation as a Secondary rock avalanche. In particular, the secondary scar presents close to the dams' frontal edge where its width is maximal, which is not favorable for slope wakening due to filtration. This scar is linked with the distinct right-side trimline (marked by T on Fig. 5.64) indicating their genetic relationship.

If the flow-like failure would occur subsequently, due to seeping water exfiltration and debris saturation, the liquefied water-rich material would sag and flow down, as it occur in the physical models presented in Okeke and Wang (2016a, b). However, the trimline

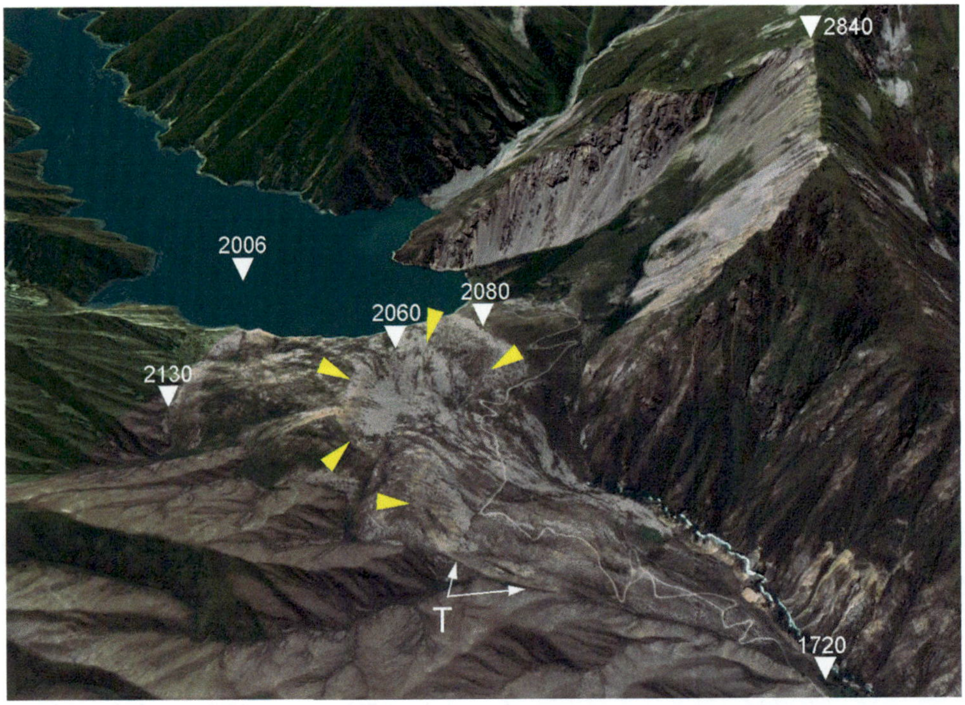

FIGURE 5.64 The Karasu Lake rockslide accompanied by the deflected secondary rock avalanche that involved ~10% of the entire rockslide and moved 1.4 km down-valley at a right angle to the direction of the initial slope failure. Its secondary scar is marked by yellow triangles. T, the distinct right-side trimline. 3D Google Earth view.

~30 m above the resultant surface shows that rockslide debris moved initially along the right slope of the valley as a thick unit. Moreover, this part was overlaid by the more flow-like material originated from the central part of the dams' crest, indicating that the secondary deformation of the frontal part of the dam somehow preceded failure of the central part of the downstream dams' face. From our point of view, it could result from the same physical process as described above, in the Kiol-say case—collision, momentum transfer, and ejection towards the free space. Much larger size of the secondary failure in the Karasu Lake case could be due to, first, overall scale of the feature and, thus, energy of the event, and, second, much larger valley width just downstream of the initial accumulation favorable for the Secondary rock avalanche motion.

One of the most impressive case studies combining the Primary and the Secondary styles of the rockslide/rock avalanche motion is the giant Big Dragon Lake rockslide in the Khalyktau Range in Xinjiang (42.6° N, 82.4° E, China 015) about 2 km^3 in volume that caved from the more than 1 km high slope of the small river valley composed of the Silurian metasediments and formed ~500 m high dam (Fig. 5.65). The initial failure exemplifies the classical Primary rock avalanche in the frontally confined conditions—the collapsed mass crossed the valley and rose about 370 m on the opposite slope above the main dams crest level. The impact with a steep obstacle confining further motion resulted not only in the run-up but also in the ejection of the 4.3-km long secondary rock avalanche from the secondary headscarp marked by 1 on Fig. 5.65. The overall final direction of the Secondary rock avalanche motion differs from the original slope failure direction for more than 90 degrees. Motion conditions of the avalanche-like part changed when it entered valley narrowing, and we can assume that the "bottleneck" effect took place (see Section 5.3.3.2) being marked by the secondary scar 2.

The uppermost 150–200 m of the rockslide body front (up to 3370 m a.s.l.) can be classified as a brandung—the enormous tongue of debris that rises above the average rockslide dam crest level for dozens and even hundreds of meters (Heim, 1932). Similar feature was described, for instance, at numerous cases in the Karakoram (Hewitt, 2002). From the mechanical point of view, the brandung origin could be very similar to the Secondary rock avalanche formation, i.e., ejection of some part of rockslide debris into free space due to momentum transfer caused by the collision with an obstacle. While in the secondary rock avalanche cases material ejects laterally, in case of the brandung formation debris moves upward, being compressed by the

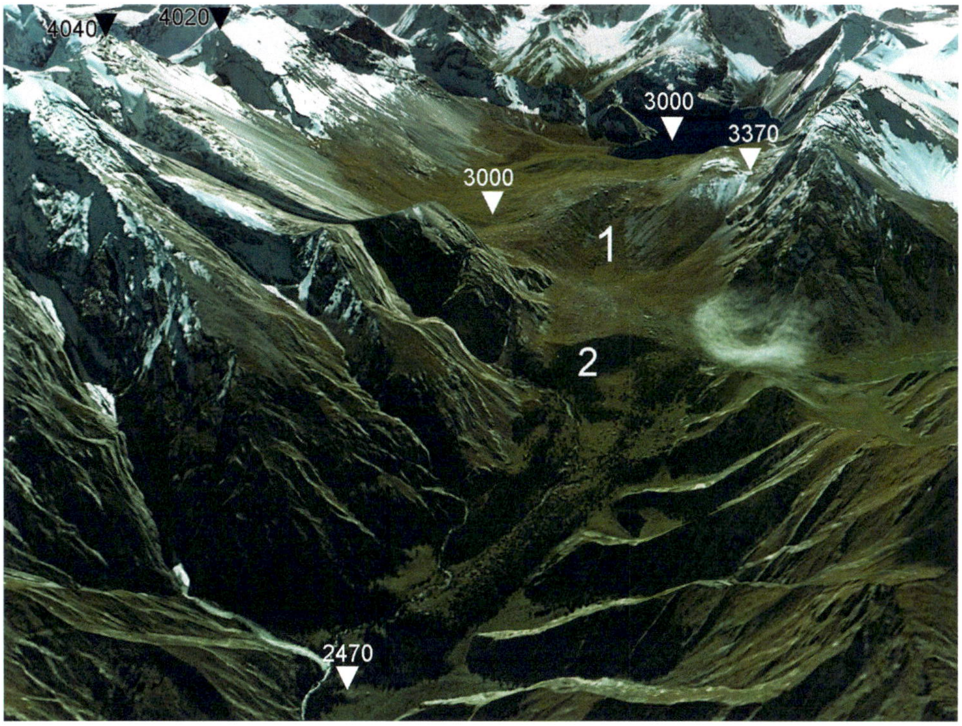

FIGURE 5.65 The Primary Big Dragon Lake rockslide about 2 km^3 in volume with 370 m run-up of its frontal part accompanied by the 4.3 km long deflected Secondary rock avalanche that originated from the secondary scar 1 and moved up to 2470 m a.s.l. with an intermediate secondary scar 2. 3D Google Earth view.

following material and extruded upslope from the rockslide front. We can also find here some similarity with the processes that take place in the Jumping rock avalanche formation, but, instead of the vertical compression and lateral extrusion (in the Jumping case), here the compression acts horizontally and the extruded material rises upslope.

One more interesting feature—the presence of an additional concave secondary scar (2 on Fig. 5.65) within the secondary rock avalanche where it had entered the valley constriction, thus demonstrating the Bottleneck effect (see Section 5.3.3.2). Just above this secondary scar, a small portion of debris entered into the left tributary valley and moved upstream along it for about 600 m. The same phenomenon—recurrent interruptions of rock avalanche motion marked by several secondary scars—was observed at some other sites and will be mentioned during regional rockslides description.

5.3.4 Some Implications of the Proposed Rock Avalanche Classification

Rock avalanche types and subtypes identified within the classification system presented above are characterized by the special mechanisms providing their high mobility—gradual undisturbed transition from the initial accelerating motion to the inertial motion for the Primary type of rock avalanches; fluidization of the frontal part of rockslide body due to dynamic loading provided by the following portions of debris and the formers' extrusion into free space for the Jumping type (see Fig. 5.42); the continuous multistage process including initial acceleration, collision with any obstacle, momentum transfer, and ejection of some portion of debris retaining possibility of further motion for Secondary type (Strom, 2006, 2010a). Formation of a certain type of rock avalanche depends neither on the collapsing rock mass volume nor on its lithology or structure of the source zone, not on the type of the initial slope failure. Analysis of the numerous rockslides and rock avalanches both from the study region and worldwide shows that the main factor predetermining the resulting emplacement mechanism is the shape of the headscarp base where the initial downslope accelerating motion of the collapsing rock mass turns into its motion due to gained momentum. It allows some qualitative prediction of the motion style in cases if any potential large-scale rock slope failure is anticipated that could provide important input data for landslide hazard and risk assessment.

One more important implication of the proposed classification system is that it is better to perform the quantitative correlation between various rock avalanche parameters (area, volume, runout, H/L ratio, etc.) for different types and subtypes separately, not mixing them. It will help to avoid hardly quantified factors such as the morphology of the transition-deposition zone and to obtain correlations that reflect physics of the phenomena in question better. It will be discussed in more details in Chapter 10 based on the analysis of the rockslide database presented in the Appendix.

CHAPTER

6

The Bedrock Landslides Database

OUTLINE

Compilation of the landslide inventories is an important and indispensable first step for revealing spatial distribution of the existing slope failures, susceptibility, landslide-related hazard and risk assessment and for understanding the qualitative and quantitative pattern of these phenomena (Soeters and Van Westen, 1996; Ercanoglu and Gokceoglu, 2004; Van Westen et al., 2008; Guzzetti et al., 2012). The Central Asian bedrock landslide inventory was compiled based on the analysis of the remote sensing data—free Google Earth, SAS Planet, and ArcGIS Earth space imagery that covers the entire Central Asia, though with variable quality of images, as well as of the old Soviet KFA-1000 stereoscopic and KFA-3000 images with a resolution of 5–8 and ~3 m correspondingly, and aerial photographs available for some parts of the study region. Soviet 1:1000,000–1:25,000 topographic maps were used as well. Besides we utilized literature data, and results of our personal field studies carried out in Kyrgyzstan, Tajikistan, and Kazakhstan since middle of 70s of the 20th century. Some preliminary results of this work were presented in Havenith et al. (2015a) and were used for the landslide susceptibility analysis of the western half of the Tien Shan (Havenith et al., 2015b).

Considering dimensions of the region in question (almost 900,000 km^2) and its rugged terrain favorable for landslide formation, it was impossible to identify and map all landslides recognizable on high-resolution space images available at present because number of such features would reach tens of thousands (see, e.g., (Havenith et al., 2015a; Golovko et al., 2017)). That is why we determined the threshold value for the features that are included in the database—1 million m^3 in volume approximately. Considering significant scatter of volume estimates, which will be discussed in more details hereafter, some smaller features were not excluded from the database if, after being identified and outlined on space images, they were found to be smaller. Besides, small rockslides that form clusters accompanying larger slope failures were included in the database and described, considering their importance for revealing possible triggering factors of large-scale rockslides formation (see Chapter 13: Seismically Induced Rockslides: Their Discrimination). Some smaller features were included in the statistical analysis of volume–mobility relationships (see Chapter 10: Quantitative Relationships of Central Asian Bedrock Landslide Parameters), though they are unrepresentative in general, as far as many of them could be just missed.

Why did not we establish the threshold landslide area that can be measured much more precisely and is widely used for the statistical analysis of landside size distribution (Hovius et al., 1997; Stark and Hovius, 2001; Malamud, et al., 2004a, b; Hovius and Stark, 2006; Guzzetti et al., 2008, 2012)? First, one of the aims of our study was to collect data suitable for the statistical analysis of large bedrock landslides' parameters for better understanding of their motion mechanism(s) and mobility. It is obvious that it is just volume, not area that determines energy and momentum released during the landslide motion. Second, the significant diversity of rockslide and rock avalanche deposits morphology resulting

DOI: https://doi.org/10.1016/B978-0-12-803204-6.00006-5

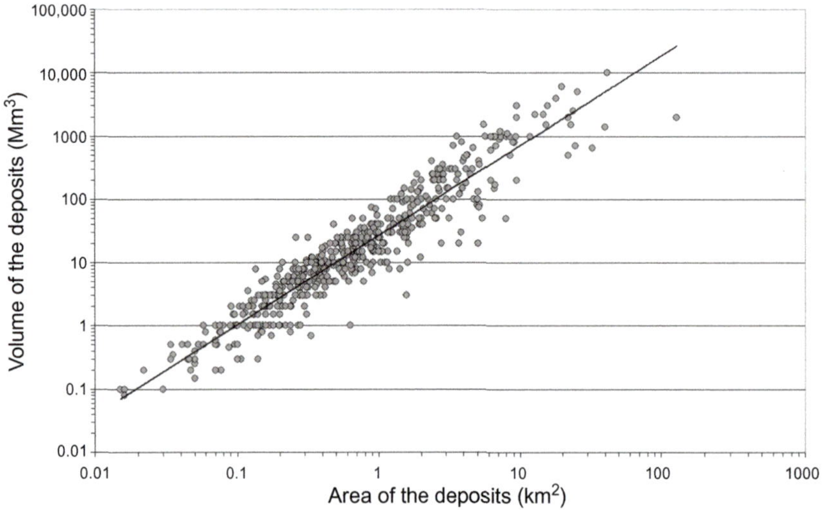

FIGURE 6.1 Landslide volume versus deposits area for 551 case studies characterized by power law distribution $V = 26.875 \times A^{1.419}$ with correlation coefficient $R^2 = 0.8933$. Notice the significant scatter of about one order of magnitude for features with the same measured area practically for the entire independent parameter (area) range.

in one order of magnitude scatter of volume of landslides measured individually (see Section 6.3) with the same depositional area (Fig. 6.1) does not allow to use the deposits' volume estimates based on area/volume statistical correlations proposed in (Simonett, 1967; Hovius et al., 1997; Korup, 2005a,b; ten Brink et al., 2006; Guzzetti et al., 2008, 2009; Hadian-Amri et al., 2014) for such analysis. We should also note that most of such regressions were derived for landslides with areas ranging from 10^2-10^6 m^2, while areas of landslides included in our database ranges from $\sim10^4$ to $\sim10^8$ m^2 (see Fig. 6.1).

Most of the area/volume correlations were derived for shallow slides in soil, large part of whose bodies did not leave their headscarps. In contrast, many of large-scale bedrock landslides included in our database had left their headscarps completely and formed variable-shaped bodies depending on their motion type (see Section 5.3 in Chapter 5: Bedrock Landslide Types and Classification Systems) and occurred in quite variable topographic conditions that resulted in such a large and statistically stable (see Fig. 6.1) scatter. That is why we came to decision to assess landslide volumes on the individual basis whose principles are described hereafter in Section 6.3.

FIGURE 6.2 The successive space images of the landslides at the southern slope of the Western Akshiyriak Range (41.263° N, 74.417° E, Kyrgyz 119). Above: from the Bing Maps (SAS Planet) with a small landslide-dammed lake (date is unknown); below: same area on Google Earth image of April 28, 2001, with this dam breached and the toe of a long runout landslide downstream partially eroded.

Because vast majority of the rockslides in the study region are prehistoric, we did not intend to analyze their modern evolution. Quite rarely, some significant change of their state has been identified on the successive space images (Fig. 6.2).

Feature contours were digitized using the Google Earth, and SAS Planet tools, saved as the *.KMZ (Google Earth) and the *.ECW (SAS Planet) files and overlaid on the 3″ (~90 m) SRTM DEM available for the entire area and on the georeferenced raster topographic and geological maps using Global Mapper software. Quantitative rockslide parameters such as area, drop height, and runout were estimated based on this DEM. Elevations of some identifiable points provided by Google Earth were used as well. We should notice that sometimes they differ up to several tens of meters from those of the 3″ SRTM DEM, when contours delineated on Google Earth are superimposed on this DEM. However, considering large size of most of the study features, such difference does not produce critical errors of landslides quantitative parameters. Free Aster DEM with nominally better (30 m) spatial resolution is also available, but its quality was found to be not sufficient because it contains too many artifacts. All data collected were stored as the Global Mapper workspace (*.GMW file). KFA space images as well as stereoscopic aerial photographs were used to check, specify, and clarify morphology of some features.

We must point out that both elevations provided by the SRTM DEM and those indicated automatically by the Google Earth software sometimes differ from those indicated on Soviet topographic maps available, in particular, from the Mapstor.com. These differences, compared for the landslide dammed lakes, whose levels are easy to fix, are quite variable and range from few meters up to several tens of meters. We did not aim to clarify, which estimates are correct. Because all local elevation differences measured and included in the database are just relative, real elevation of any particular point is not so important. That is why we used mainly the elevation values provided by the Google Earth and 3″ SRTM DEM.

The rockslide database was compiled as an Excel spreadsheet (Appendix). It includes parameters characterizing each landslide and grouped in several blocks: general information, headscarp geology and parameters, overall parameters of the rockslide deposits, rock avalanche parameters, landslide dam and dammed lake parameters. Principles of all these parameters' measurement or selection are described in the following sections. These parameters were used when compiling regional description of rockslides presented in Part 2 and for statistical analysis (see Chapter 10).

6.1 GENERAL INFORMATION

Rockslide ID	Date (if known) and name (if given)	Comments	Latitude	Longitude
1	2	3	4	5

1. **Rockslide ID** contains country name and three-figure number per country (e.g., Afghan 001, China 004, Kazakh 034, Kyrgyz 235, Tajik 030, and Uzbek 025). Numbers were given one-by-one during rockslide identification and mapping on Google Earth or SAS Planet images. However, because we checked these images repeatedly, at least for some parts of the study region, sometimes new features were found much later nearby those sites identified previously. That is why landslides located not far from each other could have ID numbers differing significantly. Sometimes, we provided IDs with additional alphabetic notations for the adjacent and, likely, interrelated features—China 006, China 006a, and China 006b. Few landslides occurred in the river valleys along which state borders pass so that their headscarps are in one state, while the main part of the deposits had crossed the river and rest in other state. For them the headscarp country acronym is used for the ID.

2. **Feature date and name**, if available. Many large rockslides in the former USSR were named by Soviet researchers according to the local site names indicated on the topographic maps (see Kurdiukov, 1964; Hodjaev, 1985; Fedorenko, 1988; Vinnichenko, 1989, 2006). Most of river-damming landslides are named by the lake name (the Karasuu, the Sarychelek, the Issyk, etc.) though in the rare cases they have special names—the Sarez Lake, for example, was formed by the1911 Usoi landslide. In this case the name of the rockslide inherits the name of the small village buried with all its inhabitants, whereas the lake inherits the name of the largest submerged village. The common names for such features in the Kyrgyz and Kazakh parts of the region such as "Kiol" or "Kiol-say" just mean "Lake" or "Valley with a lake." Some landslides first described in the literature by authors were named by ourselves (Abdrakhmatov and Strom, 2006; Strom, 1994, 1996, 1998, 2010b, 2012). For the Chinese part it was much more difficult for us because detailed topographic and geological maps are in Chinese. That is why most of the Chinese case studies have the ID only, though several sites in Xinjiang were named according to the local names on the 1:1000,000–1:500,000 Russian maps available for this part of the study region as well. Rivers', lakes', and mountain ranges' names used in this book were taken from the same maps and, sometimes, from the Google Earth, if location names are available.

Date (year) is provided for the few historical events. Due to significant scatter of the prehistoric features age estimates, which are quite rare for in Central Asia, we do not include them in the database. They will be provided for the particular case studies in Part 2 during regional description if available.

3. **Comments**. Here some most peculiar characteristics of the features in question (the extreme size, abnormal runout, special internal structure or shape of the deposits, etc.) are indicated that can be used for further

selection of case studies to be described in detail or for more logical and reasonable sampling for the statistical analysis.

4 and 5. **Coordinates (Latitude and Longitude)** provided in dd.dddd format up to four digits (~10 m accuracy). These coordinates correspond to the approximate center of the entire feature and should be used for search purpose only. In the text, we often limit coordinates up to 1–2 digits if it is accurate enough to identify the site on the external imagery resources or maps by the reader interested in some particular case study described in our book.

6.2 HEADSCARP GEOLOGY AND PARAMETERS

Rock age	Rock type	Rock sequence from top to bottom	Headscarp height (km)	Mean slope angle (degrees)	Headscarp width at its base (km)
6	7	8	9	10	11

6. **Age of rocks** involved in slope failure is provided according to the geological maps available. For the most of the former Soviet part of the study region, we have 1:200,000 State Geological maps with the detailed legends. We also used 1:500,000 geological maps of Kyrgyzstan (1978), Tajikistan (1984), and Southern Kazakhstan (1979) and the 1:1500,000 geological map of Uzbekistan (1978). Geological map of Tajikistan covers also most of the studied part of Afghanistan. For Chinese part, we used the 1:2000,000 geological map of Xinjiang (1985) and the 1:250,000 maps available on the website of the China Geological survey (http://www.ngac.org.cn/) (both in Chinese). The standard indications of the geological periods are used.

7 and 8. **Rock type(s)** of the source zone. Generally such information could be derived from the 1:200,000 State geological maps of the USSR available for large parts of the study region in Kazakhstan, Kyrgyzstan, Tajikistan, and Uzbekistan (it is included in the maps' legends). For the Chinese territory, it is more complicated for us because the 1:250,000 maps available on the website of the China Geological survey (http://geodata.ngac.cn/) are in Chinese, so we could not find data on the lithology, except for the igneous rocks marked by color are metamorphosed rocks marked by symbols. Same is for the 1:1000,000 geological map of Xinjiang and for Afghanistan. Granite is marked by "G," terrigenous sedimentary rocks are marked by "T," carbonate sedimentary rocks are marked by "C," metamorphic rocks are marked by "M," and igneous rocks other than granites are marked by "I."

For those rockslides that originated on the slopes composed of the uniform lithology or of various lithologies whose interrelation is unknown or too complex, they are listed in Column 7. For the landslides studied by the authors or described in the literature where rock types alternate from the headscarp top to its bottom, they are listed in such order in Column 8 using same symbols.

9. **The headscarp height** is measured as the elevation difference between headscarp crown and the base of the collapsed slope. While the crown elevation can be determined easily, the precise localization of the slope base is not so unequivocal. Sometimes, when all collapsed material had escaped from the source zone, its base can be defined clearly, like, for example, at the Seit rock avalanche of the Primary type in the Central Tien Shan (see Fig. 5.14) or the Imom rockslide in Pamir (see Fig. 4.14). But in many cases it is hidden under the proximal

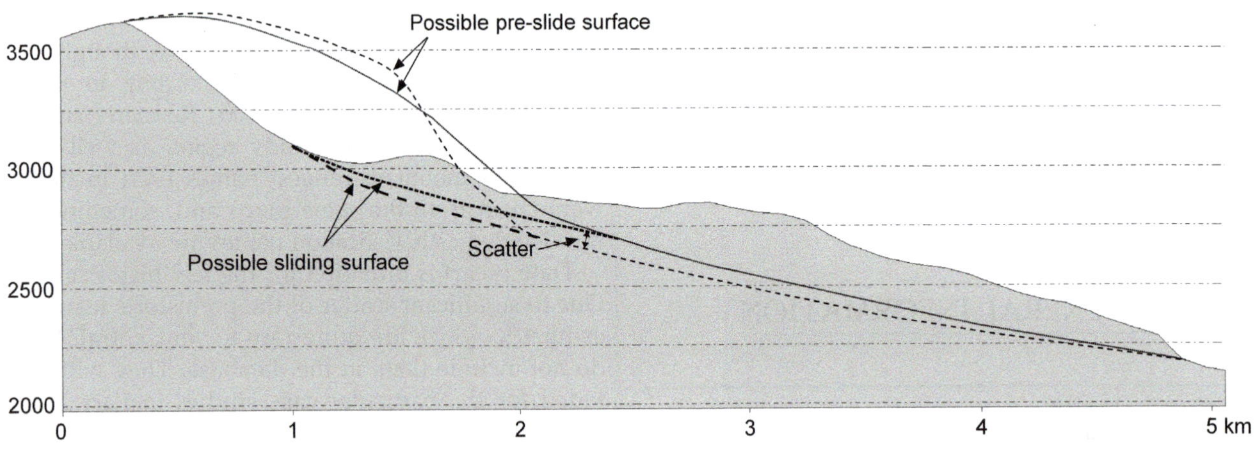

FIGURE 6.3 Cross-section of the Aidyn-Ula rockslide (China 131) shown on Fig. 4.4. Two possible variants of the preslide surface correspond to two possible position of the sliding surface. In fact real scatter of both the original day-light surface and of the sliding surface buried under the proximal part of the deposits could be even higher.

debris accumulation, most often of the unknown thickness (see the Aidyn-Ula rockslide in the Chinese Tien Shan on Fig. 4.4 and its cross-section on Fig. 6.3, the headscarp of the Usoi rockslide on Fig. 5.7 or the Karasu rockslide in the Kyrgyz Tien Shan on Fig. 5.64). We assume that in such cases the headscarp height can be determined with accuracy not better than ±10% to 15%.

10. **Mean slope angle** is derived from the 3″ SRTM DEM profiles drawn by the Global Mapper software along the straight lines connecting the headscarp crowns and feet. In most of the cases, it reflects the mean angle of sliding surface. Accuracy of these estimates ranges from ±1 degree for the simplest cases of the translational landslides that originated on slopes composed of well-bedded sedimentary rocks with constant dip angle (see Fig. 5.6) and up to ±5 degrees for those cases where the sliding surface is characterized by the significant curvature (see, e.g., Fig. 4.22) or its actual position is uncertain (see Fig. 6.3). We must point out that because profiles were plotted based on the 3″ SRTM DEM with ~90 m spacing, they provide rather rough impression of the real slope topography. However, considering size of the study features, whose linear dimensions often exceed 1 km, such accuracy seems to be more or less sufficient for the first step analysis.

11. **Width of the headscarp base** is measured as a length of a straight line connecting lowermost point of the well-defined headscarp boundaries, usually perpendicular to the direction of debris motion. This parameter, being compared with maximal width of rockslide deposits (Section 6.3), allows quantitative estimate of rock avalanche debris mobility in the transverse direction. It is quite important for landslide hazard and risk assessments due to several reasons.

First, most of destruction and casualties caused by catastrophic landslides in mountainous areas occur not within their source zones (affected slopes as for landslides in plains), but at the feet of the collapsing slopes—in the river valleys and intermontane depressions. Despite obviousness of the conclusion that, under otherwise equal conditions, the destructive effect is proportional to the area affected rather than just to the runout distance, most of the relationships derived up to now relate rockslide volume to the deposits runout or H/L ratio (A. Heims' (1932) fahrborshung) (Sheidegger, 1973; Hsü, 1975; Davies, 1982; Li, 1983; Shaller, 1991; Kobayashi, 1993, 1997; Kilburn and Søremsen, 1998; Legros, 2002), and only one that we could find — to the affected area (Li, 1983). We want to point out that conditions, typical of many parts of the study region, in the Tien Shan especially, i.e., the presence of numerous flat intermontane depressions up to dozens kilometers wide and of the trough valleys formed by old glaciers also with flat, kilometers wide bottoms, are favorable for the formation of the unconfined rock avalanches that came to halt due to basal friction and internal processes only. Analysis of such case studies shows significant difference in the plan view of their deposits (Strom, 2006). Some of them, like the ca. 5 km long Chaartash-3 rock avalanche in Kyrgyzstan (see Fig. 5.61B), the 2.5 km long rock avalanche in the Chinese Tien Shan, west of the Toksun town (see Fig. 5.32), or the 2.5 km long Imom rockslide (rock avalanche) in Pamir (Fig. 4.14) form relatively narrow stripes of debris indicating that they moved almost strictly ahead. Others, like rock avalanche in Kash River valley in Xinjiang (see Fig. 5.33) or the Atdjailau rock avalanche in the Inylchek River valley in Kyrgyzstan (42.155° N, 79.457° E, Kyrgyz 113) (Fig. 6.4) form the pancake-shape bodies, whose width is equal or even larger than their length. These features will be described in more details in Chapter 8, Tien Shan.

Another reason to collect and analyze data on the sidewise spreading of rockslide debris is their importance for the rockslide dams' stability assessment. As it was shown in Strom (2010b), evolution of natural blockages depend not only on their height and volume (Ermini and Casagli, 2003) but also on its shape that can be characterized by the ratio of the dams' height and its width (length in the along-stream direction; we use the term "rockslide dam length" to characterize its linear dimension in the direction of debris motion), which, in turn, is predetermined to a large extent by debris sidewise spreading that increases dams width, decreasing its height. We can compare quite compact blockages such as the Ak-Kiol rockslide dam in the Kokomeren River basin in Kyrgyzstan (see Fig. 4.3) or the Shiva rockslide dam in Afghan Badakhshan (see Fig. 4.22) with those like Yashilkul rockslide dam in Pamir (see Fig. 4.18) or the impressive rock avalanche (China 036) about 40 Mm^3 in volume damming a lake at the southern slope of the Sarmin-Ula Ridge in Xinjiang at 42.46° N, 85.437° E (see Fig. 4.10 at its left, western part,) that spread up- and downvalley for about 2.3 km after collision with a rocky spur at the opposite valley slope (Fig. 6.5), similar to the Yashilkul rockslide.

Besides its importance for landslide hazard assessment, analysis of the shapes of both unconfined rock avalanches and of natural blockages allows to shed light on some peculiarities of their motion mechanism if we take in consideration some additional factors that will be discussed in Chapter 12, Morphological and Structural Evidence of Rockslides' (Rock Avalanches') Motion Mechanism(s). For unconfined case studies, such additional data are the mechanical properties and

FIGURE 6.4 View on the Atdjailau rock avalanche (Kyrgyz 113) in the Inylchek River valley, whose compact proximal part about 200 m thick is surrounded by strongly eroded 10–30 m thick pancake-shape body. The promontory of the river's right bank is composed of the Inylchek rock avalanche distal part overlapping the distal part of the Atdjailau rock avalanche (see Fig. 8.45). Blue arrow indicates the direction of the Inylchek River flow; T-S R indicates the Taldy-Suu River mouth; dashed line indicates the approximate original boundary of the Atdjailau rock avalanche deposits; double-head dashed arrows indicate width of the headscarp base (1170 m) and that of the resultant deposit (2400 m); yellow dual arrows mark the site from where photograph shown on Fig. 8.45 was made; single-head yellow arrow indicates approximate location of the outcrop shown on Fig. 8.46; LS indicates Silurian limestone; Sh indicates Silurian shale underlying the limestone and outcropping in the deepest part of the headscarp. This photograph was made from the Inylchek rock avalanche headscarp (Kyrgyz 114).

FIGURE 6.5 Rockslide dam at the southern slope of the Sarmin-Ula Ridge 3.1 times wider than the headscarp outlet. The upstream part of the dam is underwater. A smaller rock avalanche (China 035), ~10 Mm^3 in volume, likely simultaneous, is visible just left from the river-damming one. Small black arrows mark the trimline 80–100 m above the deposits surface. 3D Google Earth view.

the water table level of the transitional–depositional zones (Strom, 2006); for natural blockages, it is the morphology of the opposite valley slope (the impact zone) (Strom, 2010a).

6.3 OVERALL PARAMETERS OF ROCKSLIDE DEPOSITS

Area of the deposits (km^2)	Total area affected (km^2)	Volume (10^6 m^3)	Total length (km)	Maximal deposits' width (km)	Maximal height drop (km)	Height drop to the tip (km)	Confinement	Run-up (km)
12	13	14	15	16	17	18	19	20

12. **Area of the deposits** is one of the most precisely measured parameter whose accuracy depends on the correctness of rockslide body identification and digitizing. The main ambiguity is associated with those parts of the deposits that are hidden under dammed lake water or lake sediments or have been removed by the erosion. We strove to map the area initially covered by rockslide debris, if possible (see Fig. 6.4). Sometimes the masked segment of rockslide body embraces small part of the entire deposition area, so that the latter can be estimated with reasonable confidence (see Fig. 6.5). In other cases, without detailed field check with drilling and geophysical surveys, it can be outlined just hypothetically based on the remote sensing data only (Fig. 6.6). The worst cases are those, where large part of the rockslide had been eroded. Sometimes more

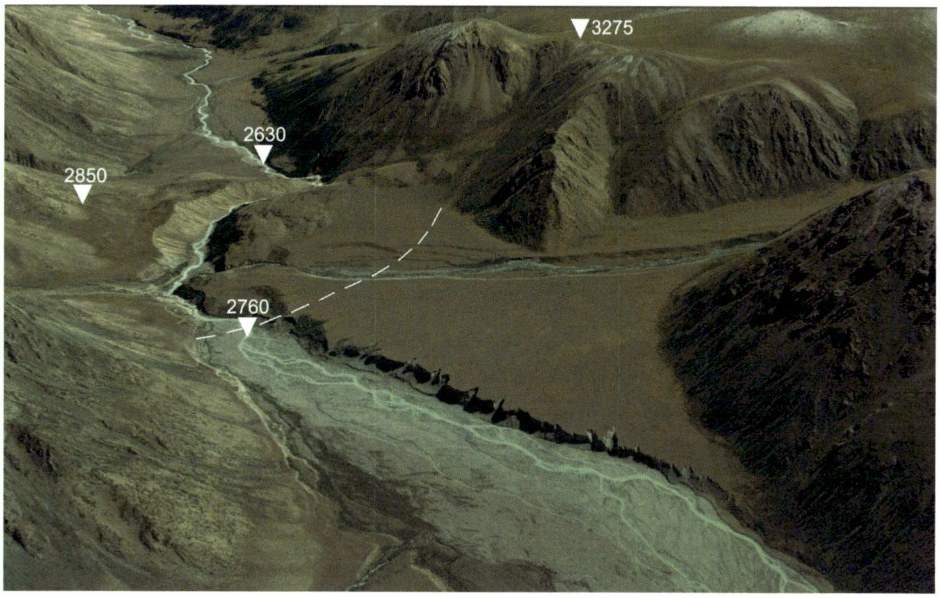

FIGURE 6.6 Breached rockslide dam about 60 Mm^3 in volume and almost 200 m high in the upper reaches of the Chilik River valley, Southern Kazakhstan (43.03° N, 77.51° E, Kazakh 020). Large part of its body (about 15% of 2.18 km^2) was eroded and/or buried by the alluvial fan of the Kairakty River that fall into the dammed lake. Dashed white line is the assumed upstream boundary of the rockslide body. Multiple irregular terraces on the right bank of this river just upstream of the dam indicate gradual multistage breach of the blockage. 3D Google Earth view.

than 50% of the area originally covered by rockslide debris has been washed out (see, e.g., Figs. 4.5, 4.8–4.12). In such cases the original deposits area estimates are rough inevitably.

13. **Total area affected** includes the headscarp, the transition zone along which moving debris had passed and the depositional area. This parameter is important for hazard assessment as far as it describes the entire area directly affected by rockslides in whatever way. It is measured by combining the headscarp and the deposits in one polygon and measuring its horizontal projection. Accuracy of this parameter estimate is the same as for the previous parameter. We should mention that in cases, where the distinct trimlines are visible, such as at the upper part of the 6.48-km long Karakystiak rock avalanche in Southern Kazakhstan (42.58° N, 73.13° E, Kazakh 001) (Fig. 6.7) or at the Seit rock avalanche in Kyrgyzstan described above (see Fig. 5.14), outer boundaries of the transition zone were delineated just along these trimlines that could remain much above the resultant deposits level.

14. **Rockslide volume** is one of the basic parameters characterizing slope failure. Most of quantitative relationships characterizing landslide mobility relate their runout (Sheidegger, 1973; Hsü, 1975; Shaller, 1991; Kilburn, Søremsen, 1998; Legros, 2002) or the affected area (Li, 1983) with volume, which, along with the drop height, predetermine the potential energy of the collapsing rock mass—the main, if not the only source of energy released during rockslide/rock avalanche motion (see Chapter 12). At the same time, this parameter is one of the most hardly determined due to drastic irregularity of both the headscarp and the deposits shape, lack of data on the prefailure topography, significant reshaping of the deposits by erosion. According to the reasons mentioned in the introductory part of this chapter, we did not use any of the statistical relationships between landslide area and volume (Simonett, 1967; Hovius et al., 1997; Korup, 2005a,b; ten Brink et al., 2006; Guzzetti et al., 2008, 2009; Hadian-Amri et al., 2014) not to increase scatter of further statistical analysis between rockslides' volume and their runout, fahrborshung, etc. Instead, volume of each particular rockslide was assessed individually, considering their headscarps and deposits as combinations of simple 3D geometrical figures—pyramids, cones, and prisms. One order scatter of the V/A ratio plotted from such data (see Fig. 6.1) from our point of view demonstrates inappropriateness of quite formal statistical approach for this purpose (see also Chapter 10).

In most of the cases, we first assessed volume of rockslide deposits and used it as a most reasonable value. The headscarp volume was sometimes measured to check reliability of the previous estimates, though we found it less grounded (see the discussion about visual difference between the deposits and the source zone of the Aidyn-Ula rockslide in the Section 4.1 and Fig. 4.4 in Chapter 4: Principles and Criteria of Landslides Identification and

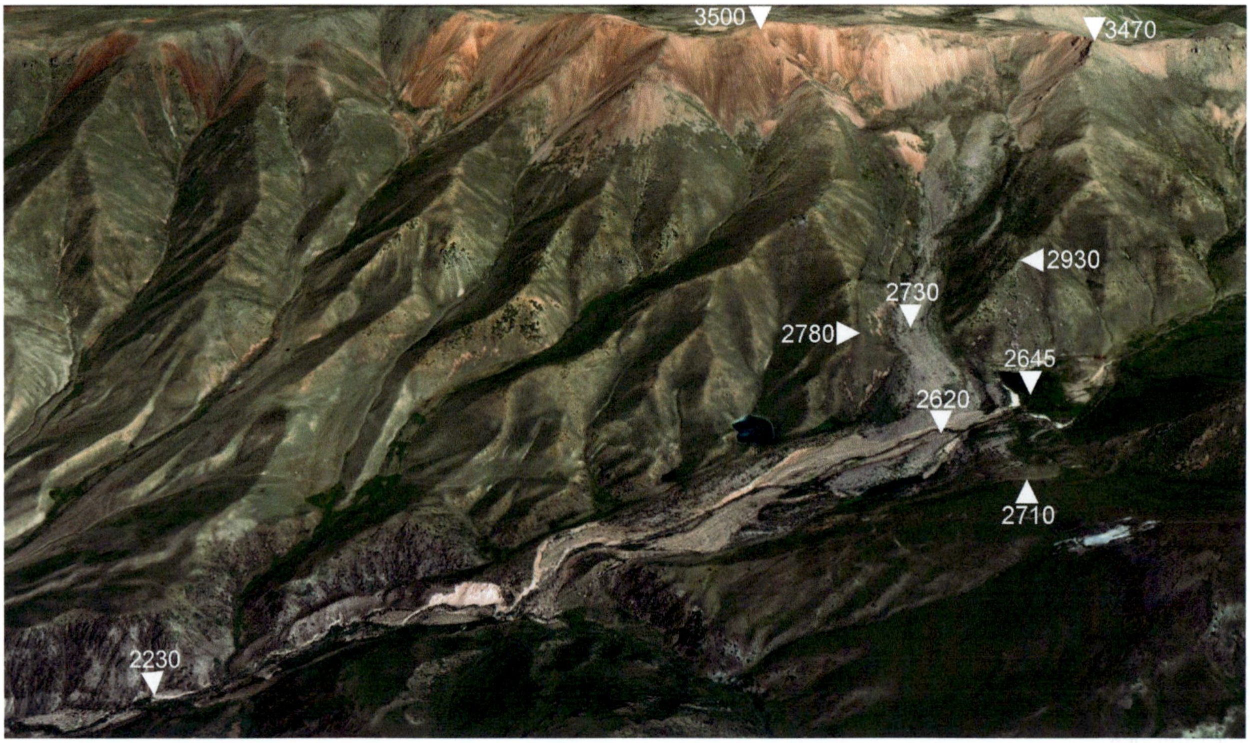

FIGURE 6.7 The 6.48 km long Karakystiak rock avalanche with funnel-shape headscarp up to 1.95 km wide (between elevation marks 3500 and 3470 m a.s.l.) and distinct trimlines along the initial part of its transition zone (marked at 2780 and 2930 m. a.s.l.) up to 200 m above the surface of the deposits. Besides the 80–90 m run-up took place where rock avalanche turned 60° right along the Busuleankiol River.

Discrimination From the Features Formed by Other Mass-Wasting Phenomena). Besides, we should take into consideration material entrainment typical of large rock avalanches (Hewitt, 2006a, b) and softening of rockslide debris relative to the initial bedrock resulting in about 18%–35% decrease of rockslide debris bulk density (Sherard et al., 1963; Hungr, 1981; Chen et al., 2006; Crosta et al., 2011; Frattini et al., 2016) and, correspondingly, in the same increase of the deposits volume relative to that of the headscarp.

Even the historical rockslides' volumes have been often estimated with the same or even larger scatter. We can mention the 1949 Khait rock avalanche (39.187 N, 70.88 E, Tajik 038, see Fig. 5.15), whose first volume estimates ranged from 400 to 500 Mm^3 (Panasenko, 1952; Leonov, 1960), while recent studies by S. Evans and his colleagues provided much smaller (and much more reliable) estimate of about 75 Mm^3 only (Evans et al., 2009b). Quite controversial data were published for the 2000 Yigong rock avalanche in Tibet for which the reported volume of the initial wedge failure ranges from 30 to 100 Mm^3 and the final volume estimates after the significant entrainment range from 105 to 300 Mm^3 (Shang et al., 2003; Wang, 2008; Xu, 2008; Yin and Xing, 2011; Evans and Delaney, 2011). The same, though with smaller scatter is for the Daguangbao landslide—the largest slope failure triggered by the 2008 Wenchuan earthquake, whose volume estimates, despite detailed topographic maps and digital elevation models available both prior and after the event, range from about 750 Mm^3 (Huang et al., 2009, 2012; Chigira et al., 2010; Wu et al., 2010; Fan et al., 2012a,b) up to 1280 Mm^3 (Chen, et al., 2014). Frattini, et al. (2016) assessed accuracy of the precise estimates of the 2004 Turwieser rock avalanche as about ±10%.

The simplest case for volume measurements is the long runout rock avalanche that moved over the unconfined flat surface forming the apron with more or less uniform thickness as the Imom rockslide (see Fig. 4.15), rock avalanches shown on Figs. 5.32 and 5.33, the Chaartash-2 and 3 rock avalanches (see Fig. 5.61), or the Yimake rock avalanche at the north-eastern foot of the Chinese Pamir (39.2° N, 76.15° E, China 148) with a gigantic 9 × 5.3 km leave-shape body about 15–20 m thick on average only (Fig. 6.8). Their deposits' volumes are calculated as a product of the deposits area and its mean thickness. Main uncertainties are caused by the unknown thickness distribution and pre-slide topography. Additional uncertainty can be introduced by the irregular hummocky surface of such rock avalanches and subsequent erosion, which are typical, for example, of the gigantic Koman and Lenin rock avalanches in the Alai Valley (Strom et al., 2014a; Robinson et al., 2015;

Reznichenko and Davies, 2015; Reznichenko et al., 2017) completely covered by 3–5 m, rarely higher hummocks (Fig. 6.9) and partially eroded by melt water from the Trans Alai Range glaciers.

To calculate the volume (V) of rockslide bodies damming rather narrow valleys, we approximated them as a combination of two pyramids or cones with the same base equal to the area of the crest cross section (ABD on Fig. 6.10) and height equal to the along-valley dams' base length (EF on Fig. 6.10). Dams' height (CD on Fig. 6.10) is measured by comparing the dams' crest elevation with its axial base level derived from the rivers' long profile. Thus $V = (AB \times CD/2) \times (EF/3) = AB \times CD \times EF/6$. Some volume underestimation of the downstream part due to stream base inclination is compensated by overestimation of the upstream part. In the cases with trapezoid-like shape of either the dam with widened crest or of the valley cross section, the central part of the rockslide body is approximated as a prism, while side parts as pyramids (Fig. 6.11). Similar approach was used to calculate volume of the rock mass removed from the source zone. An example of rockslide that can be approximated by simple geometric figure is presented on Fig. 6.12.

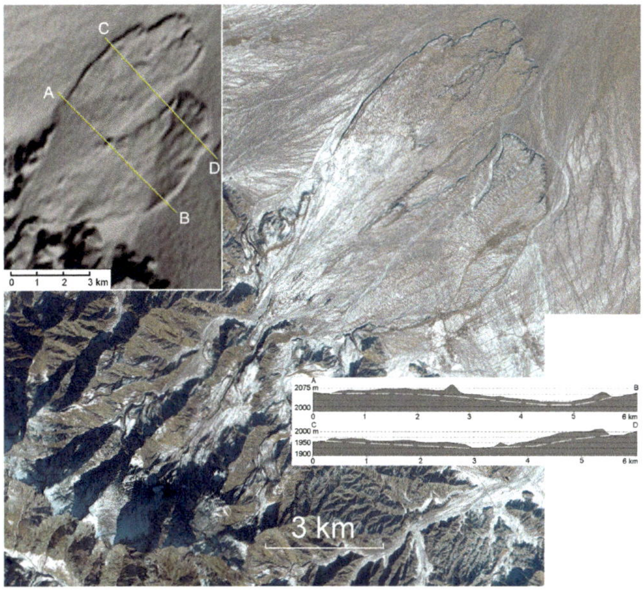

FIGURE 6.8 Space image of the Yimake rock avalanche at the north-eastern foot of the Eastern Pamir (Xinjiang). DEM of its body and the cross sections along lines A–B and C–D are shown in the insets. Dashed lines on the cross-sections are the assumed base of rock avalanche deposits.

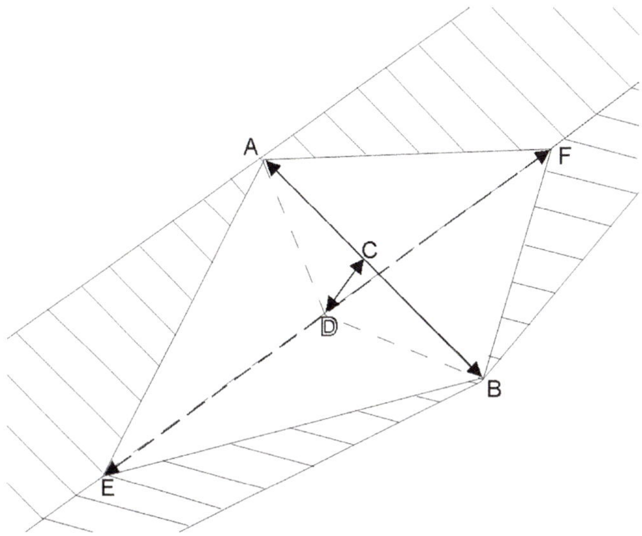

FIGURE 6.10 The schematic model of the simplest rockslide body used for its volume calculation.

FIGURE 6.9 Views of the hummocky surface of the Koman rock avalanche.

Of course the shape of many rockslide bodies is too complex (see, e.g., Figs. 4.30 or 5.65). However, they could be approximated part by part by simple geometric figures in the way described above.

Considering all the above ambiguities, we estimate general accuracy of our volume measurements to be up to ±50%, which, nevertheless, is much better that half-order scatter of the assessment based on the statistical relationships between landslide area and volume (see Fig. 6.1) and is sufficient for further quantitative analysis.

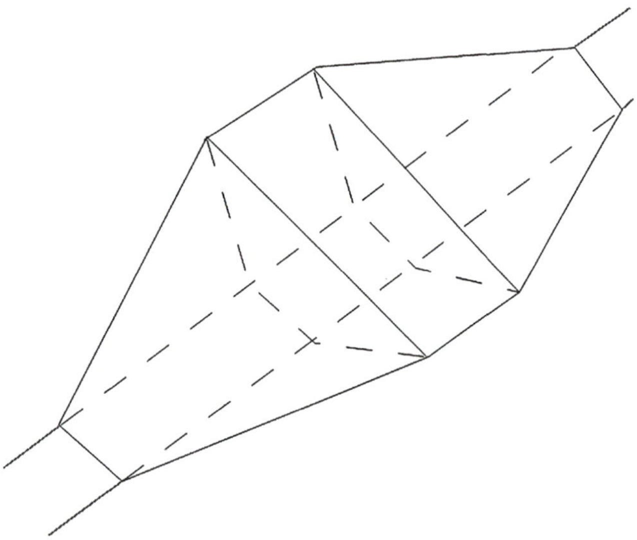

FIGURE 6.11 The schematic model of the rockslide body with widened crest resting in the trapezoid-shape valley.

FIGURE 6.12 Rockslide dam about 12 Mm^3 in volume in the Saihan River basin (42.558° N, 80.324° E, China 061) close to the Chinese–Kazakh border. AB: 1100 m, CD: 600 m, maximal dams' height ~110 m.

15. **Total length (runout)** is measured as the horizontal distance from the headscarp crown to the deposits' tip along the approximate central line of the deposits. Sometimes it is straight (as at case studies shown on Figs. 4.3, 5.9, 5.32, or 5.61), but in many cases the central line is curved or crooked alignment (Fig. 6.13). In the most complex cases, such as those described in Sections 5.3.2 and 5.3.3 in Chapter 5, where the direction of the avalanche-like motion differs significantly from that of the main part of rockslide body (see Figs. 4.30 or 5.65), we measured the runout up to the distal rim of the secondary rock avalanches (viz. up to point at 2450 m a.s.l. on Fig. 4.30 and up to point at 2470 m a.s.l. on Fig. 5.65). It was decided because rock avalanche hazard assessment depends on a large extent on the distance that can be reached by moving debris regardless of the exact motion mechanism.

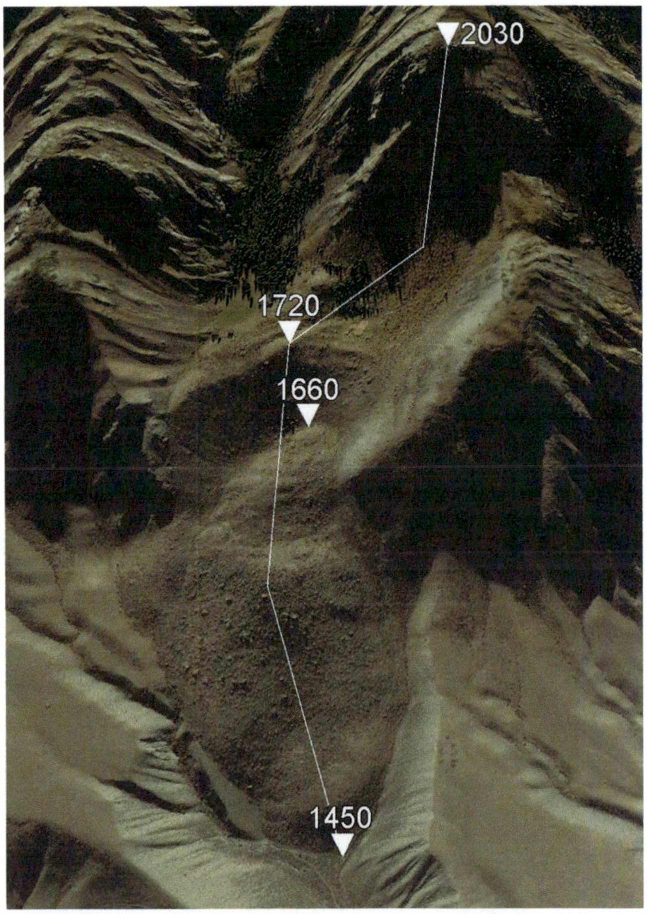

FIGURE 6.13 Measuring the 1.34 km long runout of rock avalanche in the Chinese Tien Shan at the northern slope of the Iren–Khabyrga Range (44.127° N, 84.426° E, China 033). The crown of the distinct secondary scar is at 1720 m a.s.l., and the secondary rock avalanche length was measured between elevation marks 1720 m a.s.l. and 1450 m a.s.l.

16. **Maximal deposits' width** is measured in the direction, transverse to the mean direction of debris motion (see Figs. 5.32, 5.33, or 6.4). This parameter, being compared with the headscarp base width, allows characterizing lateral debris spreading (see Section 6.2).

17. **Maximal height drop** (H_{max}) is the elevation difference between the headscarp crown and the lowermost point of the deposits. For the simplest cases where material had moved just downslope and practically did not escape from the headscarp, it is equal to the headscarp height (see parameter No 9). In more complex cases where rockslide forms the compact blockage in the valley, it is measured as the elevation difference between the headscarp crown and the lowermost point of the dam's base at its axial line.

For the events that had moved down valley (most of long runout rock avalanches including those of the Secondary deflected subtype) this parameter is defined as the elevation difference between the headscarp crown and the lowermost tip of the deposits.

18. **Height drop to the tip**. This parameter is the same as the previous one for rockslides and rock avalanches moving down valley, while for the compact dams it is measured as the elevation difference between the headscarp crown and the most distal tip of debris (see Fig. 5.34-2, Fig. 5.41—from the crown up to 1790 m a.s.l., Fig. 5.65—from 4040 m a.s.l. to 3370 m a.s.l. for the main part of its body, whose uppermost frontal part can be considered as the brandung (Heim, 1932; Hewitt, 2006a, b)).

Short comment is necessary here. We excluded such parameters as vertical descend and horizontal displacement of the center of gravity; though, strictly speaking, they have better physical meaning than the elevation differences between the headscarp crown and the deposits' tip or runout measured as the horizontal distance between these points (Adushkin, 2006). However, the irregular shape of both headscarp and the deposit, lack of data on the prefailure topography and significant (at least in many cases) subsequent erosional reworking of the deposits could result in insufficiently poor accuracy of the center of gravity determination before and after the event. Moreover, in many cases possible errors of such estimates could be hardly assessed. The additional reason of not to use these parameters is that in many long runout cases, horizontal distance between the prefailure and postfailure positions of the displaced rock mass center of gravity would be much shorter than the distance to the deposits distal rim. Thus, the relationships based on this parameter would not allow reliable hazard assessment related to the extramobility of rockslide debris, which could be underestimated significantly.

19. **Confinement** is a qualitative parameters used for the more reasonable sampling during statistical analysis of the case studies included in the database (see Chapter 10). Three main types or rockslides were selected, namely the unconfined (U) that moved without overcoming any obstacles both frontal and lateral as the Imom rockslide (see Fig. 4.14), the primary rock avalanche shown on Fig. 5.32, the Chaartash-3 rock avalanche (see Fig. 5.61-B), or the Yimake rock avalanche (Yuan et al., 2013) (see Fig. 6.8); frontally confined (F) that hit an opposite valley wall as it is shown on Fig. 5.34-2 that can be exemplified by the Ak-Kiol rockslide (see Fig. 4.3), the Yashilkul rockslide (see Fig. 4.18), the Kudara-Pasor rockslide (see Fig. 4.20), the Mini-Köfels rockslide (Fig. 5.41), and by many others; and laterally confined (L) that moved down valley being canalized by it walls as rock avalanche with headscarp crown at 3780 m a.s.l. on Fig. 6.5, the Karakystiak rock avalanche (see Fig. 6.7), or rock avalanche of the Secondary type shown on Fig. 6.13. Sometimes the confinement mode changed during rockslide motion, like at the Seit rock avalanche (see Fig. 5.38), whose motion was confined, first, laterally by steep valley slopes and, at the end, frontally, by sharp bending of the left slope (see Fig. 5.35). Such cases are marked as L, F. In contrast, many rock avalanches of the Secondary type that first hit opposite valley wall (frontal confinement) moved afterward along narrow valleys being confined laterally (see, e.g., the Karakungey rock avalanches, Figs. 5.44 and 5.55, or the Big Dragon Lake rock avalanche on Fig. 5.65). Such cases are marked as F, L.

20. **Run-up** was measured as the elevation difference between the topmost point at rockslide body front and the lowest point of its crest (between points at 3370 and 3000 m a.s.l. on Fig. 5.65; 2440 and 2415 m a.s.l. on Fig. 6.12). Sometimes it was observed at the distal parts of long runout rock avalanches as well, such as at the super-long runout (34 km) Komansu rock avalanche in the Alai valley, which front raised on the hilly foot of the Alai Range for about 100 m after crossing the lowermost part of the Alai valley bottom occupied by the Kyzylsuu River (Robinson et al., 2015). The same was observed at the terminus of the ~5.5 km long Mingteke rock avalanche (42.00° N, 74.355° E, Kyrgyz 092) in Central Kyrgyzstan with 90 m run-up (Fig. 6.14).

Run-up allows estimating debris velocity before its motion was terminated due to the impact with frontal confinement based on simple relationship between the kinetic and potential energy:

$$m \times g \times H = m \times V^2/2;\ V = (2 \times g \times H)^{0.5}$$

where m is rock mass, H is run-up, and V is debris velocity.

FIGURE 6.14 90 m (from 2575 to 2665 m a.s.l.) distal run-up of the 5.5 km long Mingteke rock avalanche. Its headscarp crown is at 3700 m a.s.l.

In the Mingteke case, it gives ~144 km/h. Same approach can be used for the velocity estimates from the debris superelevation at the pass bends, such as at the Chongsu rock avalanche in Kyrgyzstan (see Figs. 5.48 and 5.49). Similar superelevation can be observed at the Karakystiak rock avalanche in Kazakhstan (see Fig. 6.7). Such estimates, however, do not take into account possibility of some momentum transfer within rapidly moving debris. The extreme run-up, such as 600 m described for the Avalanche Lake rock avalanche (Evans et al., 1994) or 500 m for the Karivhoh rockslide in Caucasus (Strom, 2004), as well as ejection of large boulders from the main rock avalanche mass mentioned for the Avalanche Lake rock avalanche (Evans et al., 1994), and 1970 Huascaran ice-rock avalanche (Plafker and Eriksen, 1978) and described at the Chongsu rock avalanche (see Fig. 5.49) could be caused by "pushing" of the frontal part of debris by its much more voluminous following parts or by local acceleration of the separate boulders, not reflecting the bulk velocity of moving debris. Limitations in using superelevation to calculate the velocity of debris flows were demonstrated by Prochaska et al. (2008).

6.4 ROCK AVALANCHE PARAMETERS

Rock avalanche type	Rock avalanche volume (10^6 m^3)	Percentage of total rockslide volume	Rock avalanche length (km)	Rock avalanche height drop (km)	Secondary scar/jump height (km)
21	22	23	24	25	26

This group of data describes parameters of highly mobile parts of rock avalanches ascribed to the Jumping and Secondary types (see Sections 5.3.2 and 5.3.3 in Chapter 5). For the Primary rock avalanches (see Section 5.3.1), they are the same as the overall parameters listed in the previous section.

21. **Rock avalanche type** is identified according to classification system presented in Section 5.3 P: Primary, J: Jumping, SC: Secondary of the Classical subtype, and SB: Secondary of the Bottleneck subtype.

22, 23. **Rock avalanche volume** was calculated for the secondary and jumping rock avalanches following the same principles as described for the overall parameters. Besides its estimate in millions of cubic meters, the percentage of the part of debris involved in the avalanche-like motion was calculated.

24. **Rock avalanche length** was measured for rock avalanches of the Secondary type as the horizontal distance along the axial line of rock avalanche between the secondary scar crown and rock avalanche tip (see Fig. 6.13). We should point out that some rock avalanches have not one, as on Fig. 6.13, but several successive secondary scars. It was found at the Big Dragon Lake rock avalanche (marked by 1 and 2 on Fig. 5.65), at the rock avalanche at 42.467° N, 85.588° E (Fig. 6.15, China 037)—the longest one (4.21 km long) within the rockslide cluster shown on Fig. 8.11, at the Bielogorka-right rock avalanche in Northern Kyrgyzstan (42.64° N, 74.29° E, Kyrgyz 206) (Fig. 6.16) and at the central part of the 3.9 km long rock avalanche with 1.37 overall height drop in Dzungaria at the left side of the Irgaity River valley (45.65° N, 81.81° E, Kazakh 031, Fig. 6.17). In such cases, both rock avalanche length and its height drop, described as Parameter 25, are measured from the crown of the uppermost secondary scar.

For Jumping type, this parameter was measured as the distance between the foot of the compact part and the tip of the avalanche-like mobile part (see e.g. Fig. 5.43).

25. **Rock avalanche height drop** was measured for rock avalanches of the Secondary type only as an

FIGURE 6.15 The contiguous secondary scars of the 3 km long (from 3390 to 2710 m a.s.l.) secondary rock avalanche in Xinjiang (China 037). The uppermost horseshoe-shaped scar is at 3390 m a.s.l., while two other same shaped scars are visible below. This rock avalanche represents mix of the Classical and the Bottleneck subtypes of Secondary rock avalanche (see Section 5.3.3 in Chapter 5): rockslide that originated in the funnel-shaped headscarp more than 1.1 km wide passed through the 0.5 km wide valley bottom, which is typical of the Bottleneck subtype; at the same time, it has well pronounced, though low secondary scars. The additional peculiarity of this case study is a distinct 70–100 m high right-side trimline that dammed the joining gulley with a small almost completely silted pond at 3260 m a.s.l. It indicated that initially this rockslide body moved westward as up to ~100 m thick mass of crushed debris, and afterward converted into more flow-like rock avalanche with series of secondary scars at its uppermost part.

FIGURE 6.17 Two successive secondary scars at 1770 and 1630 m a.s.l. in the central part of the 3.9 km long rock avalanche in Dzungaria (Kazakh 031) that moved along steep gulley with mean channel slope of about 16 degrees. 3D Google Earth view.

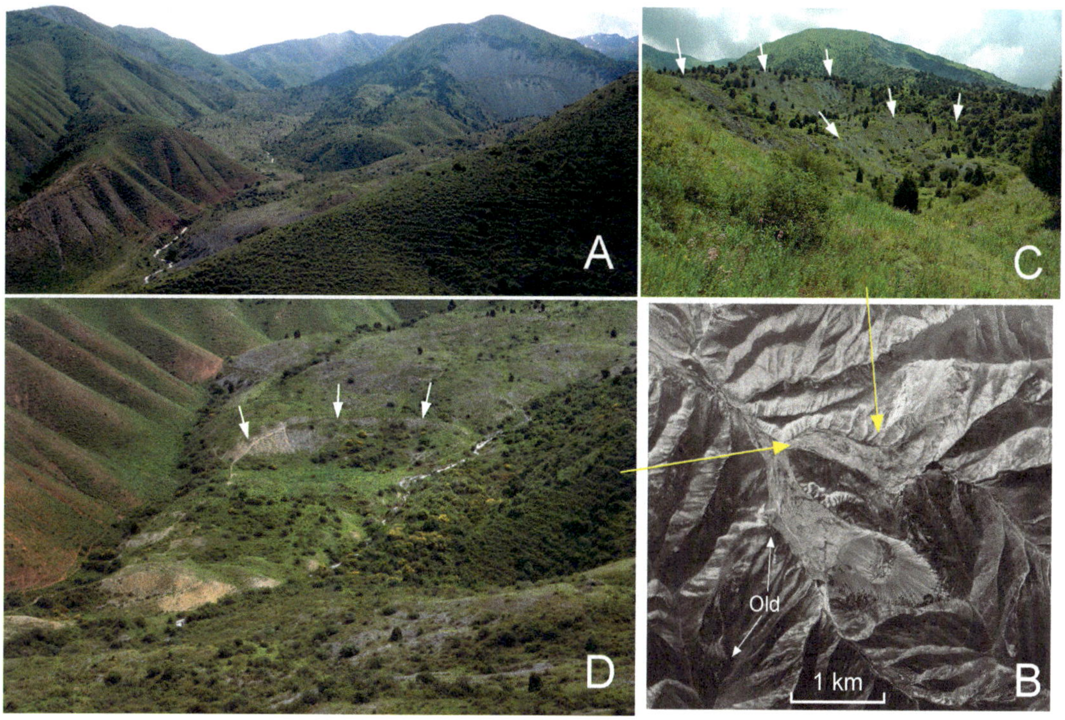

FIGURE 6.16 The successive secondary scars of the 2.4 km long Bielogorka-right rock avalanche. (A) Overview of the Bielogorka-right (Kyrgyz 206) and the Bielogorka-left (Kyrgyz 205) simultaneous rock avalanches (named the Bielogorka twins); (B) fragment of the aerial photograph. Besides the clearly visible "twins", there is a much smaller and much older rock avalanche (Kyrgyz 204) marked as "Old." Small white arrows point on its headscarp and deposits tip; (C and D) views on the concave secondary scars.

elevation difference between the secondary scar crown and rock avalanche tip (see Fig. 6.13).

25. **Height of the secondary scar** (for Secondary rock avalanches of the Classical subtype) and **jump** height (for Jumping rock avalanches). Parameters given in this column are measured in different ways. Height of the secondary scar—as the elevation difference between the highest secondary scar crown and its base (150 m for the scar on Fig. 4.30). The jump height—as the level of the "launching pad" above the valley bottom (see the model on Fig. 5.42).

6.5 ROCKSLIDE DAMS AND DAMMED LAKES PARAMETERS

Dam parameters			Lake parameters				
Effective height (m)	Dams' width (km)	The position of the lowermost part of the dams' crest	Lake state	Maximal depth (m)	Length (m)	Mean width (m)	Volume (Mm^3)
27	28	29	30	31	32	33	34

The last group of data characterizes parameters of those rockslides that formed natural blockages and of the rockslide-dammed water bodies, both existing and past, considering special hazard provided by such phenomena (Costa and Schuster, 1988, 1991; Evans et al., 2011). It is subdivided into natural dams' parameters and dammed lakes' parameters.

27. **Effective dam height**: minimal height of the blockage at its crest point where the overflow had started or could start in case of an overtopping. It is exemplified by the Upper Zhasylkol rockslide dam on the Aganakty First River in Southern Kazakhstan in Dzungaria (45.28 N, 80.62 E, Kazakh 068) that forms the Upper Zhasylkol Lake (Fig. 6.18). About 140 Mm^3 of the terrigenous rocks caved from the upper part of more than 1100 m high slope, jumped as from the launching pad ~370 m above the valley bottom and formed a dam 2.5 km wide (at its base along the river valley) with an effective height about 150 m only, while maximal thickness of the deposits exceeded 250 m and the distal run-up was up to 190 m. This rockslide demonstrates combination of features typical of the Primary rock avalanche with frontal confinement and of the Jumping rock avalanche (see Section 5.3). Significant run-up resulted in much lower effective dam height than of some other dams with similar volume but with more flat dams' crest.

Considering irregularity of the river's thalwegs quite common in the mountainous regions (Fig. 6.19), especially for rather small streams, accuracy of this parameter estimate is not very high, generally not better than 10%–20%.

28. **Dams' width** is measured along the stream at the dams' base. Its accuracy is limited by the uncertainty of measurement of its upstream part hidden underwater or buried by lake sediments. Most often it corresponds to maximal deposits' width (No 16, see Section 6.3), but sometimes differs from it significantly such as at the Bashkol rockslide that dammed the same name river—left tributary of the Uchkiol River in Eastern Kyrgyzstan (41.92° N, 78.764° E, Kyrgyz 110). Here large rockslide about 50 Mm^3 in volume filled the small gulley and just frontal part of its ~900 m wide body entered the Bashkol River valley and blocked it by the 80 m high dam at a distance of about 450 m forming a beautiful lake (Fig. 6.20).

29. **The position of the lowermost part of the dams' crest** is an important factor affecting dams' stability (Korchevskiy et al., 2011). Four types of its position are identified. D: distal that can be exemplified by the Kashkasu rock avalanche of the Jumping type (see Fig. 5.45) and by the Kiol-say rockslide (see Fig. 5.63). P: proximal, characterized by the significant run-up and exemplified by the Usoi rockslide (A. Ischuk, 2011), by the rockslide in the upper reaches of the Muzart River basin (see Fig. 5.40) or by the river damming body of the Big Dragon Lake rockslide (see Fig. 5.65). C: centered, also with the run-up, but not as impressive as in the previous cases, so that the lowermost part of the dams' crest is at its central part. It can be exemplified by the Ak-Kiol rockslide dam (Fig. 5.8) and by the Karasu rockslide dam (see Fig. 5.64). F: flat, without distinct lowering that can be seen at the Ak-Kiol rockslide dam (see Fig. 4.3) and by the Shiva rockslide dam (see Fig. 4.22).

We must note that types of the across-valley profiles presented here differ from those included in rockslide dams' classification proposed in (Hermanns et al., 2011) and shown in the upper part of Fig. 5.30.

The parameters in Columns 30–34 of the database characterize the dammed lakes.

30. **Lake's state** is a qualitative parameter describing the present-day state of the water body formed by river-damming rockslide. Three main types are selected—existing (E), breached (B), and filled (F).

Numerous natural reservoirs dammed by rockslides exist in the study region (see Figs. 4.3, 4.19, 4.22, 5.8, 5.63–5.65, and many other described in Part 2). The largest one is the 17 km^3 and ~500 m deep Sarez Lake in Pamir dammed by the Usoi rockslide (Fig. 6.21).

FIGURE 6.18 The Upper Zhasylkol rockslide dam. 3D Google Earth view.

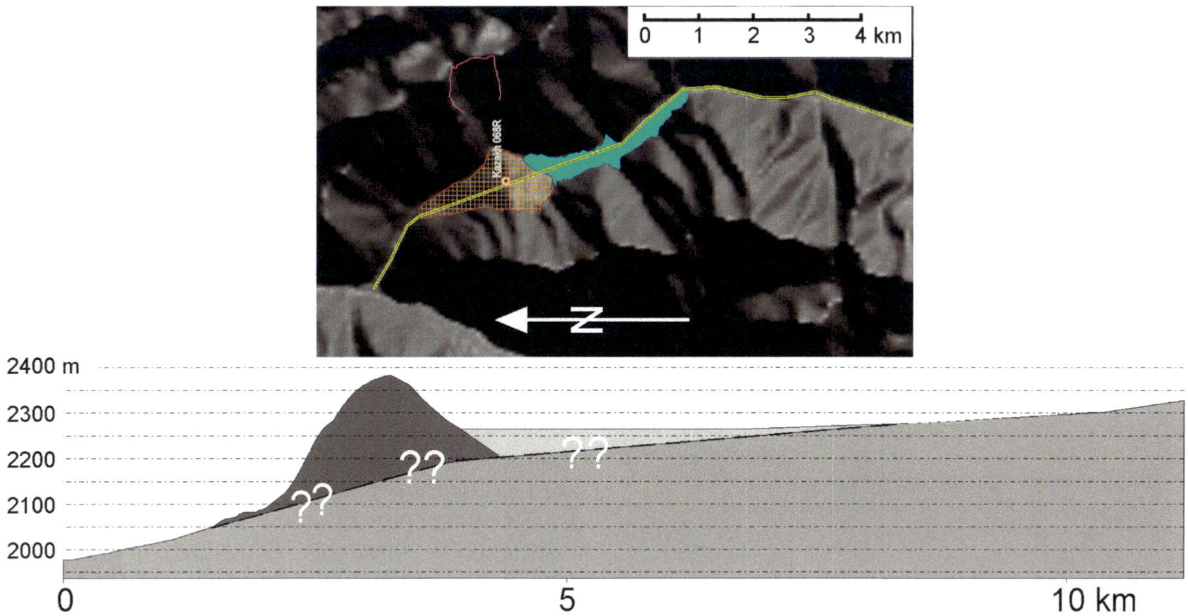

FIGURE 6.19 Assumed pre-slide thalweg profile of the Aganakty First River dammed by the Upper Zhasylkol rockslide. The uncertain position of its parts masked by the lake and rockslide body is marked by the interrogation marks.

The breached state is ascribed to those lakes where relatively small amount of lake sediments was accumulated before they were eroded either due to overtopping or piping. Some rockslide-dammed lakes that had existed for centuries and even millennia were breached recently providing catastrophic effects such as the Issyk (in 1963) and Yashinkul (in 1966) rockslide dams (see Figs. 2.1 and 2.3). The 1964 Aini rockslide dam (see Fig. 2.2) breached through the channel excavated to prevent catastrophic outburst flood

FIGURE 6.20 The Bashkol rockslide dam. Small black arrows mark recent headscarp on the left bank of the Uchkiol River valley. 3D Google Earth view.

FIGURE 6.21 Space image of the Sarez Lake—the World largest natural reservoir formed by rockslide. RB: location of the assumed Right-Bank landslide, whose catastrophic collapse proposed in Sheko and Lekhatinov (1970) and Fedorenko et al. (1981) could cause surge wave able to destroy the Usoi blockage. This point of view is opposed by N. Ischuk (2013b). *Reprinted from Strom, A., 2013a. Geological prerequisites for landslide dams' disaster assessment and mitigation in Central Asia. In: Wang F., Miyajima M., Li T., Fathani T.F. (Eds.). Progress of Geo-Disaster Mitigation Technology in Asia, Springer-Verlag Berlin, Heidelberg. pp. 17–53 (ISBN 978-3-642-29106-7), with permission of Springer.*

(see Chapter 2: Brief History of the Bedrock Landslide Studies in the Central Asia).

The last group marked by F (filled) is represented by dammed lakes that have been silted almost completely while the dam remains more or less intact. Such past lakes filled by lake sediments form flat forested or populated areas immediately upstream from rockslides. They can be exemplified by several impressive case studies—the Kashkasu rockslide (see Fig. 5.45, green forest behind the dam at 2125 m a.s.l.), the Karakul rockslide at the Eastern Karasu River mouth (Fig. 6.22; 41.633° N, 72.645° E., Kyrgyz 061), and the Lower Kulun rockslide (Fig. 6.23; 40.46° N, 74.067° E, Kyrgyz 217). Factors that predetermined such evolution of these blockages will be discussed in Chapter 14, Rockslide Dams—Origin, Evolution, and Failure.

Sometimes, after the dammed lake was silted significantly, the dam was breached (or bypassed by the newly formed gorge like in case of the giant Beshkiol landslide (Korup et al., 2006)). Such cases are marked as F, B.

31. **Maximal depth** of the dammed lake is revealed from the longitudinal profile along the blocked river or, in the rare cases, by the direct measurements if they have been published. Besides the ambiguity of the original thalweg profiling (see Fig. 6.19), accuracy of the maximal lake depth estimate is strongly affected by its siltation. Lake sediments accumulation always starts from its tail part where the stream transporting capacity decreases drastically. However, in narrow mountainous valleys with steep kilometer-high slopes, large amount of talus can be supplied to any part of the water body. In most of cases, our estimates are based on the comparison of the averaged thalweg profile with the water level (for the existing lakes) and with the lowermost point of the dams' crest for breached and filled (silted) lakes.

Measuring both depth and plan dimensions of the existing lakes, we considered that their level could change significantly depending on the season. It is well described for the Sarez Lake, whose annual level change can reach 8–11 m (see, e.g., A. Ischuk, 2011). Sometimes evidence of the water level higher than at a moment when the space image was obtained can be seen as for the small dammed lake at 45.296° N, 81.15° E (Kazakh 066) in the Attapkan River valley in Dzungaria close to Kazakhstan–China border (Fig. 6.24). In such cases maximal depth (and plan dimensions) were measured and included in the database.

32. **Lake length**: distance along the lake long axis. In case of the isometric or bifurcating shape length is measured transverse to the upstream face of the blockage up to the most distant point of the present or anticipated shoreline. Because most of the existing rockslide dammed lakes have been partially silted (see Figs. 2.1, 4.24, 6.5 and 6.20), we measured lake length including the freshest, nonvegetated part of the alluvial fan that could be submerged during flood period.

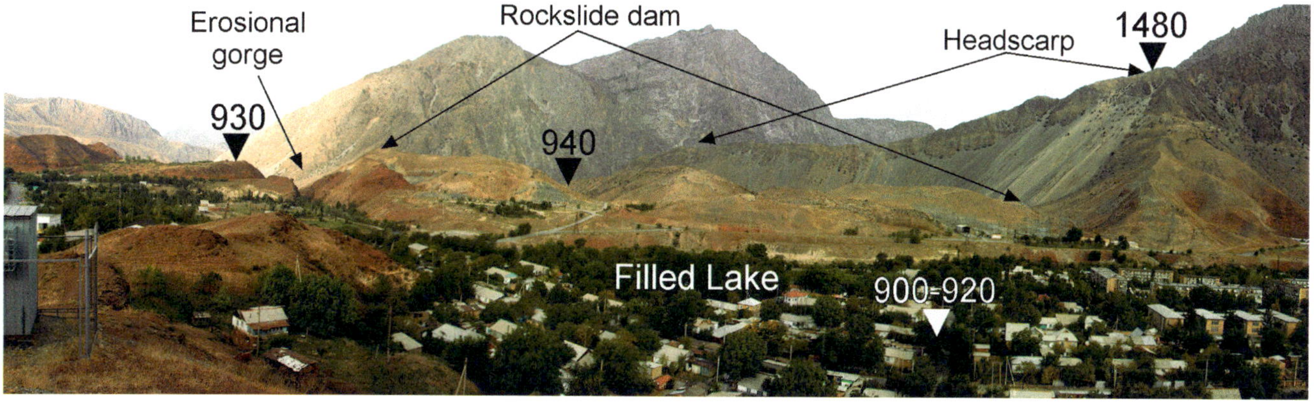

FIGURE 6.22 The silted lake dammed by the Karakul rockslide dam and occupied by the Karakul town (Kyrgyzstan). *Modified from Strom, A.L., 2010b. Landslide dams in Central Asia region. Landslides 47(6), 309–324, with permission of the Japan Landslide Society.*

FIGURE 6.23 The Lower Kulun rockslide dam and completely silted dammed lake (in the inset). Stream is forming a cut through the dams' crest and its downstream slope. Water also found its way through the left-bank contact between the rockslide front and the valley wall. White arrow shows the direction of the bedrock slope failure.

For the breached or silted lakes, their plan dimensions were outlined on space images as widened flat, often nonvegetated section of the valley bottom as for the filled lake upstream from the nearly 190 m high rockslide dam ~30 Mm^3 in volume at 42.59° N, 80.584° E, 7 km upstream from the Shaty village in the Northern Muzart River basin in Xinjiang (Fig. 6.25, China 007). Its tailing part is now above the original dams crest level that could result from gradual accumulation of the alluvium supplied by the river over the initial lake level.

33. **Lake mean width** is calculated by dividing lake area outlined on space images by its length. Values are rounded up to 10 m.

34. **Lake volume** is calculated by multiplying lake area by half of its maximal depth and dividing by 2 for the lakes in the valleys with V-shape cross-sections. For lakes in wider valleys with U-shape or trapezoid

FIGURE 6.24 Evidence of the significant (up to 12 m) seasonal change of the rockslide-dammed lake in the upper reaches of the Attapkan River valley. Large white arrow shows the direction of rockslide motion. Lower and higher lake levels were measured from Google Earth. 3D Google Earth view.

FIGURE 6.25 Silted rockslide-dammed lake 7 km upstream of the Shaty village in Xinjiang. The assumed older rockslide with well-defined spoon-shape headscarp (its crown is at 3220 m a.s.l.) and strongly reworked body extending from 2385 to 2200 m a.s.l. in the tributary valley (China 003) can be seen at the right side of the figure. 3D Google Earth view.

cross-sections lake area multiplied by half of its maximal depth was divided by 1.5 which gives more realistic estimate.

Not all positions have been filled in the Appendix up to now though for more than 500 case studies basic quntitative parameters are available. It is a permanent long-term work and we are adding new data gradually. No doubts that the detailed analysis of the particular case studies will provide more accurate estimates of their quantitative parameters. Authors will be glad if our followers will specify data collected in the database that could be used for better understanding of rockslide mechanisms and of the hazards provided by these natural phenomena.

P A R T I I

REGIONAL DESCRIPTION

In this part we will provide data on the regional bedrock landslides distribution within the Dzungaria, the Tien Shan, and the Pamir mountain systems, and brief description of the most interesting case studies. As mentioned already in Chapter 1, Introduction, it is impossible to describe all rockslides identified within the study region. Of course, the selection of case studies described individually is subjective and depends at a large extent on the availability of the high-quality space images and of the data collected during our personal field work, allowing to demonstrate their geomorphic and geological peculiarities. We strove, however, to describe case studies indicating causes and triggers of bedrock slope failures, deposits morphology, and internal structure that can help reconstructing debris motion mechanisms, etc. Features described in previous chapters will be referred as well. For each described site, we provide coordinates so that interested reader could easily find it either on space images or in the field.

We will start the description from the Dzungarian Mountains, and then will describe the Tien Shan from East to West and finally, the Pamir.

CHAPTER

7

Dzungaria

The Dzungarian mountain system stretches from east to west for 350 km and from north to south for up to 110 km being located north of the Tien Shan. Its main part, the east-west trending Dzungarian range up to 4622 m a.s.l. high is linked with the Chinese Tien Shan by the Boro-Khoro range that passes in the east-south-eastern direction separating the Ili intermontane depression from the Alakol-Ebi-Nur depression (Fig. 7.1).

This part of the study region is characterized by the extremely nonuniform spatial distribution of large-scale bedrock landslides that concentrate in the eastern part of the Dzungarian range being grouped in several distinct clusters (see Fig. 7.1). The largest and the densest one—the Northern Dzungarian cluster—consists of 13 rockslides identified within a limited area up to 12 × 15 km only in the southern part of the Jamanty River basin. One more group of three adjacent rockslides was identified about 20 km eastward in the Yrgaity River basin (Fig. 7.2). All rockslides, which volume ranges from 1–2 to 30–50 Mm^3, occurred on the slopes 300–700 m high approximately, composed of the Devonian and Carboniferous volcanic and terrigenous rocks. Most of these slopes are looking west or east and only two rockslides occurred on the north-facing slopes (see Fig. 7.2).

Most of the large-scale rock slope failures seem to be relatively young, likely Holocene. Quite often slopes failed very close (1–2 km) to each other forming chains of either existing (Fig. 7.3) or partially filled by

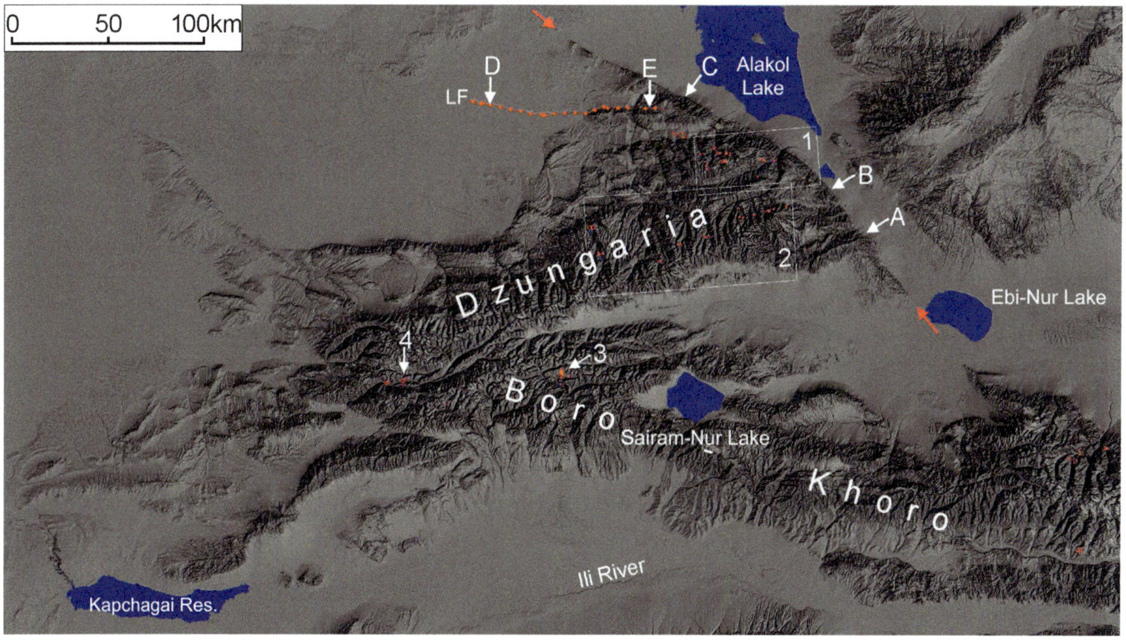

FIGURE 7.1 Shaded 3″ STRM DEM of the Dzungarian and Boro-Khoro mountain systems. Red arrows mark the Dzungarian fault; red dotted line, the Lepse fault (LF); (A–E) points with evidence of active faulting shown on Figs. 3.8 and 3.9. Marked areas and sites: 1—see Fig. 7.2; 2—see Fig. 7.5; 3—see Fig. 7.9; 4—see Fig. 7.11. Rockslides are marked by pink and brown dots; blue dots, dammed lakes.

Rockslides and Rock Avalanches of Central Asia.
DOI: https://doi.org/10.1016/B978-0-12-803204-6.00007-7

sediments and breached rockslide-dammed lakes (Fig. 7.4). Such closeness along with similar geomorphic expressiveness allows assumption that at least these adjacent slope failures occurred simultaneously. However, extensive dating is necessary to prove or disprove this assumption and, all the more so, to expand similar conclusion to the entire cluster. It can be assumed that these slope failures were triggered by strong earthquake(s). Hypothesis of their seismic triggering is supported by the position of the headscarps that embrace either upper parts of the slopes or salient rocky spurs (see Fig. 6.17, Figs. 7.3 and 7.4) and by their more or less similar orientation. Rock avalanche shown on Fig. 6.17 (Kazakh 031) has two secondary scars indicating its "interrupted" motion.

Another group of rockslides form a 60-km long "chain" stretching at the upper part of the northern slope of the main Dzungarian range close to the water divide at the elevations of about 2250–3000 m a.s.l. (Fig. 7.5). Most of these slope failures occurred within and just below the recently glaciated zone on the east- or west-facing slopes, and many of them formed

FIGURE 7.2 Map of the Northern Dzungarian rockslide cluster. 1: Two intact rockslide dams shown on Fig. 7.3; 2: two breached dams shown on Fig. 7.4; red arrows mark the Dzungarian fault. Background—The 3″ SRTM DEM visualized by Global Mapper software.

FIGURE 7.3 The Upper and Lower Kulasu rockslide-dammed lakes in the Jamanty River basin (45.654° N, 81.45° E, Kazakh 053, 054). 3D Google Earth view.

FIGURE 7.4 Two adjacent breached rockslide dams at the lower reaches of the Kok Suat River valley (45.642° N, 81.9° E, Kazakh 030, 030a). 3D Google Earth view.

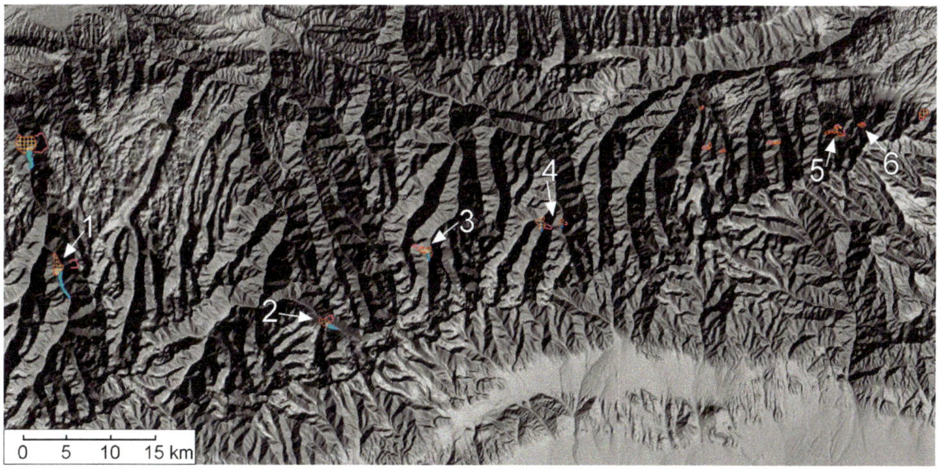

FIGURE 7.5 Location of bedrock landslides near the main Dzungarian range water divide. White arrows mark features shown on: 1: Fig. 6.18, 2: Fig. 4.24, 3: Fig. 6.24, 4: Fig. 7.8, 5: Fig. 7.7, 6: Fig. 7.6. Background—The 3″ SRTM DEM visualized by Global Mapper software.

rockslide-dammed lakes (see Figs. 4.24, 6.18, and 6.24). Most of the failed slopes are composed of the Devonian and Carboniferous terrigenous and volcanogenic rocks and only one of the granites. The latter that forms the Zhashilkiol Lake (45.4° N, 81.15° E, Kazakh 067) is the largest one in this part of the Dzungaria with volume $\sim$300 Mm^3, 1–2 orders larger than most of other rockslides that occurred on slopes composed of terrigenous and volcanic rocks.

Similar to the features forming the Northern Dzungarian cluster, some of these rockslides demonstrate evidence of simultaneous formation (Figs. 7.6 and 7.7) that, along with similar orientation of failed slopes (Fig. 7.8), supports assumption of their seismic origin.

Despite reasoning in favor of seismic origin of rockslides of both groups described above, we must point out that there is no sound spatial correlation between them and most prominent active faults of the study region—either the Dzungarian and the Lepse faults shown on Fig. 7.1 (see also Figs. 3.8 and 3.9) or with the system of neotectonic faults that form the southern boundary of the Dzungarian range.

Some rockslides in this part of the region occurred at high altitude and their deposits can be hardly distinguished from the moraines left by recent glaciation, such as at the body of the Sarybiriuk rockslide dam (Kazakh 077) in the upper reaches of the Kyzyl-Tentek River at 45.225° N, 81.00° E described in Chapter 4, Principles and Criteria of Landslides Identification and Discrimination From the Features Formed by Other Mass-Wasting Phenomena (see Fig. 4.24).

Unlike the eastern part of the Dzungarian range rich of bedrock landslides, just few similar features were found in its western parts as well as in the Boro-Khoro range despite similarly rugged terrain with numerous steep slopes up to 1–1.5 km high (see Fig. 7.1). Small group of rockslides is located in the Ulken-Kazan River basin close to the Kazakhstan–China border. One of them, looking very "fresh" blocked the junction of the Ulken-Kazan and the Kumbelsai Rivers at 44.7065° N, 80.312° E and, despite its small volume of about 3 Mm^3 only, dammed both of them forming the Kazankol Lake in the larger Ulken-Kazan River valley and one more in the Kumbelsai River valley.

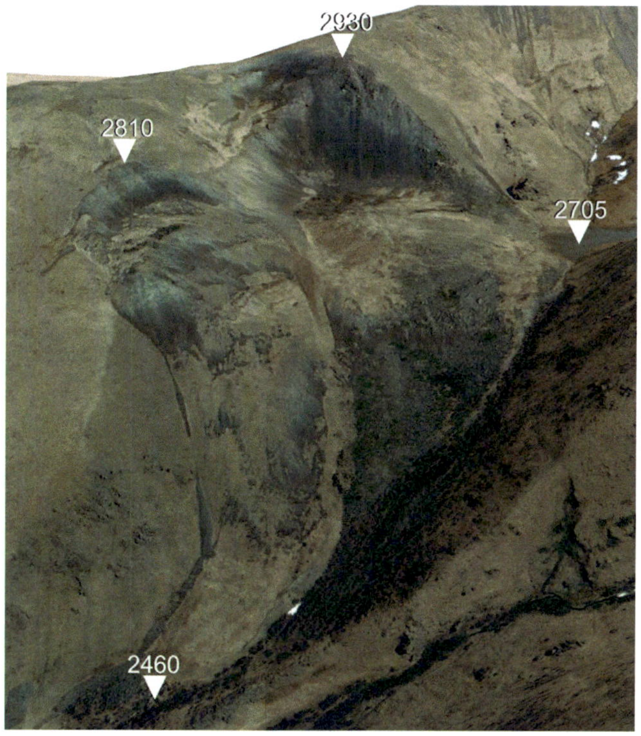

FIGURE 7.6 Twin simultaneous rockslides (Kazakh 069, 070) at 45.42° N, 81.782° E. 3D Google Earth view.

About 8 km north of it (44.77° N, 80.33° E), giant wedge failure in the Carboniferous volcanic rocks about 300 Mm^3 in volume formed a 5.4-km long rock avalanche. Its body, moving downslope, met an obstacle at 2820 m a.s.l., ~1 km bellow the headscarp crown, and split into two branches, one of which stopped shortly while another one moved further for about 1.5 km and reached the Kumbelsai River at 2550 m a.s.l. forming temporary blockage (Kazakh 085, Fig. 7.9). Besides long runout, in spite of complex confinement, this feature has two impressive secondary scars (Fig. 7.10) indicating interruptions in rockslide debris motion (see Section 5.3.3.1 in Chapter 5: Bedrock Landslide Types and Classification Systems).

FIGURE 7.7 A 2-km long rock avalanche that originated from the compound headscarp at 45.414° N, 81.743° E (Kazakh 071). 3D Google Earth view.

Several large rockslides were found in the Koksu River valley. In this part of the region, the east-west trending structures of the Dzungarian range merge with the east-south-east trending structures of the Boro-Khoro range. The area is composed of the Carboniferous terrigenous sediments intruded by granites and diorites. Most of rockslides occurred in the sedimentary rocks. One quite old Pleistocene rockslide, about 100 Mm^3 in volume, caved from the right bank of the valley, almost 3 km downstream from the Ermensai River mouth at 44,758° N, 79.196° E (Kazakh 003).

One of the most interesting sites is located 8 km east of the previous case study at 44.762° N, 79.298° E (see Fig. 7.1, No. 4). Here just upstream from the Tenteksai River mouth, the Koksu River had been blocked twice, first in Pleistocene by a giant, about 250 Mm^3 rockslide (Kazakh 012a) that caved from the left bank of the river valley (Fig. 7.11A). Much later, likely in Holocene, it was blocked again by rockslide with an impressive fresh headscarp from the right bank (Kazakh 012) (Fig. 7.11B). It can be assumed that the first rockslide had pushed the stream toward the right bank of the river, which facilitated this slope undercutting and its subsequent failure, which was, however, about one order less in volume than the first one. According to the longitudinal profile of the Koksu River, it is still dammed for about 40–50 m, but the

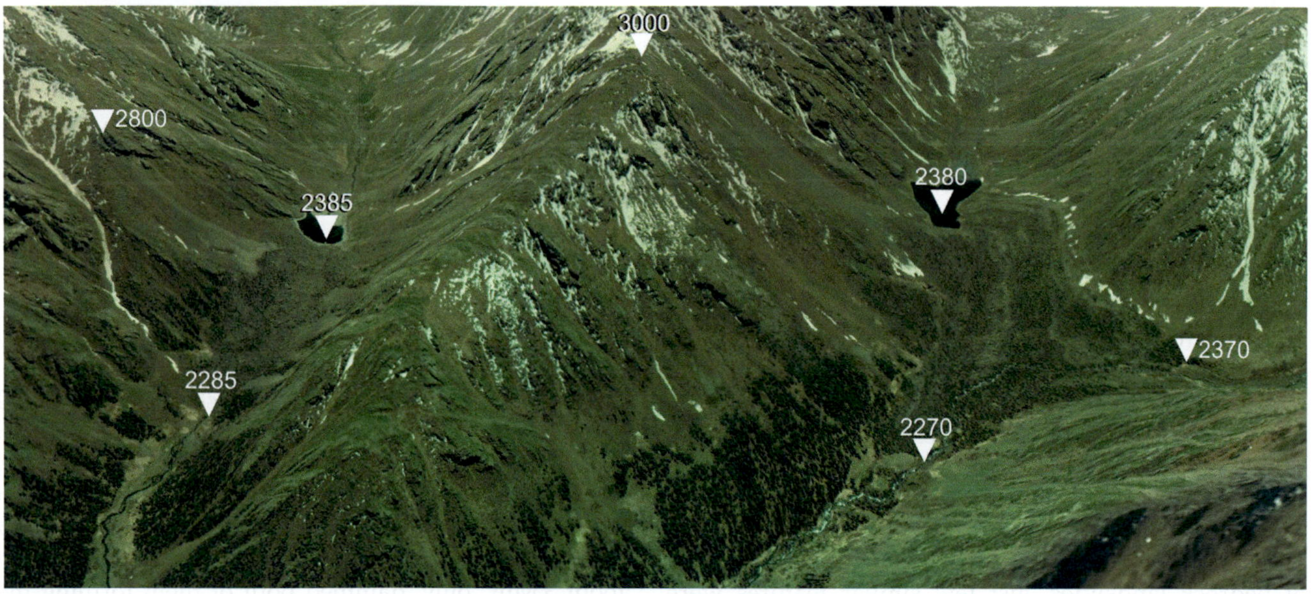

FIGURE 7.8 Similar orientation of rockslide headscarps located in the adjacent river valleys at 45.32° N, 81.33° E (Kazakh 075,076). Rockslide dams have been formed just below end moraines left by recent glaciation. 3D Google Earth view.

FIGURE 7.9 A 5.4-km long rock avalanche on the right bank of the Kumbelsai River valley. 3D Google Earth view.

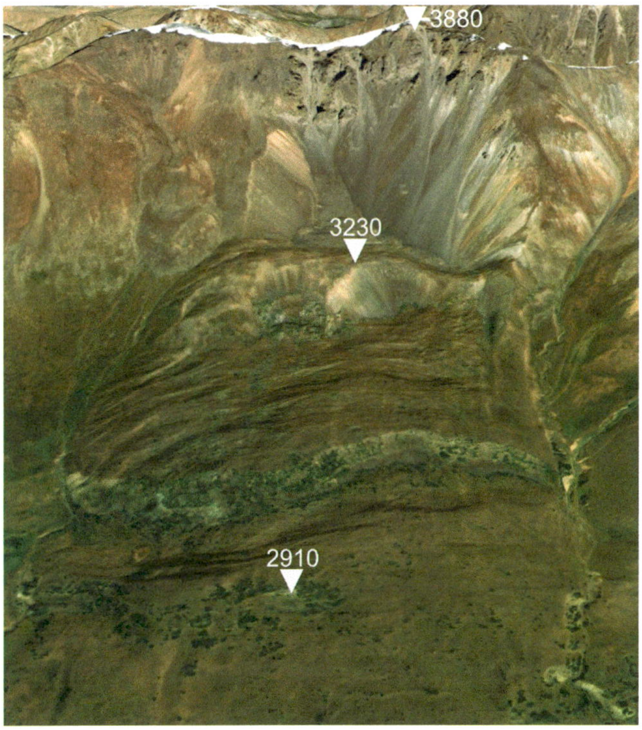

FIGURE 7.10 Front view on the headscarp and the upper part of the Kumbelsai rock avalanche with distinct secondary scars. 3D Google Earth view.

relative role of each rockslide in creation of this nick point could not be derived just by the remote sensing data analysis with confidence.

FIGURE 7.11 Large Pleistocene and Holocene rockslides and assumed landslide-related landforms on both sides of the Koksu River valley, Southern Dzungaria. Landforms marked by A, B, C, D are explained in the text. 3D Google Earth view.

Besides these two opposing features, one more old rockslide about 110 Mm^3 in volume (likely more but it was strongly eroded) has been identified presumably immediately upstream from the younger feature (Fig. 7.11C) along with a suspicious 2.3 km long and up to 210 m high step along the left-bank watershed, whose strike corresponds to the bedding and it that could be a back wall of a large subsided block (Fig. 7.11D).

We could not identify similar features along the remaining ca. 200 km long part of the Boro-Khoro range, near the Sairam-Nur lake and southeast of it.

CHAPTER

8

Tien Shan

OUTLINE

The 2500 km long Tien Shan consists of numerous mountain ranges rising from 3000–4000 to more than 7000 m a.s.l (the Victory Peak, 7439 m a.s.l.). These ranges, most of which represent neotectonic ramp anticlines, are divided by intermontane depressions often used by river valleys and, sometimes, filled by lakes.

To simplify usage of the data presented in this section we divided the Tien Shan into several conventional segments, based on the orographic criteria mainly (Fig. 8.1). We will start regional description of the Tien Shan rockslides from its eastern segment and will move westward.

8.1 EASTERN TIEN SHAN

The conventional Eastern Tien Shan includes its part east of the 82° E where the E-S-E trending Boro-Khoro Range merges the E-W trending Khabyrga Range, and up to the 88.5° E (Fig. 8.2), since no large rockslides were found east of the Bogdo-Ula Range that rises above the Urumchi City.

The easternmost group of large-scale rockslides is located about 40 km east of the Urumchi City. Here the gigantic rockslide about 1 km^3 in volume dammed the river valley that starts at the western slopes of the Bogdo-Ula Peak (5332 m a.s.l.) and created the beautiful Tianchi Lake (43.89° N, 88.11° E, China 075) used now as a resort area (Fig. 8.3). Deng and Ding (1990) provided even larger volume estimate of about 1.5 km^3.

Its headscarp destroyed the watershed at a distance of about 1.0 km similarly to the Ak-Kiol rockslide (see Section 5.3.1, Fig. 5.9) though at a much larger scale, so that its crown appeared on the opposite slope of the ridge up to ~150 m below its former crest. The northern flank of the headscarp up to 90 m high comes down almost to the lake level, while the southern one is less pronounced. Another morphological peculiarity of this rockslide is the prominent depression up to 50 m deep at its proximal part locally called the "Pan-shaped Pit" (Yi et al, 2006) and bounded from the front by a small ridge marked on Fig. 8.3 by elevation mark 2280 m a.s.l. Altogether it looks as a combination of the wedge and translational failure. This gigantic dam survived, being protected by artificial measures undertaken by Chinese authorities, despite significant erosion along its frontal part (Fig. 8.4).

The site is composed of folded Carboniferous tuffaceous siltstone and sandstone in the anticline core and andesitic volcanic breccia and tuff on both fold flanks. According to the ^{10}Be dating rockslide originated about 12 ka BP (Yi et al, 2006). (Deng and Ding, 1990) concluded that rockslide lacking loess cover should be younger than these deposits which age is determined

Rockslides and Rock Avalanches of Central Asia.
DOI: https://doi.org/10.1016/B978-0-12-803204-6.00008-9

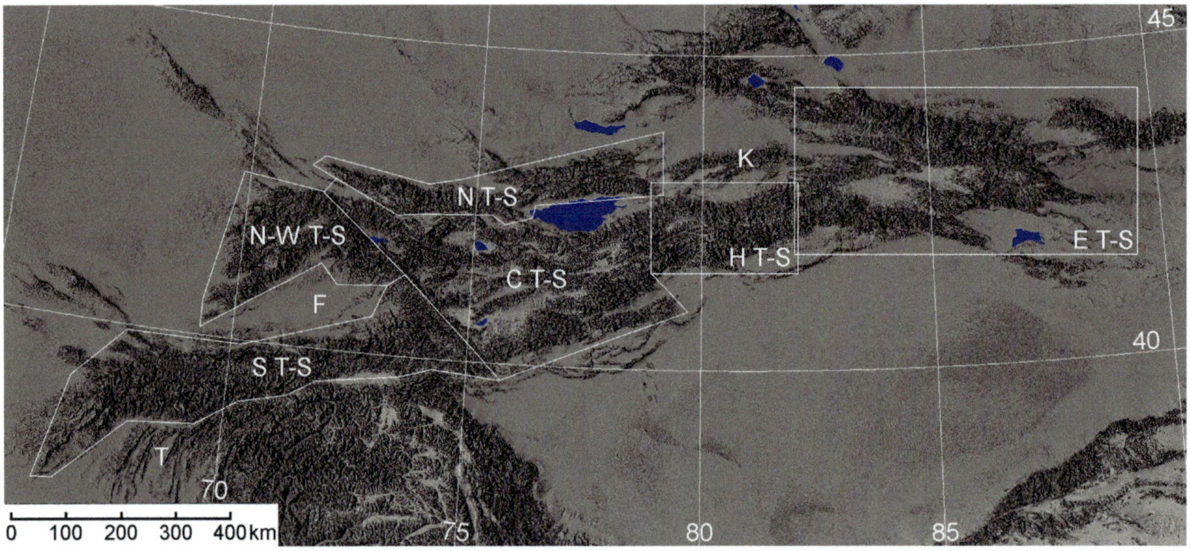

FIGURE 8.1 Schematic division of the Tien Shan mountain system. E T-S – Eastern Tien Shan (Section 8.1); H T-S – Highest Tien Shan (Section 8.2), N T-S – Northern Tien Shan (Section 8.3); C T-S – Central Tien Shan (Section 8.4); N-W T-S – North-Western Tien Shan (Section 8.5); S T-S – Southern Tien Shan (Section 8.6). K – the Ketmen Range; F – the Ferghana intermontane depression; T – the Tajik depression. Background – the 30″ SRTM DEM visualized by the Global Mapper software.

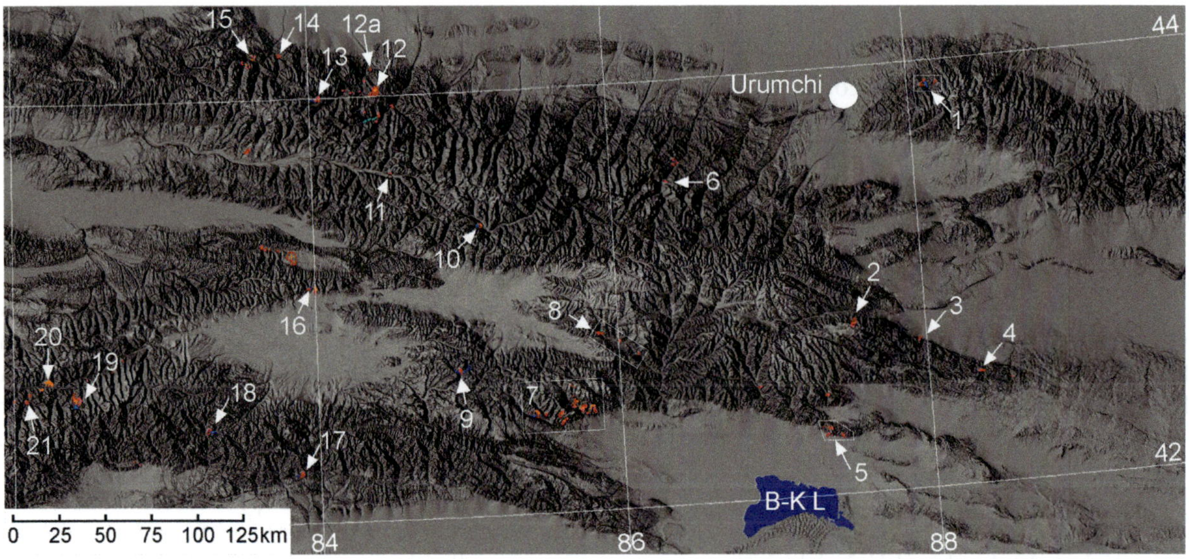

FIGURE 8.2 Shaded 3″ STRM DEM of the Eastern Tien Shan. Rockslides are marked by pink and brown dots; blue dots – dammed lakes. B-K L – the Bagrashkiol Lake. 1 – the Tianchi Lake rockslide cluster (see Figs. 8.3–8.5); 2 – the Aidyn-Ula rockslide (see Fig. 4.4); 3 – see Fig. 5.32; 4 – see Fig. 8.6; 5 – the Tsagan Giunge Mountains rockslide cluster (see Figs. 4.31, 5.2, 8.7, 8.8); 6 – see Fig. 8.9; 7 – the Sarmin-Ula rockslide cluster (see Figs. 4.9, 4.10, 6.5, 8.10–8.15); 8 – rockslides in the Arhotyn River valley (see Figs. 8.16–8.19); 9 – the Twin-lakes rock avalanche (see Fig. 8.20); 10 – see Fig. 8.21; 11 – the Kash rock avalanche (see Fig. 5.33); 12 – see Fig. 8.22; 12a – see Fig. 6.13; 13 – see Fig. 8.23; 14 – see Fig. 8.25; 15 – see Fig. 4.11; 16 – see Fig. 8.26; 17 – the Alasu rock avalanche (see Figs. 8.27, 8.28); 18 – the Keqi Kekule rockslide (see Fig. 8.29); 19 – the Big Dragon Lake rockslide (see Fig. 5.65); 20 – see Fig. 8.30; 21 – see Fig. 8.31.

as 9537 ± 114 B.P. (9244–8620 B.C. according to IntCal 13 Calibration curve) and, thus that rockslide originated in Early Holocene time.

About 5.5 km east of the Tianchi Lake dam, at 43.894° N, 88.191° E, one more bedrock slope failure of about 7 Mm^3 in volume (China 099) formed 2.15 km long rock avalanche of the Secondary type (see Section 5.3.3.1) shown on Fig. 8.5.

Another very large rock slope failure in the Eastern Tien Shan – the Aidyn-Ula rockslide about 400 Mm^3

FIGURE 8.3 The rockslide-dammed Tianchi Lake. 3D Google Earth view.

FIGURE 8.4 3D Google Earth view of the Tianchi Lake dam. Note deeply eroded canyon at its true right side now protected by the dam with a spillway. S – spring releasing seeping water.

FIGURE 8.5 Rock avalanche east of the Tianchi Lake dam. White arrows mark the secondary scar. 3D Google Earth view.

in volume described in brief in Section 4.1 (China 131) was found at 42.8° N, 87.537° E in the Borto-Ula Range, west of the deepest Tufan intermontane depression which bottom is up to -137 m bellow sea level. Interesting peculiarity of this 4.66 km long rock avalanche of the Secondary type (see Section 5.3.3) that originated on the slope composed of the Ordovician granites is that its frontal part is composed of the material, which microrelief differs significantly from that of the main body (see Fig. 4.4). It is assumed that moving rockslide debris bulldozed Quaternary, likely glacial deposits resting at the gulley bottom similar to what was described at several Pamir case studies in Section 4.2 (see, e.g. the Imom rockslide – Fig. 4.14).

About 35 km east of the Aidyn-Ula rockslide, at 42.7° N, 87.955° E, one can find a good example of the 2.5 km long Primary rock avalanche about 15 Mm^3 in volume that moved in totally unconfined conditions (China 028, see Fig. 5.32) and formed an elongated flat blade up to 730 m wide and up to 20–25 m thick approximately. This slope failure occurred on the ~500 m high slope dipping 31° and composed of the Ordovician metasediments. The bajada over which it had moved is formed by merged alluvial fans of temporary streams. Though the proximal part of this feature has been reworked at a rather large extent by temporary streams, its frontal and lateral edges are quite distinct.

One more interesting feature (China 058) was found at 42.523° N, 88.342° E, about 35 km east of the previous site. This old (Pleistocene) secondary rock avalanche about 50 Mm^3 in volume with maximal runout of 2950 m and maximal height drop of 975 m caved from the 520 m high slope composed, most likely of granites (Fig. 8.6). Rockslide body has wide proximal bench at ~1960 m a.s.l. formed by the about 170 m thick deposits and then splits into two parts being divided by a rocky spur. The longest one reached the valley bottom and, likely, blocked it, though no direct evidence of the dammed lake could be found now on the Google Earth images. We assume that distinct division of this rockslide deposits into compact proximal and highly mobile distal parts was caused by its collision with the above mentioned rocky spur that resulted in the Secondary rock avalanche formation (see Section 5.3.3).

Several interesting, though rather small features, if compared with rockslides described above, were found on the slopes of the Tsagan Giunge Mountains (No 5 on Fig. 8.2). This 5–6 km wide and up to 1.2 km high ridge is composed of the Devonian metasediments intruded by granites.

One of them – the assumed Pleistocene rockslide about 25 Mm^3 in volume at 42.31° N, 87.324° E (China 052) had dammed the small valley forcing it to cut the bypass gorge (see Fig. 4.31). Its body looks as a gigantic smoothed alluvial fan, but complex analysis of its morphology, presented in Section 4.4 allows sound conclusion on its single-event formation due to large-scale rock slope failure. One more, much smaller (~2 Mm^3) rockslide that looks much more "fresh" was found about 1.0 km downstream at 43.319° N, 87.32° E (China 051). It also dammed the river, but now only widened valley bottom filled by debris flow deposits reminds past river damming.

One of the most didactic case studies from this spatial cluster is a Primary rock avalanche at 42.28° N, 87.31° E (China 053) about 18 Mm^3 in volume that originated on the ~600 m high steep (35°) slope composed of varicolored layers of metasediments dipping in the slope. Its fan-shaped body up to 920 m both in width and length is composed of stripes of varicolored debris, which succession corresponds to that in the source zone (Fig. 8.7). Rock avalanche overlays dark-gray unit outcropping at the slope base, but its debris is absent in the rockslide body indicating that failure affected the slope above its level.

It is interesting to note that, unlike the China 028 rock avalanche that caved from the slope with more or less similar height (see Fig. 5.32), this rock avalanche with slightly larger volume spread more sidewise rather than frontwise.

Two other rockslides 18 and 10 Mm^3 in volume (China 054 and China 055, correspondingly) are located 8 and 9 km east of this feature at about 42.275° N, 87.4015° E. Here the ridge is composed of granite and both rockslides originated on the south-facing slopes 400 and 240 m high. These rockslides looking quite "fresh" are associated with an assumed surface rupture with presumably normal kinematics (Fig. 8.8). It can be hypothesized that surface rupture was formed by an earthquake. We must note that it was traced on the Google Earth image for 2.5–3.0 km only and, most likely, is not an independent structure, but just a part of a longer system of

FIGURE 8.6 About 3 km long rock avalanche at 42.523 ° N, 88.342 °E (China 058). 3D Google Earth view.

FIGURE 8.7 Fan-shaped rock avalanche body composed of varicolored stripes of debris. 3D Google Earth view.

FIGURE 8.8 Two rockslides at the southern slope of the Tsagan Giunge Mountains spatially associated with surface rupture marked by red arrows. On the left – view along this rupture strike. 3D Google Earth views.

active faults, which study, however, requires extensive field work.

Both failures occurred at the downthrown side of this active fault. The shortest distance from the rupture to the upper headscarp (China 054) is about 100 m only, while the lower headscarp is 1.5 km from it. We hypothesize that seismic strong motion, along with rapid acceleration-deceleration of the subsiding block could "dislodge" some weathered rock from the free face of the slopes. The consistence of fault kinematics and slope failure direction could be considered as one of the additional criterion of seismically triggered slope failure. Besides, "freshness" of fault scarps and rockslide bodies allows assumption of their simultaneity, which, of course, must be proved by direct dating of both.

This Tsagan Giunge Mountains cluster also includes the about 0.5 Mm^3 rockfall of the Devoniam metasediments (China 103) that caved from the ~300 m high rocky slope at 42.32° N, 87.288° E and can be considered as a small analogue of the 1991 Randa event in the Swiss Alps (Gotz and Zimmermann, 1993; Sartori et al., 2003) (see Fig. 5.2).

Small group of landslides several million cubic meters each was found in the Khutubi River valley that dissects the northern slope of the Unen Ridge – the northern chain of the Tien Shan at this section, at 43.61° N, 86.43° E. They occurred on slopes composed of the Jurassic terrigenous rocks and, most likely, represent slowly moving slope failures. Interesting case study is located about 10 km south of them at 43.526° N 86.36° E (China 130). This rock avalanche about 7 Mm^3 in volume originated in the Devonian rocks on the ~500 m high right slope of the

FIGURE 8.9 Plan view of the 2.2 km long rock avalanche that crossed the Khutubi River valley and entered the tributary valley on the opposite side of this river. SL – silted lake in the tributary valley dammed by this rock avalanche. Blue arrow shows the direction of water flow. Make notice of the significant difference of the river valley morphology up- and downstream of the blockage. Google Earth image.

Khutubi River valley and traveled 2.2 km. It crossed the river, blocked it forming the dam that was breached later on, but still forms a significant nickpoint, and finally entered for about 400 m into the left-bank tributary valley (Fig. 8.9) where one more small lake, which is totally silted now, originated. Unfortunately SRTM DEM at this part of the Unen Ridge is of poor quality that prevents quantitative assessment of this rockslide parameters.

The extremely dense cluster of 19 rockslides and rock avalanches that concentrate within a limited area of about 30 × 10 km only is located at the southern part of the Sarmin-Ula Range dissected by left tributaries of the Khaidyk-Gol River (another name – the Karashar River). The nearest similar features were mapped more than 40 km far from it

(Figs. 8.10, 8.11, Table 8.1). This cluster was described in brief in (Strom, 2015c). Most of these bedrock landslides look quite recent (Late Holocene) (see Figs. 6.5, 6.15, 8.13–8.15) and range in volume from about 0.3 to 40–60 Mm^3. One of them (China 036, for coordinates see Table 8.1) blocks the Tsagan-Gol River forming a lake (see Fig. 6.5). Besides, at some watersheds, just above or near the existing headscarp there are sackung-like features (see Fig. 8.11) reflecting either slope failures that were not released completely during rockslide formations, or the evolving DSGSD (see Section 5.1). Along with these recent and modern slope deformations, several much older, likely Pleistocene slope failures were identified within the same area (see Figs. 4.9, 4.10, 8.11). Three of them are much larger than any of the Holocene rockslide – up to 200–250 Mm^3 in volume (see Table 8.1). All rockslides occurred in granites except China 072 that affected slope composed of the Devonian metasediments.

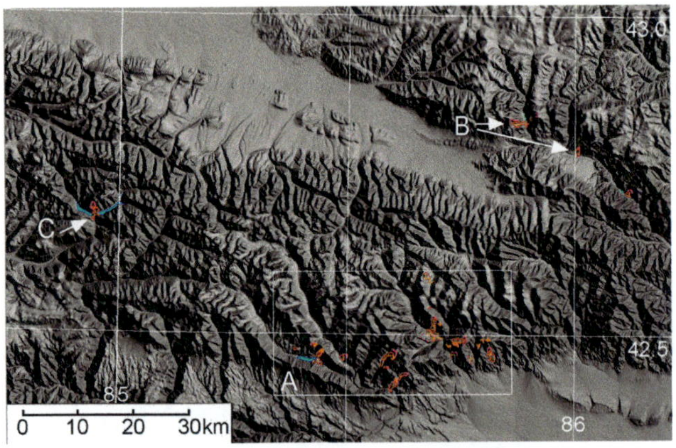

FIGURE 8.10 Uneven rockslide distribution at and around the Sarmin-Ula Range. A – the rockslide cluster shown on Fig. 8.11; B – rock avalanches shown on Figs. 8.16–8.19 (China 044, China 045, China 144); C – the Twin-Lakes rockslide dam (China 027, see Fig. 8.20). Legend is the same as for Fig. 8.11. Background – the 3″ SRTM DEM visualized by Global Mapper software.

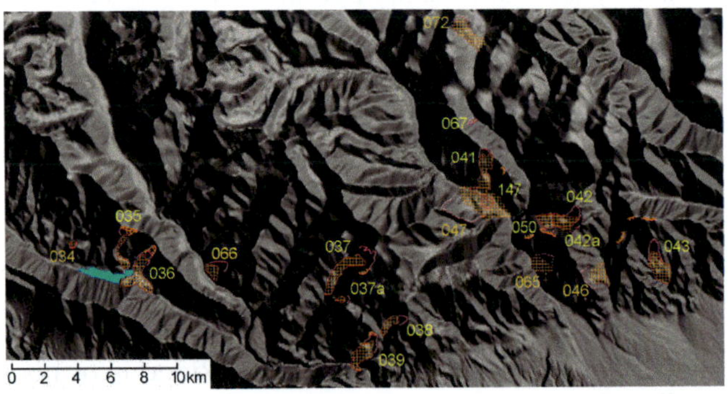

FIGURE 8.11 Detailed map of the rockslide cluster at the southern slope of the Sarmin-Ula Ridge. Cross-hatched areas with brown boundaries – confidently identified rockslide bodies; those without boundaries – old rockslide bodies; solid pink lines – recent headscarps; dashed pink lines – old headscarps; comb orange lines – sackung-like scarps; blue area – landslide-dammed lake. Numbers correspond to those in Table 8.1. Older (Pleistocene) rockslides are numbered by orange, more recent (Holocene) – by yellow. Background – the 3″ SRTM DEM visualized by the Global Mapper software.

Most of recent slope failures of this cluster demonstrate flow-like debris motion allowing classifying them as rock avalanches (see Figs. 6.5, 6.15, 8.12, 8.13), except China 050 that is a typical blockslide (Fig. 8.14). Quite often one can see distinct trimlines dozens meters above the deposits surface (see Figs. 6.5, 8.15). Rockslides can be ascribed to the Primary (see Fig. 6.5, Fig. 8.12), Jumping (see Fig. 8.13) and Secondary (see Fig. 6.15) types described in Section 5.3.

The majority of recent rockslides from this cluster occurred on the similarly-facing (towards South and South-East) slopes (see Figs. 8.11, 8.15). This fact,

TABLE 8.1 Rockslides forming dense cluster at the southern part of the Sarmin-Ula Range

ID China ...	Latitude (dd.ddddd°)	Longitude (dd.ddddd°)	Area of the deposits (km^2)	Total area affected (km^2)	Volume (10^6 m^3)
YONGER (HOLOCENE) ROCKSLIDES					
035	42.4707	85.4232	1.030	1.37	10
036	42.4609	85.4368	1.930	2.38	40
037	42.4665	85.5878	1.660	2.52	60
037a	42.4487	85.5861	0.140	0.26	0.3
038	42.4370	85.6208	0.560	0.79	9
039	42.4202	85.6010	1.245	1.68	30
041	42.5146	85.6891	1.625	1.97	20
042	42.4911	85.7497	0.910	1.22	15
067	42.5435	85.6809	0.035	0.09	0.35
043	42.4680	85.8190	1.530	2.00	40
147	42.5001	85.6909	0.120	0.36	3.5
050	42.4914	85.7369	0.663	0.87	11
042a	42.4905	85.7511	1.040	1.42	15
OLDER (LIKELY PLEISTOCENE) ROCKSLIDES					
046	42.4632	85.7757	1.000	1.70	200
047	42.5006	85.6829	3.660	5.50	250
065	42.4692	85.7320	1.200	1.60	200
066	42.4634	85.4923	0.500	1.10	15
072	42.5898	85.6759	1.610	1.80	10
034	42.4756	85.3885	0.107	0.14	0.3

Reprinted from (Strom, 2015c) with permission of the Canamaple Academia Services.

FIGURE 8.12 Primary rock avalanche (China 043) with funnel-shape confinement. 3D Google Earth view.

FIGURE 8.13 Rockslide of the Jumping type (China 041), which sliding surface daylighted at ~3300 m a.s.l., about 600 m above the valley bottom. 3D Google Earth view.

along with their distinct morphological similarity and concentration within limited area allows sound assumption on triggering of all or many of them by strong local earthquake(s).

About 35 km north of the rockslide cluster described above there are 4 rockslides in the Arhotyn River valley (see Fig. 8.10). The lowermost one (China 048), at 42.723° N, 86.118° E) forms an old deeply eroded rockslide dam looking similar to rockslide shown on

FIGURE 8.14 Conjugated bodies of rock avalanche (China 042), additional rock slope failures from the opposite slope (China 042a) and blockslide (China 050). Though rock avalanche debris of China 042 slightly overlays China 050 blockslide, it does not exclude their simultaneous formation and such mutual position could result from the bigger distance that rock avalanche debris had to pass. 3D Google Earth view.

Fig. 4.11. About 10 km upstream at 42.789° N, 86.0° E nearly 12 Mm^3 of rocks caved from the 500 m high slope, traveled ~1.7 km and reached the river, but did not cross it and stopped without any visible confinement (Fig. 8.16).

Nearly 10 km upstream the Arhotyn River two assumed rock avalanches, one of which had blocked the valley, were identified on SAS Planet imagery. One of them (China 044) at 42.83° N, 85.878° E has complex headscarp with lower (between 3610 and 3360 m a.s.l. on Fig. 8.17) and higher (with crown at 4010 m a.s.l.) concave sections. Debris from the lower part forms the left-side part of the deposits, which surface is higher than this 3.7 km long rock avalanche main body that originated from the upper section of the headscarp. This deposit about 50 Mm^3 in volume differs significantly from the Quaternary fill of the adjacent catchments both in size and morphology, similarly to the rock avalanche identified on space images in the upper reaches of the Kashkasu River valley – tributary of the Kokomeren River (see Figs. 4.43, 4.44). The proximal part of the deposit is covered by numerous conical hills (Fig. 8.18) that, most likely, are molards typical of rock avalanche deposits (see, e.g., Fig. 4.23 and also Figs. 8.143 and 8.144). Combination of all these peculiarities proves the correctness of our interpretation of this case study as a rock avalanche.

One more suspicious feature (China 144) was identified at 42.84° N, 85.905° E in the small gulley that falls into the Arhotyn River valley just at the central part of the breached dam formed by the rock avalanche

FIGURE 8.15 Two similarly directed adjacent rock avalanches (China 038 and 039). 3D Google Earth view.

FIGURE 8.16 Unconfined rock avalanche in the Arhotyn River valley (China 048).

FIGURE 8.17 Upper reaches of the Arhotyn River valley with two assumed rock avalanches, which headscarps are marked by A (China 044) and B (China 144). Outlined area is shown on Fig. 8.18. Make notice of glacial landforms north of China 044 rock avalanche with disparately smaller Quaternary fill. SAS Planet image.

described above (see Fig. 8.17). It is pronounced much worst than the previous case study (Fig. 8.19) and decision on its rock avalanche nature was based on the following considerations: 1 – the abnormally large (about 2 Mm^3) debris accumulation more than 25 m thick at the lower reaches of the gulley with sharp boundaries, 2 – freshness of the assumed headscarp, and 3 – blocking the tributary gulley at ~3300 m a.s.l. (Fig. 8.19) by the material that moved down-valley, same as at the Ak-Kiol rockslide in Kyrgyzstan (see Fig. 5.11).

Really unique rock avalanche (China 027) that had blocked two merging river valleys (Fig. 8.20) was found near the Sarmin Ula Range water divide in the upper reaches of the Khaidyk-Gol River left tributary at 42.864° N, 84.9° E. It was named the Twin-lakes rockslide in (Strom, 2010b) since the original Chinese name of this locality was not found. According to the landslide dams classification system proposed in (Hermanns et al., 2011b) it can be ascribed to the IVd subtype (see Fig. 5.29). At the same time it is a good example of the Secondary rock avalanche of the Classical subtype (see Section 5.3.3.1).

About 100 Mm^3 of the Devonian metasediments caved from the 850 m high slope just above the confluence of two river valleys, crossed the left (eastern) one and collided with its opposite slope forming about 250 m high dam, while about one half of debris retaining possibility of further motion formed an impressive

FIGURE 8.18 Multiple molards up to 30 m in diameter on top of the proximal part of the China 044 rock avalanche. SAS Planet image.

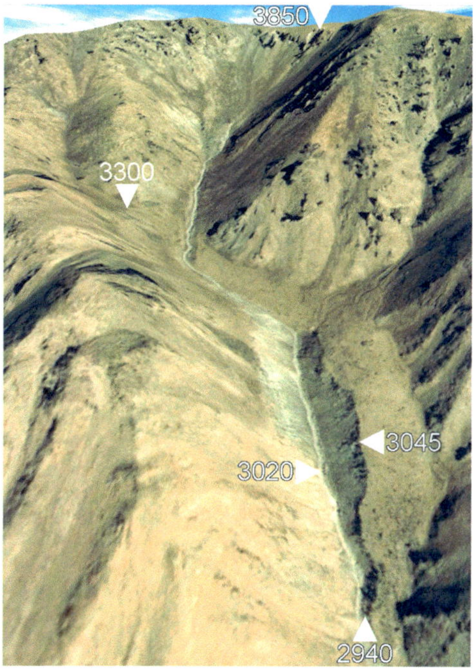

FIGURE 8.19 Assumed rock avalanche (China 144) in the left tributary of the Arhotyn River. SAS Planet image visualized in 3D by Global Mapper software.

FIGURE 8.20 Twin-lakes rock avalanche of the Secondary type that dams two merging river valleys. 3D Google Earth view.

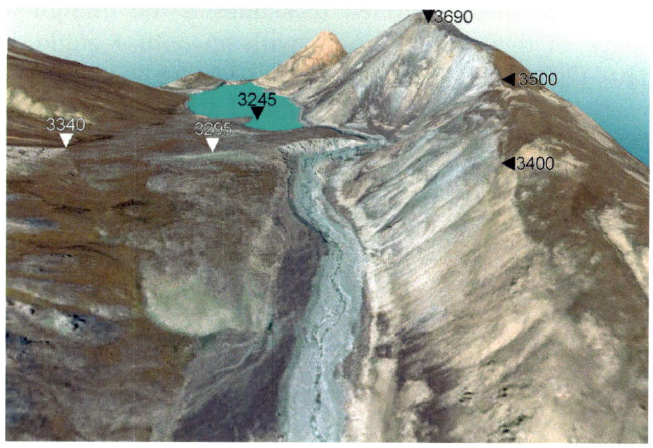

FIGURE 8.21 Rockslide dam and dammed lake in the upper reaches of the Hosutai River valley. Part of the slope between elevation marks 3500 and 3400 m a.s.l. sagged but did not collapse. SAS Planet image visualized in 3D by the Global Mapper software.

Secondary rock avalanche that traveled 2.4 km more and entered the valley downstream from the confluence. This part of the rockslide blocked the right (western) tributary by the about 100 m high dam. It resulted in formation of two lakes with 135 m difference of their water levels that makes the site quite favorable for construction of the pump-storage hydraulic power plant.

Rockslide dam up to ~120 m high that could be misinterpreted with glacial deposits was found in the upper reaches of the Hosutai River valley at 43.363° N, 85.115° E (China 125). It was formed by the fan-shaped rock avalanche about 40 Mm^3 in volume that caved from the 430 m high slope composed of Carboniferous sediments and spread over wide valley bottom with up to 60 m runup. We should notice that much larger block was detached from the host massif, but, due to the unknown reasons, did not collapse (Fig. 8.21). The dam was partially breached but still retains the lake with estimated volume of about 7 Mm^3. The shape of this dam, indicating that it moved across the valley, is very similar to that of the Rivakkul rockslide dam in Pamir (see Fig. 4.21), which also occurred in the formerly glaciated valley.

About 55 km north-east of this site, in the upper reaches of the Kash River valley at 43.622° N, 84.533° E (China 124) an interesting Kash rock avalanche about 20 Mm^3 in volume is characterized by the extreme lateral spreading of debris (see Fig. 5.33) – up to 3040 m wide, 4 times more than the headscarp outlet. Debris blanked wide river valley bottom with ~5, 12–16, and 25–30 m high terraces overlain by the alluvial fan from the tributary valley and entered the latter's mouth with up to 100 m high runup. Unlike many other cases where lateral spreading of debris was caused by collision with an opposite slope, here it reflects initial sidewise motion. Debris apron that traveled up to 1750 m from the headscarp foot is just several meters thick (see inset in Fig. 5.33). Motion mechanism of this rock avalanche requires special study.

Group of the assumed gigantic multistage slope failures in granites and Carboniferous sedimentary rocks was found at 44.02° N, 84.45° E, about 50 km south-west of the Dushanzi town (China 096, China 109, China 110, Fig. 8.22). The total area affected is about 12.6 km^2, while the deposits area reaches 10 km^2. The cumulative volume of these slope failures was estimated as ca. 800 Mm^3. Numerous younger landslides are forming now at the feet of the assumed headscarps, likely in rockslide deposits. The characteristic feature of these assumed rockslides is the triangular shape of their source zones with an obtuse angle of 110–120°. The largest rockslide deposit (China 096) was, likely, formed by the material that originated from several source zones both on its northern and north-eastern flanks (at 3500 and 3390 m a.s.l on Fig. 8.22). The hilly area at the feet of the assumed headscarps is disturbed by recent landslides with distinct horseshoe-shaped headscarps.

This group of the assumed gigantic rock slope failures marked by No 12 on Fig. 8.2 is surrounded by numerous smaller rockslides. One of the most interesting of them is located about 30 km west, at 43.99° N, 84.066° E, in the upper reaches of the Dzhergalanty River basin (China 060). Here the river valley was blocked by a large, about 150 Mm^3 rockslide that caved from the more than 1200 m high slope composed of the Devonian rocks and formed a 200 m high dam remaining almost intact (Fig. 8.23), while the dammed lake up to 5.3 km long and about 50 Mm^3 in volume was completely silted, so that at present rockslide dam forms a prominent 200 m high knick point in the river profile (Fig. 8.24). Besides this rockslide is

FIGURE 8.22 Group of several assumed gigantic rockslides with numerous subsequent landslides in the rockslide deposits. Numbers 096, 109 and 110 correspond to China 096, China 109 and China 110 case studies in the rockslide inventory. SAS Planet image visualized in 3D by the Global Mapper software.

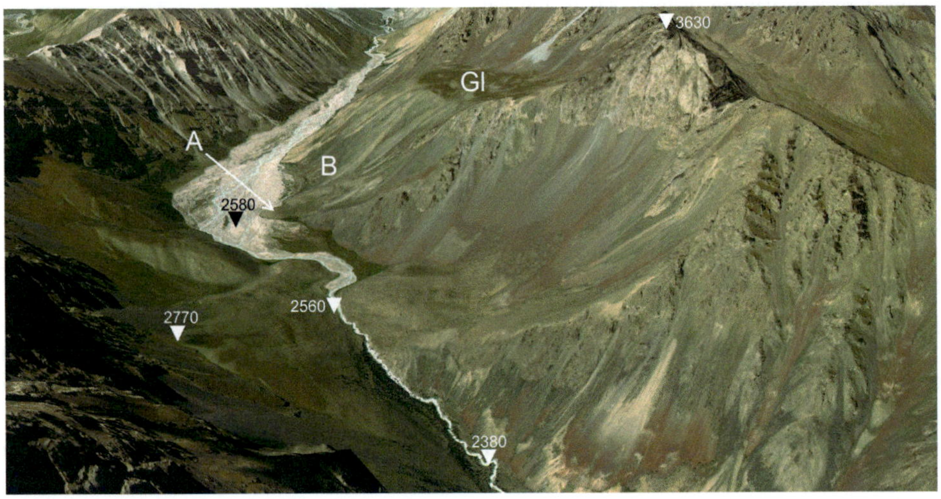

FIGURE 8.23 Rockslide dam and completely silted dammed lake in the upper reaches of the Dzhergalanty River basin. A – body of the small rockslide resting over the lacustrine sediments; B – headscarp of the rockslide, which body was buried by lake fill; Gl – assumed Pleistocene glacial landform (compare with Fig. 4.20). 3D Google Earth view.

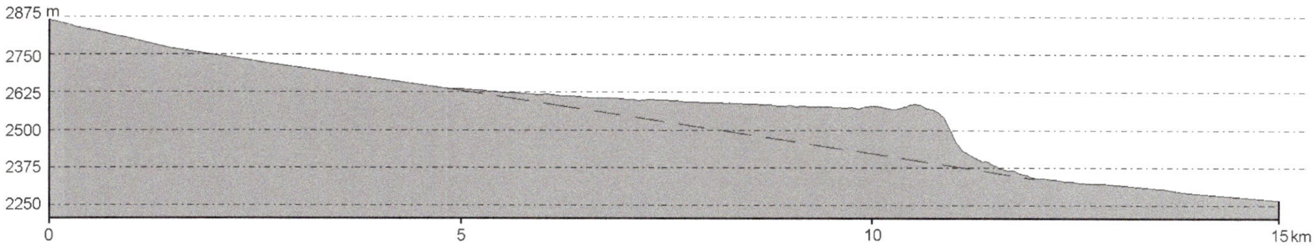

FIGURE 8.24 Prominent knick point on the river profile formed by the rockslide dam shown on Fig. 8.23. Dashed line – assumed pre-slide river profile.

characterized by a significant runup – its' deposits distal edge is up to 210 m higher than the lowermost part of the dams crest. This large rockslide was accompanied by two much smaller rock slope failures (A and B on Fig. 8.23) one of which (A), less than 1 Mm^3 in volume, rests over the lacustrine deposits, thus being much younger than the main rockslide, while the body of the second one (B), 3.5 Mm^3 in volume approximately, is buried by the lake fill, assuming that it could occur simultaneously with the "big brother".

Another didactic case study from this region is the 1.34 km long rock avalanche of the Secondary type at the northern slope of the Iren-Khabyrga Range (44.127° N, 84.426° E, China 033) (see Fig. 6.13). It originated as a wedge failure on the slope composed of well-bedded Jurasic sedimentary rocks. About 7 Mm^3 of rockslide debris accelerated descending for about 400 m and collided with a small rocky spur, which resulted in formation of the pronounced Secondary rock avalanche with distinct secondary scar up to 60 m high (see Section 5.3.3.1). The mobile part involves ~70% of the entire rockslide debris. Rockslide is located just 6 km of the range foot, less than 2 km from the road and is easily accessible.

Several interesting features could be found 25–35 km north-west of the above mentioned rockslide dam, in the Epte River basin (see Fig. 8.2, No 14 & 15). One of them (China 040) is located on the right bank of the Tsagan River valley close to its confluence with the Ashu River at 44.206° N, 83.809° E. It affected the Carboniferous rocks, which different units are characterized by the distinct lithological and color contrast (Fig. 8.25). About 15 Mm^3 of rockslide debris, approximately (perhaps more, but very irregular shape of the deposits prevents more precise volume estimate), moving downslope split into two parts being dissected by the rocky cliff. The true left one had dammed the river (the blockage was breached, likely, rather fast), while the true right one entered short steep gulley and followed it. Small portion of rockslide material that differs sharply in color from the surrounding bedrock, finally also reached the main river at 1235 m a.s.l..

FIGURE 8.25 Rockslide close to the Tsagan River mouth, which debris differs in color from the surrounding bedrock. 3D Google Earth view.

About 13 km west, in the Ashu River valley at 44.2° N, 83.644° E, there is an assumed old (Pleistocene) rockslide (China 095) briefly described in Section 4.1 (see Fig. 4.11). Its' almost 400 m thick body about 200 Mm^3 in volume is completely dissected by the river providing good opportunity to study its internal structure, similarly to the Kokomeren rockslide in Kyrgyzstan (Strom, 1994a, 2006) that will be described in Section 8.4. Two assumed large rockslides were identified on the opposite side of the Ashu River valley at 44.17° N, 83.604° E (China 107) and at 44.178° N, 83.57° E (China 108). It could not be excluded, however, that these landforms located 500–800 meters higher than the base of the feature shown on Fig. 4.11, are of glacial origin.

FIGURE 8.26 Blockslide (on the left) and rock avalanche (on the right) that caved from the same watershed at the southern slope of the Narat Ridge. 3D Google Earth northward view.

Small rockslide cluster at the southern feet of the Narat Ridge (16 on Fig. 8.2) includes several small unconfined rockslides that collapsed southward from the slope composed of granite and two rockslides at 43.107° N, 83.955° E that caved from both sides of the about 400 m high watershed, also in granites (Fig. 8.26). While the western one (China 031), about 7 Mm^3 in volume, looks as a blockslide that, practically, remains within the headscarp, the eastern one, 15 Mm^3 approximately, formed a rock avalanche with 1.9 km long runout that crossed the river valley damming a lake and, moving further, entered the tributary valley mouth for about 0.5 km (up to elevation mark 2970 m a.s.l. on Fig. 8.26).

These two rockslides are, most likely, simultaneous. We must notice that distinct fault scarps 5–10 m high can be traced along the northern boundary of the Large Yulduz intermontane depression, 4–7 km south of the Narat Ridge foot, providing sound argument in favor of seismic triggering of the above rock slope failures. However, some other rockslides that form this cluster look much older.

A very large landslide in soft Quaternary sediment with an area about 22 km^2 that was formed on a slope with mean angle 4–5° only at 43.26° N, 83.85° E in the small neotectonic depression used by the upper part of the Tsanmo River valley. It is shown on Fig. 8.2 about 15–20 km NW of the site marked as "16". It has well defined, though smoothed en-echelon headscarps and, likely, could be classified as a giant lateral spread (Varnes, 1978; Hungr et al., 2014). It is accompanied by several small rockslides downstream the Tsanmo River valley.

FIGURE 8.27 The source and the upper part of the transition-deposition zone of the Alasu rock avalanche. 3D Google Earth eastward view. Varicolored background is formed by different space images.

Gigantic, about 250 Mm^3 in volume, rock avalanche of the Secondary type (see Section 5.3.3) was found at 42.23° N, 83.87° E, in the Alasu River basin (China 087). Rockslide originated on the ~ 1 km high slope composed of Carboniferous rocks with well pronounced bedding and traveled 4.16 km with maximal height drop of 1.23 km through very irregular terrain, blocking several gullies. These small lakes have been filled by debris flows completely. The headscarp northern wall represents well-defined slightly bent sliding surface coinciding with bedding. Its upper part, up to ~2950 m a.s.l., dips south at 30°, while ~250 m high lower part is steeper – up to 39.4° (Fig. 8.27). Much smaller eastern wall of the headscarp crosses bedding, being predetermined by fracture systems. Generally, this slope failure can be classifies as a planar slide (Hungr et al., 2014) that converted into Secondary rock avalanche (Strom, 2006). Rockslide collided with an obstacle marked by 2840 m a.s.l. and its part retaining possibility of further motion rushed ahead forming 2.9 km long Secondary rock avalanche with up to 250 m high secondary scar. Same succession of the sequential emplacement stages resulting in similar morphology of the deposits took place at the Twin-Lakes rock avalanche (see Fig. 8.20).

The frontal part of this rock avalanche is covered by large isomeric angular blocks up to 20 m in size and rises gradually for ~60 m towards its distal edge (Fig. 8.28). It allows calculating debris velocity (~125 km/h in this case) and indicates retention of the rock avalanche kinetic energy before final collision with an obstacle that halted its motion far away from the source zone. Similar distal debris accumulation and runup was observed at the Mingteke rock avalanche in Kyrgyzstan (see Fig. 6.14). No evidence of overtopping or erosion can be seen on this natural dam except traces of small slope failures from the deposits' front right side near the settlement, which could be either caused by slope undercutting (which is, however, unlikely in the dammed lake environment) or, more likely, could reflect the same phenomena as of the Secondary rock avalanche formation (see Section 5.3.3) due to final collision with an obstacle. If this assumption is true, the deposit of this minor Secondary rock avalanche should be buried by lake sediments.

FIGURE 8.28 The distal part of the rock avalanche in the Alasu River basin. Blocks up to ca. 20 m in size on the rock avalanche surface at A and B are shown in the inset. Winter Google Earth image.

Another very large rockslide in this part of the Chinese Tien Shan (China 146) is located at 42.45° N, 83.27° E, in the upper reaches of the Kucha River basin that drains the southern slopes of the Khalyktau Range. It forms more than 200 m high natural dam of the Artukul Lake (site name indicated on the 1:500 000 Soviet topographic map; the alternative site name provided by Google Earth – the Keqi Kekule).

The nearly rectangular block of the Silurian metasediments 1500 × 1200 m in size and up to 150 m thick slid down from the ~1 km high right slope of the river valley along the sliding surface dipping ~37°. Bedding (or foliation) in the host rock massif is pronounced well, though not at such extent as in the previous case. Rockslide volume was estimated as 150 Mm^3. Debris blocked the valley and rose on the opposite slope practically up to the crest of a small rocky ridge (Fig. 8.29). The elevation difference between the proximal lowermost and distal highest parts of the dams' crest is up to 150–170 m, which indicates very high velocity of debris motion – more than 200 km/h. The dammed lake has never overtopped the dam which lowermost point is about 40 m above the lake level, thus water balance is maintained by filtration. Small pond on the downstream part of the dam ~55 m below the lake level indicates low gradient (~0.04) of the water seeping through the upper part of the blockage.

The next very interesting feature assigned to the Eastern Tien Shan is the gigantic Big Dragon Lake rockslide in the Khalyktau Range (42.6° N, 82.4° E, China 015) about 2 km^3 in volume described in Section 5.3.3.3 as an example of the combined Primary and Secondary rock avalanche type (see Fig. 5.65). This unique slope failure is surrounded by numerous rockslides. Most of them are much smaller except one at 42.69° N, 82.22° E, about 17 km NW of the Big Dragon Lake rockslide where the entire 1.34 km high, slightly convex slope demonstrate evidence of the ongoing slope deformations, likely deep seated, affecting area of 9.5 km^2 (China 020, Fig. 8.30). Volume of rock massif involved can exceed 1 km^3. Evidence of the ongoing slope deformations could be seen on some other slopes in the Kioksu River valley and in its tributary valleys. Since catastrophic failure of such slopes cannot be excluded, these phenomena pose a potential threat to the entire Kioksu River valley with 2 hydraulic schemes downstream and require organization of a regular monitoring.

The westernmost site described in this Section (China 010) belongs to the same group of gradually

FIGURE 8.29 The Keqi Kekule rockslide dam and dammed lake. 3D Google Earth view.

FIGURE 8.30 Right slope of the Kioksu River valley with evidence of the ongoing slope deformations. 3D Google Earth view.

moving rockslides as that shown on Fig. 8.30. It is located on the left bank of the Kioksu River left tributary at 42.608° N, 82.078° E where the entire 1.13 km high slope composed of the Silurian metasediments is affected by slope deformations. Volume of this landslide was estimated as about 25 Mm^3. The site is characterized by clear evidence of more active landslide that affects lower part of the slope up to 2700 m a.s.l. Several more similar slow moving landslides were identified nearby. One of them – the China 145 at 42.64 N, 82.079 E – is partially visible on Fig. 8.31.

FIGURE 8.31 Gradually moving rockslide, lower part of which undergoes most recent deformations. Ch 145 at the upper right corner – part of one more similar rockslide (China 145) 3.5 km north. 3D Google Earth view.

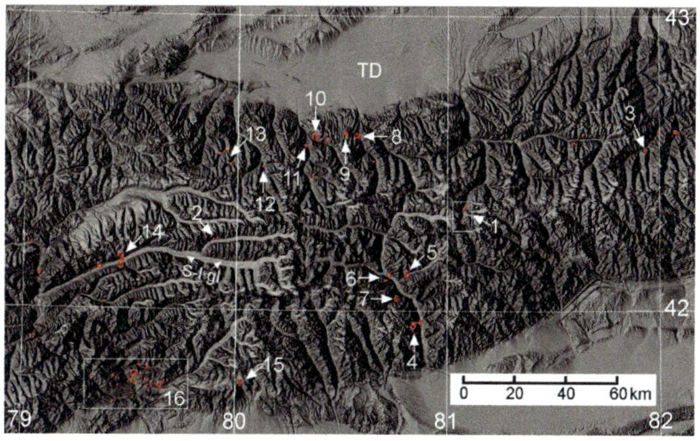

FIGURE 8.32 Shaded 3″ STRM DEM of the "Highest Tien Shan". Rockslides are marked by pink and brown dots; blue dots – dammed lakes. TD – the Tekes intermontane depression; S-I gl – the Southern Inylchek Glacier. 1 – rock avalanches at the upper part of the Koprsay glacier (see Figs. 4.32, 4.33); 2 – see Fig. 8.33; 3 – see Fig. 8.34; 4 – see Fig. 4.30; 5 – see Fig. 5.40; 6 – the Upper Muzart rock avalanche (see Figs. 8.35, 8.36); 7 – see Fig. 8.37; 8 – see Fig. 6.25; 9 – see Figs. 8.38, 8.39; 10 – see Fig. 8.40; 11 – see Fig. 6.12; 12 – see Fig. 8.41; 13 – see Fig. 8.42, 8.43; 14 – see Figs. 6.4, 8.44–8.50; 15 – see Fig. 8.51; 16 – the Sarydjaz-Aksu rockslide cluster (see Figs. 8.52, 8.53, 8.57).

8.2 HIGHEST TIEN SHAN

This part of the Tien Shan mountain system extending from 79° to 82° E (Fig. 8.32) was named "the Highest Tien Shan" because it embraces its most elevated part that reaches 7439 m a.s.l. (the Victory Peak at the Chinese-Kyrgyzian border). While its eastern limit (82° E) is just conventional, the western one coincides with the Sarydjaz-Arsu River valley (in its upper reaches – in Kyrgyzstan – the river is named the Sarydjaz and downstream – in China – the Aksu. Not to mix it with numerous other "Aksu" we use this dual name. We should notice that the 290 km long Ketmen Range (marked by "K" on Fig. 8.1) located north of this segment of the Tien Shan and divided from it by the Tekes intermontane depression (TD on Fig. 8.32), is almost free of large-scale rockslides despite high elevations – up to 3500 m a.s.l. Evidence indicating that this neotectonic feature is older than most of the nearby Tien Shan ranges can explain absence of the large-scale gravitational slope processes manifestations.

High elevations predetermine wide development of modern glaciers in this segment of the Tien Shan Mountains, the largest is the 60.5 km long Southern Inylchek Glacier that covers (at present) more than 567 km^2. Many other glaciers are up to 20–35 km long. As it was mentioned already in Section 4.5, rapidly (as compared with most of other geological processes) moving glaciers rework rockslide and rock avalanche deposits that fell on them, which can explain the deficit of identifiable rockslides in the most elevated part of the area despite quite favorable geomorphic conditions – kilometers-high steep, sometime almost vertical slopes of the trough valleys. Only most recent rock slope failures left the identifiable deposits on the glaciers (see Figs. 4.32, 4.33). On the other hand, many of the long valley glaciers, especially their lower, downstream parts are covered by moraine material almost entirely so that ice itself is practically invisible. This material is supplied both by talus and numerous minor rock falls from the steep slopes and by individual larger slope failures such as those shown on Fig. 8.33. However, identification of the deposits left by large-scale slope failures among the moraines requires special field studies and microscopic analysis of finest fractions of the material (Reznichenko et al., 2012a, b).

Spatial distribution of rockslides in the study section of the Tien Shan is characterized, similar to the Eastern Tien Shan, by the extreme unevenness (see Fig. 8.32). Most of them are grouped in several clusters (Strom, 2015c), though some features are located dozens kilometers from the nearest "neighbors".

About 13 km south-west of the site shown on Fig. 8.31, at 42.541° N, 81.936° E a small (~0.7 Mm^3 in volume only) rock avalanche caved from the upper part of a ridge up to 3400 m a.s.l. high composed of the Silurian metasediments (China 013, Fig. 8.34). Though its volume is less than the conventional

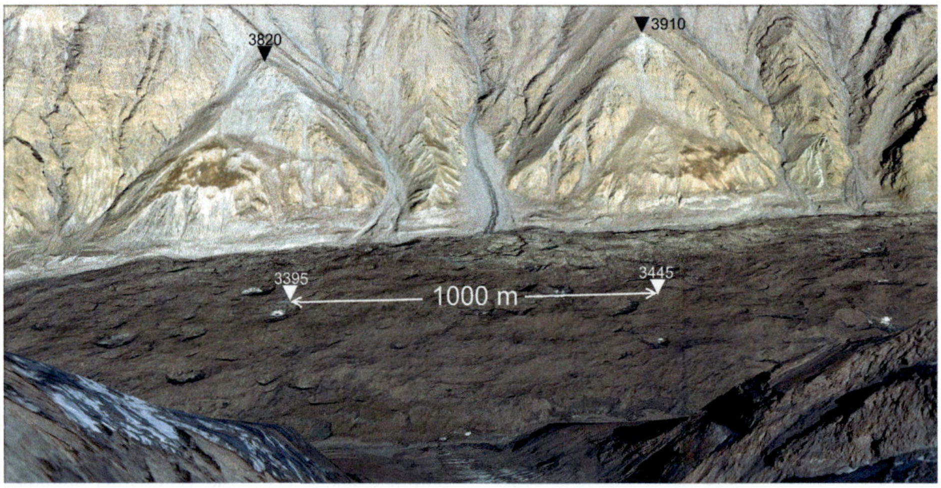

FIGURE 8.33 Relatively small (1.0–1.5 Mm^3) rockslides on the right slope of the Northern Inylchek Glacier valley, just upstream of the Mertsbaher Lake (~42.234° N, 79.883° E, Kyrgyz 198, 199). Notice the complete coverage of the glacier by moraine, so that ice can be seen in the few sinkholes only. 3D Google Earth view.

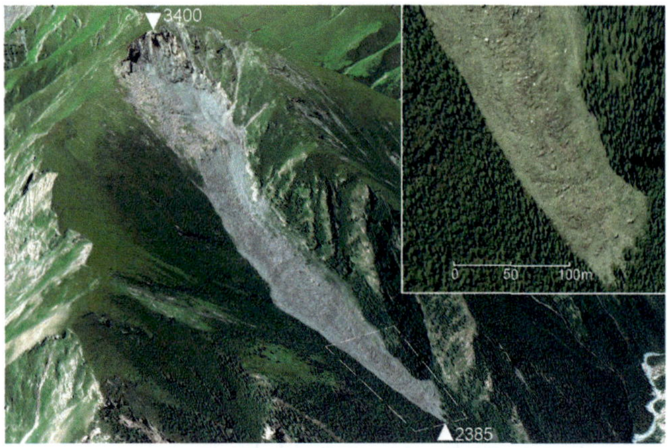

FIGURE 8.34 Small-size recent rock avalanche with 1.72 km runout. 3D Google Earth view. Plan view of its outlined distal fragment is shown in the inset.

threshold value of 1.0 Mm^3, we present its brief description due to several reasons. First, despite small volume it has rather long – 1.72 km – runout, thus representing a good example of the flow-like debris motion along the dry gulley sloping 30° that stopped by friction only, without any frontal confinement except rather thick forest. That is why this rock avalanche was included in the analysis of the relationships between rockslides' volume and parameters characterizing their mobility presented in Chapter 10. Second, since this rock avalanche is totally clear of vegetation and looks extremely fresh it can be considered as one of the most recent catastrophic events in the study region. As far as it is visible on space image obtained on August 11, 2010 provided by Google Earth, we know that failure occurred earlier, but it is definitely not older than few decades.

Group of 4 very large rockslides ranging in volume from about 60 to 400 Mm^3 was identified in the Muzart River basin draining the highest part of the Tien Shan from the east and south-east. Two of them were described in Sections 4.4 and 5.3.1. The first one (China 009) at 41.95° N, 80.87° E is a good example of the deflected Secondary rock avalanche with giant fan-shaped distal part that had blocked the Muzart River (see Fig. 4.30). It is the largest rockslide in this river basin about 400 Mm^3 in volume more than half of which – almost 250 Mm^3 was involved in the avalanche-like motion. The second one (China 002) at 42.125° N, 80.765° E, represents the Primary rock avalanche 90 Mm^3 in volume that formed ~250 m high natural dam in the narrow Akchisuu River valley – the left tributary of the Muzart River with 170 m frontal runup (see Fig. 5.40).

The Upper Muzart rock avalanche (China 133) – the uppermost feature of this group – occurred on more than 1.5 km high right slope of the Muzart River valley composed of the Paleozoic metasediments just downstream of the Karagiulsuu River and the Tuzbelchisuu River confluence (42.12° N, 80.725° E). About 60 Mm^3 of debris formed almost rectangular body 1.4 km long (across the valley) and 1.8 km wide (along it) (Fig. 8.35). This rock avalanche body somehow recalls the much larger Kudara-Pasor rock avalanche in Pamir that also had crossed wide and relatively flat bottom of the former glacial trough (see Fig. 4.20). As far as the longitudinal profile along the valley up- and downstream of this blockage is almost the same, the dam that is about 25 m high on an average only, was breached very soon after its formation, and no siltation of the dammed lake occurred. Rock avalanche deposits are composed of large angular

FIGURE 8.35 The 3D Google Earth view of the Upper Muzart rock avalanche.

blocks up to 30–60 m in size, sometimes looking as the jigsaw puzzle (Fig. 8.36-A,B), indicating that destruction of even larger blocks occurred just before motion halted. They are covered by large alluvial fans formed by debris flows both from the headscarp area and from the opposite side of the valley that are composed of much finer material (Fig. 8.36-C).

The last "member" of this group is a huge blockslide about 250 Mm^3 in volume at 42.042° N, 80.765° E (China 001) that originated on the ~1 km high slope and slumped down for about 500 m but did not convert into flow-like rock avalanche (Fig. 8.37). Likely, it had blocked the right tributary of the Muzart River, at least partially. Its headscarp differs significantly from the adjacent parts of the valley slope and landforms visible at its base provide an opportunity to identify them in the rockslide body. The site is about 3.5 km of the road that passes along the Muzart River valley.

FIGURE 8.36 The composition of the Upper Muzart rock avalanche body. Blocks up to 30 and even 60 m long are visible in the insets A and B. Sharp contrast between rock avalanche deposits and the deposits left by debris flow can be seen in the insets B and C. The jigsaw puzzle is visible in the inset B. Google Earth images.

It can be hypothesized that catastrophic breach of one of the rockslide dams in the Muzart River basin could produce large alluvial fan 27 km long and up to 7 km wide well visible at the southern foot of the Tien Shan on Goggle Earth images.

One more rockslide cluster is located in the Northern Muzart and the Lesser Muzart River basins about 10 km south of the Tekes intermontane depression southern boundary (see No 8 on Fig. 8.32). It includes almost intact 190 m high rockslide dam ~30 Mm^3 in volume at 42.59° N, 80.584° E (China 007, see Fig. 6.25) with the completely silted lake up to 4.6 km long accessible by the road. Just nearby, in the small tributary valley, another, but much older rockslide can be seen on the same image (China 003, 42.597° N, 80.567° E). It has distinct spoon-shape headscarp with a crown at about 3220 m a.s.l. and a remnant of the ~200 m high body at its foot. Both slope failures occurred on slopes composed of granite.

FIGURE 8.37 Large-scale blockslide in the Muzart River basin. 3D Google Earth view.

Just 4 km east of this site, at 42.6° N, 80.52° E, interesting group of much smaller (from less than 1 to 5 Mm^3 in volume each) slope failures occurred within the contact zone between granite intrusion and host Carboniferous metasediments. Five of them (China 004, 006, 006a, b, c) collapsed from both sides of a 250–300 m high watershed with a trace of surface rupture passing along its northern feet and clearly visible on the space image obtained in winter (Fig. 8.38). One more small (~0.5 Mm^3) rockslide occurred 1.5 km south of it (China 006d).

Three rockslides (China 006, 006a and 006b) plus one more much smaller feature not included in the database that is visible on Fig. 8.38 between China 006 and 006a have occurred, most likely, simultaneously so that their bodies merged. Such simultaneous failure of rocky slope in the arid zone is a sound argument in favor of their seismic triggering, especially considering their closeness to surface rupture. Their temporal relationships with two rockslides that originated on the northern slope (China 004 and 006c) are unclear. The latter are characterized by relatively deeper headscarps, especially the China 004, which headscarp forms up to 100 m deep bite in the north-facing slope (Fig. 8.39) and looks much younger than all the adjacent slopes. It seems that the 1.6 km long body of this rockslide overlay the trace of surface rupture (see the inset in Fig. 8.39). However, clarification of temporal relationships between all rockslides and surface ruptures in this area requires detailed field studies and dating.

About 8–12 km west of the site described above one more group of rockslides originated on slopes composed of Carboniferous sedimentary rocks. The largest of them had formed a huge dam up to 250 m high in the Urtentai River valley (the right head of the Lesser

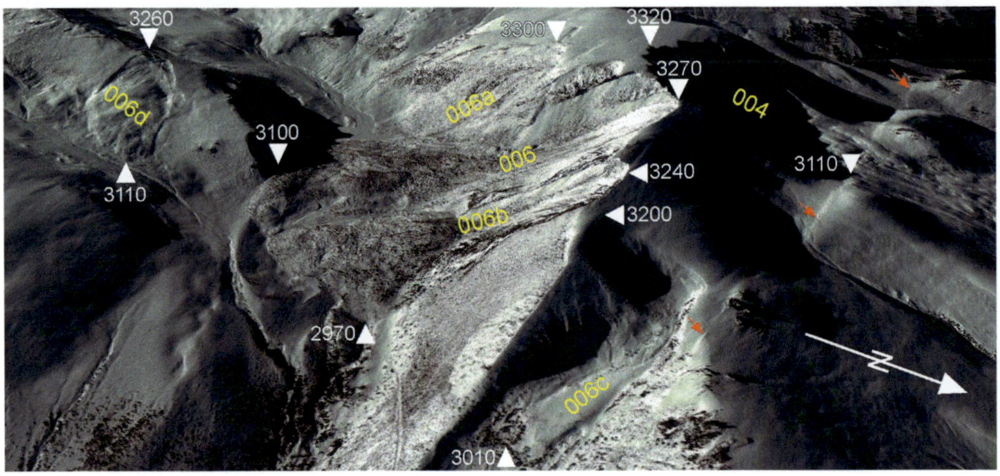

FIGURE 8.38 Group of rockslides closely associated with surface rupture marked by red arrows. 3D Google Earth view of December 28, 2010. Rockslide numbers in yellow.

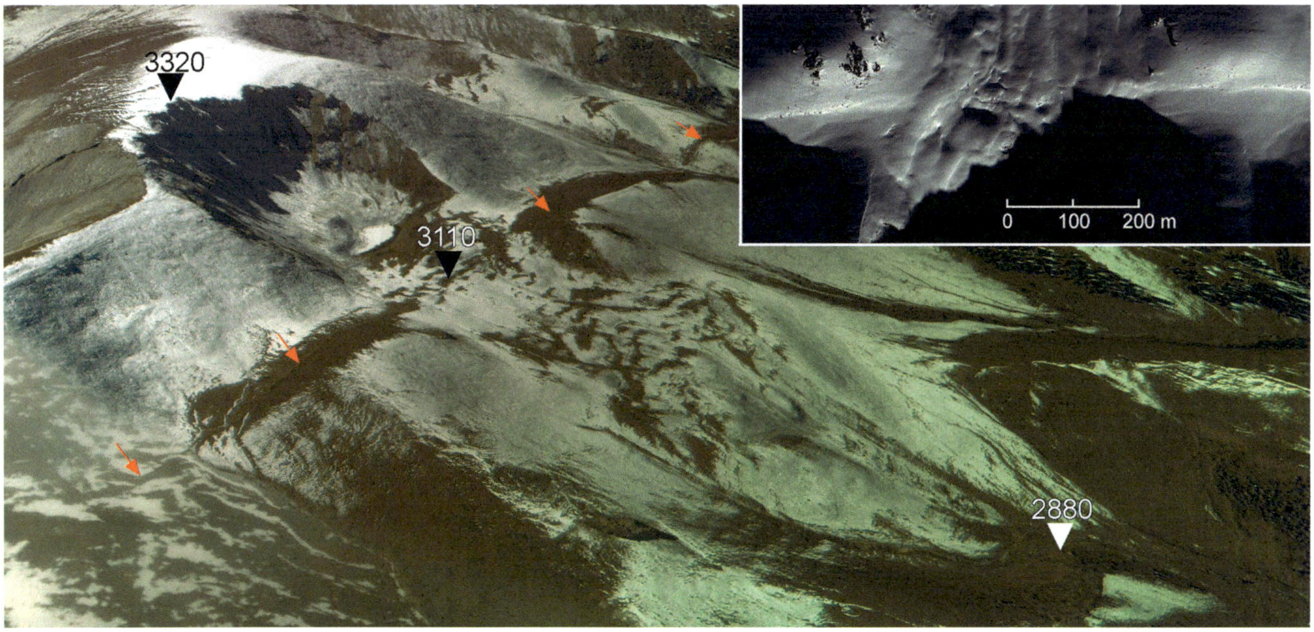

FIGURE 8.39 Rockslide (China 004) with an anomalously deep fresh headscarp and rockslide body that crosses and, likely, overlays (see the inset) surface rupture marked by red arrows on the 3D Google Earth view. Inset demonstrating its relationships with surface rupture is the zoomed fragment of winter space image, the same as on Fig. 8.38.

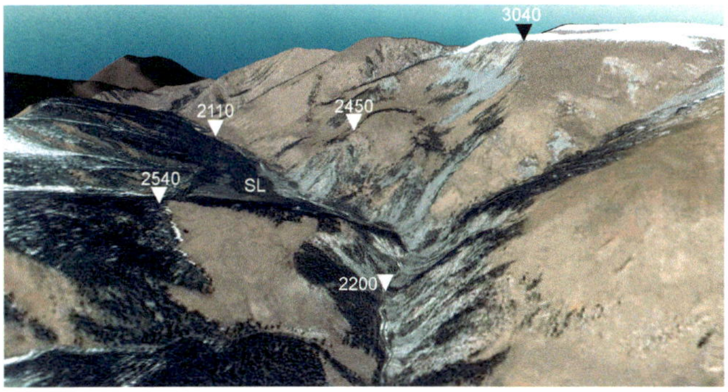

FIGURE 8.40 250 m high breached dam in the Urtentai River valley (looking downstream). Notice subsequent landslide formed in the rockslide deposits (SL). SAS Planet image visualized in 3D by Global Mapper software.

Muzart River) at 42.59° N, 80.363° E (China 005) that was breached completely so that the cross-section of its 300 Mm^3 body can be observed (Fig. 8.40). Besides, subsequent failure took place on the left side of this newly formed downstream part. Rockslide affected the entire ridge between two small gullies up to its flat top, which is considered often as being typical of the earthquake-induced slope failures (see Chapter 13: Seismically Induced Rockslides: Their Discrimination). Similar position of the headscarp – up to the ridge top – characterizes another, much smaller rockslide (China 062) ~30 Mm^3 in volume that occurred on the same right side of the Urtentai River valley about 5 km upstream (at 42.57° N, 80.423° E).

Just 4.5 km south-west, in the small tributary of the Saihan River at 42.558° N, 80.324° E one more rockslide about 12 Mm^3 in volume (China 061) forms a dam up to 110 m high with a small lake upstream. Its triangular headscarp destroyed ridge promontory just near the junction of two valleys (see Fig. 6.12)

About 18–33 km west of the Lesser Muzart River valley, on the Kazakh side of the China-Kazakh border, at the northern slope of the eastern part of the Terskey Alatau Mountain Range three rockslide in Paleozoic granites formed natural dams two of which still retain lakes, while one was silted completely. The latter one (Kazakh 004) caved from the rocky spur above the confluence of the Keskentas River and its right tributary – the Alaigyr River (42.488° N, 80.125° E). Geomorphic position and shape of its headscarp is very similar to that of the rockslide shown on Fig. 6.12 – its more than 50 m deep headscarp destroys the ridge promontory (Fig. 8.41). About 4 Mm^3 of debris blocked the valley by up to 70 m high dam that remains practically intact despite distinct overflow at its central part. The dammed lake was filled by sediments. Such erosion resistance could be explained by two main factors – presence of numerous blocks of granite up to 10 m in size in the dams' carapace (see inset in Fig. 8.41) and significant, 1.35 km along-valley debris spreading that reduced stream's gradient and its erosive power. Both prevented streams' incision.

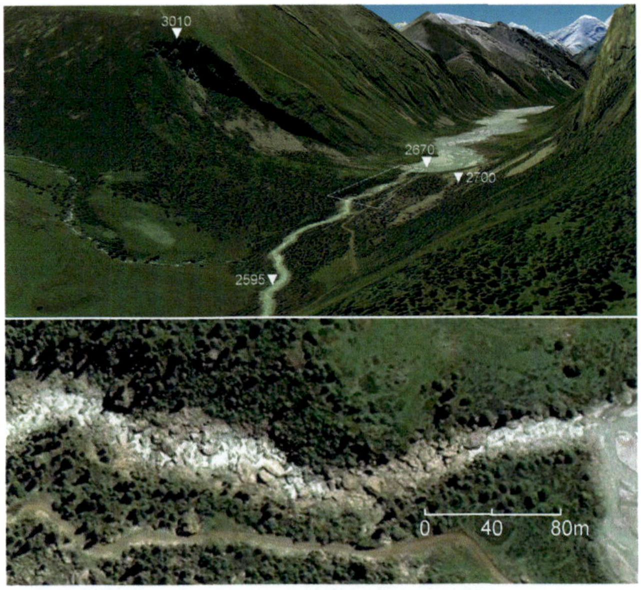

FIGURE 8.41 Above - the intact rockslide dam and silted lake at the confluence of the Keskentas and the Alaigyr Rivers. 3D Google Earth view. The zoomed outlined area shown bellow demonstrates the dam's carapace composition with angular blocks up to 10 m in size.

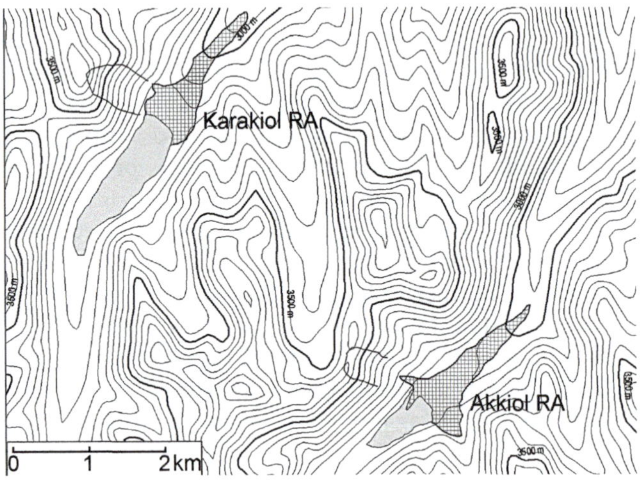

FIGURE 8.42 The schematic map of the Akkiol and Karakiol rock avalanche dams in South-Eastern Kazakhstan.

FIGURE 8.43 The Akkiol rock avalanche with long runout flow-like body. 3D Google Earth view.

Further west, in the adjacent Akkiol River and the Bazinbai River valleys, two similarely oriented rockslides 8 and 17 Mm^3 in volume correspondingly (Kazakh 037 at 42.494° N, 79.966° E and Kazakh 036 at 42.528° N, 79.925° E) form the Akkiol and the Karakiol dammed lakes (Fig. 8.42). Formation of the similarly oriented rockslides on the slopes of the adjacent river valleys was observed in the Kokomeren River valley (Southern and Northern Karakungey rock avalanches – see Fig. 5.52), at the northern slope of the main Dzungarian Range (see Fig. 7.8), and within the rockslide cluster at the southern slope of the Sarmin-Ula Ridge (see Fig. 8.15). Such similarity in the direction of slope failure, along with morphological similarity of the deposits that can indicate their simultaneity, allows assumption on seismic triggering of these "dual" slope failures.

Both the Akkiol and the Karakiol rock avalanches have deep wedge-like headscarps and distinct flow-like micro-relief of the deposits that caved from 500–600 m slopes, collided with an opposite side of the valley, and turned downstream forming about 1.5–1.7 km long tongues with linear ridges and furrows (Fig. 8.43), very similar to those of the Urmochdara rock avalanche in Pamir (see Fig. 4.41), the Upper Kashkasu rock avalanche in the Central Tien Shan (see Fig. 4.43) and, at some extent, of the Karakystiak rock avalanche in Northern Tien Shan (see Fig. 6.7). The Akkiol rock avalanche collided with a spur at the junction of the main river and its small tributary at about 3085 m a.s.l. and its debris split into two parts larger of which moved 1700 m down the main valley, while the smaller one entered the tributary valley for about 450 m up to 3135 m a.s.l. (see Fig. 8.43).

Very informative group of several multi-aged rockslides (Kyrgyz 113–117) can be found in the upper reaches of the Inylchek River valley (42.16° N, 79.457° E) at its confluence with the left tributary – the Atdjailau River and right tributary – the Taldy-Suu River, 16.5 km downstream from the present day terminus of the Southern Inylchek Glacier. One more rock avalanche (Kyrgyz 004) was found 9 km downstream at 42.143° N, 79.353° E.

Rockslides near the Atdjailau and the Taldy-Suu River mouths occurred from both sides of the Inylchek River valley (see Figs. 6.4 and 8.44) and there is distinct evidence of the recurrent slope failures at this site clearly visible in the outcrop shown on Fig. 8.45. First slope failure occurred from the 1300 m high left bank of the glacial trough composed of the Silurian limestone underline by shale of nearly the same stratigraphic age.

All succession dips steeply towards a river. The sliding surface of the compound rockslide about 50 Mm^3 in volume likely coincided, at least partially, with the stratigraphic contact between hard marble-like limestone and much weaker terrigenous sediments. The letter debris was also found in the rock avalanche deposit, most of which consist of angular limestone boulders and crushed limestone (Fig. 8.46).

Rockslide converted into highly mobile rock avalanche that practically crossed the 1.1 km wide valley and, at the same time, spread along it for 2.4 km forming a pancake-like body 10–30 m thick. We assume (Strom, 2006) that such multidirectional spreading was governed by low friction at the water-saturated alluvium of the flat bottom of the former glacier valley, which mechanical properties differ significantly from those of dry bottoms of intermontane depressions over which most of the unconfined rock avalanches described above moved. Similar multidirectional spreading of rock avalanche debris can be observed at the rock avalanche in the Kash River valley described above (see Fig. 5.33).

FIGURE 8.44 View on the right bank of the Inylchek River valley from the Atdjailau rock avalanche body (113). Photo made in June 1981. Other numbers (114–117) mark rockslide headscarps. Dashed black lines mark approximate boundaries between Carboniferous black limestone (C_2), red Paleozoic granite (γPZ) and Silurian schist (S_2), all of which are involved in the Inylchek rock avalanche (see Figs. 8.45, 8.48).

Later on this wide thin body was strongly eroded by the powerful Inylchek River with recurrent outburst floods from the glacier-dammed Mertsbaher Lake. However, its distal northernmost part survived being protected by the more recent Inylchek rock avalanche that overlaid it (see Fig. 8.45). Presence of the alluvium in the narrow erosion cut in the Atdjailau rock avalanche deposit indicates that its' overlapping by the Inylchek rock avalanche about 20 Mm^3 in volume occurred after some interval.

Besides this "intermediate" alluvium accumulation, the Atdjailau rock avalanche front appeared to be covered by loess-like loam that was abraded and deformed by rapidly moving Inylchek rock avalanche and can be visible now at the southern cliff of the right bank promontory forming the river valley narrowing (Fig. 8.47). The Inylchek rock avalanche deposits, in turn, represent the succession of debris "layers" that replicates the succession of the lithologies visible in the headscarp area (compare Figs. 8.44, 8.45 and 8.48). Notice that the Silurian shist, which debris marked by S_2 on Fig. 8.45 overlays granite debris, came not from the main belt of Silurian shale marked on Fig. 8.44, which is bellow the headscarp base, but from the schistose lens in the narrow granite belt (see Fig. 8.48). We must notice that stripe of the granite debris was identified in the field between schist and black limestone debris too, but it is too narrow to be seen on Fig. 8.45. Another interesting point is the absence of any mixing of debris from different lithologies that can be seen both in the cliff on Fig. 8.45 and at the outcrop at the base of the source zone (see Fig. 8.49).

Additional well expressed peculiarity of the Inylchek rock avalanche deposits is the contrasting

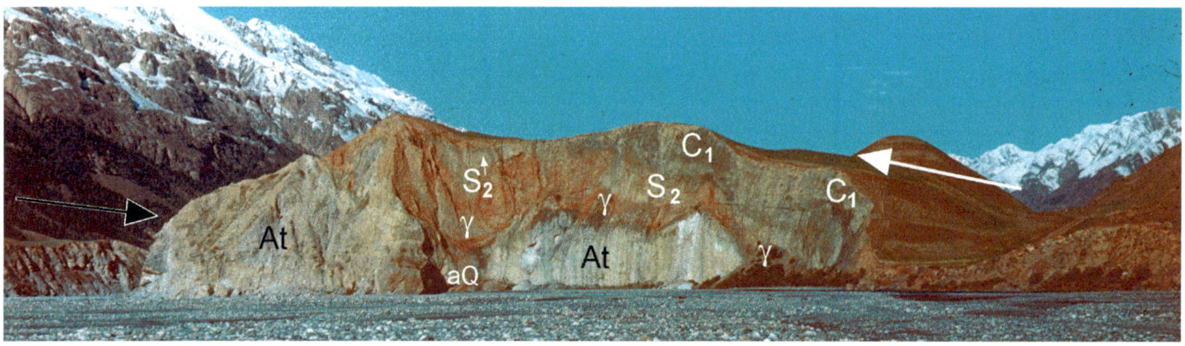

FIGURE 8.45 Distal part of the Atdjailau rock avalanche overlaid by the Inylchek rock avalanche deposits. At – crushed white limestone of the Atdjailau rock avalanche frontal part (its motion direction is indicated by black arrow; aQ – alluvial deposits remaining in the narrow erosional cut; red-gray succession – deposits of the Inylchek rock avalanche that moved in the direction indicated by white arrow: γ – Paleozoic granite, S_2 – Silurian schist, C_1 – Carboniferous black limestone, which mutual position can be seen on Figs. 8.44 and 8.48.

FIGURE 8.46 Contact of limestone (LS) and shale (Sh) debris in the "pancake" part of the Atdjailau rock avalanche.

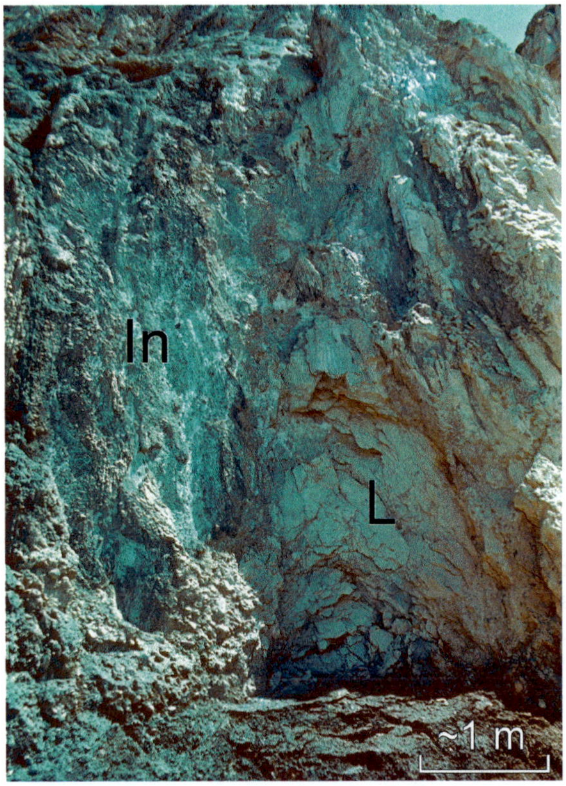

FIGURE 8.47 Contact between the intensively deformed loess-like loam (L) and the Inylchek rock avalanche debris (In).

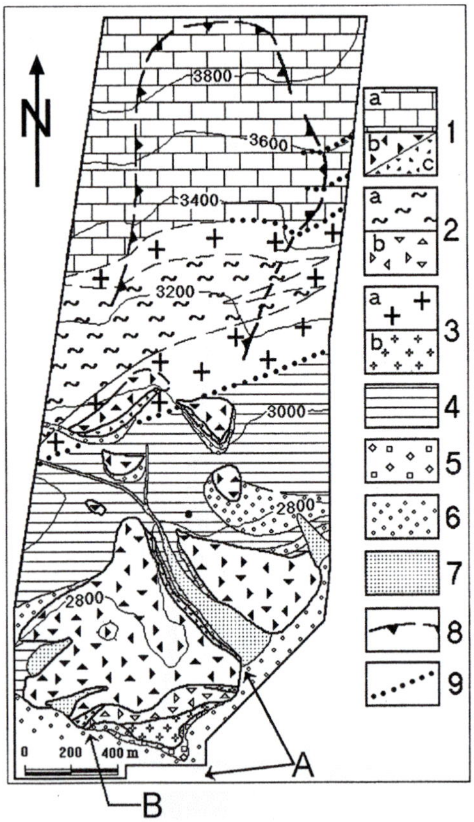

FIGURE 8.48 Schematic geological map of the Inylchek rock avalanche. Legend: 1 – black Carboniferous limestone (a – in host massif, b – big angular boulders of the carapace, c – crushed into rubble and forming a "layer" below the carapace); 2 – schist (a – in host massif, b – rubble-size debris); 3 – granite (a – in host massif, b – rubble-size debris); 4 – Silurian sandstone and shale at the slope base below the headscarp; 5 – deposits of the Atdjailau rock avalanche front; 6 – alluvium of the Inylchek River floodplain and terraces; 7 – the Taldy-suu River debris flow fan; 8 – the Inylchek rock avalanche headscarp; 9 – tectonic fault with evidence of recent motion; A – see Fig. 8.45; B – see Fig. 8.47. *Modified from Abdrakhmatov, K., and Strom, A. 2006. Dissected rockslide and rock avalanche deposits; Tien Shan Kyrgyzstan. In: Evans, S.G.; Scarascia Mugnozza, G.; Strom, A.; Hermanns, R.L. (Eds.) Landslides from Massive Rock Slope Failure. NATO Science Series: IV: Earth and Environmental Sciences, Springer, Vol. 49, 551–572 with permission of Springer.*

FIGURE 8.49 Unmixed "layers" of debris from granite (A) and black limestone (B) at the base of the source zone.

grain-size composition of most of debris "layers" and of the carapace composed of angular boulders up to several meters in size of black limestone solely that form the smooth hummocky surface of this rock avalanche body (Fig. 8.50). Similar phenomenon will be described for several case studies from the Central Tien Shan (Section 8.4).

Other rockslides on the right bank of the Inylchek River valley located close to the Inylchek rock

FIGURE 8.50 Hummocky surface of the Inylchek rock avalanche on the right bank of the Taldy-Suu River valley. The wild yak bull at the foreground serves as a scale.

avalanche are much smaller. Two of them, on the left bank of the Taldy-Suu River valley (Kyrgyz 116, 117, see Fig. 8.44) with well-defined headscarps and bodies, look rather recent, the third one (Kyrgyz 115) seems to be either of the same age as the Inylchek rock avalanche, or even older. However, it is just a preliminary estimate based on their morphological expressiveness only.

Several interesting case studies could be found also in the south-western, Chinese part of the Highest Tien Shan. One large rockslide (China 122) is located in the upper reaches of the Kokburunsu River at 41.75 N, 80.02 E. Here about 100 Mm^3 of rocks collapsed from the 1 km high slope of very complex structure well visible on the Google Earth images (Fig. 8.51). It created a dam that, despite it was breached, still forms an

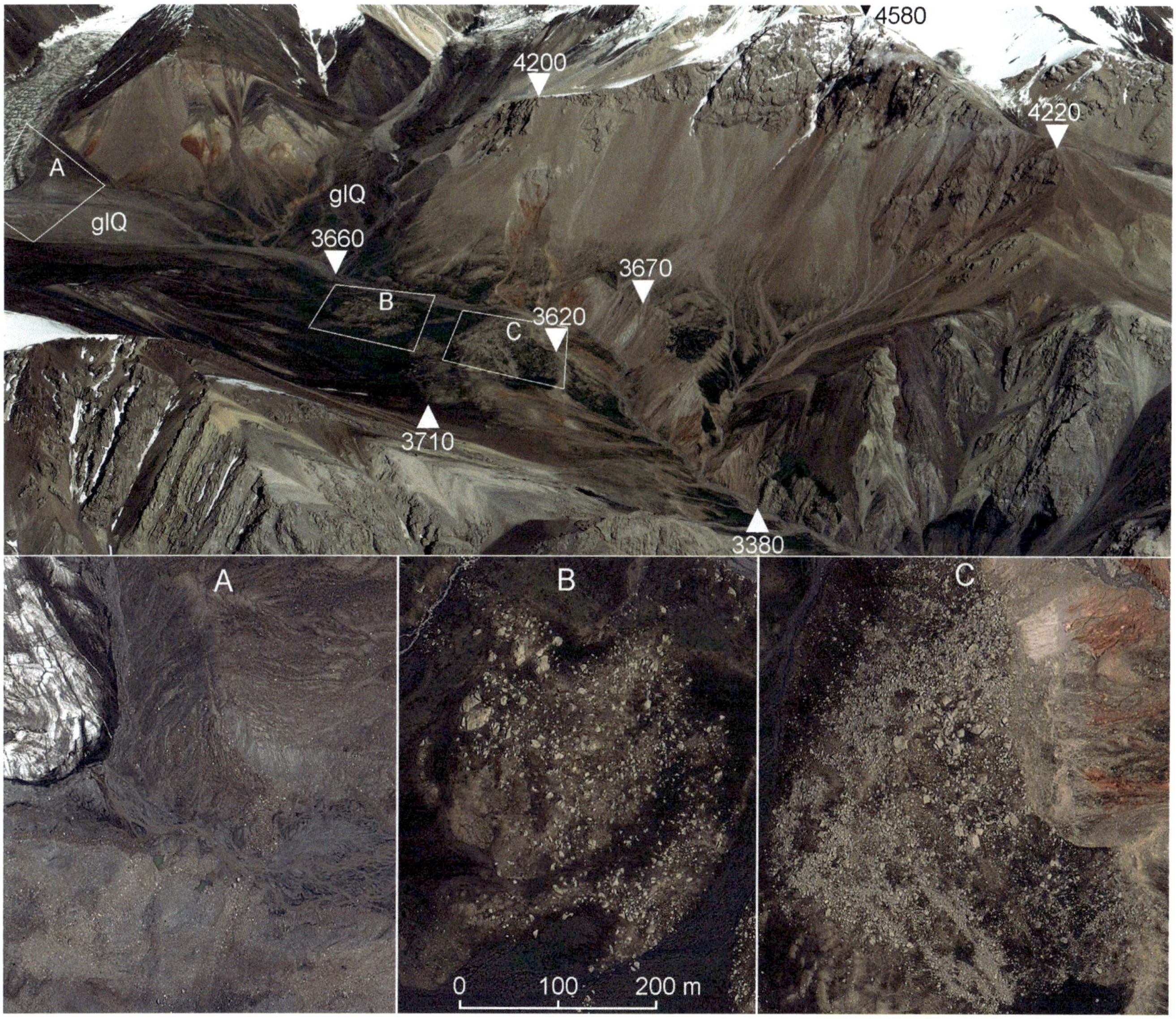

FIGURE 8.51 Deeply eroded Kokburunsu rockslide dam in the upper reaches of the Kokburunsu River with variable lithologies within its source zone. Above – 3D Google Earth view, below – glacial deposits (A) and coarse blocky rockslide carapace (B, C) visible on zoomed Google Earth images. Scale of all outlined fragments is the same.

impressive nickpoint showing that about 100 m of rockslide debris have not been incised yet. Just 1.2 – 1.5 km upstream, at 3700 m a.s.l. approximately, valleys are filled by end moraines left by small glaciers, while larger glaciers nearby terminates much lower, at 3000 and even 2600 m a.s.l. The Kokburunsu rockslide dam is covered by ~20–30 m thick carapace composed of large angular boulders, likely of granite from the headscarp crown, some of which are up to 20–30 m in size. This blocky material is much coarser than moraine material brought by both glaciers (compare insets on Fig. 8.51).

The base of the source zone is composed of red beds visible in the ravine eroded at the upstream side of the source zone (see Fig. 8.51). Rockslide debris sourced from these layers composes lower part of the dam visible in the eroded gorge. Similar succession, schematically presented on Fig. 4.17-B, can be seen in many other rockslides, such as those shown on Figs. 4.7, 6.22, 8.45 and some other that will be described in the following sections.

Just south of the Kyrgyz-China border in the lower reaches of the Sarydjaz-Aksu River valley and its tributaries one can find a cluster of at least 25 landslides concentrated within a small area of 15 × 35 km extending from 41.67° to 41.81° N and from 79.32° to 79.75° E (Fig. 8.52, Table 8.2). Unlike the Sarmin-Ula Range cluster (see Fig. 8.11, Table 8.1) where many rockslides are of the same order of size, this one can be ascribed to the "core" type (Strom, 2015C), with one giant feature accompanied by numerous landslides several orders of magnitude less than the "core member". Such distribution – one gigantic landslide surrounded by much smaller features – was described by B. Malamud and his co-authors (Malamud et al., 2004a, b; Turcotte and Malamud, 2004) as being typical of landslides triggered by highly energetic triggering events such as large earthquakes and typhoons. However, in our case no data about ages of slope failures that formed this cluster are available and, thus, such assumption is just hypothetical.

The "core" feature of the Sarydjaz-Aksu rockslide cluster is a gigantic, 300–350 m high (300 m is its assumed effective height) and about 900 Mm^3 (±30%) in volume, breached rockslide dam (China 100, 41.764° N, 79.526° E). The dammed lake was 31 km long approximately with an area of about 24 km^2 and could store up to 2.2 – 2.5 km^3 of water. Failure occurred on slope composed of the Carboniferous metasediments dark gray in color, while both its depositional zone and area downstream are composed of pink and reddish Paleozoic and Cenozoic sedimentary rocks. Such color contrast predetermines the most interesting, besides the features' size, peculiarity of this rockslide – the distinct traces of the exceptionally powerful outburst flood that had painted river terraces by black debris eroded from the dam's body (Fig. 8.53).

Immediately downstream from the dam the surge wave was up to 120 m above the riverbed and ~700 m wide (on top). About 2 km downstream outburst flood wave became wider – nearly up to 1000 m and affected river banks up to 80 m high (see Fig. 8.53). Considering generally triangular cross-section of the river valley and ~10 m/s velocity of the outburst flow (in fact it could be even higher), the peak discharge could reach 400 000 – 500 000 m^3/s – about 4 times

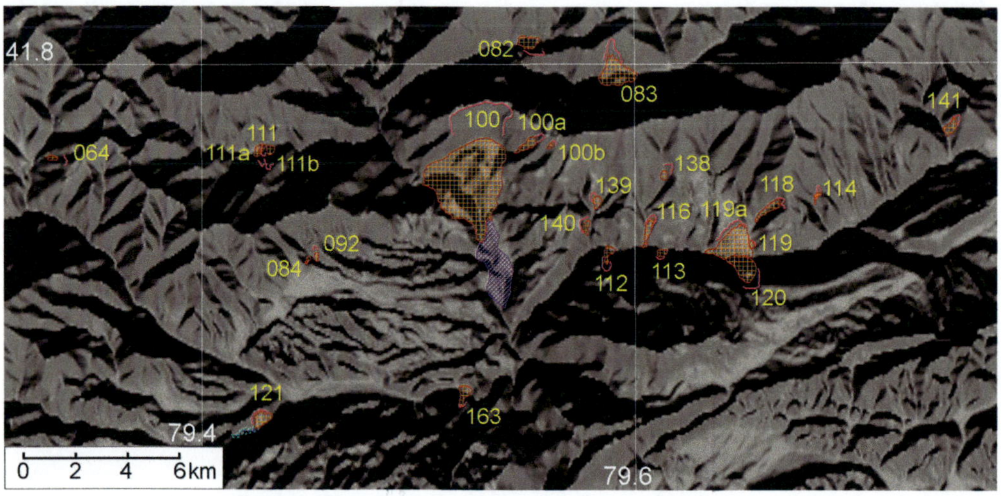

FIGURE 8.52 Detailed map of the Sarydjaz-Aksu rockslide cluster. Cross-hatched areas with brown boundaries – more recent (Holocene?) rockslide bodies; those without boundaries – older rockslide bodies; solid pink lines – more recent headscarps; dashed pink lines – older headscarps; comb orange lines – sackung-like scarps; dashed blue line – silted dammed lake; violet area – trace of the catastrophic outburst flood (see Fig. 8.53). Numbers correspond to those in Table 8.2. Background – the 3″ SRTM DEM visualized by Global Mapper software.

TABLE 8.2 Rockslides forming the Sarydjaz-Aksu rockslide cluster

ID China ...	Latitude	Longitude	Area of the deposits (km²)	Total area affected (km²)	Volume (10^6 m³)
	dd.ddddd°				
064	41.7686	79.3322	0.058	0.17	1.0
082	41.8071	79.5506	0.280	0.50	5.0
083	41.7957	79.5902	0.970	1.27	17.0
084	41.7357	79.4495	0.26	0.03	0.1
092	41.7359	79.4530	0.050	0.09	0.3
100	41.7641	79.5262	7.150	10.13[a]	900.0
100a	41.7740	79.5527	0.200	0.27	1.0
100b	41.7728	79.5611	0.030	0.05	0.1
111	41.7717	79.4283	0.050	0.09	0.3
111a	41.7717	79.4259	0.100	0.19	0.5
111b	41.7717	79.4317	0.087	0.19	0.5
112	41.7366	79.5885	0.140	0.26	1.5
113	41.7375	79.6120	0.080	0.11	1.0
114	41.7555	79.6829	0.041	0.08	0.5
116	41.7448	79.6058	0.190	0.25	3.0
118	41.7507	79.6595	0.240	0.30	4.0
119	41.7395	79.6532	0.050	0.06	0.4
119a	41.7403	79.6454	1.21	1.41	50.0
120	41.7319	79.6500	0.680	0.94	12.0
121	41.6816	79.4270	0.264	0.35	7.0
138	41.7626	79.613	0.07	0.18	0.2
139	41.7542	79.583	0.11	0.16	1.5
140	41.7457	79.5774	0.09	0.20	1.0
141	41.7787	79.7445	0.2	0.26	4.0
163	41.6899	79.5218	0.2	0.25	2.0

[a]*area affected by outburst flood not included.*

Modified from Strom, A.L. 2015c. Clustering of large bedrock landslides and recurrent slope failure: implications for landslide and seismic hazard assessment. In: Proc. of the 4th International Symposium on Mega Earthquake Induced Geo-disasters and Long Term Effects, May 9–13, 2015, Chengdu University of Technology, China with permission of the Canamaple Academia Services.

FIGURE 8.53 Evidence of the catastrophic outburst flood caused by the Sarydjaz-Aksu rockslide dam breach – terraces up to 120 m high "painted" by dark-gray debris eroded from the dam's body. 3D Google Earth view.

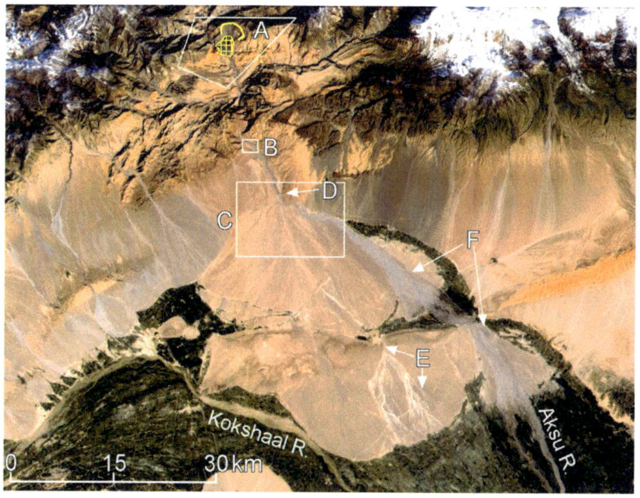

FIGURE 8.54 Overview of the multi-stage alluvial fan of the Sarydjaz-Aksu River. A – area shown on Fig. 8.53; B – flood-affected river terrace shown on Fig. 8.55; C – upper part of the alluvial fan shown on Fig. 8.56; D – the HPP dam; E – the alluvial fan that passed through a wind gap in a small ridge stretching about 25 km off the Kokshaaltau Range; F – the Sarydjaz-Aksu River modern flood plane. Google Earth image.

more than that of the 2000 Yigong outburst flood (Wang, 2008; Xu et al., 2008; Evans and Delaney, 2011) and, likely, more than peak discharge during the 1985 Bairaman rockslide dam breach in Papua New Guinea where the debris flow surge wave just bellow the breaching dam was up to 100 m high (King et al., 1989). In the Sarydjaz-Aksu rockslide case the dam was breached almost completely and do not form any distinct knick point now, unlike many other breached dams described above. Both extremely high surge wave discharge and absence of a knick point could be caused by the intensive comminution of the rockslide dam material and its erodibility.

6 km downstream from the breached rockslide dam the river enters a narrow gorge where evidence of this outburst could not be recognized distinctly (Fig. 8.54). However, they appear again 10 km further downstream at the outlet of this gorge on the 50–55 m high right bank terrace that is curved by the amygdaliform branching ravines. None of similar ravines exist on the higher terrace, where the road passes (Fig. 8.55),

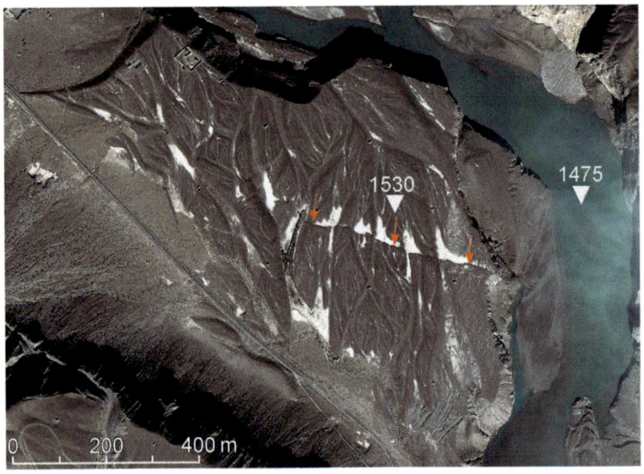

FIGURE 8.55 The amygdaliform ravines left by powerful flow on the 50-m high Sarydjaz-Aksu River terrace, later on ruptured by a small fault (marked by *red arrows*). Google Earth image.

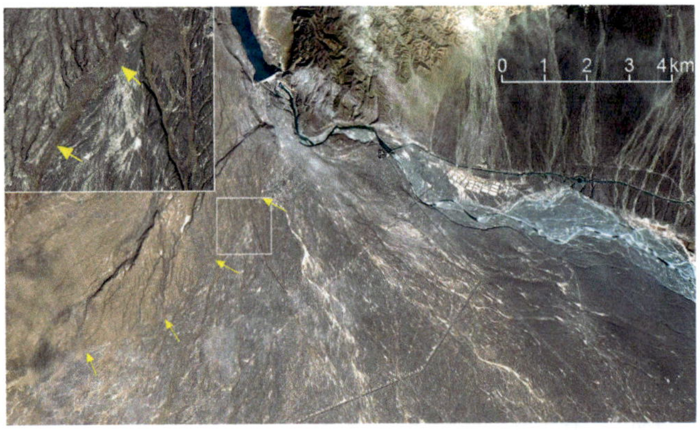

FIGURE 8.56 The dendritic amygdaliform microrelief of the alluvial fan formed by the catastrophic outburst flood contrasting from the landforms of the Sarydjaz-Aksu River older alluvial fan. Boundary between two generations of the alluvial fan landforms is marked by *yellow arrows*. Outlined area is shown in the inset. Google Earth image.

which, thus, was not affected by the flood. It is interesting that numerous minor landforms curved by this outburst flood on the terrace surface are ruptured by a recent fault, which dating should provide upper limit of the outburst flood and, thus, of the Sarydjaz-Aksu rockslide formation. We must mention, however, that if this faulting event occurred after flooding it could not be related to formation of this rockslide. Considering the 208 m^3/s mean present-day discharge of the Sarydjaz-Aksu River (Rivers and Lakes, 2012) the dam could be overtopped about 100–120 days after its formation. What time passed between the outburst flood and faulting is unknown. Opposite chronological sequence is possible too.

There is geomorphic evidence that 9 km farther downstream the surge wave spread over the gigantic older alluvial fan formed by the Sarydjaz-Aksu River at the feet of the Tien Shan Mountains. As it can be seen on Figs. 8.54 and 8.56, almost the entire fan up to ~20 km wide was affected. The preexisting microrelief at an area of about 630 km^2 was demolished and replaced by the multiple dendritic amygdaliform ravines that differ from the older landforms significantly (see inset in Fig. 8.56). It can be speculated that it was just this catastrophic event that blocked the original course of the Kokshaal River which joined that initially fall into the Sarydjaz-Aksu River immediately north of the small ridge (foreberg) stretching about 25 km off the Kokshaaltau Range (see Fig. 8.54). It forced the Kokshaal River to find a new way. The alluvial fan with the dendritic landforms spread far eastward leaving the narrow belt of unaffected low terraces on the left bank of the Sarydjaz-Aksu River that is used at present for agriculture. Later on the modern floodplain was affected by periodic outburst floods fed at a large extent

FIGURE 8.57 Erosional terraces in the lacustrine sediments upstream from the partially breached rockslide dam in the Djandjer River valley. 3D Google Earth view.

by regular floods from the Mertzbaher Lake. Part of the catastrophic outburst caused by the rockslide dam breach passed through a wind gap (E on Fig. 8.54). It is evident that this outstanding natural phenomenon requires more detail study.

Seven kilometers north-east of the Sarydjaz-Aksu rockslide at 41.796° N, 79.59° E, the Djandjer River valley was blocked by rockslide about 17 Mm^3 in volume (China 083). Rockslide fell from the right bank of the river (marked by white arrow on Fig. 8.57) and formed

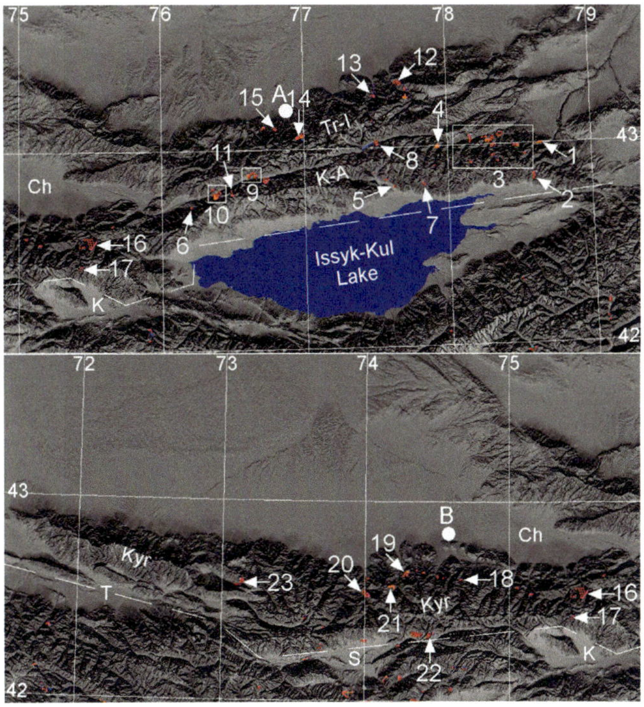

FIGURE 8.58 Shaded 3″ STRM DEM of the Northern Tien Shan. Rockslides are marked by pink and brown dots; blue dots – dammed lakes. Tr-I – the Trans-Ili Range; K-A – the Kungey-Alatau Range; Kyr – the Kyrgyz Range. Ch – the Chu; K – the Kochkor; S – the Suusamyr; T- the Talas intermontane depressions; A – Almaty City; B – Bishkek City. Dashed line – the conventional boundary between the Northern and Central Tien Shan. 1 – see Fig. 8.59; 2 – see Fig. 8.60; 3 – landslide cluster in the Chilik River basin (see Figs. 8.61–8.68); 4 – see Fig. 8.69; 5 – the Ananievo rockslide (see Figs. 8.70–8.72); 6 – the Kaindy rockslide (see Fig. 8.73); 7 – see Fig. 8.74; 8 – see Fig. 6.6; 9 – see Figs. 8.75–Fig. 8.78; 10 – see Figs. 8.75, 8.79–8.86; 11 – see Fig. 8.87; 12 – see Fig. 8.88; 13 – the Issyk rockslide (see Fig. 2.1); 14 – the Great Almaty Lake rockslide (see Fig. 8.89); 15 – the Akjar rockslide (see Fig. 8.90); 16 – the Uitabulga rock avalanche (see Figs. 8.91–8.93), 17 – see Fig. 8.94; 18 – the Arashan rockslide (see Figs. 8.95–8.96); 19 – the Bielogorka rockslides (see Figs. 8.97–8.101); 20 – the Aksu rockslide (see Figs. 8.102, 8.103); 21 – see Fig. 8.104; 22 – the Snake-Head rock avalanche (see Figs. 5.56-5.60, 8.105, 8.106); 23 – the Karakystiak rock avalanche (see Fig. 6.7)

a dam up to 170 m high. Unlike the Aksu-Sarigjaz rockslide dam that was breached completely, almost up to the original river bottom and, likely, soon after its formation, perhaps during first overtopping, this blockage remained for a much longer time. It seems that the lake was silted almost completely, up to the dams' crest level, and only after that the upper part of the dam was eroded. Such an assumption can be derived from the several terraces at the left bank of the river eroded in lacustrine sediments (see Fig. 8.57). More than 100 m of the dam is still in place forming an impressive knick point.

Most of other rockslides forming the Sarigjaz-Aksu cluster (see Fig. 8.52) are "ordinary" features and we will leave them to save more space for numerous exciting examples from the other parts of the study region.

8.3 NORTHERN TIEN SHAN

This part of the Tien Shan Mountain system includes the 200–240 km long Trans-Ili and the Kungey-Alatau Ranges north of the Issyk-Kul Lake and the 350 km long Kyrgyz Range north of the Kochkor, the Suusamyr and the Talas intermontane depressions that are considered as a conventional boundary between the Northern and the Central Tien Shan. It belongs to Kazakhstan and Kyrgyzstan (Fig. 8.58).

This part of the Tien Shan Mountain system has been affected by several historical earthquakes at the end of XIX-s – beginning of XX-s Centuries (see Chapter 2: Brief History of the Bedrock Landslide Studies in the Central Asia) that triggered large bedrock landslides and numerous landslides in soft sediments. Some of these features will be described hereafter.

The easternmost interesting site in this part of the region is the more than 3 km long sackungen associated with identifiable landslides along the crest of the 450 m high left (northern) slope of the Shet-Merke River valley in Southern Kazakhstan (43.0° N, 78.7° E, Kazakh 082) composed of Carboniferous terrigenous sediments. From the neotectonic and geomorphological viewpoint this area belongs to the northern slope of the eastern pericline of the Kungey-Alatau neotectonic anticline. This neotectonic uplift is growing actively relative both to the adjacent intermontane depressions and to the Ketmen Range located east of it. The ~250 m wide crest of the smooth ridge has a longitudinal depression with a minor lake in its lowermost part accompanied by the upslope-facing scarps at its southern side (Fig. 8.59). The latter's freshness allows assumption that this feature could be activated (or even formed) by the 1889 Chilik earthquake, though no actual data about its age are available up to now.

Very large, about 150 Mm^3 in volume, rock avalanche (Kyrgyz 186, Fig. 8.60) was found on the southern slope of the Kungey-Alatau Range about 30 km east of the Issyk-Kul Lake in the upper reaches of the Korumdy River valley (42.837° N, 78.6° E). Its headscarp in Paleozoic granites and upper part of the 4.4 km long body are at and above 3000 m a.s.l., within formerly glaciated zone. Identification of this feature as a gravitational landform was based on the same criteria as described in Section 4.5 for the Upper Kashkasu rock avalanche (see Figs. 4.43 and 4.44) – first of all on sharp morphological difference of the

FIGURE 8.59 Part of the sakungen at the top of the Shet-Merke River valley northern slope with a small lake and fresh upslope-facing scarps. SAS Planet image.

FIGURE 8.60 Assumed rock avalanche in the upper reaches of the Korumdy River valley. 3D Google Earth view.

assumed headscarp and the deposits surface from all surrounding landforms at the same altitude. Here we can also see steep nearly planar slope covered by talus that rises above the huge, about 300 m high, hill of bouldery material with a small dammed lake behind. Downstream from this compact body with poorly defined secondary scar (see Section 5.3.3.1) the valley bottom is filled at a distance of about 2.3 km by debris with striated surface. From our point of view such combination indicates that upper reaches of this valley were affected, likely in Early Holocene or in Late Pleistocene, by catastrophic rock avalanche.

Further to the north-west valleys of the Kaindy, Saty, Kurmekty, Taldy Rivers – right tributaries of the Chilik River dissecting northern slope of this part of the Kungey-Alatau Range were blocked by large bedrock landslides that occurred on slopes composed of the Carboniferous volcanic rocks (No 009, 010, 011, 014 on Fig. 8.61) or of the Paleozoic granites (No 013, 083, ibid). At their lower reaches, closer to Chilik River, there are also several large landslides in week Neogene sediments and in Quaternary deposits (No 015, 016, 045 – 050 on Fig. 8.61). One landslide (044 on the same Figure) involved both Carboniferous and Neogene deposits divided by a large fault zone. Some of these rockslides form the dammed lakes. Size distribution of rockslides forming this Southern Chilik rockslide cluster (Table 8.3) is the same as in the Northern-Dzungarian (see Fig. 7.2) and the Sarmin-Ula Range (see Fig. 8.11, Table 8.1) rockslide clusters – their volumes range within one-two orders of magnitude. Such size distribution differs from that of the Sarydjaz-Aksu cluster with one giant "core" rockslide surrounded by numerous much smaller "satellites" (see Fig. 8.52, Table 8.2).

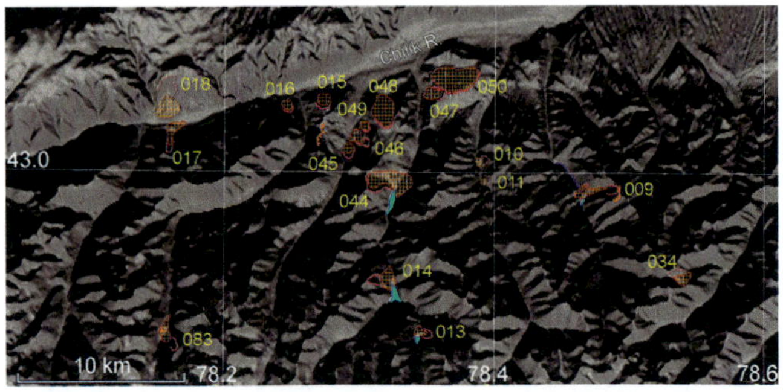

FIGURE 8.61 Landslide cluster in the Chilik River basin. Cross-hatched areas with brown boundaries – landslide bodies; those without boundaries – assumed landslide bodies; solid pink lines –headscarps; dashed pink lines – assumed headscarps; comb orange lines – sackung-like scarps; blue areas – landslide-dammed lakes. Numbers of features correspond to those in the Kazakh part of the database. Background – the 3″ SRTM DEM visualized by the Global Mapper software.

Though most of these slope failures are quite recent, likely Late Holocene, we found several geomorphic features likely formed by much older rock slope failures. Two of such sites were identified in the Saty River valley at 43.001 N, 78.388 E (Kazakh 010) and 42.995 N, 78.391 E (Kazakh 011) (Fig. 8.62). The assumed rockslide deposits form terrace-like surfaces about 50 m above the river level. No similar terraces exist both up- and downstream from these sites. The triangular facet-like concave slopes just above them are, likely, the headscarps. One more similar-shape slope rises above the river between these sites but lack of the deposits makes the assumption on its origin more speculative. Presence of such assumed old features indicate recurrent rockslides formation, well

TABLE 8.3 Rockslides and landslides forming the Southern Chilik rockslide cluster

ID Kazakh ...	Latitude dd.ddddd°	Longitude dd.ddddd°	Area of the deposits (km²)	Total area affected (km²)	Volume (10^6 m³)
009	42.9893	78.4688	0.580	1.50	8.0
009a	42.9921	78.4657	0.058	0.09	0.5
010	43.0062	78.3883	0.195	0.40	3.0
011	42.9953	78.3909	0.160	0.34	2.0
013	42.9140	78.3464	0.350	0.56	7.0
014	42.9439	78.3213	0.945	1.34	30.0
015[a]	43.0387	78.2733	0.700	0.700	7.0
016	43.0354	78.2460	0.310	0.40	4.0
017	43.0194	78.1613	0.707	0.86	8.0
018	43.0334	78.1587	1.500	2.60	35.0
034	42.9443	78.5405	0.500	0.86	5.0
044	42.9953	78.3255	2.700	3.17	135.0
045[a]	43.0124	78.2910	0.360	0.52	3.0
046[a]	43.0189	78.2971	0.500	0.500	5.0
047[a]	43.0421	78.3544	0.600	0.600	6.0
048[a]	43.0334	78.3171	2.000	2.000	20.0
049[a]	43.0244	78.3037	0.250	0.250	2.5
050[a]	43.0494	78.3711	3.089	3.089	31.0
083	42.9129	78.1577	0.650	1.01	7.0

[a]*Landslides in Quaternary deposits.*

FIGURE 8.62 Assumed old (Pleistocene?) rockslides in the Saty River valley. 3D Google Earth view.

corresponding to the recurrent surface faulting in this part of the region associated with large earthquakes (Abdrakhmatov et al., 2016).

Some of the most recent rockslides within the cluster shown on Fig. 8.61 could be triggered by the 1889 Chilik or the 1911 Kemin earthquakes, as it is written in most of popular sources like Wikipedia. However, samples of submerged trees collected by V.A. Gapich from the Kaindy Lake at 42.99° N, 78.469° E (dammed

FIGURE 8.63 Submerged trees in the Kaindy Lake (*with kind permission from Alexander Schardt; available at http://www.panoramio.com/photo/33419596?source = wapi&referrer = kh.google.com#*).

by the Kazakh 009 rockslide – Fig. 8.63) and from the Kiol-Sai Lower Lake at 42.995° N, 78.325° E (dammed by the Kazakh 044 rockslide) gave ages of 430 ± 60 BP (1406–1635 AD) and 400 ± 30 BP (1436–1625 AD) correspondingly ($\sigma = 95.4\%$), thus, indicating that they could be much older than these outstanding earthquakes. Dating was performed by L. Sullerzhitsky in the Geological Institute of Russian Academy of Sciences, Moscow (samples No GIN 6507 and GIN 6506); for calibration we used the CalPal online software (http://www.calpal-online.de).

The rockslide that formed the Kaindy Lake (Kazakh 009) originated at the upper part of the more than 1 km high right slope of the valley (Fig. 8.64), which is typical of the earthquake-induced rockslides. We should point out that only part of the disturbed rock mass formed the rockslide about 8 Mm³ in volume that had dammed the river; almost same-size, if not larger block just settled down for about 50–60 m only (see the upper left inset in Fig. 8.64).

The lake was partially breached in 1980 so that its level dropped for ~10 m. The deep erosional canyon was formed (see the lower right inset in Fig. 8.64) and traces of this outburst flood could be seen at a distance of more than 2.5 km at the valley bottom till now, more than 35 years after this catastrophic event.

Another rockslide that could originate in the beginning of the 16th Century, simultaneously with the previous case study, according to the ages of the submerged trees, forms the Kiol-Sai Lower Lake (Kazakh 044, Fig. 8.65) – the lowermost of the three rockslide dammed lakes in this river valley (the Kiol-Sai in Kazakh means – the river with lake or lakes). The peculiarity of this rockslide about 135 Mm³ in volume is that it occurred directly at the fault zone separating the Paleozoic volcanogenic and sedimentary

FIGURE 8.64 Rockslide dam in the Kaindy River valley. 3D Google Earth view.

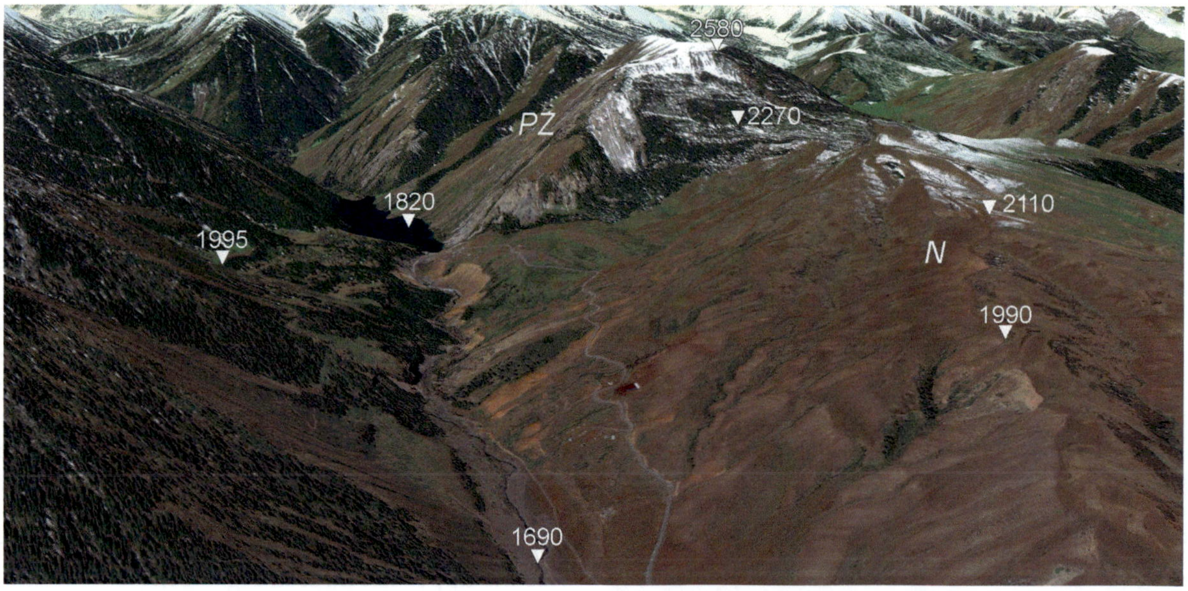

FIGURE 8.65 The Lower Kiol-Sai rockslide dam. 3D Google Earth view.

rocks from the Neogene red beds. While the upper part of its source zone is within the Paleozoic block, the main part of the displaced material that forms the dam's body is from the Neogene. The direction of its motion changed almost at a right angle from N-E ~25° in the upper part to E-S-E 110–115° at the distal one. The dam has been deeply eroded and an artificial spillway was constructed at the crest of the remaining part of the blockage to prevent its further catastrophic breach and to protect this beautiful natural feature.

Two other rockslide dams upstream, about 30 and 7 Mm^3 in volume correspondingly, form the Middle Kiol-Sai Lake at 42.944° N, 78.321° E (Kazakh 014) and the Upper Kiol-Sai Lake at 42.914° N, 78.346° E (Kazakh 013) (see Fig. 8.61). These blockages formed by debris originated from the Paleozoic volcanogenic rocks and granites are almost intact despite clear evidence of overflow during flood periods. Such drastic difference in the present-day state of these three lakes is caused, most likely, by the mechanical properties of the material and its ability to resist erosion, which is higher for debris of hard Paleozoic rocks than of that of weak Neogene red beds.

15 kilometers west of the Kiol-Sai River mouth at 43.02° N, 78.161° E the Chilik River was dammed temporarily by rock avalanche about 8 Mm^3 (Kazakh 017)

FIGURE 8.66 Photo of the Chilik rockdslide triggered by the 1911 Kemin earthquake and temporarily lake. From (Bogdanovich et al., 1914).

triggered by the 1911 Kemin earthquake (Bogdanovich et al., 1914). About 8 Mm^3 of weathered rocks collapsed from the ~800 m high slope composed of Paleozoic metasediments and granites and blocked the river. When the site was visited by the earthquake reconnaissance team led by Prof. Bogdanovich the dam was overtopped and the river started eroding it gradually (Fig. 8.66). Due to small dams' height and its large dimensions in plan view resulting in low gradient of the overflow, no catastrophic breach and outburst flood occurred and now the stream almost restored its original profile, while rock avalanche headscarp and most of the deposits are still well preserved (Fig. 8.67).

FIGURE 8.67 Modern view of the 1911 Chilik rockdslide. 3D Google Earth view.

Several other bedrock landslides looking quite "fresh" are located in the vicinity of the 1911 Chilik rockslide. Since they are located in the valleys of the Chilik River tributaries, which were not visited by the team that studied effects of the earthquake, we can not assure that they were also triggered by this earthquake. However, it is quite possible. One of such rockslides (Kazakh 083) is located in the Taldy River valley, at 42.913° N, 78.158° E, 12 km south of the Chilik rockdslide (see Fig. 8.61). About 7 Mm^3 of granite debris fills the valley bottom just downstream from the modern glacial landforms but is not affected by any processes related to glaciation (Fig. 8.68). It also looks as if it was accompanied by several minor adjacent rockslides. It is evident that its age does not exceed a century or so, thus it could be triggered either by the 1889 Chilik or the 1911 Kemin earthquakes.

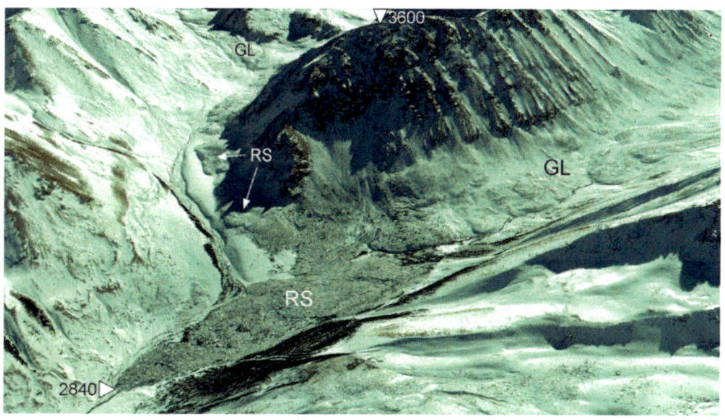

FIGURE 8.68 Recent rockslide in the Taldy River valley and its interrelation with recent glacial landforms. RS – rockslides, GL – glacial landforms (moraines and rock glaciers). 3D Google Earth view (November, 2014).

One more recent rockslide, but in totally different environment, dams a small lake in the Bolshoi (Large) Oriuktu River valley 5 km upstream from its mouth at 42.988° N, 77.935° E (Kazakh 019). At this site, composed of the Ordovician terrigenous rocks, about 50 Mm^3 of them were involved in the sagging and sackungen-like DSGSD (see Section 5.1). Besides there is a rockslide about 2 Mm^3 in volume forming the dam and minor adjacent rockslides, one of which is marked on Fig. 8.69 as RS, The most spectacular is the wedge-like ~1 km long sackungen at the watershed bounded on both sides of the ridge crest by upslope-facing scarps (see inset in Fig. 8.69).

The dam with about 80 m runup on the opposite slope of the valley was partially breached, most likely due to overtopping. Freshness of the headscarp and poor forestation of the dams' body, contrasting of that of the original valley slope indicates young age of this feature. Its location within the epicentral zone of the 1911 Kemin earthquake allows assumption that rockslide could be triggered by this event. One more much smaller "fresh" rockfall about 0.2 Mm^3 in volume caved from the slope composed of Ordovician granites in the upper reaches of the Bolshoi Oriuktu River valley at 42.902° N, 77.906° E (Kazakh 026).

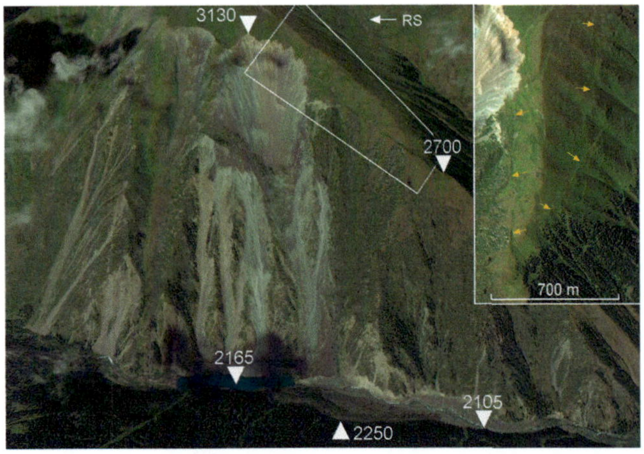

FIGURE 8.69 Rockslide dam extending downvalley up to 2105 m a.s.l. in the Bolshoi Oriuktu River valley and the DSGSD on its left watershed. RS – small rockslide body in the adjacent valley. 3D Google Earth view. Plan view of the outlined area with the sackungen is shown in the inset. Small orange arrows mark the upslope-facing scarps on both sides of the ridge crest.

Another rockslide that is definitely known to be triggered by the 1911 Kemin earthquake is located north of the Issyk-Kul Lake at the southern foot of the Kungey-Alatau Range (42.803° N, 77.629° E, Kyrgyz 184). This easily attainable feature was identified during the reconnaissance mission (Bogdanovich et al., 1914). Being located 7 km north-north-west of the Ananievo village (its previous name at the time of earthquake was the Sazanovka village) it was named in more recent publications as the Ananievo rockslide (Delvaux et al, 2001; Havenith et al, 2003).

Rockslide destroyed the rocky spur up to 450 m high and about 800 m wide (at its base) composed of granite. Just at its foot one of surface ruptures of the 1911 Kemin earthquake passes – thrust fault dipping under the ridge (Delvaux et al, 2001). It is marked by red arrows on Fig. 8.70. Failure of about 30 Mm^3 of granite affected the most prominent part of the spur. Unlike many other bedrock landslides that have spoon-shape headscarps (see e.g. Figs. 4.3, 4.14, 5.4, 5.45, 5.52, 6.5, 7.10, 8.12) here the headscarp back wall up to 200 m high and more than 600 m wide is almost planar, similar to the much bigger Aidyn-Ula rockslide in the Chinese Tien Shan (see Fig. 4.4). It looks as if the entire frontal part of the ridge was hacked away

FIGURE 8.70 View of the Ananievo rockslide. Yellowish loam visible in the ravine at the foreground is the material over which rockslide debris moved, partially bulldozing it. Red arrows – position of the Kemin earthquake surface rupture, bold wide line – the approximate position of the section that crosses the frontal bench (see Fig. 8.72).

(Fig. 8.71). In good weather this fresh nonvegetated surface can be seen even from the southern beach of the Issyk-Kul Lake.

FIGURE 8.71 Planar headscarp and proximal depression of the Ananievo rockslide. The topmost part of the spur is composed of gray granite which angular blocks form the surficial part of the rockslide body.

Rockslide formation just at this site during the M8.2 earthquake that affected severely very large area can be explained by favorable combination of several factors: 1) topographic amplification (Havenith, 2002; Havenith et al., 2003), 2) closeness to the seismogenic thrust that, besides producing intensive elastic shaking, could "knock this spur up" according to its kinematics, and 3) structure of this rock mass with weak fault zones that formed the base of the spur while its crest was composed of less fractured and less deformable granite (see Fig. 8.71). West of this spur the weak zone rises and crosses the ridge more than 100 m above its feet, while east of it weaker material does not outcrop being below the daylight surface.

Just along the frontal edge of the rockslide body there is a 7 to 10 m high bench (see Fig. 8.70) composed not of granite debris, as the entire rockslide, but of the light brown loam (Fig. 8.72). Small trenches excavated at the base of this bench (C on Fig. 8.72) and at its upper limit (B on Fig. 8.72) reveal its internal structure and relationships with the main part of the rockslide. Rockslide debris scraped the surficial part of the clayey loam and

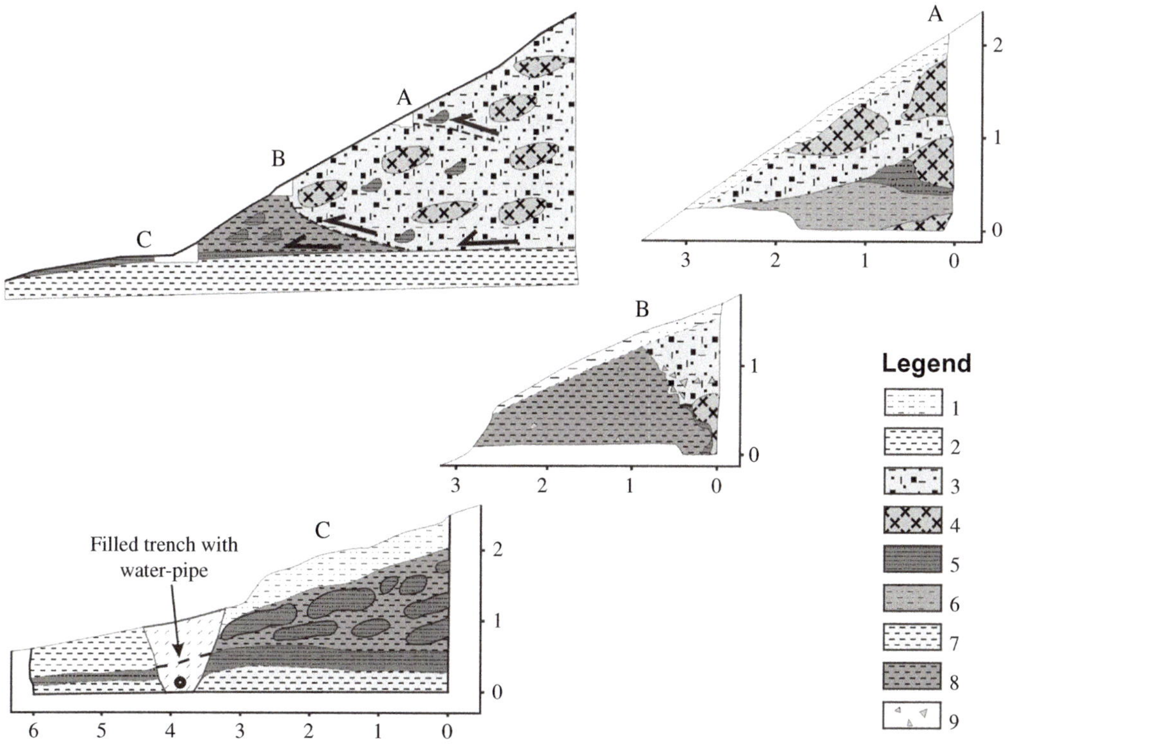

FIGURE 8.72 The internal structure of the Ananievo rockslide frontal bench. Schematic view of the toe materials with location of the trenches A, B, C, which are shown in larger scale. 1 – talus, 2 – whitish compact loam, 3 – crushed gray granite – arenites, 4 – blocks and boulders of pink granite, 5 – buried soil, 6 – brown subsoil loamy sand, 7 – whitish subsoil loam (in situ), 8 – light brown loam, 9 – small granite fragments. Axes scales are in meters. *Modified from Havenith, H.-B., Strom, A., Jongmans, D., Abdrakhmatov, K., Delvaux, D., and Trefois, P. 2003. Seismic triggering of landslides, Part A: Field evidence from the Northern Tien Shan. Natural Hazards and Earth System Sciences 3, 135–149 with permission of the European Geosciences Union and from Abdrakhmatov, K., and Strom, A. 2006. Dissected rockslide and rock avalanche deposits; Tien Shan Kyrgyzstan. In: Evans, S.G.; Scarascia Mugnozza, G.; Strom, A.; Hermanns, R.L. (Eds.) Landslides from Massive Rock Slope Failure. NATO Science Series: IV: Earth and Environmental Sciences, Springer, Vol. 49, 551–572 with permission of Springer.*

bulldozed it. Steep boundary between the granite debris and loam in front of it is visible on Fig. 8.72-B, while mixture of disrupted blocks of soil in loamy matrix moving above the buried soil layer can be seen on Fig. 8.72-C.We should point out that earthquake occurred in winter (January 4, 1911) when the soil over which rockslide moved was frozen. It could be hypothesized that such scraping could damp debris motion so that this, generally unconfined rockslide had a short runout of 1.24 km only.

Several other relatively large rockslides triggered by the 1911 Kemin earthquake were found by the reconnaissance team (Bogdanovich et al., 1914). One of them (Kyrgyz 157) – about 6 km north-west of the Ananievo rockslide in the Sutty-Bulak River valley at 42.838° N, 77.568° E, in the same granite massif. Unfortunately no good space image of this site was available either in Google Earth or in SAS Planet.

Another well known Kaindy rockslide triggered by the 1911 Kemin earthquake occurred in the western part of its epicentral zone, in the south-eastern corner of the Chon-Kemin intermontane depression (42.718° N, 76.205° E, Kyrgyz 335). About 5 Mm^3 of the Proterozoic metasediments collapsed from the 350 m high slope (Fig. 8.73) just above the surface rupture of this earthquake with predominantly left-lateral style of motion (Delvaux et al, 2001). Landslide entrained large amount of loose material at the slope foot and traveled 1.4 km being accompanied with numerous blockslides and earth flows in loose Quaternary sediments (Bogdanovich et al., 1914).

During careful examination of the space images we found one more recent rock avalanche. About 3 Mm^3 of weathered granite collapsed from the upper part of the 700 m high slope in the Chon-Uriukty River basin at 42.813° N, 77.6286° E (Kyrgyz 185) 16 km east of the Ananievo rockslide. Dry rockslide debris flowed down along rather steep gulley, and divided on the way into two parts (Fig. 8.74). The smaller one stopped at about 2410 m a.s.l., while the main rock avalanche terminated at 2200 m a.s.l. after 2.12 km runout. Freshness of its debris and closeness to the area severely affected by the 1911 Kemin earthquake allows assumption that this rock avalanche could be also triggered by this earthquake.

The 1911 Kemin earthquake was characterized by complex pattern of surface faulting – one system of surface ruptures passed in the east-north-eastern direction along the Chon-Kemin River valley and, likely, extended eastward in the upper part of the Chilik River valley. The Chilik (see Fig. 8.66) and the Kaindy (see Fig. 8.73) rockslides are associated with this zone. Another system of surface ruptures branches off it at

FIGURE 8.74 Recent rock avalanche in the Chon-Uriukty River basin. SAS Planet image combined with the 3″ SRTM DEM and visualized by Global Mapper software.

FIGURE 8.73 The 1911 Kaindy rockslide soon after its formation. From (Bogdanovich et al., 1914).

the hardly attainable watershed part of the Kungey-Alatau Range and stretches in the east-south-eastern direction towards the Issyk-Kul Lake. The Ananievo rockslide is associated with this fault zone. Besides recent rockslides definitely or presumably triggered by this earthquake, several prehistoric features were found in the Chon-Kemin and the Chilik River valleys. Some of them are much bigger than the 1911 slope failures. We must notice, however, that though these prehistoric features have been also triggered, most likely, by high magnitude past earthquakes, their larger size does not mean that these past earthquakes were larger than the 1911 Kemin one (see Chapter 13).

One of such older rockslides is located in the upper reaches of the Chilik River valley immediately downstream from the Kairakty River mouth at 43.03° N, 77.51° E (Kazakh 020, see Fig. 6.6). Failure of about 60 Mm^3 of granites formed the ~200 m high partially breached dam. Presence of multiple irregular terraces just upstream of the dam indicates gradual multistage erosion of the blockage.

Two groups of large-scale bedrock landslides were found in the central part of the Chon-Kemin River valley (Strom and Abdrakhmatov, 2004) divided by a 20 km long "gap" with only one suspicious feature that will be described later, despite similarly rugged terrain and location within the epicentral zone of the 1911 Kemin earthquake and closeness to its surface rupture (Fig. 8.75). The eastern group (see No 9 on Fig. 8.58), located close to the Djaya and the Bashi-Djaya Rivers that fall from the right in the Chon-Kemin River, consists of three rockslides – Kyrgyz 179 (42.868° N, 76.652° E), Kyrgyz 180 (42.868° N, 76.608° E) and Kyrgyz 181 (42.883° N, 76.624° E), about 20, 11 and 35 Mm^3 in volume correspondingly. One more assumed large rockslide (Kyrgyz 182) was identified ~6 km east of this group, at 42.853° N, 76.73° E.

The Bashi-Djaya rockslide (Kyrgyz 179) – one of the most interesting case studies from this cluster – collapsed from the right slope of the Chon-Kemin River valley composed of amphibolites. Internal structure of its body can be observed in the river gorge, and intensive comminution is typical of debris originated from this hard rock. The rockslide front rises about 120 m above the riverbed and overlies the terminal $Q_{III\text{-}IV}^1$ moraine (Fig. 8.76). In the road cut one can observe the distal part of rockslide debris overlying the buried soil developed on the glacial deposits (Fig. 8.77). ^{14}C AMS dating of this paleosol gave an age of 7000–7950 BP (6400–4600 BC with 95.4 % probability) that should be close to the age of this slope failure since paleosol development stopped when it was buried by rockslide debris (Abdrakhmatov and Strom, 2006). It is a rare case of dated prehistoric rockslide in the study region.

Distal part of rockslide outcropping here is composed of angular fragments about 1–3 cm in size on an average (according to visual estimate) at its base, and of coarser fragments up to 20–30 cm in size few meters higher. Study of the buried soil structure helps to understand how moving rockslide debris affected underlying surface. In this case, in particular, blocks of brown organic-rich soil (horizon "A" of the paleosol profile) were overthrusted by blocks of the horizon "B" (see Fig. 8.77) and all the succession was truncated by rockslide debris.

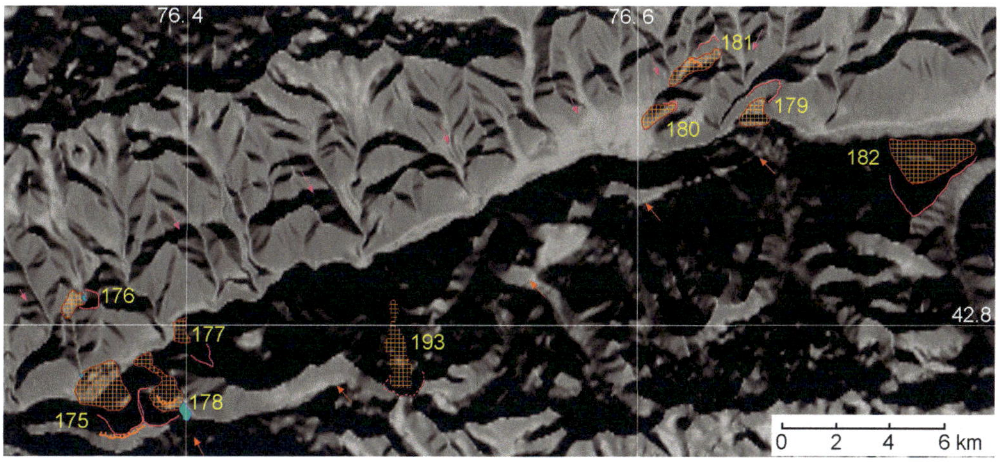

FIGURE 8.75 Rockslide clusters in the central part of the Chon-Kemin River valley. Cross-hatched areas with brown boundaries – recent rockslide bodies; those without boundaries – old rockslide bodies; solid pink lines – recent headscarps; dashed pink lines – old headscarps; comb orange lines outside rockslide bodies – sackung-like scarps; similar lines inside – secondary scars (see Section 5.3.3.1); blue areas – landslide-dammed lakes; red arrows mark approximate position of the 1911 Kemin earthquake surface rupture; purple arrows – the approximate position of recent surface rupture of unknown age on the right side of the Chon-Kemin River valley. Yellow numbers correspond to database IDs (Appendix), but without prefix "Kyrgyz". 3″ STRM DEM visualized by the Global Mapper software as a background.

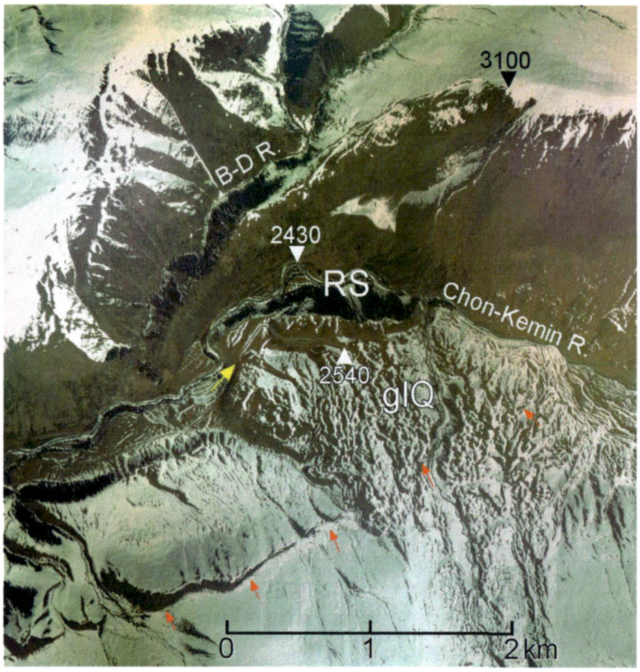

FIGURE 8.76 Google Earth Image of the Bashi-Djaya rockslide. Its headscarp crown at 3100 m a.s.l. Red arrows mark surface rupture of the 1911 Kemin earthquake; yellow arrow – outcrop shown on Fig. 8.77; B-D R. – the Bashi-Djaya River.

FIGURE 8.77 Deformed paleosol layer developed on the moraine deposits and overlaid by the frontal part of the Bashi-Djaya rockslide. Zoomed outlined fragment shown in the inset demonstrates its deformational pattern. Red lines – shears within paleosol. RS – rockslide debris; PS – paleosol; glQ – moraine deposits; A, B – fragments of the "A" and "B" horizons of the paleosol profile. Handle of the geological hummer is ~65 cm long. *Modified from Abdrakhmatov, K., and Strom, A. 2006. Dissected rockslide and rock avalanche deposits; Tien Shan Kyrgyzstan. In: Evans, S.G.; Scarascia Mugnozza, G.; Strom, A.; Hermanns, R.L. (Eds.) Landslides from Massive Rock Slope Failure. NATO Science Series: IV: Earth and Environmental Sciences, Springer, Vol. 49, 551–572 with permission of Springer.*

Another interesting case study from this cluster is the 2.6 km long Djaya rock avalanche (Kyrgyz 181) closely associated with surface rupture of the unknown age that passes just at the base of the collapsed slope (Fig. 8.78). About 35 Mm^3 of granite debris overlies the trace of this active fault, well visible both east and west of rock avalanche body. Slope failure could occur either immediately after surface faulting or later, and dating of both the latest faulting event and rock avalanche could specify their relationships. However, absence of fault trace within the rock avalanche body indicates that no permanent displacements have taken place along this fault and, thus, this surface rupture was associated, most likely, with large earthquake(s).

The western cluster in the central part of the Chon-Kemin River valley (see No 10 on Figs. 8.58 and 8.75) consists of 4 rockslides. Three of them – the Djashilkul (42.78° N, 76.362° E, Kyrgyz 175), the Kiol-Kogur (42.7795° N, 76.388° E, Kyrgyz 178) and the Upper Djashilkul (42.807 °N, 36.348° E, Kyrgyz 176) have well-defined bodies and look relatively young, while the forth one at 42.794° N, 76.398° E (Kyrgyz 177) is much older, strongly eroded and could be identified in the field only, based on the composition and internal structure of its body remnant.

The largest of these features – the Djashilkul rockslide about 200 Mm^3 in volume, collapsed from the deep semicircular headscarp on the ~800 m high left-bank side of the Chon-Kemin River valley, composed of the Proterozoic metasediments and formed the natural dam with effective height about 150 m (Fig. 8.79). This blockage was breached relatively soon after emplacement, judging from the lack of lacustrine sediments upstream. The frontal part of the dam that survived after breach has smooth wavy surface (Fig. 8.80) with distinct bow-shaped belt of large angular boulders of metasediments up to 2–3 m in size, which do not present outside this belt (marked by white arrows on Fig. 8.79). It originated from the single layer in the source zone composed of rather massive rocks that retained its integrity at some extent during rockslide motion.

Besides damming the powerful Chon-Kemin River that had filled the dammed lake in a short time resulting in dam's overtopping and breach, this rockslide blocked the mouth of its right tributary and formed a small lake that still exists (see Fig. 8.79). Its water

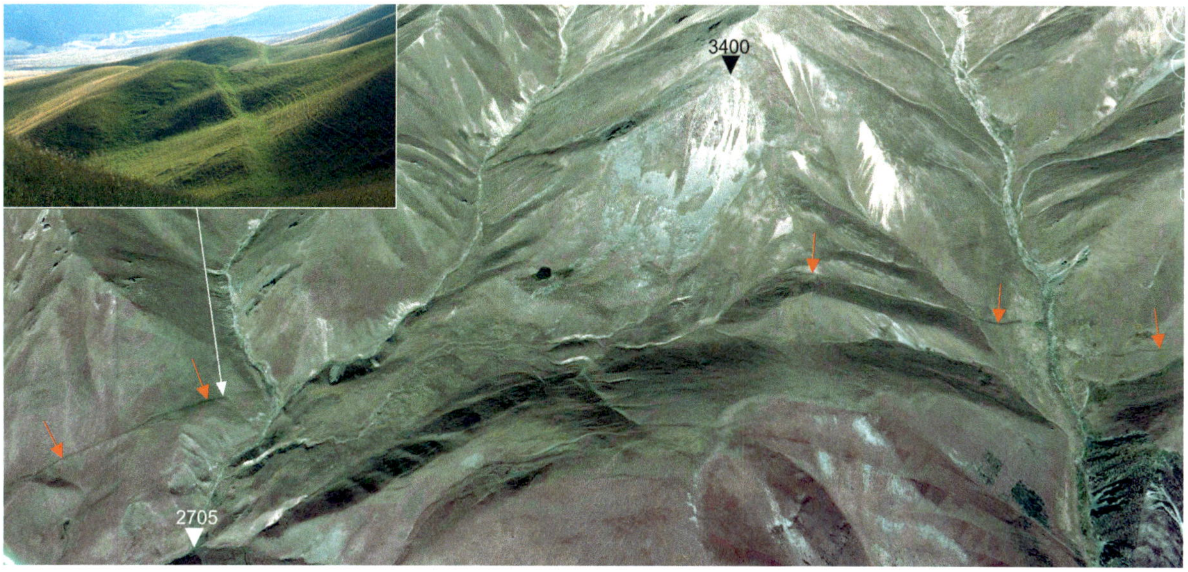

FIGURE 8.78 The Djaya rock avalanche overlying distinct trace of recent surface rupture marked by red arrows. The upslope-facing scarp of this active fault is shown in the inset.

FIGURE 8.79 Overview of the Djashilkul rockslide dam. White arrows mark bouldery belt on top of the body.

FIGURE 8.80 Wavy surface of the Djashilkul rockslide dam. Westward view; the headscarp and erosional gorge is on the left.

sometimes looks green that gave the name to this feature (Djashilkul – means Green Lake). Presence of the tributary valley mouth at the left (downstream) side of the impact zone resulted in the asymmetric along-valley profile of the blockage (Strom, 2010a) – its downstream slope, which is closer to this "trap", is steeper than the upstream one (Fig. 8.81). Causes and consequences of such asymmetry will be discussed in Chapter 12.

In the nearly 200 m-deep gorge incised in the dams' body one can find excellent outcrops allowing detail study of its internal structure (Fig. 8.82). It is clearly visible that main part of the blockage is composed of the intensively comminuted debris, which, however, is well "stratified". Boundaries between "layers" of

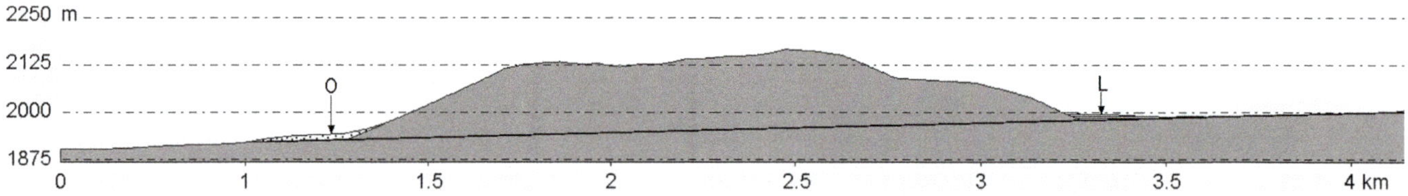

FIGURE 8.81 The along-valley profile of the Djashilkul rockslide body. L – lacustrine sediments upstream; O – deposits likely left by outburst flood during dam's breach.

FIGURE 8.82 Right bank of the gorge eroded through the Djashilkul rockslide dam where its internal structure and grain-size composition can be studied in details. Typical grain-size composition of the dams' interior can be seen in the inset. Each section of the hammer's handle is 10 cm long.

rockslide breccia composed of macadam-size fragments in the sandy-silty matrix (see inset in Fig. 8.82) are very distinct, without intermixing of adjacent lithological units. Though we could not study the detail structure of the scar area, which is rather hardly attainable (there are no bridges nearby), we assume nevertheless, that lithological sequence that can be observed within the dam's body corresponds to that in the parent rock massif. Such structure indicates that the whole descending rock mass moved entirely, being subjected to the intensive comminution during its motion, since such crushing is not observed within the host massif.

Average size of fragments (D_{50}) of the 37-kg sample of metamorphosed limestone collected at the central part of the erosional gorge about 70–80 m above the riverbed has been found to be about 1 cm (it will be discussed in more details in Chapter 11). Unlike several other deeply incised rockslide deposits such as of the Köfels (Erismann, 1979) and the Kokomeren rockslides (Strom, 1994, 2006; Abdrakhmatov and Strom, 2006), which coarse carapaces are rather thick – up to ca. ½ of the entire deposit's thickness, at the Djashilkul case the intensive shattering affected the descended rock mass almost entirely, except of its uppermost carapace that is composed, at least partially, of much larger fragments. The comminution level is comparable with that of the Kokomeren rockslide described in the next section of this Chapter (compare Tables 8.4 and 8.8) except its basal unit crushed at a higher extent (see Table 8.7). It will be discussed in Chapter 11 too.

Immediately east of the Djashilkul rockslide headscarp another large-scale rockslide (Kyrgyz 178) formed a natural dam with a small Kiol-Kogur Lake behind it (Fig. 8.83). Most likely both rockslides occurred simultaneously, which can be derived from the unified system of fractures behind their headscarp crowns (see plan view in Fig. 8.83). While main part of the Kiol-Kogur rockslide body about 120 Mm^3 in volume blocked the tributary valley and dammed the lake, about 15% of it formed the secondary rock avalanche that moved ca. 1.5 km downvalley, almost at a right angle to the direction of the initial slope failure and formed a fan-shape body at the Kiol-Kogur River mouth. It represents a minor model of the much larger (about 250 Mm^3 in volume) secondary rock avalanche in the Muzart River valley (see Fig. 4.30).

The third well preserved feature of this cluster is the Upper Djashilkul rockslide (Kyrgyz 176), 3 km north of the Djashilkul rockslide (see its coordinates above). It is located in the upper reaches of the unnamed creek that falls into the Djashilkul Lake. Rockslide about 20 Mm^3 in volume originated from the wedge-like headscarp on the slope about 350 m high composed of Paleozoic metasediments. Headscarp curvature at the level of the deposits surface reached 250–300 m – so deep was this slope failure. Debris up to 80–90 m thick filled the valley at a distance of about 1 km (Fig. 8.84). Despite relatively small runout this feature can be classified as rock avalanche that occurred in the extremely confined topographic conditions. The valley upstream was filled by sediments and only a small beautiful lake remains in the proximal depression at the headscarp foot. At present water eroded a ravine along the left boundary of the rock avalanche deposits and also infiltrates through the narrowest section of its upper part forming a series of springs. Later on small landslide marked by LS on Fig. 8.84 slid down from the headscarp back wall.

About 1.5 km upstream from the Kiol-Kogur River mouth and 3 km north of the Kiol-Kogur Lake, on the left bank of the Chong-Kemin River one can see a terrace-like shelf with slightly hilly surface at the foot

TABLE 8.4 Grain size composition of shattered limestone from the central part of the Djashilkul rockslide body about 70–80 m above the riverbed

	Fractions (mm)										
	>100	100.0-40.0	40.0-20.0	20.0-10.0	10.0-5.0	5.0-2.0	2.0-1.0	1.0-0.5	0.50-0.25	0.25-0.10	0.10-0.05
	Content (%)										
Sample 3/2003	10.75	15.59	13.44	12.1	11.55	13.47	5.77	1.3	2.12	4.62	9.29

FIGURE 8.83 Above – plan view on the unified system of fractures (marked by yellow arrows) on top of the ridge behind the Djashilkul and the Kiol-Kogur headscarp crowns. The small rockslide (RS) is visible at the top right. Below – 3D Google Earth view on the Djashilkul rockslide (to the right) and the Kiol-Kogur rockslide (to the left). Dj – the Djashilkul Lake, K-K – the Kiol-Kogur Lake.

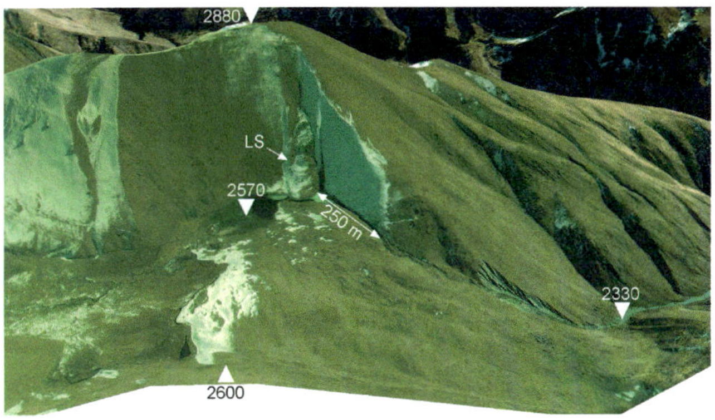

FIGURE 8.84 The Upper-Djashilkul rockslide with wedge-like headscarp up to 250–300 m deep. 3D Google Earth view.

of high densely forested concave slope (Fig. 8.85). This pseudo-terrace represents the remnants of large rockslide named the Kiol-Kogur-Northern (Kyrgyz 177). This rockslide is much older than the Djashilkul and Kiol-Kogur rockslides described above since its body had been significantly eroded and can be hardly identified from its morphology only.

However, presence of large rockslide at this site can be proved by the structure of the remaining part of its deposits with characteristic double-layer internal structure. Its upper part is composed of angular debris and boulders up to 1 m in size (Fig. 8.86-A), while the lower unit represents multi-colored succession of "layers" of comminuted red and dark-gray igneous

FIGURE 8.85 Overview of the Kiol-Kogur-Northern rockslide. White arrows mark the site where photo shown on Fig. 8.86 was made; RS – rockslide deposits; alQ – alluvium overlaid by rockslide debris.

FIGURE 8.86 Internal structure of the frontal part of the Kiol-Kogur-Northern rockslide deposits and alluvium deformed by the advancing rockslide debris. RS – rockslide deposits; alQ-1 – sandy-gravelly alluvium; alQ-2 – pebbly-cobbly alluvium; red lines – assumed thrust shears in the alluvium; A and B – zoomed views of the outlined parts of the outcrop.

rocks, most likely granite and diabase (RS on Fig. 8.86-B). Besides, this outcrop shows interaction between advancing rockslide debris and loose substrate. Several thrust shears are visible in the alluvial deposits deformed by rockslide body (see Fig. 8.86-B).

This case study indicates that large-scale bedrock landslide occur nearby repeatedly. Same recurrent slope failures were described above in the Inylchek River valley (see Figs. 6.4, 8.45), at the southern slope of the Sarmin-Ula Ridge (see Fig. 8.11) and in other parts of the study region that will be described hereafter.

As we mentioned above, almost no large-scale slope failures were identified between two rockslide clusters at the upper part of the Kungey-Alatau Range northern slope (see Fig. 8.75). The only exception is a suspicious feature at 42.795° N, 76.493° E (Kyrgyz 193). This accumulation about 35 Mm^3 in volume rests at the foot of the circus-shape scar with crown at 3800 m a.s.l., i.e. in the recently glaciated zone. However, as it can be seen on Fig. 8.87, its composition differs drastically from the deposits at the feet of similar scars at the same altitude. While outside this site slopes below the circuses are smooth, without big boulders on the surface and the only manifestations of the most recent slope movements are short rock glaciers (RG on Fig. 8.87), the body in question is the accumulation of angular clasts, blocks and boulders of the Proterozoic granites and metasediments up to ca. 20 m in size, with numerous steps, especially at its proximal part, below well-defined rock glaciers (left inset in Fig. 8.87). Foot of the assumed headscarp is covered by active rock glaciers that overlay this bouldery material.

Judging from the totally bouldery composition of material it is obvious that there was a large rock slope failure from the ridge's top below 3800 m a.s.l. Large amount of rock debris could fell either on the small glacier or on rock glacier. Stepped relief of the proximal part of this accumulation could be formed due to melting of the buried ice. The bulb-like distal part of this body, also composed of coarse material, though finer than at its proximal part, demonstrates distinct evidence of the flow-like motion somehow resembling the secondary rock avalanche of the bottleneck subtype (see Section 5.3.3.2, area marked by a circle on Fig. 5.56), though the alternative explanation is possible as well – water seeping from the melting buried ice could result in gradual earthflow of the frontal part of the this body that filled the valley up to 2660 m a.s.l. Afterward recent rock glaciers were formed in the headscarp. Further studies are necessary to reveal the real origin of this feature.

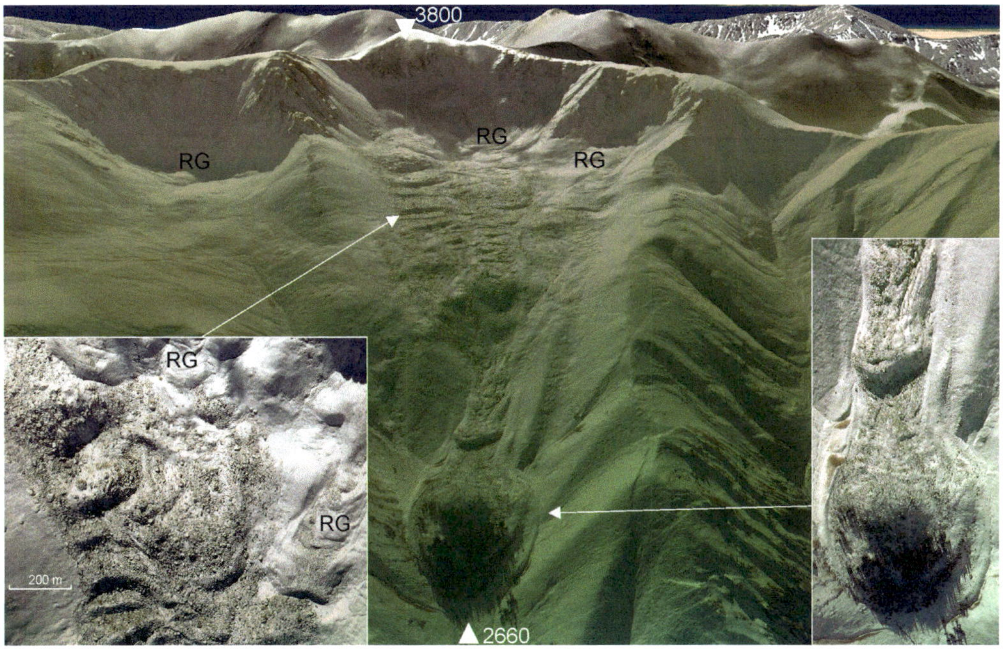

FIGURE 8.87 Strange bouldery accumulation at the foot of the headscarp at the Kungey-Alatau Range northern slope. RG – rock glaciers. Bouldery proximal part of this body with stepped relief, partially overlaid by recent rock glaciers is shown in the left inset; the bulb-like distal part – in the right inset. 3D Google Earth view.

Several interesting case studies are known in the northernmost mountain chain of the Tien Shan – the Trans-Ili Range, at its northern slopes rising above the populated Ili intermontane depression (see Fig. 8.58). The central part of this area was affected by the M ~7.3 1887 Vernyi earthquake that produced numerous slope failures studied by I. Mushketov (1890). One of them, in the Aksai River valley, will be described in brief hereafter. Two larger (M > 8.0) earthquakes – the 1889 Chilik and the M8.2 1911 Kemin that also produced intensive shaking of this area, did not cause such effects here, since their causative faults passed southward (Bogdanovich et al., 1914; Delvaux et al., 2001; Abdrakhmatov et al., 2016). Most of large-scale slope failures at the northern slope of the Trans-Ili Range are prehistoric.

Very large prehistoric rock avalanche of the Secondary type with well pronounced secondary scar occurred on the right slope of the Turgen River valley about 50 km east of the Almaty City (43.326° N, 77.658° E. Kazakh 041). After collapsing from the ~450 m high headscarp with its crown at 2250 m a.s.l., about 300 Mm^3 of granite debris moved 3 km downslope, collided with the gulley slope band, and formed Secondary rock avalanche with well-defined headscarp at 1550 m a.s.l. that traveled about 2.1 km more and stopped in the Turgen River valley at 1200 m a.s.l. approximately (Fig. 8.88). Most likely it had blocked it, though no visible evidence of the dammed lake can be derived from the space images now.

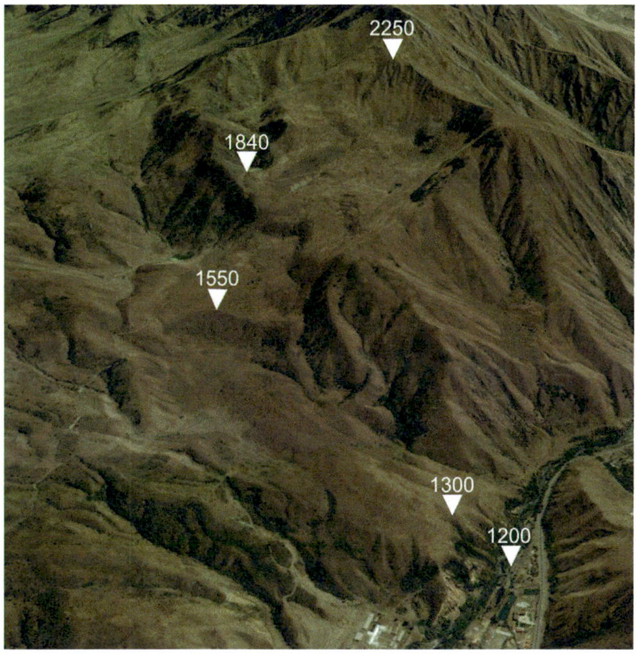

FIGURE 8.88 Large Secondary rock avalanche in the Turgen River valley. 3D Google Earth view.

Large rockslide dam forming a beautiful Issyk Lake is known in the Issyk River valley, 40 km east of Almaty City at 43.257° N, 77.487° E (Kazakh 021). It was formed by collapse of about 25 Mm^3 of diorite from the ~750 m high right slope of the valley and remained intact for centuries if not millennia. As it was described in Chapter 2, on July 7, 1963 powerful debris flow that entered the Issyk Lake caused the catastrophic breach of this dam and formation of the even more catastrophic secondary debris flow. Modern 3D Google Earth view of this site after construction of the artificial dam in the erosional gorge left by the 1963 disaster can be seen on Fig. 2.1.

Interesting peculiarity of the Issyk rockslide noticed by Eugen Gaspirovich (personal communication) is as follows: though the entire dam is composed of the diorite debris, its carapace consists of large angular blocks of hard diorite, while in the dams' interior the same rocks are crushed at a high extent, so that visually distinguishable clasts sometimes can be broken by hands. It explains the catastrophic style of the dam's breach – when blocky carapace was removed by waves triggered by debris flow surges entering the lake, underlying comminuted material was eroded rapidly.

The largest rock slope failure at the Trans Alai Range northern slope can be found about 15 km south of the Almaty City at 43.065° N, 76.986° E (Kazakh 023). About 450 Mm^3 of crushed granite blocks the valley in the upper reaches of the Bolshaya (Large) Almaatinka River valley and forms the Great Almaty Lake (Bochkariov, 2006) (Fig. 8.89). We must notice that we disagree with Bochkariov's interpretation of this blockage origin as a result of multistage slope failures from both sides of the valley accompanied by the tectonic subsidence in the lake area. From our point of view this natural dam was formed by a single-event failure from the left side of the valley composed of Paleozoic granites. Headscarp crown is marked on Fig. 8.89 by elevation marks 3310 and 2780 m a.s.l. Rockslide crossed the valley, rose 110–115 m on the opposite slope and then spread downstream almost at a right angle for about 2.5 km. Very similar kinematics of rockslide debris motion, though with better expresses secondary scar, can be observed at the Karasu Lake rockslide in Kyrgyzstan (see Fig. 5.64) and explains all observable geomorphic features typical of these rockslides.

The last site from this part of the study region that we would like to mention is the Akjar rockslide in the Aksai River basin, about 10 km south-west of the Almaty City, at 43.103° N, 76.802° E (Kazakh 006). This rockslide was the largest one triggered by the 1887 Vernyi earthquake (Mushketov, 1890; Bogdanovich et al., 1914). Failure occurred on the 150–400 m high slopes at the contact zone between granite and diorite and schist. This rockslide has very irregular and complex headscarp (Fig. 8.90). Though in the literature volume of the collapsed rock was estimated as 40 Mm^3 (Khromovskikh, 1984), according to our estimates it was much less – 7–10 Mm^3 only. As described by I. Mushketov (1890), some boulders of this rockslide were up to 20 m in size.

24 year later, during the 1911 Kemin earthquake, which intensity at the northern slopes of the Trans-Ili Range, about 25 km from the causative fault, was still very high, the deposits of the Akjar rockslide were shaken and fractured, but no new large slope failures occurred (Bogdanovich et al., 1914). Later on large amount of crushed loose rockslide debris produces recurrent debris flows, which traces are well expressed and visible on the Google Earth images.

The next group of interesting case studies is located in Northern Kyrgyzstan, on the southern and northern slopes of the Kyrgyz Range (see Fig. 8.58). Many of them are at one day trip distance from the Bishkek City.

The enormous deposits about 500 Mm^3 in volume cover about 8.53 km^2 and extend for 7.2 km from ~3500 m a.s.l. in the recently glaciated zone of the Uitabulga River valley – right tributary of the Shamci River at 42.56° N, 75.48° E (Kyrgyz 285, Fig. 8.91). Its

FIGURE 8.89 The Great Almaty Lake rockslide dam. Artificial dam on top of the rockslide body was constructed to protect the Almaty City from outburst and to use the lake for hydropower. 3D Google Earth view.

FIGURE 8.90 Complex headscarp of the 1887 Akjar rockslide triggered by M ~ 7.3 Vernyi earthquake. 3D Google Earth view.

FIGURE 8.91 3D view of the assumed Uitabulga rock avalanche. SAS Planet image visualized by Global Mapper software.

assumed bow-shaped irregular headscarp crosses bench on the northern slope of the glaciated summit more than 4000 m a.s.l. composed of the Devonian terrigenous rocks and forms a step up to 400 m high dissecting recent glacial landforms (Fig. 8.92). Tongue of the deposits bellow this escarpment reaches the zone composed of the Neogene red beds at about 2000 m a.s.l., which is much lower than the "normal" elevation of end moraines in the valleys dissecting northern slope of the Kyrgyz Range (2900–3000, rarely 2500 m a.s.l.). The lowermost part of this body (bellow ~2250 m a.s.l.) has very impressive hummocky microrelief with small conic hills up to 40–50 m in diameter (Fig. 8.93), very similar to that of rock avalanches shown on Figs. 4.9, 4.23, 4.41, 6.9, 8.18. We must notice, however, that downvalley from the point marked by 2010 m a.s.l. on Fig. 8.93, the hummocky microrelief

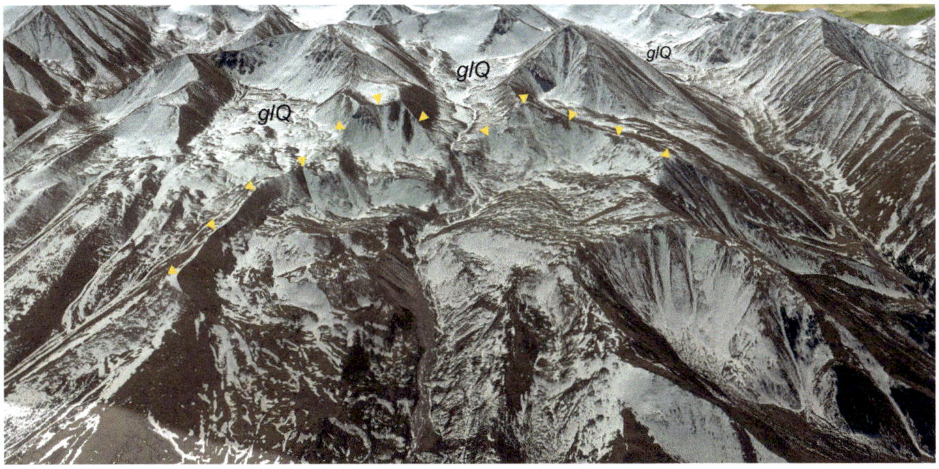

FIGURE 8.92 The assumed headscarp (marked by yellow triangles) of the Uitabulga rock avalanche. 3D Google Earth view.

FIGURE 8.93 Hummocky microrelief of the assumed Uitabulga rock avalanche distal part. 3D Google Earth view.

disappears and the lower reaches of the Uitabulga River valley up to its confluence with the Shamci River have very smooth topography. Such topographic change resembles transition from the moraine to the glaciofluvial deposits. However, abnormally low altitude of the distal part of the deposits in question, presence of conical hills, differing significantly from the so called "chukur" relief typical of the glacial deposits with multiple small drainless depressions (absent in our case) and well-defined headscarp allows sound assumption that it is rather recent rock avalanche, which, of course, had affected moraines at the valley bottom. Several other suspicious landforms could be found in the vicinity of this case study and extensive fieldwork should be carried out to reconstruct Late Quaternary history of this part of the Kyrgyz Range.

FIGURE 8.94 Natural dam formed by the Primary rock avalanche south from the Koltor pass. Headscarp crown is at the top right at 3590 m a.s.l.; the deposits tip – at 2930 m a.s.l. at the down left of the image. Google Earth image.

12 km south of the site described above, in the upper reaches of the Koltor River valley and about 2 km south of the same-name pass, at 42.419° N, 75.431° E a small dammed lake was formed by the Primary rock avalanche about 10 Mm^3 in volume (Kyrgyz 173). It originated about 700 m above, at the southern slope of a small summit composed of the Devonian terrigenous rocks. Here, similar to the Seit rock avalanche (see Fig. 5.35), almost all debris "flowed" down and formed a dam at a distal part of the transition-deposition zone (Fig. 8.94).

Relatively small rockslide about 6 Mm^3 in volume is located in the upper reaches of the Alamedin River valley (42.61° N, 74.665° E), 22 km south from the Bishkek City and just near the Arashan spa resort (Kyrgyz 159). It occurred on the right bank of the valley on the slope composed of Paleozoic granites. Failure affected only central and lower parts of the slope, approximately 2/3 of its 1-km entire height. This natural dam was dissected at its distal part where the dams' crest was lower. No remnants of rockslide debris remain on the left bank of the river (Fig. 8.95). Before damming the river channel was closer to the

FIGURE 8.95 Front view of the Arashan rockslide. A and B – location of outcrops shown on Fig. 8.96; sample 2/2003 was collected at point B, sample 1/2003 – at point C.

right bank and it can be hypothesized that valley slope undercutting by the river was one of the main causes of its failure.

According to the ^{10}Be dating this rockslide originated about 11–15 ka BP, which is consistent with radiocarbon dates of gastropod shells from reworked loess capping the landslide (Sanhueza-Pino et al., 2011). The newly formed river channel dissects both the Arashan rockslide body, the glaciofluvial deposits at its base and underlying bedrock so that the Alamedin River flows now in the canyon about 10 m deep incised in the granites. Assuming that rockslide body itself was dissected quickly, one can estimate the bedrock incision rate, which is about 1 mm/year.

Though this feature represent nothing exceptional neither by its size nor by runout, thanks to the road cut that can be seen on Fig. 8.95, it can be considered as an excellent and easily attainable site allowing detailed study of the rockslide internal structure and grain-size composition. The upper (outer) part of the rockslide body is a clast-supported non-lithified breccia of large angular granite boulders up to 1 m and even more in size with voids filled by fines (Fig. 8.96-A), while main part of the deposits form its internal "core" composed of intensively shattered granite with mean size of fragment (D_{50}) ranging from 4 mm to 1.5 cm, depending on the extent of comminution (Fig. 8.96-B). Grain size composition of its internal part based on the analysis of samples of crushed granite No 1/2003 and 2/2003 of about 155 and 65 kg correspondingly (see Fig. 8.95) is presented in Table 8.5 (see also Chapter 11). Not far from the northern (downstream) limit of the rockslide body one can observe the direct contact between bouldery glaciofluvial deposits and overlaying rockslide debris.

FIGURE 8.96 Grain-size composition typical of the outer (A) and internal (B) parts of the Arashan rockslide.

Besides intensive shattering there is one more interesting feature typical of the lower part of the Arashan rockslide body – veins of comminuted red granite inside the mass of gray granite (see, e.g. the upper right part of Fig. 8.96-B). It means that, though being completely crashed, large portions of the descending rock massif moved either as a blockslide or as a laminar flow, without mixing. Same peculiarity is typical of the Flims rockslide in the Swiss Alps (Erismann, 1979; Pollet et al, 2005), the Inylchek rockslide (see Fig. 8.45), the Djashilkul rockslide (see Fig. 8.82) and several rockslides in the Central Tien Shan that will be described in the next section of this chapter (Strom, 1994, 2006).

Another very interesting site at one-day trip distance from the Bishkek City is located at 42.638° N, 74.285 E, in the upper reaches of the Buriulu Creek – the right tributary of the Sokuluk River where two expressive recent rock avalanches, 24-25 Mm^3 in volume each, sometimes named the "Bielogorka Twins" (Kyrgyz 205, 206), most probably of the same age, and one older, much smaller rockslide, about 3 Mm^3 in volume (Kyrgyz 204) occurred at a very small area 3.5×2.3 km only (Fig. 8.97).

It was assumed that recent rock slope failures, which really look very "fresh", were triggered by the 1885 Belovodsk earthquake, which source zone was located about 20 km north of this site, just at this section of the Kyrgyz Range foothills. However, there is

TABLE 8.5 Grain size composition of shattered granite from the Arashan rockslide

	Fractions (mm)										
	>100	100.0-40.0	40.0-20.0	20.0-10.0	10.0-5.0	5.0-2.0	2.0-1.0	1.0-0.5	0.50-0.25	0.25-0.10	<0.10
	Content (%)										
Sample 1/2003	5.60	10.04	11.72	11.01	7.58	14.42	17.13	3.08	8.14	7.15	4.13
Sample 2/2003		12.92	27.54	27.23	14.41	10.02	3.42	0.81	1.13	0.84	1.68

FIGURE 8.97 Google Earth image of the Bielogorka rockslides. A – Right, B – Left, C – Old.

FIGURE 8.98 Headscarps of the Right (A) and Left (B) Bielogorka rockslides. White arrows on the view of the Right one mark affected blocks outside the main headscarp. Large block rests in the headscarp of the Left one above the distinct, up to 20 m deep, proximal lowering.

no single word in the reconnaissance report about the formation of such spectacular features visible from the populated plains (unlike the information about the adjacent Aksu River, which temporary blocking was noticed) (Ignatiev, 1886). It was mentioned only, that there were numerous slope failures in the mountains. Though one old man told that his ancestors were killed in 1885 by one of these rock avalanches (Hans-Balder Havenith, personal communication), the exact age of the "Twins" is still debatable. Our attempt to date their deposits covered by rather thick, up to 50 cm in diameter, archa trees (Central Asian juniper) by dendrochronological method failed – the thickest of them where not older than 60 years. It was assumed that older trees were used by locals for fire wood during the World War II and all present forestation is new.

Nevertheless, seismic origin of the Bielogorka rockslides is most probable (regardless of their exact age). Both headscarps have large blocks that were displaced but did not collapse completely (Fig. 8.98). The most sound argument in favor of the simultaneous origin of both rock slope failures is the confluence of distal parts of both rock avalanches without overlapping of one by another (Fig. 8.99). If they would occur successively, younger deposit should overlay an older one (as the Atjailau and the Inylchek rock avalanches in the Inylchek River valley – see Fig. 8.45).

Though being of the same order of size ($\sim 25\ Mm^3$) and originating in the similar geological formation (Left one – in the Ordovician sandstone; Right one – in the Ordovician sandstone intruded by Carboniferous Diorites), the "Bielogorka Twins" have different morphological pattern. The Right one is a typical Secondary rock avalanche (see Section 5.3.3.1), which

FIGURE 8.99 Confluence of the distal parts of the Right (A) and Left (B) Bielogorka rock avalanches.

proximal and distal portions are divided by expressive two-stepped secondary scar (its crown is at 1910 m a.s. l. on Fig. 8.97; see also Fig. 8.100) that originated when rockslide collided with an opposite slope of the valley. After that direction of rock avalanche motion changed almost at a right angle. Moreover, small bow-shaped steps that can be interpreted as secondary scars could be seen at the distal part of this body (see Fig. 8.99, just below "A").

FIGURE 8.100 The two-stepped secondary scar marked by white arrows at the downvalley side of the Right Bielogorka rockslide compact part. Headscarp crown of the Left Bielogorka rockslide at the background.

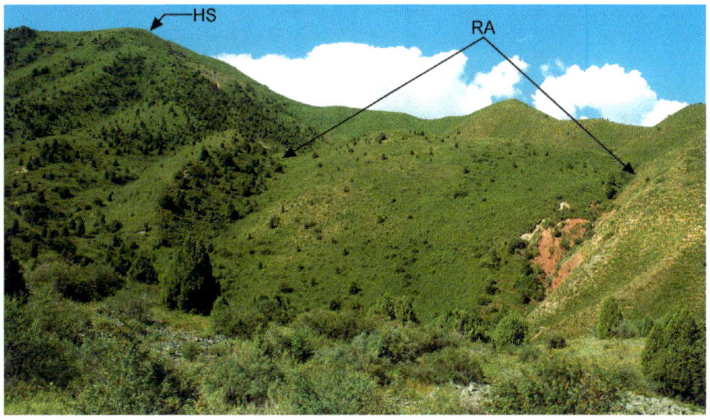

FIGURE 8.101 Body of the Old Bielogorka rock avalanche (RS). HS – headscarp crown in the upper left part of the photo.

FIGURE 8.102 View from the upstream on the Aksu rockslide dam. R – the rocky remnant of the former right bank of the river that had passed originally left of it. *Modified from Strom, A.L. 2010b. Landslide dams in Central Asia region. Landslides – Journal of the Japan Landslide Society 47(6), 309–324 with permission of the Japan Landslide Society.*

Left "Twin" is characterized by the huge block that remains inside its scar just above the prominent, about 20 m deep, proximal depression. The thickness of the rock avalanche body immediately downstream from the proximal depression seems to be maximal, up to 40–50 m (Havenith, 2002) and gradually decreases towards its toe. It is characterized also by the longitudinal levee along its central part that rises 10–15 m above the boundary zones. Near the lake dammed by the Left rockslide one can see the small levee composed of loess-like loam, likely bulldozed by rockslide during its motion (compare with the 1911 Ananievo rockslide – see Fig. 8.72). Infiltration of water in the dam's body can be observed here too. We must notice that, despite deep proximal lowering, which bottom is much lower than the lake level, no springs appears there, indicating that water table level linked to the lake decreases rapidly.

Both rock avalanches are composed (at the surface) by the macadam-size debris. According to Havenith (2002) average size of clasts is slightly less than the joint spacing observed at the bedrock outcrops. He also noted the very small content of fines on the avalanches surfaces. Though very small lakes exist in both dammed gullies, rock avalanche debris seems to be rather permeable, so that water of both dammed streams filters through it easily.

Besides the above "Twins" that are, obviously, rather recent, much older rockslide was found just west of the Left Bielogorka rock avalanche (C on Fig. 8.97). Though being much smaller it is of the came morphological type as the Left "Twin" – with big block remaining inside the headscarp and distal accumulation of diorite and sandstone debris. Unlike younger "big brothers" both the headscarp and body of this feature is almost totally grass-covered (Fig. 8.101).

Several bedrock landslides were identified in the Aksu River valley dissecting northern slope of the Kyrgyz Range west of the Sokuluk River. The largest of them, about 1 km^3 in volume (Kyrgyz 203) is located at 42.54° N, 74.0° E, 13 km south of the foot of the range. We must mention that in some previous publications (Strom, 2010b, 2013a) its volume was estimated as 1.5 km^3. However, more careful examination shows that smaller estimate presented in our earlier paper (Abdrakhmatov and Strom, 2006) seems to be more correct. As it was mentioned in Section 6.3, such accuracy of volume estimates should be considered as an acceptable. This easily attainable case study allows reconstruction of the long-term multi-stage history of river damming.

Giant rock mass collapsed from the wedge-like headscarp at the 1800 m high right slope of the Aksu River valley, which lower part, up to ~2700 m a.s.l., is

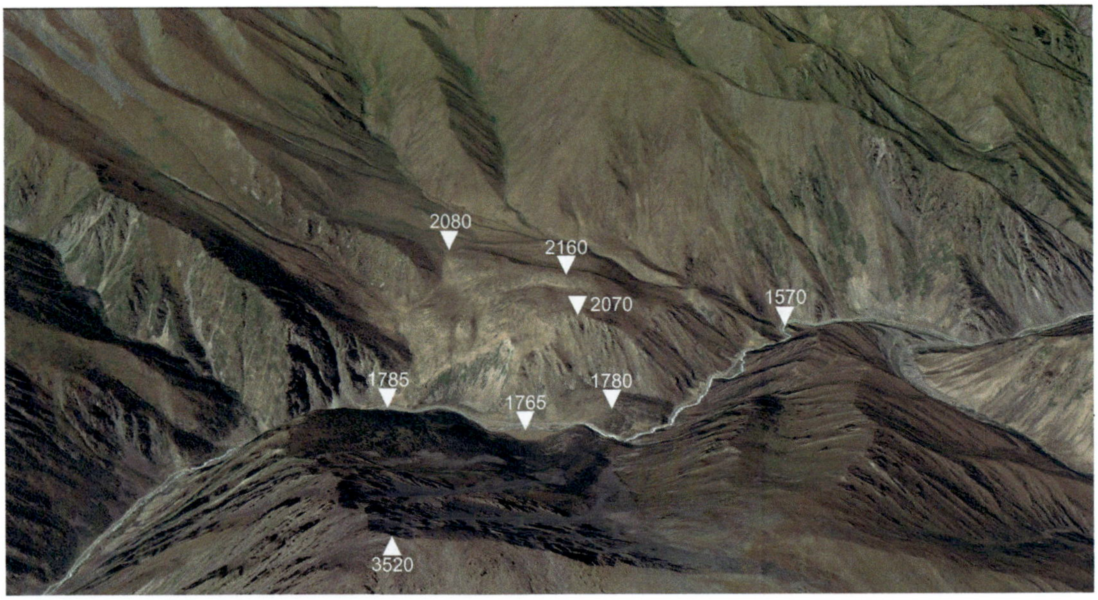

FIGURE 8.103 3D Google Earth view from the headscarp side on the Aksu rockslide dam. Distance between elevation marks 1785 and 1570 m a.s.l. is 2.5 km.

composed of granite, while upper one, up to 3500 m a.s.l. – of the Ordovician metasediments. Headscarp was up to several hundreds meters deep. It formed the dam about 400 m high (Fig. 8.102) that had blocked the valley at a distance of ~2.5 km (Fig. 8.103).

Occurrence of such a tremendous slope failure just at this site was predetermined by the presence of a deep semi-glacial landform on the opposite – left side of the valley (see Fig. 8.103). Recurrent debris flow that had originated there, inevitably forced the Aksu River towards its right bank causing gradual slope undercutting, which resulted, finally, in this slope failure that could occur, according to ^{10}Be dating, 63–67 ka BP (Sanhueza-Pino et al., 2011). The collapsing rock mass crossed the valley and entered the lowering at its other side that acted as a "trap", which led in formation of a very compact dam – 2.5 km wide only, which is quite small for such a high blockage (Strom, 2010a). This phenomenon will be discussed in more details in Chapters 12 and 14. The lowest part of the dams' crest was located at the foot of the headscarp. The frontal part of rockslide body is more than 100 m higher, which is typical of the Primary type of rock avalanches (Strom, 1996, 2006) (see Section 5.3.1).

Absence of lacustrine sediments upstream indicates that this dam, despite low DBI value (Ermini and Casagli, 2003) (1.94 according to (Strom, 2010b, 2013a) and 2.11 considering smaller rockslide volume), had been breached and dissected almost completely soon after its formation. It was hypothesized that it was dam's compactness that had played a critical role in such evolution of the blockage. Steep downstream slope of the blockage with gradient of about 0.4 was eroded easily after dam overtopping. The formation of the erosional gorge started at the proximal part of the dam, where its crest was lower and now one can see small rocky remnants of the original right bank of the river with bouldery alluvium on top of them at its left bank both at the upstream (see Fig. 8.102) and downstream limits of this blockage. The internal structure of this giant and almost completely incised dam can be seen on the left bank of the erosional gorge, though not as well as at the Inylchek (see Figs. 8.45, 8.49) and the Arashan (see Fig. 8.96) cases described above. Some outcrops can be seen on the right bank too. The topmost part of the blockage is composed of huge angular boulders and blocks up to 5–10 m in size that form the carapace similar to that at the Arashan rockslide. The structure and grain-size composition of the internal part of the dam is masked by talus, but, nevertheless, it is visible that it is much finer. We must mention that easily attainable Aksu rockslide can be used for acquisition of the geophysical methods that can be applied for study of the Usoi rockslide dam in Pamir, which is of the same order of size.

The post-failure history of the Aksu River valley can be revealed as well (Strom, 2010b). The more than 300 m deep erosional gorge incised in this Pleistocene blockage was blocked once or twice by the much smaller younger landslide(s) that originated on the steep left slope of the newly formed gorge composed of the rockslide debris. Its body is visible on Fig. 8.103 and forms a smooth surface at 1780 m a.s.l. at the foot of the bow-shaped headscarp. Actually there are two bow-shaped headscarps; the upper one is well visible on Fig. 8.102. Taking in mind the notice in the Ignatiev's description of the 1885 Belovodsk earthquake that after the earthquake water flow in the Aksu River disappeared temporarily (Ignatiev, 1886) we hypothesize that it could be the effect of such slope failure. Earthquake occurred on July 22, during the warmest period of time when glaciers melt, so the discharge of the Aksu River should be significant and the newly formed dam several tens of meters high should be overtopped shortly. It is obviously that after the earthquake a sediment yield in the river should increase due to rock falls and landslides in the catchment as it occurred after the 2008 Wenchuan earthquake in China (Fan et al, 2012a,b). It resulted in rapid siltation of this lake(s) that are represented by flat floodplains at 1765 and 1785 m a.s.l. visible on Figs. 8.102 and 8.103.

Several sackungen-like features up to several kilometers long affect watersheds between the Sokuluk and Aksu River valleys at 42.575° N, 74.067° E and at 42.575° N, 74.19° E (Fig. 8.104). It can be assumed that these slope deformations could be triggered by intensive seismic shaking either of the 1885 Belovodsk earthquake or of the preceding seismic catastrophes. In any case fresh near-watershed fractures detaching large rock massifs indicate the increased rockslide hazard at this part of the Kyrgyz Range.

The southern slopes of the Kyrgyz Range are also affected by numerous landslides, both in bedrock and

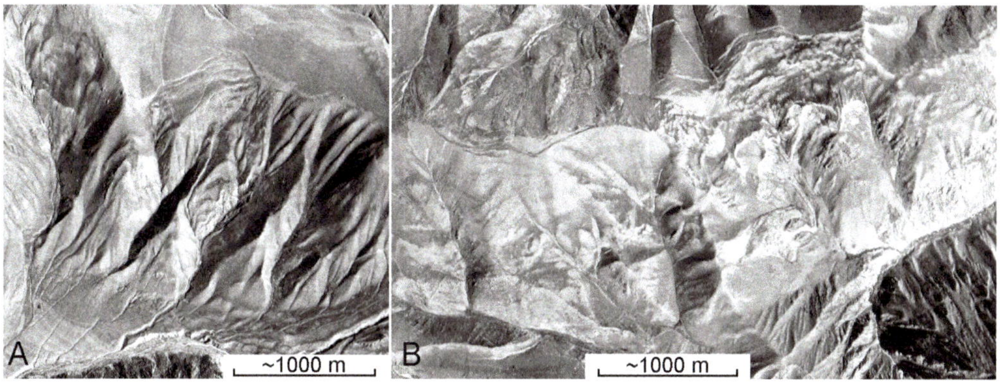

FIGURE 8.104 The sackungs between the Sokuluk and Aksu River valleys. A – at 42.575° N, 74.067° E; B – at 42.575° N, 74.19° E. Fragments of aerial photographs.

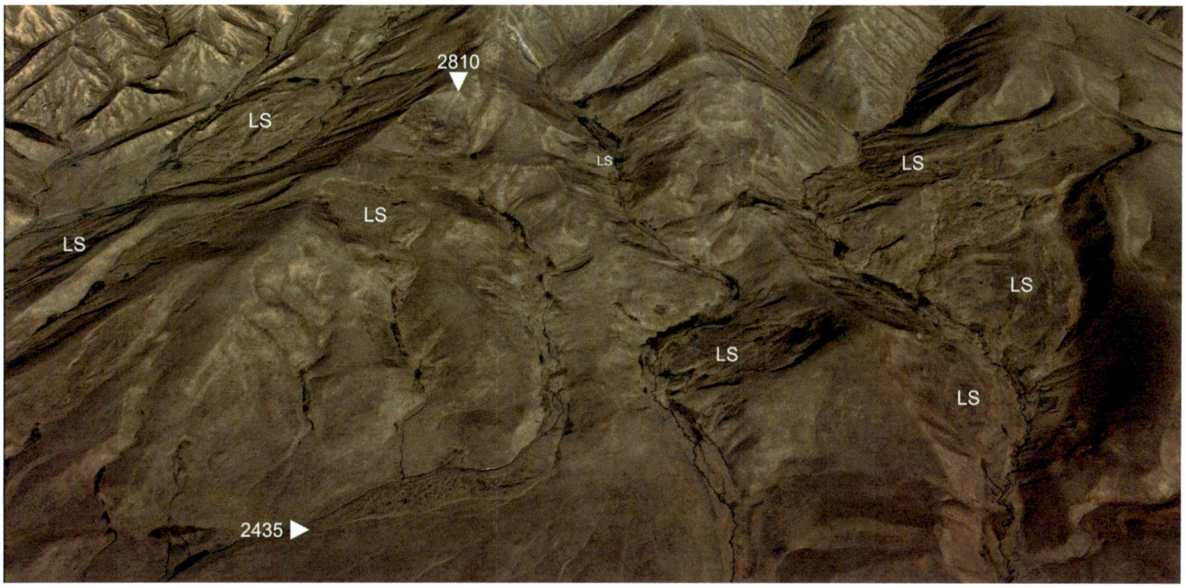

FIGURE 8.105 Numerous landslides (LS) surrounding the Snake-Head rock avalanche, which headscarp crown is at 2810 m a.s.l. and its deposits tip at 2435 m a.s.l. 3D Google Earth view.

FIGURE 8.106 Spatial distribution of landslides and recent faults in the Suusamyr intermontane depression and the surrounding ranges. Red areas – landslides; dark brown lines – assumed Late Quaternary surface ruptures. Areas marked by solid and dashed black lines are shown on Figs. 5.56 and 8.105. Soviet 1:300 000 topographic map as a background.

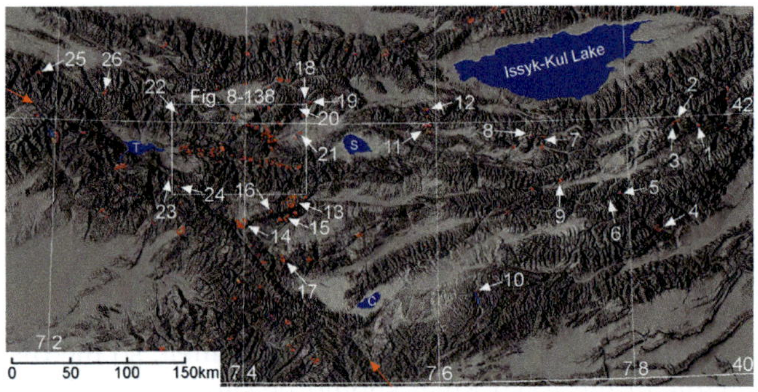

FIGURE 8.107 Shaded 3″ STRM DEM of the Central Tien Shan. Rockslides are marked by pink and brown dots; blue dots – dammed lakes. Red arrows mark the Talas-Ferghana fault; T – the Toktogul reservoir; S – the Sonkul Lake; C – the Chatyrkiol Lake. 1 – see Fig. 6.20; 2 – see Figs. 8.108, 8.109; 3 – the Djamansu rockslide (see Figs. 8.110–8.112); 4 – see Fig. 5.17, Fig. 8.113, Fig. 8.114; 5 – see Fig. 8.115; 6 – see Fig. 8.116; 7 – see Fig. 8.117; 8 – see Fig. 8.118A; 9 – see Fig. 8.118B; 10 – the Aksai-Kiolsu rock avalanche (see Fig. 8.119); 11 – the Karakudjur and Spur rockslides (see Figs. 4.23, 8.120–8.123); 12 – the Ukok rock avalanche (see Figs. 8.124, 8.125); 13 – the Beshkiol rockslide (see Figs. 4.1, 5.13, 8.126, 8.127); 14 – the Chaartash rock avalanches (see Figs. 5.61, 8.128–8.130); 15 – see Fig. 6.2; 16 – see Figs. 8.131, 8.132; 17 – the Djamantau rock avalanche (see Figs. 8.133, 8.134); 18 – the Chong-Tash rockslide (see Fig. 8.53); 19 – the Bash-Kugandy rockslide (see Fig. 8.136); 20 – the Koshbulak DSGSD (Fig. 8.137); 21 – the Chaek landslide (see Fig. 5.5); 22 – the 1992 Belaldy rockslide (see Fig. 2.4); 23 – the Karasu rockslide (see Fig. 5.64); 24 – the Kapkatash rockslide (see Fig. 8.196); 25 – see Fig. 8.197; 26 – the Beshtash rockslide (see Fig. 8.198); the detailed map of the outlined area is shown on Fig. 8.138.

in soft sediment of the Neogene and Quaternary ages. The most interesting case study is the Snake-Head Secondary rock avalanche of the bottleneck subtype (42.35° N, 74.34° E, Kyrgyz 099) described in Section 5.3.3.2 (see Figs. 5.56-5.60). This outstanding, despite its relatively small size, feature is just a member of a dense landslide cluster affecting area about 8×3.5 km only (Fig. 8.105) and separated from other groups of slope failures by large gaps almost free of similar phenomena (Fig. 8.106). It is one of the impressive examples of rockslide clustering, similar to those in other parts of the Tien Shan (Strom and Abdrakhmatov, 2004).

The westernmost spectacular rockslide of the Northern Tien Shan is the 6.48 km long Karakystiak rock avalanche in the southern Kazakhstan at 42.58° N, 73.13° E (see Fig. 6.7). About 150 Mm^3 of the Ordovician granite caved from the funnel-shape headscarp up to 1.95 km wide (between elevation marks 3500 and 3470 m a.s.l., passed through the narrow valley leaving distinct trimlines up to 200 m above the deposits' surface (marked at 2780 and 2930 m. a.s.l. on Fig. 6.7), crossed the Busuleankiol River – the right source of the Karakystiak River – with almost 100 m high runup on the opposite valley slope, turned 60° right and moved more than 3 km downvalley the Busuleankiol River. This case study is an excellent example of the flow-like long runout rock avalanche in the laterally confined conditions.

8.4 CENTRAL TIEN-SHAN

Part of the Tien Shan mountain system located south of the Issyk-Kul, the Kochkor, the Suusamyr and the Talas intermontane depressions and bounded by the Sarudjaz-Aksu River valley from the east and by the Talas-Ferghana fault zone from the west is considered as the Central Tien Shan (Fig. 8.107). Most of this part of the Tien Shan belongs to Kyrgyzstan, while the southern slopes of the Kokshaal Range that rise above the Tarim block – to China.

We must notice that most of our personal field studies on rockslides and rockslide-related phenomena were carried out just within this part of the region and many case studies describing various features typical of rockslides are presented above, in Chapters 4, 5 and 6.

Several interesting case studies could be found in the Uchkiol River valley that falls from the west in the Sarydjaz River. At 41.92° N, 78.764° E the Bashkol River – small left tributary of the Uchkiol River – was blocked by rockslide just 1 km upstream from its mouth and the 0.613 km^2 Bashkol Lake was impounded upstream (Kyrgyz 110, see also Section 6.5). About 50 Mm^3 of the Oligocene-Miocene red beds collapsed from about 800 m high slope. Rockslide debris filled the small dry gulley and its distal part formed ~100 m high dam just upstream of the deep narrow erosional gorge incised by the Bashkol

River in the ca. 300 m high and 600 m wide ridge composed of the Silurian sedimentary rocks (see Fig. 6.20). Since the Bashkol Lake level is about 230 m higher than the nearby Uchkiol River valley bottom, some seeping could occur through the bedrock that could result in relatively small-scale slope failure (its headscarp is marked on Fig. 6.20 by black arrows).

About 4.5 km east of this site, at 41.93° N, 78.82° E 1.6 km long sackungen deforms the bow-shaped crest of a short river-parallel ridge on the northern side of the Uchkiol River valley above the concave slope.

Interesting group of multi-aged rock slope failures was found in the upper reaches of the Uchkiol River valley, where it is named the Irtash River, at its confluence with the Djamansu River. About 0.3–0.8 km upstream from this confluence, at 41.963° N, 78.53° E, the 800 m wide Irtash River valley of the glacial origin is crossed by a presently breached barrier, composed of angular boulders up to 10 m in size, that differs significantly from debris flow fans and rock glaciers (Kyrgyz 215, Fig. 8.108). Right slope of the valley covered by fresh scree and steep 250 m high wall above it seems to be a headscarp back wall coinciding with large fracture in the Proterozoic metasediments. Rockslide debris formed an apron up to 15–20 m thick. Its temporal relationship with the alluvial fan of the Djamansu River is questionable and requires field inspection to be revealed. Frontal part of the deposits rest on the opposite – left bank of the Irtash River about 30 m above the riverbed and its hummocky surface differs from the surrounding terrain.

FIGURE 8.108 Rock avalanche originated as a wedge failure upstream from the confluence of the Irtash and the Djamansu Rivers. 3D Google Earth view.

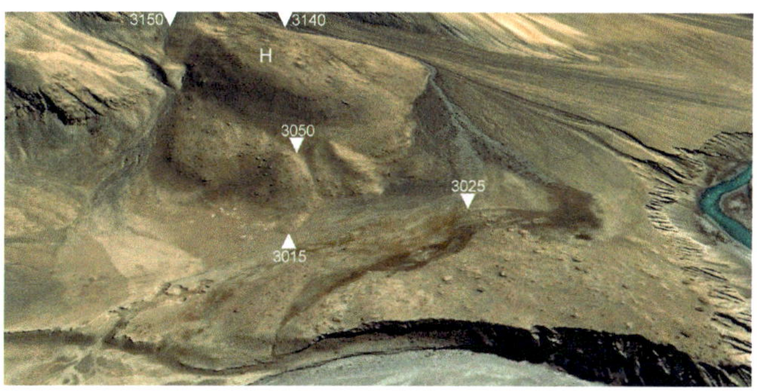

FIGURE 8.109 Suspicious hummocky deposits at the gulley mouth on the left bank of the Irtash River. Bouldery accumulation at the foreground – the frontal part of rock avalanche shown on Fig. 8.108.

One more suspicious debris accumulation up to 45–50 m thick with hummocky relief (Fig. 8.109) could be found on the opposite side of the Irtash River valley (41.96° N, 78.543° E) at the mouth of a small creek without visible traces of recent glaciation, but with concave slope of the ridge crest that could be interpreted as a headscarp (Kyrgyz 218). Unfortunately space images available for the anticipated source zone are of rather low resolution, which makes their interpretation ambiguous. However, the distal accumulation that can be seen in details is characterized by bouldery composition, though without such giant blocks as at the case study described just above, but with widespread conical hummocks on top of the deposits, likely the molards.

The most interesting feature in this area is the quite a rare case of the newly formed rockslide (Kyrgyz 111). Its fresh deposits blocking the Djamansu River valley at 41.95 N, 78.497 E and the headscarp on top of its right bank slope are absent on the Google Earth space image of September 29, 2002, but are visible on the more recent images provided by SAS Planet and ArcGIS Earth. The latter was obtained from the WorldView-1 satellite on August 6, 2009, most likely rather shortly after the event, since rock avalanche deposits have not been eroded yet (Figs. 8.110, 8.111).

About 8 Mm^3 of the metamorphic Precambrian rocks (gneiss or schist) collapsed from 1.1 km high steep slope. Almost 65000 m^2 (about 10 football pitches) of the snow-covered summit at 4200–4450 m a.s.l. disappeared (Fig. 8.112). As far as the site is located in the remote unpopulated area no eyewitness reports are available and it was unknown when this large-scale bedrock landslide occurred. We must notice that no visible indication of the ongoing instability can be recognized on the space image obtained prior to rockslide. Moreover, analysis of the earthquake catalogue for the period from September 2002, provided by the Kyrgyz Institute of Seismology, showed that no one significant earthquake had occurred in the vicinity of the Djamansu rock avalanche. According to this catalogue the nearest one with local magnitude $M_{LH} \leq 3.5$ only occurred on January 18, 2008 about 17 km north of it. Two more events, which magnitudes did not exceed 3.8, occurred on July 27, 2008 and on March 27, 2009, about 20–25 km far from the site. Even if their

FIGURE 8.110 Successive space images before and after the Djamansu rock avalanche formation. A – Google Earth image obtained on 29.09.2002; B – SAS Planet image (Nokia Digital Globe) obtained after August 6, 2009, the exact date is unknown. Outlined area is shown on Fig. 8.112.

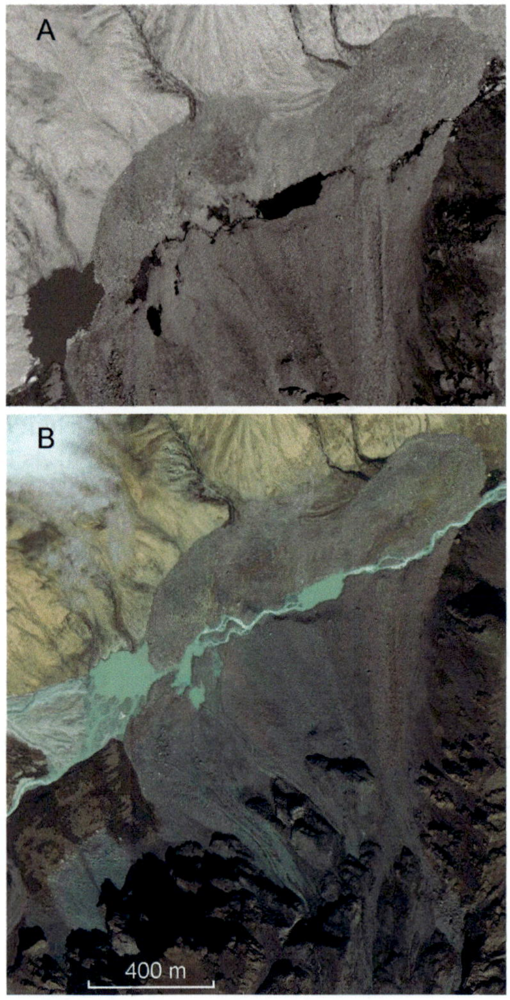

FIGURE 8.111 Space images of the Djamansu rock avalanche deposit. A – WorldView-1 image of August 6, 2009, rather soon after collapse and before dam's overtopping (provided by ArcGIS Earth); B – some time later (exact acquisition date is unknown) when levees that are undisturbed on the upper image have been eroded already (provided by SAS Planet). Small rock falls, likely simultaneous to the main failure (similarly fresh) are visible in the lower left corner.

epicenters were closer to the site, which cannot be excluded, considering low accuracy of location, it is quite unlikely that such a weak seismic shaking could play any role on detaching more than 200 m thick block from the host massif and producing such a large slope failure.

Very important information based on the records of seismic arrays of the Kazakh part of the International Monitoring System for nuclear test verification was kindly provided by the Director of Kazakhstan National Data Center Dr. Natalia Mickhailova. According to these data seismic event with $m_b = 3.1$ and $m_{pv} = 2.6$ was recorded on October 25, 2008 at 41.9347° N, 78.4832° E, about 1.5 km from the Djamansu rock avalanche headscarp, which is practically the same point, considering accuracy of epicenter determination, and at zero depth. As noticed by Dr. Mikhailova this event could be the record of the large-scale slope failure. Its occurrence in October 2008 fits well to the freshness of rock avalanche body on the WorldView-1 space image made on August 6, 2009, less than 1 year later (see Fig. 8.111-A). Considering low temperatures common of this altitude and significant Autumn daily temperature variations it can be hypothesized that slope failure could be produced by freezing and thaw of water presenting in joints. Same triggering was proposed for the 2000 Yigong rockslide in China (Shang et al., 2003; Xu et al., 2008; Evans and Delaney, 2011; Delaney and Evans, 2015) and for several rockslides in the European and New Zealand Alps that occurred without any identifiable trigger (McSaveney, 1993, 2002; McSaveney et al., 1992; Barla et al., 2000; Cox et al., 2015).

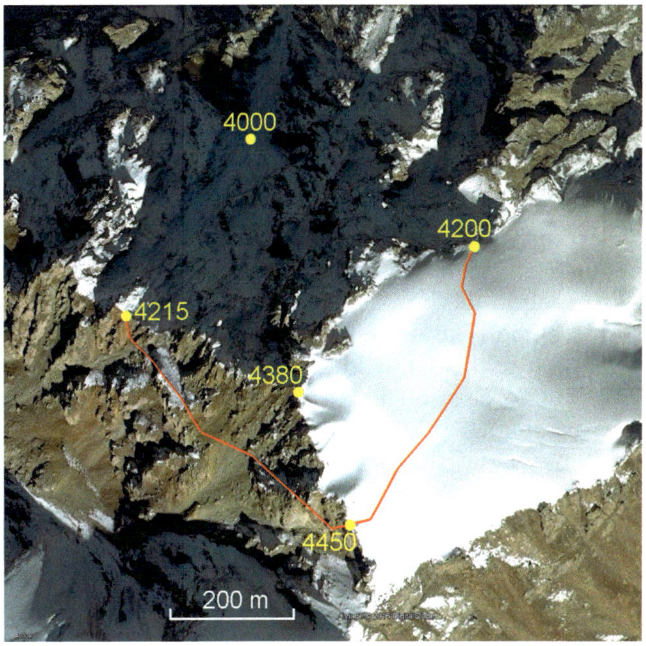

FIGURE 8.112 Position of the Djamansu rock avalanche headscarp (red line) on the pre-slide Google Earth image without any visible evidence of future instability.

FIGURE 8.113 Breached dam formed by rockslide from the right side of the Chon-Uzengiogush River valley with impressive hummocky surface. Headscarp area is in the shadow. Small white arrows mark assumed secondary scar. Whitish deposits visible on the river bank upstream (at the lower right part of the image) are, most likely, lake sediments. 3D Google Earth view.

This case study highlights the complexity of the relationships between large-scale bedrock landslides and seismicity that will be discussed further in Chapter 13.

The Djamansu rock avalanche debris composed of the intensively crushed material, though some boulders are up to 10–18 m in size, blankets about 1 km^2 of valley bottom by rather thin apron up to few tens of meters at most (at the valley bottom), but usually less, and with very sharp distal edge. Debris rose about 50 m on the opposite side of the valley and blocked there several small creeks fed by snow melt water. While rather powerful Djamansu River fed by glaciers formed a small lake upstream, small discharge of these creeks completely seeps through rock avalanche debris indicating its high permeability. Water of the lake dammed in the main valley was also seeping through the coarse debris initially as it can be seen on space image available at ArcGIS Earth (see Fig. 8.111-A), but later on it overtopped the low dam and stated eroding it (see Fig. 8.111-B).

Dense rockslides cluster that includes also several sackungen features is located at the southern part of the Kokshaaltau Range in the Chon-Uzengiogush River valley upstream from the Karatala village in Xinjiang. 5 km upstream of the village, the entire 500 m high left slope of the Betdjin Creek valley (41.146° N, 78.354° E, China 081) at a distance of about 3 km represents the single-side ridgetop spreading (Hungr et al., 2014) (see Section 5.1 and Fig. 5.17). Though, as mentioned above, this landform reflects, most likely, gradual long-term slope deformations, catastrophic collapse at this site with such secondary effects as valley damming and dam's breach cannot be excluded, especially in case of a large earthquake that could occur in this seismically active area. Site monitoring or at least its regular inspection could reduce the risk. Similar, though smaller features were found along some local watersheds at the right side of the Chon-Uzengiogush River valley.

Two kilometers upstream, at 41.134° N, 78.308° E, the Chon-Uzengiogush River had been dammed by a large rockslide about 35 Mm^3 in volume (China 102) with impressive hummocky surface. At least some of the hummocks are huge isometric angular blocks of the Devonian metasediments (most likely, metamorphosed sandstone) that form the deposit's carapace (Fig. 8.113). According to the deposit's morphology it can be assumed that below blocky carapace this body is composed of much more intensively crushed material. Some of the deposits form benches about 50–90 m high aside its proximal part. Slopes below these benches could be treated as a secondary scar (see Section 5.3.3.1). If this assumption is correct, more than 85% of the entire collapsed material was involved in the Secondary rock avalanche formation.

Upstream from this breached dam fine whitish sediments can be traced on the Chon-Uzengiogush River banks at a distance of about 2.5 km and up to 60–70 m above the present day stream level (see Fig. 8.113). Nothing similar is visible downstream from the breached dam. These definitely lacustrine sediments accumulated in the dammed lake before its breach. We must notice that siltation starts from the tail part of the dammed lake, where the stream's velocity and, thus, its transporting capacity decreases abruptly. Thus, presence of thick lake sediments just upstream from the dam indicates that the river had been blocked for a rather long time, which longevity can be derived by OSL dating of the lacustrine silt, which is overlain by rather thick debris flow fans. When the blockage breached, most of accumulated silt was eroded.

FIGURE 8.114 Recent rockslide on the left side of the Chon-Uzengiogush River valley. A – Google Earth image of its headscarp crown. White arrows mark fractures above the crown indicating potential for future failures; B – 3D Google Earth view of the entire rockslide; C – evidence of the outburst flood visible on Google Earth image.

However, the history of the Chon-Uzengiogush River damming includes at least one much more recent episode. 3 km upstream from the previous site, at 41.148° N, 78.271° E about 5 Mm^3 of rocks fell from the steep left bank of this valley and formed a dam that was, likely, overtopped and breached rather soon (China 101, Fig. 8.114-B). Rockslide may be classified as a compound slope failure with very steep back wall about 160 m high and large amount of the material still remaining in the headscarp. As far as the concave part of the slope is undermined here by the river, further activation of this landslide is quite possible. Moreover, as it can be seen on Fig. 8.114-A, large block with an area of about 16 000 m^2 had been detached from the host massif by the bended system of fractures and, thus, more than 1 Mm^3 of rocks can be involved in rocksliding in future.

At a distance of more than one kilometer downstream of the rockslide site shown on Fig. 8.114B the river banks and low terraces are covered by coarse material with striation that we believe to be deposited here by the outburst flood caused by the breach of this rockslide dam (Fig. 8.114C). Such floods could pose a threat to the hydraulic scheme build just at the confluence of the Kokshaal and the Chon-Uzengiogush Rivers. One of the most negative effects is the increase of the sediment yield resulting in rapid siltation of the reservoir.

We must notice that the nearest large-scale rockslides identified in the vicinity of this group were found more than 40 km north-west of it, in Kyrgyzstan, in the upper reaches of the Kainar River valley, that was blocked by two rockslides located about 8 km from each other at 41.39° N, 77.803° E (Kyrgyz 108) and at 41.41° N, 77.9° E (Kyrgyz 109).

The eastern one still dams the Kainar Lake. This rockslide in Carboniferous carbonate rocks is an excellent example of the wedge slope failure with one sliding surface coinciding with the steep bedding plane and another one, intersecting bedding. Similar headscarp structure has the Daguangbao rockslide more than 1 km^3 in volume – the largest one triggered by the 2008 Wenchuan earthquake that occurred, similarly, in carbonate rocks (Huang et al., 2009, 2012). In the Kainar rockslide case nearly rectangular block of the well-bedded carbonate rocks about 2000 m long, up to 800 m wide (in dip direction; dip angle is ~35°) and up to 200 m thick collapsed and blocked the river at a distance of 2500 m (Fig. 8.115). Rockslide dam is 130 m high approximately and, with a freeboard of about 40 m, seems to be quite a stable feature. Total volume of the collapsed rock mass was estimated as 130 Mm^3 (±30%). Rockslide moved not strictly downslope but in the oblique direction following the dip of the wedge hinge line marked on Fig. 8.115. During its

FIGURE 8.115 The Kainar rockslide dam. Small red arrows mark the headscarp. Dashed white arrow corresponding to the wedge hinge line shows the direction of rockslide motion. The frontal part of this Primary rock avalanche moving upstream entered the tributary valley and partially blocked it at 3550 m a.s.l., while the downstream portion that reached 3360 m a.s.l. likely moved as a Jumping rock avalanche. SAS Planet image.

FIGURE 8.116 Rockslide in the Kainar River valley with dual structure of the deposits. RG – rock glacier from the tributary valley, likely older than this rockslide. 3D Google Earth view.

motion rockslide converted into rock avalanche that can be ascribed to the Primary type (see Section 5.3.1) that "splashed" on the opposite side of the valley resulting in the about 165 m high runup. Its frontal part that moved upstream and upslope is covered with angular blocks of the limestone up to 30–40 m in size. At the same time this case study demonstrates some morphological peculiarities typical of the Jumping rock avalanches (see Section 5.3.2), since it really jumped from more than 200 m high bench at a headscarp base and its downstream avalanche-like part moved in the direction nearly opposite to that of the initial motion (see Fig. 8.115).

8 km upstream the Kainar River had been blocked by another, much smaller rockslide (Kyrgyz 108), about 7 Mm^3 in volume only, that collapsed from the steep wall of the typical trough glacial valley. The entire front of the about 500 m high rocky spur composed of the Carboniferous sedimentary rocks (likely of limestone underline by terrigenous rocks) was affected by wedge failure that formed the oval-like body 0.4 km^2 in size with two distinct concentric zones (Fig. 8.116). The outer (frontal) zone is composed of large angular blocks some of which are up to 20 m in size, while the main, inner part – of much finer material. This rockslide resembles at some extent the gigantic, more than 1 km^3 in volume, and more complex Kudara-Pasor rockslide in Pamir (see Figs. 4.20, 9.38, 9.39) that has even more impressive concentric structure.

Rockslide body has a distinct 60 m high runup on the opposite slope. We should also notice that its transverse, downvalley spreading was hampered by the rock glacier coming from the tributary valley. Rockslide had formed the lake that was partially silted and then breached, so that the distinct knick point still exists at this valley section.

Several large-scale bedrock slope failures were identified the upper reaches of the Naryn River and the Lesser Naryn River basins that were subjected to intensive glaciation in the past, with glaciers that had moved both along the main valleys and from their tributaries.

Among them there is an interesting group of three similarly (southward) oriented rockslides (Kyrgyz 226, Kyrgyz 158 and Kyrgyz 212) that occurred at the northern sides of the Burkhan, the Djilusuu and the Tekelitor River valleys correspondingly (Fig. 8.117A). The largest one (Kyrgyz 226), about 80 Mm^3 in volume, collapsed at 41.75° N, 77.104° E from the 900 m high slope composed of the Proterozoic metasediments intruded by the Ordovician granites and formed the 4.2 km long rock avalanche. Interesting peculiarity of this feature is the presence of the large subsided block resting within the headscarp that forms a 250–270 m wide step 440 m below the headscarp crown (Fig. 8.117B). It is very similar to 5.4 km long rock avalanche in the Kumbelsai River valley in Dzungarian Range shown on Fig. 7.9, the 3 km long rock avalanche in Eastern Tien Shan (see Fig. 8.6), and the Djamno rock avalanche in the Naryn River valley that will be described hereafter.

Such morphological type can be classified as a Secondary rock avalanche (see Section 5.3.3), though, unlike case studies described therein, their rapidly decelerated proximal parts are represented by large, more or less intact blocks rather than by crushed debris as at the Southern Karakungey (see Fig. 5.53) or the Chongsu (see Fig. 5.47) rock avalanches. It can be assumed that when the entire collapsing rock block

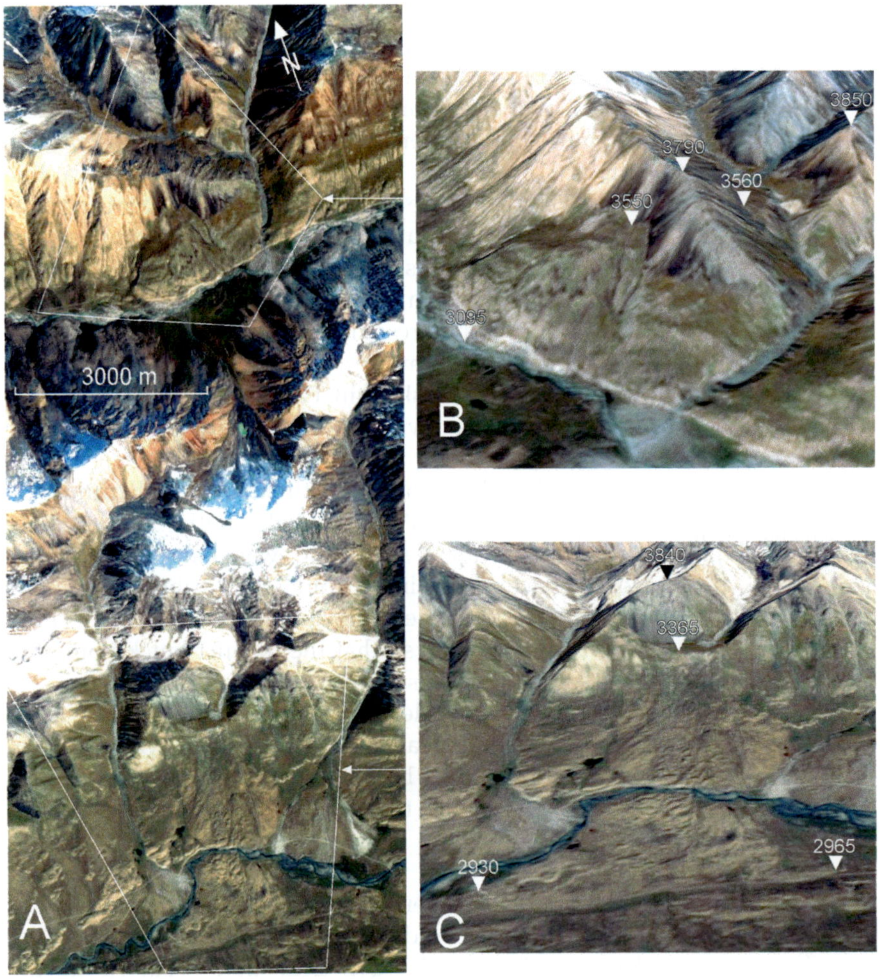

FIGURE 8.117 A – plan view of the three uni-directional rockslides in the upper reaches of the Lesser Naryn River basin. Outlined areas are shown on B and C; B – 3D view of rockslides in the Djilusuu River and the Tekelitor River valleys; C – 3D view of the 4.2 km long rock avalanche in the Burkhan River valley. Google Earth images.

suddenly decelerated, its outer (frontal) part, likely composed of more weathered rock gained some additional momentum and flung down forming long runout rock avalanche.

In this particular case rapidly moving debris crossed the Burkhan River valley and formed a very low, 15–20 m high only, and more than 1 km wide dam, presently dissected by the river, that has a knick point at this part of its long profile. Its frontal part rises on the opposite bank for about 20 m and stopped due to consume of the gained momentum, without direct collision with any pronounced obstacle. Rock avalanche surface has the specific microrelief with longitudinal furrows differing from the surrounding terrain (see Fig. 8.117C).

About 7.5 – 9 km north of its source zone two wedge-type rockslides caved from the south-facing slopes of the Djilusuu River valley at 41.88° N, 77.125° E and of its small right tributary, the Tekelitor River, at 41.893° N, 77.134° E (Fig. 8.117B). Such directed failure of the nearby slopes with the same aspect was described above in the Sarmin-Ula Range in Xinjiang (see Fig. 8.15) and at the western terminus of the Djumgal intermontain depression (see Fig. 5.52). Along with the similarity of the landslides expressiveness, likely indicating failures simultaneity, it can be considered as indirect evidence of their seismic triggering. The smaller one, about 4 Mm^3 in volume, had blocked the Tekelitor River valley, while the larger one, 30 Mm^3 approximately, just pushed the Djilusuu River stream towards its left bank.

10 km west of this group of rockslides, in the small right tributary valley of the Djilusuu River, which bottom is covered by moraine deposits, one more similarly oriented feature, likely of rockslide origin was identified at 41.88° N, 77.0° E (Kyrgyz 357). The assumed headscarp is a 1.2 km long, almost 1 km wide (on top) and up to 200 m deep wedge-like depression on the about 800 m high slope rising up to 4200 m a.s.l. and composed of the Cambrian volcanic rocks. Large fan-shaped deposit at its foot about 150 Mm^3 in volume rises 150 m on the opposite side of the valley bottom and has minor landforms indicating that debris moved across the valley (Fig. 8.118A). Such style of motion contradicts to the hypothesis of the glacial

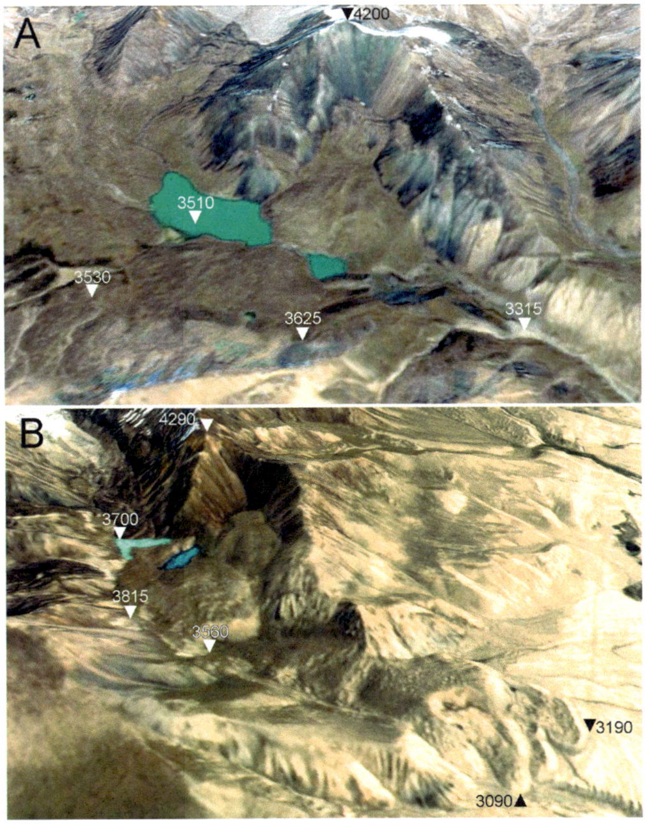

FIGURE 8.118 3D Google Earth views of two deep wedge-like negative landforms on the valley slopes at high altitude zone of the Central Tien Shan and related deposits at their feet, presumably of rockslide origin. See discussion in the text.

origin of this combination of landforms, despite its location at the altitude with well-developed traces of Late Quaternary glaciation. This body dams the valley forming a lake and one smaller lake has been formed in the depression at its central part.

Very similar negative wedge-like landform was found 45 km south of the site described above on the right side of the Naryn River valley opposite the Karakol River mouth at 41,547° N, 77.29° E (Kyrgyz 002), at the same altitude, on the slope composed of the Proterozoic metasediments. It has same dimensions in plan view but is deeper – up to 250 m deep (Fig. 8.118-B). Relationships of the accumulation that comes out from this assumed headscarp with glacial deposits are controversial, since 3 km long tongue with hummocky surface that fills the lower reaches of this valley looks similarly to the moraine deposits in the nearby valleys. At the same time recent (Holocene) end moraines terminate before it reaches a small lake dammed by this blockage. We hypothesize that the body about 180 Mm^3 in volume that comes out from this wedge-like depression could be the rock avalanche of the Secondary type (see Section 5.3.3). However,

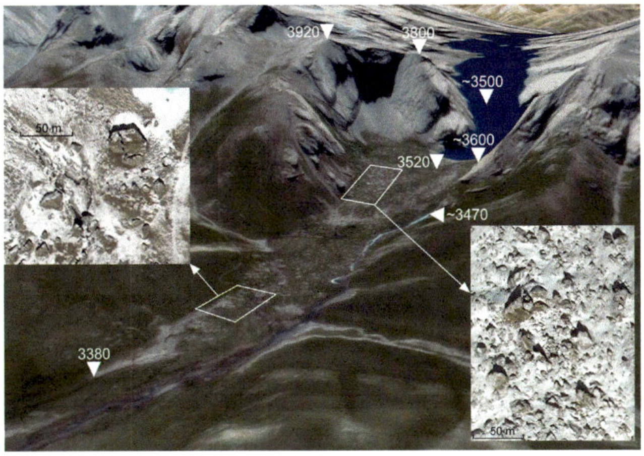

FIGURE 8.119 3D Google Earth view of the Aksai-Kiolsu rock avalanche and rockslide-dammed lake. Insets (SAS Planet images) show large angular blocks both at the proximal and distal parts of this rock avalanche.

additional studies are needed either to confirm or to disprove the assumption of rockslide origin of both features shown on Fig. 8.118.

One more rock slope failure characterized by similar deep wedge-like headscarp, this time identified more confidently, was found in the Aksai River basin, in the Kiolsu River valley (one more river named Kiolsu due to presence of the lake – the Kiol in Turkic languages) at 40.708° N, 76.39° E (Kyrgyz 200). To distinguish it from other same-name features we named it the Aksai-Kiolsu rock avalanche. Its very impressive headscarp, trapezoid-shape in cross-section is ~500 m long in the slope-dip direction, about 400 m wide (on top) and 170 m wide at its bottom, and up to 200 m deep (Fig. 8.119). About 15 Mm^3 of the Devonian sedimentary rocks formed 2 km long rock avalanche, which surface is covered by angular blocks up to 25–30 m in size (see insets on Fig. 8.119). It blocked the narrow river valley by about 100 m high dam and created a 13 km long lake. Seepage takes place through the blocky carapace of this natural dam.

Such extremely deep headscarps resemble that of the Ak-Kiol rockslide (see Section 5.3.1, Figs. 5.9, 5.10), though the latter originated due to "falling out" of the syncline core, while wedge failures at the three case studies described above seem to be associated with combinations of minor fault planes and bedding. One more possible analogue is the ~2000 Mm^3 Tamins rockslide in the Swiss Alps with very deeply incised headscarp (Calhoun et al., 2015).

Next features in the Central Tien Shan that should deserve attention of landslide researchers are the dual Pleistocene Karakudjur rockslides (Kyrgyz 106, 106a) about 1 and 5 km^3 in volume correspondingly and the approximately 10 Mm^3 Holocene Spur rock avalanche

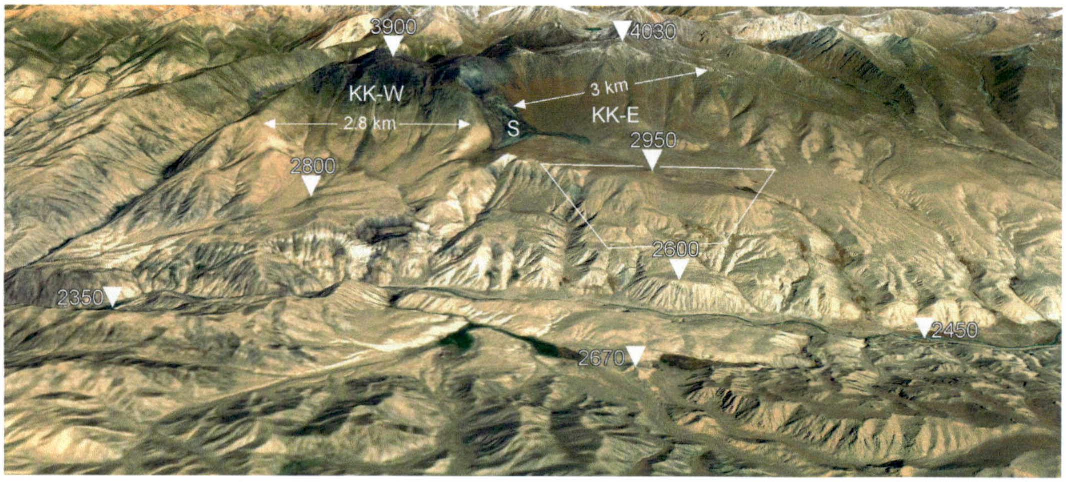

FIGURE 8.120 Rockslides at the western terminus of the Karakudjur intermontane depression. KK-W – the Western one; KK-E – the Eastern one; S – the Spur rock avalanche. 3D Google Earth view. The outlined area is shown on Fig. 8.121

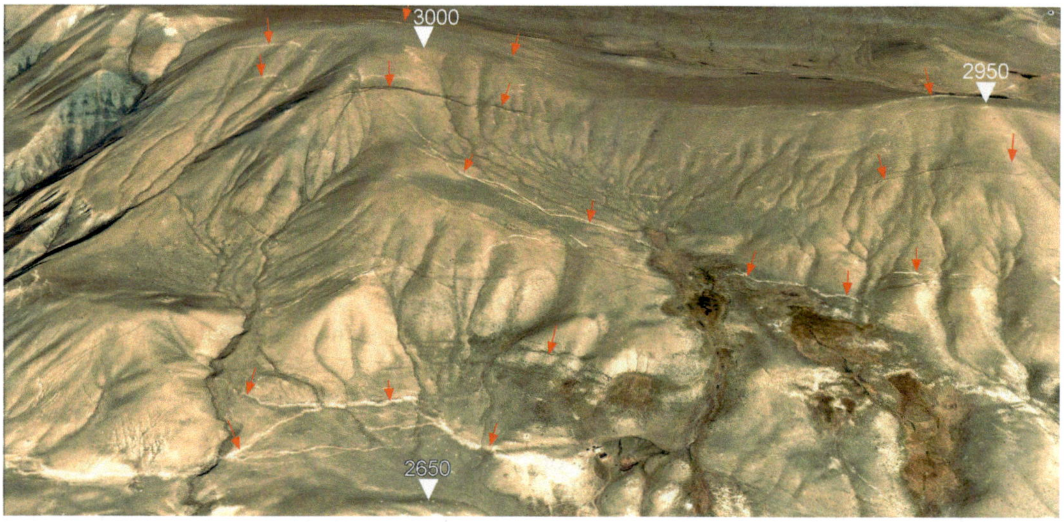

FIGURE 8.121 Fresh fissures (marked by red arrows) in the body of the Eastern Karakudjur rockslide. 3D Google Earth view.

(Kyrgyz 107) that caved from a rocky spur dividing headscarps of these gigantic rockslides.

The Karakudjur rockslides originated on the 1.5 km high right side of the Karakudjur River valley at the western terminus of the same-name intermontane depression. Their source zones expressed in the relief as smooth bow-shape lowerings about 3 km long each (Fig. 8.120) are composed of the Proterozoic metasediments intruded by granites and have very complex structure. It is assumed that these rockslides had blocked the river so that it had to erode a new valley, at least around the frontal edge of the Western Karakudjur rockslide body (41.96° N, 75.88° E) totally composed of the Proterozoic rocks. The much larger Eastern Karakudjur rockslide (41.95° N, 75.9° E) affected also the Neogene fill of the Karakudjur depression that was displaced and forms the smooth elongated hill on the left bank of the Karakudjur River valley with its tip at 2670 m a.s.l. These rockslide deposits were affected by numerous subsequent slope failures so that the internal structure of the main rockslide deposits can be observed in their headscarps. Moreover, multiple fresh fissures in this gigantic Pleistocene rockslide body indicate ongoing slope deformations affecting tens millions cubic meters that can collapse in future (Fig. 8.121).

The 700 m high rocky spur dividing the headscarps of the Karakudjur rockslides was destroyed by the Spur rock avalanche (41.975° N, 75.902° E), mentioned in brief in Section 5.1 as the case study with evidence of buckling preceding its collapse (see Fig. 5.4). It can be classified as a Secondary rock avalanche with well-expressed secondary scar at the base of its proximal part and blanket-like mobile part less than 10 m thick

FIGURE 8.122 Sharp frontal edge of the Spur rock avalanche western part.

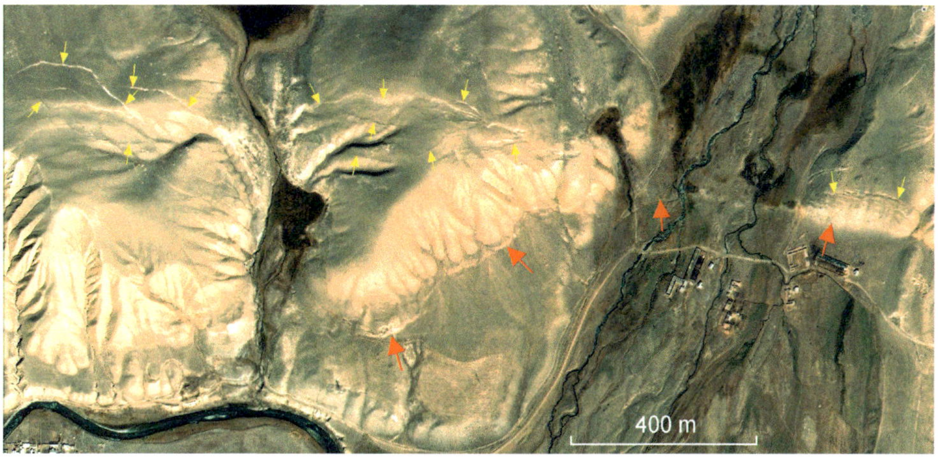

FIGURE 8.123 Surface rupture (marked by large red arrows) and associated slope deformations (marked by orange arrows pointing to the subsided side) at the right bank of the Karakudjur River. 3D Google Earth view.

on an average almost completely covered by molards (see Fig. 4.23) and with very sharp edges (Fig. 8.122). As pointed out in Section 5.1 this debris apron started expanding outside the headscarp area about 300 m above the slope base and covers the slope as it can be seen on Fig. 5.4. It is a rather rare phenomenon observed just at few case studies in the entire Central Asia region (see also the Beshtash rockslide shown on Fig. 8.198).

Multi-aged slope failures at this area could be triggered by large earthquakes associated with active fault passing along the right bank of the Karakudjur River. Recent surface ruptures that can be traced at a distance of about 30 km form several meters high scarps across low terraces of its right tributaries at the feet of the ~150 m high terrace-like landform. This surface rupture is associated with landslides and graben-like depressions on the edge of this terrace (Korjenkov and Chedia, 1986) (Fig. 8.123) resembling landforms in the hanging wall of the 1980 El-Asnam earthquake in Algeria (Philip and Meghraoui, 1983). According to (Omuraliev and Omuralieva, 2004; Djumabaeva, 2012) surface faulting in the Karakudjur depression took place 620–690 ± 45 BP, 700–800 ± 35 BP and 2815–2960 ± 35 BP. One of these events could trigger the Spur rock avalanche which exact age is still unknown.

Both slope deformations in the hanging wall of an active fault and those in the Eastern Karakudjur rockslide body (see Fig. 8.121) indicate possibility of the future large-scale slope failures, especially in case of next large earthquake and, thus, this valley must be considered as an area of the enhanced landslide hazard.

15 km north of the Karakudjur rockslides in the Ukok River valley at 42.1° N, 75.9° E about 200 Mm^3 of the Paleozoic granite and metasediments collapsed from the twofold headscarp dissected by local fault zones and formed the dam (Kyrgyz 105) with elongated downstream avalanche-like part that filled the valley bottom at a distance of ~3 km downstream from the dam's crest. Beautiful Ukok Lake was

FIGURE 8.124 Plan view of the Ukok rock avalanche. Lake is frozen. Yellow arrows mark the secondary scar; blue arrows – sharp boundary between the deposits that originated from the Western (W) and Eastern (E) headscarps (see also Fig. 8.125). SAS Planet image.

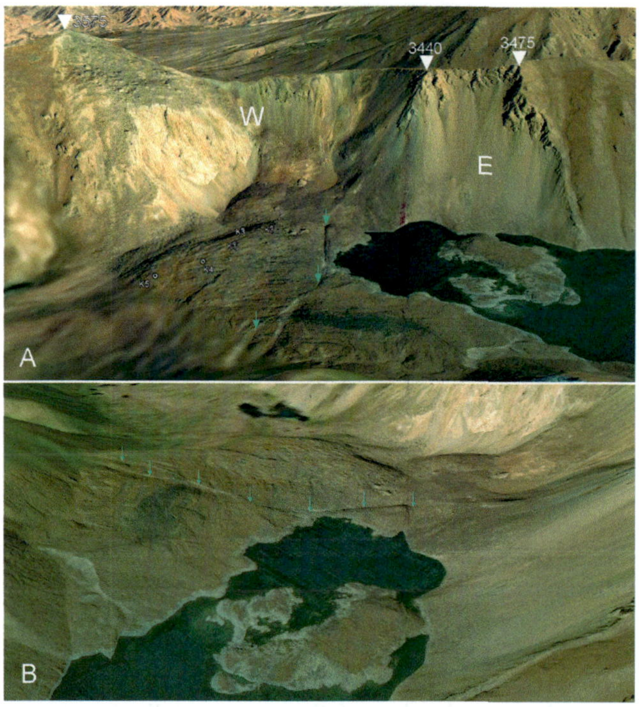

FIGURE 8.125 A – front view on the twofold headscarp and the proximal part of the Ukok rock avalanche. W & E – the same as on Fig. 8.124; K1-K5 mark location of the sampling points for ^{10}Be analyses from (Sanhueza-Pino et al., 2011). B – view downstream. Small blue arrows mark the linear furrow at the boundary between two simultaneous bodies. 3D Google Earth views.

impounded upstream of this dam. Rock avalanche has a distinct secondary scar above it (Fig. 8.124). The deposits front at its axial part on the opposite slope of the valley is up to 125 m higher than the lowermost part of the dams' crest. Another morphological peculiarity critical for understanding of its origin and motion mechanism is the distinct linear boundary between the deposits that originated from the larger Western and smaller Eastern headscarps (Fig. 8.125).

Unlike the assumption from (Sanhueza-Pino et al., 2011) who selected 4 "distinct and partly nested detachment scars" and identified "individual lobes (that) show characteristics of several superimposed catastrophic rock-avalanche deposits" we hypothesize that the Ukok Lake dam was formed by one complex event, though relatively small failures could really occur later from the highest westernmost part of the Western headscarp (see Fig. 8.125-A). From our point of view there are only 2 really individual headscarps – the much larger Western one between elevation marks of 3575 and 3440 m a.s.l. on Fig. 8.125 that embraces headscarps No 1–3 of the (Sanhueza-Pino et al., 2011) and smaller Eastern one between 3440 and 3475 m a.s.l. Moreover, we assume that main collapses from both of them had occurred simultaneously. This assumption is based on the above mentioned clear linear margin between the deposits that caved from the Western and Eastern headscarps. It recalls the relationships between the Bielogorka rock avalanches (see Fig. 8.99) and the two Ikhnachkul rock avalanches in Uzbekistan (Uzbek 019, 019a) that will be described hereafter in Section 8.5. Only simultaneously failed bodies can rest aside without overlapping.

Different ^{10}Be ages (Table 8.6) were measured from the samples that were definitely collected from the same generation of the rockslide body (see Fig. 8.125-A). Considering overall freshness of this

TABLE 8.6 ^{10}Be age of the samples from the Ukok rock avalanche body (simplified after (Sanhueza-Pino et al., 2011))

Sample No (see Fig. 8.125-A)	Lat.	Lon.	^{10}Be age (ka)	Internal 1σ error (ka)	External 1σ error (ka)
K1	42 °6.127′	75 °54.325′	1.47	0.28	0.32
K2	42 °6.143′	75 °54.443′	0.41	0.16	0.18
K3	42 °6.177′	75 °54.392′	35.3	1.2	3.2
K4	42 °6.107′	75 °54.228′	5.89	0.31	0.59
K5	42 °6.110′	75 °54.122′	8.21	0.46	0.84
K6	42 °6.300′	75 °52.928′	0.44	0.14	0.16

feature, the younger age (0.4–1.5 ka) seems to be more reasonable than larger estimates (5.9–8.2 ka). The downstream tongue of debris from which one of the youngest samples (K6) was collected, was formed, most likely, by the material from the valley bottom entrained by rock avalanche. Significant difference in the ^{10}Be ages could appear if some sampled blocks had rested on the daylight surface before rockslide occurred. Similar situation with dating of the Dead Lakes rockslide in the North-Western Tien Shan will be described in the next Section of this Chapter.

Simultaneous failure of the two adjacent rocky slopes corresponds with the assumption of its triggering by a strong earthquake, likely 0.4–1.5 ka ago (Djumabaeva, 2012). Source of this earthquake could be associated with active faults that bound the Kochkor intermontane depression from the south. It could not be excluded, however, that it was triggered by large earthquake in the Karakudjur intermontane depression (Djumabaeva, 2012).

Moving westward we should notice that large part of the Central Tien Shan between 74.7–75.8 E, i.e. the north-south trending belt about 100 km wide is almost free of large-scale bedrock landslides (see Fig. 8.107), while the area west of it is extremely rich of such features.

Several really outstanding bedrock landslides occurred at the northern slope of the 65 km long and up to 4000 m a.s.l. high Western Akshiyriak Range formed by asymmetric neotectonic anticline with well-developed south-eastern wing and northern wing ruptured by a system of active faults along which it is thrusted over the Togus-Torau intermontane depression (Chedia, 1986). Numerous relatively "small" slope failures (up to 40 Mm^3) of various ages were identified on the south-eastern slope of the range in Neogene deposits (see e.g. Fig. 6.2), but the most interesting case studies originated on the northern slope close to the eastern and western terminations of this range (Strom, 1998).

The largest slope failure of the entire Central Asia region – the Beshkiol rockslide (Kyrgyz 253) 8–12 km^3 in volume affected 55.6 km^2 of the northern slope of the Western Akshiyriak Range and had blocked the Naryn River valley at a distance of about 7 km (41.41° N, 74.48° E, see Fig. 4.1) (Strom, 1998; Strom and Korup, 2006; Korup et al., 2006). As can be seen on the schematic cross-section (see Fig. 5.13) it is a compound landslide, which proximal part caved from the 1.5 km high steep slope composed of Paleozoic rocks forming gigantic, 6 km wide and up to 600 m deep bow-shaped headscarp with a rock avalanche deposits at its foot, while the main body is in the folded Neogene red beds that formed the wide and rather gentle syncline. Here the sliding zone coincided with some weak bedding plane(s) within the gypsum-bearing Neogene sediments. Above the headscarp crown there is the 800 m wide, about 3 km long and up to 500 m high step either of the glacial origin or representing the subsided landslide block that sagged but did not collapse completely.

Landslide body has two well-pronounced steps – the upper one at about 2100 m a.s.l. and the lower one at about 1750 m a.s.l. Four small lakes fill closed depressions in the upper step of the landslide deposits. The fifth lake is at the eastern boundary of the landslide in the small gulley dammed by its body, and these five lakes – the "besh kiol" in Kyrgyz language – gave the name of the feature.

Landslide filled the Naryn River valley up to the level of the neotectonic bench that formed its right slope, so that the level of the deposits was slightly above it. The overflowing water had to incise not the erodible rockslide body composed by rather weak Neogene sediments, but the hard Paleozoic bedrock and due to it this gigantic dam had existed for about 2200–3000 years before the river eroded a new bypass gorge through the right-bank bedrock (Korup et al., 2006). The lake level could rise up to 1700 m a.s.l. that corresponds to the area of about 345 km^2 and volume of water of about 25 km^3 (Fig. 8.126) (which is slightly bigger than estimated in (Korup et al., 2006)). Judging from the level of the terraces eroded in the lake sediments, and according to the level of the deposits upstream from the dam, the lake was silted up to ~1640 m a.s.l. near the dam and up to ~ 1660 m a.s.l. at the tail of this natural reservoir by the material brought both by the main stream and by its numerous large and small tributaries (Fig. 8.127). We should notice that this abnormally horizontal terrace level was interpreted in 70th as the result of recent tectonic uplift of the Western Akshiyriak Range at its crossing with the Naryn River (Makarov, 1977). At this time, before space images became widely available, such a gigantic feature as the Beshkiol Rockslide that is much larger than area covered by a single aerial photograph or stereopair, was missed and was not shown on the

1:200,000 State Geological map. Rock avalanche deposits resting at the headscarp base were erroneously mapped as Middle Pleistocene glacial deposits.

Considering the assumed volume of the lacustrine sediments of about 11–15 km^3 (~7.5 km^3 of which have been eroded after dams' breach) and the present-day mean suspended sediment yield $Q_s = 4 \times 10^6\ m^3$/year (+20% for the bed load; A. Zhirkevitch, personal communication), lake siltation should last for 2200–3100 years. Besides overflow that was definitely the main way for water to escape from the lake, seepage could have also occurred through the Neogene debris intercalated with solution-prone gypsum beds. This assumption is based on the presence of numerous sinkholes on top of the terrace formed by the lacustrine sediments at the "entrance" of the blocked section of the Naryn River valley that are visible on the left bank of the river at 1635–1600 m a.s.l.

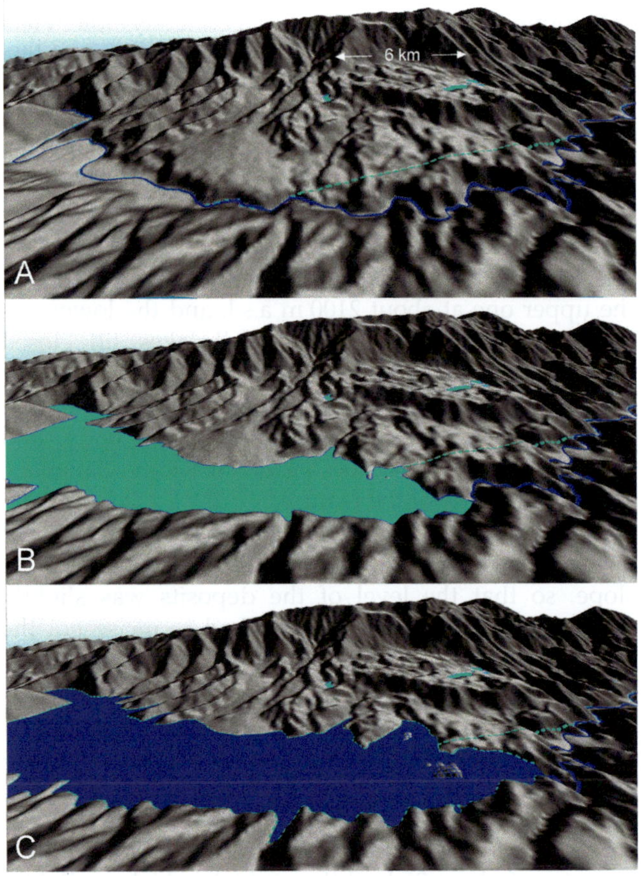

FIGURE 8.126 Shaded 3D views of the Beshkiol landslide. A – at its present state; B – at the level of the dammed lake siltation (ca. 1640 m a.s.l.); C – at maximal lake level (ca. 1700 m a.s.l.) when overflow had started. *Dark blue line* – the modern Naryn River; *dashed light blue line* – assumed valley axis before its blocking. 3″ SRTM DEM visualized by Global Mapper software.

Analysis of the Naryn River long profile (see Fig. 8.127) shows that the river eroding the bypass gorge still did not reach its initial, pre-slide level.

The triggering factor of this outstanding feature is unknown. Evidence of recent surface faulting at the fault zone passing along the northern feet of the Western Akshiyriak Range provides an argument in favor of its seismic triggering, but more studies should be performed to get more sound and univocal proof (or disproof) of this assumption.

Another group of three extraordinary rock slope failures is located at the south-western termination of the Western Akshiyriak Range at 41.22° N, 74.00° E, where it is crossed by the Kiok-Art River valley. This part of the range composed of Paleozoic granite, just north of the Makmal goldfield, is named the Chaartash Mountains, which gave the name of these long runout rock avalanches – the Chaartash-1, 2 and 3. The 9.06 km long Chaartash-2 rock avalanche was described briefly in Section 5.3.3.2 as an example of extra-mobile Secondary rock avalanche of the bottleneck subtype and was compared with the Chaartash-3 rock avalanche with almost twice smaller runout despite similarity of their volumes and height drop (see Fig. 5.61; Table 5.1). Here we describe the Chaartash rock avalanches on a more regular basis. We must notice that all these features were mapped on the 1:200 000 State Geological map as Late Pleistocene glacial deposits.

The Chaartash-1 rock avalanche (Kyrgyz 263, Fig. 8.128) originated as a compound rockslide of about 200 Mm^3 in volume on the steep walls of the deep, 1.5 km long gulley incised in Paleozoic granites and formed the distinct flow-type rock avalanche that covered the foot of the Chaartash Mountains and then split into 3 branches following the erosional valleys at

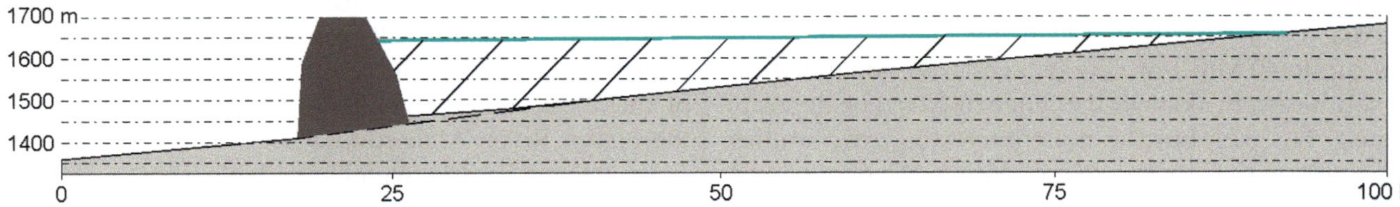

FIGURE 8.127 The Naryn River valley long profile from eastern termination of the Togus-Torau intermontane depression in the west to the Terek River mouth in the east. The original riverbed level has not been reached yet. The *light-blue line* – assumed level of lake siltation.

FIGURE 8.128 The Chaartash-1 rock avalanche. 1, 2, 3 – three branches of its long runout deposits; Ch-2 – deposits of the Chaartash-2 rock avalanche. Block outlined by *dashed line* was disturbed but did not fail. Google Earth image.

FIGURE 8.129 Microrelief of the distal part of the Chaartash-2 rock avalanche. Notice difference of the deposits surface before and after the river valley crossing. A – debris outshoot (distal edge marked by *arrows*) where the lateral confinement disappears; B – the hummocky microrelief and the debris outshoot filling the ravine (distal edge marked by *arrows*). Google Earth image.

the range feet. Large block 300 × 400 m in plan view, just south of the elevation mark 2400 m a.s.l. on Fig. 8.128 was severely disturbed but did not fail completely.

The deposits have smooth striated relief differing sharply from the dendritic erosional relief of the Neogene bedrock. The runout of its three branches (being measured from the headscarp crest) is 5.8, 4.7 and 4.7 km long correspondingly and rockslide debris had covered 6.25 km^2 in toto. We must notice that branches 2 and 3 had stopped due to cease of the momentum, without collision with any obstacle. In contract, the largest branch (No 1) had blocked the Kiok-Art River valley and, thus, collided with its left bank.

The Chaartash-2 rock avalanche (Kyrgyz 264, see A on Fig. 5.61) is one of the longest rock avalanches of the entire Central Asia. About 120 Mm^3 of the Paleozoic granite collapsed from the slopes of the gulley deeply incised in the northern slope of the ridge and, after passing its narrow funnel-shape outlet traveled 7 km more along the gently inclined bottom of the intermontane depression following shallow creek valley (see Fig. 5.61, Table 5.1). This feature is a typical case study of the Secondary rock avalanche of the bottleneck subtype (see Section 5.3.3.2).

It is interesting to compare microrelief of this flow-like feature (Fig. 8.129) with those of the Chaartash-1 (see Fig. 8.128) and Chaartash-3 (Fig. 8.130) rock avalanches. While the entire Chaartash-1 rock avalanche deposits and distal part of the Chaartash-3 have smooth surfaces with well pronounced striation, the Chaartash-2 deposits, also characterized by striated morphology, are covered by numerous hummocks (see inset B on Fig. 8.129). Besides there are several small outshoots that originated where rock avalanche debris fills small transverse ravines (ibid) or where

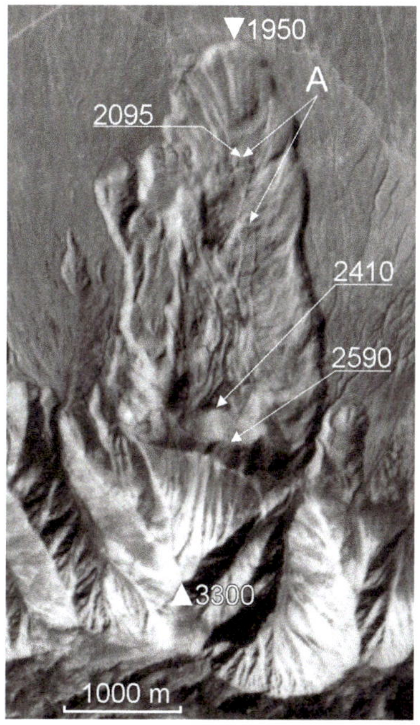

FIGURE 8.130 Morphology of the Chaartash-3 rock avalanche. A – terminations of the superficial flow-like features. KFA-1000 image.

the lateral confinement disappears so that debris can turn off from the main granular flow (see inset A in Fig. 8.129). Hummocks well pronounced on rock avalanche surface before debris reach the Kiok-Art River valley, disappears at the distal part of the deposit that had crossed the valley and rests on its left bank.

The last rock avalanche of this group – the Chaartash-3 (Kyrgyz 265, see B on Fig. 5.61 and Fig. 8.130) is of the same order of size as the two previous features (about 170 Mm^3 approximately), but unlike them originated not in the deeply incised ravine but on a slightly higher (800 m) slope rising directly above the unconfined depression bottom. According to the deposits morphology it can be ascribed to the combination of the Primary and the Secondary types (see Sections 5.3.1 and 5.3.3.1). Most of rockslide debris formed an elongated tongue from 25 to 70 m thick with nearly constant width. Similar shape of the deposits was observed at such unconfined primary rock avalanches as shown on Figs. 4.14 and 5.32, while in several other cases (see Figs. 6.4, 6.8, 8.7) debris spread over flat feet of the collapsed slopes forming fan-shaped or pancake-shape bodies.

On the other hand just at the foot of the planar triangular headscarp rockslide debris forms a bench about 180 m high and 200–300 m wide with the sinuous frontal slope representing multiple merged secondary headscarps. Two central of them rise over the strange superficial flow-like features that terminates 2 and 2.9 km from the secondary headscarp crowns, about 400 and 500 meters bellow them (their termini are marked by A on Fig. 8.130). These flow-like bodies being 1.0–1.5 Mm^3 in volume only have quite low H/L ratio of 0.2–0.24, which is typical, generally, of much larger rock avalanches. We can hardly imagine that they could reach such a long distance independently of the main rock avalanche. May be, though it is a pure speculation right now, such a long runout could take place if these secondary failures occurred when the main body was still moving and, thus, these superficial flows could move as on the vibration bed. The alternative, similarly speculative explanation is that they were ejected from the secondary scar with a crown at 2590 m a.s.l. due to momentum transfer when the proximal part of the Chaartash-3 rock avalanche body stopped abruptly.

FIGURE 8.131 140 m high step at the watershed of the Western Akshiyriak Range. *Red arrows* mark fractures on the southern side of the watershed indicating ongoing slope instability. 3D Google earth view.

One more interesting site is located in the middle of the Western Akshiyriak Range at 41.29° N, 74.23° E, where the 1300 m long and 750 m wide block likely subsided for about 140 m along the linear scarp (Kyrgyz 270, Fig. 8.131). Its flat surface is the displaced part of the Pre-Neogene planation surface that forms the southern limb of the Western Akshiyriak neotectonic anticline. Besides, numerous fractures with northern sides down can be seen along the watershed at a distance of about 4.5 km. Distinct surface ruptures passing along the northern feet of the Western Akshiyriak Range at this section (Fig. 8.132) could be associated with large past earthquake(s). It can be assumed that these earthquake(s) had triggered block sagging and future event could result in its further

FIGURE 8.132 Overview of the 10 km long central section of the Western Akshiyriak Range. Surface ruptures are marked by *red arrows*; the assumed headscarp – by *orange arrows*; the outlined part is shown on Fig. 8.131. 3D Google Earth view.

activation. If this step really reflects the ongoing slope instability, single-event collapse of this slope could affect about 2 km^3 of rocks and the resultant rock avalanche could reach the Naryn River valley that is 9 km north of it and about 2 km bellow. This site requires detailed study of both slope deformations and surface ruptures at its foot.

About 40 km south of the Western Akshiyriak Range, at the south-western corner of the Alabuga intermontane depression bounded by the Djamantau Range one more super-large rock slope failure at 40.97° N, 74.4° E displaced about 1000 Mm^3 of the Carboniferous limestone thrusted over the Neogene red beds. It formed 6.6 km long rock avalanche, which debris filled erosional landforms incised in the Neogene (Kyrgyz 039). Several small lakes were impounded within this Djamantau rock avalanche (Strom and Korup, 2006) in the closed depressions, the largest of which – the Samankiol Lake, is the one with green water at 2615 m a.s.l. on Fig. 8.133.

The important peculiarity of this case study is the evidence of the recurrent slope failure. The first generation formed the 6.6 km long rock avalanche (measured from the headscarp crown to the deposits' tip) of the Secondary type with proximal accumulation and not very distinct secondary scar that buried the preexisting rugged terrain. Its deposits are covered by widespread hummocks up to 70 m in diameter; most of them do not exceed 10, rarely 20 m (Fig. 8.134A). Though most of rock avalanche surface is composed of crushed Carboniferous limestone, locally the underlying crushed red beds can be seen through it. Presence of small but permanent lakes indicates high compaction and low permeability of rock avalanche debris below coarse blocky carapace with inevitably large voids.

FIGURE 8.133 The Djamantau rock avalanches. It is well visible how the deposits of its first generation cover the preexisting terrain. Fresh bouldery deposits free of grass cover is the body of the more recent rock avalanche. 3D Google Earth view.

The second, much younger and much smaller (ca. 20 Mm^3 only) rock avalanche has 4.5 km long runout and, unlike the first generation, forms also a branch that moved not in the main, north-western direction, but towards south-west, along the Kyz-Kurchan Creek and reached the Arpa River valley (at 2800 m a.s.l. on Fig. 8.133). The deposits of this younger rock avalanche are totally grass-free and contrasting as compared to the first generation. It has no hummocks typical of older deposits and looks coarcer than its predecessor, with some angular boulders at its distal part up to 20 m in size (Fig. 8.134-B).

The next group of bedrock landslides that deserves researchers' attention is located in the central part of the Djumgal Range, in the upper reaches of the Orto-Kugandy and Bash-Kugandy Rivers that form rockslide-dammed lakes endangering the Kaiyrma and the Bashkugandy villages in the densely populated Djumgal intermontane depression.

The first one – the Chong-Tash rockslide at 42.171° N, 74.616° E (Kyrgyz 100) about 11 Mm^3 in volume

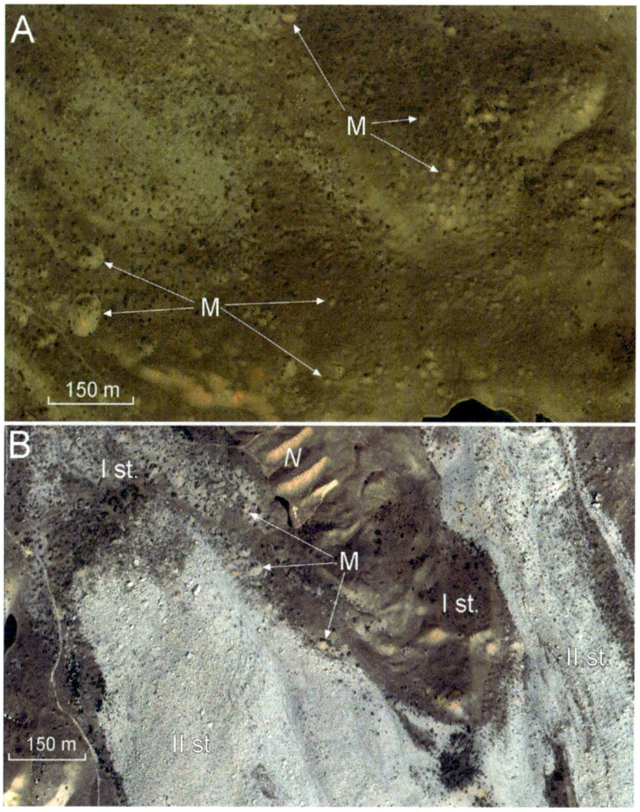

FIGURE 8.134 A – surface of the Djamantau rock avalanche first generation with widespread hummocks (molards?), some of which are marked by M; red outcrop right of the scale bar – crushed Neogene red beds underlying limestone debris. B – sharp contrast between the rock avalanche deposits of the first and the second generations. *N* – the Neogene bedrock; I st. – rockslide deposits of the 1st stage (generation); II st. – those of the 2nd stage. Google Earth images.

FIGURE 8.135 The Chong-Tash rockslide dam. 3D Google Earth view.

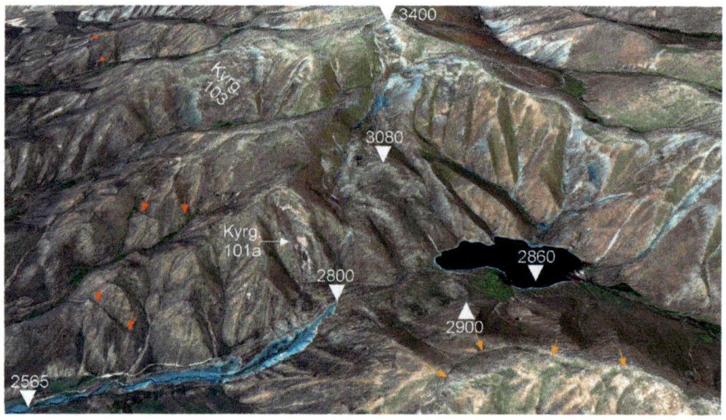

FIGURE 8.136 The Bash-Kugandy rockslide and the adjacent rock slope failures. *Small red arrows* mark upslope-facing scarps likely corresponding to surface ruptures. *Small orange arrows* – mark the horseshoe-shape headscarp of the DSGSD. Kyrgyz 101a and 103 rockslides are indicated. 3D Google Earth view.

caved from the left (eastern) slope of the valley in the upper reaches of the Orto-Kugandy River catchment composed of the Ordovician metamorphosed conglomerates and is characterized by the 110–120 m high runup, which is quite large relative to the source slope height (about 430 m only) (Fig. 8.135). It can be considered as a good example of the Primary rock avalanche in the frontally confined conditions (see Section 5.3.1). Besides the significant runup this rockslide is characterized by a compact body, which width is almost the same as that of the headscarp base. Such shape was governed by the morphology of the opposite valley slope with the pronounced lowering filled by rockslide debris. It resembles the much bigger Aksu rockslide in the Kyrgyz Range shown on Fig. 8.103. In such cases debris entering such a trap consumes its momentum for runup and friction along the side walls, and, thus, could not spread up- and downvalley.

Above the headscarp crown one can see the distinct horseshoe-shape scarp that extends further along the watershed for about 0.5 km indicating ongoing DSGSD at this part of the ridge, that affects the left bank of the dammed lake. One more interesting peculiarity of this feature is an erosional canyon dissecting the dams' downstream part. The erosion did not cause dams' breach, as at the Issyk rockslide in 1963 (see Fig. 2.1) or at the Sarydjaz-Aksu rockslide dam (see Fig. 8.53) and most of the eroded material was deposited just downstream the dam and in the lower part of the erosional canyon. However, considering significant amount of water stored in the dammed Chong-Tash Lake, and the potential instability of the lakes' left bank, this site should be considered as the hazardous one and requires monitoring or, at least, regular inspections, especially during flood periods.

About 5 km east of the Chong-Tash rockslide, in the upper reaches of the Bash-Kugandy River basin there is a group of several rockslides, one of which (Kyrgyz 101) converted in long runout rock avalanche that dams the river at 41.167° N, 74.687° E (Fig. 8.136) forming the Bash-Kugandy Lake that poses some threat to the inhabited area downstream. Two rockslides were identified nearby, one of which at 42.156° N, 74.67° E (Kyrgyz 103, ibid) is a large rocky slump about 15 Mm^3 in volume.

All these failures occurred on the slopes composed of the Ordovician conglomerates. At some rocky spurs distinct upslope-facing scarps 500–600 m long are visible. It can be assumed that they represent local surface ruptures (see Fig. 8.136).

The main feature of this group – the Bash-Kugandy rockslide about 40 Mm^3 in volume – originated on the ridge rising up to ~3400 m a.s.l., close to its watershed, moved about 2.5 km and collided with an opposite (left) slope of the river valley. The observable runup does not exceed 30 m. On the way, where slope became gentler, at about 3080 m a.s.l., one can identify well-pronounced horseshoe-shape secondary scar. It resembles the Chongsu rock avalanche of the Secondary type described in Section 5.3.3.1 (see Figs. 5.46, 5.47). However in this case, despite almost direct collision with an opposite slope, rock avalanche turned right and traveled downstream the valley for about 2.4 km so that its total runout is 4.77 km. Its distal part dammed the small tributary valley (flat green area at the lower left part of Fig. 8.136). The composition and internal structure of this rock avalanche can be observed in the canyon eroded by the overflow and, likely, seeping water along its axial part bellow 2800 m a.s.l. (see Fig. 8.136).

Besides, large, about 70 Mm^3, much older rockslide was identified on the left side of the Orto-Kugandy River valley about 3–4 km upstream from the Djumgal

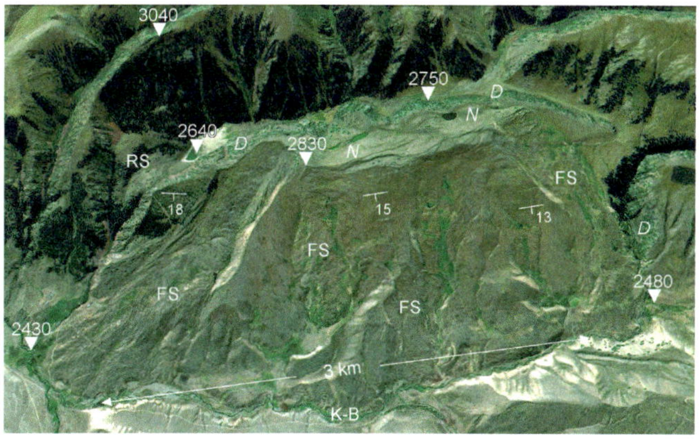

FIGURE 8.137 3D Google Earth southward view of the Koshbulak DSGSD. *D* – Devonian volcanoclastic rocks, *N* – Neogene red beds. FS – flow slides in Neogene deposits, RS – rockslide, K-B – the Koshbulak Creek.

Range southern foot at 42.062° N, 74.593° E (Kyrgyz 251).

Large complex DSGSD (see Section 5.1) accompanied by several flowslides at its sides and front is located 6 km east of the pass from the Oikaing River valley to the Orto-Kugandy River basin at 41.14° N, 74.5° E (Kyrgyz 152). Here the entire ridge south of the Koshbulak Creek is affected by slope deformations (Fig. 8.137). Its northern slope is covered by relatively shallow flow slides while complex system of bow-shaped upslope facing scarps more than 1700 m long crosses its southern slope and the watershed area composed of Neogene sediments, so that even several small lakes have been formed were these scarps dam shallow gullies. The lowermost upslope-facing scarps pass about 300 m behind the watershed and nearly 80 m bellow it, locally almost at the bottom of the adjacent valley. It means that sliding surface of this blockslide coincides with the basal layers of the Neogene sediments overlying the Pre-Neogene planation surface developed on the Paleozoic basement (Chedia, 1986). This succession was folded and tilted 13–18° northward, which formed the structural prerequisite for this DSGSD formation.

The entire volume of rock mass involved in this Koshbulak DSGSD can be estimated as ~500 Mm^3 if not more. Besides, the valley south of it was dammed at 42.133° N, 74.526° E by a ~6 Mm^3 rockslide in the Devonian volcanoclastic rocks that collapsed from the 400 m high slope and formed the partially filled Sary-Kiol Lake behind it (Kyrgyz 153, see Fig. 8.137). Several smaller landslides were identified nearby, one of which can be seen at the upper left corner of Fig. 8.137.

Central part of the Djumgal intermontane depression composed of poorly lithified Neogene and Lower Pleistocene conglomerates, sandstones and siltstones was affected by very large (about 3 km^2 in size) rotational landslides. The most expressive one (the Chaek landslide – Kyrgyz 102) on the left bank of the Djumgal River opposite the Chaek town (41.93° N, 74.57°) can be seen on Fig. 5.5. One more similar feature (Kyrgyz 328), likely older and much more reworked by erosion, is located about 3 km south-west of the Chaek town at 41.9° N, 74.52° E.

One more very large, about 15 km^2 in size, old landslide in poorly lithified Neogene and Pleistocene sediments with distinct 6.9 km long headscarp up to ca. 100 m high had affected minor neotectonic depression at the southern boundary of the Suusamyr intermontane depression at 42.04° N, 73.9° E (Kyrgyz 104). It's surface is smooth and gentle and landslide origin of this feature can be derived from the presence of the horseshoe-shape headscarp, while the entire bottom of this small depression looks as a flattened area with low levees. It can be assumed that it was affected by something like lateral spread, strongly reworked afterward by erosion and denudation.

One of the highest concentrations of the large-scale bedrock landslides all over the study region characterizes the Kokomeren River basin downstream from the Suusamyr intermontane depression, and the Naryn River valley from its junction with Kokomeren River and up to the Ketmen-Tiube intermontane depression (Fig. 8.138). Dozens of such features of all morphological types described in Section 5.3 ranging in volume from few million cubic meters and up to 1.0 – 1.5 km^3 have been identified here. Moreover, rock avalanches classification presented in Section 5.3 was proposed and elaborated on the basis of these case studies (Strom, 1996, 2006). This area can be considered as a natural rockslides museum and its eastern, more attainable part was selected for the annual International Kokomeren Summer School on Rockslides and Related Phenomena run since 2006 (Strom and Abdrakhmatov, 2009, 2017). Most interesting case studies from this region representing different morphological types of rockslides and rock avalanches were described in Chapter 5.

An impressive 5.5 km long rock avalanche about 80 Mm^3 in volume (Fig. 8.139) is located in the Mingteke River valley at 42.00° N, 74.355 ° E (Kyrgyz 092). It originated due to collapse of the 700–800 m high rocky spur about 1 km long composed of the Paleozoic deeply metamorphosed sediments, the jasper-like quartzite mainly. Its headscarp back wall that rises from ~3250 to 3700 (Fig. 8.140) is nearly planar, very similar to that of the seismically triggered 1911 Ananievo rockslide (see Fig. 8.71) and contrasting to the concave headscarps of the majority of rockslides. Considering large distance of its back wall from the pre-slide valley it is quite unlikely that such failure

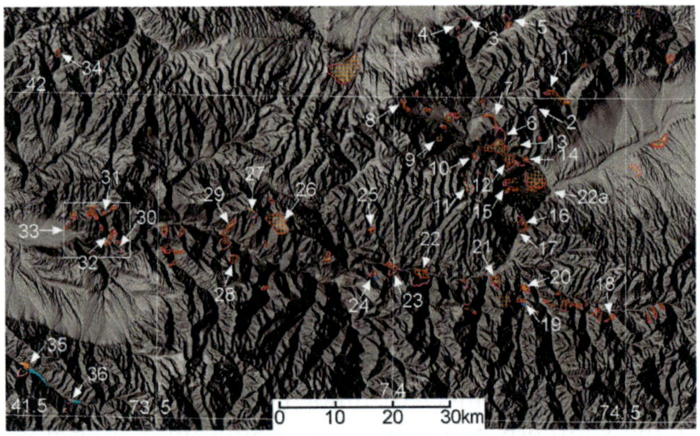

FIGURE 8.138 Rockslides in the Kokomeren River basin and the Naryn River valley from its junction with Kokomeren and up to the Ketmen-Tiube intermontane depression. 1 – the Mingteke rock avalanche (see Fig. 6.14, 8.139–8.141); 2 – see Fig. 4.39; 3 – the Seit rock avalanche (see Figs. 4.12, 5.14, 5.35-5.39, 8.142–8.144); 4 – see Fig. 4.12; 5 – the Oikaing rock avalanche (see Fig. 8.145); 6 – the Kokomeren rockslide (see Figs. 8.146–8.155); 7 – the Sarysu rock avalanche (see Figs. 8.156, 8.157); 8 – the Chongsu rock avalanche (see Figs. 5.46-5.51); 9 – see Fig. 5.3; 10 – the Kashkasu rock avalanche (see Fig. 5.45); 11 – the Upper Kashkasu rock avalanche (see Figs. 4.43, 4.44); 12 – the Displaced Peneplain rockslide (see Figs. 4.5, 4.6); 13 – see Fig. 8.158; 14 – the Mini-Köfels rockslide (see Fig. 5.41, 8.159, 8.160); 15 – the Karakungey rock avalanches (see Figs. 5.43, 5.52-5.54); 16 – the Lower Aral rockslide (see Fig. 8.161, 8.162, 8.164); 17 – see Fig. 8.163; 18 – see Fig. 8.165; 19 – the Ak-Kiol rockslide (see Figs. 5.8-5.12); 20 – the Lower Ak-Kiol rockslide (see Figs. 5.6, 5.62, 8.166); 21 – the Ornok rockslide (see Figs. 8.167–8.171); 22 – the Karachauli rockslide (see Figs. 8.172–8.176, 12.2, 13.7) and lacustrine sediments of its dammed lake – 22a (see Figs. 8.177, 8.178); 23 – the Lower Kokomeren rockslide (see Figs. 8.180–8.183, 14.3); 24 – the Aik-Kiol rockslide (see Fig. 4.3); 25 – the Kiol-say rockslide (see Fig. 5.63); 26 – the Kyzylkiol cavity (see Figs. 5.22-5.26); 27 – see Fig. 8.184; 28 – see Fig. 8.185; 29 – the Djamno rock avalanche (see Fig. 8.186); 30 – the Akterek rockslide (see Figs. 8.187, 8.188); 31 – the Djuzumdy cavity (see Fig. 5.27); 32 – the Djuzumdybulak rockslides (see Figs. 8.190–8.193); 33 – see Fig. 8.194. Besides features from this part of the Naryn River basin the following sites are indicated: 34 – the Belaldy rockslide (see Fig. 2.4); 35 – the Karasu rockslide (see Fig. 5.64); 36 – the Kapkatash rockslide (see Fig. 8.196). Background – the 3″ SRTM DEM visualized by Global Mapper software.

FIGURE 8.139 The Mingteke rock avalanche. Dark area at ~2580 m a.s.l. – completely silted dammed lake. Assemblage of aerial photographs.

FIGURE 8.140 The headscarp and proximal part of the Mingteke rock avalanche.

could be provoked by slope undercutting. Several more postglacial rockslides were identified nearby (see Fig. 4.39) along with an expressive upslope-facing scarp of the surface rupture that passes along the foot of a high slope few kilometers south of the rock avalanche headscarp. Such combination of tectonic and gravitational deformations affirms assumption of seismic origin of both features.

While part of the collapsed material accumulated at the headscarp foot (see Fig. 8.140), most of it turned 45° to 50° right (see Fig. 8.139) and was involved in the flow-like rock avalanche motion. The significant – up to 90 m – runup of its distal part (see Fig. 6.14) indicates high, up to ~152 km/h velocity of debris front motion before its halt. Rock avalanche dammed the valley. Later on this dammed lake, along with a small depression in the lowermost part of the rock avalanche, were filled by alluvium and debris flow deposits (see Fig. 8.139).

On its way rock avalanche debris undulated from one side to another (similar to the 1959 Pandemonium Creek rock avalanche (Evans et al., 1989) or the 2002 Karmadon rock-ice avalanche (Nikitin et al., 2006)) leaving trimlines up to 50–100 m above the deposits surface (Fig. 8.141). It is interesting to notice that "downstream" from the compact part the resultant surface of the rock avalanche body is much lower at its left side than at the right one, while left trimline is about 50 m higher (see Figs. 8.139 and 8.141). It allows the assumption that thickness of rock avalanche body at some stage of its motion is somehow proportional to its further mobility: where debris was thicker it became more fluid-like and passed away. Just this portion of debris jumped over about 50 m high ridge (see inset in Fig. 8.141), while main part of rock avalanche moved right from it (sides are indicated according to debris motion direction).

FIGURE 8.141 Trimlines left by moving debris at the central part of the Mingteke rock avalanche runout (marked by arrows). J – small portion of debris that jumped over the ~50 m high ridge. The frontal view on it is shown in the inset; people are passing along the foot of the main rock avalanche body.

FIGURE 8.142 Distal part of the Seit rock avalanche covered by large angular boulders.

In the upper reaches of the Kokomeren River, near the first bridge on the road connecting the Suusamyr and Djumgal intermontain depressions one can see the Seit rock avalanche (42.112° N, 74.15° E, Kyrgyz 240) – the classical example of the Primary rock avalanche with lateral confinement (see Figs. 4.12, 5.14, 5.35-5.39) which morphology was described in Section 5.3.1. It has wavy surface with stripes of coarser and finer grano-diorite and granite debris (see Fig. 5.37). The largest blocks up to 3–5 m in size concentrate in the distal part of this blocky carapace (Fig. 8.142), below which material seems to be crushed at a much larger extent (see Fig. 5.35). This rock avalanche is characterized by wide distribution of conical molards up to 1–2 m high, often with distinct jigsaw puzzle structure (Fig. 8.143) allowing sound assumption that they originated from large blocks ejected from rock avalanche interiors and "expanding" due to sudden distress. We must mention, however, that some molards look just as conical chaotic accumulation of medium-size angular boulders (Fig. 8.144).

Close to the Seit rock avalanche at 42.1° N, 74.137° E one can see the assumed remnant of much older rockslide that originated on the slope composed of granites (Kyrgyz 241, see Figs. 4.12, 4.13).

Another large rock avalanche about 35 Mm^3 in volume is located ~10 km upstream of the Oikaing River mouth, the left tributary of the Kokomeren River

FIGURE 8.143 Molard on top of the Seit rock avalanche with jigsaw puzzle structure visible in the inset. Handle of the geological hummer is 65 cm long.

FIGURE 8.144 Molard on top of the Seit rock avalanche with chaotic structure.

FIGURE 8.145 The Oikaing rock avalanche originated on the slope composed of different varicolored lithologies. 3D Google Earth view.

(42.1165° N, 74.2475° E, Kyrgyz 098). Its valley inherits narrow neotectonic depression filled by the Neogene sediments, about 18 km long and 3 km wide stretching in north-eastern direction and bounded from the south-east by steep reverse fault, while it opposite – north-western slope represents steeply inclined Pre-Neogene planation surface (Chedia, 1986). Rockslide originated on the 800 m high right-bank slope of the valley, which lower part is composed of the Ordovician granites and upper part – of the Precambrian schist and Ordovician diorites differentiating by color (Fig. 8.145). Such variable lithologies provide good chance to compare mutual position of rock avalanche debris that originated from various parts of the headscarp with the initial relationships of these lithologies, moreover, the rock avalanche body was deeply eroded by the river and its internal structure could be observed in the ~50 m high right bank outcrop. Rockslide, likely, had dammed the river, but later on its frontal part was eroded and only proximal portion extending for about 1350 m along the river and up to 500 m across it retained. However, judging from the color of the remaining part of rockslide body it could be assumed that its frontal part composed of debris of the Ordovician granites was eroded completely. Another disadvantage of this site is its relatively large (10–12 km) distance from the access road passing along the Kokomeren River.

Besides this rock avalanche several large landslides can be seen nearby in the Neogene deposits that fill the Oikaing intermontane depression. Also, just at the conjugation of the Oikaing intermontane depression and the Kokomeren River valley, at 42.043° N, 74.171° E, one can see large rockfall shown on Fig. 5.1.

One of the world-best case studies where mutual position of various lithologies in the rockslide

FIGURE 8.146 View of the Kokomeren rockslide. White arrow at the lower left part marks active fault that, presumably, could trigger this slope failure (see also Fig. 8.155). The headscarp crown is at about 3400 m a.s.l.; top of the deposits – at 2120–2280 m a.s.l.; base of the deposits – at about 1800 m a.s.l. (see also Fig. 8.147); S – much smaller "satellite" rockslide in Paleozoic granite.

FIGURE 8.147 Left bank of the Kokomeren River at the central part of the Kokomeren rockslide. γPZ – Paleozoic granite; aQ – alluvial gravels of the Middle-Late (?) Pleistocene; RS – rockslide body.

headscarp and in its body can be observed and studied in details – the Kokomeren rockslide (Abdrakhmatov and Strom, 2006; Hartvich et al., 2008; Strom, 2006; Strom and Korup, 2006) is located in the Kokomeren River valley downstream from the Kyzyl-Oi village at 41.93° N, 74.23° E (Kyrgyz 086). This gigantic, about 1.5 km^3 in volume, deeply eroded rockslide originated on the 1600 m high left slope of the river valley composed of multicolored rocks – gray and dark-gray metasediments, reddish granite, alternating red granite and dark-gray sandstone sheets, possibly with some fault zones (Fig. 8.146). The slope

FIGURE 8.148 Late Pleistocene Kokomeren River channel buried by the Kokomeren rockslide. PZ – Paleozoic metasediments; aQ_{III} – Late Pleistocene stream alluvium (boulders); RS – basal part of the rockslide body.

base visible bellow the alluvium in the high left bank strath terrace overlain by rockslide body is composed of reddish granite (Fig. 8.147).

Rockslide collapsed onto high and wide river terrace that was about 100 m above the riverbed at the time of failure (about 140–150 m above the present riverbed), crossed it and filled the river channel eroded along the contact between granites underlying the left-bank strath terrace and the dark-gray Paleozoic metasediments outcropping at the right bank. Remnants of this buried valley with bouldery alluvium overlain by rockslide debris can be found on the right bank just above the road and 40–50 m above the present-day riverbed (Fig. 8.148). It excludes river bank undercutting as a possible cause of slope instability (totally contrasting with nearly the same size Aksu rockslide in Northern Tien Shan described in Section 8.3).

Up to 400 m thick rockslide deposits resting on the left-bank terrace represent the fantastic outcrop (see part of it on Fig. 8.147) being dissected by the river that also incised for about 50 m in the bedrock bellow the riverbed that had existed at the time of failure. However, it is practically inaccessible and could be studied just visually. Fortunately, the frontal upstream part of these deposits remained on the right bank of the river, just above the road, providing excellent opportunity to investigate it in details (Figs. 8.149, 8.150). We must point out that the relationships between different lithologies, as well as grain-size characteristics of debris revealed from this easily attainable frontal part can be applied to the main left-bank part of the Kokomeren rockslide body – both are characterized by the retention of the

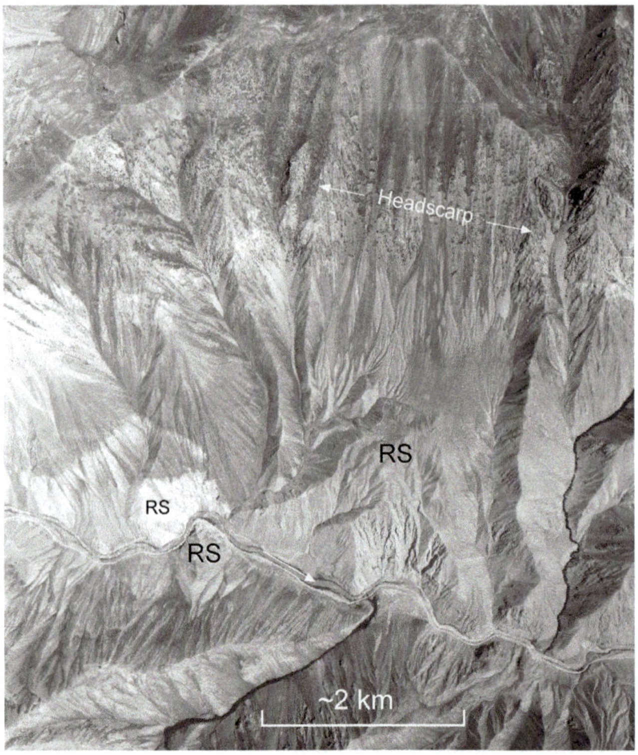

FIGURE 8.149 Aerial photograph of the Kokomeren rockslide. RS – remaining parts of the deposits.

initial structure without mixing of different lithologies, by presence of large angular boulders and even giant more or less intact blocks (in the left-bank part) in the carapace and of the intensively crushed material bellow (compare Figs. 8.150 and 8.151 with Fig. 8.152). Comparison of the initial and final position of rock types involved in this rockslide allows reconstruction of the emplacement mechanism, a model of which is presented on Fig. 8.153. Same model could be also applied to the Inylchek rock avalanche, which body also consists of "layers" of different lithology (see Fig. 8.45), to the Yashilkul rockslide (see Fig. 4.18) where bedrock debris overlays moraine bulldozed from the slope base and to many other case studies.

"Layers" of comminuted debris comprising the lower part of this giant body are inclined 20° to 30° towards the scar base and their inclination increases up to 30°–40° at the right bank of the Kokomeren River (see lower parts of Figs. 8.151 and 8.152). Though some large angular boulders could be found in the lower part as well, the extent of rock crushing in this 150–200 m thick, if not more, zone, both near its front and far away from it, is much higher than in the upper part of the body. The succession of the "layers" corresponds to the stratification in the source zone, so

FIGURE 8.150 Right-bank remnant of the Kokomeren rockslide body. Grain size composition of the blocky carapace and of the internal comminuted part of the deposits can be seen in the insets. Well-expressed varicolored badland is composed of the heavily comminuted granite (pink pyramids) and metasediments (gray and whitish pyramids).

FIGURE 8.151 Above – fault zone displaced within the alternating "layers" of crushed rockslide debris on the right bank of the Kokomeren River; below – the same outcrop from the other side before recent excavation. Crushed metasediments (gray) and granites (pink) that form unmixed units being displaced more than 2 kilometers from their original position across the river valley.

that the structure of the bedrock outcrops retains in the rockslide deposits (Fig. 8.154). Hence, this part of the sliding mass, up to 150–200-m thick, likely moved as a viscous laminar flow without any mixing. It can be assumed that the intensive shattering typical of the lower part of the rockslide body took place during the entire emplacement process, not just at a moment of its impact with the opposite valley floor as it was proposed by Erismann (1979) for the Köfels rockslide.

Outer carapace is composed of coarse angular blocks with very few, if any, fines. Similar to the comminuted debris it consists of the unmixed "layers" – the granite angular boulders just above the comminuted material that are overlain, in turn, by "layers" of metasediment debris. We must notice that extent of comminution strongly depends on the lithology (see upper part of Fig. 8.152).

The boundary between the blocky and shattered zones of the Kokomeren rockslide body is abrupt, without any pronounced transitional zone that may be interpreted as the evidence of sharp change of deformational conditions in the superficial and internal parts of the moving rockslide.

The basal unit of the right-bank part of the deposits overlaying bouldery alluvium (see Fig. 8.148) is composed of most intensively shattered granite that looks like a compact, homogeneous mass with a specific gravity of 2.67–2.68 g/cm^3, but disintegrates into fine sand being placed in water. Separate particles represent angular quartz and feldspar grain fragments with conchoidal surface texture. Most of the smallest particles are quartz scales. Its grain size composition is presented in Table 8.7. The method of sample preparation practically has no impact due to the lack of clay minerals. The latter indicates that it was formed by mechanical crushing rather than by the long-term subsequent weathering. Being compacted in the laboratory to a relative density of 1.90 g/cm^3 it has a filtration coefficient of 0.05 m/day only. Thus, we can say that this material is almost impermeable.

Grain size analysis of crushed granite and metasediment alternating "layers" shown on Figs. 8.150 and

FIGURE 8.152 Above – huge block of fractured granite overlain by debris originated from metasediments with well preserved bedding at the top of the Kokomeren rockslide left-bank body; below – heavily crushed granites (pink "layers") alternating with metasediments (gray "layers") in the basal part of the left-bank body overlaying alluvial gravels.

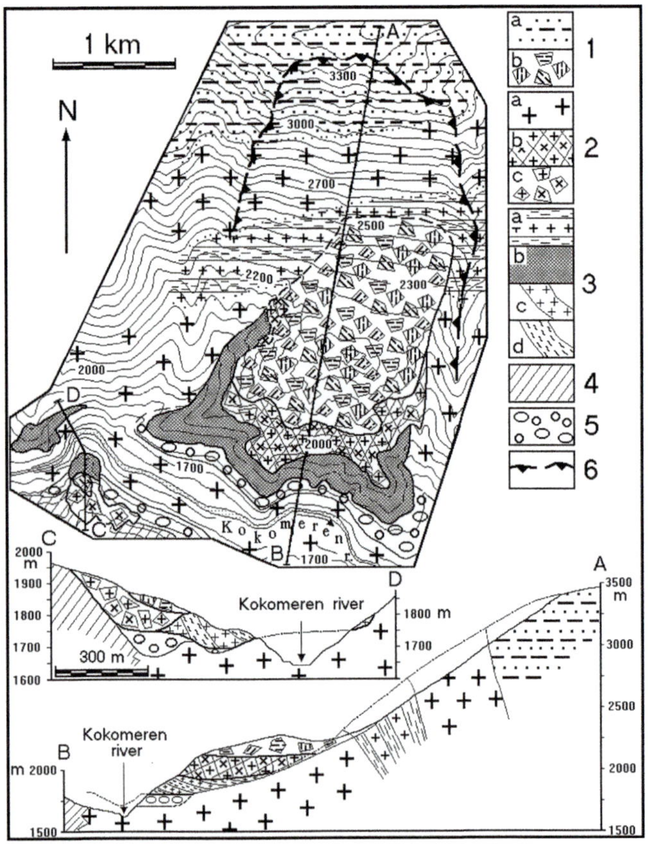

FIGURE 8.154 Schematic map and cross-sections of the Kokomeren rockslide. Legend: 1a - metasediment, 1b - metasediment fragments in rockslide deposits; 2a - granite, 2b, c - granite in rockslide deposits, forming a huge massif (b) and crushed into blocks (c); 3a – alternating granite and metasediment, 3b-d - shattered granite and metasediment in rockslide deposits (b - on the map, c, d - on the cross-sections, where c - shattered granite, d - shattered metasediment); 4 - limestone, schist and diabase alternation; 5 - alluvium; 6 - rockslide headscarp. Reprinted from *Abdrakhmatov, K., and Strom, A. 2006. Dissected rockslide and rock avalanche deposits; Tien Shan Kyrgyzstan. In: Evans, S.G.; Scarascia Mugnozza, G.; Strom, A.; Hermanns, R.L. (Eds.) Landslides from Massive Rock Slope Failure. NATO Science Series: IV: Earth and Environmental Sciences, Springer, Vol. 49, 551–572 with permission of Springer.*

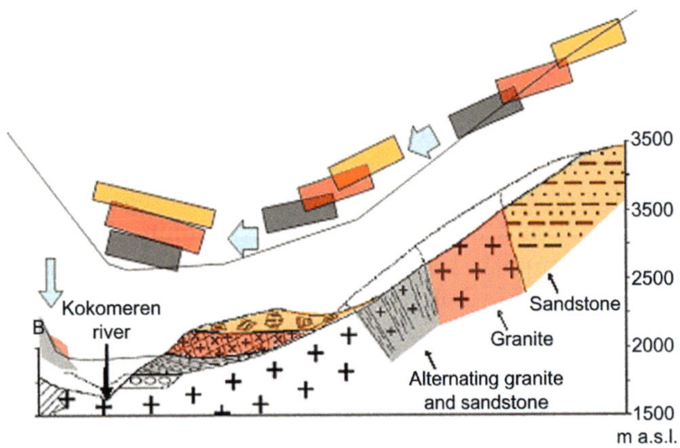

FIGURE 8.153 Reconstruction of the Kokomeren rockslide emplacement mechanism (after *Hartvich, F., Mugnai, F., Proietti, C., Smolkova, V., and Strom, A. 2008. A reconstruction of a former rockslide-dammed lake: the case of the Kokomeren River valley (Tien Shan, Kyrgyzstan). Poster presentation at the EGU conference, Vien).*

8.151 was performed in three steps – field sieving and extraction of <5-mm fraction; dry sieving and extraction of <1-mm fraction and further laboratory sieving with washing and aerometric analysis. Combined results of these three stages are presented in Table 8.8.

About 1.5 km downstream of the eastern end of the Kokomeren rockslide an expressive recent reverse fault that displaces the Holocene and older terraces can be seen on the left bank of the Kokomeren River (Fig. 8.155). Thickness of the bouldery Holocene alluvium in the hanging wall is about $^2/_3$ of that in the footwall, whereas thickness of the overlying outwash is constant within the entire outcrop. Vertical separation of the terrace base is 6.6 m, while that of the outwash base is about 2 m only. Hence it can be concluded that offsets occurred at least twice – during

TABLE 8.7 Grain size composition of shattered granite from the basal unit of the right-bank part of the Kokomeren rockslide

	spec. gravity	*prep. method	Fractions (mm) 1.0-0.5	0.5-0.25	0.25-0.1	0.1-0.05	0.05-0.01	0.01-0.005	0.005-0.002	0.002-0.001	<0.001
	g/cm³		Content (%)								
Sample 1/91	2.67	A	11.9	27.9	27.6	8.4	17.1	2.7	2.4	0.4	1.6
		B	6.9	31.6	27.6	10.5	14.1	3.7	0.4	1.6	3.6
Sample 2/91	2.68	A	6.4	28.7	25.2	11.2	21.8	3.7	–	1.4	1.6
		B	7.8	24	22.5	15.3	17.1	4.6	1.4	2.5	4.8

*Methods of sample preparation: A – to specify micro-aggregation content (shaking in water); B – to specify grain content of maximum dispersion (boiling in water with natrium pyrophosphate).

TABLE 8.8 Grain size composition of crushed granite and metasediment alternation of the Kokomeren blockage right-bank part

	Fractions (mm) >80.0	80.0-40.0	40.0-20.0	20.0-10.0	10.0-5.0	5.0-2.0	2.0-1.0	1.0-0.5	0.5-0.25	0.25-0.1	0.1-0.05	0.05-0.01	0.01-0.005	0.005-0.002	<0.002
	Content (%)														
Sample 1/2005	1.8	1.1	3.35	5.9	14.35	19.6	17.0	5.5	13.2	8.3	3.55	4.1	0.8	0.3	1.2
Sample 2/2005	2.9	8.0	12.4	14.3	15.0	16.7	12.2	1.1	4.35	2.7	2.0	3.7	1.75	1.3	1.6

Sample 1/2005 – 44.7 kg of crushed granite; Sample 2/2005 – 42.7 kg of crushed metasediment; spec. gravity = 2.66 g/cm³.

alluvium accumulation and after outwash accumulation.

Both events should be younger than the failure of the Kokomeren rockslide, since at that time the Kokomeren riverbed was much higher than the displaced 15–18 m high terrace. However, the displaced alluvium of the high river terrace (approximately the same that underlays the left-bank part of the Kokomeren rockslide) indicates earlier offsets. The amount of its vertical separation (see upper inset in Fig. 8.155) is not less than 14 m, according to laser measurements.

The north-western continuation of this fault stretches directly towards the Kokomeren rockslide scar (see Fig. 8.146) and we assume that one of the rupturing episodes on this fault, most likely associated with a strong earthquake, could have triggered this gigantic slope failure (Strom and Stepanchikova, 2008).

Rockslide had formed the dam, which minimal height could be estimated from the altitude of lacustrine deposits. The rockslide deposits base can be found on the right bank of Kokomeren River at a level of about 1690 m a.s.l., while the top of deposits that filled the past river channel was at least at ~1900 m a. s.l. Lacustrine deposits upstream can be found at a level up to 1880–1890 m. a.s.l. Thus, we can expect that the blockage's effective height was 200 m at least (Hartvich et al., 2008). Later on the river had dissected the landslide dam material and 50 m of bedrock below its bed. Locally, it eroded a new gorge about 70 m deep that bypassed the blocked channel. Such incision in the bedrock was the long-tern gradual process. We have no data to conclude if the breach of the upper part of the dam was catastrophic or not.

This blockage had caused inundation of the Kokomeren River valley at a distance of not less than 15 km (the outermost site where fine lacustrine sediments can be observed is at 42° N, 74.16° E, in the road cut opposite the Burundu River mouth). Best outcrops of the lacustrine sediments are located in the southern part of the Kyzyl-Oi intermontane depression downstream of the Kyzyl-Oi village also at the road cut. Here whitish laminated silt up to 5–8 m thick is underline by 1–2 m thick basal layer composed of centimeter-size dark-gray clasts of Paleozoic metasediments from the left bank of the river valley mixed with matrix and clasts of Neogene red beds brought from the right bank of the river, from the western part of the Kyzyl-Oi intermontane depression. Opposite this village, at the left bank of the river, lake sediments at a distance of about 2 km are represented by merged alluvial fans (debris flows deposits) several tens meters thick that rest over fluvial gravels ~30 m above the present day riverbed not eroding them. It can be hypothesized that debris flows terminated in the former lake, thus loosing their erodibility.

The lake's volume can be roughly estimated as 10^9 m³. The powerful Kokomeren River, which present-day mean annual discharge is 2.5×10^9 m³/year (State

FIGURE 8.155 Recent fault displacing river terraces downstream from the Kokomeren rockslide (see Fig. 8.146). *Red arrows* mark rupture planes; *white triangles* mark base of the Holocene (?) alluvium; *yellow triangles* – contact between alluvium and overlaying slope outwash; *light blue triangles* – base of much older alluvium that forms a stripe on the slope. Its non-horizontality is just visual due to oblique view.

Water Inventory, 1987c) should impound this lake in less than one year. However, judging from the altitude and thickness of thin-bedded silty lacustrine sediments upstream, the dam should have existed for a rather long time, especially considering time necessary to incise the about 70 m deep bypass gorge through the bedrock (Hartvich et al., 2008). As far as these lacustrine deposits are not higher than 100 m above the present-day riverbed level, the dammed lake was silted just partially before its breach. It differs significantly from the lake in the middle part of the Naryn River basin dammed by the gigantic (~10 km^3) Beshkiol landslide that was silted almost completely (see Figs. 8.126, 8.127) (Korup et al., 2006).

The total volume of lacustrine sediments accumulated in the dammed Kokomeren River valley can be roughly estimated as 10^8 m^3 ($\sim 1/10$ of the lake volume). This allows the determination of the longevity of river damming (Korup et. al., 2006). According to the State Water Inventory (1987C), present-day mean sediment yield of the Kokomeren River is 140,000 tons/year, which roughly corresponds to 7×10^4 m^3/year. Assuming the same rate of sedimentation at the end of the Pleistocene when river damming took place, we counted that the time necessary to accumulate the above-mentioned amount of sediments was about 1500 years. Palinological analysis of two samples collected from lake sediments indicates that dam formation most likely took place in the Late Pleistocene, during a warm and rather wet epoch. This, in turn, allows the assumption that both river discharge and sedimentary yield in the period of river damming were larger than at present and, thus, the longevity of the rockslide dammed lake should be shorter, likely not more than few centuries.

Close to the gigantic Kokomeren rockslide there are several much smaller rock slope failures. One of these rockslides about 3 Mm^3 in volume descended from the left bank of the Kokomeren River valley about 2 km upstream (41.92° N, 74.19° E, Kyrgyz 246). Remnants of its deposits can be found on the river's right bank, where a large amount of angular boulders of the same material that forms the upper part of the scar area rests on the terrace-like surface several dozens of meters above the riverbed. The geomorphic position of the rockslide deposits on the slope of the Kokomeren River valley is very similar to the position of the right-bank part of the Kokomeren rockslide body, thus, it is likely that both failures occurred simultaneously.

One more rockslide (Kyrgyz 249) less than 1 Mm^3 had blocked the small gully east of the Kokomeren rockslide at 41.92° N, 74.25° E (marked by "S" on Fig. 8.146). Since it forms a rather well-preserved dam it can be assumed that this slope failure is much younger and could be triggered by a strong earthquake associated with the nearby active fault shown on Fig. 8.155.

Long runout Sarysu rock avalanche (41.97° N, 74.21° E, Kyrgyz 089) caved from the other side of the same watershed as the Kokomeren rockslide and traveled 4.5-km along the dry gully at a mean angle of about 11°. Its deposits rest in the eastern part of the Kyzyl-Oi depression crossed by the Sarysu River. The initial descent of the huge block of metasediments was $\sim$200 m only (Fig. 8.156), but it resulted in quite a long runout, presumably due to the bottleneck effect described in Section 5.3.3.2.

The rock avalanche body resting in the rather narrow valley has a smooth, grassy surface with expressive trimlines rising 10–20 m above the main part of the rock avalanche debris and visible, first, along the right side of the travel path, and then along its left side (Fig. 8.157). It indicates rapid flow-like motion of debris accompanied by its "swinging" from one side to another, similar to the Mingteke rock avalanche described above (see Fig. 8.141) or the 1959 Pandemonium Creek rock avalanche (Evans et al., 1989), though not so intensive.

FIGURE 8.156 The funnel-shape headscarp of the Sarysu rock avalanche and proximal part of its body.

FIGURE 8.157 Trimlines (marked by *white arrows*) left by Sarysu rock avalanche first on the right side and then on the left side of its path.

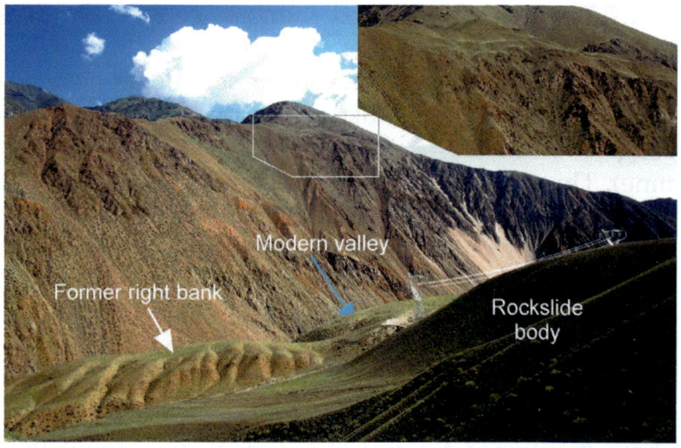

FIGURE 8.158 Modern valley of the Kokomeren River undercutting left bank and probable evidence of ongoing slope deformations – smooth steps parallel to the slope edge on its top (see zoomed fragment shown in the inset).

Most of its surface is composed of intensely crushed macadam-size debris, locally with larger angular boulders up to 0.5 m in size. The more intensive crushing of the rock avalanche carapace than at many other rockslides described above is, most likely, due to the much lower strength of the fillite and sandstone involved in this slope failure in comparison with granite or gneiss of many other rock avalanches from this region.

Several landslides and rockslides were found in the western part of the Kyzyl-Oi intermontane depression (see Fig. 8.138). The most interesting features here are the Chongsu rock avalanche (Kyrgyz 094) described in Section 5.3.3.1 (see Figs. 5.46-5.51) – one of the Secondary rock avalanches of the classical subtype, and the unstable slope at 41.94° N, 74.095° E (Kyrgyz 080) with distinct evidence of the flexural toppling (see Fig. 5.3).

Bedrock landslides are widely developed in the valleys of the Kokomeren right tributaries (see Fig. 8.138). Most interesting of them can be found in the Kashkasu River valley. Here, 1.5 km upstream from its mouth at 41.91° N, 74.17° E is an impressive Kashkasu Jumping rock avalanche (Kyrgyz 245, see Fig. 5.45), which morphology was described in Section 5.3.2. We should add that along the path, well visible on Fig. 5.45, that crosses upper part of this body, composed of gneiss-like granite mainly, one can see rather narrow "belts" of angular boulders of dark gray to black mafic igneous rocks stretching parallel to the direction of debris motion. These boulders practically do not mix with those of granite, demonstrating the laminar style of motion even within coarse blocky carapace. It will be shown hereafter, on Fig. 11.1.

About 2 km upstream one can reach the tongue of the 5.7 km long Upper Kashkasu rock avalanche (Kyrgyz 314, see Figs. 4.43, 4.44) that originated in the recently glaciated zone but can be distinguished with confidence from the glacial landform and deposits (see Section 4.5).

Few kilometers downstream from the Kokomeren rockslide the river had been blocked once more by another rockslide about 400 Mm^3 in volume, this time from the right side of the Kokomeren River valley. This feature, named the Displaced Peneplain rockslide (Kyrgyz 087, see Figs. 4.5 and 4.6) was described in Section 4.1 as an example of old strongly reworked rockslide that can be hardly identified on the space or aerial images only. The Kokomeren River, being blocked at a distance of about 1 km, eroded a bypass gorge through granites of its left bank, as it is shown on Fig. 4.5, which resulted in slope undercutting and, presumably, could cause formation of large rockslide here in future. Indirect evidence of the potential instability could be seen on top of the left slope of the valley (Fig. 8.158).

The temporal relationships between these two river damming rockslides – the Kokomeren and the Displaced Peneplain – are still unknown. Level of the abandoned valley riverbed that was blocked by the Kokomeren rockslide above the modern riverbed (see Fig. 8.148) is higher than that of the Displaced Peneplain one (see inset in Fig. 4.6). However, the first one is located in the hanging, uplifted side of the active fault shown on Fig. 8.155, while the second one – within its footwall. Considering possibility of ongoing neotectonic folding and location of the Kokomeren rockslide closer to the axis of the Djumgal anticline we can only say that both blockages occurred, most likely, in Late Pleistocene. More precise reconstruction of the Kokomeren River valley evolution that strictly corresponds to the "Disturbance regime" concept proposed by K. Hewitt (2006a) still awaits its researcher.

FIGURE 8.159 Frontal view on the Mini-Köfels rockslide body from its headscarp. Most of the deposits are red (gneiss debris) and only downstream part has gray color (metasediments debris).

About 2.5 km downstream from the Displaced Peneplain rockslide at 41.906° N, 74.28° E, the Kokomeren River had been blocked again, this time just temporarily, by the Mini-Köfels rockslide dam (Kyrgyz 085, see Fig. 5.41) exemplifying the Primary type of rock avalanche in frontally confined conditions. Besides impressive morphology of the deposits described in Section 5.3.1, it is characterized by the inherited relationships between rock types in its source zone that was on the right bank of the Kokomeren River valley and in the body, which frontal part that survived after dams' breach rests on the opposite – left bank of the river (Fig. 8.159). Right side of the 800 m high and 500–600 m wide headscarp is in dark-gray gneiss-like metasediments, while its central and left side – in dark-red gneiss. Same debris distribution can be seen in the rockslide body, which right (downstream) part is composed of dark-gray debris (seen on Fig. 5.41, and at the right part of Fig. 8.159), while most of the rockslide debris is composed of crushed red gneiss, well visible also on Fig. 8.160. Similar relationship of rock types was described above for the Kashkasu rockslide.

Small portion of the intensively crushed gneiss that formed upstream "wing" of the Mini-Köfels rockslide overlies low alluvial terrace on the right bank just above the road (Fig. 8.160). It is a homogenous matrix-supported material with predominant sand-silt size fines and rare angular clasts up to 10–20 cm. Visually similar material forms the main part of the left-bank body, which is blanket by relatively thin blocky carapace.

Further south, just above western termination of the Djumgal intermontane depression, two very

FIGURE 8.160 Crushed gneiss of the upstream part of the Mini-Köfels rockslide (RS) overlying alluvium (aQ) of ~10 m high terrace on the right bank of the Kokomeren River.

impressive rockslides – the ~10 Mm^3 Northern Karakungey (41.87° N, 74.24° E, Kyrgyz 321) and ~20 Mm^3 Southern Karakungey (41.86° N, 74.25° E, Kyrgyz 323) caved from the slopes of the nearly same height, steepness and aspect composed of Paleozoic granite (see Fig. 5.52). While the first one is an excellent example of the Jumping rock avalanche (see Fig. 5.43 in Section 5.3.2), the second one transformed in the Secondary deflected rock avalanche (see Figs. 5.53 and 5.54 in Section 5.3.3.1). Both features were described in the corresponding sections. One more much older headscarp could be identified here presumably, though its deposits could be hardly distinguished from the bouldery semi-glacial accumulations (ibid.). Several kilometers west of them large blockslide in Neogene

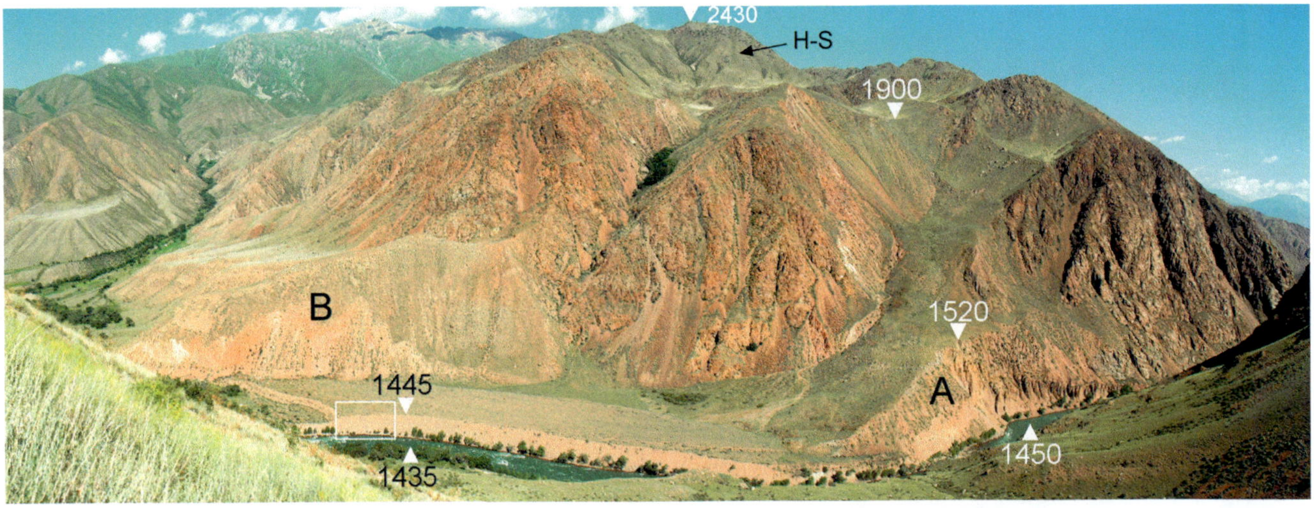

FIGURE 8.161 Lower Aral rockslide. A and B – the upstream (upper) and the downstream (lower) pats of the deposits; H-S –the headscarp area. Outlined outcrop of the terrace-like surface left by the outburst flood is shown on Fig. 8.164.

and Quaternary deposits affected almost entire western termination of the Djumgal intermontane depression just north of the confluence of the Kokomeren and Djumgal Rivers (see Fig. 8.138).

Further downstream, between the Djumgal and Minkush-Kokomeren intermontaine neotectonic depressions the Kokomeren River had been dammed at 41.798° N, 74.288° E by the Lower Aral rock avalanche (Kyrgyz 327) about 25 Mm^3 in volume that originated at the crest of the ~1 km high ridge forming right slope of the valley and composed of the same Paleozoic granite as the Karakungey rock avalanches. Such position of the headscarp excluding slope undercutting by erosion can be considered as the indirect evidence of seismic triggering of this rockslide, though uncertainty of this criterion was demonstrated by the Djamansu rockslide described above (see Figs. 8.110 and 8.112). The collapsing debris split on its way into two parts that formed two separate bodies ~800 m from each other (Fig. 8.161). The upper one is composed of heavily comminuted granite with a coarse carapace without any transitional zone; the lower one has the same internal structure, though without such a sharp contrast between main comminuted body and carapace (Fig. 8.162).

The upper body formed the 70 m high dam, whereas the lower one presumably caused just partial blockage of the river (see Fig. 8.161). Up to 250 Mm^3 of water (may be even more) could be accumulated in the dammed lake that extended for at least 9–10 km upstream. Considering relatively small size of the blockage and high extent of debris comminution, most likely that this dam had been breached just after first overtopping. This case study provides excellent evidence of catastrophic outburst flood downstream

FIGURE 8.162 Above – sharp contrast between comminuted material and blocky carapace of the Lower Aral rockslide upper body (just above A on Fig. 8.161); bellow – internal structure of its lower body (B on Fig. 8.161).

caused by dam's breach (Strom and Zhirkevich, 2013). About 2 km downstream of the dam there is a small, ~37 m high cliff on the right bank of the river covered by an apron of angular granite clasts and with accumulation of large angular boulders of granite up to 1 m in size at its downstream side (Fig. 8.163). It can

FIGURE 8.163 Debris left by outburst flood after breach of the Lower Aral rock avalanche dam on the isolated 37 m high cliff. See explanations in the text.

be explained as follows: the first, most powerful surge wave of the outburst flood carrying boulders eroded from the dams' carapace (see Fig. 8.162) was so high that it could not pass just through the stream and, thus, overrode this cliff. Those boulders that passed along the stream were brought far away, but those that passed just above the cliff were deposited in the 'shadow' behind it. The following portion of the outburst debris flow left smaller fragments on top of the cliff. The imbrication of flattened fragments visible in the upper inset in Fig. 8.163 clearly indicates the direction of debris flow motion. This debris flow was thick enough to leave about 2-m of debris nearly 40 m above the riverbed level. Rough calculations gave peak discharge of this debris flow of about 28,000–30,000 m^3/sec (Strom and Zhirkevich, 2013) – almost 50 times more than the maximum value ever recorded here ($\sim$600 m^3/sec) (State Water Inventory, 1987c).

Deposits left by the final stages of this catastrophic outburst could be seen on the right bank of the Kokomeren River between two bodies of the Lower Aral rock avalanche, where they form the terrace-like surface about 14 m above the modern riverbed level (see Fig. 8.161). While morphologically it looks similar to the alluvial river terrace, it is composed of layers of fines alternating with layers of angular clasts (Fig. 8.164). The latter correspond to the episodes of more powerful flow caused by temporal blocking of the erosional canal incised in the blockage.

FIGURE 8.164 Typical section of the outburst flood deposits about 14 m thick, forming terrace-like surface just downstream of the Lower Aral breached dam.

Downstream of the Lower Aral rock avalanche the Kokomeren River sharply turns to the west and follows the Minkush-Kokomeren intermontane depression. Unlike most of other neotectonic depressions of the Central Tien Shan corresponding to the asymmetric synclines bounded from one side by reverse faults, this one represents a narrow ramp graben bounded by reverse faults both from north and south (Sadybakasov, 1972, 1990; Makarov, 1977; Chedia, 1986). It is composed of the intensively folded Neogene conglomerates mainly, with subordinate sandstone, siltstone and mudstone, sometimes with gypsum. Locally Neogene deposits alternate with the Jurassic coal-bearing sediments. This neotectonic zone that stretches for about 120 km from the Sonkul

FIGURE 8.165 Rockslide at the left bank of the Tuyuksu River above the radioactive waste disposal (WD). Distinct evidence of deformations can be visible at the upper part of the slope. Its presently active part is on the slope left from the WD. 3D Google Earth view.

depression in the east to the Ketmen-Tiube depression in the west is characterized by the highest concentration of rockslides within the entire study region (see Fig. 8.107, Fig. 8.138). One of rockslides on the right bank of the Tuyuksu River near the Minkush town at (41.662° N, 74.47° E. Kyrgyz 303) endangers radioactive waste disposal left after uranium mining (Torgoev et al., 2013). Though its presently moving part is about 700,000 m^3 in volume only, it seems that much larger volume had been active in the past and could be activated again (Fig. 8.165).

Very impressive rockslides had dammed the Unkursai River valley – left tributary of the Kokomeren River. Both were described in Section 5.1. The upstream one – the Ak-Kiol rockslide still damming the valley – as an example of the very deep wedge failure (see Figs. 5.8-5.12) that formed Primary rock avalanche in frontally confined conditions, while the Lower Ak-Kiol rockslide – as a planar slide where huge block of Neogene conglomerates about 30 Mm^3 in volume slid down along the inclined bedding plane (see Fig. 5.6). This rock slope failure has very distinct, though rather short secondary rock avalanche (see Fig. 5.62). Rockslide dam had been breached, likely recently, though no historical data about this event are available. We hypothesize that this breach could be triggered by partial breach of the Ak-Kiol rockslide dam evident from the presence of the deep erosional canyon dissecting the latter's body (see Fig. 5.12) when large amount of water had entered in the lake dammed by the Lower Ak-Kiol rockslide. Such a domino effect is similar to the well-known breach of the Issyk rockslide dam in 1963 mentioned in Chapter 2. It is interesting that on the banks of the emptied lake one can see the cover of the dried and calcified algae marking the water level in the dammed lake (Fig. 8.166).

FIGURE 8.166 View on the emptied lake from the Lower Ak-Kiol rockslide dam. Blue arrows mark the level of the calcified algae shown in the lower inset; outlet of the erosional canyon dissecting the Ak-Kiol rockslide dam can be seen in the upper inset.

One of the most interesting case studies in the Kokomeren River basin – the Ornok rockslide about 200 Mm^3 in volume (Kyrgyz 010) is located on the left bank of the Kokomeren River valley at 41.716° N, 74.223° E. It is just the frontal part of the gigantic, about 17 km^2 DSGSD evolving between the Unkursai and the Tabylgaty River valleys, in the hanging wall of the neotectonic reverse fault bounding the Minkush-Kokomeren intermontane depression from the south that brings the greenish and violet Ordovician shale and siltstone over the Neogene red beds (Fig. 8.167). Besides this boundary fault one more south-dipping reverse fault passes parallel to it about 1 km to the north, through the folded Jurassic and Neogene deposits. Such fault is shown on the 1:200 000 State Geological map west of the Tabylgaty River valley.

The Ornok rockslide most likely represents a unique combination of the rotational blockslide and rock

FIGURE 8.167 Above – 3D Google Earth view of the giant DSGSD on the left side of the Kokomeren River valley and the Ornok rockslide. Below – hand-made photo of the same area. White arrows mark the crown of the DSGSD, yellow arrows – crown of the Ornok rockslide headscarp, orange horizontal arrows – front of the rock avalanche deposits, red arrows – position of neotectonic fault, pink arrows – fault zones displaced by rockslide; that marked by A is shown on Fig. 8.170; blue arrow marked by L – lake sediments accumulated in the proximal depression of the rotational blockslide; yellow line – position of the cross-section shown on Fig. 8.171; N – Neogene red beds; PZ Paleozoic metasediments; Unk. – the Unkyrsay River valley; Tab. – the Tabylgaty River valley.

FIGURE 8.168 Frontal part of the Ornok rockslide – rock avalanche. View from west. Pink arrows – fault zones displaced within the rockslide body; one marked by A is shown on Fig. 8.170; B – outcrop of rock avalanche debris overlying fluvial pebbles (ibid); 1 – proximal rotated block of Paleozoic metasediments; 2 – well-bedded Neogene rotated and displaced together with metasediments; 3 – horizontal multi-color "layers" – frontal avalanche-like part of this complex rockslide.

avalanche at its frontal part. The proximal part of this body is a huge block of the Paleozoic metasediments thrusted over Neogene red beds that slid down from the 1400 m wide and 400 m high concave headscarp. Reverse fault dividing metasediments and Neogene red beds is shifted forward by the rockslide for about 600–700 m (see upper image on Fig. 8.167 and Fig. 8.170). The inclined Neogene strata that outcrop just in front of the displaced Paleozoic block and retain the original bedding (Fig. 8.168) gradually

FIGURE 8.169 Contact between rotated block (RS) and overlying avalanche-like part (RA) of the Ornok rockslide. Boundary between these two parts is marked by dashed white line. Notice gradual shift from well-bedded Neogene on the right to the blocks in matrix on the left. Arrows marked by L – lake sediments accumulated in the depression on top of the avalanche-like part.

disintegrates northward converting into large harder blocks in the uniform matrix (Fig. 8.169) that belongs to the frontal avalanche-like part of the rockslide body. It is overlain by the punctuated deformed "layer" of scree. The approximate boundary between this upper part and the lower/proximal rotational part is marked by the dashed white line.

The frontal part of this rockslide – rock avalanche body that rests over fluvial gravels (Fig. 8.170) includes the displaced fault zone marked by 'A' on Fig. 8.167. It is either the same fault zone dividing Ordovician rocks from the Neogene (between 1 and 2 on Figure 8.168) that was displsaced once more being involved in the avalanche-like motion of the frontal part of the Ornok rockslide that collapsed from the frontal side of the blockslide, moved ahead and formed the frontal portion of its body, or another fault displaced within the frontal part of the rotational slide. The latter model is shown on Fig. 8.171. The avalanche-like part is formed by the deposits with typical pseudo-stratified internal structure similar to that observed at the Inylchek and Kokomeren rockslides (see Figs. 8.45, 8.153 and 14.1).

The lacustrine sediments – whitish laminated silt overlain by debris flow deposits had accumulated in the lowermost part of the proximal depression at the foot of the headscarp (see 'L' on Fig. 8.167) and on top of the avalanche-like part in the depression that had resulted from uplift of the body front typical of the rotational slides (see 'L' on Figs. 8.169 and 8.171). Age of their basal layers should correspond to the age of the Ornok rockslide formation.

Among numerous rockslides in the downstream part of the Kokomeren River valley west of the Ornok village the most interesting case studies are the Karachauli (Kyrgyz 077) and the Lower Kokomeren (Kyrgyz 011) rockslides (see Fig. 8.138).

The Karachauli rockslide (41.725° N, 74.06° E, Fig. 8.172) is the gigantic feature strongly reworked by erosion. Its volume could be roughly estimated as about 650 Mm^3. Unlike the Kokomeren rockslide which Late Pleistocene age could be derived from the position of the blocked river channed about 50 m above the present-day riverbed (see Fig. 8.148), the sliding surface of the Karachauli rockslide body could be even below the modern Kokomeren riverbed. Nevertheless its age seems to be Late Pleistocene as well. We should point out that the Kokomeren rockslide occurred within the axial part of the uplifting neotectonic anticline, while the rockslides at the east-west trending section of the Kokomeren River valley are within the Minkush-Kokomeren neotectonic

FIGURE 8.170 Above – view of the frontal part of the Ornok rockslide; below left – fluvial gravels (aQ) overlain by scree of Paleozoic metasediments (S), which is the lowermost unit of the "layered" avalanche-like portion of debris (RA); below right – fault zone composed of mylonite along which Paleozoic metasediments had been thrusted over the scree and the entire succession was probably moved within the rotational part of the Ornok rockslide. Its front was truncated by the overlying avalanche-like part (see Fig. 8.171).

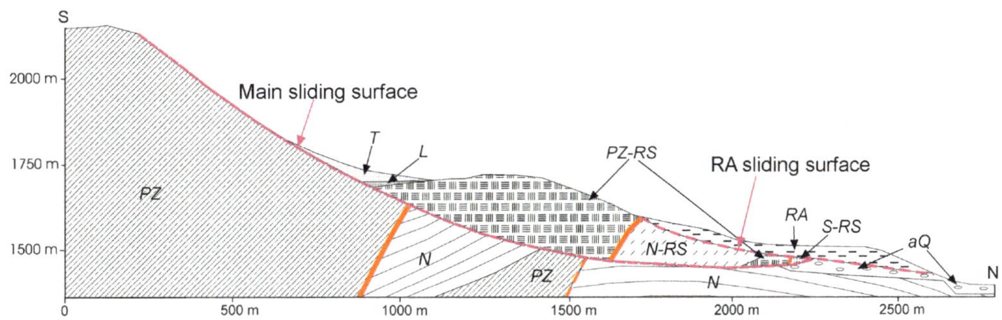

FIGURE 8.171 Schematic cross-section of the Ornok rockslide along the line marked on Fig. 8.167. *PZ* and *N* – the Paleozoic and Neogene bedrock; *PZ-RS* and *N-RS* – same units displaced by the rockslide; *RA* – rock avalanche deposits; *S-RS* – Quaternary scree overthrusted by Paleozoic rocks and displaced by rockslide (see Fig. 8.170, below right); *aQ* – Quaternary gravels overlaid by rockslide (see Fig. 8.170, below left); *L* – post-rockslide lake sediment in the proximal depression; *T* – talus accumulated at the foot of the headscarp.

depression where the river is characterized by some overdeepening, demonstrating long profile typical of the disturbance regime proposed by K. Hewitt (2006a).

Similarly to the Ornok rockslide the Karachauli one originated on the slope composed of Paleozoic metasediments (Ordovician volcano-clastic rocks) thrusted over the Neogene red beds, but in the much more narrow section of the river valley (compare Figs. 8.172 and 8.167). In both cases the Paleozoic rocks that could be found within the rockslide bodies have been shifted for several hundreds meters forward from their original position. Moreover, the frontal part of the Karachauli rockslide body composed of the displaced Neogene red beds mainly filled the mouth of the right tributary valley and could be seen on the opposite – right bank of the Kokomeren River. More or less

FIGURE 8.172 Above – 3D Google Earth view of the Karachauli rockslide, below – front view from the opposite bank of the Kokomeren River. Red arrows mark the position of the fault zone dividing Paleozoic and Neogene rocks outside the rockslide body, which part between elevation marks 1315–1680 m a.s.l. is composed of Paleozoic rocks, while its frontal part up to ~1600 m a.s.l. – of the displaced and crushed Neogene conglomerates (see Fig. 8.173).

FIGURE 8.173 Inclusion of rounded granite boulders (aQ) in crushed Neogene red beds (Cr. N) at the frontal part of the Karachauli rockslide body. Notice difference between Neogene bedrock (N) and more homogenous crushed Neogene conglomerates of the rockslide frontal part (Cr. N).

unidorm mass of Neogene conglomerates has inclusions of rounded granite boulders that can be nothing else but the Kokomeren stream or terrace alluvium (Fig. 8.173; see also Fig. 12.2) carried here when the entire valley bottom composed of Neogene was bulldozed by this gigantic slope failure. Besides, angular clasts of metasediments can be seen on top of the right-bank portion of rockslide debris (Fig. 8.174). Few hundreds meters westward one can distinguish frontal part of rockslide deposits composed of crushed Neogene conglomerate and sandstone from the overlying compact scree composed of the same material that fell from the steep right bank of the valley after (may be immediately after) rockslide formation (Fig. 8.175).

The simplified schematic cross-section of the Karachauli rockslide is shown on Fig. 8.176 (compare with that of the Ornok rockslide on Fig. 8.171).

Before being dissected by the erosion, the Karachauli rockslide had formed natural dam, presumably up to 300 m high, that had existed for rather long time judging from the presence of laminated silt about 10 m thick in the road cut at the junction of the Kokomeren and Djumgal Rivers near the Aral village (Fig. 8.177), about 30 km upstream from the blockage. This lake more than 40 km long with an area of about 85 km^2 could store up to 7 km^3 of water. We have no data about the style of its breach. However, considering that river had eroded through the dam's body, it could be assumed that dams breach was catastrophic.

Our attempt to date lake sediments using OSL method gave Late Pleistocene age but with inconsistent numerical values ranging from 18 to 61 Ka (Swenja Rosenwinkel personal communication). It was assumed previously that these lacustrine deposits were accumulated in the lake formed by the Lower Aral rockslide about 8–9 km downstream from this site (Strom, 2013b). However, more careful analysis shows that this assumption was wrong. First, the level of the Lower Aral rockslide dam crest is slightly lower than the top of the laminated silt succession; second, as we mentioned above, size of this dam was too small to block powerful Kokomeren River for a rather long time necessary to accumulate these deposits; and third, the Lower Aral rockslide looks too "fresh" to be Late Pleistocene feature, while relatively large age of the lacustrine deposits is confirmed by OSL dates and by the presence of debris flow deposits eroding and overlying silt succession (see Fig. 8.177).

We must point out that lacustrine sedimentation was interrupted by intensive liquefaction episode that took place soon after main river damming event (Fig. 8.178) (unfortunately this outcrop was significantly reworked in 2015 during road reconstruction). The lowermost silt laminae sank into the liquefied underlying basal sand that accumulated due to initial reworking of the slope by raising water (similar succession in the lacustrine sediments was observed upstream of the Kokomeren rockslide) and the ejected sand spread over the lake bottom forming the so called "event horizon" (McCalpin, 2009), while overlying silt have not been subjected to such deformations. Judging from the small thickness of the affected silt, liquefaction should take place, likely, within first years after the lake formation.

FIGURE 8.174 Angular clasts of Paleozoic rocks mixed with fluvial boulders (FB) on top of the right-bank part of the Karachauli rockslide composed of crushed Neogene conglomerates mainly.

Such an intensive liquefaction is a typical effect of strong seismic shaking (see, e.g. Chapter 7 in McCalpin, 2009), though limited extent of the outcrop were it is visible (about 100 m only) does not allow such sound conclusions on its seismic origin as, for example, in the Issyk-Kul basin, where liquefaction-induced features were observed in the numerous kilometers-long exposures of Quaternary lacustrine sediments (Korjenkov, 2000).

Strom (2013b) hypothesized that liquefaction could be caused by strong aftershock of an earthquake which main shock triggered the river-damming slope failure. As mentioned above it was assumed erroneously that the latter was the Lower Aral rockslide, while, more

FIGURE 8.175 Crushed Neogene conglomerates of the frontal part of the Karachauli rockslide body (RS) overlain by layered talus of the same conglomerated that fell from the right-bank valley slope (T). Original photo and the same image in false colors.

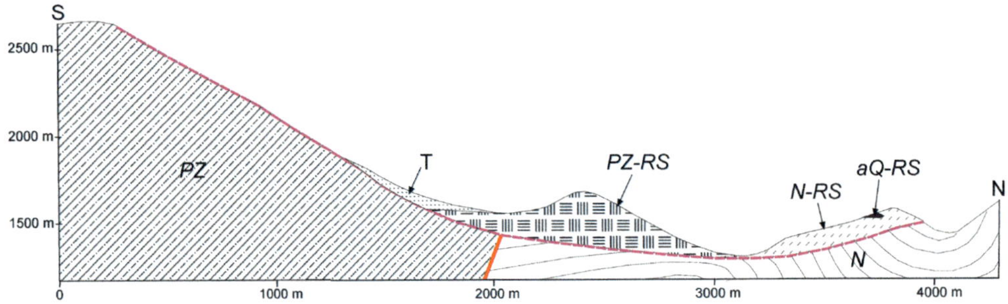

FIGURE 8.176 The schematic cross-section of the Karachauli rockslide. *PZ* and *N* – the Paleozoic and Neogene bedrock; *PZ-RS* and *N-RS* – same units displaced by the rockslide; *aQ-RS* – Quaternary gravels displaced by rockslide (see Fig. 8.173); *T* – talus accumulated at the foot of the headscarp.

FIGURE 8.177 Lacustrine sediments at the junction of the Kokomeren and Djumgal Rivers (whitish laminated silt - L) eroded and overlain by thick debris flow deposits (DF).

FIGURE 8.178 Lacustrine sediment accumulated at the tail-end of the past lake dammed, most likely, by the Karachauli rockslide and evidence of liquefaction. 1 – basal massive coarse arkose sand liquefied and ejected (1′ – the event horizon); 2 – lower part of silt unit that sank in the basal sand; 3 – upper non-deformed part of silt unit represented by nearly horizontal laminae.

likely, lacustrine sediments accumulated in the lake dammed by much bigger Karachauli rockslide, about 30 km downstream. Nevertheless, the assumption that liquefaction was cased by an aftershock of an earthquake that had triggered this large-scale rockslide seems to be reasonable, though its proof or disproof requires more accurate dating both of rockslide and of lake sediments. Further discussion on this topic will be presented in Chapter 13.

Additional, though indirect evidence of river damming is the presence of thick debris flow accumulations upstream from the Karachauli rockslide (Fig. 8.179-B, C). Unlike deposits of most common debris flow fans with chaotic structure that erode underlying deposits (see, e.g., Fig. 8.177) these accumulations up to 40–50 m thick are well-bedded and relatively fine (see Fig. 8.179-B, C). We assume that they were deposited underwater, when debris flows had entered the dammed lake. Similar deposits were found upstream of the Kokomeren rockslide in the Kyzyl-Oi depression opposite the same-name village (see Fig. 8.179-A)

We must notice, however, that debris flow deposits shown on Fig. 8.179-B, C could be accumulated in the lake dammed not by the Karachauli rockslide, but by the another, likely much younger Lower Kokomeren rockslide (Kyrgyz 011) that had blocked the valley 5 km downstream at 41.736° N, 74.0° E.

Similar to the Karachauli rockslide, this slope failure originated on the ~900 m high slope composed of varicolored Paleozoic metasediments thrusted over Neogene conglomerates. However, in contrast, it practically did not affect the Neogene deposits themselves (their debris could be seen at the eastern edge of rockslide body only), but, instead, displaced mylonite and cataclastic rocks several dozens meters thick forming the fault zone that passes here almost along the stream, and brought them on the right bank of the river.

FIGURE 8.179 Debris flow deposits (DF) accumulated in the lakes formed by large-scale rockslide dams. A – upstream of the Kokomeren rockslide (aQ – underlying fluvial gravels); B, C – upstream from the Karachauli rockslide.

Rockslide body about 20 Mm^3 in volume had crossed the Kokomeren River valley and blocked it raising more than 130 m on the opposite – right bank where it had dammed the dry valley by the levee about 15 m high at present (Fig. 8.180). It can be classified as a Primary rock avalanche with the extreme frontal confinement (see Fig. 5.31–1b), since the opposite valley wall was much steeper than in most of similar cases described above (compare with Figs. 4.20, 5.40, 5.41, 6.24, 8.23, 8.103, 8.116).

Thanks to the construction of the new road that crosses the rockslide body, its internal structure could be studied in detail in fresh outcrops in the road cut. Despite sharply bended trajectory of rockslide debris that descended several hundreds meters downslope, then moved across the valley bottom, and finally rose more than 130 m upslope – different lithological units demonstrate retention of their mutual position without any mixing. It is typical both of coarse blocky carapace (Fig. 8.181), locally with jigsaw puzzle structure of large fractured blocks (Fig. 8.182), and of inner part composed of tectonically comminuted rocks of the fault zone visible in the eastern part of the road cut (Fig. 8.183). Outside of the dammed tributary valley (see Fig. 8.180) rockslide body collided with an opposite valley wall that could cause rockfall of steeply dipping Neogene conglomerates visible just east of the

FIGURE 8.180 Front view of the Lower Kokomeren rockslide from the dammed dry valley. Headscarp is at the background.

FIGURE 8.181 Unmixed varicolored debris of Paleozoic metasediments on top of the western part of the Lower Kokomeren rockslide body.

FIGURE 8.182 Block of metasediments with jigsaw puzzle structure.

FIGURE 8.183 Unmixed lithologies in the eastern part of the Lower Kokomeren rockslide body. Ms – coarse metasediments debris; FZ – tectonically comminuted rocks of the fault zone; N – debris of Neogene conglomerates; S – scree displaced from the base of the source zone; N-T – talus of Neogene conglomerates from the right bank of the Kokomeren River valley overlying rockslide deposits; RF – rockfall of Neogene conglomerates.

rockslide and gradually replaced by talus overlying its body (see Fig. 8.183).

Two more rockslides in the tributary valleys at the lower part of the Kokomeren River basin have been described above in Chapters 4 and 5. The Aik-Kiol rockslide (Kyrgyz 076, see Fig. 4.3) is an example of well-defined rockslide with nearly planar top of the deposits without any prominent lowering or runup (Korchevskiy et al., 2011). The Kiol-say rockslide (Kyrgyz 078, see Fig. 5.63) represents the deflected Secondary rock avalanche with small but quite mobile avalanche-like portion of debris.

Further to the west this east-west-trending zone enriched by large-scale rock slope failures follows the western narrow part of the Minkush-Kokomeren neotectonic depression inherited by the Naryn River valley and reaches the Ketmen-Tiube intermontane depression filled by the Toktogul reservoir (see Fig. 8.107). Besides numerous bedrock landslides two unique caldera-like cavities described in Section 5.1 are located within this zone at 41.8° N 73.75° E (see Figs. 5.22-5.26) and at 41.814° N, 73.385° E (see Fig. 5.27) on top of the ranges bounding the Naryn River valley from the north. Their formation model is shown in Fig. 5.28.

These unique features are accompanied by numerous rockslides of different types and ages varying from Pleistocene, strongly reworked features to the

FIGURE 8.184 Upslope-facing scarps west of the Kyzylkiol cavity at 41.812° N, 73.717° E. The lowermost scarp is highlighted by snow patches remaining in the dammed gullies. Google Earth image and zoomed fragment of the July, 1980 aerial photograph in the inset.

FIGURE 8.185 Aerial photograph of the sackungen on the Aksu – Konchakiol watershed at the left bank of the Naryn River valley. Ak. – the Aksu Creek, Kon. – the Konchakiol Creek.

ongoing sackungen-like slope deformations. The latter form well-expressed upslope-facing scarps that stretch at a distance of more than 1.5 km along the watershed

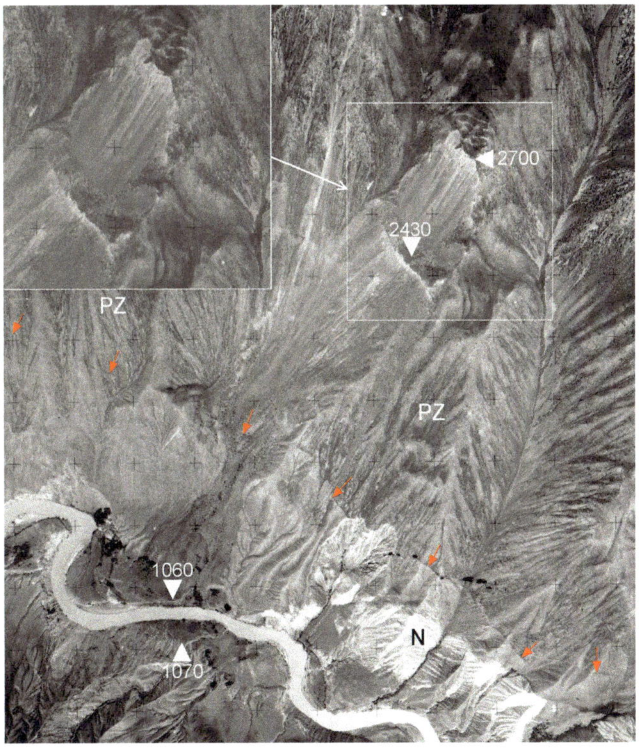

FIGURE 8.186 Aerial photograph of the Djamno rock avalanche. Red arrows mark surface rupture associated with the fault zone between Ordovician metasediments (PZ) and Neogene red beds (N). Darker patch of debris on the left bank of the Naryn River at 1070 m a.s.l. could represent the frontal part of rock avalanche body.

at the elevation above 3000 m a.s.l. east and, especially, west of the Kyzylkiol cavity (Fig. 8.184).

One more site with evidence of the ongoing slope deformations is located on the left bank of the Naryn River valley on the 600 m high ridge between small Aksu and Konchakiol Creeks at 41.747° N, 73.662° E (Kyrgyz 282). Its eastern 35°-36° slope differs drastically from the western one. The latter is grass-covered with relatively smooth minor landforms, while the former one is unstable, covered by talus (Fig. 8.185). It indicates that, despite presence of the upslope-facing scarps on both sides of the ridge, the entire rock massif about 1.5 km long, almost 0.5 km thick and ca. 150 Mm^3 in volume composed of the Late Precambrian metasediments is moving eastward gradually.

One more impressive, likely Holocene rockslide in this part of the Naryn River valley occurred at 41.8° N, 73.654° E on its right, more than 1700 m high slope, composed of the Ordovician metasediments near the ruins of the Djamno hamlet (Kyrgyz 281). It represents combination of the blockslide that descended ~270 m, and of the mobile rock avalanche about 3 km long, likely of the Secondary type

FIGURE 8.187 The Akterek rockslide downstream of the Kambarata-1 dam site. A – Source zone on the left bank slope of the Naryn River valley. H-S – headscarp; γS – the Silurian granites thrusted over the Neogene conglomerates (N); red arrows mark the Akterek fault zone. B – rockslide body (RS) resting on the right bank of the river on the ~50 m high strath terrace with Neogene conglomerates at its bed (N) overlain by alluvial gravels (aQ).

(see Section 5.3.3) (Fig. 8.186). Slope above the headscarp crown about 70 m high is fractured (see the inset in Fig. 8.186). Blockslide forms a bench about 450 m long and 100 m wide at ca. 2430 m a.s.l. Rock avalanche originated on its frontal part and reached the valley bottom. Moreover, there are some indications that rock avalanche had crossed the Naryn River and stopped on its left bank at about 1070 m a.s.l. It can be assumed that the rapidly descending block had stopped abruptly and its near-surface more weathered part was detached and rushed down along the 35° slope. Presence of the distinct surface rupture well visible outside rock avalanche body and slightly pronounced within it allows assumption of seismic triggering of this rock slope failure.

It should be noticed that features shown on Figs. 8.184–8.186 are located on the slopes of that part of the Naryn River valley that will be submerged by the planned Kambarata-1 HPP reservoir.

Another group of large-scale rockslides is located downstream from the Kambarata-1 damsite on the banks of the existing Kambarata-2 reservoir. Three of them had crossed the valley so that deposits rest on the river banks opposite to those where failure had occurred. One of such rockslides – the Akterek ~10 Mm^3 in volume (Kyrgyz 016) originated at 41.76° N, 73.42° E on the about 750 m high left slope of the Naryn River valley composed of the (presumably) Silurian granites thrusted over Neogene coarse conglomerates. The Akterek fault zone tens of meters thick composed of the tectonically alternated and weathered granite passes along the slope at about 1/4 of its height (Fig. 8.187-A). Up to 170 m thick deposits of this rockslide rest over the ~50 m high right-bank strath terrace (Fig. 8.187-B). As in many cases described above, debris of the lower part of this rockslide body was heavily comminuted while overlaying carapace is much coarser (Fig. 8.188).

The Akterek rockslide had dammed the Naryn River, though no evidence of either lacustrine sediments or outburst flood have been identified yet. Morphologically it is very similar to the Mini-Köfels rockslide (see Fig. 5.41) that caved from the slope composed of the Proterozoic gneiss and metasediments and also rose on the opposite slope after crossing the river valley. Main parameters of these case studies are also nearly the same (Table 8.9). Both can be classified as Primary rock avalanches with frontal confinement.

FIGURE 8.188 The bipartite internal structure of the Akterek rockslide deposits. Intensively cruched main body is blanked by coarse carapace shown in more details in the inset.

TABLE 8.9 Comparison of the Mini-Köfels and Akterek rockslides parameters

	Height of the collapsed slope (m)	Length (m)	Headscarp / deposits width ratio	Volume (Mm^3)	Runup (m)	H/L ratio
Mini-Köfels	800	2200	0.53	12	170	0.37
Akterek	680	2000	0.55	10	170	0.34

Though no direct data on the age of this rockslide are available, it can be determined from the ages of the strath terraced of the Naryn River about 2.5 km upstream derived by Prof. Yeong Bae Seong from the Korea University (Personal communication). Ages of the terraces just downstream of the proposed Kambarata-1 dam site (Fig. 8.189) were obtained by ^{10}Be method that showed good correlation with

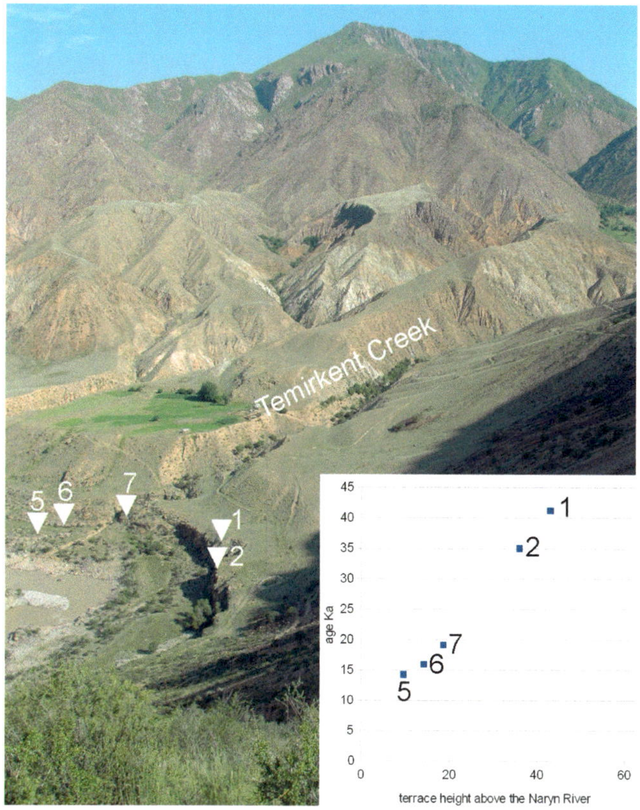

FIGURE 8.189 Position of quartz samples collected from the surfaces of strath terrace basements on the right bank of the Naryn River close to the Temirkent Creek mouth. Relationship between terraces age and height above the river is shown in the inset.

terraces height (see insert in Fig. 8.189). According to this plot the 50 m high terrace overlain by the Akterek rockslide deposits should be about 40–50 ka old. Considering that the dated terraces at the Temirkent Creek mouth are located within the hanging wall of the Akterek reverse fault with cumulative Late Pleistocene offset of about 15–20 m and the terrace overlain by the Akterek rockslide body – in its footwall, the latter could be even older. However, it means only that rockslide is not older than this terrace, while its real age could be much younger, as far as debris crossed the river channel that could be incised much less at that time than at present.

1.5 – 2 km downstream from the Akterek rockslide, opposite the Djuzumdybulak Creek mouth one can see the deposits of the 2-stage rockslide (Kyrgyz 279 & 279a) resting on the left bank of the Naryn River valley at its sharp bend (41.775° N, 73.4° E, Fig. 8.190). Both slope failures occurred on the high right slope of the valley composed of the about 1 km wide tectonically deformed zone with two opposing reverse faults bounding narrow ramp graben of folded weak Neogene conglomerate, sandstone and claystone from Paleozoic granite (Fig. 8.191). Faults demonstrate

FIGURE 8.190 Above – 3D Google Earth view of the Djuzumdybulak rockslides. Below – view on their headscarp and multi-age deposits from upstream. PZ, γPZ – Paleozoic metasediments and granites; N – Neogene sedimentary rocks; RS1 & RS2 – rockslide deposits of the first (older) and the second (younger) generations correspondingly; white dashed lines – approximate fronts of these rockslides; red arrows mark active fault; A – site where 6-m likely single-event vertical offset was dated as 111 BC – 240 AD.

FIGURE 8.191 Neogene ramp graben bounded by the opposing fault zones that crosses headscarp of the older Djuzumdybulak rockslide. Red arrows mark fault plains.

evidence of recent tectonic activity (Belousov et al., 1994) (see Fig. 8.190). The deposits of the older rockslide 30 Mm^3 in volume approximately (the body was eroded at a large extent) blanket series of strath river terraces from ca. 45 to 100 m high (Fig. 8.192) and extend up to 1.5–1.8 km from the foot of its source

FIGURE 8.192 Closer view on the deposits of the Djuzumdybulak rockslides showing their relationships with strath terraces. Most of indications are the same as on Fig. 8.190; aQ – alluvium of the strath terraces of different height; F – fresh fracture with vertical offset on top of loose deposits.

zone. It looks as if this body climbed over these terraces as over giant stairs (see left part of the photograph on Fig. 8.190). Surface of this body is smooth, grass-covered and really looks old that differs it significantly from the younger Djuzumdybulak rockslide that fills the younger valley and, partially, overlies older rockslide deposits.

This younger body up to 70–80 m thick and about 10 Mm^3 in volume has hummocky surface with classical badland on its steep slopes (see Fig. 8.192). It rests on the terraces about 15–20 m high, may be even less, and its front rises 50–60 m up to ~1000 m a.s.l. Judging from the ages of the Naryn River strath terraces upstream (see Fig. 8.189) it can be assumed that older rockslide should occur about 35–40 Ka BP, while the younger one – about 14 Ka BP or later.

Both bodies are composed of the intensively crushed material with distinct stripes of debris of different color. Now most of the younger body has been submerged by the Kokomeren-2 reservoir and cannot be observed directly.

It is still questionable if both failures occurred from the same source zone or not. First assumption is supported by the presence of thick loose deposits at the base of the headscarp with a crown at 1620 m a.s.l., at the level, corresponding to the base level of the younger rockslide (see Fig. 8.190). Several years ago large fissure with some vertical offset started opening at the top of this loose accumulation (it is marked by F on Fig. 8.192). There is unverified information that sliding of loose material at this slope was triggered by the 1992 M7.3 Suusamyr earthquake which epicenter was about 30–35 km north of this site. However, these recent landslides could evolve in the headscarp of the older catastrophic rockslide just gradually, not being

FIGURE 8.193 Distinct eastward inclination of the "layers" of debris of the yonger Djuzumdybulak rockslide (RS2) indicating westward direction of its motion. View from the right bank of the Naryn River near the Djuzumdybulak Creek mouth. Rockslide debris overlies debris flow deposits (pQ) brought by the Djuzumdybulak Creek, which, in turn, rest over the Naryn River alluvium (aQ).

directly related to the second (younger) catastrophic slope failure that could originate from the separate headscarp with a crown at 1470 m a.s.l. visible on Fig. 8.190.

The latter assumption is supported by the observations on the internal structure of the younger rockslide deposits visible from the opposite bank of the Naryn River (Fig. 8.193). They differ significantly from the underlying thick brownish debris flow deposits that had been sourced from the Neogene sediments shown on Fig. 8.191. Younger rockslide body itself, in contrast, is composed of the tectonically crushed granites of the fault zone only. Besides, as one can see on Fig. 8.193, "layers" in rockslide debris dip towards the east. Such orientation along with evidence that slightly upstream base of rockslide body is lower than at the outcrop shown on Fig. 8.193, and absence of Neogene debris in the deposits indicate that this rockslide originated not from the same headscarp as an older one.

It was hypothesized that the younger Djuzumdybulak rockslide could be triggered by an earthquake sourced at active fault, visible on Fig. 8.190, which latest ~6 m vertical displacement occurred between 111 BC and 240 AD (with 95.4% confidence). The attempt to date this rockslide by lichenometric analysis revealed that it was not younger than ca. 1000 years, which does not contradict this assumption (Belousov et al., 1994).

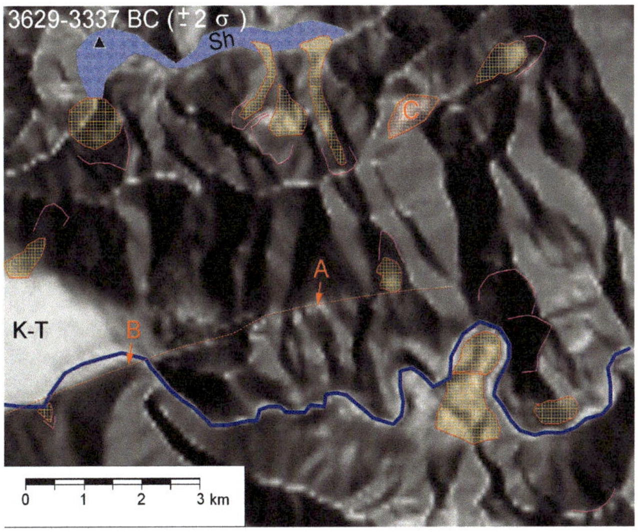

FIGURE 8.194 Rockslides (light brown hatched areas) north of the Naryn River valley where it enters the Ketmen-Tiube intermontane depression (K-T). C – the Djuzumdy cavity; dashed red line –fault trace with evidence of recent activity; A – site where 6-m offset occurred about 2000 years ago; B – site where evidence of several recurrent rupturing events were found out; blue hatched area – dated lacustrine sediments accumulated in the landslide-dammed lake in the Shilenhana intermontane depression (Sh). *Modified from Belousov T.P., Skobelev S.F., and Strom A.L. 1994. On estimation of the recurrence period of strong earthquakes of the central Tien Shan (according to the data of absolute geochronology). Journal of Earthquake Prediction Research, 3, 226–236.*

Several more rockslides, though of much smaller volume could be found in the vicinity of this site (Fig. 8.194). Nearby, at 41.814° N, 73.385° E, there is also the Djuzumdy caldera-like cavity (Strom and Groshev, 2009) (see Fig. 5.27). It can be speculated that formation of this cavity could be simultaneous with the slope failure that formed the older Djuzumdybulak rockslide, likely triggered by large earthquake, but more studies are necessary to clarify if it is true or not.

More grounded assumption on seismic origin of the younger Djuzumdybulak rockslide and some other nearby slope failures (see Fig. 8.194) can be made based on several indirect indicators. As we mentioned

FIGURE 8.195 View from the downstream on the Uch-Terek blastfill dam few hours after its creation by the explosion (upper-right inset). The lower-left inset presents the front view on this slope failure.

already, age of this rockslide is not less than 1000 years. The latest rupturing event at the nearby active fault occurred ca. 2000 years ago. 2–3 km north of it and just west of the Djuzumdy caldera-like cavity three small rockslides in Paleozoic granites occurred on the northern slope of the ridge bounding small Shilenhana intermontane depression from south. One of them formed a compact dam, while two other converted into rock avalanches 1.5–2.0 km long which distal parts entered flat bottom of the intermontane depression formed by debris flow deposits gradually transplaced by the fine-graded whitish silt and sand with gravel. These sediments likely of the lacustrine origin could be accumulated in the lake dammed by older rockslide that had blocked the valley draining the Shilenhana intermontane depression. Dated charcoal sample collected in these deposits gave their radiocarbon age of 4640 ± 50 years BP (GIN 4724) that corresponds to the calibrated age of 3629–3141 years BC (OxCal 4.2). Since rock avalanches mentioned above overly these sediments, they should be younger. Thus we could propose noncontradictory hypothesis that both younger Djuzumdybulak rockslide (older than 1000 years) and rock avalanches in the Shilenhana intermontane depression (younger than ca. 5400 years) could be triggered by the earthquake that occurred between 111 BC and 240 AD (about 2100 year ago) (Belousov et al., 1994).

South of this section of the Naryn River valley, at 41.722° N, 72.245° E, in the ca.150 m deep gorge of the Uch-Terek River one can see the very "fresh" 50–55 m high rockslide dam (Fig. 8.195) that, in fact, was created on June 11, 1989 by the experimental blast of 2,000 tons of the crude dispersive igdanite performed in laminated steeply dipping siltstone within the frames of the Kambarata-1 HPP project to test the possibility of dam construction by the artificially induced rockslide (Adushkin, 2011).

One more artificial blastfill blockage about 50 m high was created on February 8, 1975 in the Burlykia River valley at 41.882° N, 73.239° E by blasting of 703 tons of grained granulite in granites [ibid]. We mention these features to avoid their mixing with natural rockslide dams by next generations of landslide researchers.

About 30 km north of the Naryn River in the upper reaches of its small right-bank tributary – the Belaldy River valley at 42.06° N, 73.279° E there is the same-name rockslide about 20 Mm^3 in volume – the only one large-scale bedrock landslide triggered by the 1992 M7.3 Suusamyr earthquake (Kyrgyz 254, see Fig. 2.4) that was described in brief in Chapter 2.

Two rockslides are associated directly with the Talas-Ferghana fault that is considered as the southwestern boundary of this part of the Tien Shan dividing it from the North-Western and Southern parts (see Fig. 8.1). They are located in the Eastern Karasu River valley – left tributary of the Naryn River. Both rockslides dam the river forming lakes. Some morphological peculiarities of the larger one – the 200–250 Mm^3 Karasu Lake rockslide at 41.57° N, 73.22° E (Kyrgyz 238) were described in brief in Section 5.3.3.3 (see Fig. 5.64). Here we should notice that, unlike the interpretation of this rockslide dam as a result of several successive slope failures with significant time intervals between them, presented in (Matveev, 1968) and

FIGURE 8.196 The Kapkatash rockslide and the assumed position of its source zone at 3240 m a.s.l.

FIGURE 8.197 Deeply eroded rockslide dam in the Chong-Kongur-Debe River valley. 3D Google Earth view.

(Belousov et al., 1994) we consider the Karasu Lake rockslide as a single-event catastrophic slope failure. Difference in the morphology and grain-size composition of its different parts visible well on Fig. 5.64 resulted from the mechanism of its motion that can be classified as a deflected Secondary rock avalanche of the Classical subtype (see Section 5.3.3.1).

10 km upstream from the Karasu Lake rockslide at 41.524° N, 73.322° E the Eastern Karasu River is dammed by another rockslide about 20 Mm^3 in volume (Kyrgyz 239) that forms the Kapkatash Lake (Fig. 8.196). The dam up to 120 m high is composed of the same carbonate debris as the Karasu Lake dam. Unlike the former one that had never been overtopped and were first springs appear about 1 km from the lake and 140–150 m below its level (see Fig. 5.64), the much smaller Kapkatash Lake dam demonstrates evidence of the overflow during flood seasons and springs appear in the erosional canyon about 10 m below lake level at the low-water season.

From the first glance it looks as if the Kapkatash rockslide originated from the double-headed headscarp with a crown at 2520 m a.s.l. (see Fig. 8.196). It is, however, unlikely that such failure could form the south-eastern part of the blockage upstream from the dams' crest since it would require upward (upstream) debris motion in the direction transverse to the initial failure vector. Much more likely that rockslide was sourced about 1 km above the valley floor at the spoon-shape lowering with a crown at 3240 m a.s.l. also marked in this Figure. Huge amount of crushed limestone rushed down along the steep tributary gulley, crossed the river valley, collided with a right-bank rocky spur between two gullies with up to 70 m runup and split into two parts. The upstream one is now hidden underwater, while the downstream one spread for about 1 km blocking on the way the small right bank gulley.

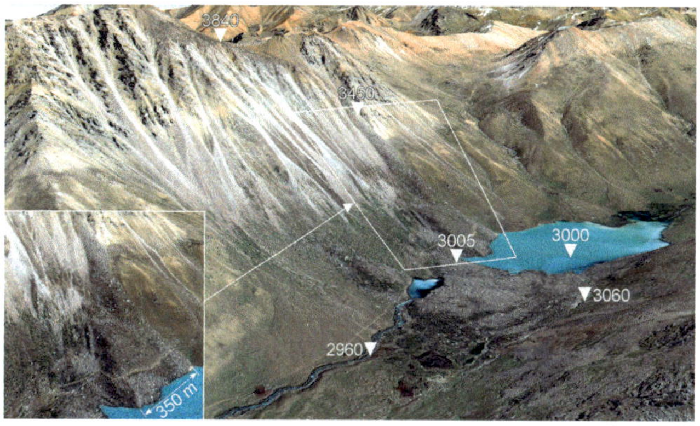

FIGURE 8.198 The Beshtash rockslide dam. Debris apron outside the headscarp limit is shown in the inset. 3D Google Earth view.

Based on the similarity of landscapes of both the Karasu Lake and the Kapkatash Lake dams it can be assumed that both rockslides occurred simultaneously, being triggered, most likely, by strong earthquake associated with the Talas-Ferghana fault.

Finishing the description of the Central Tien Shan rockslides we cannot miss two interesting case studies in its north-western part – on the northern slopes of the Talas Range. One of these features (Kyrgyz 071) is located at 42.353° N, 71.798° E and represents old rockslide dam completely dissected by erosion in the Chong-Kongur-Debe River valley – right tributary of the Kiumush-Too River, which, in turn, falls from the south in the Talas River. Wedge failure of about 20 Mm^3 of the Precambrian terrigenous and carbonate rocks occurred on the ca. 950 m high slope and produced a compact, ~1 km wide dam about 140–150 m high at its proximal, lowermost, part and with significant ~150 m runup (Fig. 8.197). This Primary rock avalanche with pronounced frontal confinement (see Section 5.3.1) somehow resembles the much bigger

Aksu rockslide near Bishkek (see Fig. 8.103), though being much smaller and much more reworked by the erosion. Similar to the Aksu case study no traces of valley siltation can be seen upstream, indicating that it could be breached soon after its formation.

Another rockslide in this part of the Central Tien Shan (Kyrgyz 360) seems to be much younger and still dams the Beshtash River at 42.207 °N 72.486 °E forming the Beshtash Lake. This river falls in the Talas River just in the same-name city. Failure of ca. 15 Mm^3 of weathered rocks occurred on 1 km high slope composed of Paleozoic granites and affected almost 1 km long section of the slope (Fig. 8.198). One of the most interesting peculiarities of this case study is rockslide debris spreading over the slope outside the headscarp boundary. Same phenomenon was described at the Spur rock avalanche in the Karakudjur River valley (see Fig. 5.4). In the Beshtash case debris started expanding over the headscarp eastern boundary about 400 m above slope foot and covers the slope by thin apron up to 310–320 m wide at its base (see Fig. 8.198). From our point of view it could result from some prestressing of the deforming rock mass before its failure that provides not only vertical (downslope) motion of debris under gravity force, but also its lateral, sidewise motion.

Presence of the concentric belts of debris in the frontal part of the low (ca. 40 m only) and 1400 m wide body with about 55 m runup indicates that this coarse material moved across the valley, overlying older fluvial-glacial deposits. Similar, though much more pronounced belts could be observed, for example, at the Kudara-Pasor rockslide dam in Pamir (see Fig. 4.20).

8.5 NORTH-WESTERN TIEN SHAN

North-Western Tien Shan embraces system of northeast-trending Ugam, Pskem, Chandalash and Chatkal ranges drained by the Pskem, the Chatkal and the

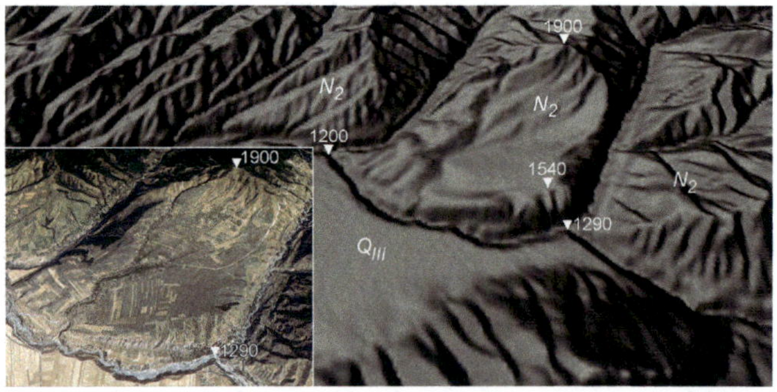

FIGURE 8.200 The shaded 3D relief of the assumed Kugart landslide and the adjacent area (3″ SRTM DEM visualized by Global Mapper Software). N_2 – Pliocene sediments; Q_{III} – Late Pleistocene fluvial deposits. 3D Google Earth view of this area is shown in the inset.

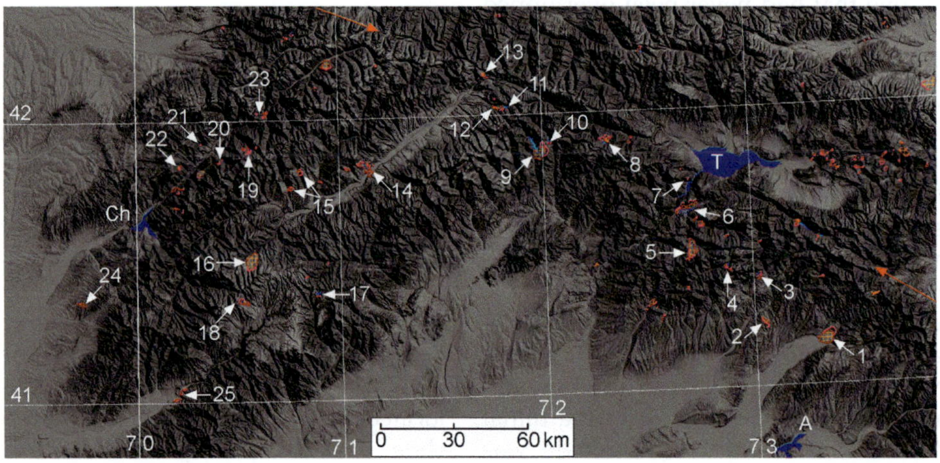

FIGURE 8.199 Shaded 3″ STRM DEM of the North-Western Tien Shan. Rockslides are marked by pink and brown dots; blue dots – dammed lakes. Red arrows mark the Talas-Ferghana fault; T – the Toktogul reservoir; Ch – the Charvak reservoir; A – the Andizhan reservoir. 1 – the assumed Kugart landslide (see Fig. 8.200); 2 – see Fig. 8.201; 3 – see Fig. 8.202; 4 – the Kutmankul and the Baubashata rockslides (see Figs. 8.203 and 8.204); 5 – the Dead Lakes rockslide and the nearby slope failures (see Figs. 8.205–8.207); 6 – rockslides in the lower reaches of the Eastern Karasu River (see Figs. 8.208–8.212); 7 – the 1946 Sarykamysh rockslide (see Fig. 8.213); 8 – see Fig. 8.214; 9 – the Sarychelek rockslide (see Figs. 8.215, 8.216); 10 – see Fig. 8.217; 11 – the Karatoko Lake dam (see Fig. 8.218); 12 – see Fig. 8.219; 13 – the Ak-Tash rock avalanche (see Fig. 8.220); 14 - the Chukurchak rock avalanche (see Figs. 8.221, 8.222); 15 – see Fig. 8.223; 16 – the Ters rockslide (see Figs. 4.2, 5.16); 17 – the Kiok-Ala rockslide (see Fig. 8.224); 18 – the Arashansai rock avalanche (see Fig. 8.225); 19 – the Ikhnachkul rockslide cluster (see Figs. 8.226, 8.227); 20 – the Badak Lake rockslide (see Fig. 8.228); 21 – see Fig. 8.229; 22 – the Kharamdazasai rock avalanche (see Fig. 8.230); 23 – the Sarydjayak rock avalanche and adjacent rockslides (see Figs. 8.231–8.233); 24 – see Fig. 8.234; 25 – the Atcha landslide (see Fig. 8.235).

FIGURE 8.201 Stepped relief with small lakes in the depressions formed by landslide in the Cretaceous sedimentary rocks dipping south-west at the Kara-Unkiur – Kugart watershed. 3D Google Earth view.

Angren Rivers mainly and located north of the Ferghana intermontane depression. Pskem and Chatkal Rivers merge about 60 km north-east of the Tashkent City and form the Chirchik River dammed in its upper reaches by the Charvak HPP with large reservoir. In this Section we will describe also landslides of the north-western part of the Ferghana range south-west of the Talas-Ferghana fault and north-west of the Kara-Daria River with Andishan reservoir at its middle reaches (Fig. 8.199).

Evidence of a very large old landslide (Kyrgyz 044) was found in the Kugart (Kiok-Art) River valley about 35–40 km upstream of the Djalalabad City at 41.17° N, 73.35° E. The smooth concave surface of ca. 30 km^2 on its right bank composed of the gently dipping Neogene terrigenous rocks differs drastically from the surrounding slopes despite many of them are composed of the same Neogene deposits (Fig. 8.200). Such morphology is the main reason to identify this 8 km long landslide. The affected slope is 700 m high only, thus providing very low H/L ratio of ~0.09 that corresponds to an angle of about 5–6° (Strom and Korup, 2006). However, such a low angle of sliding surface is not surprising if we take in mind that the largest terrestrial Baga Bogdo landside in central Mongolia with an area about 300 km^2 and volume of ca. 50 km^3 originated on a slope with mean angle ~3° only (Philip and Ritz, 1999).

Despite very large area affected, we estimated volume of this assumed Kugart landslide to be about 2–2.5 km^3 only due to the relatively thin displaced mass. Its sliding surface at least partially could coincide with bedding planes in the Neogene (Pliocene) deposits. It was speculated in (Strom and Korup, 2006) that such slope failure could be triggered by a strong earthquake. We realize, however, that this assumption is based on the features' size and very low angle of the possible sliding surface and that more studies are needed both to interpret this geomorphic anomaly unequivocally and to explain its origin.

We must notice that the eastern rim of the Ferghana valley – one of the largest intermontane depressions of the entire Tien Shan, composed of rather weak poorly lithified Mesozoic and Cenozoic sedimentary rocks is extremely rich of landslides. Thousands of them have been identified in this part of the region (Roessner et al., 2005; Havenith et al., 2006b, 2015a, b; Schlögel et al., 2011; Torgoev et al., 2005, 2013; Golovko et al., 2014, 2017). Since many landslides are either permanently active or activate almost every year affecting densely populated areas they have attracted researchers' attention much more than most of large-scale bedrock landslides in the basement rocks. As we mentioned already in the Introduction, we do not intend to analyze soil slides. That is why we will describe here in brief only few largest landslides in MZ-KZ sediments.

Besides the assumed Kugart landslide, about 25 km west of it, at 41.23° N, 73.05° E the north-facing slope of the Kara-Unkiur – Kugart watershed composed of weak Cretaceous claystone and sandstone dipping 15–20° south-west, locally with gypsum and limestone, is affected by large landslide at a distance of more than 3 km with several steps on the north-east facing slope and with small water bodies in the local depressions (Kyrgyz 342) (Fig. 8.201). Unlike the assumed Kugart landslide that slid down conformably with bedding, here the sliding surface crosses bedding. In contrast, several shallow landslides affect the opposite – southern slope of this gentle ridge, likely following bedding. It resembles the landslides in monoclinal flish in the Czech (Carpathians) (Břežný and Pánek, 2017) some of which are cataclinal (like the Kugart landslide) while other are anaclinal – similar to the Kyrgyz 342 case study. Similar relationships between rock structure and relief governing rockslides

FIGURE 8.202 Rockslides on both sides of a small ridge west of the Kara-Unkiur River. White arrows marked by K 312 and K 343 (Kyrgyz 312 and 343) indicate the directions of both slope failures. 3D Google Earth view.

FIGURE 8.204 Relationships between coarse Baubashata rockslide debris (RS) and modern terminal moraines of the glaciers (glQ_{IV}^2), south of the Baubashata Mountain. 3D Google Earth view.

FIGURE 8.203 3D Google Earth view of the Kutmankul and the Baubashata rockslides originated from the slopes of the Baubashata Mountain (4428 m a.s.l.). The jigsaw puzzle structure of angular boulders on top of the Kutmankul dam is shown in the inset. Area outlined at the right part is shown on Figure 8.204.

formation were described by Murge (1965) at the Sawtooth Ridge, Montana. Since the slope is a built-up area of the Karakiol village, detailed study and real-time monitoring of this feature would be of high practical importance.

Moving upstream the Kara-Unkiur River just at the boundary between the steeply dipping well bedded Cretaceous sediments and the Paleozoic basement, at 41.394° N, 73.03° E one can find a small, 400 – 500 m high north-east trending ridge composed of the Devonian limestone with rockslides on its both sides (Kyrgyz 312 and 343, Fig. 8.202). Debris of the southern one (Kyrgyz 312) about 25 Mm^3 in volume fills the upper reaches of a gulley (its tip is at 1450 m a.s.l.) and overlaid the boundary between the Paleogene – Upper Cretaceous terrigenous succession and the Devonian limestone. The northern rockslide also ca. 25 Mm^3 in volume (Kyrgyz 343) formed 150 m high dam with more or less planar crest that blocks the steep valley. The dammed lake was completely silted. If it will be found that both rockslides occurred simultaneously, it would be a sound argument in favor of their seismic origin (compare with rockslides shown on Figs. 8.26 and 8.38).

One more pair of large rockslides in the eastern part of the North-Western Tien Shan occurred on the north-western and south-western slopes of the

FIGURE 8.205 KFA-1000 space image of the Dead Lakes rockslide. White circles – elevation marks, black quadrate at 2180 m a.s.l. – point where quartz veins in large boulders were sampled for the cosmogenic dating. *Modified from Strom, A.L., and Korup, O. 2006. Extremely Large Rockslides and Rock Avalanches in the Tien Shan, Kyrgyzstan. Landslides, 3, 125-136 with permission of Springer.*

4448 m a.s.l. Baubashata Mountain – the topmost peak of this area composed of the Devonian volcanic rocks (Fig. 8.203). Rockslide about 55 Mm^3 in volume caved at 41.432° N, 72.868° E from the ridge extending west of this peak (headscarp crown is between elevation marks 3275 and 3300 m a.s.l. on Fig. 8.203) and formed the ca. 180 m high Kutmankul Lake dam in the Kerey-Zan River valley – one of the sources of the Maili-Suu River (Kyrgyz 055). Another, much smaller rock slope failure about 10 Mm^3 in volume (Kyrgyz 055a) originated just on the southern slope of the Baubashata Mountain and filled the bottom of the glacial valley at 41.418° N, 72.887° E. Surface of this Baubashata rockslide is about 500–600 m higher than the Kutmankul rockslide dam crest. Both are covered by huge angular boulders up to 20–30 m in size, sometimes with the jigsaw puzzle structure (see inset in Fig. 8.203).

The Kutmankul rockslide with proximal lowering and up to 160 m runup represents a good example of the primary rock avalanche with sharp frontal confinement. The dam looks quite "fresh". It has never been overtopped and seeping water appears at the base of its downstream slope without any visible springs on its downstream face above. It can be hypothesized that seeping could take place through the debris that had rested at the valley bottom before the rockslide occurred.

The Baubashata rockslide body has been affected directly by modern glaciers, which terminal moraines with drastically finer composition overly coarse rockslide debris (Fig. 8.204). Sharp difference between their grain-size composition proves different origin of these Holocene accumulations.

FIGURE 8.206 The internal structure of the Dead Lakes rockslide visible in the ca. 250 m long outcrop in the right bank of the canyon eroded through its frontal part. Contrast between the blocky carapace and heavily comminuted interior can be seen in the inset.

TABLE 8.10 ^{10}Be ages of quartz samples collected at the Dead Lakes rockslide

Sample No	Age (yr)	Error (yr)
DLR-1	41284	19947
DLR-2	44848	2189
DLR-3	44837	2158
DLR-4	54055	2122
DLR-5	71850	3460

FIGURE 8.207 Headscarp of the Dead Lakes rockslide (DL RS) and the adjacent rockslide, which deposits (RS) form the convex body in the upper reaches of the unnamed gulley between 2910 and 2400 m a.s.l. 3D Google Earth view.

Similarly to the previous site, if the simultaneous age of both rockslides that originated on the opposite slopes of the same ridge will be proved by any dating method, it would be an argument in favor of their seismic triggering.

One of the largest rockslides in this part of the study region was found at 41.5° N, 72.71° E on the left side of the Kara-Kiol River valley that falls into the Eastern Karasu River. Several small lakes on its surface lacking fish are called Dead Lakes by locals. We named this outstanding feature – the Dead Lakes rockslide (Kyrgyz 060) (Strom and Korup, 2006). Failure occurred on the 1.7 km high and more than 7 km long slope composed of the Paleozoic schist and ultramafic igneous rocks. The gigantic horseshoe-shape headscarp 1.9 km wide and more than 2 km long has a steep back wall up to ca. 500 m high (Fig. 8.205). Rockslide body covers 14.65 km^2. Considering its mean thickness of about 150 m we get volume estimate of about 2.2 km^3 (previous more rough estimates gave 2.5 km^3 (Strom and Korup, 2006)). H/L ratio of the 7.5-km long-runout rock avalanche is 0.21.

Debris motion had to be quite rapid because its portion crossed a local interfluve and entered an adjacent gully (see Fig. 8.205). Finally, rockslide formed a 2.65 km wide dam and came to rest against the opposite slope without visible evidence of run-up, but with some upstream and downstream spreading. This natural blockage has been dissected by the river and the internal structure of its frontal part can be observed in the ca. 250 m long canyon (Fig. 8.206). While the deposits surface all over the rockslide body is covered by large angular boulders more than 1 m in size it is likely that they form just a rather thin carapace, while most of underlying debris is crushed at a large extent. Indirect evidence of the intensive comminution is the presence of numerous small lakes in the closed depressions on the rockslide surface – water could not be stored if most of rockslide would consist of coarse bouldery material with inevitably high permeability. Same phenomena was observed at the Djamantau rock avalanche in the Central Tien Shan (see Section 8.4, Fig. 8.133)

In 2009 together with Prof. Yeong Bae Seong from the Korea University we took 5 samples for ^{10}Be dating from quartz veins in the large angular boulders of schist in the central part of the Dead Lakes rockslide body at 2180 m a.s.l. (see Fig. 8.205). Though all samples were collected from the nearby boulders they returned rather variable ages (Table 8.10, Yeong Bae Seong personal communication). If there was no processing error it can be hypothesized that quartz veins from which samples with older ages (54–71 ka) were collected could be on the daylight surface before rockslide occurrence, while younger dates provide real age of this slope failure – about 42–46 ka. Similar discrepancy appeared with dating of the Ukok rock avalanche in Central Tien Shan (Sanhueza-Pino et al., 2011). It was described in Section 8.4 (see Fig. 8.125-A, Table 8.6).

According to the geomorphic data one more rockslide body (Kyrgyz 060b) about 40 Mm^3 in volume rests at 41.463° N, 72.683° E in the upper reaches of the circus-like depression just west of the Dead Lakes rockslide headscarp (Fig. 8.207). This body looks similar to moraine deposits but no similar geomorphic features exist at the same altitude around this site, which excludes its glacial origin (see also Section 4.2). Evidence of one more smaller slope failure in the vicinity of the Dead Lakes rockslide (Kyrgyz 060a) were found east of it at 41.475° N, 72.718° E. Similar to the case studies described above, such clustering could indicate seismic triggering, but additional studies are needed to substantiate this assumption.

Group of the large-scale heterochronous bedrock landslides is located in the lower reaches of the Eastern Karasu River (see Fig. 8.199), 3–5 km south of the Toktogul dam. The most recent one – the Karakul rockslide (Kyrgyz 061) – dams the mouth of Eastern Karasu River valley at 41.633° N, 72.645° E and the Karakul town has been built on the deposits that had filled this dammed lake completely (see Fig. 6.22) (Strom 2010b). Rockslide had dammed the Naryn

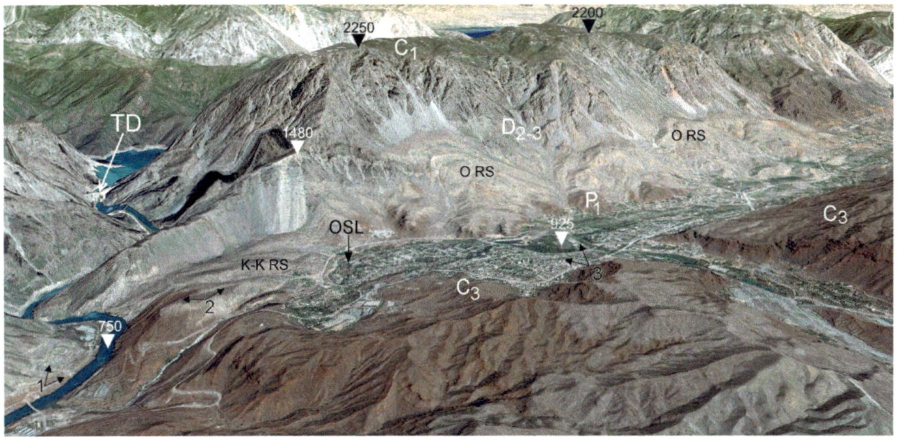

FIGURE 8.208 The heterochronous rockslides near the Karakul town. TD – the Toktogul dam; K-K RS – the Karakul rockslide; O RS – older (Pleistocene) rockslides, which headscarp crowns are at 2250–2200 m a.s.l.; D_{2-3}, C_1 – the Devonian and Lower Carboniferous limestone and dolomite; C_3, P_1 – Upper Carboniferous and Lower Permian red beds; Doubled black arrows marked by 1–3 – points from where photographs shown on Figs. 8.209–8.211 were made; OSL – point where lacustrine silt was sampled for OSL dating. 3D Google Earth view.

FIGURE 8.209 View of the Karakul rockslide from the right bank of the Naryn River valley (near the checkpoint on the road to the Toktogul dam). The lower part of the valley is submerged by the tailing part of the Kurpsay reservoir. Water seeping through steeply dipping bedrock is shown in the inset. D_{2-3}, C_1, C_3 – see caption of Fig. 8.208; RS – rockslide body represented by the carbonate rock debris where it blocks the Eastern Karasu River mouth and by debris of the Permian (?) red beds on top of the former left bank of this river.

FIGURE 8.210 Front view of the Karakul rockslide. This picture was made from the hill in the frontal part of the rockslide body marked by 2 on Fig. 8.208 and composed of the carbonate debris that filled the erosional gulley on the former slope of the Naryn River valley. Red beds at the central part of the image are, most likely, the Permian bedrock.

River as well, but this powerful stream had eroded the blockage and only small remnant of this body could be seen at the base of the right bank terrace.

Right slope of the Eastern Karasu River valley up to ca. 1500 m high is composed (from top to bottom) by the Lower Carboniferous and Devonian carbonate rocks (limestone, dolomite) thrusted over the Lower Permian and Upper Carboniferous terrigenous and volcano-clastic red beds (Fig. 8.208) being divided by large fault zone. The Eastern Karasu River close to its mouth had passed just at the base of this slope, likely following the fault zone, and undercut the slope creating the prerequisites of its failure, though direct trigger is unknown. Rockslide body is formed by varicolored "layers" of debris that originated from both Permian red beds and Devonian carbonate rocks crushed at different extent that fills the Eastern Karasu River mouth (Figs. 8.209, 8.210). Most of rockslide debris, which total volume was estimated as ca. 200 Mm^3, originated from the Devonian carbonate rocks. It buried rugged topography of that time left bank of the Eastern Karasu River (Fig. 8.210), while debris of the Permian red beds forms relatively thin "interbeds" at the base of the rockslide body (see Figs. 6.22, 8.209).

We should point out that several springs fed by the underground water of the Eastern Karasu River valley appear not on the modern slope composed of rockslide debris but at its part composed of the steeply dipping bedrock layers that had stretched along its left bank before its blocking. It shows that the lower part of the rockslide body is less permeable than the bedding planes in the bedrock.

The blockage was so large that rather small Eastern Karasu River could not erode it unlike the powerful Naryn River that washed out the western part of the rockslide body. Despite rather high – 2.85 DBI value (Strom, 2010b) the Karakul rockslide dam has not been breached and the lake formed by this blockage was silted completely. River found its way through the bedrock of its former left bank gradually eroding the new bypass gorge with nearly 180 m high rapids.

In 60 s of the last Century this rockslide was investigated by the researchers from Moscow State University (Matveev, 1968) who proposed that this natural blockage was formed, similarly to the Karasu Lake rockslide, mentioned in the previous Section, by the recurrent slope failures that they dated conventionally as the Middle and Late Pleistocene (Matveev, 1968). We disagree both with such multi-event slope failures and with age estimates. The "pseudo-stratified" internal structure of this rockslide body is absolutely similar to that of the Inylchek, the Kokomeren, the Ornok rockslides described in Sections 8.2 and 8.4 and typical of many other rockslides worldwide (Heim, 1932; Johnson, 1978; Hewitt, 1998, 1999, 2001,

FIGURE 8.211 Western old rockslide (O RS) on the slope above the Karakul town.

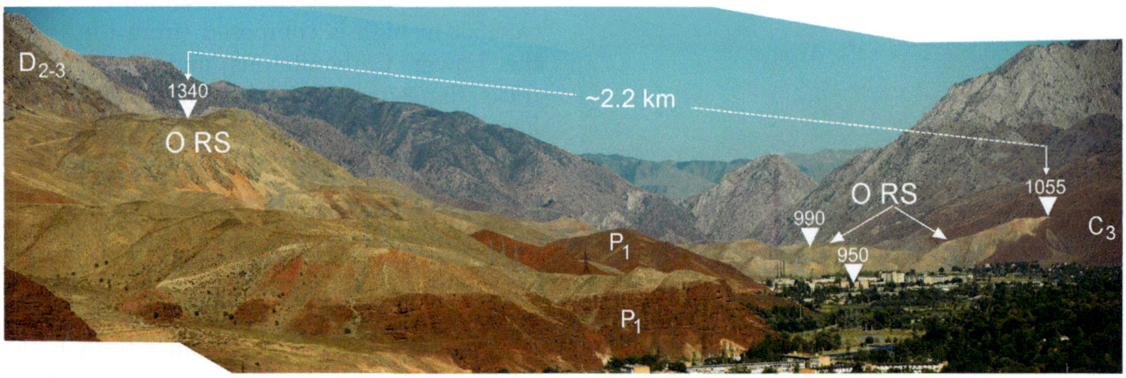

FIGURE 8.212 Eastern old rockslide (O RS) east of the Karakul town that had crossed the Eastern Karasu River valley and rose up to ca. 65 m on the opposite valley slope.

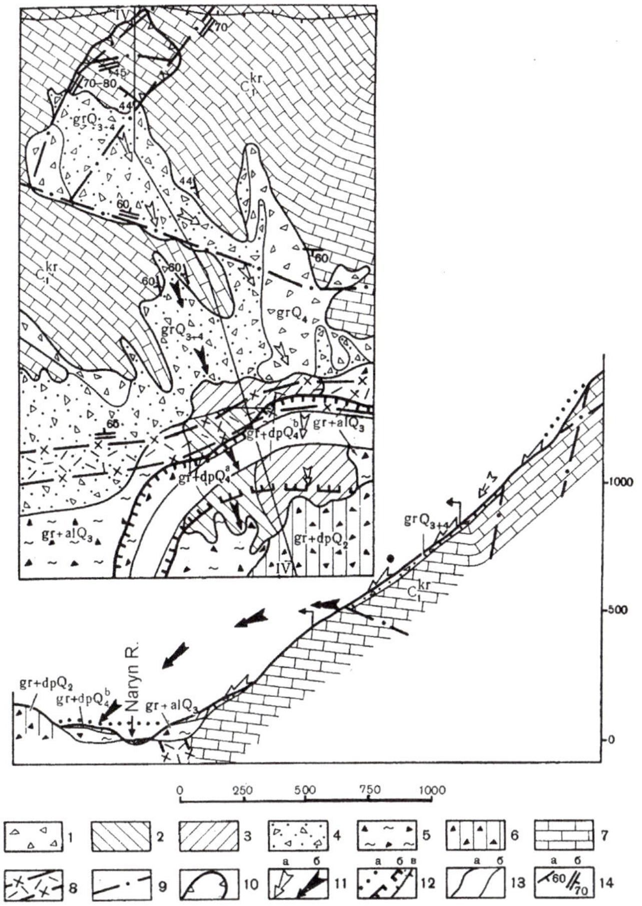

FIGURE 8.213 The geological scheme and cross-section of the 1946 Sarykamysh rockslide site (after *Matveev, Yu. D. 1968. Regulationg of the colossal rock falls and landslides in the Toktogul dam construction region. In: Problems of engineering geology and soil science. Issue 2. Moscow, 245–258 (in Russian)* with minor modifications). 1 – recent talus (gr Q_4); 2 – rock avalanche deposits that jumped over the river from the rocky bench (gr + dp Q_4^a); 3 – material that moved like a slump (gr + dp Q_4^b); 4 – not defined Holocene and Late Pleistocene talus (gr Q_{3+4}); 5 – Late Pleistocene fluvial and gravitational deposits (gr + al Q_3); 6 – Middle Pleistocene slump deposits (gr + dp Q_2); 7 – the Lower Carboniferous limestone of the Kerney Series (C_1^{kr}); 8 – Large fault zone; 9 – minor faults; 10 – headscarp of the 1946 bedrock landslide; 11 - directions of motion of slump (white arrows) and of the Jumping rock avalanche (black arrows); 12 – a: the daylight slope surface before its collapse and the original surface of the rockslide dam, b – terrace edges, c – the watershed; 13 – the geological and lithological boundaries; 14 – dip and strike of the bedding (a) and faults (b).

2006a, b; Abdrakhmatov and Strom, 2006) and can be explained within the simpler one-stage failure model, which will be discussed in Chapters 11 and 12.

Age of this natural barrier could be determined by dating of the lacustrine sediments that filled the landslide dammed lake almost completely. OSL dating of two silt samples collected from the upper, most recent part of these sediments (see Fig. 8.209) returned

FIGURE 8.214 Large-scale wedge failure about 5 km from the Talas-Ferghana fault damming the small valley and assumed smaller rockslides nearby. 3D Google Earth view.

Holocene ages from 4.1 to 2.9 (±0.1) Ka BP (Yeong Bae Seong personal communication). Considering about 1 m difference of the sampling points altitude, we can estimate the deposition rate of ca. 1 mm/year. We think, however, that simple multiplication of this rate by approximate thickness of lake sediments that could be about 150–160 m close to the dam will provide significant overestimation of the real age of this river-damming event. Reconstructing history of its evolution we must take into account that both rivers that brought sediments to this natural lake have been blocked by other large rockslides – the Karasu Lake rockslide with an intact dam (see Fig. 5.64) and the Dead Lakes rockslide that was breached, at least partially (see Figs. 8.205, 8.206). If the Karakul rockslide was formed before the Karasu Lake formation, its siltation rate should be much higher than if it occurred later than the Eastern Karasu River was blocked upstream. We assume that the Karakul rockslide was formed either in the beginning of the Holocene or at the end of Late Pleistocene. General influence of the cascade blocking of river valleys on the evolution of natural dams was described in (Strom, 2010b) and will be discussed hereafter in Chapter 14.

Besides this relatively recent rock slope failure, at least two much older large rockslides could be identified in the lower reaches of the Eastern Karasu River valley at 41.648° N, 72.673° E (Fig. 8.211, Kyrgyz 062) and at 41.648° N, 72.716° E (Fig. 8.212, Kyrgyz 063). Their position is shown also on Fig. 8.208. Deposits of these rockslides which volumes range from 100 Mm^3 to ca. 700 Mm^3 are composed of the carbonate debris sharply contrasting with the Carboniferous and Permian red beds forming the valley bottom. The second one had crossed the valley and its body demonstrates up to 50–65 m runup on the opposite valley slope.

On November 2, 1946 the rockslide triggered by the M7.5 Chatkal earthquake occurred in the widened part of the Naryn River canyon (the so called Sarykamysh intermontane depression) at 41.745° N, 72.701° E (Kyrgyz 035). The formation mechanism and main parameters of this Sarykamysh rockslide were described in (Matveev, 1968). According to his description about 1.2–1.8 million cubic meters of the Carboniferous limestone caved from the more than 1 km high right slope of the Naryn River valley at 41.745° N, 72.701° E (Fig. 8.213), fell on talus accumulations at the upper part of the slope and rushed downslope entraining these loose material saturated by the Autumn rains. Part of it jumped from the rocky bench at the central part of the slope, while other part affected talus accumulations at the foot of this slope. Finally it formed the dam up to 50–60 m high with estimated volume of about 4–5 Mm^3 that had blocked the river and caused valley inundation (see also Chapter 2). To prevent most adverse effects the dam was blasted and, later on, eroded by the river. Now remnants of this rockslide body are submerged by the Toktogul Reservoir.

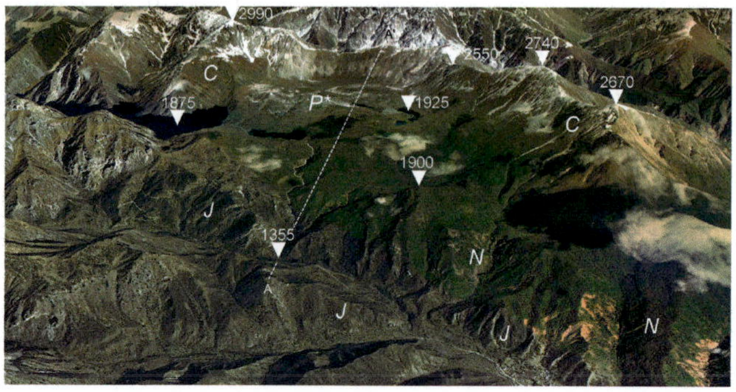

FIGURE 8.215 3D Google Earth view of the Sarychelek rockslide. Distance between summits at the headscarp crown with elevation marks 2990 m a.s.l. and 2550 m a.s.l. is about 5 km. N – the Neogene red beds, J – Jurassic coal-bearing rocks, P* – block of Permian carbonate and terrigenous rocks within the rockslide body, C – Carboniferous carbonate, volcanic and silica rocks, dashed line A-A' – position of the profile shown on Fig. 8.216. The diagonal boundary between more green upper right part and more gray lower left part is due to different space images available on Google Earth.

Further north-west along the Talas-Ferghana fault that passes here along the Karakuldja River valley very large rockslide about 350 Mm^3 in volume (Kyrgyz 064) dams its small right tributary at 41.893° N, 72.313° E. Rockslide originated on the 1200 m high slope composed of the Devonian and Carboniferous carbonate and volcanic rocks. Its wedge-shape headscarp is about 300 m deep (Fig. 8.214).

This rockslide forms the 250–300 m high dam with steep upstream slope and more gentle elongated downstream side that can be interpreted as a deflected Secondary rock avalanche (see Section 5.3.3.1) that moved about 1.5 km downstream. Crest of this dam with not very well defined downstream-facing secondary scar is almost even with insignificant (relative to the size of this feature), 30–40 m high only, distal runup.

Besides this giant slope failure evidence of two much smaller rockslides with their headscarp crowns at 2850 and 2140 m a.s.l. could be found nearby (see Fig. 8.214). All three rockslides occurred on slopes with similar, SE aspect. It could be speculated that such similarity was caused by strong motion, similar to the Karakungey rockslides described above (see Fig. 5.52). Two rockslides identified at the mouth of the Chong Kuraaryk River that falls into Karakuldja River about 7 km south-east at 41.905° N, 72.431° E (Kyrgyz 075 and 088) are much smaller – about 8–12 Mme only each. All these nearby rock slope failures form rather compact rockslide cluster located close to the Talas-Ferghana fault zone and, likely, associated with recent activity of this major neotectonic structure.

The largest and one of the most interesting features in this part of the North-Western Tien Shan is the tremendous Sarychelek rockslide (41.85° N, 71.98° E, Kyrgyz 066) that dams the same-name lake up to 234 m deep fed by the Khodja-Ata River. Detailed geological description of this blockage was presented in (Fedorenko, 1968) who, however, denied its landslide origin and proposed, following Gerasimov and Rantsman, (1964), formation of the lake due to

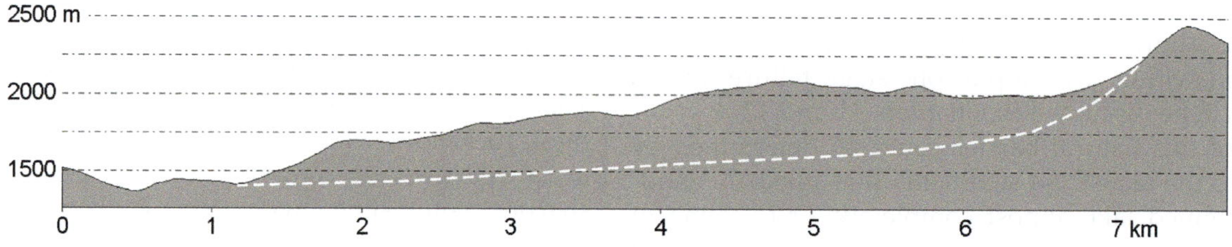

FIGURE 8.216 Profile across the Sarycheleck rockslide and anticipated position of its sliding surface.

intensive recent tectonic block movements. We totally disagree with such interpretation. First, the geomorphic analysis of the area shows that steep 600–800 m high horseshoe-shape wall bounding from the east the hummocky plateau with small lakes on it, represents a typical, though very large headscarp, while the plateau itself – giant rockslide body with an area of ca. 18 km^2 and assumed volume of about 4 km^3. Second, the formation of the natural dam was, in any case, a catastrophic phenomenon. Considering the Sarychelek Lake depth (more than 200 m) and lack of any evidence of water flow over the dam, it means that single-event vertical offsets along local faults that do not extend out of the Fedorenko's "blocky field" (rockslide body) should reach dozens if not hundreds meters and that these movements were, basically, upward directed rather than subsidence. However, it is well known from the worldwide statistics of seismic surface faulting that single-event vertical tectonic offsets do not exceed ca. 10–12 m and, moreover, are well correlated with rupture length so that such large displacements occur along surface ruptures tens and hundreds kilometers long only (Wells and Coppersmith 1994; Strom and Nikonov, 1997, 2000; Strom 1998; Chipizubov, 1998). Thus, such localized neotectonic deformations seem to be impossible.

Working on this book we reanalyzed and correct the Sarychelek rockslide parameters provided in (Strom and Korup, 2006). While area affected by this rockslide (about 40 km^2) was overestimated significantly (new estimates are 18 km^2 for the deposits area and 24.3 km^2 for the total area affected), its volume appears to be much larger than it was assumed (4 km^3 instead of 2.0–2.5 km^3 in this paper). Previously we assumed that the headscarp visible on Fig. 8.215 extends from elevation mark 2990 m a.s.l. up to mark 2670 m a.s.l. (about 8 km long). It was also assumed that distal part of the deposits had reached the Kichikul River valley at the ca. 4 km wide front. However, more careful examination of the new space images with better resolution shows that the southern end of the headscarp was, likely, close to the elevation mark 2550 m a.s.l. (see Fig. 8.215) and that rockslide body had blocked the mouth of the Khodja-Ata River at 1355 m a.s.l. not extending further west.

In contrast, previous estimates of the deposits thickness underestimated this important parameter and, thus, rockslide volume. Of course these estimates are not grounded either by boreholes or by geophysics and are much more speculative than area assessment. However, based on geomorphic profiles an example of which is shown on Fig. 8.216 we can say that maximal thickness of rockslide body exceeds 400 m, while inclination of the sliding surface does not exceed 4–5°, except its part close to the headscarp back wall. It allows estimating rockslide volume to be about 4 km^3, which makes the Sarychelek rockslide one of the largest in the entire Central Asia region. Of course additional studies of this outstanding feature are necessary

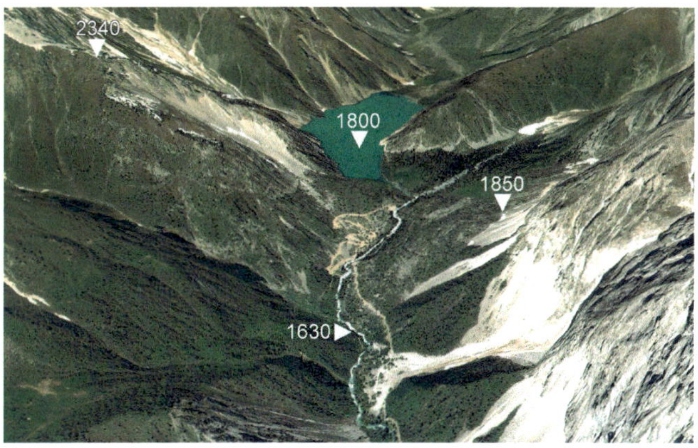

FIGURE 8.217 Rockslide dam in the upper reaches of the Western Karasu River valley. 3D Google Earth view.

FIGURE 8.218 The Karatoko Lake natural dam. Above – view towards west; below – view towards south. 3D Google Earth views.

FIGURE 8.219 The deflected Secondary rock avalanche with nearly 90 m initial runup 3.5 km west of the Karatoko Lake. 3D Google Earth view.

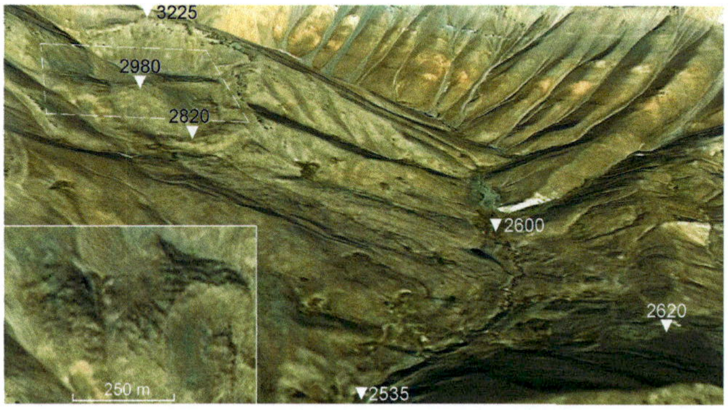

FIGURE 8.220 The Ak-Tash rock avalanche in the upper reaches of the Chatkal River basin. 3D Google Earth view. The plan view of the outlined proximal compact part with hummocky relief is shown in the inset.

FIGURE 8.221 The Chukurchak rock avalanche in the central part of the Chatkal intermantane depression. Above – the headscarp and proximal part of the rock avalanche deposits with tongues of bouldery debris that splashed over the true right side of the headscarp at 2355 m a.s.l. and 2275 m a.s.l. Below – general view with right-side and left side branches of rock avalanche shown in the insets. 3D Google Earth views.

to understand its origin, motion mechanism, possible trigger and age.

The Sarychelek rockslide is located within the epicentral zone of the 1946 M7.5 Chatkal earthquake that produced numerous surface disturbances (Fedorenko, 1968; Leonov, 1970; Kuchai, 1971). Unfortunately they have not been mapped in detail (or these data remain unpublished). Of course it is quite likely that the Sarychelek rockslide was triggered by a large earthquake as well (Kuchai, 1971). However, we must point out that the enormous size of this rockslide several orders of magnitude larger than any slope failure associated with the 1946 earthquake, does not mean that the prehistoric earthquake, which definitely could trigger it, was proportionally larger than the known historical earthquake. It will be discussed further in Chapter 13.

Just west of the ridge between the Khodja-Ata and the Western Karasu River valleys from where the gigantic Sarychelek rockslide had caved, one more much smaller failure of about 15 Mm^3 of the Carboniferous carbonate, volcanic and silica rocks from the ca. 500 m high rocky spur formed the ~200 m high dam in the Western Karasu River valley at 41.896° N, 72.035° E (Kyrgyz 065, Fig. 8.217). It can be classified as a Primary rock avalanche with frontal confinement (see Section 5.3.1, Fig. 5.31–1b) and significant, about 50–60 m high runup. Water flows over the dam but does not form any significant erosional canyon yet, which could indicate that this feature is relatively young, likely younger than its giant neighbor.

Further north-west, in the upper reaches of the Chatkal River basin there are several natural blockages in the Karatoko River valley. The largest one forms the ~1 km^2 Karatoko Lake. Its twofold dam at 42.03° N, 71.813° E (Kyrgyz 068) consists of the 1.6 km long bar stretching along the river and of the conic debris accumulation at the mouth of the steep gulley merging the Karatoko River from the left side. The latter comes from the very deep glacial circus (Fig. 8.218), which supports its glacial origin. At the same time several distinct wedge-like lowerings on the 900 m high left slope of the valley composed of the Carboniferous limestone (west of the point with elevation mark 3680 m a.s.l.) could be the source zone of large-scale slope failures. This controversy – rockslide vs. moraine – cannot be defined unambiguously in this particular case based on the remote sensing data only (Fig. 8.218) and requires field investigations to be solved.

3.5 – 4.0 km downstream from this dam, at 42.03° N, 71.77° E, the Karatoko River has been blocked once more, this time by the univocally identified rockslide (rock avalanche) dam (Kyrgyz 069, Fig. 8.219). This rock slope failure looking very fresh, and, thus, quite recent, originated on the ca. 400 m high slope as a translation slide along the steeply dipping bedding plane in Carboniferous limestone. The initial failure of about 5 Mm^3 of rocks formed the dam with very impressive ~90 m high runup at its front where rockslide hit the opposite slope directly. Left part of the sliding block just fell from the steep cliff as a rockfall. After collision with an opposite slope the secondary scar formed (in the shadow below 2805 m a.s.l. on Fig. 8.219) from which about 40% of the collapsed rock mass moved 1.1 km downstream, entered the Karatoko River valley and blocked it by up to 30 m high dam at a distance of about 400 m. Evidence of the intermediate secondary scar can be seen at 2670 m a.s.l.

One more impressive Secondary rock avalanche (Kyrgyz 070) is located in the Ak-Tash River valley – the small right tributary of the Chatkal River, at 42.14° N, 71.73° E (Fig. 8.220). Unlike case studies described above it originated on the slope composed of the Paleozoic granite. The ca. 200 m high and 800 m long back wall of its headscarp rises above the about 600 × 150 m hummocky bench (see inset in Fig. 8.220) that corresponds to the top of the proximal subsided block, which frontal slope is the up to ~160 m high secondary scar. Main part of this Ak-Tash rock avalanche is formed by the 1.8 km long and up to ~700 m wide body with furrows and levees on its surface indicating the flow-like style of motion. We must notice that on the 1:200 000 State Geological map published in 1964 this body was marked as glacial deposits. Rock avalanche had dammed the Ak-Tash River, likely for a short time, and rose up to ca. 2620 m a.s.l. on the gentle opposite slope of the valley that corresponds to the ~20 m runup. This case study demonstrates quite well the abrupt change of motion style during the Secondary rock avalanche formation – the slump-like subsidence of the proximal compact part and the flow-like motion of its mobile portion.

One of the most interesting case studies in the Chatkal River basin is the Chukurchak rock avalanche (Kyrgyz 074) described in (Guschenko and Leonov, 1972; Chedia and Lemzin, 1980). It involves about 450 Mm^3 of debris and had blocked the Chatkal River valley at 41.805° N, 71.15° E (Fig. 8.221). Its 2.6 km wide headscarp lies within the active Southern Chandalash fault zone that bounds the Chatkal

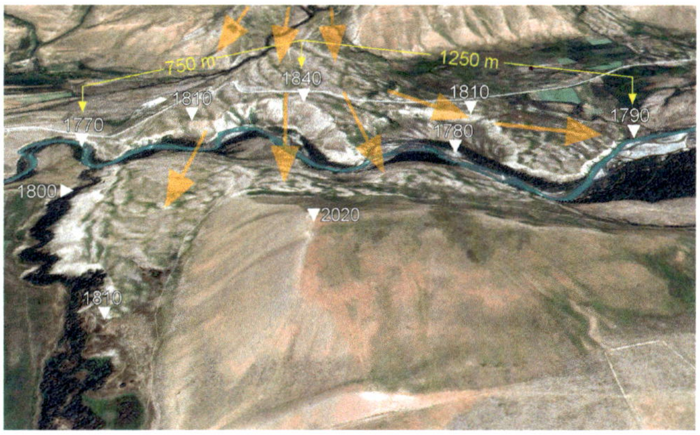

FIGURE 8.222 Zoomed view of the distal part of the Chukurchak rock avalanche in the Chatkal River valley. Notice asymmetric along-river profile of the deposits with steeper and higher downstream slope and more gentle and elongated upstream slope. Brown transparent arrows show direction of debris motion. 3D Google Earth view.

intermontane depression from the north-west (Chedia and Lemzin, 1980; Lemzin, 1988). The main part of this rock avalanche body is composed of debris originated from the Pliocene conglomerate and breccia forming the footwall of this large fault zone. Angular boulders of Carboniferous limestone from its hanging wall prevail at the proximal part of rock avalanche deposits. Chedia and Lemzin (1980) proposed that the proximal part of rock avalanche deposits composed of the Carboniferous limestone debris only originated later, as the 2nd phase of this slope failure. We assume, however, that such composition of the deposits reflects just motion mechanism of the one-stage catastrophic slope failure. It is difficult to explain how rocks of the hanging wall that would be inevitably unstable after collapse of the tremendous amount of the footwall rocks could remain on the slope. Landscape visible on the space image (see Fig. 8.221) does not show such a contrast as one can see at the really recurrent slope failures as the Djamantau rock avalanches (see Figs. 8.133, 8.134) or the Djuzumdybulak rockslides (see Fig. 8.190).

When rock avalanche reached small local ridges marked on Fig. 8.221 at 2040 m a.s.l. and 2135 m a.s.l. it divided into three parts. The true right part shown in the left inset in Fig. 8.221 is rather short and wide with distinct longitudinal stripes of coarse angular boulders on its surface. Left branch reached deep valley aside the main body, partially "flowed" into it and came out on the low terrace of the Chatkal River up to elevation mark 1800 m a.s.l. near the road (see right inset in Fig. 8.221).

The main central branch is the largest one with 7.5 km long runout and H/L ratio of ~0.16. It reached the Chatkal River valley, crossed it and hit its opposite bank resulting in spreading of rock avalanche debris for 1200 m upstream, while left part of this body moved straight ahead, entered the tributary valley of the Kara-Terek River and followed it for about 700 m up to 1810 m a.s.l. (see Fig. 8.221). It resulted in the significant asymmetry of the 70–80 m thick deposits blocking the Chartkal River valley that have rather straight and steep downstream slope with 70 m elevation difference at a distance of about 750 m and more gentle and elongated upstream one with only 50 m difference at a distance of 1250 m (Fig. 8.222). Similar asymmetry with steeper downstream side was described at the Djashilkul rockslide dam (see Fig. 8.81). Process resulting in such asymmetry was proposed in (Strom, 2010a) and will be discussed later in Chapter 12.

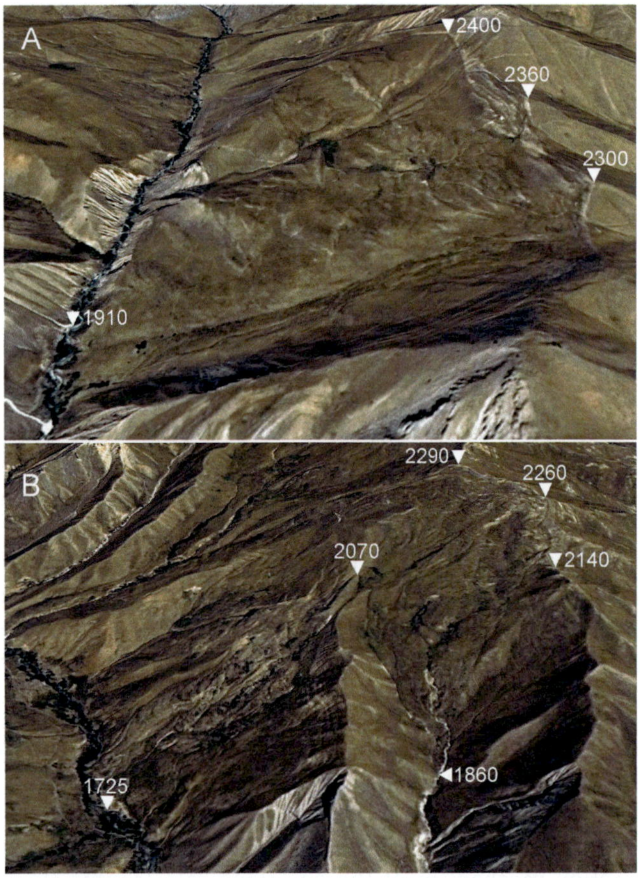

FIGURE 8.223 Landslides in the Neogene deposits west of the Sandalsh River valley on ca. 400 m high slopes, which headscarps embrace local watersheds. A, Kyrgyz 138; B, Kyrgyz 359. 3D Google Earth views.

Another interesting peculiarity of this rock avalanche is its rather low-lying source area, especially considering 7.5 km long runout. Slope failure affected the lower ~450 m of the 1400 m high slope only. It can be assumed, however, that the horizontal "depth" of its headscarp could be very large, up to several hundreds meters.

According to the geomorphic data this rock avalanche could occur in Holocene, after formation of the 20–12 m high terraces of the Chatkal River and before the formation of the 6-m high terrace incised in the rock avalanche deposits (Chedia and Lemzin, 1980).

Further to the west, several slow moving landslides affect slopes up to several hundreds meters high composed of Neogene red beds in the Chandalash (Sandalash) River basin – right tributary of the Chatkal River. They can be exemplified by two large landslides at 41.814° N, 70.81° E (Kyrgyz 138, Fig. 8.223-A) and at 41.758° N, 70.764° E (Kyrgyz 359, Fig. 8.223-B). Landslides affect both lower and the uppermost parts of slopes, often embracing local watersheds. As we mentioned above, it may indicate, though not always, that these slope failures could be

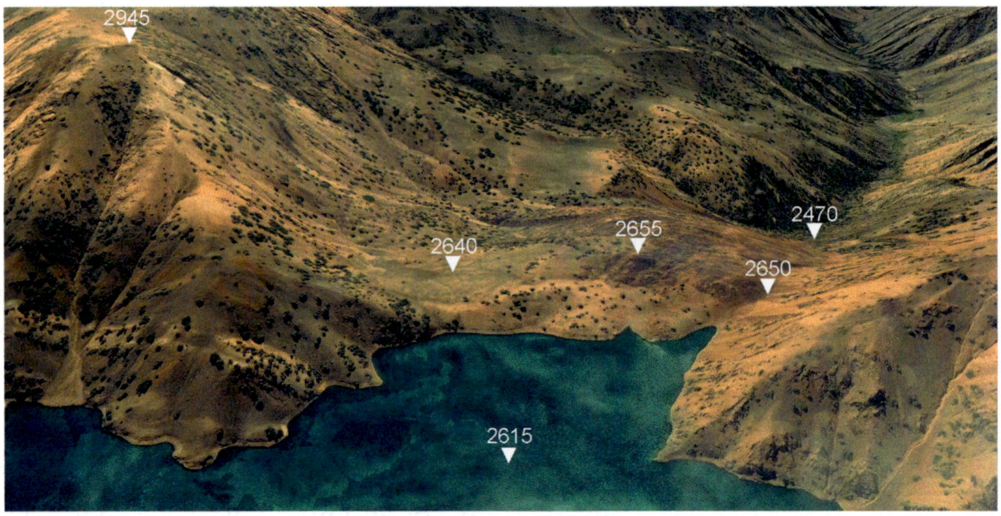

FIGURE 8.224 The Kiok-Ala rockslide dam. 3D Google Earth view.

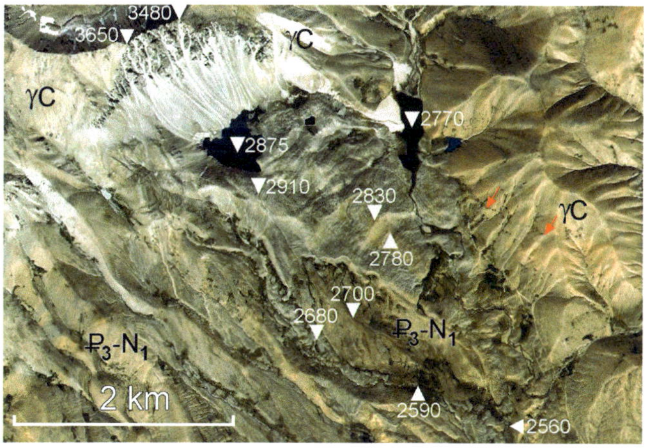

FIGURE 8.225 The Arashansai rock avalanche. γC – the Carboniferous granites, ₽$_3$-N$_1$ – the Oligocene-Miocene terrigenous rocks; red arrows mark assumed surface rupture. Notice varicolored belts of debris near elevation marks 2830 and 2780 m a.s.l. ArcGIS Earth image.

triggered by an earthquake, which intensity increases upslope (Havenith, 2002; Havenith et al., 2003).

Moving further south-west, on the right bank of the Ters River valley – the left tributary of the Chatkal River, at 41.5° N, 72.53° E one can find another unique feature – the Ters landslide (Kyrgyz 313) affecting about 25 km^2 of the gentle slope composed of granite with thin apron of Quaternary deposits. This feature was mentioned above in Section 4.2 and described briefly in Section 5.1 (see Figs. 4.2, 5.16). As mentioned in Section 5.1, slope process that caused its formation could be classified as a specific case of lateral spread (Hungr et al., 2014).

Several interesting case studies could be found on the south-eastern slopes of the Chatkal Range facing towards the Feghana Valley. Some of them were described by A. Hodjaev (1974, 1985) who treated these bedrock landslides as evidence of large prehistoric earthquakes. One of them – the Kiok-Ala rockslide (Kyrgyz 190) is located in the upper reaches of the Gava-Sai River valley at 41.373° N, 70.885° E (Fig. 8.224). Here collapse of about 25 Mm^3 of the Permian volcanic and igneous rocks formed the ca. 160 m high rockslide dam of a beautiful Kiok-Ala Lake. Judging from the deposits morphology this rockslide can be classified as a typical Primary rock avalanche with frontal confinement (see Fig. 5.31–1b). The dam has never been overtopped and two lithological units can be distinguished in its body – the frontal-downstream and the proximal-upstream differing significantly on space image. Likely this differentiation of rockslide body corresponds to the different lithology of the source zone.

One more interesting rockslide (Uzbek 031) is located in the uppermost reaches of the Akhangaran River valley at 41.356 N, 70.51 E (Fig. 8.225) just near the Chatkal Range watershed. This 5.3 km long Arashansai rock avalanche about 130 Mm^3 in volume caved from the 750–850 m high slope composed of the Carboniferous granites thrusted over the Oligocene-Miocene weak terrigenous rocks. Its body dams small creeks forming several lakes largest of which fills the ca. 40 m deep proximal depression resembling the morphological pattern of the Lower Ak-Kiol rockslide (see Fig. 5.6) and of the Bielogorka-left rock avalanche (see Fig. 8.98-B) with similar proximal lowerings.

FIGURE 8.226 Plan view of the Ikhnachkul rockslide cluster. Numbers correspond to the features' IDs in the database (see Appendix) without country (Uzbekistan) name. Google Earth image.

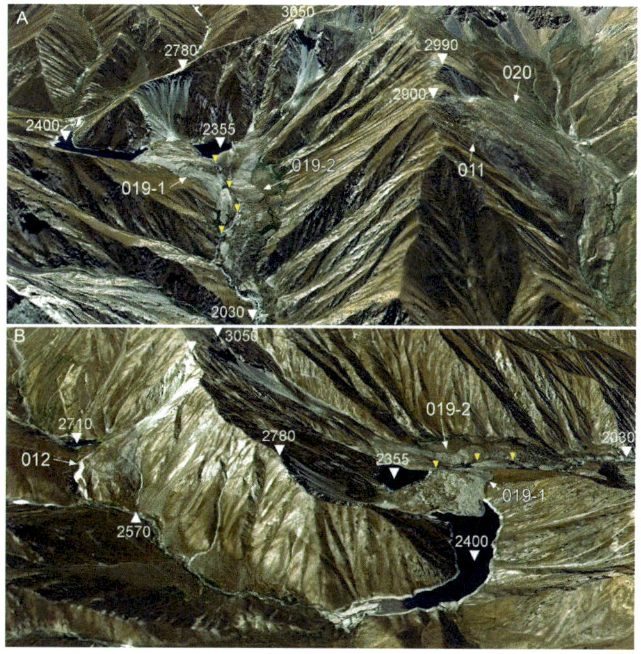

FIGURE 8.227 3D Google Earth views on the Ikhnachkul rockslide cluster. A – from the north; B – from the east. Numbers are the same as on the plan view. *Yellow triangles mark* the boundary between two synchronous Ikhnachkul rockslides.

FIGURE 8.228 The Badak Lake rockslide dam. Closer view on the dam is shown in the inset. 3D Google Earth views.

FIGURE 8.229 Natural blockage in the Urungach River valley. 3D Google Earth view.

Main part of rock avalanche body is crossed by transverse steps up to ~50 m high that can be treated as secondary scars (see Section 5.3.3), though not as distinct as in many other cases classified as Secondary rock avalanches. Besides, such a distinct proximal depression is also typical of the Primary rather than of the Secondary type of rock avalanche (see Section 5.3.1).

In the central part of rock avalanche deposits several belts of varicolored debris with sharp boundaries could be delineated along its path indicating absence of mixing of different lithologies. Similar phenomenon was described in many cases above. It also looks that the true right (western) branch of this rock avalanche fills the small valley separated from the main body by the low ridge (see Fig. 8.225).

Close association of this large-scale bedrock slope failure with neotectonic fault allows assumption of its seismic triggering, though evidence of recent surface ruptures along this fault zone are not so impressive as those shown for example, on Fig. 8.78. We should also mention that on the 1:200 000 State geological map published in 1974 deposits at the southern feet of this section of the Chatkal Range were classified erroneously as fluvial.

Another zone rich of bedrock landslides is located in Uzbekistan in the Pskem River basin (see Fig. 8.199) composed of the Devonian and Carboniferous carbonate rocks mainly. Dense cluster of at least 6 rockslides two of which formed the landslide-dammed Ikhnachkul Lakes was found at 41.89 °N, 70.55 °E in the Ikhnachsay River valley – left tributary of the Pskem River within a small area 6 × 3 km only (Fig. 8.226).

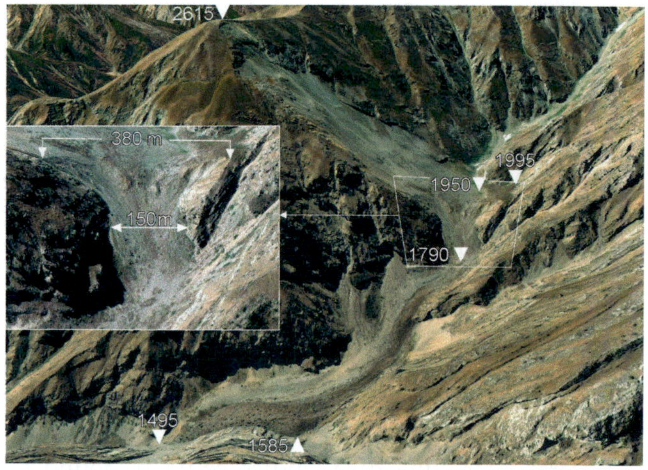

FIGURE 8.230 The 3 km long Kharamdazasai rock avalanche. The transition zone from the compact to the mobile part of the body is shown in the inset. 3D Google Earth view.

Their volumes range within one order of magnitude from 2–3 Mm^3 to 20–25 Mm^3, so that, according to size distribution, this rockslide cluster can be classified as the "uniform" one – same as the Northern-Dzungarian (see Fig. 7.2), the Sarmin-Ula Range (see Fig. 8.11, Table 8.1) and the Southern Chilik (see Fig. 8.61, Table 8.3) rockslide clusters (Strom, 2015c).

If it would be found that all these rockslides occurred simultaneously it would be a sound argument in favor of their triggering by a strong earthquake. Well grounded conclusion on real-time simultaneity of two of these rockslides can be done based on the confluence of rock avalanche bodies damming the Ikhnachkul Lakes (Fig. 8.227) without overlapping of one by another. Similar relationships of two rock slope failures sourced at different slopes were

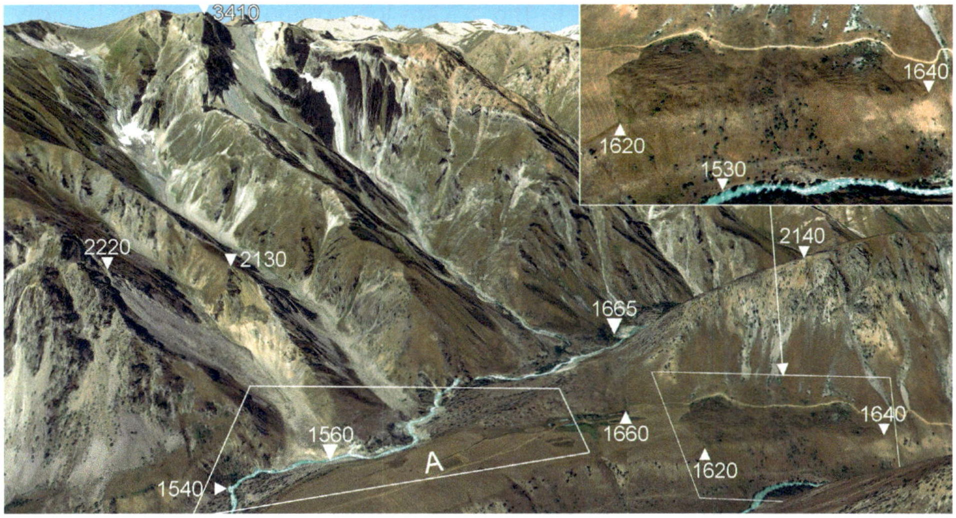

FIGURE 8.231 Rockslides the Maidantal and the Oigaing River valleys junction. 3D Google Earth view. More contrast view of the Uzbek 035 rockslide body is shown in the inset; the outlined area marked by A is shown on Fig. 8.233.

found at the Bielogorka rock avalanches in Northern Kyrgyzstan (see Fig. 8.99).

About 12 km south-west of the center of the Ikhnachkul rockslide cluster at 41.861° N, 70.419° E one more rockslide in carbonate rocks forms the Badak Lake dam (Uzbek 014, Fig. 8.228). About 13 Mm^3 of limestone or dolomite collapsed from the 900 m high slope with the unusually steep upper part (up to 60°) and formed the compact dam about 80 m high (at its axial section). The lake is characterized by the significant, ~10 m seasonal variations of its level. Judging from the appearance of the filtering water about 200 m downstream from the dams' base (marked by blue arrow on Fig. 8.228) it can be assumed that seepage takes place through the talus that had been accumulated at the valley bottom prior to this rockslide formation, similar to the Kutmakul rockslide dam (see Fig. 8.203).

One more small lake dammed by rock slope failure (Uzbek 009) can be seen at the opposite side of the Pskem River valley, in the Urungach River valley, 5.5 km upstream from its mouth at 41.923° N, 70.314° E. Failure of about 3 Mm^3 of the Carboniferous limestone occurred on much lower slope, ~300 m high only, and formed the conical body gradually lowering towards the opposite slope of the valley and damming the river by up to 100 m high blockage (Fig. 8.229). This feature resembles the famous 1991 Randa rockfall in Swiss Alps (Gotz and Zimmermann, 1993; Sartori et al., 2003). The main difference is that at Randa the affected valley was much wider, so that failure of about 40 Mm^3 of rocks did not block it completely, while in our case the narrow U-shape valley has been dammed. The dammed

FIGURE 8.232 Deposits of the Sarydjayak rock avalanche on the left bank of the Maidantal River. Transparent orange arrow indicates direction of debris motion. 3D Google Earth view.

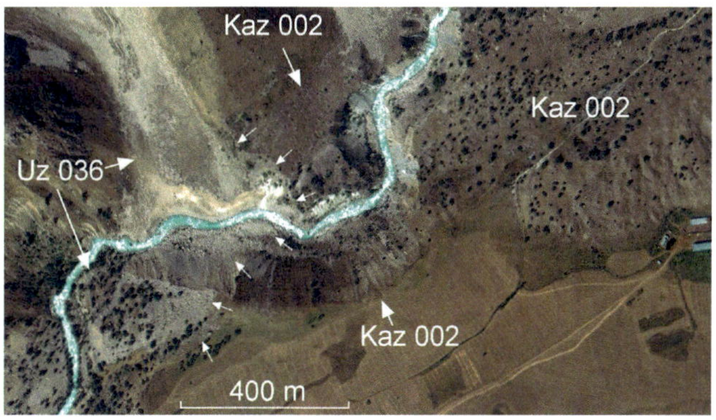

FIGURE 8.233 Relationships between the deposits of the Sarydjayak rock avalanche (Kaz 002) and of the Uzbek 036 (UZ 036) rockslide. Boundary between them is marked by white arrows. Google Earth image.

lake has been silted at a large extent. Similarly to the Badak rockslide dam, no overflow occurred and the seeping water appears much lower than the downstream tip of the blockage indicating that the underground flow passes most likely through the pre-slide talus accumulation, similar to several case studies described above.

One more attractive slope failure is in the Kharamdazasai Creek valley – one of the right tributaries of the Pskem River at 41.841 °N, 70.207 °E (Uzbek 008). Though lithology of this site is similar to that at the Badak rockslide area, this case study is characterized by totally different style of debris motion. Failure of about 12 Mm^3 of rocks from the about 650 m high slope with steeper (up to 38°) upper part and more gentle (~ 21°) lower part resulted in 3 km long runout (Fig. 8.230). Rapidly descending rock mass that slid, at least partially, along steep bedding planes, collided with opposite slope at a right angle and nearly 5 Mm^3 of debris deflected and traveled through the narrow, steep valley as a 1.6 km long dry granular flow. This Secondary rock avalanche demonstrates features typical of both Classical and Bottleneck subtypes (see Sections 5.3.3.1 and 5.3.3.2) – rather distinct 380 m wide secondary scar at the downstream face of the blockage and sharp narrowing up to 150 m just below it (see inset in Fig. 8.230). Freshness of both the headscarp and of the deposits indicates recent age of this feature.

FIGURE 8.234 About 5 km long landslide in Neogene deposits damming the Achi-Kul Lake. 3D Google Earth view.

Several rockslides originated on high slopes composed of carbonate rocks mainly near the junction of the Maidantal and the Oigaing Rivers. This site is located practically at the Kazakh-Uzbek border. The largest of them – the Sarydjayak rock avalanche about 30 Mm^3 in volume (Kazakh 002) caved from the topmost part of the 1.6 km high ridge at 3410 m a.s.l. (Fig. 8.231), and after passing about 4 km down the steep gulley crossed the Maidantal River and formed the hummocky apron on its left-bank terrace at 42.022° N, 70.637° E that covers about 1 km^2 and rose for several tens of meters on the opposite slope (Fig. 8.232). Rock avalanche debris with numerous large angular boulders has very sharp front indicating abrupt halt of its motion. It spread up- and down-valley for about 2 km (from 1665 m a.s.l. to 1560 m a.s.l. on Fig. 8.231) forming ca. 40–50 m high dam. Dammed lake was partially filled and then breached.

One more rockslide (Uzbek 035) that originated on the 500 m high right slope of the Oigaing River valley also formed the blanket-like body on the ~100 m high river terrace at 42.028° N, 70.649° E (see inset in Fig. 8.231). The third rock slope failure (Uzbek 036) with an impressive double-headed scarp (its crowns at

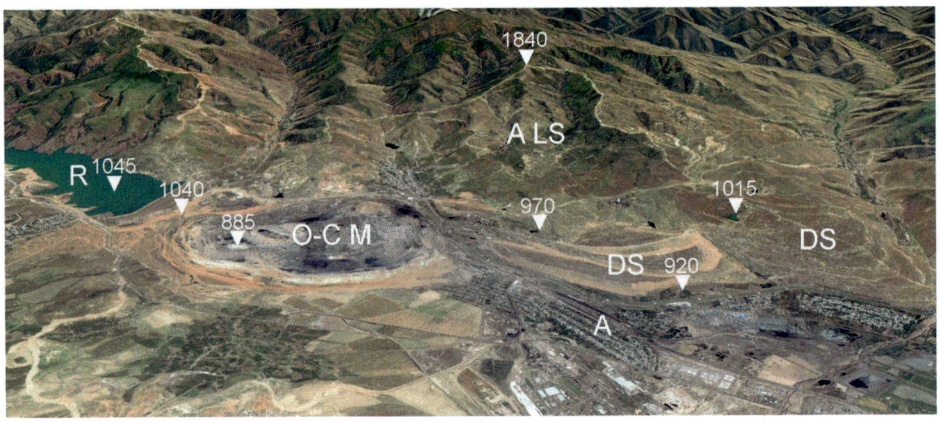

FIGURE 8.235 The Atcha landslide and damp deposits placed at the slope base to stabilize it. R – the Akhangaran Reservoir; O-C M – open-cast coal mine excavated at the drained Akhangaran River valley bottom (river was diverted through the tunnel); A – the Angren town; ALS – the Atcha landslide; DS – dump storage. 3D Google Earth view.

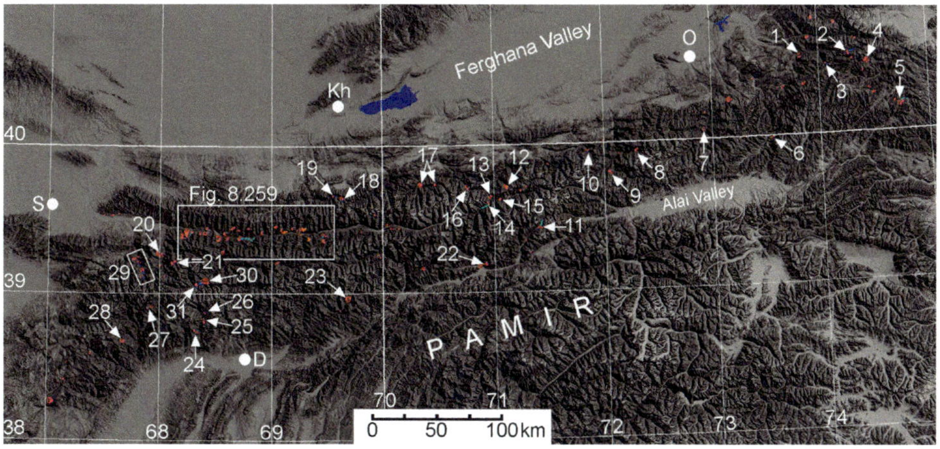

FIGURE 8.236 Shaded 3″ STRM DEM of the Southern Tien Shan. Rockslides are marked by pink and brown dots; blue dots – dammed lakes. D – Dushanbe; Kh – Khudjand; O – Osh; S – Samarkand. 1 – the Terek-Suu rockslide (see Figs. 4.7, 4.8); 2 – the Kulun rockslide (see Fig. 8.237); 3 – the Lower Kulun rockslide (see Figs. 6.23, 8.238); 4 – see Fig. 8.239; 5 – see Fig. 8.240; 6 – the Kolduk rockslide (see Fig. 8.241); 7 – see Fig. 8.242; 8 – the Yashinkul rockslide dam breached in 1966 (see Fig. 2.3); 9 – the Ming-Djilki rockslide (see Fig. 8.243); 10 – the Kurbankul rockslide (see Figs. 8.244, 8.245); 11 – the Kuturgansu-mouth rockslide (see Figs. 8.245, 8.246); 12 – the Kyzyl-Katta rockslide (see Figs. 8.248–8.250); 13 – the Korgon rock avalanche (see Figs. 8.251, 8.252); 14 – the Ak-Terek rockslide (see Fig. 8.151); 15 – see Fig. 8.254; 16 – the Kara-Kiol-Katta rockslide (see Fig. 8.255); 17 – rockslides in the Karavshin River basin (see Fig. 8.256); 18 – the Ai-Kul rock avalanche (see Fig. 8.257); 19 – see Fig. 8.258; 20 – the Khushtarita rock avalanches (see Fig. 8.268); 21 – the Big Allo rock avalanche and rock avalanche nearby (see Figs. 8.269, 8.270); 22 – the 1949 Khait rock avalanche and older rockslide nearby (see Figs. 5.15, 8.271–8.273); 23 – the Kafirnihon rockslide (see Fig. 8.274); 24 – see Fig. 8.275; 25 – the Timurdara rockslide (see Fig. 8.276); 26 – the Pairon rockslide (see Fig. 8.277); 27 – the Aigyrkul rock avalanche (see Figs. 8.278, 8.279); 28 – see Fig. 8.280; 29 – rockslide cluster in the Shing River valley (see Figs. 8.281–8.286); 30 – the Iskanderkul blockage (see Figs. 8.287, 8.289, 8.290); 31 – the Sarytag blockage (see Fig. 8.288); the large outlined area – the Zeravshan River basin rich of large-scale slope failures (see Fig. 8.259).

2200 and 2130 m a.s.l. are marked on Fig. 8.231) also blocked the Maidantal River temporarily at a distance of about 500 m. Its relationship with much larger Sarydjayak rock avalanche can be seen on Fig. 8.233. The Uzbek 036 deposits truncate distal downstream part of the Kazakh 002 body indicating that the latter should originate earlier. The same can be also derived from much less vegetated bouldery surface of the Uzbek 036 rockslide body.

Numerous landslides in poorly lithified Oligocene and Neogene sediments are known in the western foothills of the Chatkal Range east of the densely populated Tashkent City (Niyazov, 2009). One of the largest of them (Uzbek 006, 41.357° N, 69.725° E) that moved for about 5 km from 1300 m a.s.l. to 770 m a.s.l and dammed the Achi-Kul salty Lake can be seen on Fig. 8.234. Sliding occurred on the gentle slope following bedding dip direction.

Finishing the description of rockslides in the North-Western Tien Shan we have to mention one of the largest rock slope failure triggered by human activity – the Atcha landslide – the largest one, about 800 Mm^3 in volume (41.012° N, 70.184° E), among numerous slope instabilities near the Angren coal-mining field about 60 km south-east of the Tashkent City (Fig. 8.235). Unlike vast majority of the historical and prehistoric case studies described above that failed catastrophically (their description and analysis is the main goal of our book), this gigantic landslide that originated in 1972 being triggered by the in situ coal gasification, moves gradually. However, the enormous amount of rocks involved in sliding doesn't allow ignoring this case study, which detailed description and analysis of both triggering factors and hazard mitigation activities can be found in (Niyazov, 2009). Maximal sliding rate of the Atcha landslide reached 6 mm/day in late 70-s and cumulative displacement reached ca. 35 m in 2005. Special measures were undertaken to stabilize the slope, including, in particular, dump storage at the base of the unstable slope (see Fig. 8.235). Volume of this abutment exceeded 110 Mm^3 that helped to decrease the deformation rate significantly [ibid.].

8.6 SOUTHERN TIEN SHAN

Part of the mountain system ascribed to the Southern Tien Shan includes east-west trending Turkestan, Zeravshan, Gissar, Karategin, Alai and Kichi-Alai ranges located between the Ferghana intermontane depression in the north and the Tajik

FIGURE 8.237 The Kulun rockslide dam, the Kulun Lake and the dammed lake downstream of it. 3D Google Earth view.

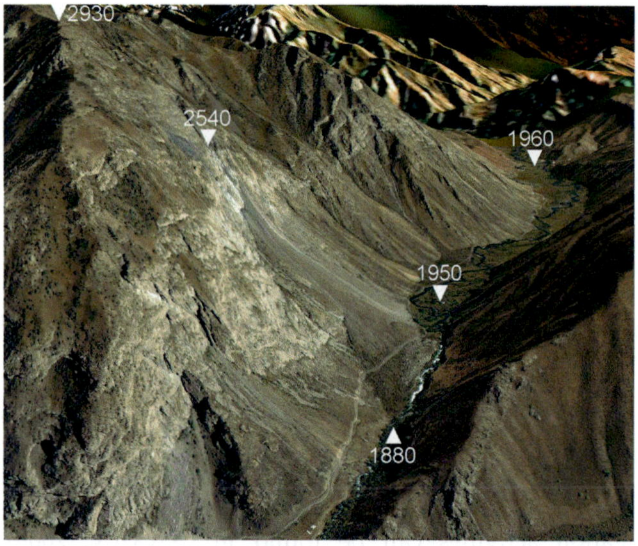

FIGURE 8.238 The Lower Kulun rockslide dam and its completely silted lake. 3D Google Earth view.

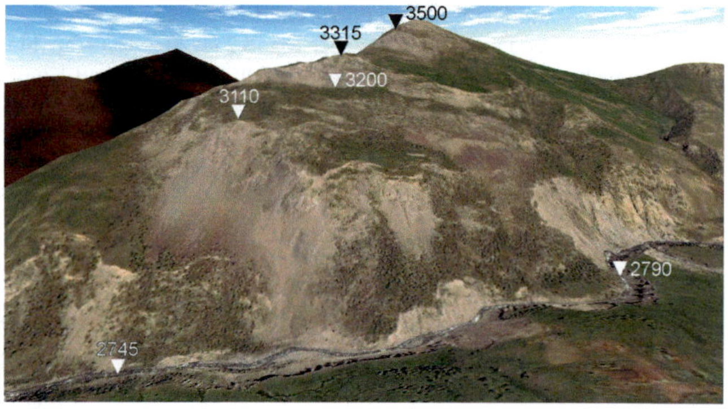

FIGURE 8.239 Blockslide on the right bank of the Terek River. SAS Planet image visualized in 3D by the Global Mapper software.

depression and Northern chains of the Pamir in the south (see Fig. 8.1). Besides it includes the part of the Ferghana Range south-east of the Kara-Daria River and the so-called South-Western Gissar bounding the Tajik depression from the west (Fig. 8.236).

Several large, likely slowly moving landslides could be found in the Ferghana Range on the slopes composed of relatively weak Jurassic coal-bearing terrigenous rocks. Such landslides affect slopes at 40.79° N, 73.94° E (Kyrgyz 045) and 40.746° N, 74.166° E (Kyrgyz 046) in the Yassy River basin.

Very large Pleistocene Terek-Suu rockslide in the Tar River valley at 40.53° N, 73.825° E (Kyrgyz 235) was described in Chapter 4. Didactic outcrops of its deposits can be seen just above the road passing along its right bank (see Figs. 4.7 and 4.8).

Interesting chain of rockslide dams is in the Kulun River valley – the right tributary of the Tar River. The upper reach of this river is blocked at 44.537° N, 74.285° E by the Kulun rockslide about 130 Mm^3 in volume (Kyrgyz 050) that forms the 280 m high dam of the Kulun Lake (Fig. 8.237) with about 50 m freeboard. One more much smaller rockslide (Kyrgyz 051) about 3 Mm^3 in volume forms the small lake 4 km downstream at 40.519° N, 74.253° E. This water body is protected from the siltation since all sediment yield has been captured by the large natural reservoir upstream. About ~23 km downstream, almost at the Kulun River mouth, at 40.46° N, 74.067° E, its valley was blocked by the third rockslide also of the small volume (ca. 2 Mm^3 only) that forms ~60 m high dam (see Figs. 6.23, 8.238). Despite its relatively small size and large catchment area resulting in high (3.61) DBI value (Strom, 2010b), this Lower Kulun rockslide

FIGURE 8.240 Slowly moving landslides in Jurassic terrigenous rocks in the Tahta-Korumsu River valley. 3D Google Earth view.

dam (Kyrgyz 217) survived and rather large, more than 3 km long dammed lake has been silted with time completely (see Figs. 6.23, 8.238). As it was proposed in (Strom, 2010b, 2013a) such an abnormal survivability of this blockage is caused by regulation of peak discharge by the voluminous Kulun Lake that accumulates most of water during the flood period releasing it gradually by filtration. It decreases the erosional power of the stream significantly so that it could not dissect coarse blocky carapace of the Lower Kulun rockslide body. Similar effect could be responsible for the preservation of the Karakul rockslide dam at the mouth of the Eastern Karasu River, which peak flood discharge could be reduced by the rockslide dammed Karasu Lake in the upper reaches of the same valley (see Section 8.4).

All these rockslides occurred on slopes composed of the Paleozoic sedimentary rocks. The Lower Kulun rockslide originated at the middle part of the slope composed of coarse-bedded limestone and carbonate sandstone steeply dipping towards the valley but not as steep as the valley slope.

The much larger Kulun rockslide affected the 1300 m high slope so that the crown of its up to 250 m deep headscarp crossed the watershed (Fig. 8.237). Considering depth of the affected slope it resembles the Ak-Kiol rockslide (see Fig. 5.9). One more peculiarity of this rockslide is that just opposite the collapsed slope there was a mouth of a rather deep gulley in which rockslide body entered as in the trap, similar to the Aksu rockslide described in Section 8.3 (see Fig. 8.103). It prevented debris up- and downstream spreading and predetermined formation of such a high blockage. This dam which front is about 60 m higher than the lowermost proximal part resembles other Primary rock avalanches with pronounced frontal

FIGURE 8.241 The Kolduk rockslide in the Gulcha River basin. C – Carboniferous units; K-Pg – Cretaceous – Paleogene units; yellow triangles –outer boundary of the rockslide frontal part. LS – smaller landslides near this "core" feature. Slightly darker color of the lower right part is due to availability of different space images. 3D Google Earth view.

confinement such as the Mini-Köfels rockslide (see Fig. 5.41).

Large-scale blockslide affects more than 800 m high right slope of the Terek River valley just upstream of the Chakarak River mouth at 40.47° N, 74.44° E (Kyrgyz 053). Slope composed of the Jurassic terrigenous rocks is crossed by several sub-parallel scarps up to 100–150 m high dividing sagging blocks up to 200–470 m wide (Fig. 8.239). The lower part of this blockslide is affected by numerous shallow slope failures at a distance of about 1.5 km. No evidence of any long-term river blocking by this complex rockslide have been found on the space images available. Besides this feature, several other gradually developing landslides affect slopes composed of the same rocks in the vicinity.

FIGURE 8.242 Rock avalanche dam in the left tributary of the Akbura River. 1, 2, 3 – different lithological units on the slope identifiable on space image. 3D Google Earth view.

One more group of large, up to four square kilometers in plan view, gradually moving landslides is located in the Chinese Tien Shan, west of the Talas-Ferghana fault, on the 500–700 m high left slope of the Tahta-Korumsu River valley at 40.18° N, 74.74° E. Two of them – China 091 and China 085 can be seen on Fig. 8.240. Despite their gradual, generally non-catastrophic style of motion these landslides caused temporary river damming that can be revealed from the presence of the distinct, 15 to 20 m deep V-shaped erosional canyons 2.4 and 1.4 km long at the feet of landslide-affected slopes differing drastically from the 200 m wide flat valley bottom upstream.

The next site that deserves attention is located in the Gulcha River basin near the Kolduk village at 39.98 N, 73.54 E (Kyrgyz 122). Failure occurred just at the fault zone dividing the Carboniferous carbonate and terrigenous rocks from the Upper Cretaceous and Paleogene gypsum-bearing claystone and limestone. About 80 Mm^3 of the latter unit mainly collapsed from the ca. 700 m high slope leaving at least the 80–90 m deep headscarp and filled the valley at a distance of about 1.5 km. It formed a dam up to 150 m thick with a small lake behind it partially filled by the alluvial fan deposits (Fig. 8.241). One more much smaller water body accumulated in the more than 40 m deep proximal depression of the rockslide body. This Kolduk rockslide resembles the Arashansai rock avalanche in North-Western Tien Shan (see Fig. 8.225) that originated at the slope crossed by the fault zone between Carboniferous granite and Cenozoic sedimentary rocks and also has a lake in the proximal depression. Similar morphology has the Bielogorka-Left rock avalanche (see Fig. 8.97), though there the proximal depression about 20 m deep is dry.

FIGURE 8.243 3D Google Earth view of the Ming-Djilki rockslide in the same name river valley (central image). Above – the syncline structure of the host rock massif (plan view); below – the brandung in front of the main debris accumulation rising from 3060 to 3130 m a.s.l.

Besides large-scale Kolduk rockslide several much smaller slope failures occurred in its vicinity. Sliding surfaces of most of them coincide with steeply dipping bedding planes of the Upper Cretaceous and Paleogene sedimentary rocks so that sliding started in the form of blockslides but transformed into flowslides. This situation recalls the Sarydjaz-Aksu rockslide cluster (see Table 8.2) with a gigantic "core" feature and numerous much smaller "satellites".

Rock avalanche that originated at the topmost part of the ca. 900 m high ridge composed (from top to bottom) of the Silurian-Devonian metasediments, Permian granosyenite and Devonian metasediments (Fig. 8.242) and blocked the valley was found in the Akbura River basin at 40.07° N, 72.93° E (Kyrgyz 275). About 5 Mm^3 of, likely, metasediments and granosyenite debris traveled along the steep slope with mean angle of 27° and accumulated at its foot forming a compact dam up to 80–90 m high that blocked the steep gully with gradient of about 10°. Almost no material remained along the 700 m long transit zone indicating Primary style of rock avalanche motion (see Section 5.3.1). This dam has not been breached and the dammed lake was silted completely.

One of the most disastrous events related to the rockslides in Central Asia region was the June 18, 1966 catastrophic breach of the Yashinkul rockslide dam in the Tegermach River valley in Kyrgyzstan (39.927° N, 72.297° E, Kyrgyz 129) described in brief in Chapter 2 based on the detailed descriptions from (Rezvoi and Rezvoi, 1969; Pushkarenko and Nikitin, 1988). Rockslide about 25 Mm^3 in volume occurred on 1 km high slope composed of the Silurian shale and sandstone on top and the Carboniferous terrigenous rocks and limestone at the central and lower part and affected its upper part mainly up to watershed (see Fig. 2.3). The headscarp reaches its maximal visible depth (ca. 80 m) at the central part of the slope at 3020 m a.s.l. Some part of debris accumulated on the slope on a bench at is central part at 2800 m a.s.l. Thickness of this proximal part of the rockslide body reaches ~200 m. Frontal part of the collapsing rock mass most likely jumped from this bench and had dammed two merging rivers. The dam's crest level slightly decreased towards it distal edge, which is typical of the Jumping rock avalanches (see Section 5.3.2). While the main dammed lake was emptied by the 1966 breach completely, the much smaller one with water level at ~2700 m a.s.l. – about 100 m higher than the bottom of the breached lake, still exist in the right tributary valley (see Fig. 2.3). One more rockslide about 8 Mm^3 in volume (Kyrgyz 128) is located 2 km northeast of the Yashinkul rockslide dam.

Large-scale wedge failure occurred on the eastern side of the ridge dividing the Isfairamsay River valley from the valley of its large left tributary – the Surmentash River, in the Ming-Djilki River valley at 39.789° N, 72.064° E (Kyrgyz 131). Slope failure affected the entire slope up to 1200 m high, composed of the Carboniferous terrigenous rocks intruded by dykes and destroyed about 1 km of its watershed. Bedrock strata here form the syncline with hinge line steeply dipping towards the Ming-Djilki River (Fig. 8.243), which up to 250 m thick core collapsed forming the rockslide that recalls the Ak-Kiol rockslide described above (see Fig. 5.12). Volume estimates of this Ming-Djilki rockslide gave mean value of about 90 Mm^3. Rockslide debris up to 200 m thick filled the valley at a distance of 2100 m and has a pronounced brandung with up to 70 m runup. The gulley bottom upstream at a distance of ca. 600 m is filled by sediments.

As we mentioned in Chapter 2, catastrophic outburst floods could be caused by partial breach of the

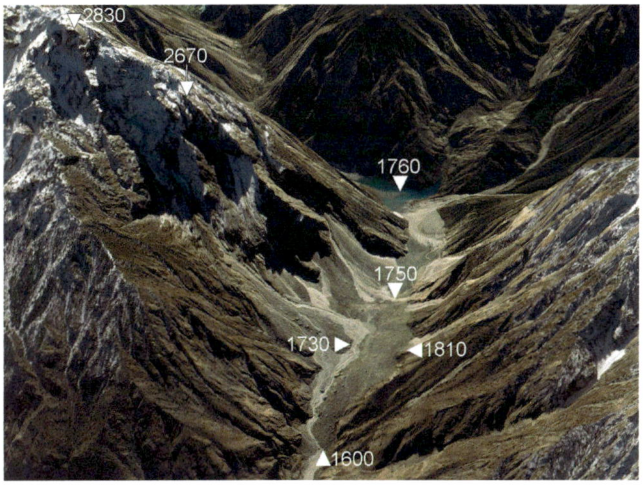

FIGURE 8.244 The Kurbankul rockslide dam and dammed lake partially filled by debris flow deposits up to elevation mark 1750 m a.s.l. 3D Google Earth view.

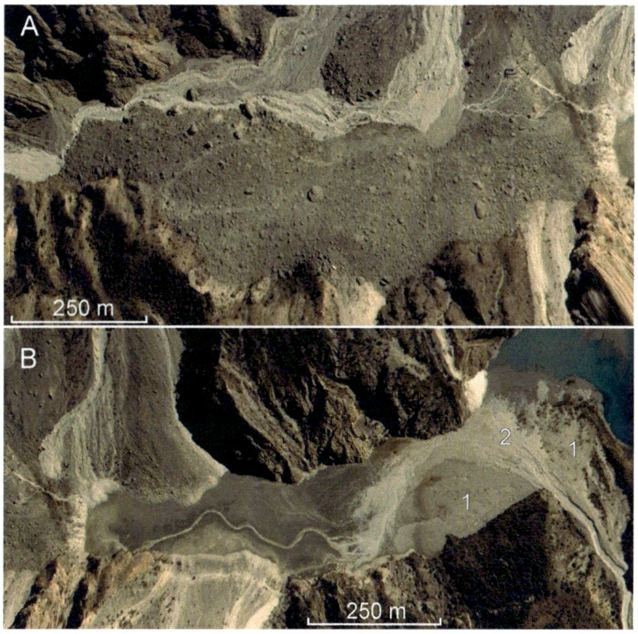

FIGURE 8.245 A – plan view of the Kurbankul rockslide dam. Largest boulders are up to 30 m in size. Notice that there is no evidence of the dams' crest erosion. B – the Kurbankul Lake most of which is filled by debris flow deposits among which two generations marked by 1 and 2 can be identified. Google Earth images.

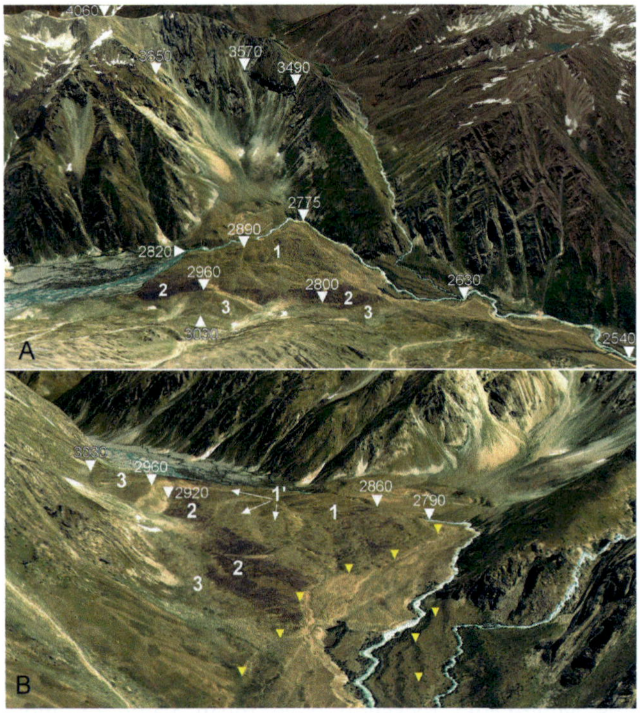

FIGURE 8.246 3D Goggle Earth views of the Kuturgansu-mouth rockslide dam. A – front view, B – view from downstream. 1, 2, 3 – lithological zones of the deposits; 1′ – belt of coarse boulders in zone 1; yellow triangles – outer limit of the erosional canyon and of the debris re-deposited by outburst flood.

FIGURE 8.247 4.75 km long zone of rockslide debris re-deposition downstream of the breached Kuturgansu-mouth rockslide dam. 3D Goggle Earth view.

Kurbankul Lake (39.936° N, 71.845° E, Kyrgyz 132) in the Shahimardan River basin in 1977 and in 1998 (Ecological state, 2006). This lake was formed by the rockslide about 15 Mm^3 in volume that caved from the ca. 1100 m high right-bank slope of the valley composed of the Carboniferous limestone. The headscarp shape is very irregular and represents compound wedge failure with the assumed headscarp crowns at 2670 and 2830 m a.s.l. (Fig. 8.244). Though in many popular sources such as Wikipedia it is mentioned that rockslide "occurred in 1766 being triggered by the series of strong earthquakes" none of them is mentioned in the earthquake catalogues (Mushketov and Orlov, 1893; Semionov, 1958; Kondorskaya and Shebalin, 1982). Thus, origin of this feature as well as the cause of the outbust floods require additional geological and historical investigations.

The Kurbankul rockslide dam is about 1200 m wide (along the river) and up to 120 m high with 80 m runup of its front over the opposite slope of the valley. Its surface is covered by coarse blocks largest of which are up to 30 m in size (Fig. 8.245-A).

Area of the upstream catchment is 127 km^2 and the corresponding DBI value (Ermini and Casagli, 2003) is 3.0, which put this blockage in the uncertainty domain. Despite information that dam was partially beached twice at the end of XX-s Century (Ecological state, 2006) no evidence of its erosion similar to many other breached dams can be seen on space images (see Fig. 8.245-A). The dam was eroded by several debris flows well visible on Fig. 8.245-A but they have no relation to the lake outburst or overflow. Thus, this case study differs significantly from other well known breaches of rockslide dams such as the 1963 Issyk (see Fig. 2.1) or 1966 Yashinkul (see Fig. 2.3) disasters when rockslide dams were really breached and deeply eroded.

Some similarity with the 1963 Issyk catastrophe is that in both cases powerful debris flows entered in rockslide-dammed lakes. In the Kurbankul case, however, even if some overflow took place it did not result in stream incision in the dam's body. In the Issyk case debris flow entered the lake from its upstream side, so that generated surge waves hit the dam directly, removing coarse blocky carapace and eroding the comminuted rockslide body interior. In the Kurbankul case, in contrast, debris flows entered the lake from its left bank, practically increasing the dams' width so that at present the lake is dammed by debris flow deposits not by rockslide body. At least two generations of debris flow accumulations can be identified. They are marked by 1 and 2 on Fig. 8.245-B.

Next feature that definitely deserves attention is the gigantic, about 300 Mm^3 breached Kuturgansu-mouth rockslide dam (Tajik 049) that had blocked the Ptovkul River valley just upstream of the Kuturgansu River mouth at 39.44° N, 71.416° E (Fig. 8.246). The triangular fan-shape body 2 km long (across the river) and with 2 km wide front caved from the 1.2 km high slope composed of granites. Unlike many other case studies, its 200 m deep headscarp did not affect the entire slope up to its top but has the crown about 400–500 m below the watershed at 3650–3490 m a.s.l. (see Fig. 8.246-A). It can be compared with the 1911 Usoi rockslide (see Fig. 5.7) and the Kudara-Pasor rockslide (see Fig. 4.20) in Pamir.

FIGURE 8.248 Front view of the Kyzyl-Katta rockslide. 3D Google Earth view.

Three distinct zones can be delineated in the rockslide body. Its main part has slightly hummocky grass-covered surface with several belts of coarser angular boulders stretching transverse to rockslide motion and, most likely, representing lithological unit which destruction produced larger boulders. Similar belt was found on top of the Djashilkul rockslide in Northern Tien Shan (see Fig. 8.79). This zone is bounded from its front by the 100–370 m wide belt of boulders covered with desert varnish. The distal zone demonstrating significant, more than 100 m high runup (see Fig. 8.246-B) is composed, most likely of the Quaternary valley fill (moraine?) bulldozed by rockslide body in front of its main part composed of crushed bedrock. Similar bulldozing was found at many case studies described above (see Figs. 4.4, 4.14, 4.15, 4.16, 4.18). Most impressively such transverse zoning is visible on the above mentioned gigantic Kudara-Pasor blockage in Pamir (see Fig. 4.20 and Figs. 9.38, 9.39).

Another interesting peculiarity of this rockslide is more than 4 km long belt of debris eroded from this rockslide dam during its breach and re-deposited downstream (see Fig. 8.246-B, Fig. 8.247). The downstream face of the blockage was eroded by outburst that formed a 500 m wide and about 50 m deep breach. Deposits left by this flow are up to 25–30 m thick. We can assume that several erosional terraces on the left bank of the river just downstream of the blockage at the promontory of the Ptovkul and the Kuturgansu Rivers confluence (see the center-bottom part of Fig. 8.246-B) were also eroded by this flow. Total volume of debris eroded from the dam and re-deposited downstream could reach ca. 30 Mm^3. Part of it has been reworked by the river later on.

However, unlike the Lower Aral (see Fig. 8.163) and the Sarydjaz-Aksu (see Fig. 8.53) case studies where

FIGURE 8.249 Side view on the Kyzyl-Katta rockslide from the east. Bedding in the bedrock is well visible. C - carbonate rocks; T - terrigenous rocks. 3D Google Earth view.

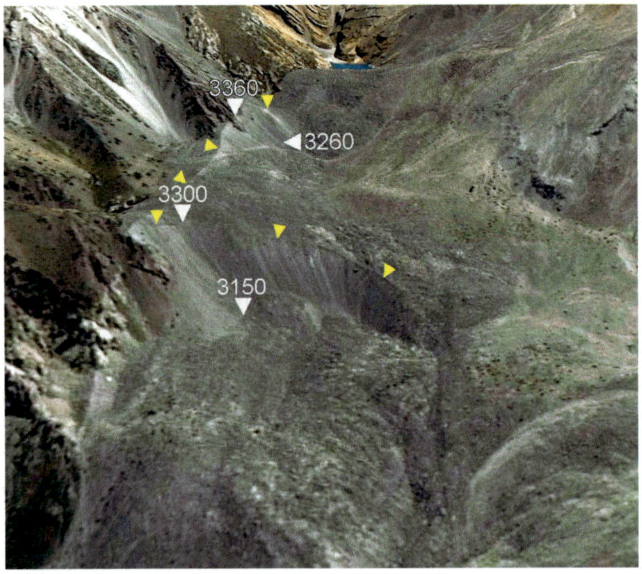

FIGURE 8.250 100–150 m high secondary scars (marked by yellow triangles) at the right upper part of the Kyzyl-Katta rockslide body. 3D Google Earth view.

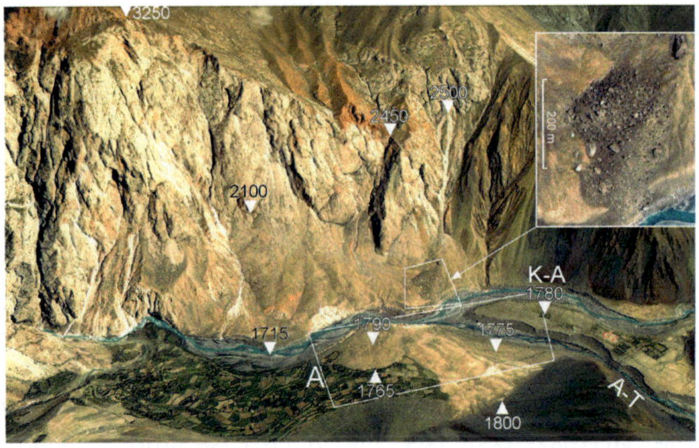

FIGURE 8.251 The Korgon rock avalanche. A-T – Ak-Terek River, K-A – Khodja-Achkan River. More recent coarse accumulation is shown in the inset; A – area shown on Fig. 8.252. 3D Google Earth view.

we have more distinct markers of the outburst flow peak depth, here it is more difficult to estimate surge wave height and peak discharge since downstream part of rockslide body was eroded first and later on debris washed out from the upper part of the water gap was re-deposited here. But in any case we can hold that it was a catastrophic process.

Several large-scale rockslides in the western part of the Alai Range (Kyrgyz 144, 146) and eastern part of the Turkestan Range (Kyrgyz 073) demonstrate certain similarity of failure type, motion mechanism and deposits shape. All these rockslides originated on steep rocky slopes composed of carbonate and terrigenous

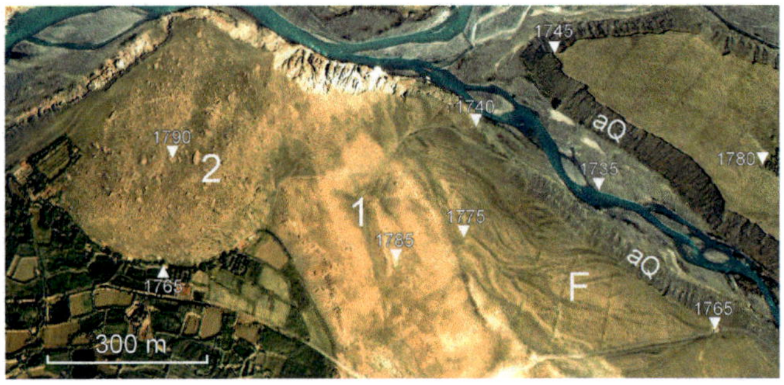

FIGURE 8.252 Possible two generations of the Korgon rock avalanche body and evidence of the outburst flood from upstream. 1 – older generation of debris with smooth surface composed of finer material; 2 – younger generation with much coarser material; F – the specific landforms left by outburst flood; aQ – fluvial deposits. Google Earth image.

sedimentary rocks of the Devonian and Carboniferous age, and their sliding surfaces coincided, at least partially, with steeply dipping bedding planes resembling the 2008 Daguangbao rockslide in China (Huang et al., 2009, 2012).

The largest of them – the Kyzyl-Katta rockslide (Kyrgyz 146) about 500 Mm^3 in volume occurred at 39.71° N, 71.115° E at the upper reaches of the Augul River – right tributary of the Sokh River and dams the small Kyzyl-Katta Lake (Kyzyl-Katta-Kiol in Kyrgyz language). One more minor lake is at the eastern boundary of the rockslide body (Fig. 8.248).

Failure occurred on the rocky spur that rose above the confluence of two rivers for about 1.1 km being composed of carbonate rocks underline by terrigenous unit (Fig. 8.249). The sliding surface is slightly steeper than bedding on top of the ridge and, likely coincides with bedding downslope. As it can be derived from rockslide body color, most of it is composed of carbonate debris. Considering large volume of the deposits, almost entire spur several hundreds meters long composed of carbonate rocks had collapsed, similar to the 1911 Ananievo rockslide (see Fig. 8.70) or the Mingteke rock avalanche (see Fig. 8.140).

Rockslide debris filled the junction of two valleys damming small lakes on both sides of the more than 3.5 km long body. Its surface is quite irregular with the distinct bow-shape secondary scar up to 150–160 m high above the Secondary rock avalanche ~50 Mm^3 in volume (Fig. 8.250). One more secondary scar up to 100 m high, likely caused by the elastic rebound after rockslide mass collision with the opposite valley wall, passes along the true right boundary of the deposits.

Three large rockslides had dammed rivers south of the Kyrgyz 146 feature in the Sokh River basin. Two of them had blocked its sources, the Ak-Terek River at 39.60° N, 70.95° E (Kyrgyz 145) and the

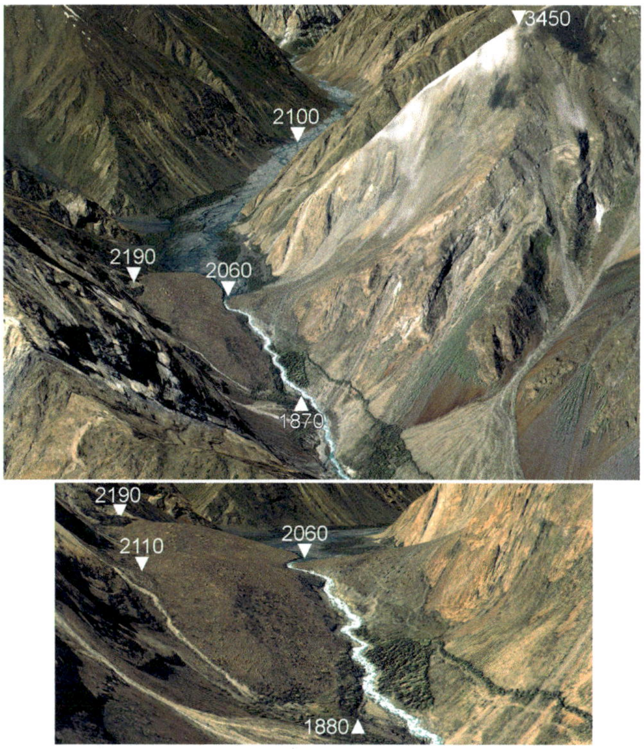

FIGURE 8.253 Breached rockslide dam in the Ak-Terek River valley. Above – overall view; below – closer view on the blockage with well-pronounced runup. 3D Google Earth view.

FIGURE 8.254 Breached rockslide dam in the Khodja-Achkan River valley. Some evidence of the outburst flood (branching ravines) can be seen on the ca. 20 m high left-bank terrace at 1930 m a.s.l. 3D Google Earth view.

Khodja-Achkan River at 39.625° N, 71.05° E (Kyrgyz 147), while the third one (Kyrgyz 333) had formed the dam just downstream of their confluence at 39.64° N, 70.966° E. This rock avalanche about 30 Mm^3 in volume caved from the ca. 1500 m high steep cliff with slope angle more than 50° composed of the Carboniferous carbonate and siliceous rocks. It formed up to 40–50 m thick fan-shaped body immediately

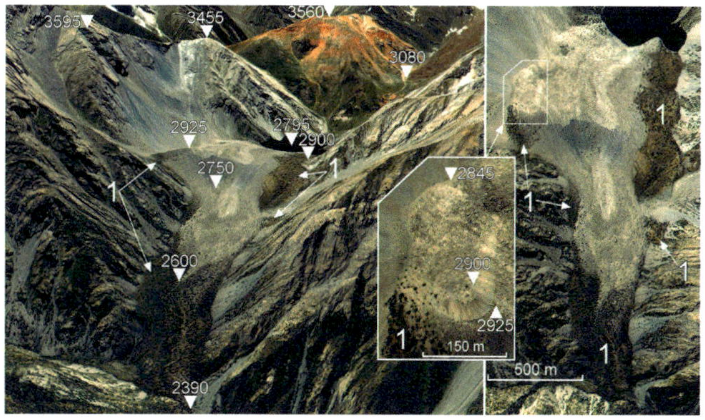

FIGURE 8.255 Left – 3D Google Earth view of the Kara-Kiol-Katta rockslide. Right – plan Google Earth view of its deposits. 1 – rockslide deposits of 1st generation. Rock glacier at the upper right corner terminates at 3080 m a.s.l. Backward sliding towards 80 m deep proximal depression is shown in the inset.

upstream from the Korgon village (on Google Earth it is marked as the Zardaly village). Some parts of this body are composed of coarse – up to 30–40 m angular boulders (Fig. 8.251). It can not be excluded that coarse accumulations were formed later than the main fan-shape body with smooth surface, being originated from different parts of the cliff (Fig. 8.252). Such multiple failures are quite possible considering the extreme topography of the right-bank slope of the Sokh River valley. Most mobile part of this rock avalanche rose towards opposite valley slope with ca. 20 m runup. Downstream of the rivers' confluence rockslide debris had dammed the valley.

One of the most interesting peculiarities of this Korgon rock avalanche is the evidence of the powerful outburst flood that left the amygdaliform branching ravines on the 40 m high left bank terrace and partially affected rock avalanche body. Similar ravines are well expressed downstream of the breached Sarydjaz-Aksu blockage (see Fig. 8.55). At the same time absence of such landforms on the slightly higher (47–50 m high) right bank terrace provides proxies of the flood wave height estimates. Considering ~400 m width of the valley at this level and shape of the river channel, its passage is about 7000 m^2. If the velocity of the flood was of the order of 10 m/s, which is quite realistic, peak discharge of this flood could be about 70 000 m^3/s.

Most probable source of such a powerful outburst flood was the breach of the ~150 m high rock avalanche dam about 40 Mm^3 in volume that had blocked the Ak-Terek River valley 4.5 km upstream (Kyrgyz 145, Fig. 8.253). We use the hyphen in the name of this river to distinguish this feature from the Akterek rockslide in Central Tien Shan (see Figs. 8.187, 8.188) The Ak-Terek rockslide originated from the ca. 1550 m high and very steep (more than 50°) left slope of the valley composed of the metamorphosed terrigenous Silurian rocks. We should notice that its source zone coincides with a fault that, according to the 1:200000 Stage geological map, crosses the river valley. It was a typical Primary rock avalanche in frontally confined conditions (see Section 5.3.1) with nearly 110 m runup (measured with due regard to the subsequent incision). The lake was silted at a large extent but upper part of the dam was eroded by overflow that produced flood wave, which traces are visible on Fig. 8.252. See also the discussion in Section 14.3.

In contrast with the blockage in the Ak-Terek River valley, natural dam produced by rockslide that caved from 1200 m high slope had blocked the Khodja-Achkan River, was breached completely without evidence of the dammed lake siltation (Fig. 8.254). Besides its smaller volume ($\sim$25 Mm^3 only), such difference in the blockage evolution could be explained by comparison of the valleys' shape and gradient (the Khodja-Achkan River valley upstream of this site is much narrower and steeper than the Ak-Terek one) and of the grain-size composition of these dams. Here debris that originated from lithologically more variable and thin-bedded Carboniferous rocks is much finer and, thus, much more erodible. At the same time the Khodja-Achkan River catchment ($\sim$800 km^2) is about 1.5 times larger than that of the Ak-Terek River ($\sim$525 km^2) and we can assume that its discharge is larger in the same proportion. All these factors that will be discussed further in Chapter 14, resulted in a short- term blocking of the Khodja-Achkan River.

Rockslide had dammed not only this river, but also its small right tributary gulley, likely without permanent flow. This part of the blockage remains more intact and large amount of debris flow deposits had accumulated here before it was partially incised allowing transit release of sediments.

Morphologically this rockslide resembles the Basi-Djaya rockslide in the Chong-Kemin River valley in Northern Tien Shan (see Fig. 8.76), though the latter has much more pronounced frontal runup.

Large-scale rockslide resembling the Kyzyl-Katta case study (see Fig. 8.248) was found at 39.71° N, 70.75° E, in the upper reaches of the small Nurlau River – the right tributary of the Kshemysh River that passes through the Vorukh enclave (Tajikistan territory in Kyrgyzstan). It dams the small Kara-Kiol-Katta Lake that gave the name of this feature (Kyrgyz 144). The Kara-Kiol-Katta rockslide originated on more than 600 m high slope composed of the Devonian, predominantly carbonate rocks dipping towards the valley (Fig. 8.255). Similarly to many other case studies rock mass, which volume can be estimated as ca. 95 Mm^3, had collapsed from the prominence on the slope. Slope failure can be classified as the irregular slide (Hungr et al., 2014) (see Section 5.1), though large depth of the headscarp (more than 200 m) allows classifying it as the "irregular wedge failure" as well.

This rockslide can be also classified as a deflected Secondary rock avalanche of the classical subtype (see Section 5.3.3.1). It has very impressive secondary scar up to 120–130 m high from which tongue of debris moved

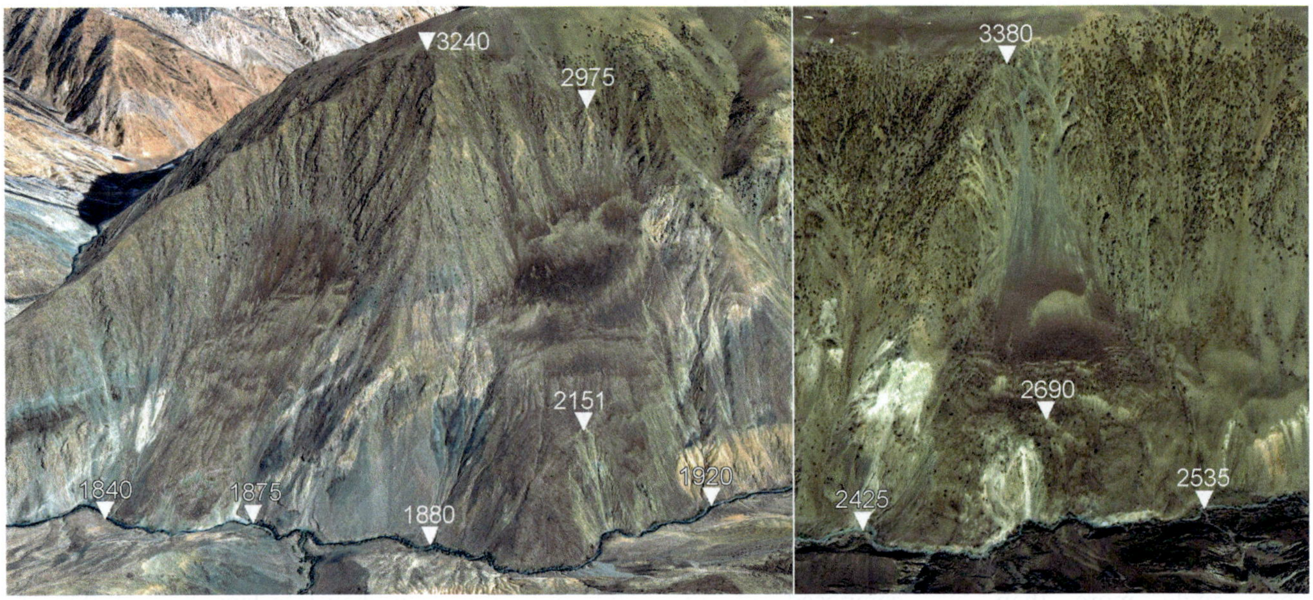

FIGURE 8.256 Three gradually evolving landslides (Kyrgyz 141, 142 and 143, from left to right) in the Karavshin River basin. 3D Google Earth views.

1.9 km downvalley. Interesting peculiarity of this body is its distinct division into two parts – the outer, likely older, with well vegetated surface and the inner one practically without any vegetation on it (see Fig. 8.255). The outer one forms the front of the river-damming blockage. Similar vegetated surface could be found at the right side of the body just at the headscarp base and also formed the ca. 950 m long lower part of rock avalanche. Very contrasting non-vegetated part overlays an older one and demonstrates similar bipartition with compact body and mobile deflected avalanche. Such an inheritance of the successive failures that originated from the same source and has similar mechanism of debris motion resulting in morphological similarity is rather rare phenomenon. In the study region it can be compared with the Djamantau rock avalanche (see Fig. 8.133). The Djuzumdybulak rockslides in the Naryn River valley described above in Section 8.4 (see Figs. 8.190 and 8.192) where such inheritance could be anticipated, most likely originated from the adjacent, but different source zones. Recurrent large-scale failures from the same source zone were described, for example, by Reginald Hermanns and his co-authors (Hermanns and Strecker, 1999; Hermanns et al., 2000, 2001) in Argentina.

FIGURE 8.257 The Ai-Kul rock avalanche in the upper reaches of the Sardala River. 3D Google Earth view.

Considering thickness of debris of the second generation the older compact body had to have a significant frontal runup. We cannot divide the entire volume of both failures into 1st and 2nd generations with confidence and estimated deposits volume of about 95 Mm^3 is the sum of both.

Another interesting and didactic peculiarity of the younger part of this feature is the distinct concentric belts of varicolored debris visible both on top of the compact body and at its avalanche-like part that indicate flow-like stile of motion of both parts (see right part of Fig. 8.255). Besides there is evidence of backward sliding of debris, likely towards the proximal depression at the headscarp base (see the inset in Fig. 8.255).

Close to this complex rockslide the sackung-like feature evolves on top of the red mountain with elevation mark 3560 m a.s.l. visible on Fig. 8.255 at the background.

Three large gradually moving landslides about 15–30 Mm^3 each evolve in the Karavshin River basin that drains north-eastern part of the Turkestan Range. Two of them (Kyrgyz 141 and 142) – on the 1200–1400 m high right-bank slope of the Kyndyk River valley at 39.728° N, 70.337° E, just at a confluence

FIGURE 8.258 Small rockslide dam and its completely silted dammed lake in the Sardala River valley 5 km downstream of the Ai-Kul rock avalanche. 3D Google Earth view.

with the Karavshin River, and the third one (Kyrgyz 143) – on the 900 m high right-bank slope of the Djiptyk River valley at 39.734° N, 70.476° E (Fig. 8.256). Landslide processes on these slopes composed of the Upper Carboniferous – Lower Permian terrigenous rocks, likely with carbonate interbeds, differ drastically from the catastrophic failures widely developed in the surrounding areas where limestone and dolostone prevail. The landslides in question most likely caused just temporary small-scale blocking of the streams and did not pose a significant threat to the downstream populated part of the Karavshin River valley. In (Fedorenko, 1988) volume of the upstream one named the Karavshin rockslide was erroneously estimated as 200 Mm^3 which seems to be one order larger than the actual one.

Further to the west, in the Sarkat River basin, its primary source – the Sardala River is dammed at 39.65° N, 69.63° E by the extremely impressive Ai-Kul rock avalanche (Kyrgyz 073) that blocks the valley creating the Ai-Kul Lake (Fig. 8.257). The entire rocky spur composed of the Carboniferous limestone that rises for about 1 km above the confluence of the two glacial valleys collapsed, blocking both of them and forming more than 3 km long rock avalanche. It resembles rockslides that occurred in the same geological conditions and can be seen on Figs. 8.248 and 8.255 and could be classified as a wedge failure (Hungr et al., 2012, 2014) with steep main sliding surface coinciding with bedding plain that dips up to 52°. Rockslide body collided with a promontory on the opposite slope forming the 100 m runup (from 2970 to 3070 m a.s.l.) and split into two parts. The smaller one moved upstream, while the larger followed the valley and filled it at a distance of about 2 km at a depth up to 300 m. Similar splitting was observed at rockslide shown on Fig. 6.5 and at the Spur rock avalanche marked by S on Fig. 8.120. The Ai-Kul rock avalanche can be ascribed to the Primary type (see Section 5.3.1), first with frontal and, afterward, with lateral confinement.

Besides this main wedge failure, which volume can be estimated as ca. 75 Mm^3, additional rockslide, likely of the translational type occurred at its true right side (its headscarp crown at 3050 m a.s.l. is marked on Fig. 8.257) and blocked the tributary valley between elevation marks 2825 and 2745 m a.s.l.

These slope failures look very "fresh" without any vegetation on top of their deposits composed, from the surface, by coarse angular clasts and blocks up to 20–30 m in size.

One more much smaller rockslide about 3 Mm^3 only (Kyrgyz 072) had blocked the same Sardala River at 39.66° N, 69.56° E, 5 km downstream of the Ai-Kul

FIGURE 8.260 Landslides in the Rivut River valley. Headscarp crowns at 3605, 3750 and 3345 m a.s.l. One of them is shown in the left inset. Right inset – breached landslide dam and former lake. 3D Google Earth view.

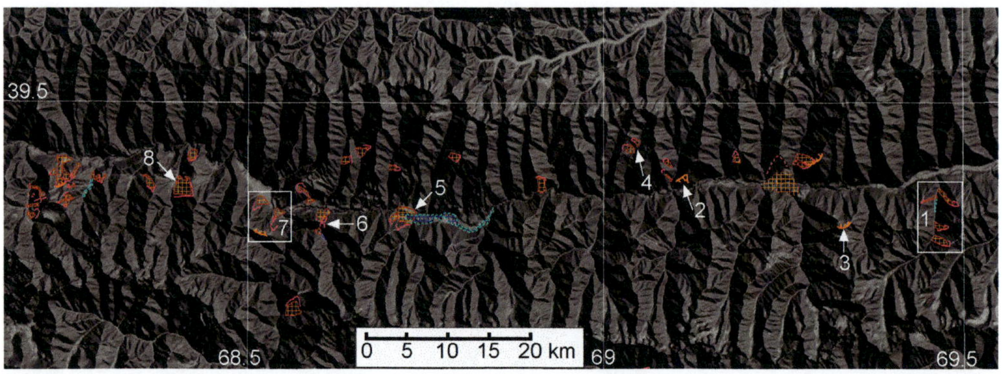

FIGURE 8.259 Landslides in the Zeravshan River valley. Cross-hatched areas with brown boundaries – recent rockslide bodies; those without boundaries – assumed rockslide bodies; solid pink lines – headscarps; dashed pink lines – assumed headscarps; comb orange lines – sackung-like scarps; blue cross-hatched areas – largest breached lakes. 1 – rock avalanches in the Rivut River valley (see Fig. 8.260); 2 – see Fig. 8.261; 3 – see Fig. 5.18; 4 – see Fig. 8.262; 5 – Rarz rockslide (see Figs. 8.263, 8.264); 6 – see Fig. 8.265; 7 – the 1964 Aini rockslide and adjacent slope deformations (see Figs. 2.2, 8.266); 8 – the Kum rockslide (see Fig. 8.267). Background – the 3″ SRTM DEM visualized by Global Mapper software.

rock avalanche deposits tip. This failure occurred on the ca. 450 m high slope promontory composed of the Devonian carbonate rocks and formed a ~30 m high dam. Despite relatively small volume of the collapsed rock mass and the headscarp height rockslide body has a prominent – up to 55 m runup, allowing its classifying as Primary rock avalanche in frontally confined conditions (see Section 5.3.1).

The dammed lake was silted almost completely before the dam was cut through by the erosional channel (Fig. 8.258). Both rockslide body and the terrace-like surface of the former dammed lake bottom are covered by grass and bushes indicating that this feature could be much older than the nearby Ai-Kul rock avalanche described above.

One of the most prominent geomorphic features of the western part of the Southern Tien Shan is the Zeravshan River valley that stretches for about 280 km from east to west between the Turkestan Range at the north and the Zeravshan Range at the south. At a long distance this river valley inherits the neotectonic fault zone with narrow stripe of the Jurassic, Cretaceous and Paleogene-Neogene weak sedimentary rocks overthrusted by much stronger Silurian sedimentary rock of quite variable lithology. From the structural point of view this zone somehow resembles the Minkush-Kokomeren intermontaine neotectonic depression in the Central Tien Shan (see Section 8.4) and, similar to it, is enriched by numerous bedrock landslides and sackungen-like features (Fig. 8.259). We must mention that rockslides in the Zeravshan River basin were studied on a regular basis in 60s-80s of the past Century by researchers from the Moscow State University. Results of these studies that had started after the 1964 Aini rockslide (see Fig. 2.2 and brief description of this event in Chapter 2) were summarized in (Fedorenko, 1988). Unfortunately, due to formal restrictions, coordinates of the numerous sites described in this book were not indicated. Most of landslides occurred on slopes composed of the Silurian terrigenous rocks. Hereafter we present the description of several most prominent case studies from this part of the region.

Several long narrow landslides originated on top of 1.0–1.5 km high ridges on both sides of the Rivut River valley – the left tributary of the Zeravshan River and moved downslope along the steep dry gullies up to the valley bottom (Fig. 8.260). Moving for a long (up to 3.5 km) distance along the steep (27–34°) slopes these rock masses 5 to 10 Mm^3 in volume converted into debris avalanches or earth flows and finally dammed the river partially or completely. One of these landslides (Tajik 120) formed a rather long-living dam about 50 m high at 39.406° N, 69.466° E. Dammed lake was partially silted and then breached. Judging from sharp boundaries of the former lake bottom, not reworked and buried by talus, breach had to occur recently, may be within few decades, though we have no data about such event. We have to notice that larger landslide (Tajik 047) that originated on the opposite slope at 3750 m a.s.l., reached the valley bottom downstream (at 2125 m a.s.l. on Fig. 8.260, 39.407° N, 69.466° E) and formed more than 100 m high dam, has no evidence of valley inundation. It can be explained as follows: if both landslides occurred simultaneously, the upper one had blocked the stream preventing both downstream water flow and sediment transport. When the upper dam was breached the resultant outburst flood eroded both dams together.

The impressive evidence of toppling can be seen on the right-bank slope of the Zeravshan River valley at 39.42° N, 69.11° E, just upstream of the Dashtiobburdon village (Tajik 031, Fig. 8.261). More than 1 km high slope with slightly convex profile is

FIGURE 8.261 Toppling on the right-bank slope of the Zeravshan River valley upstream of the Dashtiobburdon village. 3D Google Earth view.

FIGURE 8.262 Adjacent rockslides in the Shamtuch River valley. 3D Google Earth view.

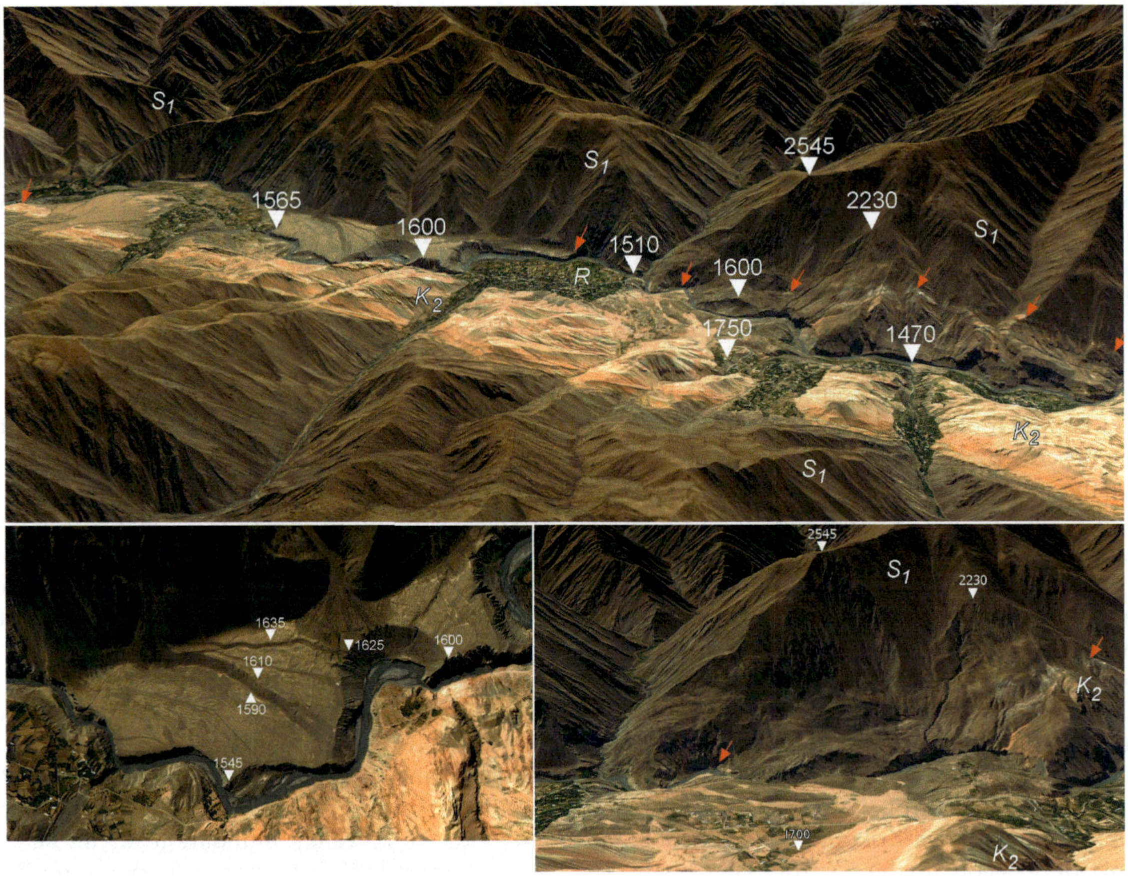

FIGURE 8.263 Above – 3D Google Earth view of Rarz rockslide and the upstream section of the Zeravshan River valley. Rarz rockslide headscarp crown is at 2545 m a.s.l.; rockslide deposits front – at 1750–1770 m a.s.l.; *R* – Rarz village, S_1 – Lower Silurian units; K_2 – Upper Cretaceous layers; *red arrows mark* fault zone between the Silurian and Cretaceous units. Below left – series of the erosional terraces in lacustrine deposits overlain by much younger debris flow fans. Below right – closer view of the Rarz rockslide.

composed of the Silurian terrigenous rocks dipping in the slope. Such combination of the geological structure and landform provides preconditions favorable for toppling (see Section 5.1). The crown of the affected rock massif seems to be identified at ca. 2820 m a.s.l. If it will fail in one large-scale event the assumed volume of such rockslide could exceed 50 Mm^3 and it would provide major hazard for both upstream and downstream parts of the Zeravshan River valley. Regular monitoring of this site should be performed and if evidence of ongoing deformations will be found out, special measures should be undertaken to prevent most adverse effects.

One more site where large-scale slope failure can be anticipated was found on the left side of the Zeravshan River valley above the Rogiv village at 39.374 °N, 69.335 °E (Tajik 078, see Fig. 5.18). This impressive example of the DSGSG was described in Section 5.1 of Chapter 5.

Slope failures in the Zeravshan River valley and in its tributary valleys often occur nearby from each other. It could be seen in the Rivut River valley (see Fig. 8.260) and can be also exemplified by rockslides on the right-bank slope of the Shamtuch River valley (Tajik 045, 046). These failures of the order of 10–30 Mm^3 in volume occurred at 39.46° N, 69.046° E and at 39.45° N, 69.035° E on the 600–800 m high slopes, at the spurs on both sides of the dry gulley (Fig. 8.262). Slopes dipping 27–28° are composed of the Silurian terrigenous rocks. Both headscarps are up to 50–70 m deep. Absence of any evidence of valley inundation upstream from these landslides allows assumption that they are moving gradually, though initial failure could occur catastrophically.

One of the largest rockslide in the Zeravshan River valley (Tajik 028) is located at 39.38° N, 68.72° E just downstream of the Rarz village that gave the name to this feature described first by V. Fedorenko (1988). Failure occurred on the kilometer-high slope composed of the predominantly terrigenous Silurian rocks thrusted over the weak Cretaceous gypsum bearing clays, sandstones and conglomerates (Fig. 8.263). This tectonic setting is somehow similar to that of the Minkush-Kokomeren intermontane depression in the

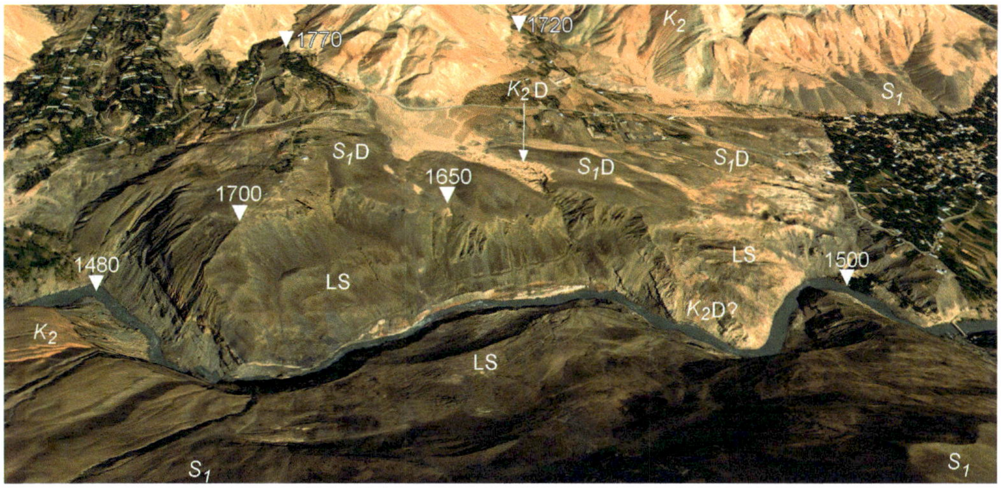

FIGURE 8.264 Internal structure of the Rarz rockslide body. S_1 – Lower Silurian rocks; S_1D – corresponding rockslide debris; K_2 – Upper Cretaceous rocks; K_2D – corresponding rockslide debris; LS – modern rockslides. 3D Google Earth view.

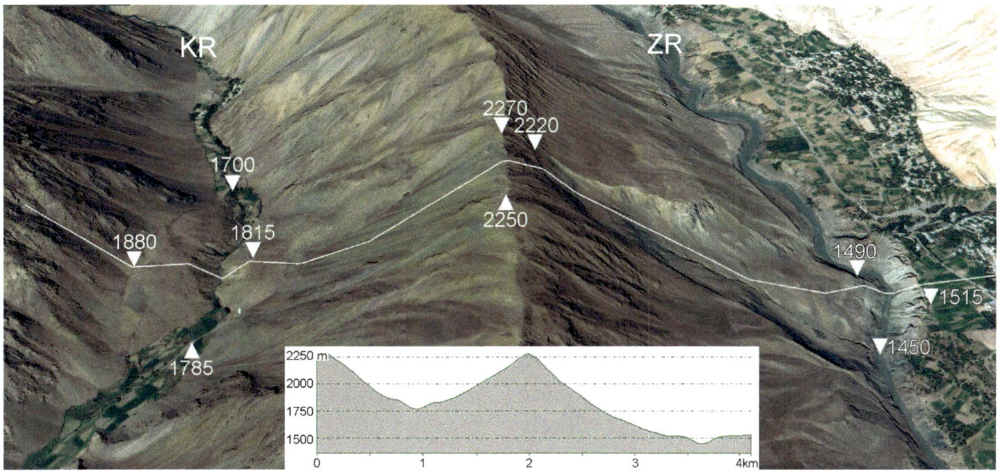

FIGURE 8.265 Two rockslides on both sides of the ridge 6 km east of the Aini town. ZR – Zeravshan River, KR – Kumar River. 3D Google Earth view. The schematic topographic profile along the line crossing both features is shown in in the inset.

Central Tien Shan where the Karachauli and the Lower Kokomeren rockslides are located (see Section 8.4, Figs. 8.172, 8.180). Their headscarps are within hard Paleozoic rocks thrusted over much weaker Neogene conglomerates. However, unlike the Minkush-Kokomeren depression bounded by the reverse faults both from north and south, Cretaceous sediments developed in the Zeravshan River valley have the tectonic contact with Paleozoic units from the south only while from the north they overlay the Paleozoic basement with angular discontinuity. Just west of the Rarz rockslide one more rockslide about 10 Mm^3 in volume occurred (its headscarp crown is marked by 2230 m a.s.l. on Fig. 8.263), likely simultaneous with its "big brother".

Rarz rockslide body formed the dam up to 200 m thick with up to 60–70 m high runup of the frontal part and can be classified as the Primary rock avalanche with frontal confinement (see Section 5.3.1). It covers about 3.6 km^2 of the rather wide valley bottom and extends from the headscarp foot for ca. 2.5 km. Volume of the deposits is estimated as 200 Mm^3 (Fedorenko, 1988). This 2.6 km wide blockage formed a lake that had been silted for about half of its maximal depth before it was breached and incised by the Zeravshan River almost up to its original level (compare with the Beshkiol rockslide dam where the original river level has not been reached yet – see Fig. 8.127). Though banks of this erosional canyon are affected by numerous modern landslides, the internal structure of the deposits can be investigated in canyon walls. "Layers" of debris dipping toward the source zone can be observed in the right bank of the river canyon where debris that originated from the Silurian rocks overlays debris of the Cretaceous sediments

(Fig. 8.264). Contact between these debris units with contrasting colors is sharp, similar to such boundaries at the Inylchek (see Fig. 8.45), the Kokomeren (see Fig. 8.153) and the Ornok (see Fig. 8.168) rockslides described above.

Geomorphological analysis allows reconstructing the post-failure evolution of this part of the Zeravshan River valley. About 4 km upstream of the blockage several terraces were eroded in thick lacustrine sediments that accumulated in the dammed lake (see Fig. 8.263). Presence of the series of such terraces indicates that the dam failed not in a single catastrophic event but it in several steps, similar to case studies shown on Figs. 6.6 and 8.57. One more important observation is the presence of the small debris flow fans that overlay these erosional terraces instead of forming the polygenetic surface with the fluvial deposits (see Fig. 8.263). Such relationships indicate that debris flow fans are much younger and started forming when the blockage was breached and lake sediment were eroded. It provides some chance to date the dam's breach.

Next interesting case study in the Zeravshan River valley could be found 6 km east of the Aini town at 39.372° N, 68.607° E where two rockslides about 10 Mm^3 each (Tajik 025 and Tajik 027) collapsed from both sides of the narrow ridge composed of the Silurian and Carboniferous terrigenous rocks. It divides the Zeravshan River valley from the Kumar River valley that falls into Fandaria River 3 km upstream from its confluence with the Zeravshan (see Figs. 8.259, 8.265). Northern rockslide originated on the 800 m high slope and likely dammed the Zeravshan River by the ca. 50 m high dam that had blocked the erosional canal incised in the 50–55 m high terrace. Powerful Zeravshan should overtop and erode this dam soon after its formation. It can be assumed that this failure occurred on the eastern side of the larger and older landslide, which can be derived from slope morphology.

The southern slope of the ridge is much lower – about 530 m only. The dam that had blocked the Kumar River valley later was also incised almost completely but, despite the incision, its body with an impressive ca. 70 m runup retaines its shape that allows classifying it as a Primary rock avalanche in frontally confined conditions (see Section 5.3.1).

Such association of large-scale slope failures that affected both sides of the same ridge can be considered as an indication of their seismic triggering (see also rockslides shown on Figs. 8.26 and 8.38 and the Kokomeren and the Sarysu rockslides described in Section 8.4). Same assumption for rockslides in the study area was proposed in (Fedorenko, 1988). It will be discussed further in Chapter 13, Seismically induced rockslides: their discrimination.

As we mentioned already, the increased awareness on landslide hazard in the Zeravshan River basin was stipulated at a large extent by the 1964 Aini rockslide disaster described in brief in Chapter 2 (see also (Neshikhovskiy, 1988; Strom, 2010b, 2013a)). This slope failure occurred within the source area of much larger Pleistocene rockslide with headscarp crown at 2720 m a.s.l. (see Figs. 2.2, 8.266) and there are sound geomorphic indications that large-scale rock slope failures at this site could occur in future. Besides arcuate scarp just above the headscarp crown of the older landslide marked on Fig. 2.2, there is a 1.9 km long sackungen at the ridge watershed just west of it (see Fig. 8.266) indicating ongoing large-scale slope deformations on top of the ridge composed of the Silurian and

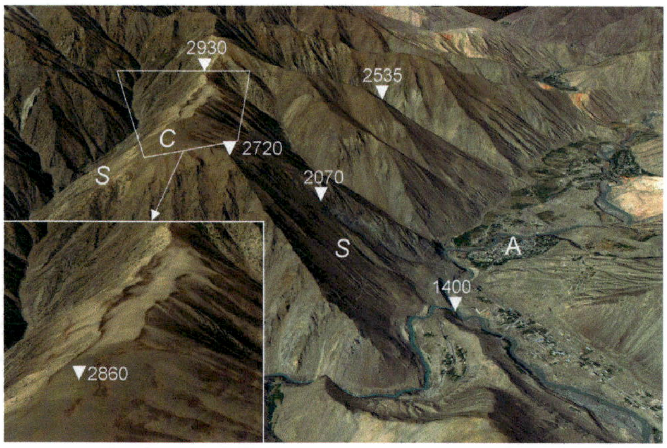

FIGURE 8.266 Sackungen along the watershed of the ridge south-west of the 1964 Aini rockslide. Its closer view is in the inset. S – Silurian units; C – Carboniferous unit; A – the Aini town. 3D Google Earth view.

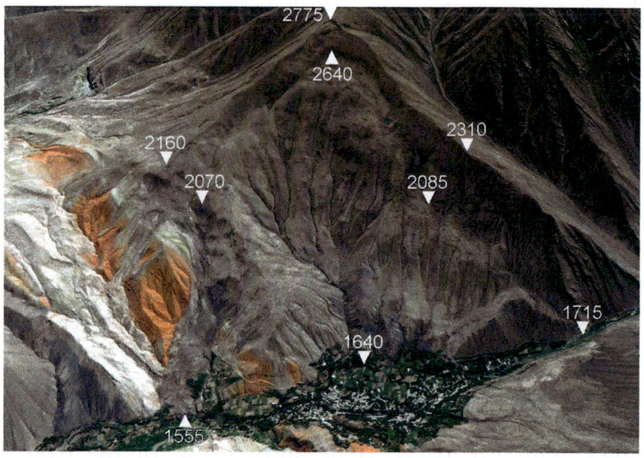

FIGURE 8.267 The Kum rockslide. 3D Google Earth view.

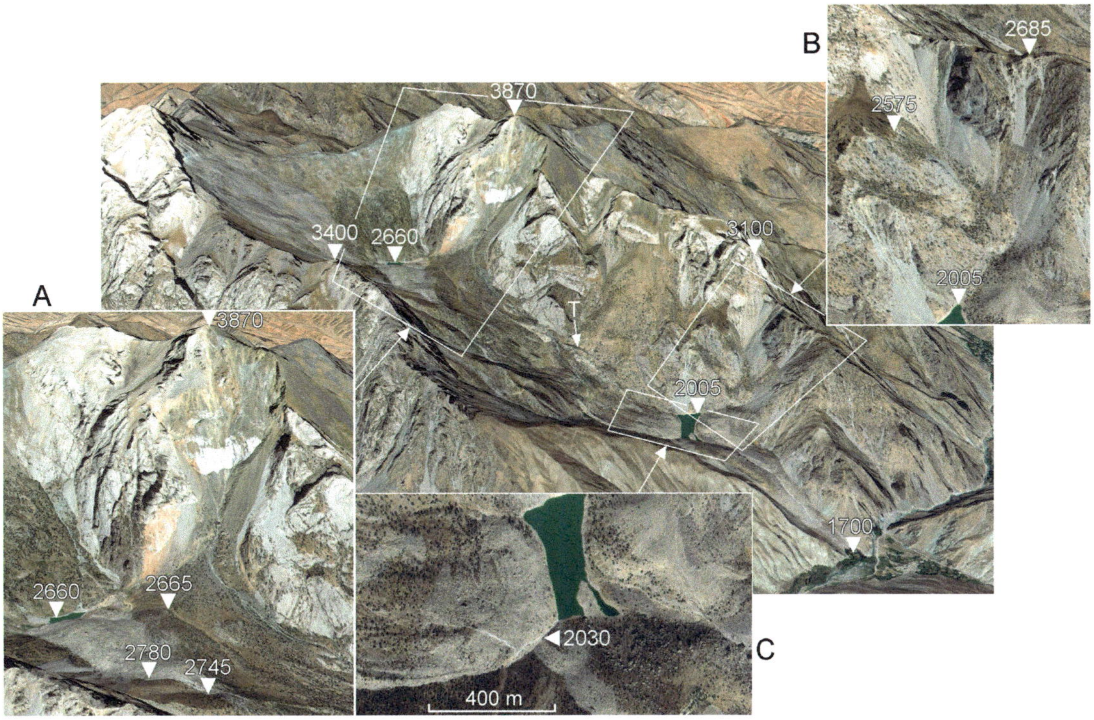

FIGURE 8.268 The Khushtarita rock avalanches. A – headscarp of the Khushtarita Western rock avalanche; B – that of the Khushtarita Eastern rock avalanche; C – overlapping of the eastern rock avalanche body by the terminal past of the western one; T – the trimline left by the western rock avalanche. 3D Google Earth views.

Carboniferous terrigenous rocks. The scarp likely coincides with the uncomfortable boundary between the Silurian and Carboniferous units. Considering the location of this suspicious feature just above the Aini town, it should be studied in more details and monitored if necessary to understand its nature and behavior better.

Remnants of one older rockslide can be seen on the left-bank slope of the Zeravshan River valley about 2 km from the 1946 Aini roclslide. Its headscarp crown at 2535 m a.s.l. is marked on Fig. 8.266.

About 12 km west of the Aini town at 39.408° N, 68.407° E large rockslide nearly 200 Mm^3 in volume (Tajik 086) originated on the 1100 m high right-bank slope of the Kum Creek, just above the Kum village, composed of the Silurian terrigenous rocks thrusted over the clay, sandstone, conglomerate and marl of the Paleogene age. It has a well-pronounced headscarp with more than 100 m high back wall and rather smooth surface partially reworked by erosion. Most likely it evolves gradually, not resulting in river damming. Interesting peculiarity of this feature is the transformation of its right-side part into earthflows up to 1.0–1.2 km long that cover Paleogene marls and red beds (Fig. 8.267). One of them moved from 2070 to 1555 m a.s.l.

FIGURE 8.269 Large-scale rockslides in the vicinity of the Big Allo Lake. Notice color contrast between the main part of the rock avalanche damming this lake and its proximal part. 3D Google Earth view.

Dense cluster of rockslides could be found about 25 km west of the Aini town on the southern side of the Zeravshan River valley between its left tributaries – the Madm and the Rivat Creeks (39.42° N, 68.25° E, see Fig. 8.259). Most of them occurred on slopes composed of weak Jurassic, Cretaceous and Paleogene coal-bearing and gypsum-bearing terrigenous and, rarely, carbonate sediments or on slopes composed of the Silurian terrigenous sediments thrusted over Mezo-Cenozoic units.

Several interesting case studies could be found in the highly elevated western part of the Zeravshan

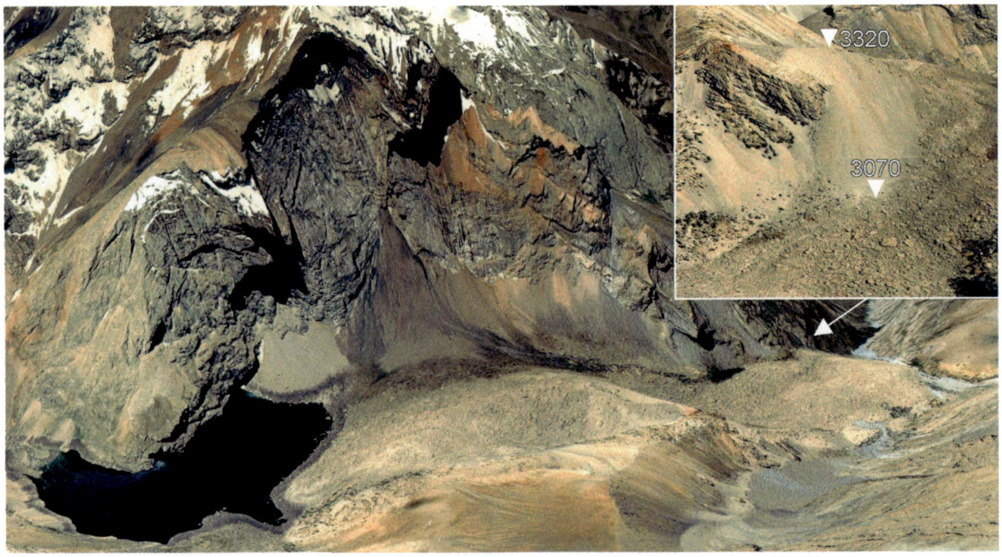

FIGURE 8.270 Front view on the Big Allo rock avalanche. Sharp difference between coarse carapace of rock avalanche body and its intensively comminuted interior can be seen in the inset. 3D Google Earth view.

Range composed of Paleozoic sedimentary and metamorphic rocks with complex geological structure, in the Kshtut River basin.

Two large-scale rock slope failures occurred on the south-facing slope 3.5 km from each other and the small Khushtarita Lake accumulated in the closed depression between their bodies (Fig. 8.268). The bigger one – the Khushtarita Western rock avalanche (39.27° N, 67.98° E, Tajik 040) originated as a ca. 250 m deep wedge failure on the more than 1200 m high slope dipping 39°. It crossed the valley with ca. 120-140 m runup, turned left and traveled downstream with total runout of 5.23 km. On the way it left the trimline on the left slope of the valley about 80 m above the final debris surface (T on Fig. 8.268). Total volume of these deposits is about 150 Mm^3.

The second one – the Khushtarita Eastern rock avalanche (39.26° N, 68.02° E, Tajik 040a) – is smaller – about 40 Mm^3 only. It also originated as a ca. 200 m deep wedge failure on the 750–1000 m high slope and moved similarly to the western one – crossed the valley with almost 140 m runup, turned left and moved downvalley practically up to the Kshtut River at 1700 m a.s.l. Total runout of this feature is 3.15 km.

Just south from the Khushtarita Lake one can see that the distal part of the western rock avalanche overlays the deposit of the eastern one (see inset C Fig. 8.268). However, judging from the similarity of the deposits' landscape of both Khushtarita rock avalanches, most likely they occurred simultaneously, but, due to significantly different distance from the headscarps to the confluence of rock avalanche bodies, the frontal part of the western one overlays the eastern one. Similar relationships could be observed at some sites of the Sarmin-Ula rockslide cluster shown on Fig. 8.14. A the same time it differs from the conjugated bodies of some other doubled rock slope failures such as the Bielogorka rock avalanches in Northern Tien Shan (see Fig. 8.99) and the Ikhnachkul rock avalanches in North-Western Tien Shan (see Figs. 8.226 and 8.227).

FIGURE 8.271 The 1949 Khait rock avalanche and older rockslide dam in the Obi-Khauz-Dara River valley. Google Earth image.

One more couple of large rockslides is located about 12 km west of the Chimtarga Peak (5490 m a.s.l.) – one of the highest summits of the Southern Tien Shan. One of them – the Big Allo rock avalanche (39.21° N, 68.115° E, Tajik 061) dams the Zindan River and forms the Big Allo Lake; one more (Tajik 061a) – blocks it right tributary just upstream from the lake at 39.216°

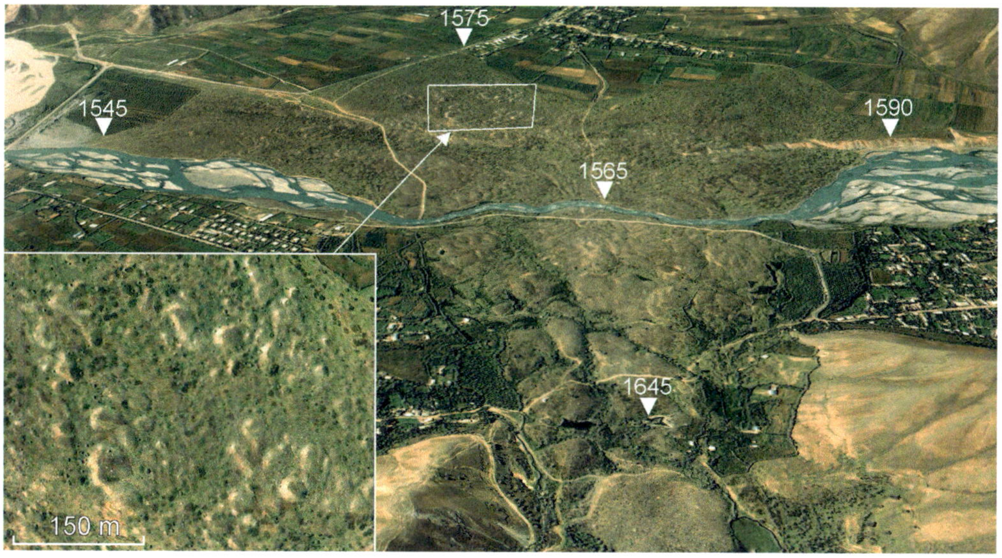

FIGURE 8.272 3D Google Earth view of the distal part of the Khait rock avalanche. Conical hummocks on its surface are shown in the inset.

N, 68.137° E (Fig. 8.269). There is a significant likelihood that both slope failures occurred simultaneously but contact between their bodies is hidden under the lake and only dating of both features could help to prove or disprove this assumption.

Both failures occurred on ca. 1 km high and very steep – up to 56° – slopes composed of well bedded schist and can be classified either as irregular or, even, as planar slides (see Figs. 8.269, 8.270). Though debris of the Big Allo rock avalanche filled the valley bottom with a stream gradient of about 7° at a distance of 2900 m, it can be classified as a Primary rock avalanche with frontal confinement (see Section 5.3.1) and high – up to 140 m runup. Looking from downstream on this rock avalanche deposits ca. 100 Mm^3 in volume one can see sharp difference between thin blocky carapace with numerous blocks up to 20–30 m in size and the deposits interior composed of much finer material (see inset in Fig. 8.270).

One more peculiarity of this case study that can be derived from the remote sensing analysis is the significant – up to 10–15 m – seasonal variability of the Big Allo Lake level. Seeping water appears just at the downstream tip of the deposits.

The second rockslide in the vicinity of the Big Allo Lake is smaller – about 30 Mm^3 in volume, and of the same morphological type as the Big Allo rock avalanche. Its debris also filled the bottom of the narrow glacial valley with 8° gradient and rose on the opposite slope for about 80 m spreading along the valley so that its downstream tip is hidden under the lake (see Fig. 8.269). The small dammed lake formed in the deep trough valley with kilometer high steep walls was silted, most likely, very soon and the small river eroded the upstream section of the dam at its proximal lowering and then disappears seeping through the rock avalanche deposits towards the Big Allo Lake.

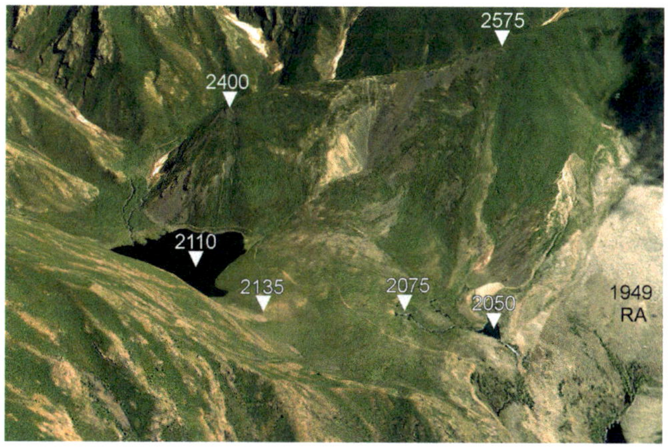

FIGURE 8.273 The Pre-Khait rockslide dam in the Obi-Khauz-Dara River valley. 1949 RA – deposits of the 1949 Khait rock avalanche. 3D Google Earth view.

One of the most famous and, definitely most disastrous (see Chapter 2) rockslide of the entire Central Asia region was the 1949 Khait rock avalanche (39.19° N, 70.9° E, Tajik 038) that collapsed from the more than 1 km high north-western slope of the Chokrak mountain leaving the impressive irregular headscarp up to ~300 m deep (see Fig. 5.15). According to (Leonov, 1960) failure occurred along the tectonic contact between gneiss, schist and granites. Besides the tremendous amount of fractured bedrock seismic shaking mobilized large amount of loess-like loam widely developed on the slopes in this part of the Tien Shan that was mixed with rock avalanche debris.

FIGURE 8.274 The Kafirnihon rockslide dam. 3D Google Earth view.

According to personal communication of N. Leonov the Khait earthquake was preceded by several days long rainy period that resulted in the excessive saturation of the loess-like cover and facilitated its sliding. Mixture of rock debris and saturated loess-like loam rushed along the Obi-Khauz-Dara River valley, entered wide and flat valley of the Yarhych River – the right tributary of the Surkhob River, blocked its flood plane and spread over the right-bank terrace forming the 1.5 km long and 2.6 km wide fan-shape body up to ca. 20 m thick (Fig. 8.271). Later on this low dam was eroded and incised by the Yarhych River.

Surface of this fan-shaped body is hummocky and covered by numerous low conical hills (Fig. 8.272) resembling rock avalanches shown on Figs. 4.22, 4.23, 6.9, 8.18, 8.50, 8.93, 8.134-A, 8.232. However, large proportion of loess-like loam in this material makes its surface smoother than that of many other rock avalanches with carapace composed of coarse bouldery material only (see, e.g. Figs. 5.37, 8.50).

The total runout of the Khait rock avalanche was 7.41 km and the total area covered by the deposits – 5.045 km^2. Published estimates of the volume of this extraordinary historical event are controversial (see Section 6.3). Earlier researchers (Leonov, 1960, Nikonov and Shebalina,1986), following the first description of G. Panasenko (1952) estimated the total volume as 400–500 Mm^3, which seems to be exaggerated significantly. More reliable estimate – ca. 75 Mm^3 – was provided in (Evans et al. 2009b). From our point of view, considering area of the deposits (5 km^2) and their mean thickness of about 20 m, its volume could be of the order of 100 Mm^3.

As it can be seen on Fig. 5.15, the left-bank slope of the Obi-Khauz-Dara River valley was affected, aside of the main deep headscarp, by numerous landslides, some of which are shallow, while others could be quite deep-seated and create prerequisites for future slope failures. Most likely they were formed by the same earthquake. Besides, at 39.201 °N, 70.913 °E this river valley is dammed by an older rockslide just upstream from the site where the Khait rock avalanche turned on its way from the headscarp (Tajik 037, Fig. 8.273). This natural dam existed before the 1949 earthquake, but its ca. 500 m high headscarp was rejuvenated by this event, along with small landslide on the same slope just upstream from the lake (its headscarp crown is marked by 2400 m a.s.l. on Fig. 8.273). According to lichenometric data (Nikonov and Shebalina, 1986) this natural dam was

FIGURE 8.275 Long runout rock avalanche on the right-bank slope of the Karatag River valley. Deep erosional canyon incised in the dam's body with distinct dual internal structure of the deposits is shown in the inset. 3D Google Earth view. Notice that space image overlays the DEM with some distortion.

FIGURE 8.276 3D Google Earth view of the Timurdara rockslide and rock avalanche in the Karatag River valley (see its plan view in the inset).

formed between 1610 and 1750 A.D. Some grooves on the same slope were dated by the same method as "before 1300 A.D." [ibid]. If these age estimates are correct and if the assumption that all these features were triggered by past earthquakes, their recurrence interval could be about 300 years. With a freeboard of ~25 m this rockslide dam has never been overtopped and water seeps through its body with ~3° water-table gradient.

One of the largest rock slope failures in the entire Southern Tien Shan was found in the upper reaches of the Kafirnihon River valley that divides the Gissar and Karategin Ranges (Tajik 052). Failure of about 800 Mm^3 of Paleozoic granites and metasediments took place at 38.96° N, 69.69° E on the ca. 1400 m high right-bank slope of this valley (Fig. 8.274). Upper part of the sliding surface, within metasediments, dips 32°, while dip angle of its lower part in granites increases up to 39°. Its giant triangular headscarp is nearly planar, with poorly pronounced concavity, similar as sliding surfaces of the 1911 Ananievo rockslide (see Fig. 8.71) or the prehistoric Mingteke rock avalanche (see Fig. 8.140).

This old deeply incised rockslide dam up to 450 m high that had blocked river valley at a distance of 3000 m resembles the Aksu rockslide in Northern Tien Shan (see Fig. 8.103). Both can be classified as Primary rock avalanche in the frontally confined conditions (see Fig. 5.31-1b) with post-failure erosional canyon at the proximal lowermost part of the dams. We expect that the original river valley was more straight and passed closer to the left side of the valley, presumably along the lower part of the dams' body at 2375 m a.s.l. (see Fig. 8.274) The internal structure of the rockslide body can be studied in the ~200 m deep erosional gorge.

Four river-damming rockslides are in the Karatag River valley and in the valley of its left tributary – the Pairon River that dissect the southern slope of the Gissar Range north-west of the Dushanbe City.

FIGURE 8.277 The Pairon rockslide dam. 3D Google Earth view.

The southern one (Tajik 012) is the 5 km long rock avalanche that originated on top of the 1.7 km high slope composed of Carboniferous volcanic rocks, traveled along the steep gulley leaving almost empty headscarp and finally had blocked the Karatag River valley at 38.736 °N, 68.358 °E by the ca. 50–60 m high dam with 70–75 m frontal runup (Fig. 8.275). This case study slightly resembles rock avalanche in the Chon-Uriukty River basin north of the Issyk-Kul Lake shown on Fig. 8.74. This dam survived likely due to rather low gradient of its surface so that small dammed lake was silted almost completely. At present the dam is dissected by the erosional canyon up to ~40 m deep at the dams' axis, while at the upstream 250 m wide part of the blockage depth of the incision does not exceed 10 m. In the canyon one can see that below thin carapace composed of angular boulders up to 10–20 m in size most of the dams' body consists of heavily comminuted material (see the inset in Fig. 8.275).

About 10 km north of the site described above, at 38.82° N, 68.4° E large rockslide – the largest one in this area (Tajik 053) – blocked two small merging valleys that fall from the east in the Karatag River and formed the small Timurdara Lake (Fig. 8.276). One more, much smaller rock avalanche (Tajik 053a) had blocked the Karatag River valley 2 km of it at 38.826° N, 68.38° E.

The Timurdara rockslide collapsed from the upper part of the ca. 1 km high slope composed of Paleozoic granites. The true right part of this body, which total volume estimate is of the order of 40 Mm^3 reached the slope base at the confluence of two valleys, climbed up to 90 m on the slope of the ridge dividing two valleys

FIGURE 8.278 The Aigyrkul rock avalanche body and its assumed source zone at 3710 m a.s.l. 3D Google Earth view.

and formed a lake in the right one, while the dammed part of the left one at 2110 m a.s.l. was silted completely. At present both parts of the dam have surficial outflows, at least in the flood period, but they do not erode the dam body, unlike the vast majority of the case studies described above. The true left part (west of the point marked by 2180 m a.s.l. on Fig. 8.276) accumulated on the slope bench about 300 m above the valley bottom. Perhaps this case study could be classified as Jumping rock avalanche (see Section 5.3.2), but it is not so well pronounced as the Kashkasu (see Fig. 5.45) or Northern Karakungey (see Fig. 5.43) cases.

3 km north of the Timurdara rockslide headscarp about 6 Mm^3 of the Paleozoic metasediments collapsed from the 800 m left-bank slope of the Karatag River valley. This rock avalanche is similar to that shown on Fig. 8.275 – in both cases failures occurred on top of the ridge and almost all material "flowed" down along steep dry gulley and formed a compact dam in the river valley, thus representing good examples of Primary rock avalanches confined, first, laterally, and at the end – frontally. In this case the dam is very compact, because rock avalanche entered in the gulley mouth as in a trap, similarly to the Aksu rockslide shown on Fig. 8.103. Most likely it was breached soon after the overtopping and the Karatag River cut the dam practically to its base.

About 6 km north of the Timurdara rockslide, at 38.87° N, 68.41° E one more rockslide (Tajik 054) originated on the ca. 1100 m high slope with the same aspect, also composed of the Paleozoic granite. It blocked the Pairon River – left tributary of the Karatag River and formed the Pairon Lake, slightly larger than the Timurdara Lake. Such uniform orientation of the nearby rockslides was described in Central Tien Shan (see Karakungey rock avalanches shown on Fig. 5.52), in the Sarmin-Ula cluster in the Eastern Tien Shan (see Figs. 8.11, 8.15) or in Dzungaria (see Figs. 7.3, 7.4, 7.8). It could be considered as an argument in favor of their seismic triggering, if simultaneity of such adjacent rock slope failures could be substantiated.

FIGURE 8.279 Closer 3D Google Earth view of the Aigyrkul rock avalanche body.

This slope failure of either wedge or irregular type in Paleozoic granites (see Section 5.1) with estimated volume of about 10 Mm^3 formed up to 150 m high dam with distinct distal lowering (Fig. 8.277), that differs it from other rockslide dams in the Karatag River basin. The effective dam's height is about 100 m. Lake has been partially silted by sediments transported by the river. Water is released both by surface overflow along the dams' front and by filtration. Seeping water

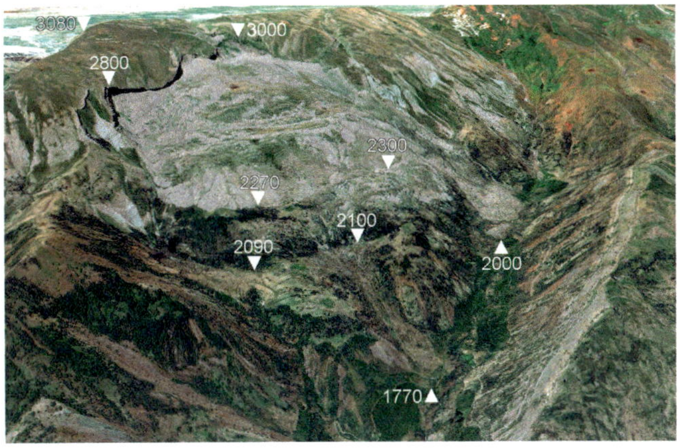

FIGURE 8.280 Translational slide in the Mesozoic sedimentary rocks south of Chash River. 3D Google Earth view.

FIGURE 8.281 Part of the Shing River valley dammed by numerous blockages. 1 – the Nezhigon, 2 – the Soya, 3 – the Gushor, 4 – the Nofin, 5 – the Hurdak, 6 – the Marguzor, 7 – the Hazorchashma Lakes; 8 – silted lake; 9 – rockslide in the tributary valley. Google Earth image.

comes out on the downstream face of the dam about 45 m below lake level, which corresponds to the water table gradient of about 10° that means that filtration takes place through the upper part of the dams body only. Same phenomenon has been described at the Usoi dam in Pamir (Gaziev, 1984; Fedorenko, 1988; Alford and Schuster, 2000; A. Ischuk, 2011).

One more interesting case study (Uzbek 013) is located at 38.91° N, 67.93° E in the Tamarhut Creek valley (the Tupolangdaria River basin). We named it the Aigyrkul rock avalanche after the name of a small lake dammed by rock avalanche deposits. It is a strange feature composed of clastic material that fills one square kilometer of the valley bottom at a distance of about 2.3 km. The problem is that there is no such a distinct headscarp on the surrounding slopes from which this material could collapse (Fig. 8.278). From the first glance it could be assumed that headscarp rises just above the Aigyrkul Lake (slope with elevation mark 2700 m a.s.l. at the upper left part of Fig. 8.279). However, part of the body between elevation marks 2570 and 2620 m a.s.l. could not originate here since it had to cross the ridge stretching from 2620 m a.s.l. towards the elevation mark 2700 m a.s.l. From our point of view the most reliable source of this material is the up to 200 meter deep circus-like depression with steep back wall near the ridge crest more than 3 km south-west of this body and 1200 m above it (see Fig. 8.278). Glacial origin of this circus is unlike due to its uniqueness (see Section 4.2; compare with Upper Kashkasu rock avalanche headscarp shown on Fig. 4.44). If our assumption is true, about 30 Mm^3 of the Paleozoic granite had passed along the gulley with mean angle of about 21° (from 26° on top to 15° below), collided with a ridge shown on Fig. 8.279, split into two parts and the main one reached altitude 2220 m a.s.l. with 5.4 km total runout, thus being one of the longest rock avalanches in this part of the Tien Shan mountains.

Another interesting case study in the Tupolangdaria River basin (Uzbek 001) was found on top of the 1.5 km high ridge south of the Chash River at 38.674 °N, 67.675 °E. Ridge is composed of the Jurassic, Cretaceous and Paleogene sedimentary rocks dipping 20–21° towards north-west and overlying the Paleozoic basement with angular unconformity. This rock slope failure is a typical planar (translational) slide ~4.5 km^2 in size, which sliding surface coincides with bedding plane (see Section 5.1). Its 2 km long back wall up to 50–80 m high at 2800–3000 m a.s.l. is very steep, locally vertical (Fig. 8.280). Downslope advance of rockslide body was rather limited, except two frontal appendices that reached the creek bottom at 2000 and 1770 m a.s.l. Despite the fact that about 250 Mm^3 of well-layered sedimentary rocks slid along the bedding plane, rockslide body crushed at a rather large extent so that it could be classified as flowslide rather than a blockslide, especially considering its distal appendices.

Judging from the vegetation development on rockslide body it can be assumed that there were two main stages of sliding. The deposits of the first one formed the lower, more vegetated belt between 2090–2100 m a.s.l. and 2270–2300 m a.s.l. marked on Fig. 8.280, while the younger one, above this level, is almost free of grass and bushes.

Finishing this Section we want to describe two groups of river-damming blockages in the western part of the Zeravshan Range forming the lakes that are among the most well known water bodies in the entire Southern Tien Shan – the Iskanderkul Lake and 7 lakes in the Shing River valley which level ranges from 1600 to 2400 m a.s.l. Origin of these blockages is still debatable and two opposing points of view have been proposed (since most of their descriptions are in unpublished technical reports hardly available now,

FIGURE 8.282 3D Google Earth views of the three downstream blockages: A – that forms the silted lake, B – of the Nezhigon Lake, C – of the Soya Lake.

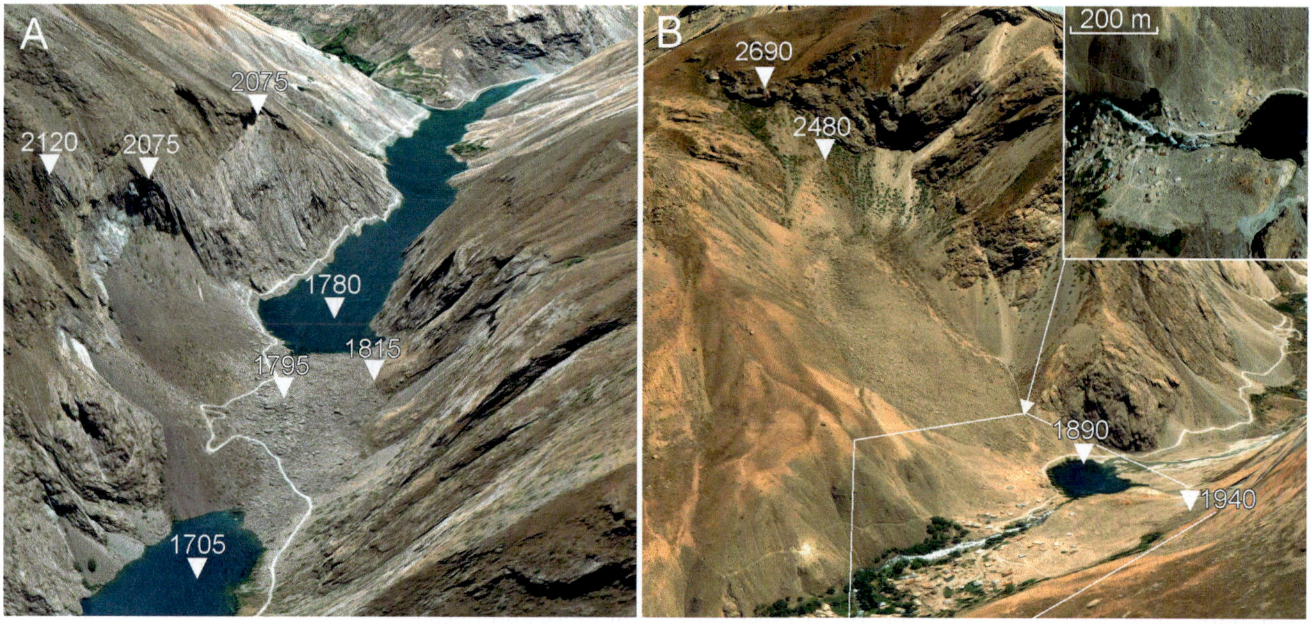

FIGURE 8.283 3D Google Earth views of the Nofin Lake (A) and the Hurdak Lake (B) dams. Water body at the foreground at 1705 m a.s.l. is the Gushor Lake separated from the Soya Lake by debris flow dam. Plan view of the Hurdak Lake dam with distinct evidence of the across-valley debris motion is shown in the inset.

we do not include all of them in the References). One of these viewpoints treats these blockages as large-scale, likely seismically triggered rockslides (Tadjibekov and Tokarev, 1990; Tadjibekov, 1991), while the opposite one considers all of them as end moraines left by Late Pleistocene – early Holocene glaciation (N. Ischuk, 2016). Both interpretations are based on field observations and sometimes their geomorphic and geological peculiarities are treated in different way to prove these opposing viewpoints. We must say that from our point of view rockslide origin of most of these blockages seems to be more reliable.

FIGURE 8.284 A – the Marguzor Lake dam and its possible source zones on both sides of the valley; B – view on the Shing River valley left side at the Marguzor Lake dam. 3D Google Earth views.

Some arguments were discussed in Sections 4.1 and 4.2. Here we will present geomorphic and sedimentological arguments mainly to substantiate our position.

Seven blockages interrupt the flow of the Shing River – left tributary of the Zeravshan River at a distance of about 15 km forming 7 beautiful lakes (also named the Marguzor Lakes) (Fig. 8.281). The valley is deep and narrow with nearly vertical walls several hundreds meters high at their lower parts. Four of the seven blockages (1, 2, 4 and 5 on Fig. 8.281) have distinct headscarps at the right-bank slope of the valley (Fig. 8.282-B, C, Fig. 8.283). One more landslide downstream from the 1st lake dammed at 39.225° N, 67.802 E° (Tajik 060a), also with headscarp on the right side of the valley, had dammed the valley too at 39.248° N, 67.798° E (No 8 on Fig. 8.281), but this lake was silted and does not exist at present (No 8 on Fig. 8.281, Fig. 8.282-A). One more rockslide (Tajik 125, No 9 on Fig. 8.281) was identified in the tributary valley at 39.235° N, 67.78° E).

The Gushor Lake is divided from the Soya Lake by debris flow fan from the left bank tributary valley and has nearly the same level, may be ~5 m higher. Initially it was a single water body dammed by the

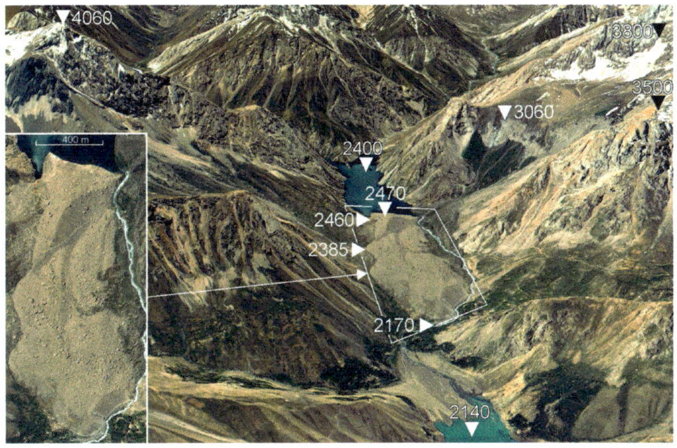

FIGURE 8.285 3D Google Earth view of the Hazorchashma Lake dam. Plan view of the dam's body in the inset.

rockslide. This valley represents an excellent example of the "Disturbance regime landscape" proposed by K. Hewitt (2006a). We must notice that different sources provide different values of lake levels. Those indicated on Figs. 8.282–8.285 are from the 1:50 000 State topographic maps of the USSR rounded to 5 m.

All three downstream blockages were formed by rather shallow rockslides that originated on ca. 700 m high right-bank slope of the Shing River valley composed of Paleozoic metasediments with complex geological structure (see Fig. 8.282). Dam of the Soya Lake (Tajik 060, 39.214° N, 67.804° E) demonstrates evidence of two-phase slope failure. Its main part is formed by gray-color debris that, likely, originated from the headscarp with its crown at 2160 m a.s.l., while upper part, coarser and more brownish in color – from the slope above it (up to 2370 m a.s.l.).

Rockslide that forms the Nofin Lake dam (Tajik 059, 39.198° N, 67.818° E) originated as a rather deep planar slide on the right side of the valley and its debris has a ~20 m runup on the opposite bank (see Fig. 8.283-A). Similar across-valley motion of debris with up to 50 m runup formed the dam of the smallest Hurdak Lake (Tajik 058, 39.167° N, 67.84° E, see Fig. 8.283-B). Younger gradually moving landslide with its headscarp crown at 2480 m a.s.l. can be seen above the lake. We should notice that the presence of distinct headscarps above all these blockages is an important argument in favor of their rockslide origin.

The Marguzor Lake dam (Tajik 002, 39.154° N, 67.847° E) has morphologically expressed concave lowerings on both sides of more than kilometer-deep valley (Fig. 8.284-A), more expressive than those above the Nofin Lake dam (see Fig. 8.283). Coarse bouldery accumulation that rests on top of the blockage close to the left bank definitely originated from the left-bank headscarp (Fig. 8.284-B). However, the main, older

FIGURE 8.286 Slope failures from the banks of the Nezhigon (A) and the Nofin (B) Lakes and debris accumulation at their feet. 3D Google Earth views.

river-damming body was sourced either from the right-bank headscarp, or could be the end moraine of the valley glacier as it was proposed by N. Ischuk (2016). Judging from the deposits morphology with distinct asymmetry (left-bank side is about 60 m higher than the right one) we tend to its rockslide origin. If it is true, this rockslide about 40 Mm^3 in volume, the largest of all Shing blockages described above that forms the ~170 m high dam, could be classified as rock avalanche of Jumping type (see Section 5.3.2, Fig. 8.284-B).

FIGURE 8.287 The Iskanderkul Lake dam and its assumed headscarp. 3D Google Earth view.

FIGURE 8.288 The Sarytag blockage upstream of the Iskanderkul Lake. 3D Google Earth view.

Origin of the uppermost Hazorchashma Lake fed by the Hazorchashma and Chapdara Rivers seems to be more complex and this blockage (Tajik 003, 39.12° N, 67.86° E) could be really formed by combination of glacial and gravitational landforms. Similar to the Marguzor Lake dam it has dual structure – its upper part is composed of angular clastic accumulation with largest boulders up to 20–30 m in size with very sharp eastern boundary, while the lower one visible in the arcuate scarp at the upper part of the inset in Fig. 8.285 – of much finer clastic material (Fig. 8.285). Such composition along with the orientation of the arcuate scarps up to 80 m high on top of the deposit indicate that coarse bouldery material on top of this dam originated from the left-bank tributary valley with elevation marks 3800 and 3060 m a.s.l., most likely as rock avalanche about 20 Mm^3 in volume. Such grain-size distribution is typical of rock avalanches, while glacial till is much more uniform throughout its cross-section. Possible source of such failure could be at the northern part of this gulley on the southern side of the ridge with elevation mark 3500 m a.s.l. Lower part of the Hazorchashma blockage, however, could be formed by end moraine either of the main valley glacier or of the hanging glaciers that had existed in the past on both sides of the Shing River valley. Additional field studies are necessary to clarify the evolution of this interesting site.

One of the sound arguments in favor of glacial origin of these blockages used by N. Ishuk (2016) is the much smaller depth of the Marguzor Lakes that it could be expected for the water bodies dammed by rockslides (see Fig. 4.25). However, as we mentioned already in Section 4.2, lakes' depth could be reduced significantly by slope failures from the steep valley walls that took place along the entire lakes' length. Evidence of such failures are well expressed at the banks of the Nezhigon and the Nofin Lakes (Fig. 8.286). At the right bank of the smaller Nezhigon Lake talus cones rise above the water level. Though

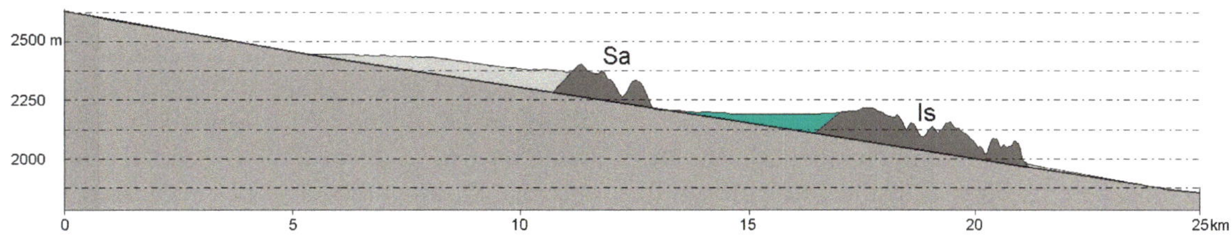

FIGURE 8.289 Schematic profile along the Sarytag and Iskanderdaria Rivers illustrating the disturbance regime landscape. Is – the Iskanderkul Lake dam; Sa – the Sarytag blockage.

FIGURE 8.290 Evidence of high stand of water (blue arrows) in the Iskanderkul Lake. 3D Google Earth view.

we do not see them at the banks of the deeper Nofin Lake, curved rocky cliffs above it must produce significant amount of such debris too.

Another group of river-damming bodies, which origin is still controversial is located near the Iskanderkul Lake in the upper reaches of the Fandaria River basin. Lake is located at the junction of several rivers flowing trough glacial valleys. N. Ischuk (2016) who takes a strong stand on the glacial origin of this blockage relies on the sedimentological observations mainly. However, most of such evidence (coarse carapace, badland relief widely developed on the deposits, internal deformations, etc., could be interpreted in both ways – see Sections 4.1 and 4.2). Since Authors did not perform their own fieldwork at this site, our interpretation in based on geomorphic criteria mainly (Strom, 2010b) along with some considerations on the evolution of the relief in the vicinity of the Iskanderkul Lake.

Besides the blockage that forms the Iskanderkul Lake itself (Tajik 014, 39.09° N, 68.39° E, Fig. 8.287), one more river-damming feature (Tajik 013) had blocked the confluence of the Sarytag and Kanchoch Rivers at 39.05° N., 68.33° E (Fig. 8.288). Slope failure just at this site could be facilitated by the Kanchoch River alluvial fan that could push the Sarytag River towards its left bank, which undercutting decreased its stability (compare with the Aksu rockslide in Northern Tien Shan, see Fig. 8.103). Unlike the much bigger Iskanderkul Lake two dammed lakes in these valleys were silted completely, which, likely, facilitated preservation of the water body downstream, since large amount of sediments was stored upstream of the Sarytag blockage. On the left bank of the Sarytag River valley there is a 950 m high triangular slope covered by scree that can be interpreted as a rockslide headscarp. Lower part of the slope up to 2750–2800 m a.s.l. is composed of pink limestone, while its upper part – of the gray-color sedimentary rocks (see Fig. 8.288). Same succession – gray debris above pink debris – can be seen in the blockage interiors visible in the canyon incised by the Sarytag River in up to ca. 180 m high dam. Such preservation of the lithologies' order is typical of rockslides and was described in numerous case studies above (see e.g. Fig. 8.153).

Both blockages produced the "Disturbance regime landscape" (Hewitt, 2006a) that can be well expressed by the schematic profile along the Sarytag and Iskanderdaria Rivers (Fig. 8.289).

Coming back to the Iskanderkul Lake dam (see Fig. 8.287) we want to highlight some peculiarities of the terrain evolution, which, from our point of view, contradict the hypothesis of the glacial origin of this feature.

As one can see in the canyon incised in the Iskanderkul blockage, its body files the valley down to its bottom at ca. 1970–2000 m a.s.l. It means that at a moment of the blockage formation glacial valley was incised up to this level. The topmost part of the glacier, according to N. Ischuk (2016), is marked by talus accumulations at 2820- 2940–2780 m a.s.l. entering the right-bank slope of the Iskanderdaria River valley above the dam. Thus thickness of this glacier here should be about 1 km. How so thick plastic ice body could stop just here? On the other hand, if the base of the glacial valley was higher than at present, how could the material left by this glacier rest below this level? From our point of view it is impossible.

The nearly 2 km high concave slope rising above the dam, from out point of view is the headscarp of rather old (Pleistocene?) rockslide about 700 Mm^3 in volume (Strom, 2010b) main part of which formed ca. 150 m high dam, while its true right portion crossed the local watershed and moved down the adjacent gulley from 3020 to 1970 m a.s.l. (see Fig. 8.287) (compare with the Dead Lakes rockslide in the Central Tien Shan shown on Fig. 8.205).

Temporal relationships between the Sarytag and the Iskanderkul blockages are unknown and more studies are necessary to reveal them. We can just speculate that they could be formed at the same time and that sediments transported by the Sarytag and Kanchoch Rivers did not reach the Iskanderkul Lake. Originally this lake should be much deeper than at present and high stand of water left distinct trace at about 2305 m a.s.l. ($\sim$110 m above the present day lake level) visible on Fig. 8.290.

CHAPTER

9

Pamir

OUTLINE

The Pamir Mountains ("Roof of the World") extending for about 550 km from east to west and for 300 km from north to south include the highest ranges of the entire Central Asia. Five of six summits exceeding 7 km a.s.l. are located here. Most of rockslides are in the Eastern (Chinese) and Western (Tajik and Afghan) parts of the Pamir with deep tectonic and erosional valleys and at the Trans Alai Range in Kyrgyzstan that bounds Pamir from the north raising above the Alai intermontane depression (Fig. 9.1) (see Chapter 3: Overview of the Study Region). Central Pamir (eastern part of the Gorno-Badakhshan Autonomous Region of Tajikistan) with much gentler plateau-like relief has just few features of this type. To make search of the case studies described hereafter easier for the reader, we divide this chapter into three sections—the first one devoted to the Eastern (Chinese) and Northern (Kyrgyz) parts of the Pamir, the second one to the Western and South-Western Pamir in Tajikistan, and the third one to the Pianj River valley dividing the Tajik and Afghan parts of this mountain system and to its westernmost (Afghan) part.

9.1 EASTERN AND NORTHERN PAMIR

In the north-eastern part of the Pamir composed of intensively folded Cretaceous and Cenozoic terrigenous rocks, most of the rockslides represent wedge failures whose main sliding surfaces coincide with steeply dipping bedding planes. They can be exemplified by several case studies from the Markansu River basin—at 39.55° N, 74.417° E (China 076, Fig. 9.2A) and at 39.562° N., 74.327° E (China 078, Fig. 9.2B). Besides these rockslides, 2.2 km long sackungen passes along the watershed at 39.537° N, 74.309° E (China 077, Fig. 9.3). Concentration of such features within the limited area about 10 × 6 km in seismically active area (see Fig. 3.10) allows assumption that they could be triggered by strong local earthquake.

Interesting 3.95 km long Archabel rock avalanche (China 080) collapsed from the southern slope of more than 1300 m high Archabel Ridge and blocked the Beretokaisu River up- and downstream of its right tributary—the Giulchineksu River mouth at 39.4° N, 74.55° E (Fig. 9.4). Rock mass about 30 Mm^3 in volume slid down along the 620 m high and steep (up to 53 degrees) sliding surface, likely along large fracture, and rushed ahead along much gentler—ca. 20-degree slope. After ca. 800 m descend it collided with the small ridge that split moving rockslide debris into two parts nearly of the same size, both of which traveled afterward more than 2.5 km until they crossed the Beretokaisu River and collided with its right bank. This collision produced not only significant—about 50 m run-up, but also splash of much finer material that covered the right bank of the valley 40 m above the tips of rock avalanche debris (see insets A and, especially, B in Fig. 9.4). We assume that it could be fluvial gravels and silt from the valley bottom mobilized by rock avalanche and differing significantly

Rockslides and Rock Avalanches of Central Asia.
DOI: https://doi.org/10.1016/B978-0-12-803204-6.00009-0

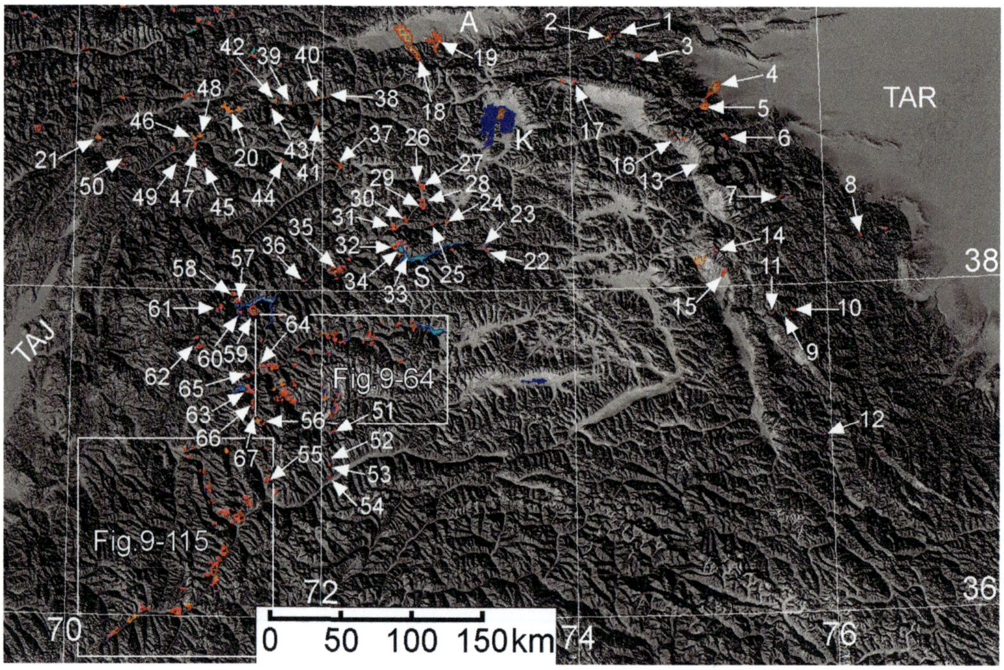

FIGURE 9.1 Shaded 3″ STRM DEM of the Pamir Mountains. S—Sarez Lake; K—Karakul Lake; TAR—Tarim depression; A—Alai depression; TAJ—Tajik depression. 1—see Fig. 9.2; 2—see Fig. 9.3; 3—the Archabel rock avalanche (see Fig. 9.4); 4—the Yimake rock avalanche (see Figs. 6.8, 9.5, and 9.6); 5—the Ke-Keka'er rockslide (see Fig. 9.7); 6—the Aerpa Aigezi rock avalanche and adjacent rockslides (see Fig. 9.8); 7—see Fig. 9.9; 8—the Ye Xiliha rockslide (see Fig. 9.10); 9—the Bile Jiyi rockslide (see Fig. 9.11A); 10—see Fig. 9.11B; 11—see Fig. 9.11C; 12—see Fig. 9.12; 13—see Fig. 9.13; 14—the debatable Muztagh rock avalanche (see Figs. 9.14–9.16); 15—the Taheman rock avalanche (see Fig. 9.17); 16—see Fig. 9.18; 17—see Fig. 9.19; 18—the Koman rock avalanche (see Figs. 9.20–9.23); 19—the Lenin rock avalanche (see Figs. 9.21, 9.24, and 9.25); 20—see Fig. 9.26; 21—see Fig. 9.27; 22—see Fig. 9.28A; 23—see Fig. 9.28B; 24—see Fig. 9.29; 25—see Fig. 9.30; 26—see Fig. 9.31; 27—see Figs. 9.32 and 9.33; 28—see Figs. 9.34 and 9.35; 29—the Damamatsay rockslide (see Fig. 9.36); 30—the Tanymas Mouth rockslide (see Fig. 9.37); 31—the Kudara-Pasor rockslide (see Figs. 4.20, 9.38, and 9.39); 32—the Usoi rockslide (see Figs. 5.7, 9.40–9.47, and 9.49); 33—see Fig. 9.48; 34—the assumed Murgab rockslide (see Fig. 9.49); 35—see Figs. 9.50 and 9.51; 36—see Fig. 9.52; 37—the RGO Glacier rock avalanche (see Figs. 4.34 and 4.35); 38—the Kyzylsu rockslide (see Fig. 9.53A); 39—the Shirvosa rockslide (see Fig. 9.53B); 40—the Satsu rockslide (see Fig. 9.54); 41—the Valter Glacier rock avalanche (see Figs. 4.36-4.38); 42—see Fig. 9.55; 43—see Fig. 9.56; 44—see Fig. 9.57; 45—see Fig. 9.58; 46—the Ragnow-mouth rockslide (see Figs. 9.59 and 9.60); 47—see Fig. 9.61; 48—see Fig. 9.62; 49—see Fig. 9.63A; 50—see Fig. 9.63B; 51—the Zardiv rockslide (see Fig. 9.95); 52—see Fig. 9.96A; 53—see Fig. 9.96B; 54—see Fig. 9.96C; 55—the Sumdjin rockslides (see Fig. 9.97); 56—rockslides in the Garmchashma River valley; 57—the Shidz rockslide (see Figs. 9.98–9.101, and 9.103); 58—the Voznavd rockslide (see Figs. 9.99, 9.101, and 9.102); 59—the Padjvar rockslide (see Figs. 9.103–9.105); 60—the Yorkhdara rockslide (see Figs. 9.103 and 9.106); 61—the Panjai-Mard and Khudardjak rockslides (see Figs. 9.103, 9.107, and 9.108); 62—see Fig. 9.109; 63—the Shiva rockslide (see Figs. 4.22 and 9.110); 64—see Fig. 9.111; 65—see Fig. 9.112; 66—see Fig. 9.113; 67—see Fig. 9.114. Background—the 3″ SRTM DEM visualized by Global Mapper software.

from coarse bouldery debris of the "original" rock avalanche. Sediments transported by both rivers silted dammed lakes. Because rock avalanche deposits and lake sediments look very fresh, this rock avalanche could occur few millennia if not few centuries ago.

One more, much smaller—about 1 Mm^3 only—rock avalanche that originated nearby (China 086, see inset C in Fig. 9.4) passed 1.0 km only and stopped 630 m below its headscarp crown.

Two outstanding rockslides are located at the north-eastern boundary of the Pamir, where this more than 7 km high mountain system merges with the Tarim depression, which bottom is at ca. 2000 m a.s.l. One of them at 39.2° N, 75.15° E is the Yimake rock avalanche (China 148) with 16.36 runout (see Fig. 6.8)—the third longest one after the Koman and Achiktash rock avalanches in the Alai valley that will be described hereafter. Volume of this fan-shape rock avalanche deposit that occupies 46.4 km^2 at the eastern foot of the Pamir was estimated as 1.4 km^3 (Yuan et al., 2013) and it originated 7.1 ± 0.6 ka BP according to ^{10}Be dating [ibid].

Rockslide collapsed from the top of the ca. 1500 m high ridge composed of chlorite schist and limestone and moved over the gently sloping depression bottom. Last 4 km of its travel path is the depression surface with mean angle of 2 degrees only (Fig. 9.5). We disagree with the assumption that its source zone consists of three small separate headscarps (Figs. 8 and 9A from Yuan et al., 2013). As one can see on Fig. 9.6 it originated from the single headscarp (the south-eastern one of three visible at the bottom left corner of

FIGURE 9.2 Wedge failures in steeply dipping red beds. (A) In the Markansu River valley with permanent water flow; angular boulders up to 20 m in size forming rockslide body are shown in the inset. (B) In the small gulley with completely silted lake. 3D Google Earth views.

FIGURE 9.3 2.2 km long sackungen at the crest of the ridge. 3D Google Earth view.

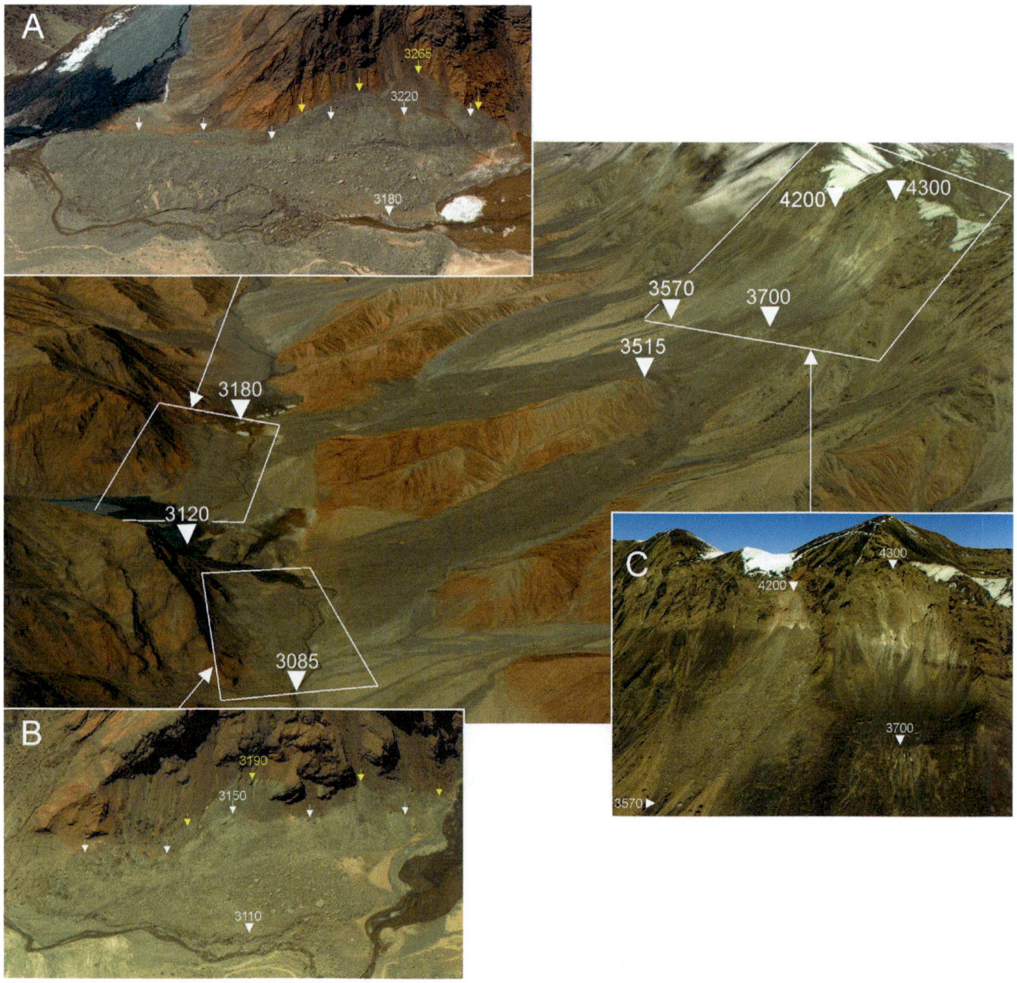

FIGURE 9.4 The bicorn Archabel rock avalanche. A and B-front of both branches of rock avalanche. White arrows mark the run-up of rock avalanche debris; yellow arrows—splash of much finer entrained mud; C-adjacent headscarps of two rock avalanches, smaller of which remains on the slope. 3D Google Earth view.

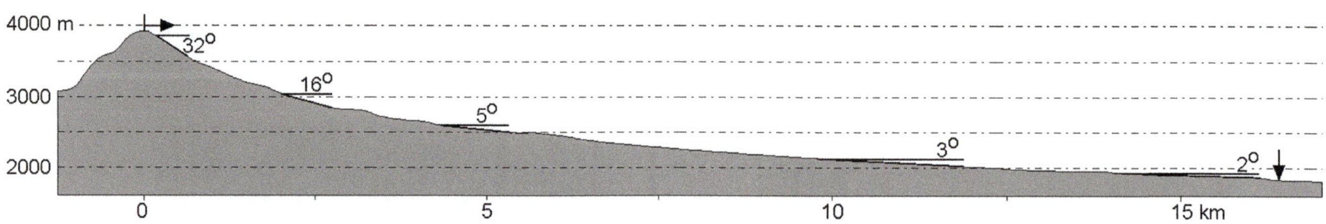

FIGURE 9.5 Longitudinal profile of the Yimake rock avalanche travel path. Arrows mark the position of headscarp crown and deposits' tip.

Fig. 9.6). Besides clear geomorphic evidence it is rather difficult to imagine simultaneous failure from three separate headscarps that would form such a uniform and definitely "single-event" body. Reader can compare it with the Ukok rock avalanche (see Fig. 8.125) that really originated from two adjacent headscarps and whose body has a distinct boundary between parts sourced from either of them.

The Yimake rock avalanche has a hummocky surface and very sharp steep edges up to 15–30 m high (Yuan et al., 2013) indicating abrupt halt of motion. It represents an excellent example of the Primary rock avalanche in unconfined conditions (see Section 5.3.1 in Chapter 5: Bedrock Landslide Types and Classification Systems) that stopped due to basal friction and energy dissipation caused by the internal processes in moving debris—material crushing, internal friction, etc.—only. Similar characteristics have much smaller Chaartash 3 rock avalanche (see Fig. 8.130) and rock avalanche in Xinjiang shown on Fig. 5.32. As it was mentioned in Yuan et al. (2013), at the eastern margin of rock avalanche body it can be seen that its deposits have coarse and blocky carapace underline by fine and cemented basal unit about 8 m thick. Such internal structure is typical of many other case studies described above both in the confined conditions (see Figs. 4.7B, 8.82, 8.96, 8.147, and 8.150) and without pronounced confinement (see Figs. 8.49 and 8.168). It was also described at the well-known Blackhawk landslide (Johnson, 1978).

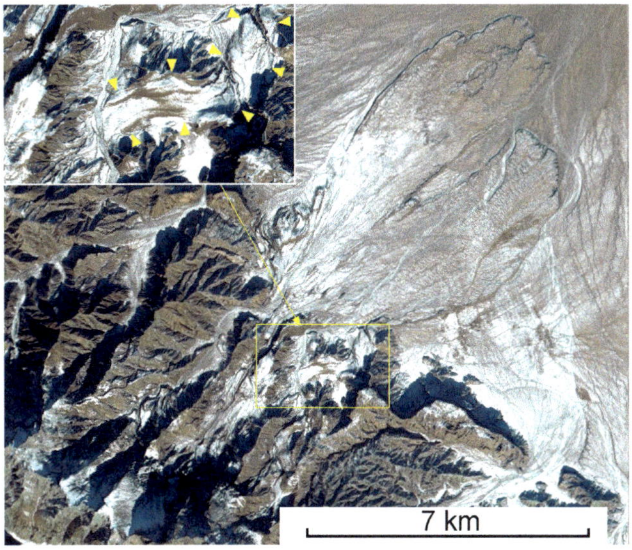

FIGURE 9.6 Right-side branch of the Yimake rock avalanche deposit. Flow-like belt of debris is marked by yellow triangles in the inset. SAS Planet image.

It looks that on its way part of debris crossed the local watershed, entered in the adjacent valley right of the main body, and formed there about 3 km long branch with evidence of flow-like motion (Fig. 9.6). Similar phenomenon was observed at the Dead Lakes rockslide in the Central Tien Shan (see Fig. 8.205) and, likely, at the Iskanderkul rockslide in Southern Tien Shan (see Fig. 8.287).

5 km south of the Yimake rock avalanche headscarp, near the Ke-Keka'er village, the gigantic block of the Silurian metasediments about 3 km^3 in volume slid down from the right-bank slope of the Yelchimnaksu River valley at 39.1° N, 75.075° E leaving the 4.5 km long and 300–450 m high back wall (Fig. 9.7). This compound Ke-Keka'er rockslide (China 149) moved about 3.5 km ahead and blocked the Yelchimnaksu River valley. Remnants of its frontal part can be seen on the left bank of the river and its front is marked on Fig. 9.7 by elevation marks 2940, 3060, and 3100 m a.s. l. Despite its extraordinary size, this 200–250 m high blockage had been eroded, likely, soon after its formation judging from the absence of thick lacustrine

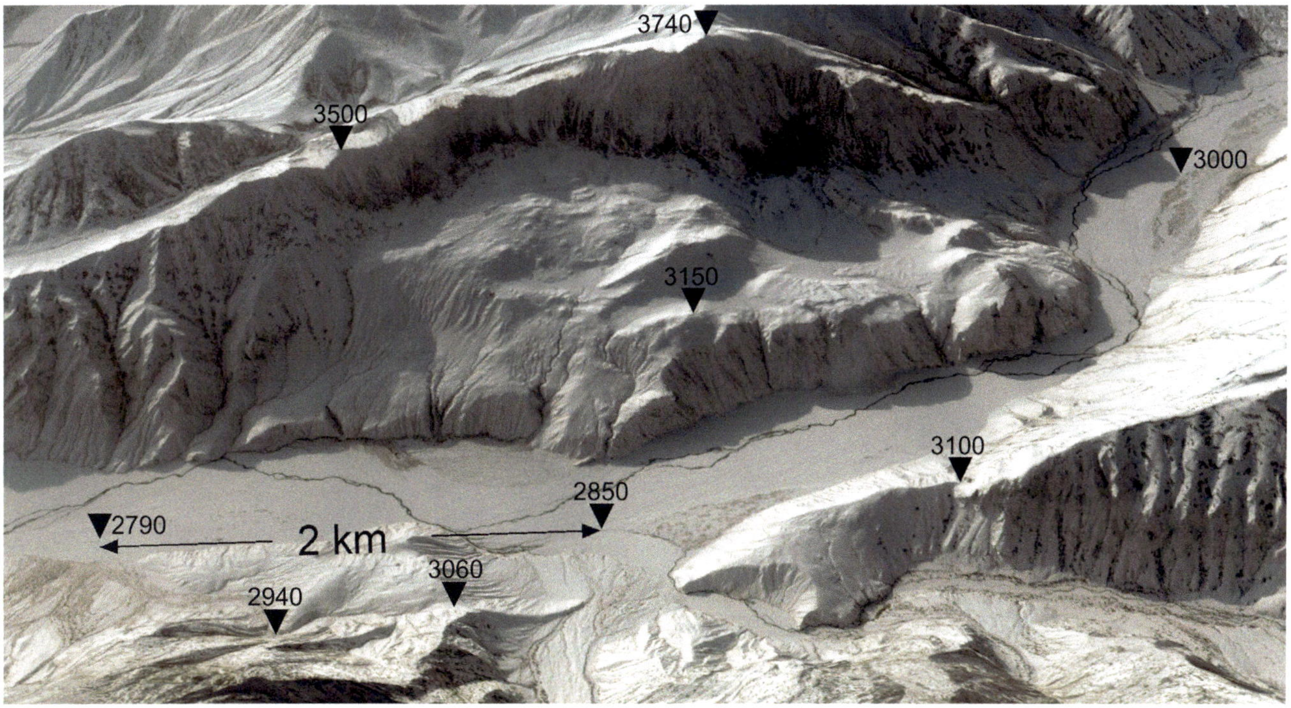

FIGURE 9.7 The Ke-Keka'er rockslide about 3 km^3 in volume. Winter 3D Google Earth view.

deposits upstream and was dissected completely, so that its present-day thalweg is almost straight, without any nick points. We must notice that the direction of this rock slope failure is almost the same as of the Yimake rock avalanche. If the age of the Ke-Keka'er rockslide would be found to be similar, such uniform directivity could be an argument of seismic triggering of both outstanding rockslides.

Group of three bedrock landslides is located in the Oitagsu River basin near the Aoyi-Take township. One of them—the Aerpa Aigezi rock avalanche (38.91° N, 75.239° E, China 151) of the Primary type (see Section 5.3.1 in Chapter 5) was described in Yuan et al. (2013). Our estimates of its area and volume differs from those provided in this paper (they measured area of $\sim 6.2 \times 10^5$ m^2 and volume of ~ 16 Mm^3 while according to our measurements area of the deposits is $\sim 5.8 \times 10^5$ m^2 and its volume is about 8 Mm^3). Such discrepancy, however, is not critical. Rockslide collapsed from the steep slope about 770 m high composed of the Devonian metagreywacke (Robinson et al., 2007) and its debris spread over the valley bottom as a thin apron with extremely high—up to 130 m—run-up (Fig. 9.8). Judging from the thalweg profile this low but wide dam was dissected by the river almost completely.

Besides this primary rock avalanche, two more rockslides were identified nearby (see Fig. 9.8). One of them (China 150 marked by B on Fig. 9.8) occurred at 38.9° N, 75.25° E on the ca. 800 m high slope south of the Aerpa Aigezi rock avalanche. This rockslide differs significantly from its highly mobile neighbor. Despite comparable slope height and much bigger volume (~ 15 Mm^3 vs ~ 8 Mm^3), most of its body still rests in the headscarp. The third one (China 154—C on Fig. 9.8) is much smaller, about 0.5 Mm^3 only, and can be classified as rock or debris avalanche. All three features look quite "fresh" and could be triggered by the same event, likely strong earthquake. The convex prominence on the right slope of the valley immediately downstream of the "C" deposit (see Fig. 9.8) could be a body of much older rockslide.

Another low but wide rockslide dam (China 152) had blocked the Khantereksu River valley at 38.536° N, 75.658° E, about 2 km downstream of the junction of the Chimgesu and the Merkesu Rivers (Fig. 9.9). Failure of about 8 Mm^3 of the Devonian metamorphosed rocks occurred from more than 2.2 km high and very steep slope. Some part of debris could remain on the slope more than 1 km below the headscarp crown (between 3850 and 3010 m a.s.l.), while most of it spread 1.38 km along the valley forming the 20–40 m high dam that later on was breached and incised by the stream. When rock avalanche reached the valley bottom, it produce up to 25 m high run-up on the opposite bank and spread down- and, mainly, up valley (see Fig. 9.9).

Another large rockslide in the easternmost chain of the Pamir Mountains originated near the Ye Xiliha village (38.281° N, 76.275° E, China 143), on the left side of a deep gulley eroded in the Devonian-Carboniferous sedimentary rocks. The 850 m wide and 250–420 m high headscarp of the Ye Xiliha rockslide between elevation marks 3580–3230 m a.s.l. (Fig. 9.10) looks much more fresh than the adjacent slopes of the gulley. It can be assumed that large rocky prominence about 300 m thick if not more collapsed along the nearly planar sliding surface cross-cutting bedding

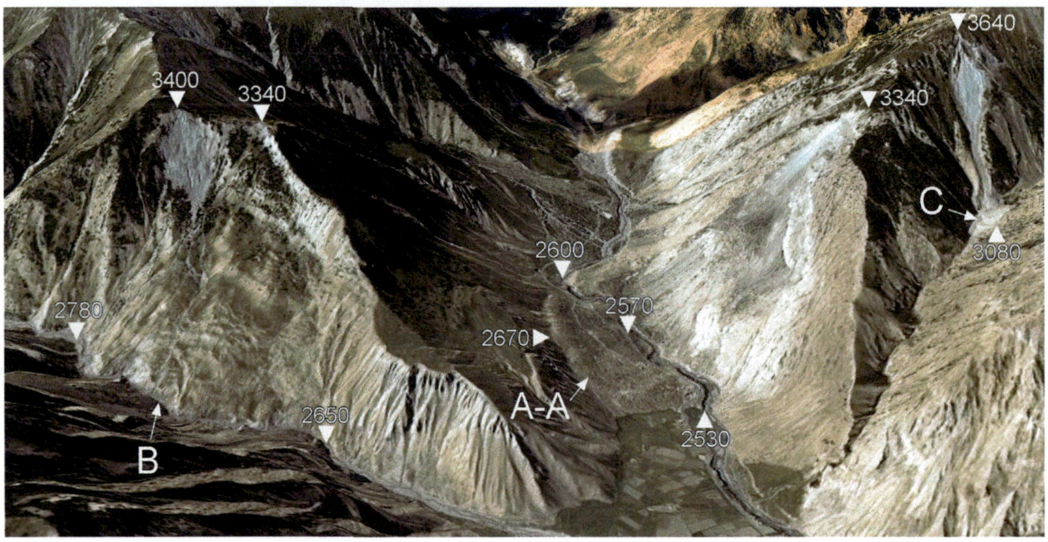

FIGURE 9.8 The Aerpa Aigezi rock avalanche (A-A) and two other, likely simultaneous, rockslides (B) and (C). 3D Google Earth view.

FIGURE 9.9 Above—the right bank of Khantereksu River valley from where rockslide that formed low and wide dam had collapsed. Assumed headscarp crown is at 4925–4780 m a.s.l. Below—closer view on the dam from the headscarp side. 3D Google Earth views.

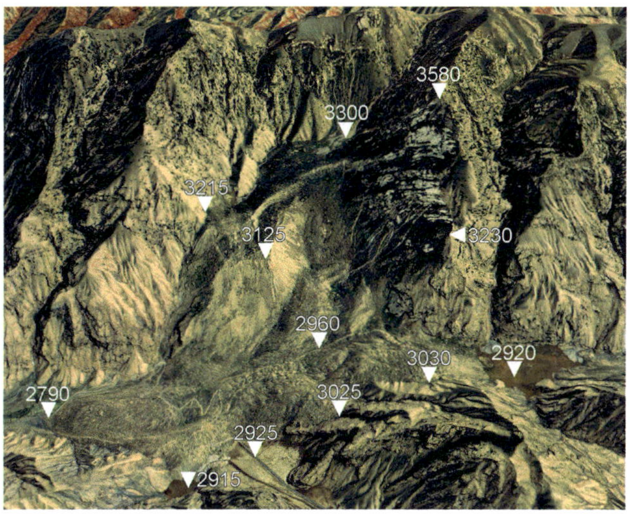

FIGURE 9.10 The Ye Xiliha rockslide. 3D Google Earth view.

planes. Rockslide about 100 Mm^3 in volume moved first across the gulley, collided with its opposite slope, and then turned left. Part of it halted but other part that retained its mobility moved ahead and formed up to 130 m high dam with ~65–70 m run-up that spread along the main valley for about 2 km and blocked both the main river and its left tributaries. Small dammed lakes have been silted completely. Boundary between these parts of rockslide body is marked by well-defined secondary scarps more than 100 m high allowing classifying the Ye Xiliha rockslide as a deflected Secondary rock avalanche (see Section 5.3.3.1 in Chapter 5).

Several bedrock landslides have been identified in the Tashkurgan River valley—the left tributary of the Yarkand River that drains southern intermontane depressions of Eastern Pamir and dissects its outer chain. One of them is the Bile Jiyi rockslide (China 155, 37.815° N, 75.679° E) described in brief in (Yuan et al., 2013). According to our estimates its debris covers about 4×10^5 m^2 (more than it was mentioned in this paper) and volume could be about 6 Mm^3. The left-bank slope of the Tashkurgan River valley from which it originated is ~930 m high and up to 38-degree steep. Rockslide had blocked the valley by coarse debris more than 20 m thick with ~80 m high run-up on the opposite slope (Fig. 9.11A). Largest block is about 30×30 m in size. Dammed valley was inundated and partially silted before the dam was breached.

More or less similar rockslides were found in the Tashkurgan River valley 15 km upstream (China 164, 37.825° N, 75.57° E) and 5.7 km downstream of the Bile Jiyi rockslide (China 165, 37.838°, N 75.732° E). All three of them originated on the left-bank slope of the valley composed of granite and gneiss and caused temporary river damming. The downstream rockslide ~15 Mm^3 in volume formed the dam whose crest composed of coarse gneiss debris rises up to 200 m above the riverbed with distinct proximal lowering through which it was eroded (Fig. 9.11B). Its' body front is overlain by more recent rock fall from the right-bank slope of the valley.

The upstream rockslide about 10 Mm^3 in volume originated on the ca. 550 m high slope composed of granite and formed a cone at the base of the collapsed slope that resembles the 1991 Randa rockslide (Gotz and Zimmermann, 1993; Sartori et al., 2003). It caused only partial damming of the Tashkurgan River (Fig. 9.11C).

One more Randa-looking rockslide (China 161) is located at a sharp bend of a rather small left tributary of the Yarkand River at 37.072 N, 75.941 E. Almost entire slope up to 460 m high composed of the Proterozoic gneiss is covered by angular boulders of the rockslide body (Fig. 9.12). Judging from the

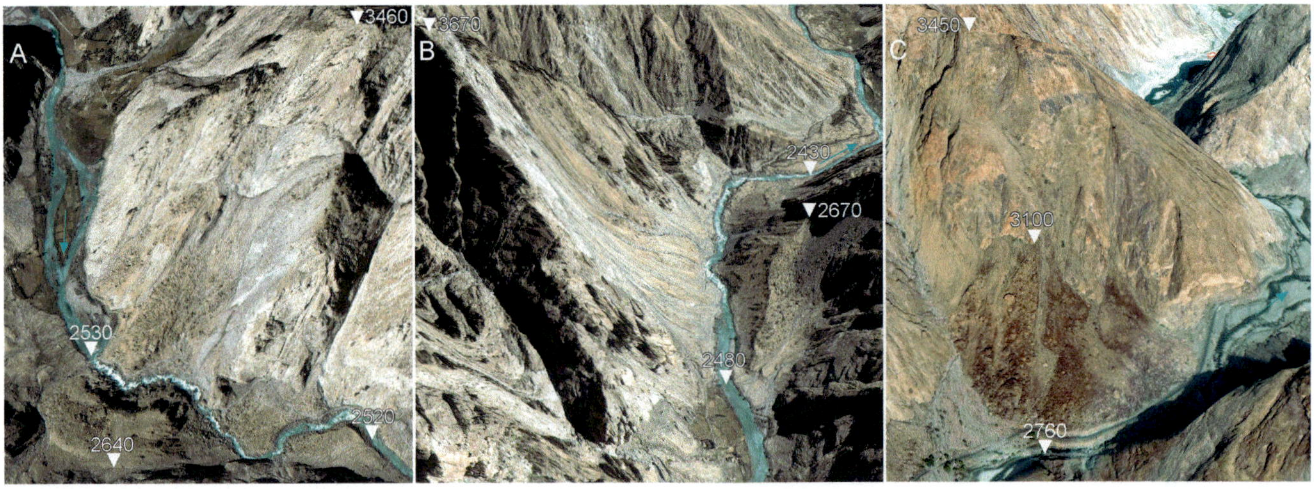

FIGURE 9.11 Rockslides in the Tashkurgan River valley. (A) The Bile Jiyi rockslide; (B) China 165 rockslide; (C) China 164 rockslide. Blue arrows indicate direction of water flow. 3D Google Earth views.

FIGURE 9.12 Rockslide in the left tributary of the Yarkand River valley and the dammed lake filled by lacustrine sediments. 3D Google Earth views.

thalweg profile thickness of rockslide debris at the valley bottom could be about 100 m, and the entire volume of this rock slope failure could reach 30 Mm^3. Dammed lake was silted almost completely and river upstream is now meandering through the flat forested area (compare with the Lower Kulun rockslide, Fig. 8.238). Rockslide in the Tashkurgan River valley shown on Fig. 9.11C and rockslides in the Tien Shan shown on Figs. 5.2 and 8.229 can be ascribed to the same type of rock slope failure.

Several bedrock landslides originated on the high slopes rising above the intermontane neotectonic depressions that divide highly elevated outer chain of Eastern Pamir with the Kongur and Muztagh-Ata peaks exceeding 7000 m a.s.l. from the more plateau-like Central Pamir. Three of them were described in Yuan et al. (2013). However, analysis of high-resolution space images raised doubts on the correctness of the interpretation of some of these case studies.

Fig. 9.13 presents the 3D Google Earth view of the site at the southernmost end of Muji Valley (38.775° N, 75.027° E) where the Bulunkou landslide ~17 Mm^3 in volume described in Yuan et al. (2013) was identified. These researchers identified the 8.6×10^5 m^2 headscarp on a SW-facing slope composed of quartz-feldspathic gneisses (Robinson et al., 2004) between ~4500 and ~5100 m a.s.l. and pointed out that its sliding surface is nearly parallel to strike of foliation in the bedrock. They estimated the deposits area of ca. 1.7×10^6 m^2 (including the buried part), volume of more than 17 Mm^3, and dated these deposits by the ^{10}Be method that returned age of 2.0 ± 0.1 ka. There is, however, some controversy because the alluvial fan from the assumed headscarp area does not overlay these deposits but truncates and erode them (see Fig. 9.13). It is difficult, however, to imagine that streams draining such a small catchment could erode most of the deposits that collapsed so recently. Besides, the assumed detachment zone looks more as the erosional feature rather than landslide headscarp. From our point of view, it is most likely that coarse bouldery hummocky deposits at the bottom of the valley incised in old moraine accumulations marked as OM on Fig. 9.13 are the terminal part of the Late Holocene moraine left by the degrading glacier still visible several kilometers north.

One of the largest rock avalanches in the Eastern Pamir was identified at the south-western feet of the Muztagh-Ata Massif at 38.15° N, 74.98° E (China 166) (Fort and Peulvast, 1995; Seong et al., 2009; Yuan et al., 2013). According to these researchers, this Muztagh

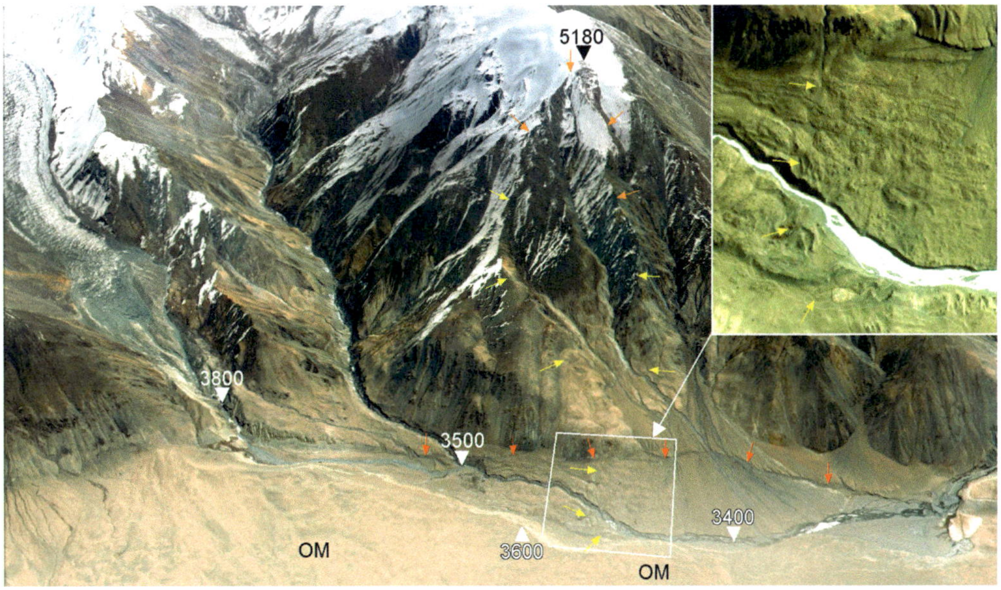

FIGURE 9.13 3D Google Earth view of the Bulunkou site (view from the west). OM—old moraine deposits. Orange and yellow arrows mark the detachment zone and inferred boundary of rockslide as it was proposed by Yuan et al. (2013); red arrows—active fault passing at the range foot with up to 9 m vertical offset [ibid]. From our point of view, the geomorphic pattern of the deposits on both sides from the boundary marked by yellow arrows in the inset does not demonstrate any significant difference.

(Kokoser) rock avalanche with about 18 km long runout covers more than 40 km^2 being composed of the Triassic quartz-feldspathic gneiss debris (Robinson et al., 2012). Its volume was estimated as ca. 470 Mm^3 if not more (Yuan et al., 2013). Rock avalanche had crossed the entire intermontane depression and individual lobes at its frontal part rose up to 120 m (according to Google Earth) an up to 170 m (according to 3″ SRTM DEM) on the opposite slope (Fig. 9.14). According to the ^{10}Be method, this giant slope failure had occurred 14.3 ± 0.8 ka BP (Yuan et al., 2013).

Fort and Peulvast (1995) interpreted these deposits as ice-rock avalanche. Similar assumption for the even more mobile Koman rock avalanche that will be described hereafter was proposed in Strom (1914a) but this assumption was argued in Reznichenko et al. (2017).

We must notice that, unlike other extra-large and extra-mobile rock avalanches in the Eastern and Northern Pamir that can be interpreted unequivocally, such as the Yimake rock avalanche (see Figs. 6.8 and 9.6) and the Koman and Lenin rock avalanches that were mentioned already in Chapters 2 and 4, Brief History of the Bedrock Landslide Studies in the Central Asia, and Principles and Criteria of Landslides Identification and Discrimination from the Features Formed By Other Mass-Wasting Phenomena, respectively, and will be described hereafter, the Muztagh rock avalanche with total descend of 3000 m has some characteristics that makes its interpretation more controversial. First, more than half of its body—the entire proximal part — is overlain by moraine deposits of the Olimde glacial stage identified in Seong et al. (2009) (Fig. 9.15), while above-mentioned rock avalanches are younger than the surrounding moraines, except those of modern glaciers that do not extend outside presently glaciated circuses. Second, no distinct headscarp differing from the adjacent glacial landforms can be identified in the ~9 km long U-shape valley from which both more recent glacial deposits and older rock avalanche (?) deposits came out (see inset in Fig. 9.16). Slightly leaping ahead we must notice that it differs significantly from much more distinct headscarps of the Koman and Lenin rock avalanches shown on Fig. 9.21. From our point of view (though, based on the remote sensing data analysis only), it cannot be excluded that lobes composed of bouldery deposits that rest on the opposite slope of the depression up to 120 m above its bottom could be left by old glacier that had crossed the valley and whose thickness (before it started melting) was large enough to leave material high on the opposite slope. Sound arguments in favor or against this assumption could be provided by the sedimentological analysis proposed in Reznichenko et al. (2012a) and utilized successfully for identification of the Koman and Lenin rock avalanches (Reznichenko et al., 2017).

The undoubtful Taheman rock avalanche (Seong et al., 2009; Yuan et al., 2013) was identified about 16 km southwest of the previous site at the southern foot of the Muztagh Ata massif at 38.09° N, 75.19° E

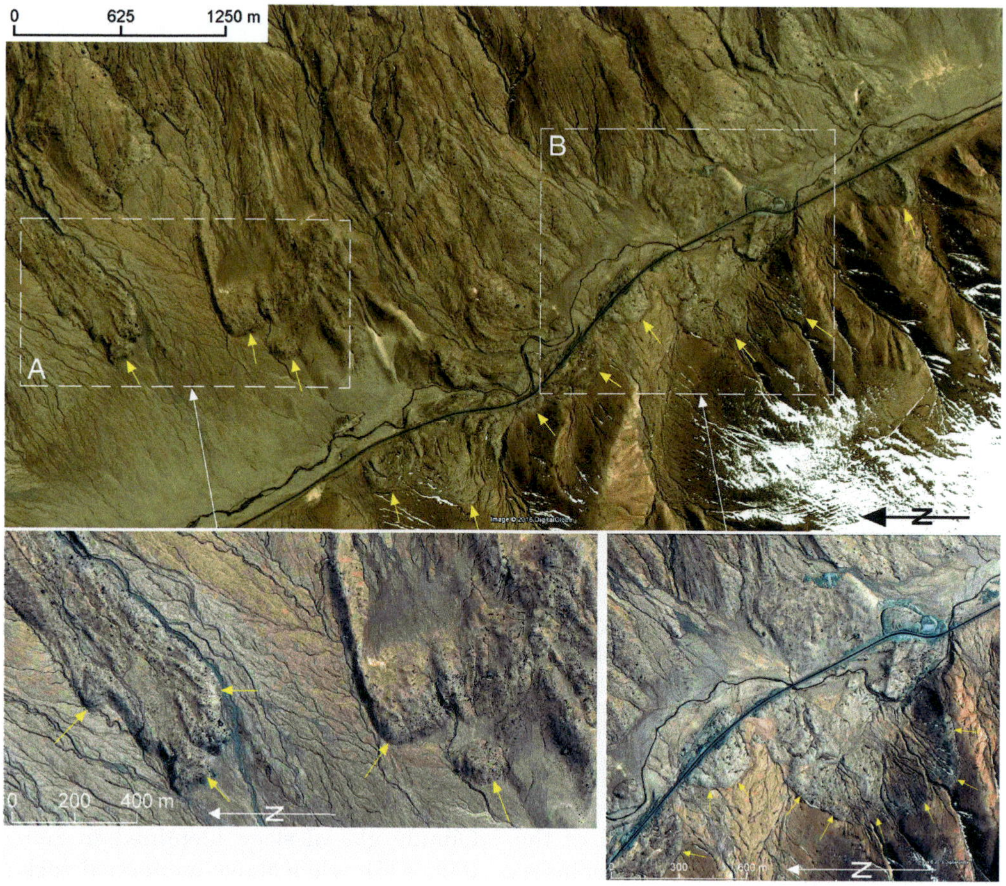

FIGURE 9.14 Frontal part of the assumed Muztagh rock avalanche. Its front marked by yellow arrows on the left side of the valley is sharp and steep (A) and rises on the opposite slope where it crosses the main valley (B). Google Earth images.

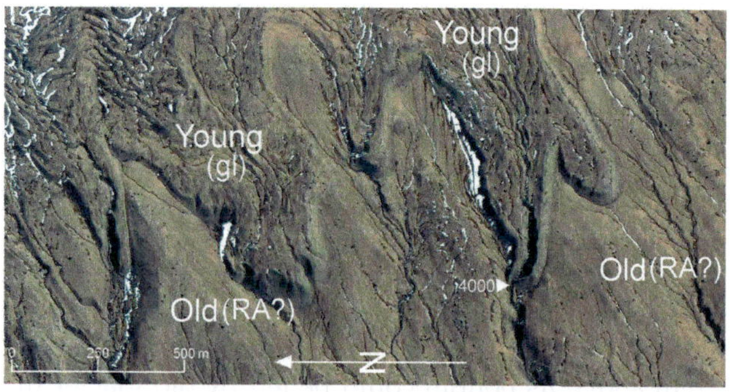

FIGURE 9.15 Young glacial deposits (gl) overlying older accumulations left either by rock avalanche (RA?) or, may be, by older glacier at the foot of the Muztagh-Ata massif. This site is about 3 km far from the front of the deposits shown on Fig. 9.14. Google Earth image.

(China 153, Fig. 9.17). It is a typical unconfined Primary rock avalanche (see Figs. 5.31-1A, 5.34-1). Most of debris accumulated at the distal part of its 5.3 km long runout, and its motion halted due to basal friction and internal energy dissipation only. Failure affected the ca. 1000 m high slope composed of gneiss with foliation dipping ~60° SW, i.e., in the same direction as the landslide moved (Yuan et al., 2013). However, dip angle is much higher than that of the slope. Total drop height of this feature is about 1470 m. Distal part of rock avalanche body consists of the concentric belts well visible on Fig. 9.17 that indicates flow-like style of debris motion.

Rock avalanche body which surface is composed of angular boulders up to 20 m in size [ibid] is crossed by surface rupture passing along the range feet and marked by red arrows on Fig. 9.17, though rupture trace where it crosses rock avalanche debris is less distinct than southeast of it. One more surface rupture likely disappears before it reaches debris accumulation.

We should notice that volume estimates provided in Yuan et al. (2012) (155 Mm^3) and, moreover, in Yuan et al. (2013) (~260 Mm^3) seems to be exaggerated significantly. The latter value was obtained by multiplying of the deposits area (5.1×10^6 m^2) by 50 m that roughly corresponds to maximal thickness of the distal part of this body. However, considering that such thickness can be ascribed to ca. 1.8 km long distal part only, while the remaining more than 2 km long part is

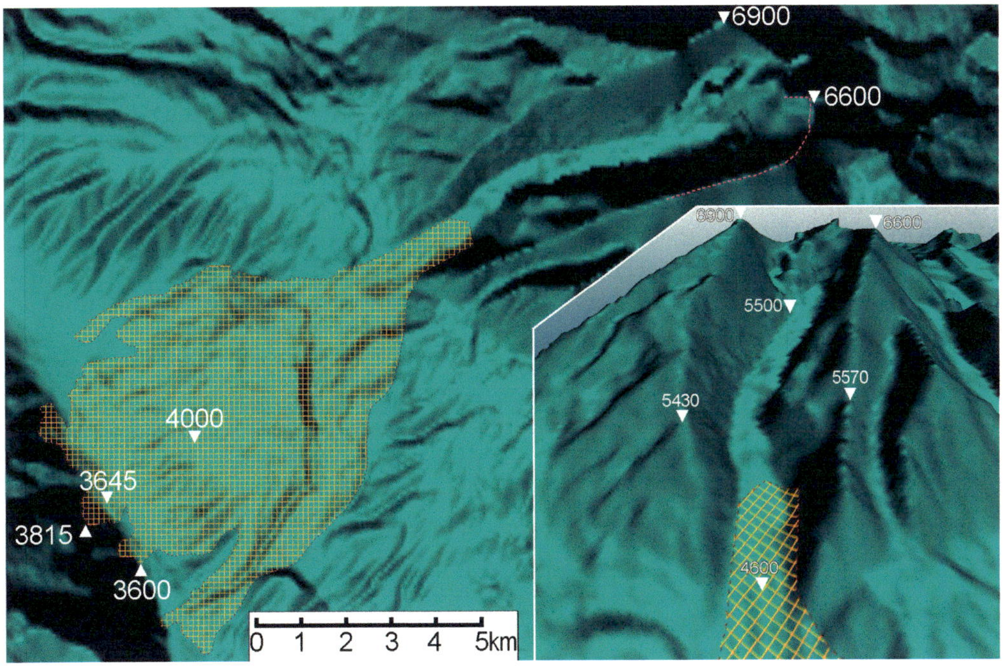

FIGURE 9.16 Overview of the assumed Muztagh rock avalanche. Cross-hatched area—inferred rock avalanche large part of which is buried now by recent moraine. Dashed pink line—possible position of rock avalanche headscarp. 3D view of its source zone is shown in the inset. 3″ SRTM DEM visualized by Global Mapper software.

FIGURE 9.17 The Taheman rock avalanche. Red arrows mark surface ruptures. 3D Google Earth view.

much thinner, the entire volume of debris could hardly exceed 100 Mm^3. Moreover, if we divide volume provided in Yuan et al. (2013) by the headscarp area of $\sim 4.6 \times 10^5$ m^2 [ibid], mean thickness of the bedrock detached from the slope will be 565 m, which is several times more than reliable estimate of its maximal thickness that could hardly exceed 200 m. Thus we assume that reasonable volume estimate of the Taheman rock avalanche is about 80 Mm^3.

The assumption that this bedrock landslide was triggered by the M7.5 1895 Tashkurgan earthquake (Seong et al., 2009) was opposed in (Yuan et al., 2012, 2013) based on the ^{10}Be dating method that returned its early Holocene age (6.8 ± 0.2 ka).

Despite sharp topographic gradient and the presence of the impressive surface rupture along the eastern–north-eastern boundaries of the Eastern-Pamir intermontane depressions (see Fig. 3.6) large-scale bedrock landslides are rather rare here, likely due to intensive glaciation at this high-altitude zone. Next group of several bedrock landslides was found about 80 km north of the Mustag-Ata massif. One of them at 38.913° N, 74.842° E (China 159) likely represents rather rare case of the repeated failure from the same headscarp. Both rockslides converted into unconfined Primary rock avalanches resembling the Taheman case study, but of much smaller size (Fig. 9.18). Distinct headscarp crown crosses the slope composed of the Devonian metamorphic rocks at about 4515 m a.s.l., though it can be hypothesized that it is the scar of younger event while older rockslide could originate higher at a bench at 4620 m a.s.l. marked on Fig. 9.18. Older rock avalanche about 5 Mm^3 in volume passed ~1 km up to 4070 m a.s.l. It has light color except its tip, where debris is dark, likely being covered with desert varnish. Similar dark debris covers the proximal and central part of this body, and this unit has sharp boundary with the underlying debris (see inset in Fig. 9.18). Volume of this, likely younger, rock slope failure is about 1 Mm^3.

FIGURE 9.18 Rock avalanches likely representing repeated failure. 3D Google Earth view.

One more rockslide (China 160) that formed compact bubble-shape body at the slope base was found at the northern terminus of the Sarykol intermontane depression, on the left bank of the Kumbelsu River valley (39.27 N, 74.031 E, Fig. 9.19). Similar to other large-scale rock slope failures near the eastern boundary of the Pamir intermontane depressions, it originated on the slope undercut by active fault, though directly at this site the latter is not expressed as clearly as ~2 km west or east of it. Such spatial coincidence allows assumption that this rockslide was triggered by large earthquake.

Two really outstanding rock avalanches with extreme mobility—the Koman (Komansuu) rock avalanche (39.55° N, 72.68° E; Kyrgyz 345) and the Lenin rock avalanche (39.52° N, 72.93° E; Kyrgyz 346) were identified in the Alai valley dividing Northern Pamir from the Southern Tien Shan. As mentioned in

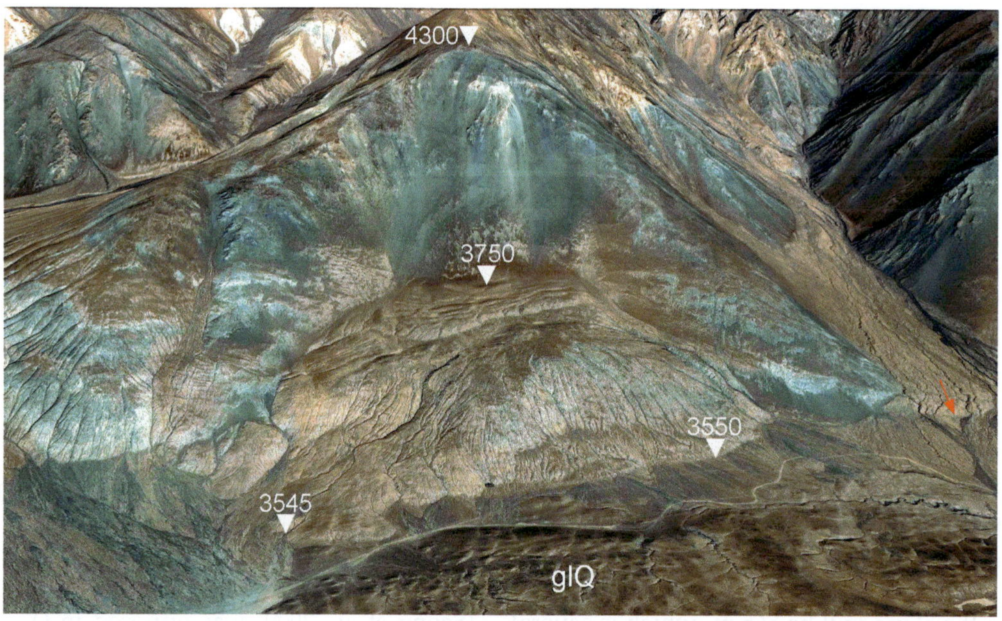

FIGURE 9.19 Rockslide at the northern terminus of the Sarykol intermontane depression. Red arrow marks active fault scarp. 3D Google Earth view.

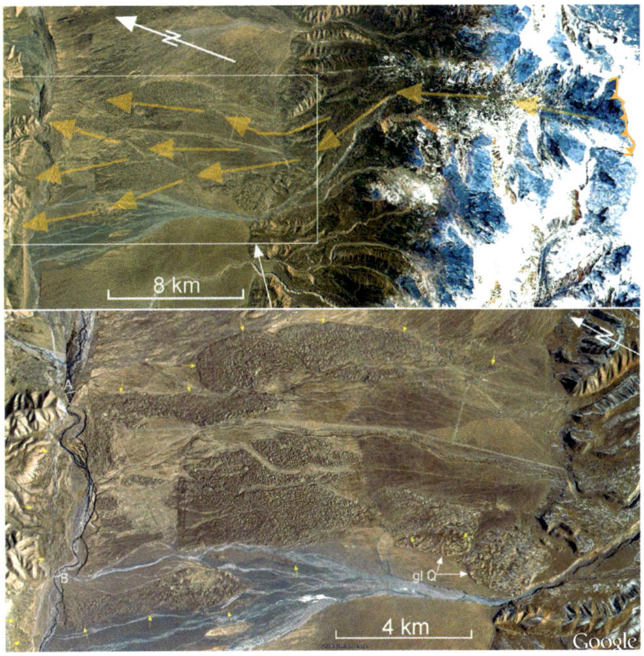

FIGURE 9.20 The Koman rock avalanche. Above—general view. Orange line—assumed headscarp; transparent arrows show direction of debris motion. Below—closer view of the deposits, whose external limits are marked by yellow arrows; gl Q—preslide glacial deposits; A and B—sites shown on Figs. 9.22 and 9.23. Google Earth images.

Chapter 2, Brief History of the Bedrock Landslide Studies in the Central Asia, they were identified as the extremely mobile gravitational slope deformations first by K.V. Kurdiukov (1950, 1964) and, after several decades of debates, correctness of his interpretation was proved in Strom et al. (2014a); Robinson et al. (2015); Reznichenko and Davies (2015); Reznichenko et al. (2017). Here we present general description of these features and will refer to Reznichenko et al. (2017) for more details.

The largest of two—the Koman rock avalanche (Fig. 9.20) demonstrates extraordinary mobility. Rock mass about 1.5–2.0 km^3 in volume (we could not provide more accurate estimate due to quite irregular surface of the deposits; besides this body was eroded significantly) collapsed from the axial part of the Zaalai (Trans Alai) Range and left the headscarp with 1500–2400 m high back wall (Fig. 9.21). Rock avalanche traveled first about 16 km along the trough valley with mean angle of about 4°, entered the Alai valley and crossed it at a distance of about 15 km with mean angle at this section of the travel path of ~1.5° only. At the end it crossed the Kyzylsu River and rose up to ca. 50 m above the former valley bottom that corresponds to the velocity of about 113 km/h (according to (Robinson et al., 2015) maximal frontal run-up was up to 100 m, but it seems to be exaggerated). Considering the entire height drop of about 3300–3500 m and ca. 34 km long runout, it corresponds to H/L ratio of ~0.1 only (mean angle is 5.5°–6°). Discussion about factors supporting such an extreme mobility could be found in Robinson et al. (2015); Reznichenko et al. (2017). We will revert to it in Chapter 12, Morphological and Structural Evidence of Rockslides' (Rock Avalanches') Motion Mechanism(s).

The Koman rock avalanche had covered about 130 km^2 of the Alai valley bottom by thin apron 15–20 m thick on average that has an impressive hummocky surface (see Fig. 6.9). Its morphology, relationships with adjacent glacial deposits and possible origin were described in details in Reznichenko et al. (2017).

Rock avalanche had blocked the Kyzylsu River valley at a distance of ca. 7 km. Though the present-day profile of the river thalweg is almost linear, with a mean angle of 0.37° and without any distinct nick point, one can see that at its upstream part river is still incising through the blockage (Fig. 9.22), while at the downstream portion modern valley have cut it completely, so that its base with impressive shear basal zone is observable (Fig. 9.23).

The Lenin rock avalanche first identified by K.V. Kurduikov (1964) is located about 20 km east of the Koman rock avalanche, where the bottom of the Alai valley is ~200 higher. Here, at the eastern part of this intermontane depression, end moraines of past glaciers had covered the valley bottom extending up to 10–15 km from the Zaalai Range foot, different from the Koman site. Rock avalanche that originated east from the Lenin Peak and left the headscarp with more than 1200 m high back wall traveled up to 24 km and covers at present the area of 32.5 km^2, while at the time of deposition its area could reach 54 km^2 (Reznichenko et al., 2017). Unlike the Koman rock avalanche that moved ahead over the flat and smooth valley bottom, debris of the Lenin rock avalanche had passed over hummocky end moraine incised by streams from melting glacier and had to adjust to the preexisting topography. It resulted in its splitting into three lobes (Fig. 9.24), which resembles the Chaartash-1 rock avalanche in Central Tien Shan (see Fig. 8.128). However, unlike that feature with well-expressed linear striation, the Lenin rock avalanche deposits have hummocky surface. As it was described in Reznichenko et al. (2017), these hummocks differ significantly from much larger hummocks on the glacial deposits that can be seen on Fig. 9.25. Volume of this rock slope failure was estimated as of the order of 1 km^3 [ibid], though in fact it could be even less—about 650 Mm^3.

West of the Altyn-Dara River where the Alai intermontane depression terminates, style of neotectonic deformations changes abruptly, and it results in the

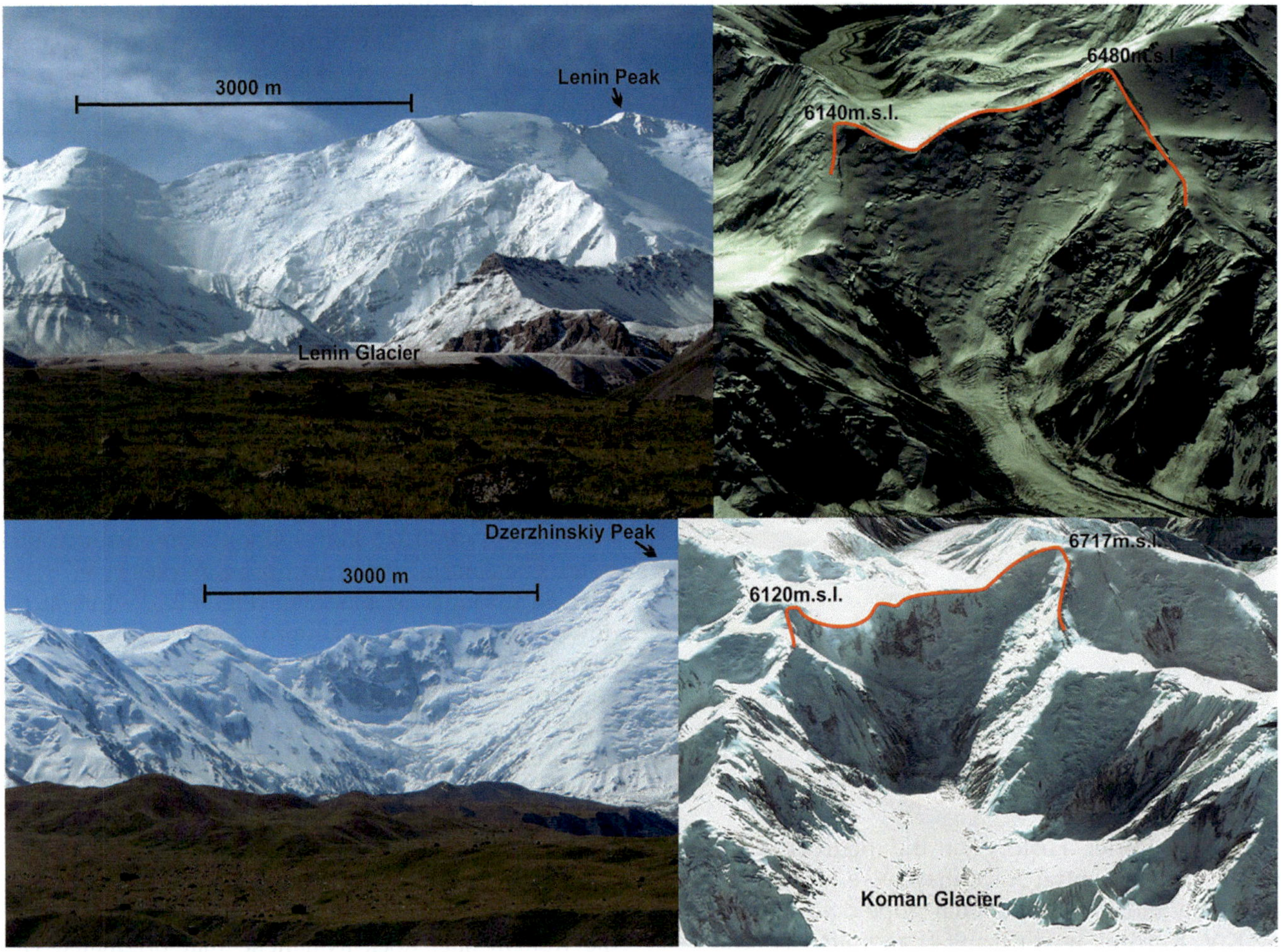

FIGURE 9.21 Most probable headscarps of the Lenin (above) and the Koman (below) rock avalanches. *From Reznichenko, N.V., Andrews, G. R., Geater R.E., and Strom, A., 2017. Origin of large hummock deposits "chukuryi" in Alai Valley, Northern Pamir: geomorphological and sedimentological investigation. Geomorphology, 285, 347–362, with permission of Elsevier.*

FIGURE 9.22 The Koman rock avalanche body (RA) at its upstream limit on the left bank of the Kyzylsu River overlaid by laminated red colored silt likely of lacustrine origin (L(?) Q).

FIGURE 9.23 Above—the Koman rock avalanche body (RA) at its downstream limit on the right bank of the Kyzylsu River resting over the alluvial gravels (aQ). Below—shear zone at the base of rock avalanche deposits (this outcrop is indicated by orange arrow above).

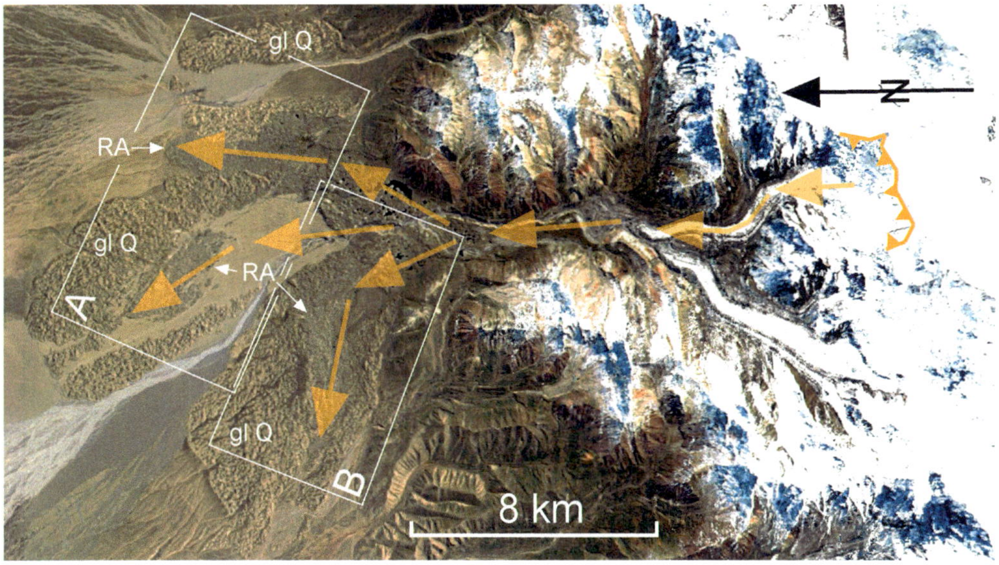

FIGURE 9.24 The Lenin rock avalanche. Orange line—assumed headscarp; transparent arrows show direction of rock avalanche (RA) debris motion; gl Q—preslide glacial deposits; outlined areas (A) and (B) are shown on Fig. 9.25. Google Earth image.

significant change of the slope deformation pattern. The Alai valley is bounded from the south by active reverse fault with distinct traces of surface ruptures (see Fig. 3.7A), indicating recurrent occurrence of large earthquakes (Arrowsmith and Strecker, 1999). It can be hypothesized that these earthquakes had triggered the enormous catastrophic Koman and Lenin rock avalanches described above. In contrast, northern slopes of the Zaalai Range westernmost part and of the Peter the First Range represent the front of the giant thrust, whose hanging wall is affected almost entirely by the DSGSD (see Figs. 5.20 and 5.21). We do not identify all

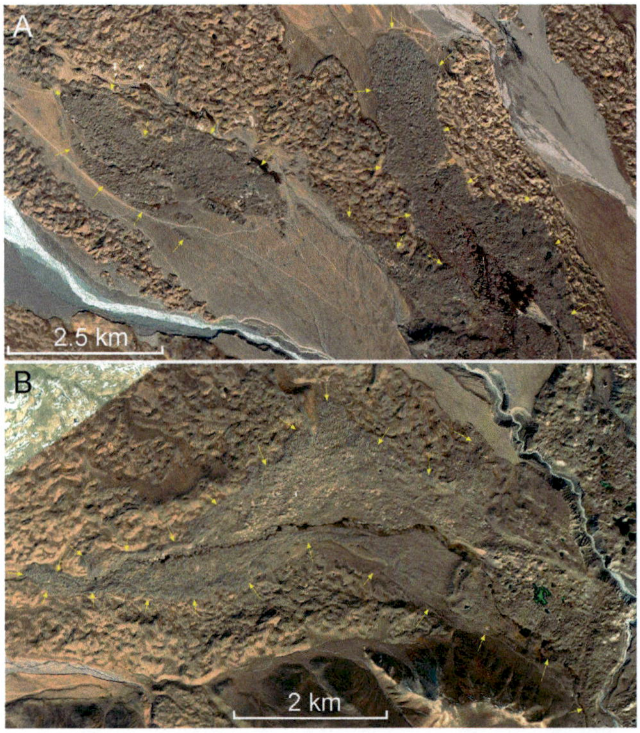

FIGURE 9.25 Closer views of the areas, marked by (A) and (B) on Fig. 9.24. The external limits of the rock avalanche lobes surrounded by the hummocky end moraine are marked by yellow arrows. Google Earth images.

individual landslides at this part of the area except a few examples of most impressive slope deformations at 39.11° N, 71.3° E (Tajik 117) and 39.12° N, 71.2° E (Tajik 118) shown on Fig. 9.26. Besides large arcuate scarps up to several hundred meters high with rather smooth slopes, one can see fresh scarps indicating ongoing deformations. Locally they could be intensified by karst processes (see sinkholes on Fig. 9.26A). Similar evidence of the DSGSD was found near the westernmost termination of the Zaalai Range. They will be described briefly hereafter.

One more similar 2 km long graben-like subsidence could be found at the western part of the Peter the First Range at 38.92° N, 70.18° E, about 15 km northeast of the confluence of the Surkhob and Obikhingou Rivers (Tajik 119, Fig. 9.27).

9.2 WESTERN AND SOUTH-WESTERN PAMIR

Central part of the Pamir mountains embracing the plateau-like highlands between the Tajikistan–China border in the east and ca. 73° E in the west is almost free of large-scale rockslides. We will start the description of the Western Pamir case studies from the Murgab–Bartang River basin. It is the largest river

FIGURE 9.26 DSGSD at the northern slope of the Peter the First Range opposite the Djirgital town. (A) Tajik 117, (B) Tajik 118. Orange arrows mark fresh scarps, S on (A)—sinkholes. 3D Google Earth views.

FIGURE 9.27 Graben-like DSGSD at the western termination of the Peter the First Range. Orange arrows mark scarps on both sides of the ridge. 3D Google Earth view.

system of the Pamir that starts from the Murgab River draining large part of the Central Pamir and bringing its water to the Sarez Lake dammed by the Usoi rockslide. 12 km downstream of the dam, it merges with the Kudara River forming the Bartang River—the major tributary of the Pianj River that flows along the Tajik–Afghan border.

Several rockslides were found in the tributary valleys of the Murgab River about 15 km upstream from the tail part of the Sarez Lake. Two of them at 38.244° N, 73.331° E (Tajik 123, Fig. 9.28A) and at 38.268° N, 73.308° E (Tajik 124, Fig. 9.28B) originated on the steep walls of glacial trough valleys 620 and 1150 m high correspondingly composed of the Triassic and Paleogene terrigenous and volcanic rocks. Both rockslides about 25 Mm^3 in volume can be ascribed to the Primary type with frontal confinement (see Fig. 5.31-1B). They blocked the valleys and have significant, ~120 and ~80 m high run-up, thus resembling rockslides in the Tashkurgan River valley in the Chinese Pamir shown on Fig. 9.11, rockslide in the Muzart River basin in the Chinese Tien Shan (see Fig. 5.40), and the Mini-Köfels rockslide in the Central Tien Shan (see Fig. 5.41). Rockslides blocked valleys close to their mouths, and these lakes have been filled by sediments almost completely due to high sediment yield. It is interesting that rockslide that caved from the much lower slope (see Fig. 9.28A) produced much more compact and higher dam with higher frontal run-up than its neighbor with almost two times larger descend. We hypothesize that it could result from more smoothed curve of the former's travel path that provided smaller resistance for its motion.

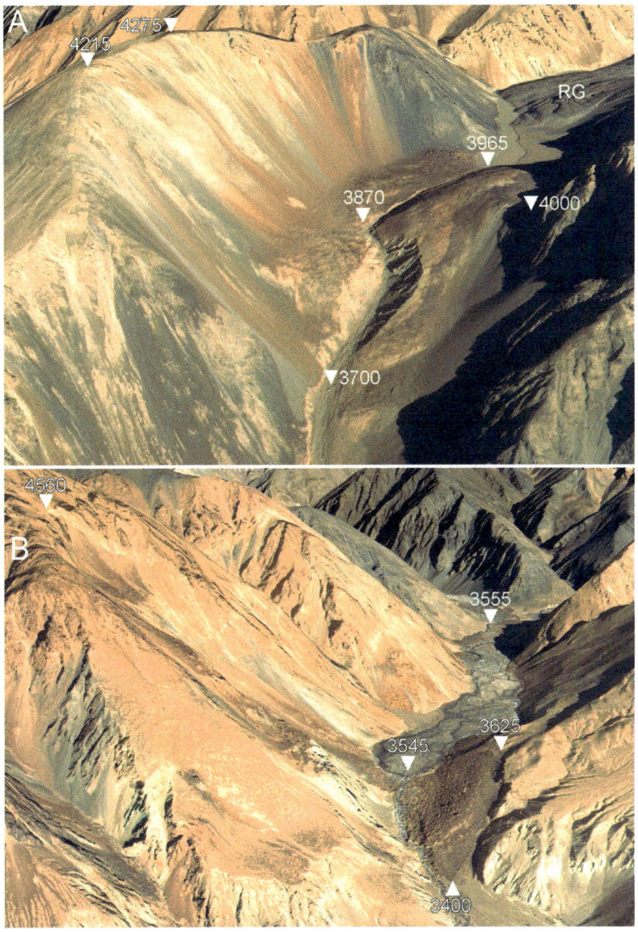

FIGURE 9.28 Rockslide dams ~250 m high (A) and ~130 m high (B) with ~130 and ~80 m high run-up correspondingly and with dammed lakes completely filled by sediments in the Murgab River tributaries upstream of the Sarez Lake. RG—rock glacier. 3D Google Earth views.

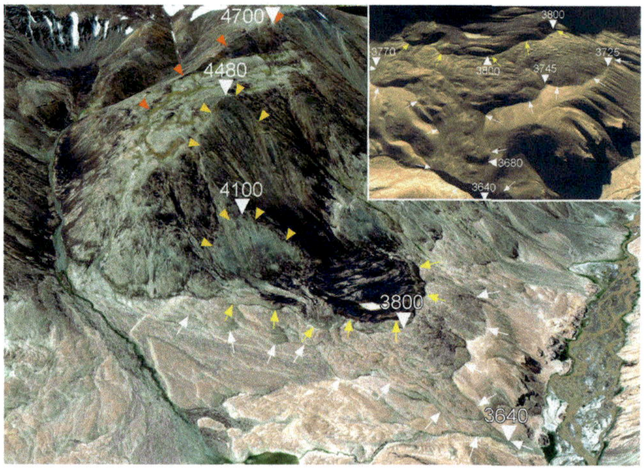

FIGURE 9.29 DSGSD and assumed recurrent rockslides on the left-bank slope of the Kokuibelsu River valley. Red triangles—the back scarp of the DSGSD; orange triangles—headscarps of the more recent flow-like rockslides whose front is marked by yellow arrows; white arrows mark front of the older deposits with hummocky surface well visible in the inset. 3D Google Earth views.

FIGURE 9.30 Breached rockslide dam with ca. 60 m run-up in Kokuibelsu River valley. 3D Google earth view.

Evidence of the recurrent past slope failures and of the DSGSD that can convert into catastrophic large-scale slope failure in future was found at 38.42° N, 73.00° E on the ca, 900 m high left-bank slope of the Kokuibelsu River valley, 3.5 km downstream of the Bozbaital River mouth (Tajik 127, Fig. 9.29). The DSGSD represents large block about 1100 m long and 350 m wide that subsided up to 100 m. We assume that light-color flow-like accumulation about 25 Mm^3 in volume with hummocky surface that covers the slope foot could represent rock avalanche that collapsed from the free face of the subsided block. This unit is overlaid by much darker accumulation about 6 Mm^3 in volume that originated from the spoon-shape headscarp on the slope. One more, smaller flow slide, originated from the adjacent headscarp, which crown is at 4100 m a.s.l. (see Fig. 9.29). Slope is composed of the Carboniferous terrigenous and metamorphic rocks with bedding/foliation dipping in the slope. Flow-like accumulations at the slope foot recalls at some extent the strange flow-like bouldery accumulation at the northern slope of the Kungey-Alatau Range shown on Fig. 8.87.

Moving west along the Kokuibelsu River valley, nearly 10 km downstream of the previous site, at 38.41° N, 72.886° E, there is one more rockslide about 15 Mm^3 in volume (Tajik 128) that caved from ca. 1200 m high right-bank slope of the valley composed of the Cretaceous or Paleogene granite (Fig. 9.30). Similar to the rockslides shown on Figs. 9.11 and 9.28, it crossed the valley and came to rest on the opposite bank mainly, thus representing the primary style of motion. Dam that was breached at its proximal part forms a 40 m high nick point in the valley and the dam's front is ~120 m higher than the stream level in the erosional gorge. Considering the incision depth of about 60 m, it corresponds to ~60 m run-up.

Further downstream the Kokuibelsu River merges with the Tanymas River. The latter flows along trough valley about 1500 m deep with ca. 1000 m wide flat bottom. Its slopes have been affected by numerous bedrock landslides, some of which move gradually, while other, most likely collapsed catastrophically.

One group of rockslides was identified on both sides of the Tanymas River valley near the Kok-Djar River mouth. This part of the area is composed of the Carboniferous-Permian sandstone, shale, and quartzite. One of these rockslides (Tajik 132) originated on the ca. 700 m high right-bank slope of the valley at 38.661° N, 72.773° E and has a distinct spoon-shape headscarp and convex body about 11 Mm^3 in volume (Fig. 9.31). Rockslide affects 800 m long section of the slope base being just slightly eroded by the river. Despite the absence of any frontal confinement, it did not convert in the highly mobile rock avalanche that allows assumption on the long-term noncatastrophic gradual formation of this rock slope failure.

Much larger, also gradually evolving complex rockslide (Tajik 133) was identified at 38.636° N, 72.8° E, on the ca. 1400 m high left-bank slope of the Tanymas River valley, 5–7.5 km downstream of the Kok-Djar River mouth. It has giant ~3700 long and more than 300 m deep bow-shape headscarp and landslide body deeply incised and reworked by erosion (Fig. 9.32). Similar to the previous case study this landslide about 200 Mm^3 in volume if not more comes out on the Tanymas River floodplain, without any frontal confinement but also does not demonstrate any evidence of long runout and river blocking. Some narrowing of the flood plane is due to debris flow fans. Fresh fractures and relatively small-size flow-like landslides visible at some parts of this complex feature indicate ongoing slope deformations (Fig. 9.33). It can be assumed that formation of this very large rockslide had started when the valley was still glaciated, and part of the displaced rock mass was removed not by water erosion, but by glacier, according to the model proposed in (McCall et al., 2017).

FIGURE 9.31 Rockslide on the right side of the Tanymas River valley opposite the Kok-Djar River mouth. 3D Google Earth view.

One more group of landslides affects both sides of the Tanymas River valley just downstream of the Kyzyltokoisay River mouth. Two adjacent concave lowerings with convex accumulations below them are visible on the left-bank slope of the valley (Fig. 9.34) composed of the Jurassic terrigenous rocks at 38.558° N, 72.81° E. The smaller one (Tajik 134a) with the headscarp crown at 3950 m a.s.l. looks older. It has more smooth surface, and some evidence of the most recent debris motion (may be rock glacier) can be seen at its upper part above 3490 m a.s.l. The southern, much larger body about 20 Mm^3 in volume (Tajik 134) with small rock glacier at its upper part that terminates

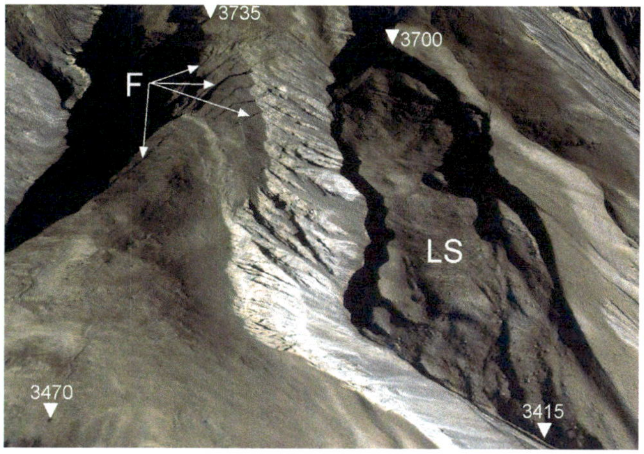

FIGURE 9.33 Evidence of recent and ongoing slope deformations within complex landslide shown on Fig. 9.32. LS—landslide, F—fissures. 3D Google Earth view.

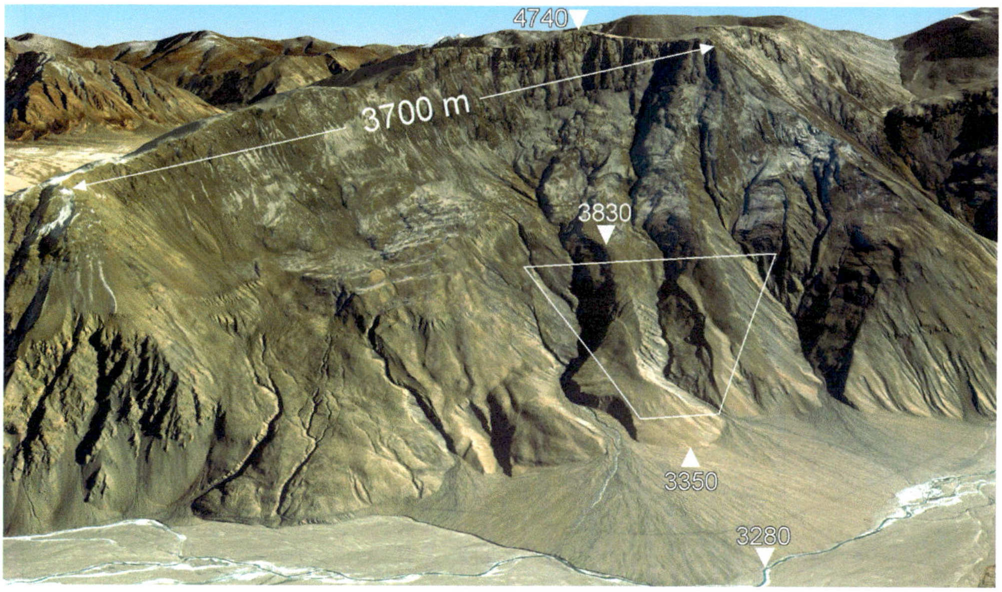

FIGURE 9.32 Rockslide on the left side of the Tanymas River valley downstream of the Kok-Djar River mouth. Zoomed view of the outlined area is shown on Fig. 9.33. 3D Google Earth view.

at 3730 m a.s.l. represents ongoing slope deformations within the rockslide deposits. This modern landslide of the flow type affects slope below 3625 m a.s.l. and splits on the way into two parts that reach the slope base. The triangular smooth surface between these flow slides is the preserved slope of the "original" rockslide body.

Two rockslides, whose distinct headscarps are much below than the same age glacial circuses typical of this part of the Pamir, are located on the right bank of the valley just opposite the site described above. The affected slope is composed of the Triassic terrigenous rocks. The larger rockslide about 8 Mm^3 in volume (Tajik 131) that destroyed the ca. 1000 m high rocky spur at 38.56° N, 72.79° E had advanced from the slope base in the flood plane for about 300 m only, despite its significant height drop (Fig. 9.35). Insignificant advance characterizes smaller rockslide just south of it (Tajik 130). It can be assumed that all rockslides on the slopes of the Tanymas River valley that did not move far from the slopes' base originated due to the postglacial debuttressing, and this could be a long-term gradual process.

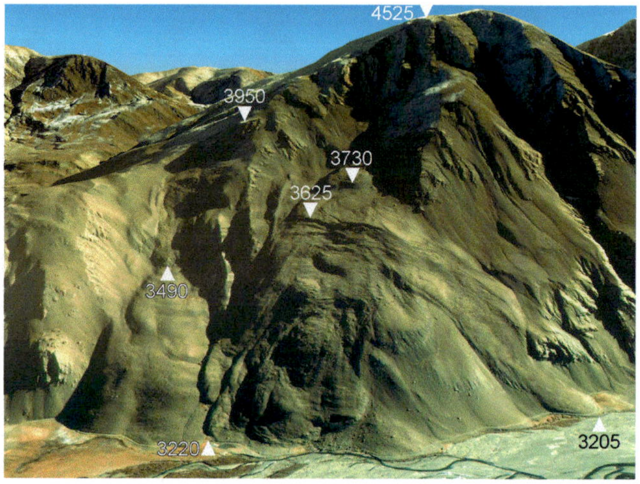

FIGURE 9.34 Landslides on the eastern side of the Tanymas River valley downstream of the Kyzyltokoisay River mouth. 3D Google Earth view.

However, there is also evidence of slope failures in the Tanymas River valley that occurred, most likely, catastrophically. It can be exemplified by very large Damamatsay rockslide (Tajik 129, 38.518° N, 72.808° E) that destroyed the rocky spur composed of the Jurassic sandstone and shale with interbeds of conglomerate, marl, and limestone. Its up to 300 m high back wall rises above the 6.22 km^2 body that moved north and collided with the right bank of the Damamatsay Creek at a front of about 2.7 km (Fig. 9.36). Rockslide about 600 Mm^3 in volume moved, likely, as a block slide rather than as a rock avalanche. Its central part is affected by the 2.6 km long secondary rock avalanche with distinct horseshoe-shape secondary headscarp with its crown at 4120 m a.s.l. and flow-like style of motion that can be derived from the microrelief of its surface differing drastically from that of the main body (see inset in Fig. 9.36). We assume that the secondary rock avalanche originated due to the collision of the main rockslide body with an opposite slope of the creek valley and resulting momentum transfer (similar to other secondary rock avalanches—see Section 5.3.3 in

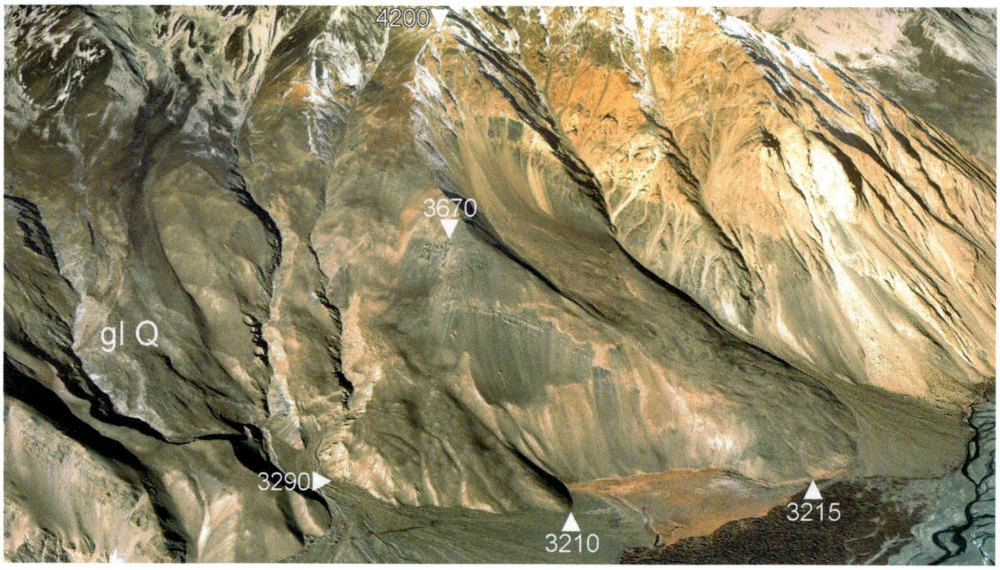

FIGURE 9.35 Rockslides on the western side of the Tanymas River valley downstream of the Kyzyltokoisay River mouth. gl Q—glacial deposits. 3D Google Earth view.

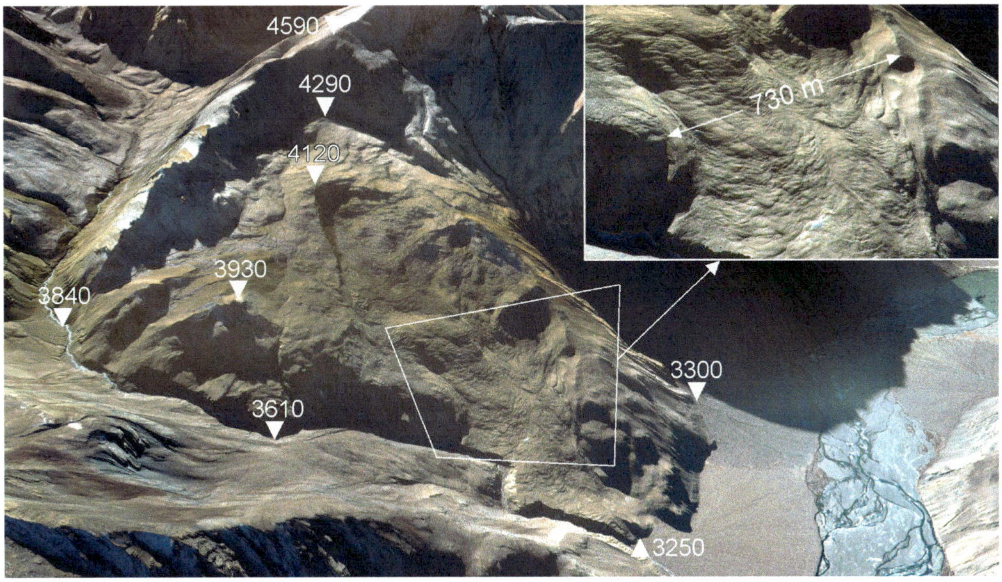

FIGURE 9.36 The giant Damamatsay rockslide in the Tanymas River valley. Difference between surface morphology of the initial rockslide body and secondary rock avalanche can be seen in the inset. 3D Google Earth view.

Chapter 5 and Chapter 12). It cannot be excluded, however, that in this case such secondary failure could originate later not being associated directly with the formation of the gigantic initial rockslide.

We want to notice that this giant rock slope failure somehow resembles the Kyzyl-Katta and the Ai-Kul rockslides in the Southern Tien Shan that also affected prominences of mountain ridges rising above the conjugating river valleys (see Figs. 8.248 and 8.257). One more, even better analogue, very similar to this one, was found in the Afghan part of the Pamir and will be described in Section 9.3 (it will be shown on Fig. 9.135).

One more, likely catastrophic slope failure (Tajik 135) occurred on the right-bank slope of the Tanymas River valley 3.5 km upstream from its confluence with the Kokuibelsu River and formed the dam at 38.431° N, 72.682° E that has been breached afterward. It is hypothesized that this Tanymas Mouth rockslide originated close to the headscarp of older rockslide, whose remnants can be seen on the slope (Fig. 9.37). Younger rockslide originated, most likely, as a wedge failure from the top of the ridge (see inset in Fig. 9.37). Block of the Triassic terrigenous rocks 400 × 180 m in plan view and about 270 m in vertical direction (~6 Mm^3) collapsed from the 2 km high slope with mean angle of ca. 34 ° whose large part was covered by the deposits of the assumed older rockslide and talus. Due to the entrainment of this loose material, volume of the resultant debris that formed the blockage about 1.6 km^2 in size increased up to ~30 Mm^3 approximately. Debris spread along the river for 1.9 km. The dammed lake was partially filled and then breached. At present, dam's remnant forms the ca. 50 m high nick point in the river profile that corresponds to the ca. 30 m thick deposits that have not been incised yet.

Remaining part of the assumed older rockslide body can be seen on the right-bank slope of the Tanymas River valley immediately upstream of the younger rockslide and covers the slope up to 3700 m a. s.l. (see Fig. 9.37). It seems that this slope failure was similar to those shown on Figs. 9.31 and 9.34 and did not result in the complete blocking of the valley, unlike its younger successor.

At 6 km downstream from the confluence of the Tanymas and the Kokuibelsu Rivers, at 38.39° N, 72.58° E, the valley of the Kudara River was blocked by the gigantic Kudara-Pasor rockslide dam (Tajik 092, see Fig. 4.20) described in Section 4.2 of Chapter 4, Principles and Criteria of Landslides Identification and Discrimination from the Features Formed By Other Mass-Wasting Phenomena, as the example of the feature erroneously interpreted as end moraine. Similarly to the famous 1911 Usoi rockslide that will be described in brief hereafter, it originated not from the top of the ridge but from the prominent bench about 700 m below the watershed (at 3835 m a.s.l. marked on Fig. 4.20) representing the bottom level of the old glacial valley. This rockslide can be classified as the Primary rock avalanche (see Section 5.3.1 in Chapter 5) with frontal confinement—after crossing the 2 km wide valley bottom the frontal part of rapidly moving rock avalanche climbed on the opposite slope for about

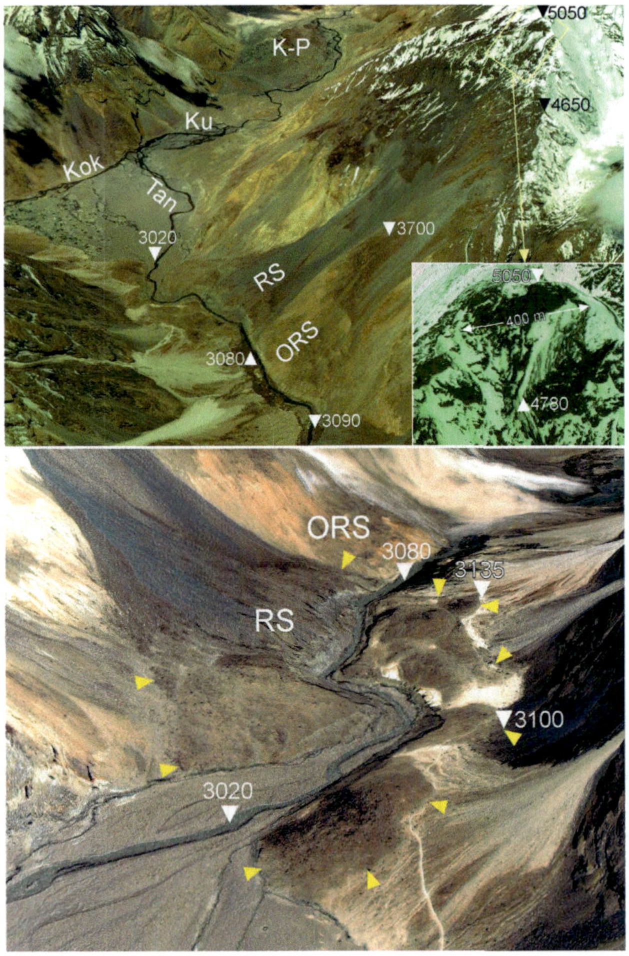

FIGURE 9.37 Two generations of rock slope failures on the right-bank slope of the Tanymas River valley near its mouth. RS—younger rockslide; ORS—remnants of older rockslide; Tan—the Tanymas River; Kok—the Kokuibelsu River; Ku—the Kudara River; K-P—the Kudara-Pasor rockslide dam (see Fig. 4.20 and the description hereafter). Oblique views from upstream (above) and downstream (below). Yellow triangles—outer limits of the younger rockslide body. Assumed headscarp of the younger rockslide is shown in the inset. 3D Google Earth views.

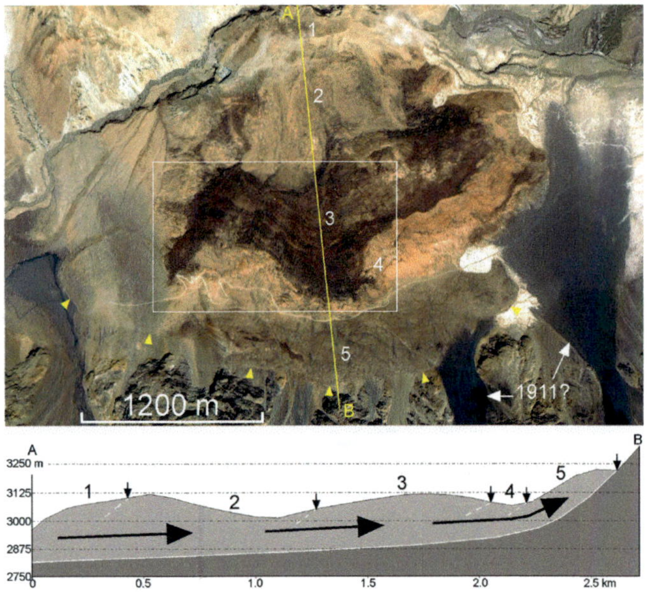

FIGURE 9.38 Plan view of the Kudara-Pasor rockslide deposits with inherited lithological zonality, indicated also on the profile along line (A–B) (Units 1–5). Boundaries between them are marked on the profile by black arrows and dashed white lines indicating their anticipated dip. Yellow triangles mark deposits front. Outlined area is shown on Fig. 9.39. Google Earth image.

FIGURE 9.39 Zoomed image of the part of the Kudara-Pasor rockslide deposits with retained succession of separate thin layers within Unit 3. Google Earth image.

155 m (being measured at the deposits surface – see the profile on Fig. 9.38).

Deposits cover almost 8.06 km^2 of the valley bottom and, despite the dam had been breached, they form the ~250 m high nick point in the valley profile that corresponds to the ca. 100 m thick rockslide deposits remaining below the stream level (similar to the Tanymas Mouth rockslide, but at much larger scale). Volume of this blockage is of the order of 1 km^3.

One more peculiarity of the Kudara-Pasor rockslide is the retention of the unmixed belts of debris of different lithology, which is expressed extremely clear (Fig. 9.38), similar to many other case studies throughout the entire Central Asian region (compare with Figs. 8.7, 8.45, 8.146, 8.151, 8.168, and 8.170). We assume that these units successively overthrust each other, which is indicated on the profile on Fig. 9.38. Besides general zonality well visible on Figs. 4.20 and 9.38, even separate layers can be traced at a long distance at some parts of its body (Fig. 9.39), despite significant crushing of the material differing this case study from the Usoi rockslide that will be described hereafter. We assume that Units 2–4 represent crushed bedrock (Carboniferous sandstone, schist, and quartzite with interbeds of marble, amphibolite, and andesite), Unit 1—glacial deposits from the old trough valley whose remnants are seen above the headscarp crown on Fig. 4.20, and Unit 5—Late Quaternary fluvial and glacial deposits bulldozed in front of moving bedrock debris.

Dark gray fans visible on Fig. 9.38 at the upstream-frontal edge of this blockage are, most likely, the traces of a snow—ice—debris avalanche produced by the collapse of hanging glacier from upper part of the Muzkol Range dividing the Murgab and Kudara River valleys that occurred during the 1911 Sarez earthquake, which was mentioned by V.S. Kolesnikov (1929).

The most well-known bedrock landslide in the Central Asian region is the famous Usoi rockslide (Tajik 090) triggered by the 1911 M7.3 Sarez earthquake (Golitzin, 1915; Ambraseys and Bilham, 2012; Kulikova et al., 2016) that blocks the Murgab River valley at 38.27° N, 72.6° E and forms the presently world deepest natural reservoir—the 500 m deep and 55.8 km long Sarez Lake that stores about 16–17 km^3 of water (A. Ischuk, 2011). Considering the potential threat that this tremendous water body poses to the population living in the valleys downstream, this outstanding natural phenomenon has been a subject of extensive scientific and technical investigations for more than a century since its origin (see Chapter 2). Besides papers on the Sarez Lake problem referred in Chapters 2 and 3, we should mention several other publications (Lange, 1926, 1929; Kolesnikov, 1929; Rodionov, 1931; Selivanov and Andreev, 1959; Fedorenko, 1988) describing state and evolution of the dam and lake. We must point out that, nevertheless, many aspects of the Usoi blockage structure, composition, and further evolution still remain unknown.

As it was mentioned in Chapter 5, rockslide originated as a wedge failure on the ca. 2.5 km high right-bank slope of the Murgab River valley and affected about two-third of its entire height below the prominent bench in the relief (see Fig. 5.7). Upper part of the southern slope of the Muzkol Range rising above this bench up to 5500 m a.s.l. has been dissected by glacial valleys whose frontal parts were truncated by the headscarp back wall and these moraines provide material for debris flows fed by melting glaciers that accumulates in the ca. 180 m deep proximal lowering of this giant blockage. Besides, these debris flows had eroded the downstream canyon. Headscarp walls are composed of the sandstone, schist, and quartzite of the Carboniferous Sarez Formation overlain by the Permo-Triassic dolomite and limestone with gypsum and anhydrite in its upper part and divided by the Usoi Thrust (Fig. 9.40).

According to I.A. Preobrajensky, who presented one of the most detailed and accurate descriptions of the Usoi blockage, volume of the displaced rock mass was about 2.2 km^3 while volume of the rockslide body was estimated as ca. 2.7 km^3 (Preobrajensky, 1920). Such increase was explained as the effect of the decompaction. He counted the height drop of the center of mass as 725 m.

FIGURE 9.40 Upper part of the Usoi rockslide headscarp. C_{sr}—Carboniferous terrigenous Sarez Formation, P-T—the Permo-Triassic dolomite and limestone with gypsum; glQ—glacial deposits.

Despite the fact that uppermost part of the Usoi rockslide represents a giant block of the terrigenous rocks of the Sarez series about 1 km^2 in plan view with well-preserved northwest striking bedding and intensive northeast fracturing originated during its motion (Figs. 9.41 and 9.42), this feature can be classified as a primary rock avalanche with prominent proximal lowering and predominantly distal accumulation. One of the most impressive morphological peculiarities of the Usoi rockslide is the system of the arcuate downslope facing scarps up to 100–150 m high that cross the entire blockage from its front up to proximal lowering filled by more loose debris (see Figs. 9.41 and 9.43). Because largest of these scarps is well expressed on the first maps of the Usoi blockage compiled by I.A. Preobrajensky in 1915 (Preobrajensky, 1920) and by V.S. Kolesnikov in 1925 (Kolesnikov, 1929) (Fig. 9.44), it can be assumed that these landforms originated during the blockage formation. We must notice that intensive deformations affect the downstream, gentler side of the blockage while its upstream slope submerged now by the Sarez Lake is much steeper. It can be explained by the significant difference of the morphology of the left bank of the Murgab River—the rockslide impact zone. Eastern part of moving rockslide entered the mouth of the Shadau River valley that acted as a trap, while its central and western parts collided with the steep valley slope directly producing intensive deformation in the transverse direction. Same phenomenon was described for the Djashilkul rockslide dam in

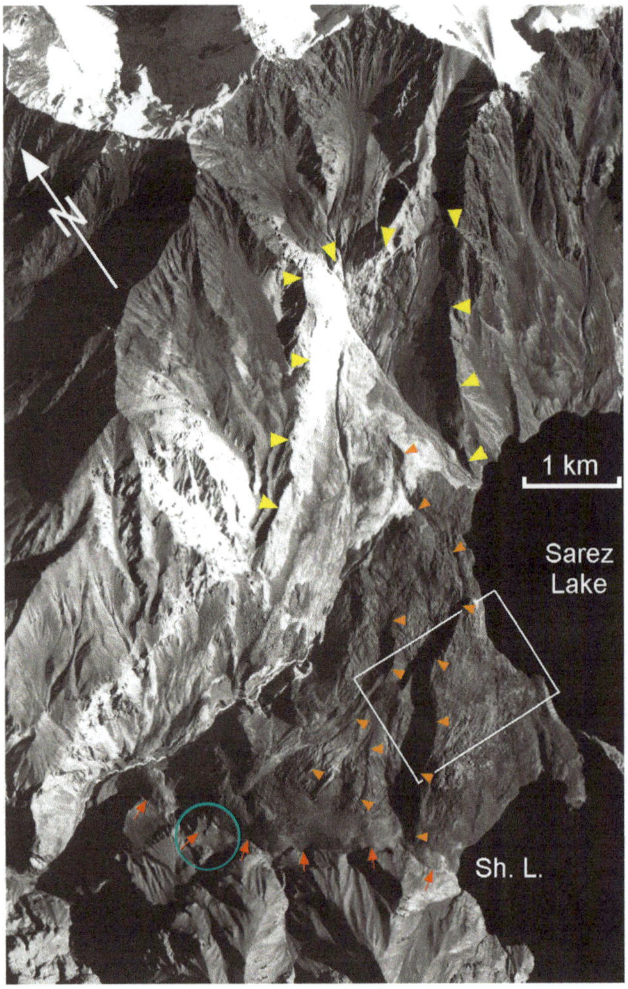

FIGURE 9.41 KFA-3000 space image of the Usoi rockslide. Yellow triangles mark headscarp; red arrows—rockslide front; orange triangles—prominent secondary scarps on the rockslide surface shown on Fig. 9.43; outlined rectangle—area shown on Fig. 9.42; blue circle—site with mud splash in front of the rockslide debris shown on Fig. 9.45; Sh. L.—the Shadau Lake.

FIGURE 9.42 Displaced block of rocks with preserved northwest striking bedding and intensive northeast striking fracturing. Google Earth image. White arrow marks the building of the Monitoring and Early Warning Station—gift from the Government and People of Japan.

FIGURE 9.43 Secondary scarps on the surface of the Usoi blockage (marked by white arrows). Photograph made in 1984 from the helicopter. Sh. L.—the Shadau Lake. *Modified from Strom, A.L., 2010b. Landslide dams in Central Asia region. Landslides 47(6), 309–324, with permission of the Japan Landslide Society.*

the Northern Tien Shan (see Fig. 8.81). Its mechanical aspects will be discussed in Chapter 12.

On the left-bank slope of the Murgab River valley very close to the path from the Barchdiv village at the junction of the Murgab and Kudara River valleys to the dam's crest, at about 3115–3200 m a.s.l., one can see the trace of the mud splash in front of the rockslide debris (Fig. 9.45).The most probable source of this material was the valley bottom about 300 m below this site. The narrow isthmus between the Sarez and the Shadau Lakes is also composed of well-rounded pebbles and boulders most likely bulldozed here by rockslide from the valley bottom.

No seeping through the dam was observed for about 3 years after dam formation, thus the Murgab River was blocked completely and the lake level increased rapidly. During this period the large Sarez village, whose buildings survived the 1911 earthquake, was submerged and the Sarez Lake inherits its name. Only in spring 1914 when the lake level rised up to 3050 m a.s.l. (being about 290 m deep) first springs appeared releasing ca. 2.0 m^3/s. At present, the lake level fluctuates at about 3260 m a.s.l. $\pm$6 m with maximum measured outflow of 47.1 m^3/s. Seeping water appears in the erosional canyon mainly at its ~850 m long head portion about 150 m below lake level (Fig. 9.46). Minimal freeboard of the natural dam above the Sarez Lake is 38 m. According to the long-term

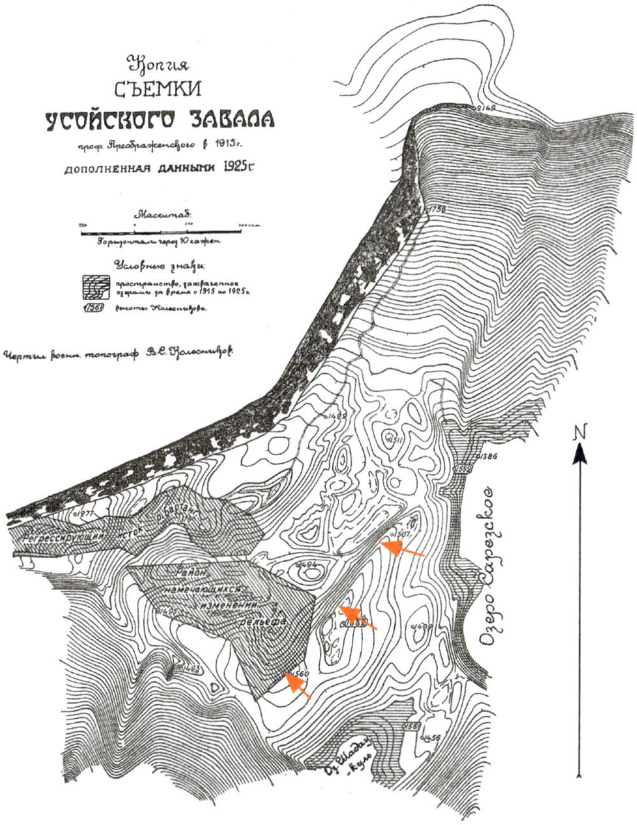

FIGURE 9.44 The topographic map of the Usoi blockage compiled by Prof. Preobrajensky in 1915 with additions made by the army surveyor V.S. Kolesnikov in 1925. Contour lines are in 10 Russian fathoms (21.3 m); horizontally hatched areas were submerged between 1915 and 1925; cross-hatched areas experienced ongoing deformations in 1925; elevation marks in Russian fathoms. We added red arrows to highlight the prominent scarp shown on Fig. 9.43. *After Kolesnikov, V. S., 1929. Brief description of the trip to the Sarez Lake in 1925. Proceedings of the Central Asian Geographical Society, XIX, Tashkent (in Russian).*

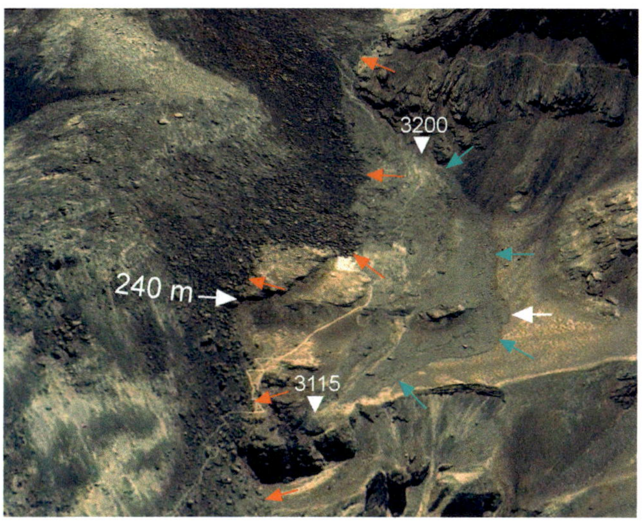

FIGURE 9.45 Traces of the assumed mud splash in front of the rockslide body. Red arrows—front of rockslide debris; blue arrows—front of the mud splash. Distance between white arrows is 240 m. 3D Google Earth view.

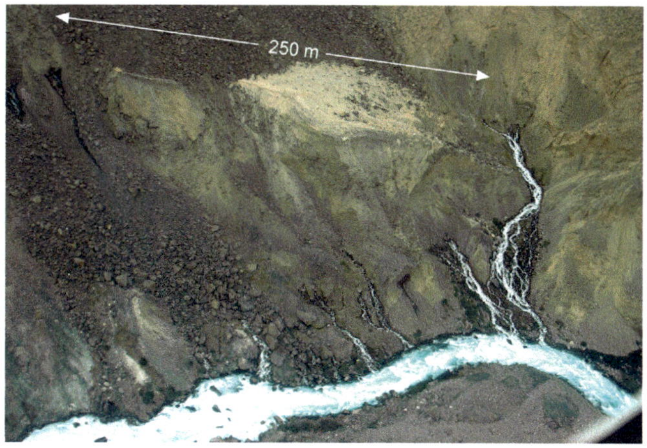

FIGURE 9.46 Springs in the erosional canyon about 800 m from the canyon head. Whitish patch on the canyon wall—remnant of debris flow sediments brought here from the headscarp area that survived from the subsequent erosion. Photograph made in September 2005.

observations, mean lake level increases gradually (see A. Ischuk, 2011, for more details).

On December 7, 2015, the M7.2 earthquake occurred near the Sarez Lake with its source associated with fault zone that crosses the eastern part of the lake (Kulikova et al., 2016). Though this earthquake did not produce any significant slope failures on the lake banks, ca. 20 cm rise of the water level was recorded directly at the blockage and at the Irkht meteorological station about 9 km upstream. It was hypothesized that such variation of the lake level was caused by the tectonic crustal deformation of the area associated with the earthquake that resulted in the overall tilt of the area with western side down that led to redistribution of water body (Ischuk and Strom, 2016).

What is very important is that the seepage takes place through the upper 140–150 m of the dam body below lake level. Thus, the lower ca. 350 m of its body is almost impermeable. We will discuss this phenomenon in Chapter 11, Internal Structure and Grain-Size Composition of Large-Scale Bedrock Landslides. The gigantic almost intact (apart from the intensive fracturing) block that forms the topmost part of the blockage (see Fig. 9.42) is surrounded by the same terrigenous rocks crushed at a rather high extent with impressive hummocky surface, likely formed by the contiguous molards (Fig. 9.47). Proximal lowermost part of rockslide dam is composed of crushed fragments of lime- and dolostone with gypsum that collapsed from the headscarp walls. It was hypothesized that some amount of ice from the displaced frontal parts of the glacial valleys (see Fig. 9.40) could be buried here too. Its melting could provide some deformations of this part of the dam that is considered as the weakest

FIGURE 9.47 Several meters high hummocks on the surface of the Usoi rockslide near the Monitoring and Early Warning Station (see Fig. 9.42) close to the Sarez Lake bank.

zone of the entire blockage both due to its altitude and material composition. It could be eroded if overtopping would occur due to any reason.

As we mentioned already neither structure nor grain-size composition of the Usoi blockage interiors are known. Judging from the numerous deeply dissected rockslides both from the Central Asian region (the Kokomeren rockslide—see Figs. 8.147, 8.150, 8.152, the Djashilkul rockslide—see Fig. 8.82, the Aksu rockslide—see Fig. 8.102) and from other parts of the world (e.g., Köfels rockslide in the Austrian Alps (Erismann, 1979; Brückl and Heuberger, 2001), rockslide dams in Indus River basin (Hewitt, 1988, 1998, 2002, 2006a, b, 2011)), we hypothesize that below coarse blocky carapace the internal almost impermeable part of the Usoi rockslide is composed of heavily comminuted and compacted rockslide debris.

We did not intend to discuss here all aspects of the Sarez Lake problem. Interested reader can find such analysis in Alford and Schuster (2000); A. Ischuk (2011); N. Ischuk (2014); Strom (2014b). Here we just describe several slope failures on the Sarez Lake banks, both past and still active, some of which may influence lake and dam safety.

The most widely discussed is the so-called Right-Bank landslide that was supposed to affect the lake bank of the Sarez Lake ~5 km of the dam (see RB on Fig. 6.21) (Sheko, 1968; Sheko and Lekhatinov, 1970; Fedorenko, 1988; Alford and Schuster, 2000). Though both the existence of this feature and its volume estimates have been debatable, in the worst case scenario it was assumed that up to 1–2 km^3 of rocks could collapse in the deepest part of the lake creating the surge wave that could erode the lowermost part of the Usoi blockage resulting in its catastrophic breach. This point of view is opposed by N. Ischuk (2008, 2013b, 2014) who explains the suspicious landforms as glacial features. Leaving this discussion beyond the scope of our book, we should notice that about 5 km of the Usoi blockage at 38.248° N, 72.66° E the right bank of the lake up to 340–400 m above the mean lake level is really affected by the block slide about 50 Mm^3 in volume (considering its submerged part), eastern portion of which had collapsed already (Tajik 088, Fig. 9.48). It can be hypothesized that this fresh slope failure could occur simultaneously with the Usoi rockslide.

V.S. Fedorenko (1988) proposed that both the 1911 Usoi rockslide and the assumed (large) Right-bank rockslide could originate on both sides of the much larger Pleistocene Murgab rockslide (Tajik 089, 38.27° N, 72.65° E) that had collapsed from the more than 2.5 km high and nearly 6 km long southern flank of the Muzkol Range and had displaced 700–800 m thick massif of the Carboniferous terrigenous rocks. This assumption was supported by the unusual topography of the left bank of the Sarez Lake between the Shadau and Irkht River valleys—the so-called Marjanai Plateau that rises 450–600 m above the lake. It is considered to be the body of this gigantic Pleistocene rockslide and, at its eastern part, of another giant Pleistocene slope failure—the Irkht rockslide (Fig. 9.49). Cumulative volume of these assumed slope failures could exceed 10 km^3. However, more investigations are necessary to make final decision on the correctness of interpretation of the Marjanai Plateau as the body of the gigantic Pleistocene rockslide(s). We must mention that

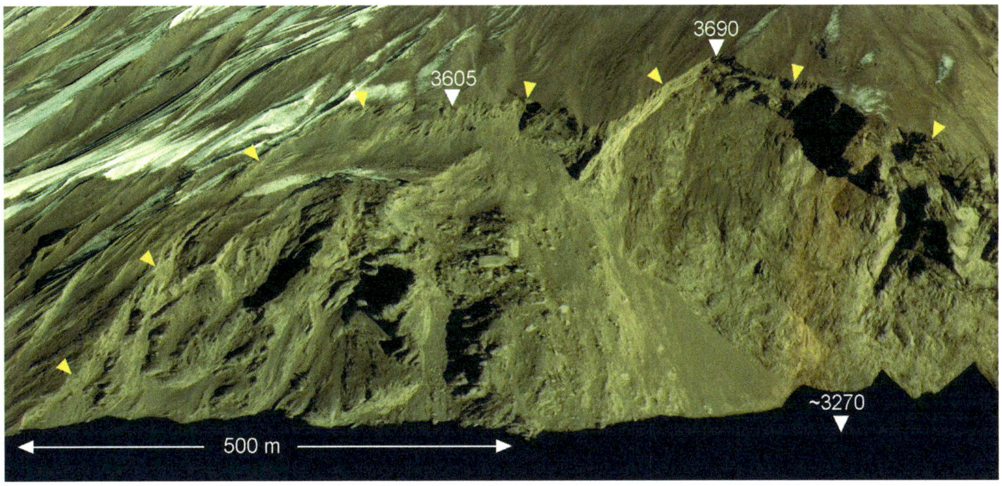

FIGURE 9.48 Block slide on the right bank of the Sarez Lake about 5 km east of the blockage. Orange triangles mark the headscarp. 3D Google Earth view.

rockslide origin of the Marjanai Plateau was first proposed by Prof. Preobrajensky (1920). Presence of the pre-Usoi rockslides in this part of the Murgab River valley was also proposed by Sh. Denikaev (1970).

Northern front of the Marjanai Plateau was affected by shallow slope deformations during the 2015 earthquake (Ischuk and Strom, 2016). Most likely that this concave slope is the headscarp of the landslide, whose deposits rest on the bottom of the Sarez Lake.

Surprisingly, we could find very few similar large-scale bedrock landslides in the deeply incised Bartang River valley downstream of the confluence of the Murgab and Kudara Rivers, and almost none of them in the Yazgulem and Vanch River valleys, except the dense cluster of about 10 landslides located in the Bartang River valley just downstream of the Basid village, on the right-bank side of the valley mainly. Two largest of these landslides (Tajik 081, 081a), first identified by Svetlana Vinnichenko (personal communication) at 38.13° N, 72.14° E on the more than 1.5 km high slope, are shown on Fig. 9.50. All of these slope failures originated on slopes composed of the Triassic terrigenous rocks cut by several fault zones. Most of these landslides rest on slopes or in the steep gullies and, likely, did not cause streams damming. That is why, despite Svetlana Vinnichenko assumption of their seismic triggering, it seems that most of these landslides are gradually evolving features, though, of course, earthquake as a triggering factor could not be excluded. But lithology and local tectonic structure play the critical role in their formation.

Besides numerous landslides, evidence of the DSGSD—series of the upslope facing scarps (Tajik 082) passing along the crest of the ca. 1 km high ridge at a distance of about 2 km (Fig. 9.51) was found at the

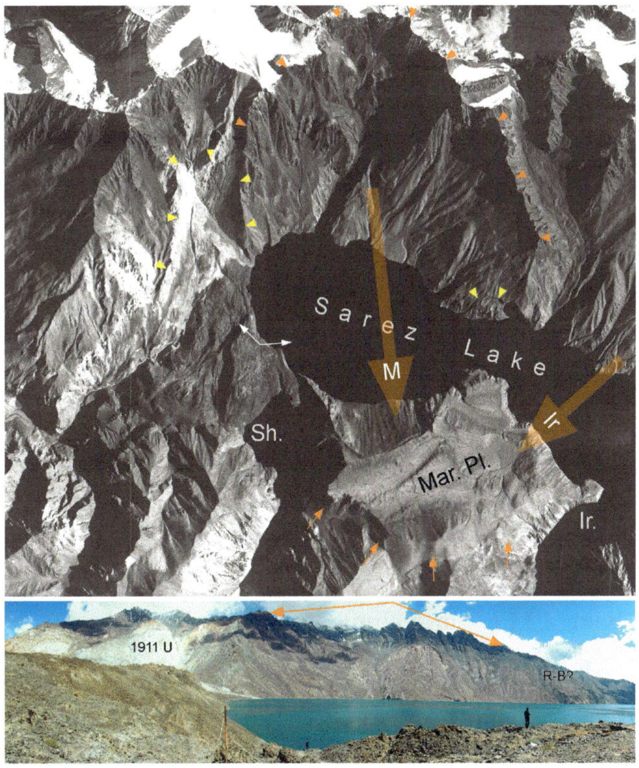

FIGURE 9.49 Above—the KFA-3000 space image of the head part of the Sarez Lake. Yellow triangles mark the headscarp of the 1911 Usoi rockslide and of the rockslide shown on Fig. 9.48; orange triangles—the assumed headscarp of the Murgab rockslide and orange arrows—front of the assumed combined rockslide body (the Marjanai Plateau) formed by the Murgab (brown arrow with M) and the Irkht (brown arrow with Ir) rockslides. Sh.—the Shadau Lake, Ir.—the Irkht bay. Below—the panorama of the Muzkol Range between the 1911 Usoi rockslide (1911 U) and the assumed Right-Bank rockslide (R-B?) made from the site marked above by the double-head white arrow. The double-head orange arrow shows the limits of the assumed Murgab rockslide headscarp.

FIGURE 9.50 Panorama of two landslides on the right-bank slope of the Bartang River valley opposite the Basid village. *Photo courtesy: Svetlana Vinnichenko.*

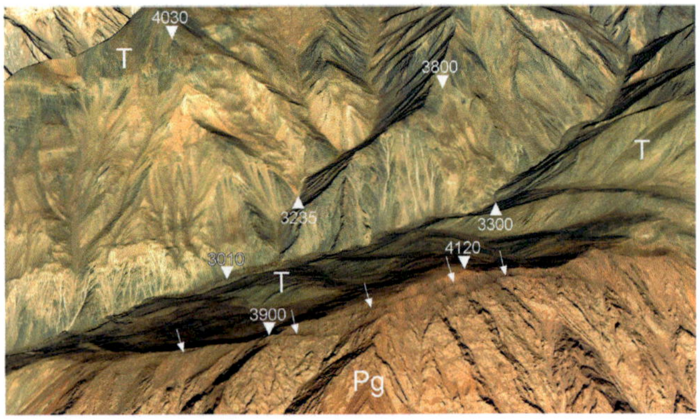

FIGURE 9.51 Upslope facing scarps (marked by arrows) along the local watershed. Elevation marks 4030 and 3800 m a.s.l mark headscarp crowns of landslides on the opposite side of the valley. T—Triassic terrigenous rocks, Pg—Paleogene volcano-clastic rocks. 3D Google Earth view.

local watershed at 38.124° N, 72.1° E, just west of the landslides shown on Fig. 9.50.

One more relatively small rockslide (Tajik 109) about 1 Mm^3 in volume was found at 39.059° N, 71.856° E near the mouth of the Khodorjiodara River—the small right tributary of the Bartang River. It caved from the slope composed of the Paleogene volcanic rocks and formed the blockage about 80 m high, but due to extreme steepness of the Khodorjiodara River thalved (about 21 degrees), the blocked part of the valley was filled by fluvial and slope deposits in a short time (Fig. 9.52). Dam's surface is covered by angular blocks up to 20 m in size while inner part (below the irrigation canal that crosses the downstream slope of the blockage)—of much finer material.

In the Vanch River valley, rock slope failure occurred in its upper reaches at the RGO glacier (Tajik 100, 38.761° N, 72.143° E) most likely in August–September 2001 (Kotliakov et al., 2003; Stark

FIGURE 9.52 Rockslide at the Khodorjiodara River mouth. 3D Google Earth view. Blocks up to 20 m in size on top of the dam are shown in the inset.

et al., 2012). It was described in Section 4.5 of Chapter 4 as one of the rare examples of rockslides that could be identified in the presently glaciated parts of the study region (see Figs. 4.34 and 4.35). As mentioned in Section 4.5 its volume was estimated as 20 ± 5 Mm^3. According to personal communication of A. Ischuk the RGO Glacier rock avalanche could be triggered directly or indirectly by one of the deep Afghan earthquakes. We have to notice, however, that none of such earthquakes that occurred within this time interval had magnitude more than 6.5 and epicenter location close to the RGO glacier.

Unlike Vanch and Yazgulem River valleys, numerous large rockslides, many of which are still evolving, were identified in the Muksu and the Obikhingou River valleys draining the southern slopes of the Zaalai and Peter the First Ranges and the large part of the most elevated "knot" of the Central Pamir with the Ismail Somoni and the Korjenevskaya Peaks exceeding 7 km a.s.l. Some of their landslides were described by Pyzhyanov and Presnuhin (1987).

In the Muksu River trough valley, whose upper reaches are filled by the Fedchenko glacier—the world-longest valley glacier outside the polar areas, large-scale bedrock landslides were identified on right-bank slope of the middle reaches of this valley mainly. One of them (Tajik 136) occurred at 39.19° N, 72.041° E, upstream of the mouth of the Kyzylsu River—right

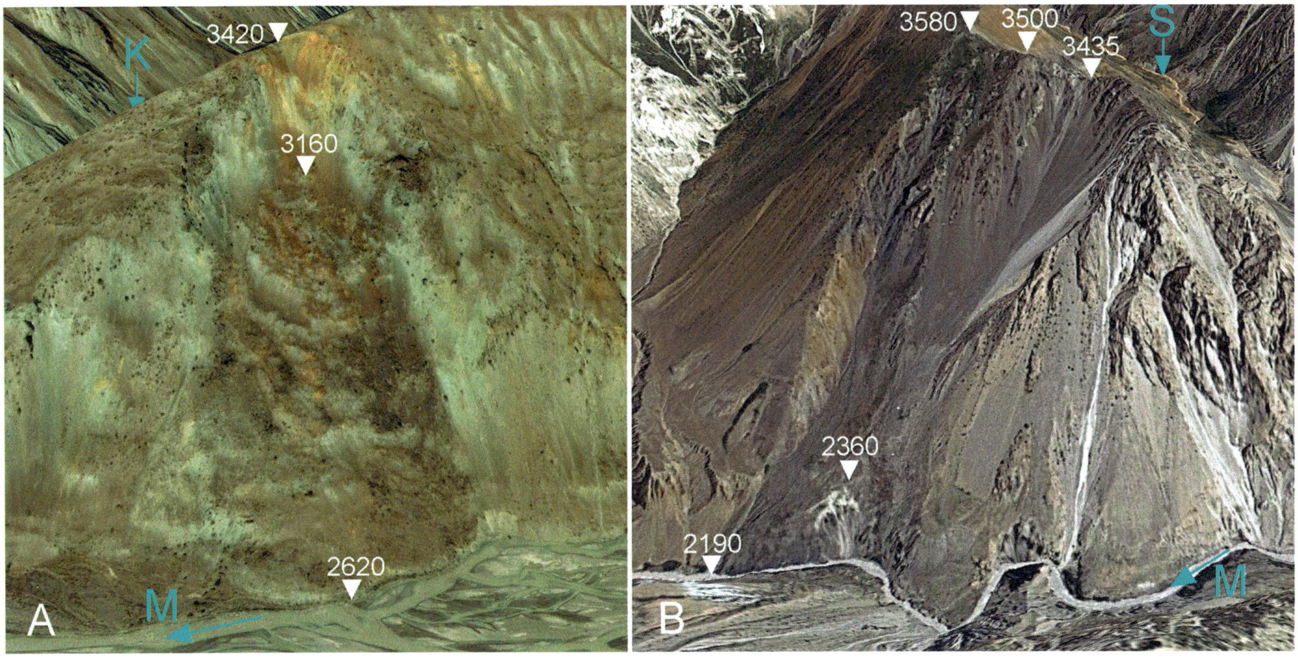

FIGURE 9.53 Rockslides on the right-bank slope of the Muksu River valley. (A) The Kyzylsu rockslide; (B) the Shirvosa rockslide; M—Muksu River; K—Kyzylsu River; S—Shirvosa River. 3D Google Earth views.

tributary of the Muksu River (do not mix with the same name river draining the Alai intermontane depression) on the 800 m high slope dipping 33–30 degrees and composed of the volcaniclastic, locally ferruginated carboniferous rocks. Landslide likely affected most weathered superficial rocks up to 30–50 m thick. Despite height and steepness of the slope and absence of frontal confinement, most of the landslide body about 8 Mm^3 is still on the slope. Though its toe is eroded gradually by the Muksu River (Fig. 9.53A), it does not demonstrate evidence of present-day activity unlike landslide on the same right-bank slope of the Muksu River valley just upstream of the Satsu River mouth (Tajik 137; 39.174° N., 71.962° E).

The Satsu rockslide originated on the slope composed of the same volcaniclastic rocks, also ferruginated. Lower part of the slope about 350 m high (up to 2880 m a.s.l. on Fig. 9.54) is affected by the pronounced active slide, while flat bench on the slope above its headscarp up to 3030 m a.s.l. is crossed by numerous fissures and minor scarps demonstrating ongoing deep-seated deformations (see the inset in Fig. 9.54). Generally this landslide could be classified as the retrogressive slope failure that initiated due to slope undercutting by the river and evolves gradually. Pyzhyanov and Presnuhin (1987) identified displacements of its body between 1962, when it was mapped first time, and mid-1980s, when the referred paper was written. Besides they assumed that the abnormal climatic event or earthquake could cause catastrophic

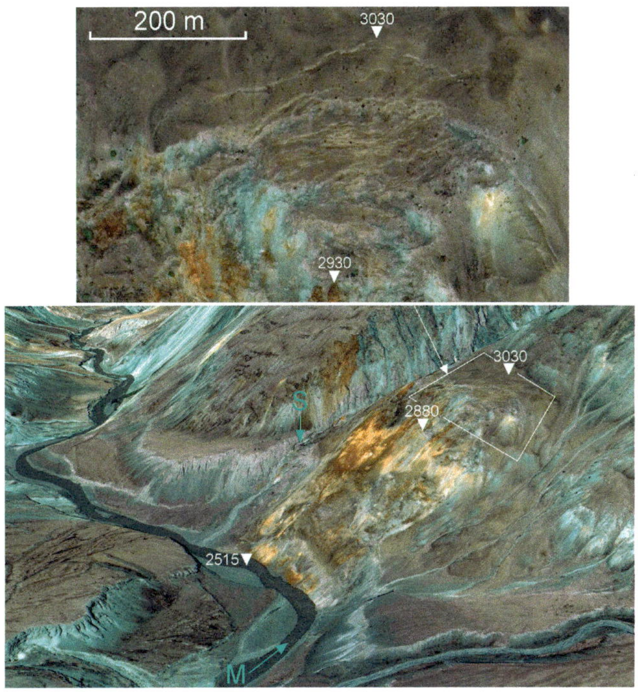

FIGURE 9.54 The Satsu rockslide. M—Muksu River; S—Satsu River. 3D Google Earth view. Scarps and fissures above the well-defined headscarp can be seen in the inset.

failure of this rockslide that could result in the Muksu River blocking for up to 200 and even 400 m high. However, size (900 m along the slope base) and volume (53.5—83.3 Mm^3) estimates of the Satsu rockslide provided by Pyzhyanov and Presnuhin (1987) seems to be exaggerated. According to our measurements its front is 670 m wide and volume (including the upper part just prepared for sliding) is about 25 Mm^3 only. Significant (up to 1 km) width of the Muksu River valley at the level of its terraces makes the formation of the dam here more that 100–120 m high almost impossible.

Multiple scarps on the right-bank slope of the Muksu River valley nearby, most likely, are the glacial landforms left by several stages of the Pleistocene glaciation.

The third rockslide mentioned in Pyzhyanov and Presnuhin (1987) is located at 39.135° N, 71.73° E, 2 km downstream of the Shirvosa River mouth (Tajik 138). This catastrophic slope failure about 12 Mm^3 in volume occurred on the ca.1300 m high slope composed of the Jurassic volcanic clastic rocks. It likely caused temporary damming of the Muksu River, whose valley here is much narrower than upstream. Headscarp of this rockslide was delineated between elevation marks 3500 and 3435 m a.s.l. based on its slightly brighter color that the adjacent parts of this local watershed (Fig. 9.53B).

In the left-side part of the Muksu River basin with well-pronounced modern glaciation, large resent rock slope failure occurred between 1990 and September 2000 (see Fig. 4.37) at 39.03° N, 71.963° E, and its deposits still could be seen near the terminus of the Valter glacier (Tajik 101). This case study was described in Section 4.5 in Chapter 4 (see Figs. 4.36-4.38).

We must notice that large parts of some glaciers in this part of the Pamir are almost completely covered by debris, so that ice itself is invisible. It can be assumed that part of this debris was supplied by more or less large rockslides and rock avalanches, However, except the Valter glacier case study, none of them could be identified at present as individual rock slope failures.

Downstream part of the Muksu River valley is incised in the Jurassic and Cretaceous sedimentary rocks typical of the Tajik depression (see Chapter 3) that differ significantly from the metamorphic, igneous, and volcaniclastic rocks of the main part of the Pamir. Slope deformations here are of the same types as at the Peter the First Range northern slope (see Figs. 5.20, 5.21, 9.26, and 9.27).

Features typical of the DSGSD were found on top of the right-bank slope of the Muksu River valley at 39.165° N, 71.614° E (Tajik 139). Both slopes of the ridge, whose crest is displaced by more than 1 km long up-slope facing scarp, are affected by landslides (Fig. 9.55). This feature strikes across bedding of the Jurassic sandstone, siltstone, and mudstone. Several local ridges nearby demonstrate similar deformations.

One more recent rock slope failure in the Cretaceous red beds (Tajik 140) affects the right-bank slope of the Shagazy River valley (left tributary of the Muksu River) at 39.128° N, 71.571° E. Despite its rather large size (about 5 Mm^3 in volume) and 250–280 m high back wall of the headscarp, there is no evidence of river damming (Fig. 9.56). Thus this feature seems to be rather slow moving landslide whose toe has been eroded gradually.

Several rockslides were identified in the Obikhingou River basin. Valleys of its upstream part are incised in Paleozoic metamorphic and igneous rocks of the Darvaz Range—the outer mountain chain of the Pamir, while the downstream part of the river basin—in the folded Jurassic, Cretaceous, and Cenozoic sedimentary rocks of the Tajik depression (see Chapter 3).

The uppermost part of its basin subjected to intensive glaciation in the past is almost free of such features except one site (Tajik 141, 39.795° N, 71.67° E) on the southern slope of the Pashimgar Ridge composed

FIGURE 9.55 Up-slope facing scarp (DSGSD) on top of the right-bank slope of the Muksu River valley (marked by yellow arrows). LS—landslides. 3D Google Earth view.

FIGURE 9.56 Rockslide in the Cretaceous red beds in the Shagazy River valley. 3D Google Earth view.

FIGURE 9.57 Rockslide on the southern slope of the Pashimgar Ridge. 3D Google Earth view.

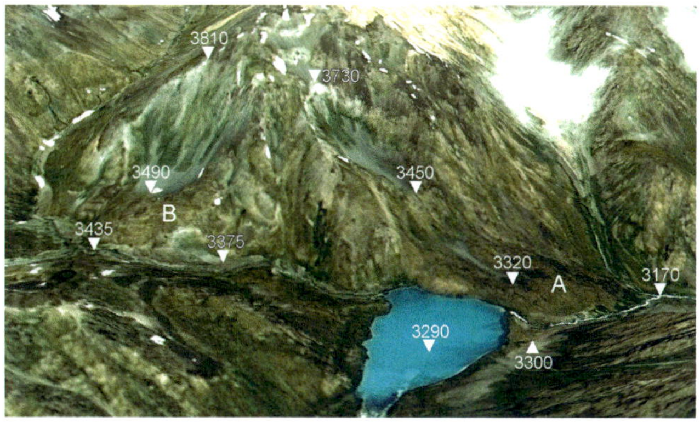

FIGURE 9.58 Natural dams at 38.766° N, 71.053° E (Tajik 142—A) and 38.759° N 71.041° E (Tajik 143—B). 3D Google Earth view.

of Permian terrigenous rocks. It has a well-defined headscarp with its crown at ~4280 m a.s.l. and 800 m long body about 6 Mm^3 in volume that filled the ravine on the valley slope and terminated about 350 m above the Bahud River valley bottom (Fig. 9.57). Rockslide body practically remains in the scar. Gradual destruction of its front supplies material for the large debris flow fan at the valley bottom.

Several natural blockages, some of which with lakes upstream, were found in the unnamed river valley that falls into Obikhingou from the left, 9 km downstream of the Obimazor River mouth (Tajik 142–144). The footpath from the Obikhingou to the Vanch River valley passes along this stream. Valley is incised in Paleozoic granite and granodiorite. Though all of these accumulations of coarse material at the valley bottom have spoon-shape lowerings on the slopes just above them (their crowns are marked on Fig. 9.58 at 3810 and 3730 m a.s.l.), we cannot exclude their glacial origin completely, considering high altitude of the area, though shape of these blockages indicates that debris moved across the valley rather than along it excluding interpretation of these bodies as end moraine of the along-valley glacier. However, if one assumes that these lowerings were formed by slope glaciers, it would be difficult to explain how this long-term gradually evolving process of slope destruction could form so high (up to 50—100 m high) and compact accumulations at the valley bottom affected by along-valley streams fed by much larger glaciers in the upper reaches of the valley. Nevertheless, detail field study is necessary to clarify origin of these landforms and deposits.

Potentially hazardous site was identified near the junction of the Obikhingou River with its large right tributary—the Ragnow River (on some maps it is named the Shaklysu River). 3–4 km upstream of its mouth, at 38.921° N, 70.964° E, the Ragnow River valley had been blocked quite recently by rockslide from its 820 m high right-bank slope that formed 30–35 m high dam (Tajik 145, Fig. 9.59). Smaller recent rockslide has its crown at 2660 m a.s.l. Both rockslides are visible on the KFA-1000 space images obtained in the middle of 1970s, about 40 years ago. Thus they should be older, but, judging from the freshness of the dam's remnant shown on Fig. 9.59, unlikely that several decades older. Considering that this site is located 28 km south of the Khait rock avalanche (see Figs. 5.15, 8.271, and 8.272) triggered by the 1949 M7.4 Khait earthquake just south of its epicentral zone, it can be assumed that slope deformations at this site could be also triggered by this earthquake. We believe that time of this failure occurrence could be specified by interviewing of local villagers. We also want to mention that these large-scale rock slope failures have almost opposite orientation—direction of the initial motion of the Khait rockslide body that later converted into long runout rock avalanche was NW-305°, while that of the Ragnow-mouth DSGSD, whose evidence can be observed on the slope above recent landslide (Fig. 9.60) is SE-120°. Failure of the frontal part of this unstable slope occurred between 1956 (it is not visible on aerial photograps made that year) and 1970s, since it is visible on KFA-1000 space image of these years (it will be shown on Fig. 15.1 in Chapter 15, Evidence of Slope Instabilities – Potential of Future Large-Scale Slope Failures).

Right-bank slope of the valley is composed of the folded Upper Cretaceous sedimentary rocks (limestone, sandstone, siltstone, and gypsum) with relatively low strength striking along the slope and dipping 40–45 degrees in the slope. According to the 1:200,000 State Geological map, river here passes just along the fault zone. Smooth hilly area above the headscarp of this recent slope failure is crossed by numerous normal scarps and upslope-facing scarps reflecting

FIGURE 9.59 The Ragnow-mouth rockslide. Scarps and fractures are developed above the recent headscarp crown up to ca. 3100 m a. s.l. Remnants of the breached rockslide dam can be seen in the inset below. 3D Google Earth views compiled from the space image obtained on August 13, 2008.

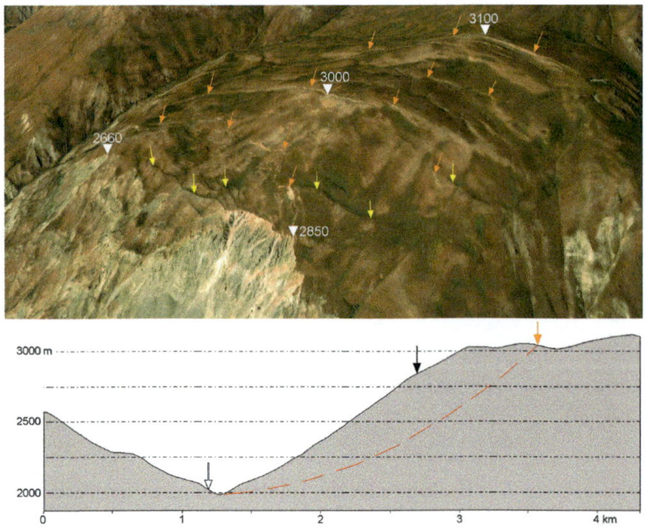

FIGURE 9.60 Above—evidence of the ongoing slope deformations above the headscarp crown of the recent Ragnow-mouth rockslide. Yellow arrows—upslope-facing scarps; orange arrows—downslope-facing scarps. 3D Google Earth view. Below—schematic cross-section of this slope and position of the assumed sliding surface (red dashed line); orange arrow—the outermost downslope-facing scarp; black arrow—crown of the recent rockslide headscarp; white arrow—tip of recent rockslide deposits. Profile compiled by Global Mapper software based on the 3″ SRTM DEM.

intensive ongoing toppling. These landforms extend at a distance up to 830 m upslope from the recent headscarp crown (see Fig. 9.60) indicating the possibility of quite voluminous, up to 300 Mm^3, future slope failure that, if it would occur catastrophically, "in one gulp," could create really a large dam several hundred meters high that could impound the lake up to ca. 0.4 km^3 in volume. Considering weak mechanical properties of the sedimentary rocks that could be involved in this failure, such dam would be breached inevitably and, likely, catastrophically. Thus, this site requires special attention taking in mind several hydraulic schemes located downstream in the Vakhsh River, including the existing Nurek HPP with 300 m high rock-filled dam and the Rogun HPP with even higher dam under construction.

We should notice that this Ragnow-mouth rockslide resembles the Satsu rockslide in the Muksu River valley described above (see Fig. 9.54) and the evolving slope deformations at the Aksu—Konchakiol watershed at the left bank of the Naryn River valley (see Fig. 8.185).

Another very large landslide affects the ~1 km high and rather gentle (~16° only) left-bank slope of the Obikhingou River valley, composed of the gypsum and salt-bearing weak sedimentary rocks of the Jurassic age, immediately upstream of its confluence with the Ragnow River (Tajik 146, 38.89° N, 70.96° E). This ca. 4 km^2 landslide evolves more or less parallel to the strike of the Jurassic layers and comes out to the Obikhingou River stream at a 1700 m long and more than 150 m high toe (Fig. 9.61). Unlike case study described above with very attractive contemporaneous slope deformations really endangering the Obikhingou River valley, this landslide about 200 Mm^3 in volume does not demonstrate evidence of present-day activity. Nevertheless, its activation due to abnormal precipitation or earthquake cannot be excluded and, considering its very large volume and composition, could be hazardous as well.

One more site in this part of the Obikhingou River basin that deserves landslide researchers' attention is located on the crest of the local ridge that forms the left-bank slope of the Ragnow River valley at 38.94° N, 71.0° E (Tajik 147), about 3.5 km northeast of the Ragnow-mouth rockslide. It is a typical DSGSD (sackungen) with 2.6 km upslope-facing scarp ~80 m below the ridge crest at its other side and with evidence of slope deformations on its river-facing side (Fig. 9.62). Ridge is composed of the Upper Cretaceous and

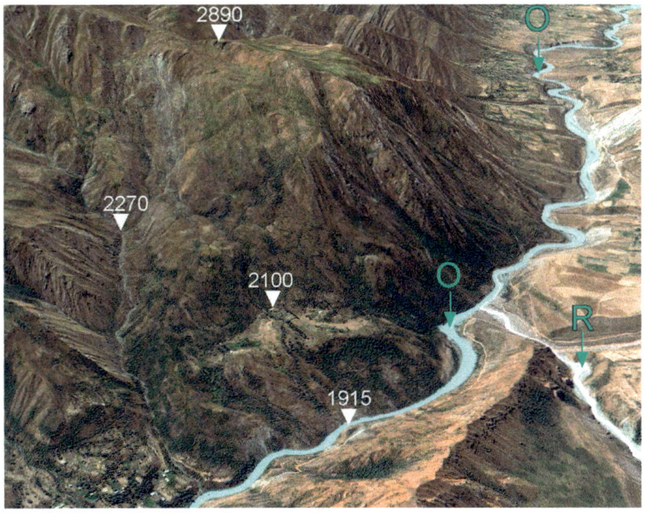

FIGURE 9.61 Landslide in the Jurassic sedimentary rocks near the confluence of the Obikhingou (O) and the Ragnow (R) Rivers. 3D Google Earth view.

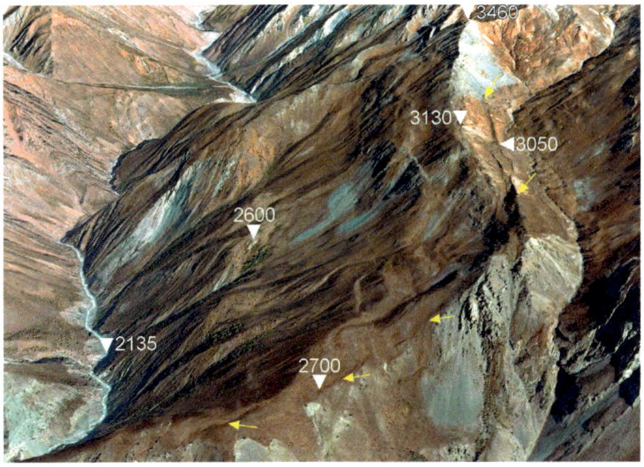

FIGURE 9.62 DSGSD on top of the left-bank slope of the Ragnow River valley. Upslope-facing scarps are marked by yellow arrows. 3D Google Earth view.

Paleogene weak gypsum-bearing sedimentary rocks folded in the syncline. It can be hypothesized that the sackungen scarps inherit bedding.

In its lower reaches (at a distance of more than 60 km) the Obikhingou River passes through folded Neogene conglomerates, sandstones, and mudstones, locally through the Paleogene and Upper Cretaceous sedimentary rocks outcropping at the axial parts of anticlines. Landslides forming in these weak rocks in the tributary valleys with steep thalwegs provide large amount of loose material that is reworked by debris flows. The latter sometimes are so powerful and voluminous that they cause temporary damming of the Obikhingou River, which is one of the major hazards in this part of Tajikistan. Two examples of such landslides can be seen on Fig. 9.63. Very recent rockslide at 38.7885° N, 70.821° E, near the Gundarabolo village (Tajik 150) collapsed from the 350 m high slope composed of the Upper Cretaceous carbonate gypsum-bearing rocks. Its deposit overlays low dam formed by the previous larger failure from the same slope. Deposits of this older landslide fill the valley bottom between 2450 and 2430 m a.s.l. (Fig. 9.63A). Rockslide shown on Fig. 9.63B (Tajik 151) occurred at 38.789° N, 70.406° E, in the upper reaches of the Rabatkoli Creek incised in Neogene conglomerates and formed easily erodible body that serves as a source of material for debris flows.

Very high concentration of large-scale bedrock landslides in the Western Pamir is in the Gunt River valley and in the lower reaches of its largest tributary—the Shakhdara River valley (Fig. 9.64). It is one of the most attainable parts of the Tajik Pamir due to the Osh-Khorog road passing here. The Gunt River originates from the Yashilkul Lake dammed at 37.78° N, 72.735° E by the Yashilkul rockslide about 50 Mm^3 in volume (Tajik 091) described in brief in Section 4.2 of Chapter 4 as an example of bedrock landslide that bulldozed moraine material resting at the foot of the collapsed slope, threw it into the stream channel, and overlaid this pseudo-moraine dam by the carapace of angular gneiss boulders (see Fig. 4.19). Due to frontal confinement, rockslide debris rised ca. 140 m on the opposite slope (Fig. 9.65) and spread along the valley for more than 4 km (see Fig. 4.18). Such spreading resulted in low gradient of both overflowing and seeping water. Being relatively low—its effective height is about 50 m only—this natural dam blocks a large catchment of 5166 km^2. However, despite high (3.71) DBI value (Strom, 2010b) that place this feature far in the instability domain (Ermini and Casagli, 2002, 2003), this blockage seems to be absolutely stable. Low-gradient water flow passing through the coarse blocky carapace of the dam with large voids between angular gneiss boulders (see Fig. 4.19) could neither remove these boulders nor incise through them.

Gunt River basin at its upper reaches, before its confluence with the Toguzbulak River, is quite asymmetric—left-bank slope of the valley rises sharply towards the ca. 5 km high local watershed of the Bachor Ridge, while its right side is dissected by numerous ~20 km long tributary valleys originating from the Rushan Range with summits up to 5.5—6 km a.s.l. Most of the left-side rockslides in the upper part of the Gunt River valley up to the Toguzbulak River mouth and even further, up to the Pathur River mouth collapsed directly into the main valley, while most of right-side rockslides occurred in the tributary valleys (see Fig. 9.64).

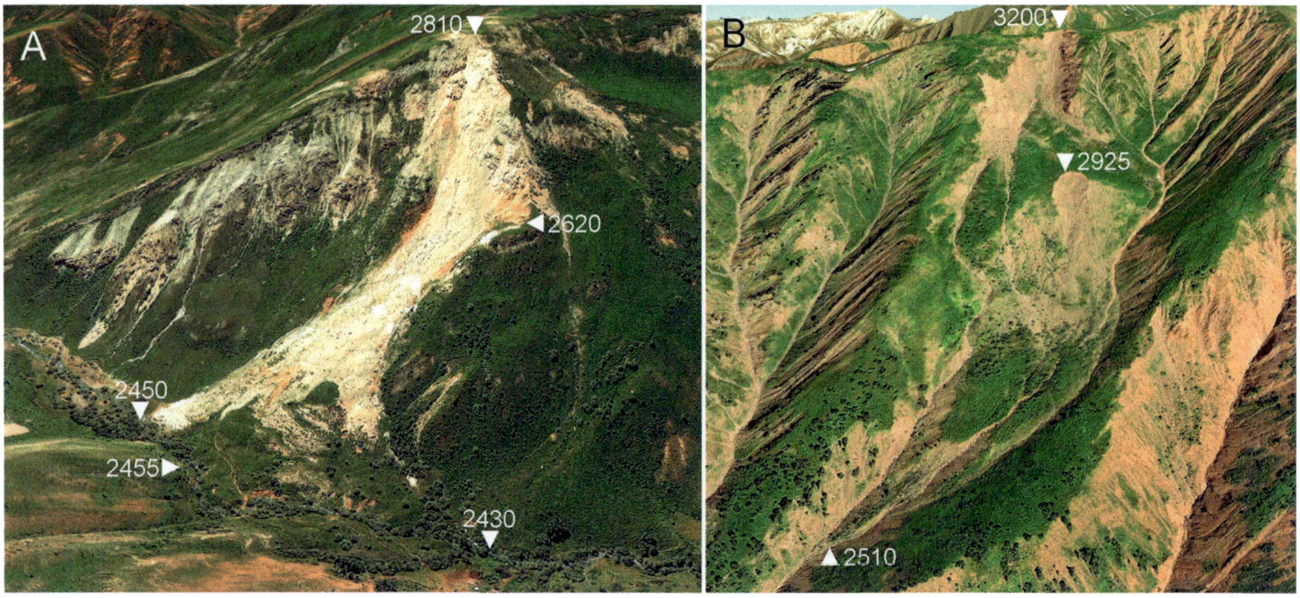

FIGURE 9.63 Landslides near the Gundarabolo village (A) and in the Rabatkoli Creek valley (B). 3D Google Earth views.

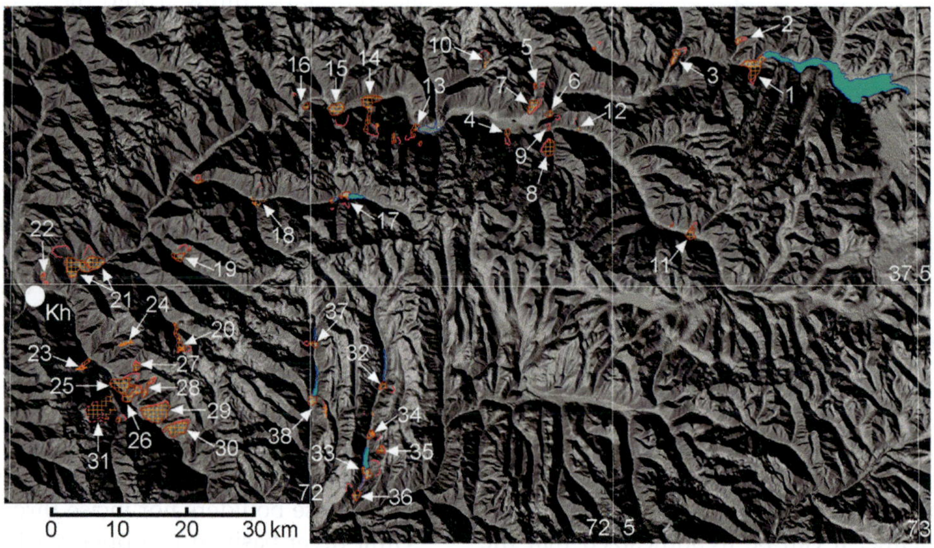

FIGURE 9.64 Rockslides in the Gunt River valley and its tributaries. Background—the 3″ SRTM DEM visualized by Global Mapper software. Rockslides are marked by pink and brown dots; blue dots—dammed lakes. Kh—the Khorog town. 1—the Yashilkul rockslide (see Figs. 4.18, 4.19, and 9.65); 2—the Liangar rockslide (see Fig. 4.16); 3—the Andravdj rock avalanche (see Fig. 9.66); 4—the Imom rockslide (see Figs. 4.14 and 4.15); 5—the Upper Shazuddara rockslide (see Fig. 9.67); 6—see Fig. 9.67; 7—the Shazud rockslide (see Fig. 9.67); 8—the Duzakhdara rockslide (see Fig. 9.68); 9—the Lower Toguzbulak rockslide (see Fig. 9.69); 10—see Fig. 9.70; 11—the Upper Toguzbulak rockslide (see Fig. 9.71); 12—see Fig. 9.72; 13—the Chartym rock avalanche (see Figs. 9.73, 9.74); 14—the Vir-Tangiv rock avalanche (see Figures 4.26–4.28, 9.75); 15—the Zuvor rockslide (see Figs. 4.26 and 4.42); 16—the Ruzhak rockslide (see Fig. 9.76); 17—the Rivakkul rockslide (see Fig. 4.21); 18—see Fig. 9.77; 19—the Shoripdara rockslide (see Fig. 9.78); 20—the Urmochdara rock avalanche (see Fig. 4.40 and 4.41); 21—the Suchan rockslides (see Fig. 9.79); 22—see Fig. 9.80; 23—see Fig. 9.81; 24—the Ridjistdara DSGSD (see Fig. 9.82); 25—the Tusen rockslide (see Fig. 9.83); 26—the Chagev rockslide (see Figs. 9.83, 9.84); 27—see Fig. 9.83; 28—see Figs. 9.83 and 9.85; 29—see Fig. 9.85; 30—see Fig. 9.86; 31—see Fig. 9.87; 32—the Upper Dirumkul rockslide (see Fig. 9.90); 33—the Dirumkul Lake dam (see Fig. 9.91); 34—see Fig. 9.91; 35—see Fig. 9.91; 36—the Lower Dirumkul rockslide (see Fig. 9.92); 37—see Fig. 9.93; 38—the Oikul rockslide and DSGSD (see Fig. 9.94)

FIGURE 9.65 View on the Yashilkul rockslide dam from the lake. Splitting of rock avalanche debris part of which rose on the opposite slope up to 3860 m a.s.l., while other part spread along the valley can be seen in the inset.

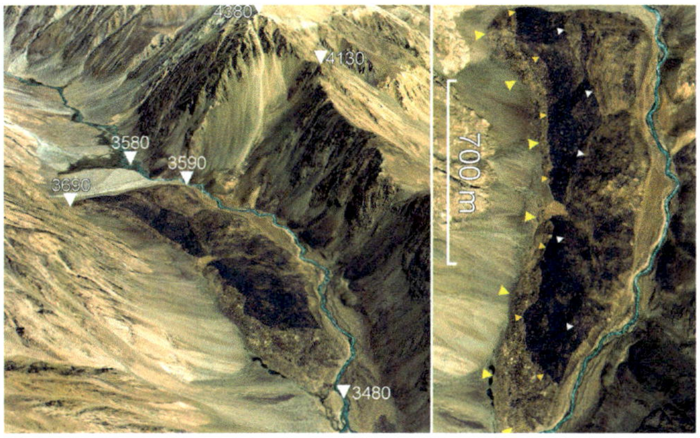

FIGURE 9.66 The Andravdj rock avalanche. Left—3D Google Earth view; right—plan view of the deposits. Yellow triangles—the outer front of rock avalanche body; smaller orange and white triangles mark boundaries between lithological zones.

About 4 km north of the Yashilkul rockslide dam, at 37.814° N, 72.71° E, the Liangar River—first right tributary of the Gunt River — was blocked by the Liangar rockslide about 8 Mm^3 in volume that bulldozed moraine material in front of it (see Fig. 4.16) similar to the Yashilkul rockslide. Rockslide originated on the about 900 m high slope composed of Mesozoic granite, but the headscarp crown is ca. 670 m above the valley bottom only. Rockslide dam crest level gradually decreased towards the right bank of the valley where its 15–20 m thick toe was incised by the river.

Large rockslide had collapsed at 37.795° N, 72.604° E from the 800 m high slope of the Andravdj River valley (the second right tributary of the Gunt River) composed of the Proterozoic gneiss and covered about 1.5 km^2 of this valley bottom (Tajik 106, Fig. 9.66). Rockslide about 30 Mm^3 in volume moved aslant to the valley orientation and, due to that, it converted into rock avalanche that spread for more than 2 km along the valley being 25–70 m thick and rose 100 m on the opposite slope at the upstream part of the blockage. Interesting peculiarity of this rock avalanche is the presence of three distinct zones on its surface differing in color and, likely, in the lithology (see right part of Fig. 9.66). It makes this case study very similar to the Kuturgansu-mouth rock avalanche (see Fig. 8.246) and the Kudara-Pasor rock avalanche (see Figs. 9.38 and 9.39). It can be hypothesized that the outer zone is composed of the moraine material from the slope base and/or valley floor bulldozed by main part of this rockslide. The Andravdj rock avalanche can be ascribed to the Primary type (see Section 5.3.1 in Chapter 5), and the blockage was eroded at its proximal left-bank side. Dammed lake was partially filled by fluvial deposits transported by the Andravdj River and by debris flows from the gullies dissecting valley slopes.

Dense cluster of bedrock landslides could be identified at the confluence of the Gunt, Toguzbulak, and Shazuddara Rivers (see Fig. 9.64). Besides the Imom rockslide at 37.697 N, 72.326 E (Tajik 093) described in Chapter 4 as an example of bedrock landside bulldozing moraine material in front of it (see Figs. 4.14 and 4.15), it consists of two catastrophic river-damming rockslides in the Shazuddara River valley (Tajik 111 at 37.734° N, 72.371° E and Tajik 112 at 37.757° N, 72.372° E), two rockslides on both sides of the Bachor Ridge dividing Gunt and Toguzbulak River valleys (Tajik 107 at 37.72° N, 72.397° E and Tajik 115 at 37.703° N, 72.395° E) and of large gradually evolving rockslide on the left-bank slope of the Duzakhdara River valley at 37.675° N, 72.398° E (Tajik 116). All of them originated on the slopes composed of granites and Proterozoic schist and gneiss. Rockslides in the Shazuddara River valley had blocked it by the presently breached dams, similarly to the rockslide in the Gunt River valley 3.7 km upstream of its confluence with the Toguzbulak River (Fig. 9.67). The Upper Shazuddara rockslide

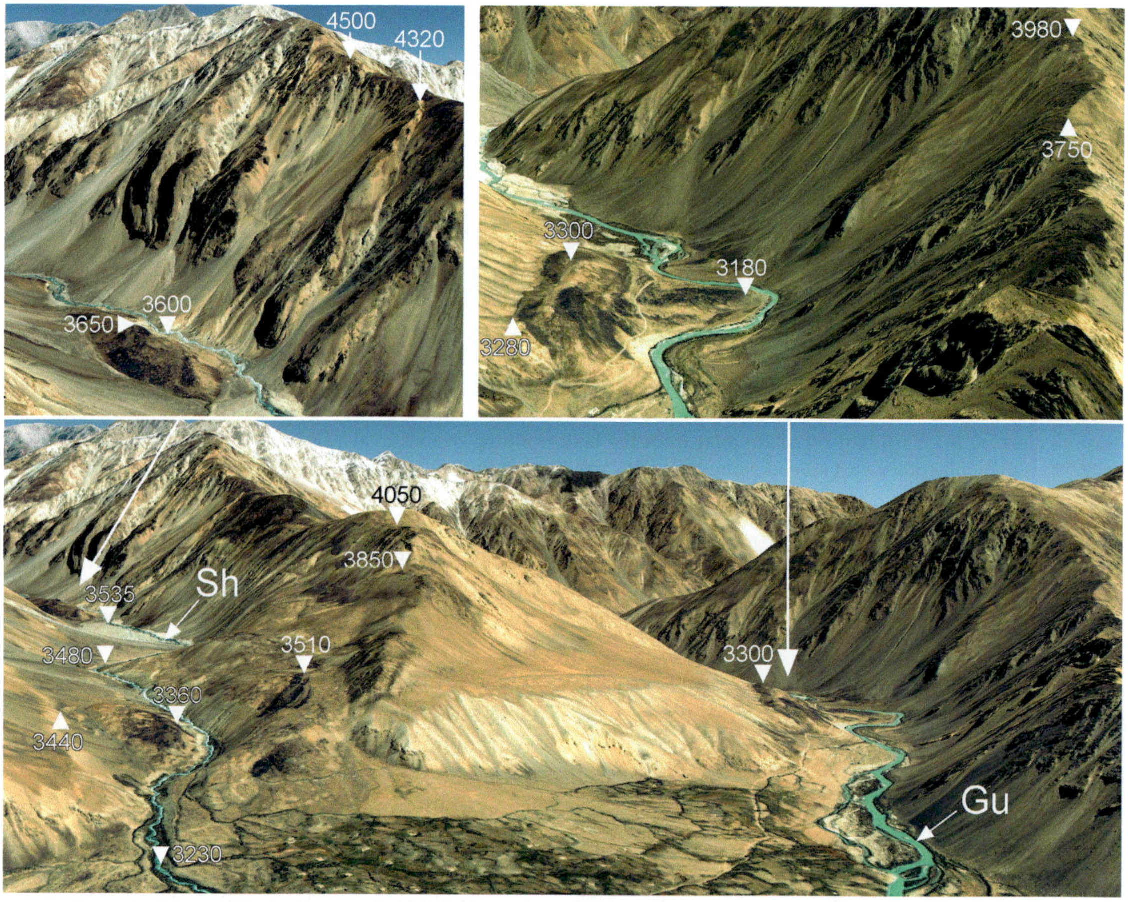

FIGURE 9.67 Rockslides at the confluence of the Gunt (Gu) and the Shazuddara (Sh) Rivers. 3D Google Earth views.

(Tajik 112) about 5 Mm^3 in volume (upper-left inset in Fig. 9.67) resembles slightly smaller rockslide in the Safedobdara River at 37.804° N, 72.471° E (Tajik 153) that originated on the slope composed of granites with more or less same height (750–900 m) and same aspect.

Rockslides shown in more detail in the insets in Fig. 9.67 are typical primary rock avalanches in frontally confined conditions (see Section 5.3.1 in Chapter 5), whose debris accumulated at their distal parts with an impressive run-up (ca. 120 m in case of the rockslide in the Gunt River valley about 9 Mm^3 in volume). Both rivers eroded these dams at their proximal parts where dams' crests were at the lowest level.

Unlike these case studies the Shazud rockslide (Tajik 111) about 90 Mm^3 in volume can be classified as a block slide whose frontal part underwent more flow-like style of motion (see Fig. 9.67). Just this frontal part of rockslide body had crossed the river valley and blocked it before being incised by the stream. Rockslide originated on more than 600 m high slope and displaced rock mass about 150 m deep if not more. It recalls much larger compound Ke-Keka'er rockslide in the Chinese part of Pamir—the gigantic block slide about 3 km^3 in volume, whose frontal part also had crossed the river valley (see Fig. 9.7). However, partially breached Shazuddara rockslide dam forms about 240 m high nick point in the Shazuddara River thalweg allowing assumption that

FIGURE 9.68 Large Duzakhdara rockslide on the left-bank slope of the Duzakhdara River valley. 3D Google Earth view.

rockslide debris up to 90–100 m thick still rests below the present-day valley bottom.

The largest rockslide of this cluster, up to 300 Mm^3 in volume, affects the 2 km long and up to 950 m high left-bank slope of the Duzakhdara River valley composed of granite. Unlike other features from this part of the Gunt River basin it moves, most likely, gradually. River erodes its toe forming the V-shape narrow gorge up to 150 m deep (Fig. 9.68). Rockslide has a very distinct, 50–100 m high headscarp that affected local watershed. This rockslide could accelerate resulting in river damming and, thus, this site requires monitoring or at least regular inspection.

The smallest member of the same cluster—the Lower Toguzbulak rockslide (37.703° N, 72.395° E, Tajik 115) about 1 Mm^3 only. It collapsed from the 400 m slope composed of the Proterozoic gneiss and schist and forms the fan-shape body ~400 × 500 m in size and up to ca. 20–25 m thick whose frontal part has been eroded by the river. Despite its relatively small size, it has the distinct bilayered internal structure typical of the majority of the flow-like bedrock landslides (rock avalanches) (Fig. 9.69). This site is easily attainable being located just aside the Osh-Khorog road.

One more feature presumably of rockslide origin was identified at the presently glaciated part of the Rushan Range in the upper reaches of the Shazuddara River valley (Tajik 114, 37.789° N, 72.293° E). Coarse bouldery accumulation about 15 Mm^3 in volume covers about 730,000 m^2 of the valley bottom at

FIGURE 9.69 The bilayered internal structure of the Lower Toguzbulak rockslide frontal part. Yellow triangles mark the boundary between more compact matrix-supported debris of the lower unit and much coarser clast-supported debris of the upper unit. Position of this outcrop is indicated by yellow arrow downstream of the bridge (see Google Earth image at the right part of this figure).

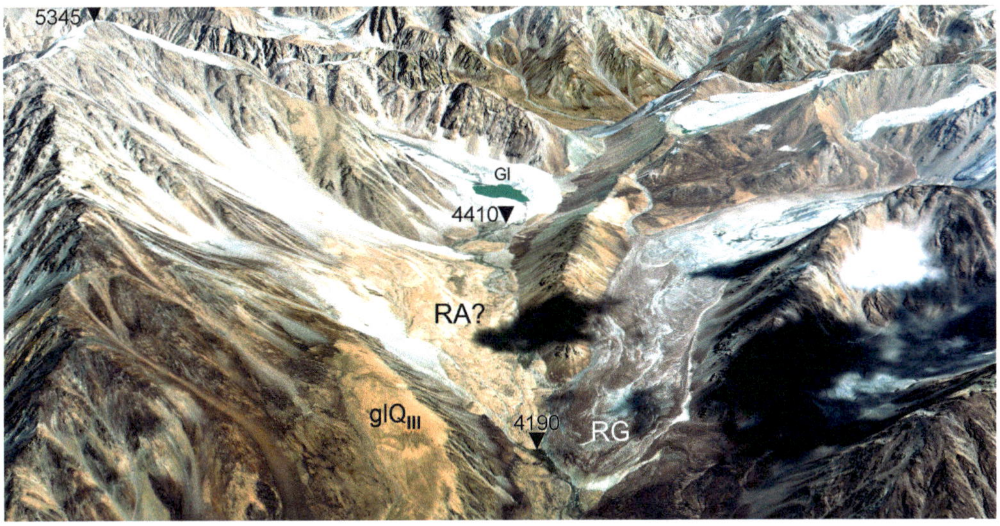

FIGURE 9.70 Assumed rock avalanche (RA?) in the high altitude glaciated part of the Rushan Range. Gl—glacier; RG—rock glacier; glQ_{III}—Late Pleistocene glacial deposits. 3D Google Earth view.

4180–4570 m a.s.l. below the circus-like source zone with its crown at ca. 5345 m a.s.l. Ridge is composed of the Proterozoic migmatites. This body differs significantly both from end moraines of the surrounding glaciers and from gradually moving rock glaciers (Fig. 9.70). Significant amount of bouldery deposits that definitely originated from the single source zone just above it, not being transported by the valley glacier, allows assumption that it was formed by the single-event or few successive but, nevertheless, large-scale rock slope failures. We should notice that the Pathur-Nemos active fault, whose most impressive fragments are shown on Fig. 3.13, passes just at the foot of the assumed headscarp, though at this particular site it is masked by recent deposits.

Besides the Lower Toguzbulak rockslide described above, rockslides were found upstream the Toguzbulak River valley, along the Osh-Khorog road, 6.5, 27, and 30 km upstream from its confluence with the Gunt River. This valley was heavily glaciated in the past and lower parts of its slopes are covered by the Pleistocene moraine deposits. It provided conditions favorable for formation of rockslides whose bodies have dual-zone structure described in Section 4.2 of Chapter 4 and exemplified by the Imom rockslide (see Figs. 4.14 and 4.15). Its frontal (outer) part is composed of moraine material with rounded and semirounded boulders bulldozed by rockslide, whereas inner part—of crushed angular bedrock debris. Just such structure is well expressed in the deposits of the Upper Toguzbulak rockslide (Tajik 161) that collapsed from the right-bank slope at 37.566° N, 72.63° E, 3.5 km upstream of the Djilandy village well known for its hot springs. A 2.6 km long rock avalanche about 30 Mm^3 in volume collapsed from the distinct spoon-shape headscarp at the upper part of the ca. 800 m high slope composed of Mesozoic granodiorite and blanketed by moraine material almost up to half of its height (Fig. 9.71). The moraine material forms the 300 m wide frontal belt of its body resembling such case studies as the Kuturgansu-mouth rock avalanche (see Fig. 8.246), the Kudara-Pasor rock avalanche (see Figs. 4.20 and 9.38), and the Andravdj rock avalanche (see Fig. 9.66) described above. Inner part of the deposits on the left bank of the Toguzbulak River has several "belts" of debris similar to the much larger Kudara-Pasor rock avalanche (see Fig. 9.39). The Upper Toguzbulak fan-shape body up to 1240 m wide is three times wider than the outlet of its source zone that makes it somehow similar to the Atdjailau rock avalanche in the Inylchek River valley (see Fig. 6.4) that also moved over wide flat bottom of the formerly glaciated trough valley.

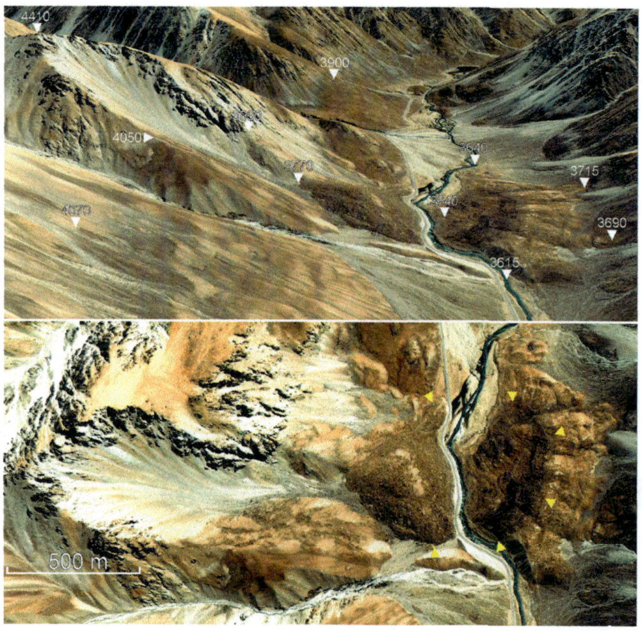

FIGURE 9.71 The Upper Toguzbulak rockslide. Above—3D Google Earth view. Old moraine deposits rest on the slope up to 3900–4070 m a.s.l. Below—plan view. Yellow triangles mark the boundary between the outer and inner parts of rock avalanche body.

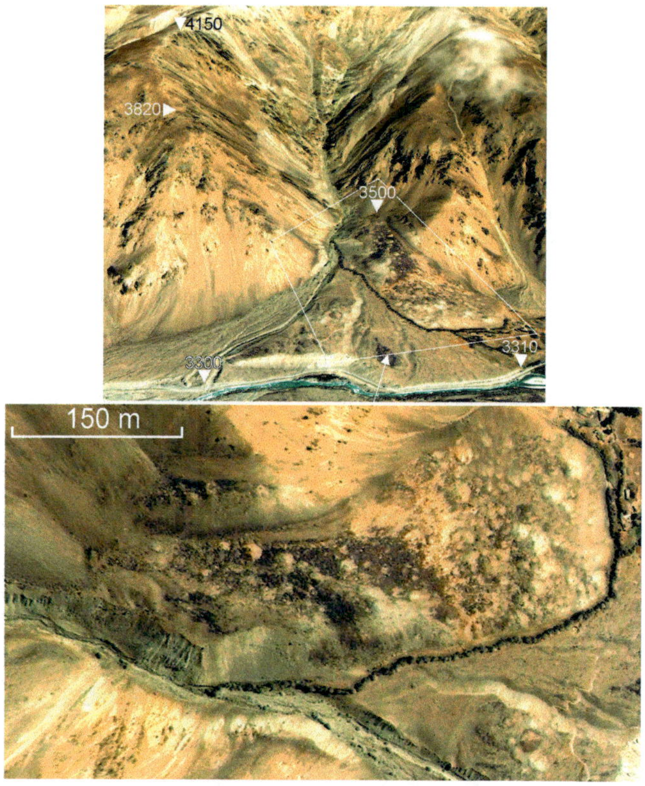

FIGURE 9.72 Remnant of the assumed rock avalanche near the Mordj village with hummocky microrelief of its body. 3D Google Earth view and zoomed plan view of the deposits.

FIGURE 9.73 the Chartym rock avalanche. Ch—the Charsem (Chartym) village. 3D Google Earth view.

One more much smaller rockslide (Tajik 162) about 0.9 Mm^3 in volume is located 3 km west of it at 37.578° N, 72.597° E, just near the Djilandy village.

Suspicious strongly eroded hummocky deposits 25–30 m thick were found at the foot of the right-bank slope of the Toguzbulak River valley at the mouth of the unnamed creek near the Mordj village (Tajik 163, 37.702° N, 72.444° E). They form the fan-shape body (Fig. 9.72) reworked both by the main stream and by debris flows from the creek, whose V-shape valley does not look as a glacial landform. Moreover, similar deposits are absent at the mouths of other creeks dissecting the right-bank slope of the Toguzbulak River valley nearby. Such uniqueness allows the assumption that the deposits in question were left by rock avalanche whose headscarp could be placed on the true right side of the creek valley with its crown between elevation marks 3820–4150 m a.s.l. (see Fig. 9.72). The hummocky surface of this body resembles those of rock avalanches shown on Figs. 4.41, 8.18, 8.93, 8.134, and 8.272.

About 17 km downstream of the Toguzbulak River mouth, near the Charsem (Chartym) village at 37.703° N, 72.174° E the kilometer-wide bottom of the Gunt River trough valley has been blocked by the Chartym rock avalanche—15 Mm^3 failure of the Archean gneiss (Tajik 096). Dammed lake about 2.4 km^2 in size and up to 50 m deep was filled by sediments through which Gunt is meandering forming numerous temporary islands (Figs. 9.73 and 9.74). It can be hypothesized that this natural dam rests on the lacustrine deposits of the older lake dammed by the much larger Vir-Tangiv rock avalanche (see Fig. 4.27).

The Chartym rock avalanche caved from the upper part of more than 1200 m high left-bank slope of the Gunt River valley dipping 31–32 degrees towards the river, slightly steeper than foliation in the gneiss. The 350–400 m high lower part of this slope is much steeper—about 40 degrees. Such morphology facilitates formation of Jumping rock avalanches (see Section 5.3.2 in Chapter 5). In this case, however, it is not so obvious as in cases of the Northern Karakungey (see Figs. 5.43 and 5.44) or the Kashkasu (see Fig. 5.45) rockslides, though thickness of the Chartym rock avalanche debris decreases towards its frontal part too (see Fig. 9.73). Similar to many other case studies described above, the Chartym rock avalanche body consists of the two main facies—coarser and finer. However, coarse carapace with numerous angular blocks up to 20–30 m in size (about 10,000 tons) of gneiss covers this fan-shape body only partially, at its southern, upstream part, while northern part of the deposits is much finer, more smooth, and visually looks much older (Fig. 9.74). Because coarser unit had crossed the valley and rests now on the right bank of the erosional canal, its mobility was higher than that of the underlying finer unit, while Jumping rock avalanches have more mobile lower parts (see Section 5.3.2 in Chapter 5, Fig. 5.42). It facilitates the assumption that coarser upper unit really represents the deposits of younger rock slope failure from the same source zone. We can just speculate on it now, because no dating of these deposits has been performed yet, though gneiss is suitable for the cosmogenic isotopes dating (Walker, 2005).

Finishing the description of this case study, we want to mention that it occurred at a very short distance (~2 km only) from the Pathur-Nemos active fault whose traces can be identified both north of the Gunt River valley (see Fig. 3.13), and also south of it, at 37.708 N, 72.1232 E, where rupture reaches the headscarp of the gigantic Vir-Tangiv rock avalanche (Fig. 9.75).

This rock avalanche (Tajik 097, 37.73° N, 72.097° E) described in Chapter 4 as an example of river damming feature is one of the largest of the entire Gunt River basin—about 350 Mm^3 in volume and, definitely, the longest one—about 6.5 km long (see Figs. 4.26 and 4.28). Crown of its enormous, 1.1 km high and ~200 m deep headscarp rises for about 2 km above the Gunt valley bottom. The affected slope is composed of the Mesozoic granite and Archean gneiss divided by the above-mentioned fault. It can be hypothesized that this slope failure could be triggered by the earthquake associated with this fault zone. We must point out that judging from the Gunt River long profile (see Fig. 4.27), the Vir-Tangiv rockslide should be much older than the Chartym one.

Rock avalanche debris that reached the Gunt River valley had formed a dam up to 300 m high composed of heavily crushed material overline by coarse blocky

FIGURE 9.74 Frontal part of the Chartym rock avalanche and silted lake with meandering river at the background. Yellow triangles in the inset mark boundary between finer (at the foreground) and coarser units likely representing two generations of rock avalanche debris.

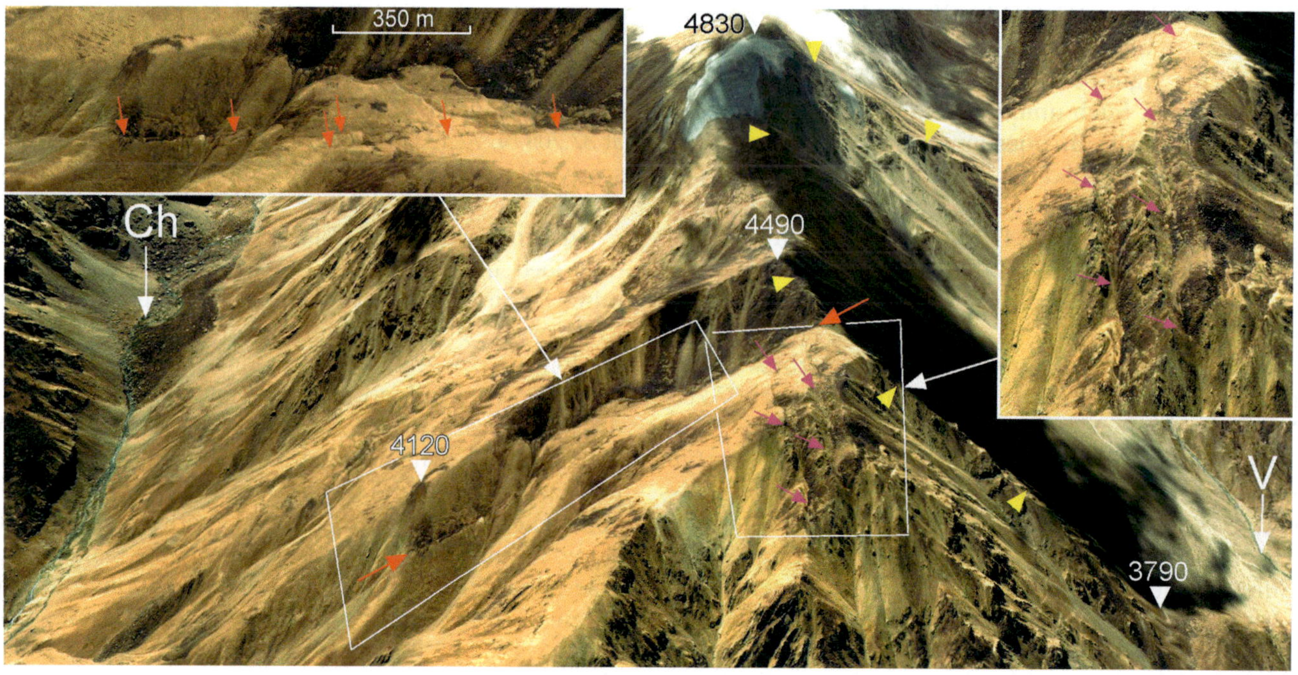

FIGURE 9.75 South-western extension of the Pathur-Nemos active fault near the headscarp of the Vir-Tangiv rock avalanche (partially in the shadow, marked by yellow triangles). V—Virdara River; Ch—Charsemdara River. Left inset shows trace of the recent surface rupture (marked by red arrows). Right inset—scarps near the headscarp of the Vir-Tangiv rock avalanche indicating possibility of future large-scale rock slope failures (purple arrows). 3D Google Earth view.

carapace (visible on Fig. 4.28 as a dark patch above elevation mark 3150 m a.s.l.). Coarse angular boulders also fill the Virdara River valley above the Vir village. As we mentioned in Section 4.2 of Chapter 4, thickness of lake sediments just upstream of the Vir-Tangive blockage could exceed 50–60 m (see Fig. 4.27).

This site should be considered as the potentially hazardous one where large-scale slope failure could occur in future. It is indicated by the arcuate scarps about 500–600 m long that affect the ridge crest at ca. 4400–4300 m a.s.l. just north of the existing headscarp (see right inset in Fig. 9.75). Such failure from the ca. 900 m high slope, if it would occur catastrophically, could produce rock avalanche that could reach the Vir village. Its closeness to the active fault enhances reliability of such scenario.

One more breached blockage formed by the Zuvor rockslide about 200 Mm^3 in volume was identified ~3 km downstream of the Vir-Tangiv rock avalanche, at 37.726° N, 74.045° E (Tajik 098) where the Zuvordara and the Shitamdara Rivers merge Gunt River (see Fig. 4.42). We have to notice that on Soviet topographic maps one more river falling from the right in the Gunt about 5.5 km downstream has the same name—the Zuvordara River.

Rockslide originated on the ca. 1.1 km high slope composed of the Mesozoic granite and phyllonite very similar to those forming the source zone of the Vir-Tangiv rock avalanche. Most likely that the Zuvor rockslide could be ascribed to the Primary type (see Section 5.3.1 in Chapter 5) because it accumulated at the right bank of the Gunt River mainly where thickness of the intensively crushed deposits exceeds 200 m (see Fig. 4.42). This feature is a rather rare case of a rockslide that was overlaid by end moraine from the tributary valley—in our case from the Zuvordara River valley that can be seen at the lower left part of the Fig. 4.42. Some considerations on temporal relationships between the Zuvor and the Vir-Tangiv rockslides were presented in Section 4.2 in Chapter 4 (see Fig. 4.27) and will be discussed in Section 9.3.

About 4.5 km downstream of the Zuvor rockslide, opposite the small Ruzhak village, on the left bank of Gunt River at 37.371° N, 71.992° E, one can see the compact body up to 200–250 m thick (Tajik 099) composed of angular clasts of light color granite that rest on the fluvial gravels of the ca. 10 m high terrace (Fig. 9.76). Uniform lithology of this debris indicates its rockslide origin and, because the entire left-bank slope of the Gunt River valley is composed of gneiss, this slope failure could not be sourced here. Failure of about 13 Mm^3 of granite occurred on the opposite—right (northern) slope of the valley ~1300 m high composed entirely of the similar light color granite. Headscarp of this Ruzhak rockslide occupies upper

FIGURE 9.76 The Ruzhak rockslide in Gunt River valley. Sh—Shitamdara River; Z—Zuvordara River. 3D Google Earth view.

part of the slope between 3300 and 4030 m a.s.l. (see Fig. 9.76), and collapsing debris should block the river temporarily. Despite the location of the source zone at the upper part of the slope typical of Jumping rock avalanches (see Section 5.3.2 in Chapter 5), the Ruzhak rockslide study can be ascribed to the Primary type in the frontally confined conditions (see Section 5.3.1 in Chapter 5) with a significant, about 200 m high runup. It can be hypothesized that descending rock mass moved along the smoothly curved surface of the Gunt valley bottom that facilitates the Primary style of rock avalanche motion. Relationships between the alluvial gravels and overlying rockslide debris could be seen in the Osh-Khorog road cut.

Lower part of the Gunt River catchment, similarly to the upper one is asymmetric, but here its left tributaries are larger—up to 20–35 km long. They dissect northern slope of more than 5 km high Shugnan Range, glaciated at its watershed zone, while right side of the catchment that occupies the south-eastern slope of the Rushan Range is 7–8 km wide only, and end moraines in steep valleys, dissecting it, terminate 2.2–3.5 km from the their mouths.

Most of the rock slope failures within this part of the region are in the valleys of the lower Gunt left tributaries—the Rivakdara, Shoripdara, and Urmochdara Rivers (see Fig. 9.64). Some of these case studies were described in Part 1. Several rockslides were found in the Rivakdara trough valley. The largest of them (Tajik 095) at 37.617° N, 72.06° E dams the Rivakkul Lake, being accompanied by two smaller rock slope failures downstream (see Fig. 4.21). Some researchers consider this blockage as end moraine (N. Ischuk, 2011). Moreover, on the 1:200,000 State geological map central part of the dam was mapped as the Holocene deposits, while its frontal part—as Late Pleistocene glacial deposits, despite obvious uniformity of this body visible on Fig. 4.21. Rockslide about

30 Mm^3 in volume originated on the ca. 850 m high slope composed of granites and forms a pancake-shape body up to ca. 140 m thick with ~40 m frontal run-up covered by numerous conical hills (molards?) about 10 m in diameter. Seeping water releases in a powerful spring at the proximal part of the blockage, probably through the scree that had accumulated at the slope foot prior to its collapse. Volumes of the satellite rockslides do not exceed 1.0–1.5 Mm^3.

One more interesting site was found at the middle reaches of the Rivak River valley at 37.605° N, 71.911° E, just at its confluence with its left tributary—the Zirgazdara River (Tajik 164). 1.34 km^2 at the latter's mouth and at the adjoining section of the Rivak valley bottom are covered by coarse debris up to ~20–30 m thick with angular boulders up to 15–20 m in size (Fig. 9.77). Evidence of temporary damming of both valleys, along with the absence of similar coarse material at the bottom of the Zirgazdara River valley indicate that blocking of both valleys was caused not by end moraine from this valley but by rock slope failure of the Mesozoic granite about 15 Mm^3 in volume from the ca. 1300 m high right-bank slope of the Rivakdara River valley just opposite the Zirgazdara River mouth. Besides moving 1 km ahead into the Zirgazdara River valley with total run-up of 140 m, rock avalanche debris splashed over the Rivakdara valley bottom at a distance of 2.3 km even with some evidence of the reversive motion, especially towards the downstream section of the Rivakdara River valley (from the elevation mark 3390 m a.s.l. towards elevation mark 3180 m a.s.l on Fig. 9.77). We assume that such reversive motion could start after collision of rock avalanche with the steep left slope of the Zirgazdara River valley. Similar evidence of the backward motion was found at the Kara-Kiol-Katta rockslide in Southern Tien Shan (see Fig. 8.255).

FIGURE 9.77 Assumed rock avalanche at the Zirgazdara River mouth. Above—3D Google Earth view. R—the Rivakdara River; Z—the Zirgazdara River. Below—SAS Planet plan view of the outlined area.

Interesting rockslide about 55 Mm^3 in volume (Tajik 104) had blocked the Shoripdara River valley—left tributary of the Gunt River, at 37.536° N, 71.78° E (Fig. 9.78). It originated on the ca. 900 m high slope of the valley composed of the Mesozoic granite, whose base was overline by Quaternary glacial deposits—rounded and subrounded boulders in sandy matrix. Headscarp crown extends for ca. 1300 m between elevation marks 3500 and 3640 m a.s.l. on Fig. 9.78. The upstream-frontal portion of this Shoripdara rockslide marked on this figure as gr-glQ is composed just of glacial material displaced and bulldozed by main part of the rockslide body composed of crushed bedrock. This bulldozed part rises on the left bank of the valley with ca. 60 m high run-up and forms the elongated hill about 1 km long and 350 m wide with rather smooth surface and distinct sharp frontal edge. Rockslide surface is covered by numerous hummocks up to 5 m high, which is typical of rockslide/rock avalanche deposits (see Figs. 4.41, 8.18, 8.93, 8.134, and 8.272). The Shoripdara rockslide resembles numerous case

FIGURE 9.78 The Shoripdara rockslide whose frontal part is composed of the displaced moraine deposits (gr-glQ) bulldozed by bedrock debris (gr). 3D Google Earth view.

studies from formerly glaciated parts of the Pamir (the Imom, Liangar, and Yashilkul rockslides—see Figs. 4.14, 4.18, and 4.19) and of the Tien Shan (the Aidyn-Ula rockslide—see Fig. 4.4). It formed the dam that was incised by the stream fed by multiple small glaciers at the watershed part of the Shugnan Range.

One more large-scale slope failure about 50 Mm^3 in volume had occurred at 37.423° N, 71.782° E in the Urmochdara River valley dissecting northern slope of the Shugnan Range (Tajik 103). It was described in Chapter 4, as an example of the relatively young rock avalanche in the recently glaciated high altitude zone (see Figs. 4.40 and 4.41). Failure occurred on the 730 m high slope composed of the Mesozoic granite and formed highly mobile, up to 5.5 km long granular flow with the striated microrelief (see Fig. 4.41) and numerous conical-shape molards. It can be assumed that this slope failure could be triggered by the shallow-focus earthquake associated with surface rupture that, besides all, had dammed the small glacial lake and is traceable on space images 3–5 km north-east of the Urmochdara rock avalanche (see Fig. 4.40).

The lower reaches of the Gunt River valley 3–8 km upstream of the Khorog town had been affected by a couple of very large rockslides (Fig. 9.79) that originated on its 1.7–2.2 km high right-bank slope composed of the Archean and Triassic gneiss and migmatites with foliation dipping 20–30 degrees towards the river valley. Deposits of the upstream one—the Suchan-1 rockslide (Tajik 166; 37.53° N, 71.637° E) form the ca. 4 km^2 terrace-like surface about 350 m above the valley bottom, occupied by several villages, one of which—the Suchan village—gave the name to these features. The Suchan-2 rockslide (Tajik 167, 37.52° N, 71.6° E) caved from the higher slope rising 2.2 km above the sharp bend of the Gunt River that undercuts this slope at present. It seems that at the time of these rockslides origin the river had passed southward, closer to the left side of the valley, but this U-shaped canyon, being filled by rockslide debris either of the Suchan-1 or of the Suchan-2 rockslide had been abandoned and Gunt River had eroded a new bypass gorge through the proximal part of the blockage. We must notice that, despite very large size of the Suchan rockslides (roughly estimated as 1800 Mm^3 for both of them), which is comparable with the 1911 Usoi rockslide, we could not find lacustrine sediments accumulated in the dammed lake. It can be hypothesized that the blockage was eroded rather soon after its overtopping. The Suchan rockslides originated, definitely, in Pleistocene, but special studies (dating by the cosmogenic isotopes method (Walker, 2005)) should be carried out to determine it more precisely.

Potentially quite hazardous site was found at 37.501° N, 71.558° E, on the right-bank slope of the Gunt River valley (Tajik 169). Here about 2 Mm^3 of the Triassic migmatites are involved in sliding bounded from top by the distinct headscarp with up to ca. 50 m high back wall. Landslide can be seen on the slope about 400–500 m just above the Khorog town (Fig. 9.80). At present, this body is a source zone of rock falls that form 200–250 m talus cones. However, if, due to any reasons (earthquake, rainstorm, etc.), it would be activated entirely or even partially, it could devastate densely populated terraces at the foot of this slope. This site requires detailed study and monitoring.

Numerous large-scale rockslides, both those that failed catastrophically, and those representing long-term slope deformations, have been identified in the deep gorges of the Shakhdara River basin (see Fig. 9.1)

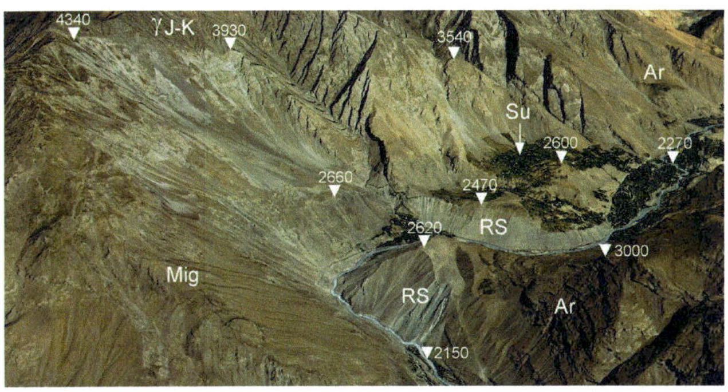

FIGURE 9.79 The Suchan rockslides. Whitish homogenous-looking outcrops marked by RS—rockslide deposits; γ J-K—Mesozoic granite; Mig—migmatites; AR—Archean gneiss; Su—the Suchan village. 3D Google Earth view.

FIGURE 9.80 Rockslide above the Khorog town. SAS Planet image.

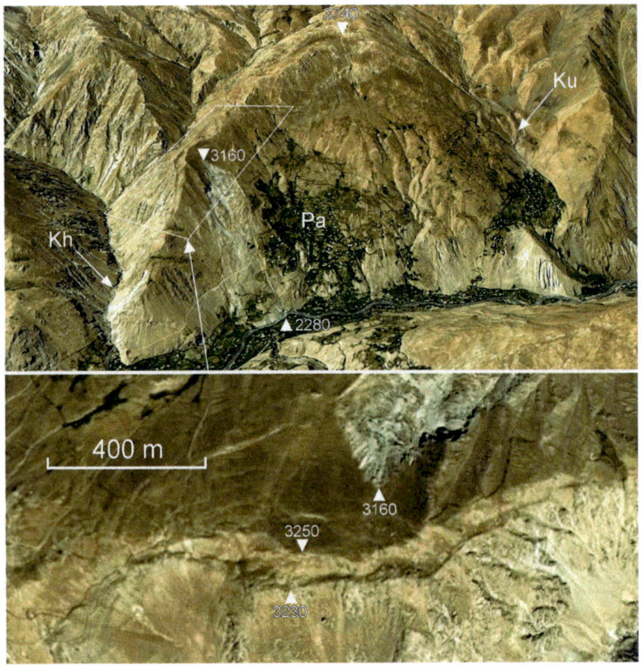

FIGURE 9.81 Rockslide and DSGSD near the Parshed village (Pa); Kh—the Khidordjavdara River; Ku—the Kurtsak River. Above—3D Google Earth view; below—plan view of the outlined area with well-defined upslope-facing scarp.

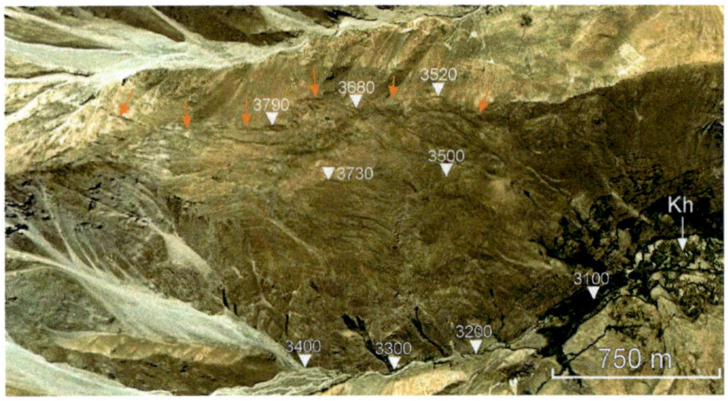

FIGURE 9.82 The Ridjistdara DSGSD. Kh—the Khut village. Orange arrows mark the outermost scarps bounding the DSGSD. Google Earth image.

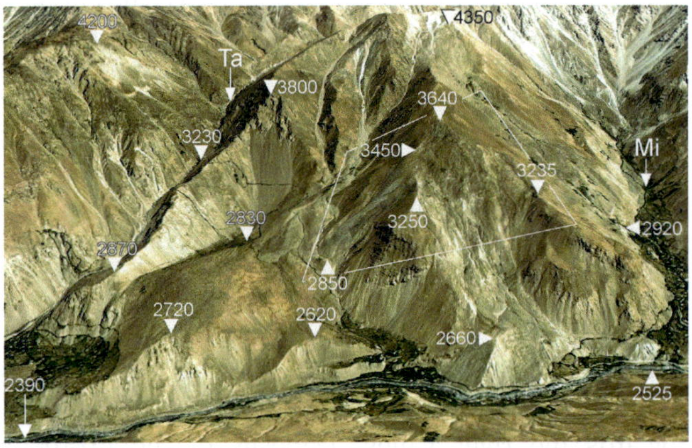

FIGURE 9.83 Rockslides at the right side of the Shakhdara River valley and in the Tavdemdara (Ta) and Midensharv (Mi) river valleys. Outlined area is shown on Fig. 9.84. 3D Google Earth view.

incised in the Archean gneisses intruded by the Mesozoic granites (see Fig. 3.1).

Despite hard rocks (gneiss and granite predominantly), many of the high slopes in this part of the region are affected by gradually evolving landslides combined with well pronounced DSGSDs. Evidence of such a large-scale slope instability could be found in the Shakhdara River valley, at it left-bank slope above the Parshed village (Tajik 170, 37.4° N, 71.626° E). Here besides well-pronounced rockslide about 5 Mm^3 in volume that originated on the ca. 1 km high slope composed of granite, 1.4 km long upslope-facing scarp passes along the crest of the ridge dividing the Shakhdara and the Khidordjavdara River valleys (Fig. 9.81). Moreover, the entire concave slope of the Shakhdara River valley between the Khidordjavdara and the Kurtsak River mouths, also in granite and migmatite, demonstrates some long-term instability, being much smoother and "loose" if compared with the adjacent rocky slopes (see Fig. 9.81).

Much larger DSGSD (Tajik 171) affects the entire left-bank slope of the Ridjistdara River valley upstream of the Khut village, at 37.43° N, 71.683° E. Here up- and downslope facing scarps can be traced at a distance of 2.2 km along the valley slope and up to 1.5 km across it (Fig. 9.82) with estimated volume of the affected rocks (Mesozoic granite) up to ca. 350 Mm^3. Deep-seated deformations affect the entire 300–500 m high north-facing convex slope of the local ridge, its watershed and extend towards the opposite slope, so that the ridge crest up to 300–400 m wide became almost flat, contrasting to its sections not affected by the DSGSDs.

Numerous large-scale bedrock landslides where identified in the Shakhdara River valley and its tributary valleys at a distance of about 15 km upstream of the Shakhdara River junction with its large left tributary—the Sharfdara River, within the granite massif. One of the largest of them—the Tusen rockslide about 300 Mm^3 in volume (Tajik 172, 37.372° N, 71.685° E) forms the giant drop-shaped body up to ca. 300 m high at the foot of more than 2 km long and 1 km wide (in plan view) wedge-like headscarp upstream of the Tavdemdara River mouth (Fig. 9.83). Rockslide originated on the 1.3 km high slope and, likely, destroyed the entire rocky spur, because volume of the deposits exceeds that of the headscarp depression. It is obvious that slope failure occurred catastrophically, because its more than 200 m deep headscarp is almost debris-free. Deposits filled the Shakhdara River valley at a distance of about 2.75 km and entered the mouth of the Sharfdara River valley. Judging from the Shakhdara River thalweg profile the blockage has been deeply incised but not up to the pre-slide level yet.

The adjacent Chagev rockslide (Tajik 173, 37.356° N, 71.697° E), about 100 Mm^3 in volume originated on the 1.15 km high slope and, most likely, also destroyed the ridge promontory between elevation marks 3250 and 3235 m a.s.l. on Fig. 9.83. Its interesting peculiarity is that part of rockslide debris 2–3 Mm^3 in volume overpassed the headscarp true right back wall between 3450 and 3250 m a.s.l. and moved for more than 1 km towards the Tusen rockslide, to 2850 m a.s.l. (see Figs. 9.83, and 9.84). Similar phenomenon was observed at the Dead Lakes rockslide in Central Tien Shan (see Fig. 8.205), at the Iskanderkul rockslide in Southern Tien Shan (see Fig. 8.287) at the unnamed rock avalanche in the Eastern Tien Shan (see Fig. 8.6) and at the Bashanabad rockslide in the Afghan Pamir that will be described in the next Section of this Chapter (see Fig. 9.120). Main part of rockslide debris had blocked the Shakhdara River valley and, similarly to the Tusen rockslide, was subsequently incised by the river, but, again, not to the original level. Same was observed, for example, at the Beshkiol rockslide in Central Tien Shan (see Fig. 8.127) and at Vir-Tangiv rockslide in Gunt River valley (see Fig. 4.27).

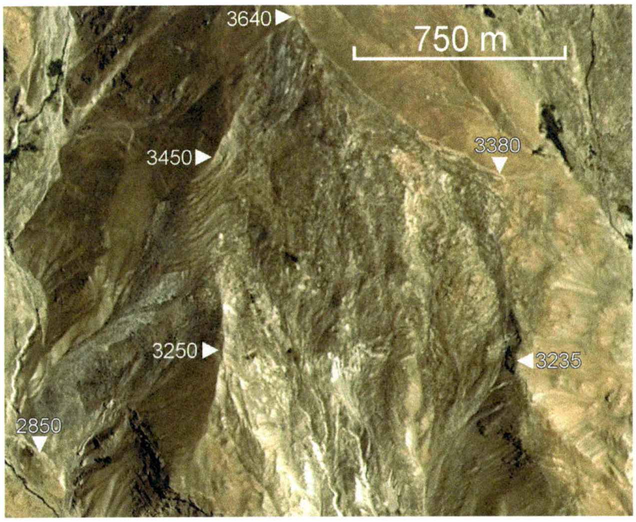

FIGURE 9.84 Headscarp of the Chagev rockslide and branching part of its deposits. Google Earth image.

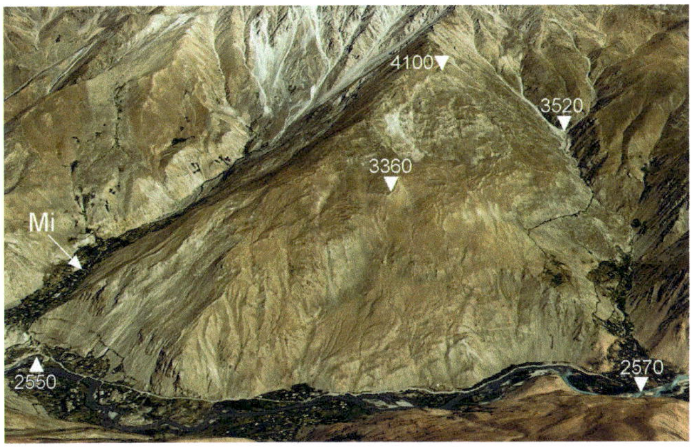

FIGURE 9.85 Very large slow moving upstream of the Midensharv River (Mi) mouth. Rockslide on the right-bank slope of the Midensharv River valley can be seen at the upper-left part of the image. 3D Google Earth view.

Two other rockslides visible on Fig. 9.83, one in the Tavdemdara River valley (Tajik 174, 37.395° N, 71.71° E), and another one in the Midensharv River valley (Tajik 175, 37.366° N, 71.724° E) are much smaller and did not produced such distinct blockages in these steep valleys as the previously described rockslides that had blocked the Shakhdara River valley temporarily. Their volumes can be estimated as 17 and 13 Mm^3 correspondingly. While the first one can be classified as the block slide, the second one with 3.7 km long runout is a flow slide (rock avalanche?) that filled the gulley bottom being distinguished from the surrounding "original" slopes by color (see Figs. 9.83 and 9.85).

Upstream of the Midensharv River mouth the entire 1.6 km high right-bank slope of the Shakhdara River valley composed of the Mesozoic granite is affected by very large gradually moving rockslide (Tajik 176, 37.34° N, 71.74° E). More than 8 km^2 is dissected by numerous upslope- and downslope-facing scarps, small- and medium-size landslides (Fig. 9.85). Assuming that thickness of the affected rock mass exceeds 100 m we can roughly estimate volume of this complex rockslide as ca. 1 km^3. Taking in mind that its activation could cause blocking of the Shakhdara River with quite adverse secondary effects—submergence of the valley upstream and devastating outburst flood downstream, this site requires regular monitoring. We must also notice very low stream gradient at this section of the Shakhdara River, likely due to its damming by the Tusen and the Chagev rockslides shown on Fig. 9.83.

One more high slope with distinct evidence of the ongoing large-scale deformations is located further upstream the Shakhdara River, just above the Barodj village (Tajik 177, 37.315° N, 71.78° E). This ca. 1.5 km high slope is composed of the Mesozoic granites underlined by the Precambrian migmatites. Its left (Fig. 9.86) part has a well-pronounced headscarp with its crown at 4040 m a.s.l. The 400 m wide and 800 m long step about 450 m below is crossed by numerous upslope-facing scarps (see inset in Fig. 9.86). Lower part of the slope undercut by the river is affected by several landslides. These slope deformations endanger

FIGURE 9.86 Rockslide-prone slope on the right side of the Shakhdara River valley above the Barodj village (Bar). Series of upslope-facing scarps is shown in the inset. 3D Google Earth view.

FIGURE 9.87 Assumed landslides on the left-bank slope of the Sharfdara River valley. 3D Google Earth view.

the road passing along the Shakhdara River and pose a threat to the Barodj village located at the base of the destroyed slope. This site, along with the slope shown on Fig. 9.85, deserves special study and monitoring.

Two large adjacent negative landforms (Tajik 178 at 37.34° N, 71.65° E and Tajik 179 at 37.34° N, 71.67° E) on the rather gentle (16° -21°) left-bank slope of the Sharfdara River valley rising for 1000–1300 m above its thalweg can be seen on Fig. 9.87. Similarly, to the most of other sites in the Shakhdara River basin, this slope is composed of the Mesozoic granites. Loose deposits with flat, slightly convex surfaces fill elongated circus-like lowerings 4.9 and 3.1 km long (Fig. 9.87). They seem to be more or less stable at present, judging from the absence of fresh minor landforms (upslope- and downslope-facing scarps, minor landslides, etc.) typical of active landslides developing on more steep (>23 degrees) slopes described above (see Figs. 9.85 and 9.86). Unlikely that these are glacial landforms—they are lower than most of unequivocally glacial landforms with similar freshness of their source zones (see also Section 4.2 in Chapter 4). On the other hand, volume of these lowerings is much larger than volume of the deposits filling their

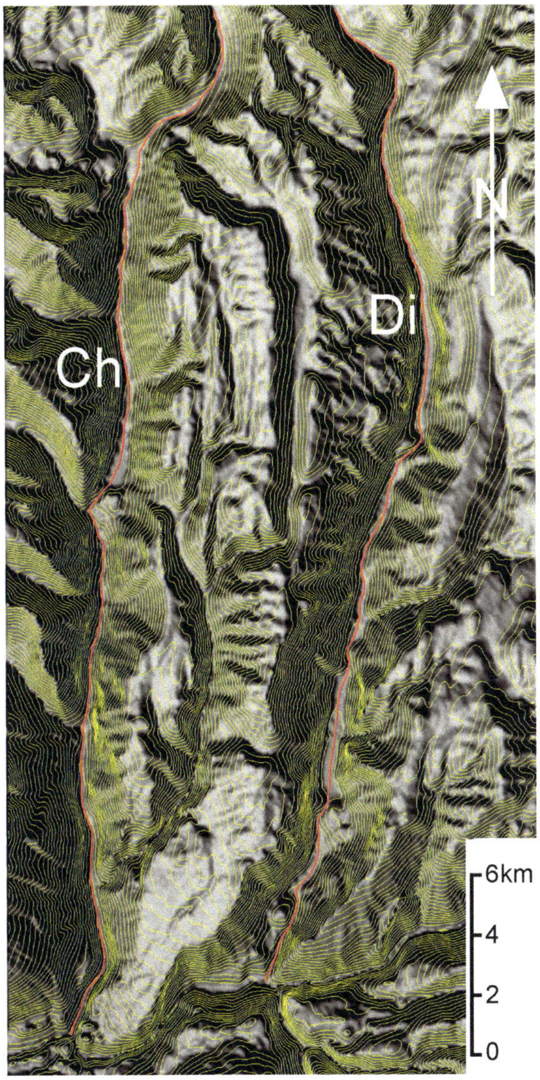

FIGURE 9.88 Shaded relief highlighted by contour lines with 50 m spacing of the Dirumdara (Di) and the Chandindara (Ch) River valleys. Thalweg profiles of these rivers (red lines) are shown on Fig. 9.89.

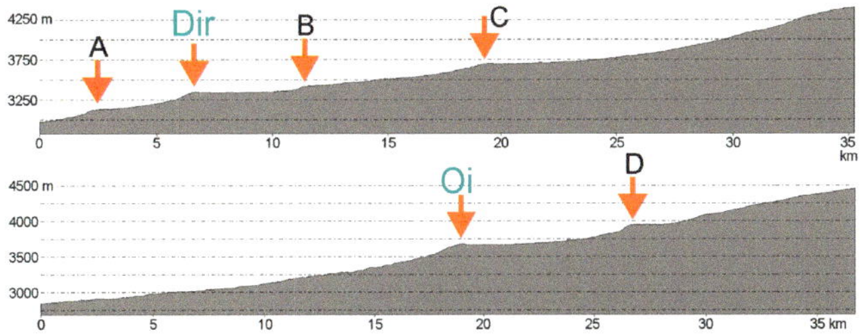

FIGURE 9.89 Schematic long profiles of the Dirumdara (above) and Chandindara (below) Rivers. Red arrows—nick points created by rockslide blockages. Oi—the Oikul Lake dam—see Fig. 9.94; Dir—the Dirumkul Lake dam — see this blockage and dam marked by (B) on Fig. 9.91; (A) see Fig. 9.92; (C) see Fig. 9.90; (D) see Fig. 9.93.

bottoms (~250 and ~75 Mm^3 correspondingly), and additional studies should be performed to explain this discrepancy.

Several rockslides were found in the trough valleys of the Dirumdara, the Chandindara, and the Sazhddara River valleys—right and left tributaries of the Shakhdara River incised 500–1000 m in the ca. 40 km wide plateau-like planation surface between the watersheds of the Shugnan Range at the north and the Vakhan Range at the south, locally covered by moraines more than 100 m thick left by the Late Pleistocene plateau glacier (age according to the 1:200,000 State geological map of the USSR). Moraine overlays the Archean gneiss. Part of this plateau is shown on Fig. 9.88. Several more rockslides were found further south, in the Saryshishtharv and Rostodjirav River valleys dissecting the southern slope of the Vakhan Range. All of them are located within the narrow north-south trending zone between 72.0° E and 71.1° E (see Figs. 9.1 and 9.64).

The Dirumdara and the Chandindara River valleys have gently sopping thalwegs with mean inclination of about 2–2.3 degrees, with several distinct nick points (Fig. 9.89). These are excellent examples of the so-called "Disturbance regime" proposed by K. Hewitt (2006a). "Normal" valley thalweg profiles are disturbed by the existing Dirumkul and the Oikul Lakes and by several other partially filled and breached dammed lakes. We should notice that landslide origin of these blockages is opposed by N. Ischuk (2013), who provided reasonings in favor of their moraine origin. However, some of their peculiarities considered by N. Ischuk are indicators of the glacial origin of these blockages ("coarse layering," "intact alternating lighter and darker sedimentary strata" (N. Ischuk, 2013, p. 15)) are typical of rockslide deposits as well; moreover, we anticipate that they are typical just of rockslide deposits—see Section 4.2 in Chapter 4, and Chapter 11) and, thus, cannot prove his concept. All blockages have the distinct concave lowerings on the valley slopes above them that are the headscarps from which rockslides had caved. Besides there are indications that debris moved across river valleys rather than along them.

One more important argument in favor of rockslide origin of these blockages is their temporal relationships. Though these features have not been dated yet by any absolute dating method (^{14}C, ^{10}Be, OSL), it is obvious that if the blockages were left by the degrading valley glaciers, those upstream must be younger than downstream, otherwise the former would not survive next glacier advance that must overcome them (same argument could be applied to the Shing River valley in Southern Tien Shan—see Section 8.6 in Chapter 8: Tien Shan). However, visually most of the

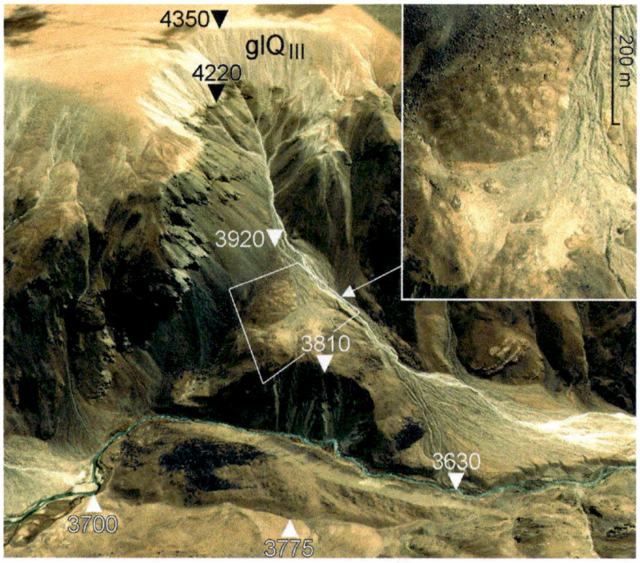

FIGURE 9.90 The Upper Dirumkul rockslide dam (marked by C on Fig. 9.89). glQ_{III}—Late Pleistocene moraine overlying Archean gneiss. 3D Google Earth view. See the hummocky surface in the inset.

FIGURE 9.91 The Dirumkul Lake dam (Dir on Fig. 9.89) and rockslides upstream and nearby. 3D Google Earth views.

FIGURE 9.92 The Lower Dirumkul rockslide dam (marked by A on Fig. 9.89). 3D Google Earth view. Plan view of the deposits is shown in the inset.

blockages look more or less as of the same age (Figs. 9.90–9.92).

The Upper Dirumkul rockslide (Tajik 186, 37.373° N, 72.12° E) originated as a wedge failure on ca. 700 m high left-bank slope of the trough valley in the Archean gneiss overlain by thick moraine deposits (see Fig. 9.90). About 17 Mm^3 of debris formed a triangular in plan view dam up to ca. 100 m high (the effective height) that blocked the river valley and formed the lake more than 5 km long that was partially silted before its, likely catastrophic, breach. The hummocky relief of the deposits surface, typical of rockslide bodies (see, e.g., Figs. 4.23 and 8.18) can be seen in the inset.

Natural dam of the Dirumkul Lake (Tajik 183, 37.26° N, 72.093° E) is formed, from our point of view, by rock slope failures from two adjoining headscarps, which crowns can be seen on Fig. 9.91 as 4240 and 4290 m a.s.l. Total volume of the deposit was estimated as ca. 50 Mm^3. One more block slide (Tajik 183a, 37.29° N, 32.119° E) originated about 2 km north at the edge of the trough valley (see upper-right inset in Fig. 9.91). The presence of this univocal rockslide can be considered as an addition argument in favor of similar origin of the Dirumkul Lake dam. The dam was partially breached and this breach occurred, most likely, catastrophically, which can be derived from the presence of the deposits with amygdaliform microrelief downstream of the erosional canyon (compare with similar erosional landforms visible on Figs. 8.55, 8.56, 8.135, and 8.252).

Upstream of the Dirumkul Lake and downstream of the Upper Dirumkul rockslide dam at least two (may be even more) rockslide bodies could be identified within the 8.5 km long section of the Dirumkul River valley (Tajik 184, 37.31° N, 72.10° E and Tajik 185, 37.337° N, 72.104° E, see upper-left inset in Fig. 9.91). Both are characterized by the significant, 75 to 95 m run-up of debris at the frontal edges of rockslide bodies, which is impossible either for moraines of the hanging niche glaciers or for rock glaciers. The larger one, about 10 Mm^3 in volume, forms the nick point marked by Fig. 9.89B. The smaller one, about 2 Mm^3 in volume had crossed the river and entered the tributary valley mouth, so that its body visible on the upper-left inset in Fig. 9.91 between elevation marks 3505 and 3580 m a.s.l. overlies the alluvial fan on the opposite slope and did not create real dam. One more slope failure from the right-bank slope of the valley probably occurred between these sites (it is visible in the upper-left inset in Fig. 9.91 opposite the mouth of a tributary valley).

The distinct Lower Dirumkul rockslide had blocked the Dirumkul River valley 1.5 - 3.0 km upstream of its confluence with the Shakhdara River (Tajik 182, 37.233° N, 72.0777° E). It originated from the ~1 km wide and more than 150 m deep funnel-shaped headscarp that destroyed upper two-fifth of the ca. 1000 m high right-bank slope of the valley (Fig. 9.92). Rockslide can be ascribed to the Jumping type (see Section 5.3.2 in Chapter 5). About 30 Mm^3 of rockslide debris rest, most likely, on the ~40 m high terrace visible on the right bank of the river downstream of the blockage. The concentric belts of varicolored debris similar to many other rockslides (see, e.g., Figs. 9.39, 9.66, 9.67, and 9.71) can be seen on top of this body (see inset in Fig. 9.92).

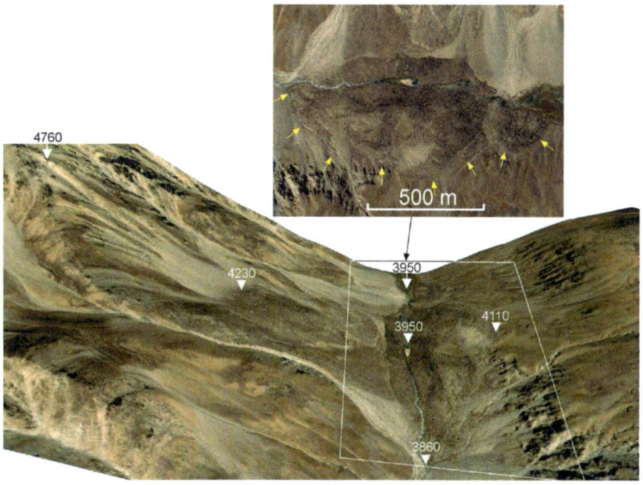

FIGURE 9.93 The upstream landslide dam in the Chandindara River valley. Plan view of the deposits is shown in the inset. Yellow arrows mark its front. SAS Planet image visualized by Global Mapper Software.

Two rockslide dams were identified in the Chandindara River valley. The upstream one (Tajik 198, 37.426° N, 72.007° E) about 12 Mm^3 in volume originated at a rocky spur about 900 m high and created the ca. 70 m high dam whose distal edge rose 160 m on the opposite bank of the river (Fig. 9.93). Dammed lake was silted completely and now this rockslide forms the distinct nick point in the river thalweg (marked by D on Fig. 9.89). Surface of rockslide body is covered by coarse angular boulders that differ significantly from the much finer moraine material of the valley bottom.

The Oikul rockslide (Tajik 187, 37.354° N, 72.003° E) about 50 Mm^3 in volume collapsed from the ~850 m high slope at the northern end of the much larger DSGSD, which scarps up to 30–40 m high pass for about 1.5 km along the local watershed (Fig. 9.94). It can be assumed that some other parts of this left-bank slope of the trough valley within the DSGSD and further south of it had collapsed in the past, but no related deposits could be found at the valley bottom. It can be hypothesized that slope failures could occur soon after or during glacier retreat and their deposits were eroded or even carried out by the glacier. The

FIGURE 9.94 The Oikul rockslide dam and the DSGSD south of the rockslide headscarp. Orange arrows mark the scarps of the DSGSD. Part of this view is from the winter space image of lower quality. 3D Google Earth view.

Oikul Lake dam was breached only partially and small Oikul Lake still exist behind it. Initially this dammed lake was much larger, more than 4 km long.

One more rockslide (Tajik 181) within the aforementioned north-south trending zone blocks the Sazhddara River—the left tributary of the Shakhdara River at 37.14° N, 72.089° E forming the Zardiv Lake (Fig. 9.95). Wedge failure of about 30 Mm^3 of the Archean-Proterozoic gneiss occurred on the ~1500 m high right-bank slope of the trough valley leaving deep "fresh" headscarp and created the 180 m high dam. The Zardiv rockslide body has well-pronounced zonal structure (see inset in Fig. 9.95) resembling that of many other Pamir rockslides (e.g., the Andravdj rockslide—see Fig. 9.66) and indicating that the material that forms this blockage moved across the river valley, not along it.

Interesting peculiarity of this case study is the evidence, though rather controversial, of the one more slope failure at this site (Tajik 181a), preceding the Zardiv rockslide formation. The triangular ca. 20 m thick body that rests on the left-bank slope of the Shakhdara River valley came out from the 1.7 km wide funnel-shaped circus about 200 m deep with curved back wall indicating its much older age than that of the Zardiv rockslide headscarp. However, it has smooth surface similar to moraine accumulations at the adjacent parts of the valley slopes that differs significantly from the recent Zardiv rockslide body composed (from the surface) of much coarser bouldery material (see inset in Fig. 9.95). Additional field studies are necessary to clarify its origin.

Further southward, at the southern side of the Vakhan Range composed of the well-bedded Archean and Proterozoic gneiss with well-pronounced foliation and within the same north-south trending zone, three impressive rockslides were found in the Saryshishtharv, Rostoshitkharv, and Rostodjirav River valleys (Fig. 9.96). That in the Saryshishtharv River valley (Tajik 188, 36.97° N, 72.055° E) forms a small lake; lakes dammed by two other (Tajik 189, 36.93° N, 72.06° E and Tajik 190, 36.86° N, 72.061° E) have been silted completely.

All these rockslides originated as wedge or compound slides on the 500–900 m high slopes. The characteristic peculiarity of all these three case studies is the southward directivity of slope failures, so that debris accumulations are slightly shifted southward relative to the headscarps feet (see Fig. 9.96). It can be speculated that such directivity could be caused by the southward directivity of the triggering factor—earthquake strong motion, though, of course, more

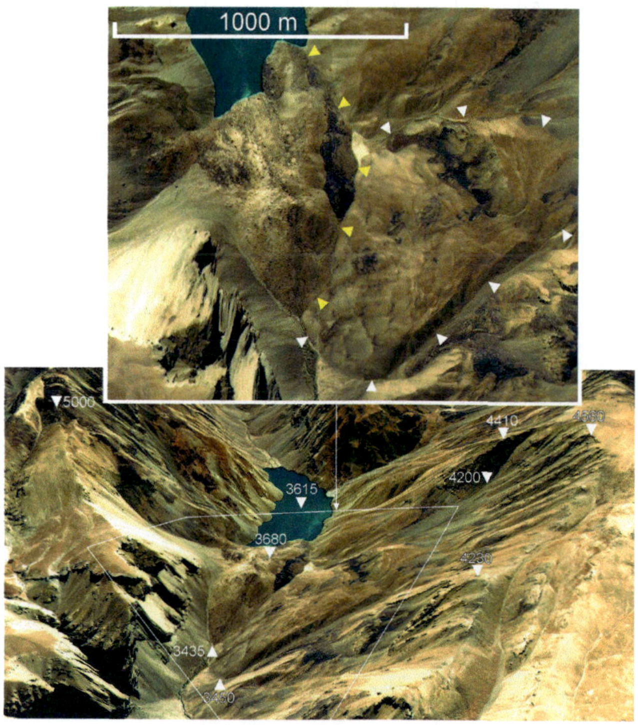

FIGURE 9.95 3D Google Earth view of the Zardiv rockslide dam and of the assumed preceding slope failure from the opposite side of the valley. Plan view of both bodies can be seen in the inset. Yellow triangles—front of the recent Zardiv rockslide body; white triangles—boundaries of the suspicious deposits.

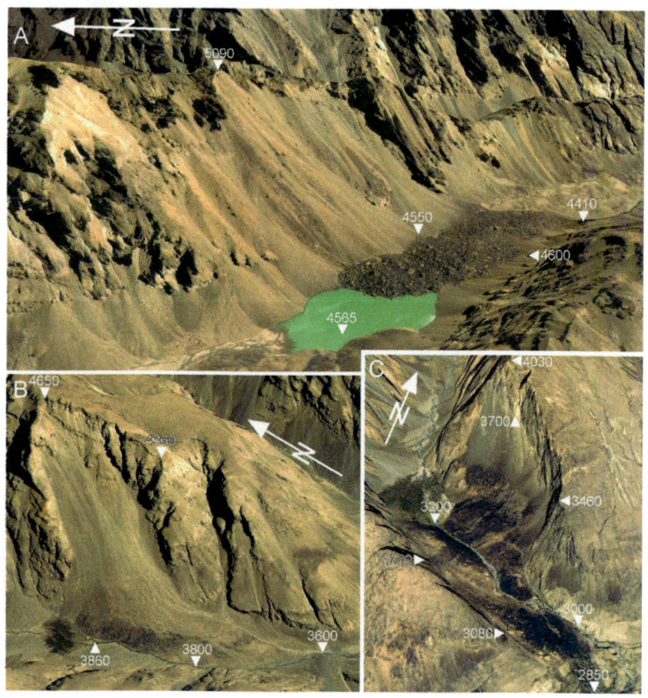

FIGURE 9.96 Rockslides with similar southward directivity of slope failure and debris motion in the Saryshishtharv (A), Rostoshitkharv (B), and Rostodjirav (C) River valleys. 3D Google Earth views.

detailed analysis of the host massifs' structure is required to substantiate this assumption.

Rockslide dam in the Saryshishtharv River valley about 8 Mm^3 in volume and up to 100 m high fills the bottom of the trough valley with thick talus cones at the feet of its steep walls. It is composed (from the surface) by very coarse angular blocks up to 30 m in size. Frontal edge of this body is up to 50 m higher than its proximal part. It can be assumed that below this blocky carapace dam is composed of less permeable material with higher content of fines. However, no overflow occurs and water balance is maintained by seeping only, probably through the more permeable talus at the valley bottom. This assumption is based on the fact that water appears just below the downstream tip of the blockage at 4410 m a.s.l. (see Fig. 9.96A).

Rockslide in the Rostoshitkharv River valley about 10 Mm^3 in volume with well-defined deep wedge-like headscarp also formed the ca. 100 m high dam, but, unlike the previous case study, with distal lowering and being composed of finer material. Large part of this dam is covered by the deposits of small debris flows originating within the headscarp. Also unlike the previous case study, the small dammed lake has been silted completely and the river flows over the frontal edge of the blockage (see Fig. 9.96B).

The southernmost rockslide that had dammed the Rostodjirav River practically at the mouth where it enters 3.5 km wide Pianj River valley slightly resembles the Ai-Kul rockslide in the Southern Tien Shan (see Fig. 8.257). In both cases headscarps destroyed rocky prominences at the confluence of two deep valleys, here—of the Rostodjirav and the Pianj Rivers (see Fig. 9.96C). Rockslide about 30 Mm^3 in volume and with almost 100 m high frontal run-up formed a dam up to 160–170 m high. The small dammed lake was silted completely and river flows along the lowermost central part of the blockage, so that downstream of the elevation mark 3000 m a.s.l. on Fig. 9.96C, the Rostodjira River had to erode a new gorge through the bedrock of its left bank to bypass the old buried valley.

One more interesting site is located at the right (Tajik) side of the Pianj River valley near the Sumdjin village. Two nearby rock slope failures occurred here in the Archean gneiss (Fig. 9.97). One at 36.828° N, 71.554° E (Tajik 192) formed a small rock avalanche ca.

FIGURE 9.97 3D Google Earth view of two rockslides near the Sumdjin village at the right side of the Pianj River valley. S—the Sumdjin village. Yellow triangles at the left inset mark the boundary between the bedrock and rockslide body.

6 Mm^3 in volume that moved without any confinement over the ~5 m high right-bank terrace of the Pianj River (see right inset in Fig. 9.97) and formed up to 25 m thick blade with concentric lithological belts resembling much larger Taheman rock avalanche in the Chinese Pamir (see Fig. 9.17). Because this body does not demonstrate any evidence of the erosion, it occurred after formation of this low terrace, likely during the last Millennium.

Another rockslide shown on Fig. 9.97 (Tajik 193, 36.84° N, 71.56° E) originated on more than 1 km high left-bank slope of the small tributary valley that fell at that time into the Pianj River south of the present-day mouth of the Sumdjin River, closer to rock avalanche described above. Rockslide body about 300 Mm^3 in volume filled the short valley completely, forcing the small Sumdjin River to erode the new bypass gorge through the bedrock Archean gneiss with very distinct layering (see left inset in Fig. 9.97). Maximal thickness of rockslide body in the buried valley could exceed 300–400 m. Judging from the relationships of rockslide bodies with the alluvial fans, we assume that this very large rockslide is much older than its small neighbor.

The last group of rockslides in the Tajik part of the Pamir that deserves to be mentioned, is located in the lower reaches of the Garmchashma River valley on its left-bank slopes above the hot sulfur springs (37.2033° N, 71.5381° E) that gave the name to this valley. These slopes, featuring evidence of gradual sliding, resemble the left-bank slopes of the Shakhdara River valley shown on Fig. 9.81.

9.3 PIANJ RIVER VALLEY AND THE AFGHAN BADAKHSHAN

In this section, we will describe rockslides in the westernmost—Afghan part of the Pamir and those case studies in the Pianj River valley whose headscarps are located at one side of the valley coinciding with the Afghanistan–Tajikistan state border, while the deposits rest "abroad"—at the opposite side of the valley. The latter cases got their IDs according to the rule described in Section 6.1. Two such rockslides have been found ca. 31 and 43 km downstream of the Bartang River's mouth close to Shidz village (37.93° N, 71.31° E, Tajik 108) and Voznavd village (37.97° N, 71.27° E, Tajik 110). Both originated from the 1400–1500 m high right-bank slope of the Pianj River valley composed of granites intruded in the Proterozoic migmatites and had blocked this major stream.

The Shidz rockslide ~700 Mm^3 in volume (considering its part buried by fluvial deposits) created the dam with the effective height more than 450 m. Its part that survived erosion forms now the 300–400 m high

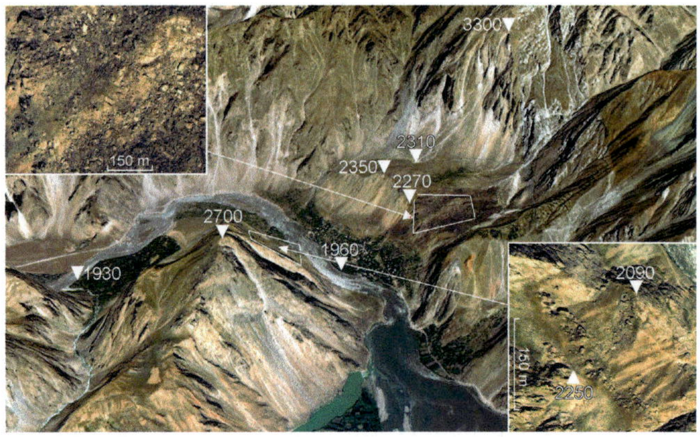

FIGURE 9.98 3D Google Earth view of the Shidz rockslide. Coarse angular boulders forming rockslide carapace are shown in the insets.

slightly convex terrace-like surface at the foot of the tremendous, about 2 km wide and 800 m deep headscarp with a very steep (~50°) back wall up to 1 km high (Fig. 9.98). Remnants of the rockslide distal part could be found on the Afghan side of the valley at 2090–2250 m a.s.l. (see lower inset in this figure). Rockslide still partially dams the Pianj River, and judging from its thalweg profile (Fig. 9.99), this blockage has not been incised up to its base yet. Up to 150 m of its body is buried from the upstream by fluvial deposits accumulated in the dammed lake (unit A on Fig. 9.99), and from the downstream—by debris eroded from the dam during its breach (unit B on Fig. 9.99) that protected the lowermost part of the blockage from further incision.

Debris accumulation about 100 m thick composed of rather fine material can be seen on the left bank of the river downstream of the former blockage (Fig. 9.100). We assume that it was also eroded from the dams' body. It cannot be excluded, however, that these deposits are the part of the original rockslide body protected from the erosion by the ca. 400 m high rocky promontory, so that it was just smoothed by the surge wave.

Unlike the Sarez Lake that survived due to lucky relationship of relatively small mean annual discharge of the Murgab River (about 50 m^3/s) and very large size of the uppermost permeable zone of the blockage balancing the inflow and filtration through the dam (though gradual long-term increase of the lake level takes place; A. Ischuk, 2011), the Shidz rockslide had dammed the much more powerful Pianj River with present-day mean annual discharge of about 800 m^3/s. Such inflow could not be released by filtration and dam's overtopping should occur inevitably.

Considering the lowermost level of the dam's crest at ca. 2200 m a.s.l., which seems to be the minimal value, the impounded lake should extend for more

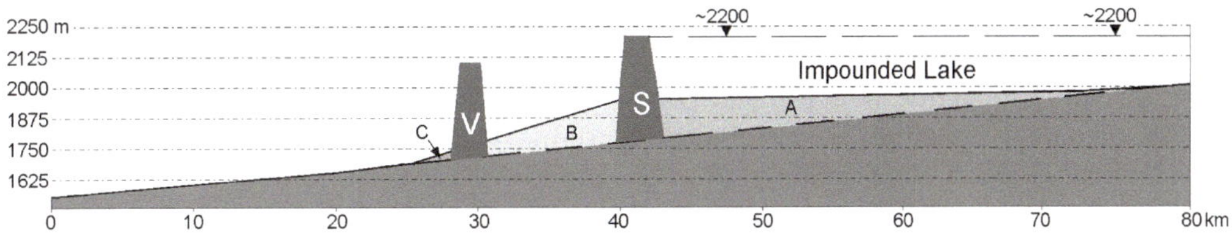

FIGURE 9.99 Thalweg profile of the Pianj River between the Bartang and Yazgulem River mouths. A: sediments accumulated in the dammed lake; B: deposits eroded from the Shidz blockage; C: deposits accumulated downstream of the Voznavd rockslide dam.

FIGURE 9.100 Debris accumulation presumably left by the outburst flood after Shidz rockslide dam breach. 3D Google Earth view.

than 100 km along the Pianj River and for more than 40 km along the Bartang River (Fig. 9.101) and could store up to 31 km^3 of water. As we mentioned already, the lowermost part of this water body has been silted so that about 43 km of the Pianj River valley up to the Badzhuvdara River mouth and about 10 km of the Bartang River valley have flattened bottoms and rivers meander through lacustrine sediments. Total volume of these deposits up to 150 m thick close to the blockage can reach ca. 2.5–3.0 km^3. Considering the annual sediment yield of the Pianj River of the order of $5-10 \times 10^6\ m^3$ (no direct measurements of this parameter are available for the river along which state border passes; State Water Inventory, 1990), it took about 300–600 years to accumulate. However, this blockage could hardly survive the first overtopping, if one takes in mind significant discharge of the Pianj River. When the dam was breached, about 200–250 Mm^3 of debris had been eroded from its body. This material filled the bottom of the valley at a distance of about 10 km (see Unit B on Fig. 9.99) and protected the lowermost part of the blockage from further erosion so that the siltation of the remaining part of the dammed lake could continue for a long time. It can explain the controversy between the large amount of accumulated lacustrine sediments that required long-term river damming and assumed short-term existence of the high-level impoundment.

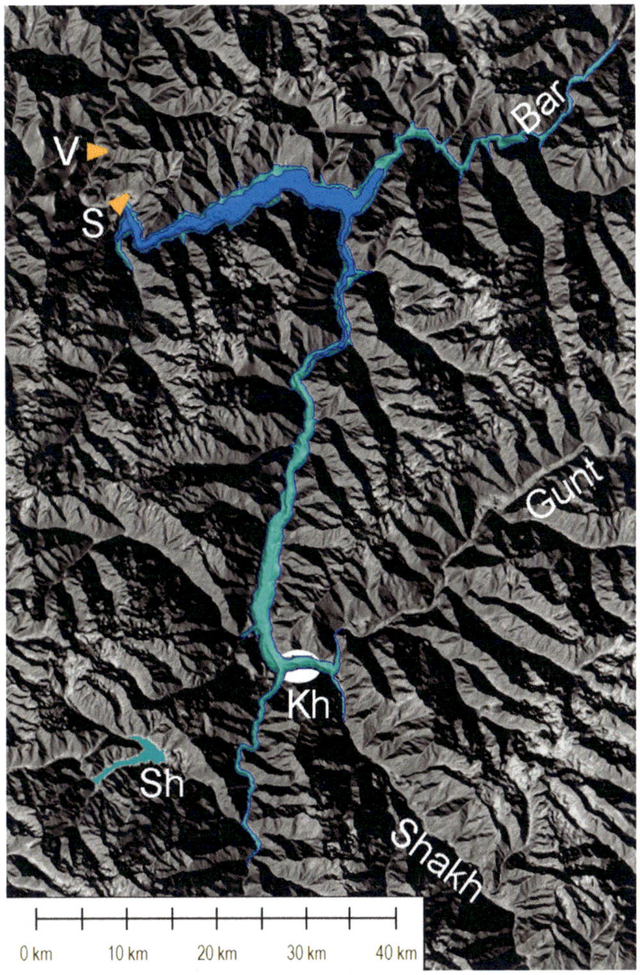

FIGURE 9.101 The schematic map of the lake dammed by the Shidz rockslide (S). V: the Voznadj rockslide dam (shown on Fig. 9.102). Bar: the Bartang River, Gunt: the Gunt River, Shakh: the Shakhdara River, Sh: the Shiva Lake (see Fig. 4.22), Kh: the Khorog town. Dark blue: silted part of the lake; light blue: the assumed lake at 2200 m a.s.l. Background: shaded SRTM DEM visualized by the Global Mapper software.

Ten kilometers downstream of the Shidz rockslide, the Pianj River valley was blocked by another very large Voznavd rockslide about 250 Mm^3 in volume (Fig. 9.102). It originated from more than 1.5 km high slope and its headscarp back wall about 1 km high is even steeper than that of the Shidz rockslide—up to

56°. Minor rock fall about 0.7 Mm^3 in volume occurred recently (we have no data about is exact age) at its northern side. Unlike Shidz rockslide which distal part was lower that the proximal one, the Voznavd rockslide moved as a typical primary rock avalanche and its toe can be identified about 450 m above the riverbed on the opposite (Afghan) bank of the Pianj River, so that the dam was eroded at its proximal section. Because the left-bank slope of the valley had some lowering, the latter acted as a trap, preventing sidewise debris spreading, same as at several other case studies described above such as the Aksu (see Fig. 8.103) and the Kulun rockslides (see Fig. 8.237).

Temporal relationships between the Shidz rockslide and the Voznavd rockslide are unknown. From our point of view, most likely that both of them had originated simultaneously. This assumption is based on the character of the valley fill visible on Fig. 9.99. If the Voznavd rockslide would block the valley earlier or later than the Shidz one, the valley upstream should be silted in the same way as it occurred upstream of the Shidz blockage so that the Pianj River thalweg upstream of the Voznavd rockslide should be similarly horizontal, which is not observed. If, however, both dams were formed, as we assume, at the same time, initially the upstream dam had protected the valley section between these two blockages from siltation. When the Shidz rockslide dam was overtopped and its catastrophic breach started—presence of the next—the Voznavd rockslide dam had prevented free transport of the eroded material downstream the valley, at least before the latter was overtopped, being about 100 m lower than the Shidz one (see Fig. 9.99). After that, it had been breached in turn. The material eroded from the upper blockage filled this section of the valley between these two blockages and formed the deposits with inclined surface (unit B on Fig. 9.99). After breach of the Voznavd blockage, nothing hampered free transport of the material eroded from both dams downstream the Pianj River and we can see only minor deposition downstream of the Voznavd rockslide (Unit C on Fig. 9.99).

FIGURE 9.102 3D Google Earth view of the Voznavd rockslide. 3D Google Earth view. Bouldery carapace of the distal part of rockslide body is shown in the inset.

Same relationships have been found in the Gunt River valley that had been dammed three times at a short section about 20 km long by the Zuvor, the Vir-Tangiv, and the Chartym rockslides (see Fig. 4.27). The Gunt thalweg between the Zuvor and the Vir-Tangiv breached rockslide dams is inclined (same as in the Voznavd–Shidz case) while upstream of the Vir-Tangiv and the Chartym rockslides river thalweg is almost horizontal, indicating temporally independent occurrence of these blockages.

Rockslide dams described above that had blocked the Pianj River are just members of the cluster of several extraordinary rock slope failures (Table 9.1) one of which—the gigantic Padjvar rockslide belongs to the largest of the entire Central Asia region. Despite the enormous size of the Shidz and the Voznavd rockslides, this cluster can be ascribed to the core type similarly to the Sarydjaz-Aksu cluster (see Table 8.2)

TABLE 9.1 Basic Parameters of Rockslides Forming Rockslide Cluster Between Pianj and Tangshev River Valleys

	dd.ddddd°				
ID	**Latitude**	**Longitude**	**Area of the deposits (km^2)**	**Total area affected (km^2)**	**Volume (10^6 m^3)**
YOUNGER (HOLOCENE) ROCKSLIDES					
Tajik 108	37.9272	71.3119	3.382	5.12	700
Tajik 110	37.9737	71.2738	2.420	3.27	250
Afghan 001	37.8700	71.4500	19.72	26.94	6000
Afghan 005	37.8603	71.3199	0.79	2.15	35
Afghan 008	37.8744	71.1578	1.28	2.06	100
Afghan 009	37.8931	71.1883	1.353	2.43	100

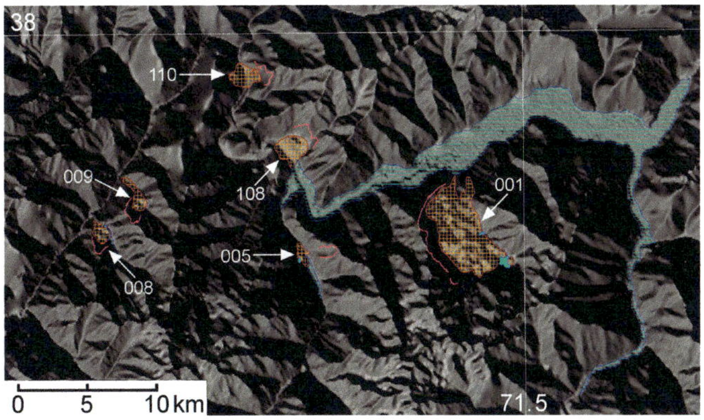

FIGURE 9.103 Rockslide cluster between Pianj and Tangshev River valleys. Cross-hatched brown areas: landslide bodies; pink lines: headscarps; bright blue spots: landslide dammed lakes; hatched blue areas: landslide dammed lakes filled by deposits. Numbers correspond to those in Table 9.1 without country names. Background: the 3″ SRTM DEM visualized by the Global Mapper software.

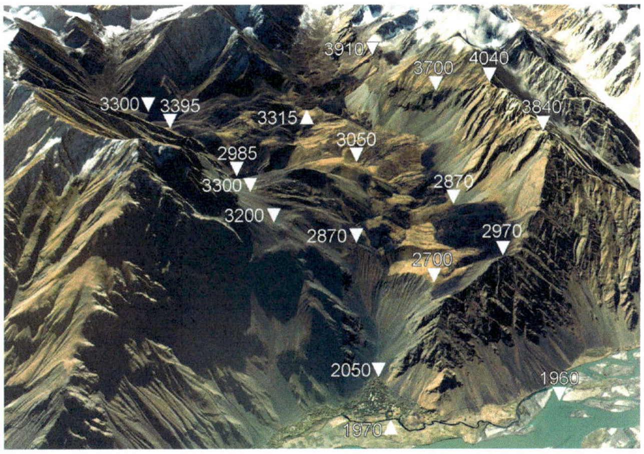

FIGURE 9.104 3D Google Earth view of the gigantic Padjvar rockslide.

because the Padjvar rockslide is much larger than most of other rockslides in this area (Fig. 9.103).

The Padjvar rockslide (37.87° N, 71,45° E, Afghan 001) is the rare case when the entire ridge about 6.5 km long slid down in the adjacent valley and filled it almost completely (Fig. 9.104). Similar phenomena can be exemplified by the Sarychelek rockslide in North-Western Tien Shan (see Figs. 8.215 and 8.216), by the Green Lake rockslide in New Zealand (Hancox and Perrin, 2009) and by the Mahu rockslide in the Sichuan Province—one of the largest in China (Cui et al., in press). Rockslide deposits up to 650–700 m thick cover 19.72 km^2 (the total area affected is about 27 km^2) and its volume was estimated as ca. 6 km^3, assuming the V-shape profile of the valley filled by its body.

According to the Geological map of Tajik SSR and adjacent territories (1984), the collapsed ridge raising up to more than 4 km a.s.l. was composed of the Paleogene granodiorite intruded in the Triassic slate. As it can be derived from the space images, most of the arcuate back wall of its headscarp ca. 600–950 m high is composed of slate while the topmost part of the ridge — of the granodiorite. It can be assumed that rock massif more than 1 km thick collapsed into the ~2 km deep valley incised up to ca. 2050–2100 m a.s.l. Surface of the deposits that fill the valley is extremely irregular with closed depressions, some of which are filled by small lakes, and with hills raising above them up to 300–400 m (see Fig. 9.104). Frontal edge of rockslide body (Fig. 9.105) is 250–350 m higher than its proximal part indicating the primary style of motion (see Section 5.3.1 in Chapter 5).

Age of this extraordinary event as well as its possible trigger are unknown. We can just anticipate that it could be triggered by a strong earthquake, either by local crustal or by deep-foci from the Hindu Kush focal zone that passes nearby in the north-eastern direction (Sippl et al., 2013) (see Fig. 3.10). At its highest—southern part rockslide body is covered by end moraine and rock glaciers' deposits from the adjacent presently glaciated valleys (see upper left inset in Fig. 9.105) that can be considered as some proxy for its age assessment—nearly 1 km advance of rock glacier over the rockslide body could last from about 1 to more than 10 millennia if its mean velocity was within the same range as it was measured at some Austrian rock glaciers—from a few centimeters to ~1 m per year (Kellerer-Pirklbauer and Kaufmann, 2012). Surface of this body is practically nonvegetated, which, however, is typical of most of the Pamir mountains at such altitude and cannot be used for quantitative age assessment. It is evident that this outstanding feature deserves special research project to learn more about its structure and origin.

About 10 km west of the Padjvar rockslide, at 37.86° N, 71.32° E, an impressive primary rock avalanche about 50 Mm^3 in volume (Afghan 005) caved from the ca. 1400 m high right bank of the Yorkhdara River valley composed of Paleogene granite. It blocked the river by 100 m high dam and the dammed lake was silted almost completely. Interesting peculiarity of the Yorkhdara rockslide is that its body that moved as a typical primary rock avalanche in frontally confined conditions (see Section 5.3.1 in Chapter 5) dammed also the small tributary valley (Fig. 9.106) and, due to the significant—more than 100 m high run-up, formed a small lake there, about 100 m above the crest of this blockage in the Yorkhdara River valley. However, unlike most of other case studies where rockslide debris motion was captured by the trap at the opposite

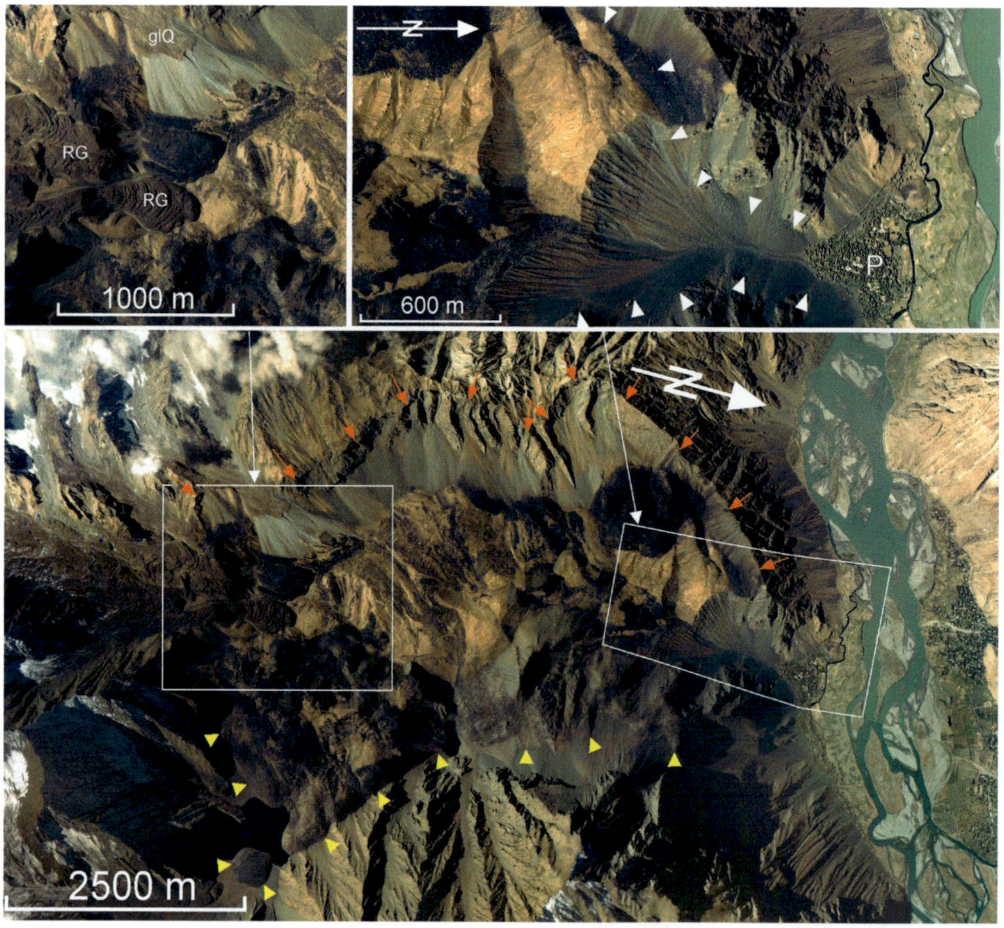

FIGURE 9.105 Plan view of the Padjvar rockslide. P—the Padjvar village; red arrows mark the headscarp, yellow triangles indicate the frontal edge of the deposits, white triangles in the upper right inset indicate shape of the deep V-shaped valley filled by rockslide. RG: rock glaciers, glQ: recent moraine deposits.

FIGURE 9.106 The Yorkhdara rockslide. SAS Planet image visualized in 3D by Global Mapper software. Google Earth image of the body is in the inset. Transparent brown arrow shows direction of the rockslide initial motion.

side of the valley (see Fig. 8.103 or 8.237), it produced a significant 1700 m long transverse (up- and down valley) debris spreading, most likely because the tributary mouth was located not just opposite the headscarp axis but aside of the main impact zone so that the debris moved there after collision with an opposite slope of the valley (see inset in Fig. 9.106).

The dammed lake was silted almost completely and the river passes over the proximal part of the dam without significant incision, presumably due to its rather low gradient caused by the along-valley debris spreading. That is why this blockage still forms a distinct 170 m high nick point in the river thalweg.

Two large-scale rock slope failures—the Panjai-Mard rockslide at 37.87° N, 71.16° E (Afghan 008) and the Khudardjak rockslide at 37.89° N, 71.19° E (Afghan 009) were found in the lower reaches of the Tangshev River valley that falls in the Pianj River 2 km downstream of the Voznavd rockslide. Both occurred on the north-facing slopes of its short right tributaries composed of the Proterozoic metamorphic rocks and could be

FIGURE 9.107 3D Google Earth views of the Panjai-Mard rockslide (lower inset) and of the Khudardjak rockslide (upper inset) in the Tangshev River valley.

classified as wedge failures (Fig. 9.107). Similar north-eastern orientation of both rockslides along with large, up to 200—250 m, depth of their headscarps, allows assumption that they could be triggered by a strong earthquake. Without any age determination, we can just speculate if it could be the same event that could trigger the Padjvar rockslide about 20 km east of the Tangshev River or not, but concentration of so large bedrock landslides at such a limited area deserves special attention.

The Panjai-Mard rockslide about 100 Mm^3 in volume represents the promontory failure similar to those that formed the Kolduk (see Fig. 8.241) and the Ai-Kul rockslides (see Fig. 8.257) or the rockslide in the Rostodjirav River valley (see Fig. 9.96C). The affected rock massif could be up to 500 m deep if not more. Rockslide formed the compact body up to 250 m thick that dammed two creeks; the lake in the larger one has been silted almost completely, while the smaller one still exist. Surface of the blockage is nearly horizontal, gradually decreasing towards its distal edge, where the water from the dammed creeks found its way out. The funnel-shape downstream slope of the dam below the elevation mark 2380 m a.s.l. at the lower inset in Fig. 9.107 that could not be formed by the erosion resembles the Kiol-say rockslide in the Kokomeren River basin (see Fig. 5.63). However, here the assumed secondary rock avalanche deposits could not be identified so distinctly.

About 3 km north, nearly 1.5 km^2 of the northern slope of the adjacent local ridge collapsed leaving up to a 250–280 m deep headscarp and forming the Khudardjak rockslide also ca. 100 Mm^3 in volume. Its body can be divided into two parts—the compact one about 250 m thick with ca. 130 m run-up that filled the creek bottom at the base of the headscarp and ~1.4 km long secondary avalanche about 4 Mm^3 and up to 50 m thick (at its distal part) that spread along the creek almost up its mouth (see Figs. 9.107 and 9.108). It originated from the very distinct secondary scar about 300 m high and can be classified as the Secondary rock avalanche of the deflected type (see Section 5.3.3 in Chapter 5), because the direction of its motion differs from that of the initial slope failure for about 55 degrees. It slightly recalls secondary rock avalanche of the much bigger Big Dragon rockslide in the Chinese Tien Shan (see Fig. 5.65).

About 25–30 km upstream, in the valley of the Tangshev River right tributary—the Golomdara River—at 37.65°–37.71° N, 71.0° E, several landslides (Afghan 010–012) originate on the slopes composed of the terrigenous and carbonate Jurassic rocks (Fig. 9.109). Unlike case studies from this region described above, they look more as gradually moving landslides rather than catastrophic rock slope failures. Largest of them (Afghan 010) that affects 1300 m high and more than 4 km long slope with an affected area of ca. 7 km^2 reaches the Golomdara River at a distance of about 1 km and, likely, caused it short-term blocking. Volume of the displaced rock massif could reach 1 km^3.

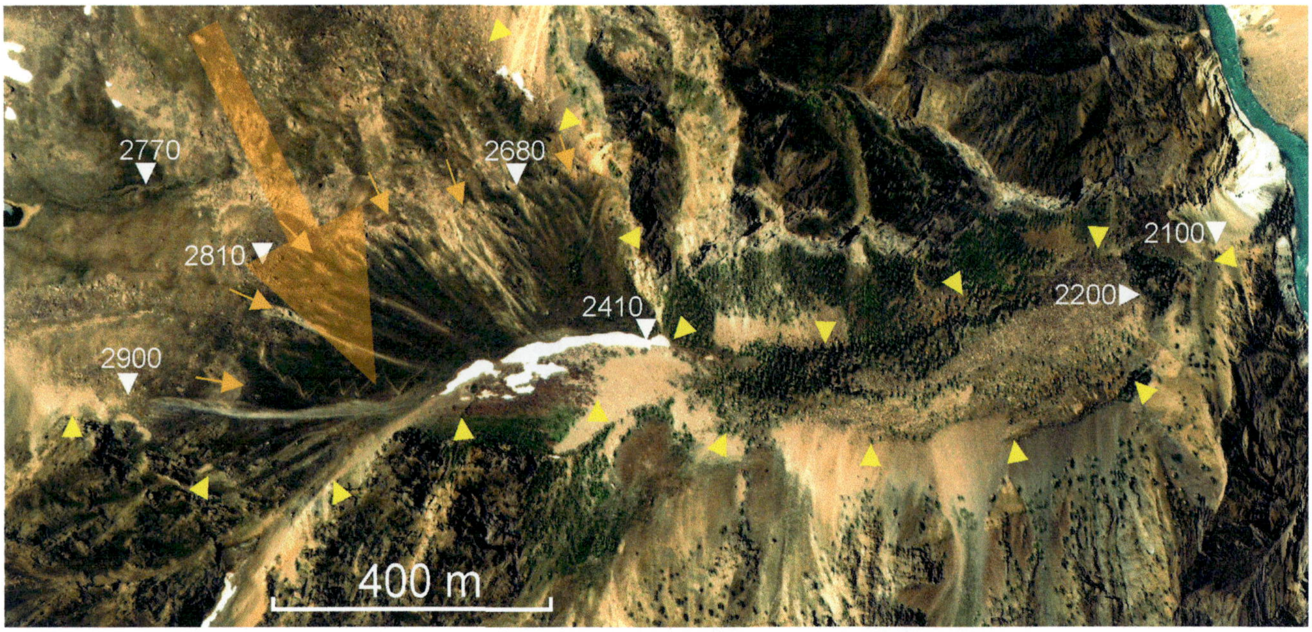

FIGURE 9.108 Secondary rock avalanche of the Khudardjak rockslide. Yellow triangles: outer boundary of rockslide debris; orange arrows: secondary scar crown; the transparent brown arrow: direction of the initial rockslide motion. Google Earth image.

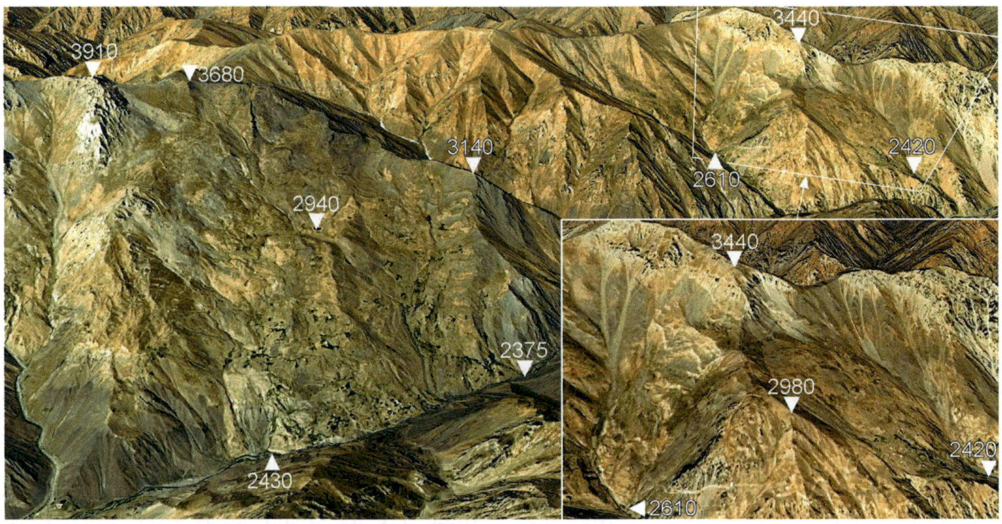

FIGURE 9.109 Landslides in the Golomdara River valley. 2D Google Earth view.

Another part of the Afghanistan territory and the Pianj River valley where numerous large-scale rockslides are widespread is near the Shiva Lake and opposite the Khorog town (see Fig. 9.1). They are distributed within the ca. 40 km long north-south trending belt that extends up to lower reaches of the Garmchashma River valley mentioned at the end of Section 9.2. The largest one is the Shiva rockslide dam at 37.39° N, 71.4° E (Afghan 007) up to 450–480 m high and about 1100 Mm^3 in volume. We must mention that in Strom (2010b) volume of this rockslide was measured as 2.0 km^3, nearly the same as that of the Usoi rockslide (2.7 km^3 according to Preobrajensky (1920)). However, comparison of its area (~8.18 km^2) and height (450–480 m) with those of the Usoi rockslide (~12.76 km^2 and ~600 m) shows that it was exaggerated significantly and that the present estimate is much more reliable.

This case study was mentioned in Section 4.2 of Chapter 4 (see Fig. 4.22) as an example of the erroneous interpretation of rockslide dam as the glacial landform (N. Ischuk, 2011, 2013a). The Shiva rockslide originated from the more than 300 m deep and up to 2370 m wide headscarp incised in the ca. 1500 m high valley slope composed of the Archean gneiss. Its nearly horizontal surface is covered by multiple oval hummocks from 20 to 100 m in diameter (Fig. 9.110) resembling the surface of the long runout Koman rock avalanche in the Alai valley (see Figs. 6.9 and 9.20). Material at the frontal part of its surface that forms these hummocks is finer than coarse angular blocks at its proximal part. The latter has been partially reworked by the evolving rock glacier (see Fig. 4.22).

The Shiva rockslides form the permanent dam that seems to be a stable and safe feature. It blocks rather small catchments of 232 km^2 only, and, with DBI value of 1.98, falls into the "stability domain" (Ermini and Casagli, 2002, 2003). Water from the Shiva Lake filters towards the minor pond on the downstream side of the blockage at its proximal part, which is about 100 m below the lake level and from which it flows down towards the Arakht River. It can be hypothesized that seeping takes place through coarser material at the proximal part of rockslide carapace that recalls the Usoi rockslide dam. One more small water body fills the closed depression at the dam's front which is ca. 50 m above the lake level (see Fig. 4.22). Breach of this dam whose crest is about 100 m above the Shiva Lake level is almost impossible.

FIGURE 9.110 The hummocky surface of the Shiva rockslide. Google Earth image.

About 2 km downstream of the Shiva rockslide dam at 37.37° N, 71.44° E, there is a ca. 50 m high nick point in the Arakht River thalweg likely formed by the deeply eroded blockage created by rockslide in the Archean gneiss, about 30 Mm^3 in volume (Afghan 015).

The suspicious set of landforms was found in the Pianj River valley at 37.535° N, 71.48°–71.52° E, 6.0 km downstream of the Gunt River mouth (Fig. 9.111). We hypothesize that the spoon-shaped lowering on the ca. 1700 m high right (Tajik) side of the valley corresponds to the headscarp in the Archean migmatites thrusted over the Triassic terrigenous rocks, while 50–70 m thick deposits resting at the foot of this headscarp on the ~550 m high terrace-like surface (likely, the bottom of an old trough valley)—small proximal portion of its body. Because the source of this assumed rockslide is at the Tajik side of the valley it was included in the database as the Tajik 202 case study. Just opposite this site, at the foot of the Pianj River valley left-bank slope, there is an elongated hill ~2200 m long, up to 600 m wide, and up to 140 m high composed of whitish material resting on the 80–100 m high river terrace formed by gray fluvial gravels (see inset in Fig. 9.111). We hypothesize that it could be the distal part of the rockslide that caved from the right-bank slope of the valley, crossed it, and reached its opposite bank, which corresponds to 5.42 km long runout. Assumed volume of this rockslide is about 50 Mm^3. Additional field studies should be carried out either to prove or to discard this assumption.

Several rockslides, whose bodies also rest on the Quaternary fill of the valley, this time likely of the glacial origin, were found in the Khidardjakdara River valley that falls in the Pianj River just opposite the Khorog town. They occurred on the slopes composed of the Archean metamorphic rocks, though of different lithology, judging from the different color of these sites on space images (Fig. 9.112). The upstream one at 37.47° N, 71.414° E (Afghan 016) is a typical block slide with short runout and compact body about 170 m thick resting on the 150–200 m high terrace-like surface used for agriculture—likely on the moraine fill of this formerly glaciated valley. Sliding affected the entire rocky prominence about 500 m high between two valleys. This rockslide about 40 Mm^3 in volume with its triangular back wall of the headscarp and compact

FIGURE 9.111 Assumed rockslide deposits (RS?) in the Pianj River valley 6 km downstream of the Gunt River mouth. Blue arrow marked by G: the Gunt River mouth; alQ: fluvial gravels. 3D Google Earth view.

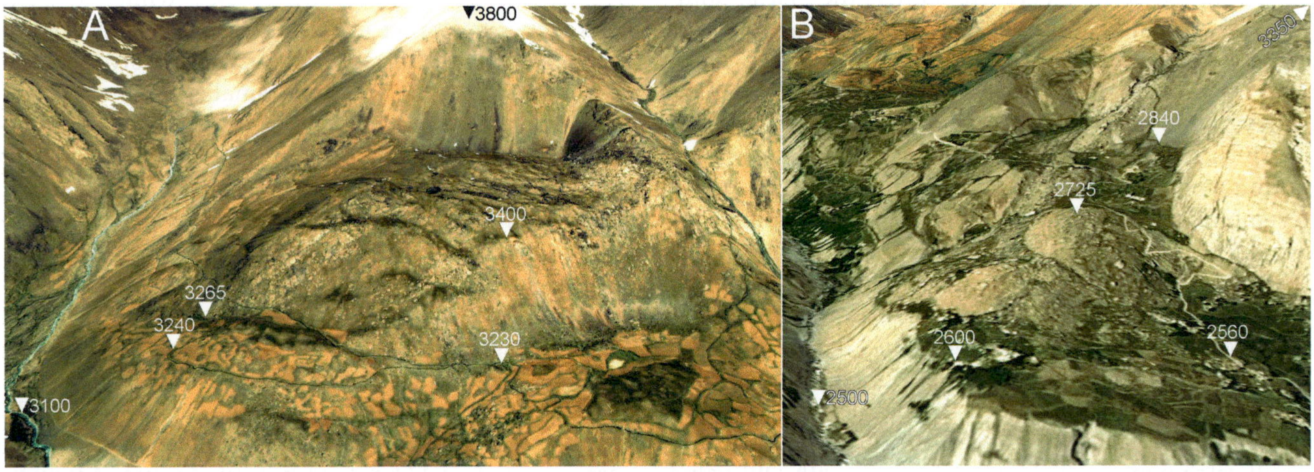

FIGURE 9.112 Rockslides in the Khidardjakdara River valley resting on thick Quaternary deposits. (A) Afghan 016, (B) Afghan 017. 3D Google Earth views.

body (see Fig. 9.112A) looks very similar to the 1911 seismically triggered Ananievo rockslide in Northern Tien Shan (see Figs. 8.70 and 8.71).

The second one (Aghan 017) also originated on the same left-bank slope of the Khidardjakdara River valley, about 4 km downstream, at 37.49° N, 71.46° E. This feature has a poorly defined headscarp about 550 m high, but very distinct deposits composed of angular boulders up to 20–30 m in size that cover 765 000 m^2 of the left bank river terrace raising 90–120 m above the present-day valley bottom. This body about 15 Mm^3 in volume differs drastically from the underlying Quaternary (fluvial or fluvioglacial) deposits, which are much finer (see Fig. 9.112B). Smaller assumed rockslide body overlays the same terrace-like landform 1 km downstream.

Further south of the Shiva Lake, two unidirected rockslides had occurred at 37.316° N, 71.446° E (Afghan 018) and at 37.384° N, 71.45° E (Afghan 019). Both failures had affected north-facing slopes of the short valleys of the Pianj River left tributaries composed of the Archean metamorphic rocks (Fig. 9.113). Such directivity of rock slope failures on the nearby slopes was described at numerous sites in the Central Asia region (see Figs. 5.52, 5.61, 6.5, 7.8, 8.8, 8.15, 8.38, 8.42, 8.117, 8.268, 9.67, 9.96, and 9.107). It could be governed by the host rock massif structure (similar orientation of bedding/foliation/fractures) and slopes'

FIGURE 9.113 Unidirected rockslides 9–12 km south of the Shiva Lake. 3D Google Earth views.

FIGURE 9.114 Rockslide on the left bank of the Pianj River ~1 km upstream of the Garmchashma River mouth. 3D Google Earth view.

aspect on the one hand, or by the same trigger, i.e. the directed earthquake strong motion. Here failures affected the entire slopes up to 970–1300 m high sloping 20–22 degrees and are more pronounced at their uppermost parts (see insets in Fig. 9.113). Destruction of the near-watershed zones can be also considered as an argument in favor of their seismic triggering (Havenith, 2002; Havenith et al., 2003), though the Djamansu rockslide (see Figs. 8.110-8.112) demonstrates that this indicator cannot be considered as a univocal.

One more rock slope failure in this part of the Pianj River valley (Afghan 014) was identified on its 720 m high left-bank slope at 37.231° N, 71.466° E, about 1 km upstream of the Garmchashma River mouth. This slope is also composed of Archean metamorphic rocks. Most of rockslide body about 25 Mm^3 in volume still rests on the slope inclined 26 degrees on average (Fig. 9.114) and, thus, it poses a potential threat to the settlements and infrastructure downstream, because its possible activation could cause temporary river damming and subsequent inundation and outburst flood.

Very high concentration of rockslides was detected at the south-western part of the Afghan Badakhshan drained by numerous deep valleys of the Kokcha River basin. This area coincides with the Hindu Kush upper mantle focal zone (Sippl et al., 2013) (see Fig. 3.10) where about 15 earthquakes with magnitudes from 7.0 to 7.8 have occurred during the 20th–21st centuries along with numerous smaller earthquakes (Fig. 9.115). The possible role of earthquakes in triggering these rockslides will be discussed further in Chapter 13, Seismically Induced Rockslides: Their Discrimination.

In the northern part of the Kokcha River basin, two interesting case studies were identified near the confluence of the major right tributaries of the Kokcha River—the Zardiv and the Vardudj Rivers, 5.5 km upstream from their junction. Near the Bakharak village at 37.009° N, 70.915° E, the Zardiv River had been blocked by the secondary rock avalanche (Afghan 036) that had caved from the ca. 1000 m high right-bank slope of the valley (Fig. 9.116) composed of the Archean metamorphic rocks.

This slope failure (the Bakharak rock avalanche) originated as a block slide about 50 Mm^3 in volume that subsided for about 300 m and stopped (see the inset in Fig. 9.116). However its frontal, likely most weathered part about 25 Mm^3 in volume moved further producing secondary rock avalanche that traveled ~2.7 km downslope, crossed the river and accumulated on its opposite left bank. The dam was breached at its proximal right-bank part but still forms ~50 m high nick point in the river profile.

Similar transformation of the block slides that originated at the upper parts of the high slopes and stop after 200–300 m descend "throwing out" their frontal parts that convert into long runout secondary rock avalanches (see Section 5.3.3 in Chapter 5) was observed at several other case studies from the Dzungaria and Tien Shan. In Dzungaria, same bipartition took place at rockslide that originated on a much gentler slope (see Figs. 7.9 and 7.10). In Tien Shan, very similar rock avalanches were found in China

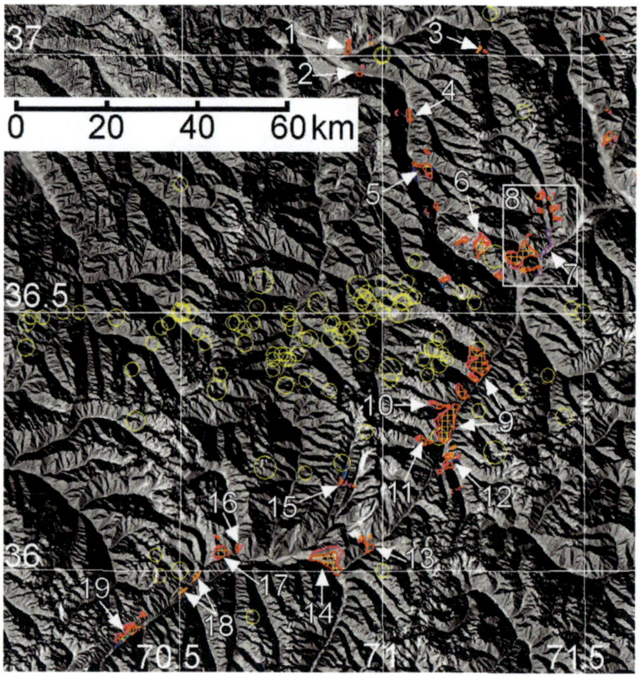

FIGURE 9.115 Rockslide in the upper reaches of the Kokcha River basin. Cross-hatched brown areas: landslide bodies; pink lines: headscarps; bright blue spots: landslide dammed lakes; hatched blue areas: landslide dammed lakes filled by deposits or breached; violet cross-hatched areas: traces of outburst floods; yellow circles: epicenters of the deep-foci earthquakes with magnitudes 6.0–6.4, 6.5–6.9, 7.0–7.4, and ≥7.5. 1: the Bakharak rock avalanche (see Fig. 9.116); 2; the Ardar rockslide (see Fig. 9.117); 3: see Fig. 9.118; 4: the Aqshireh rocks avalanche (see Fig. 9.119); 5: the Bashanabad rockslide (see Fig. 9.120); 6: the Rabati-Chikhiltan rockslide (see Fig. 9.121); 7: see Fig. 9.124; 8: see Fig. 9.125; 9: see Fig. 9.129; 10: see Fig. 9.130; 11: see Fig. 9.131; 12: see Figs. 9.132 and 9.133; 13: the Maghnsavol rockslide (see Fig. 9.134); 14: the Pazhuk rockslide (see Fig. 9.135); 15: see Fig. 9.136; 16: the Rakhoi Mouth rockslide (see Fig. 9.137); 17: see Fig. 9.138; 18: see Fig. 9.139; 19: see Fig. 9.140. Background: the 3″ SRTM DEM visualized by the Global Mapper software.

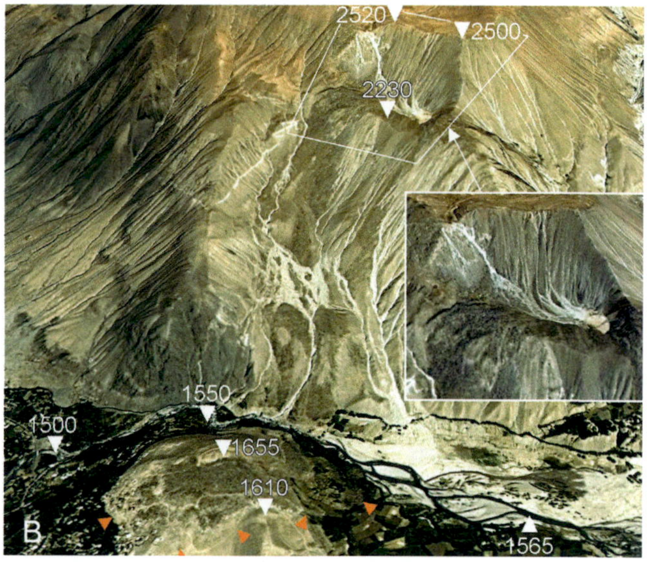

FIGURE 9.116 Block slide on the right-bank slope of the Zardiv River valley that converted into long runout Bakharak rock avalanche, whose front is marked by orange triangles. B: The Bakharak village. Zoomed headscarp is shown in the inset. 3D Google Earth view.

(see Fig. 8.6) and in Kyrgyzstan (the Djamno rock avalanche—see Fig. 8.186). Some evidence of similar bipartition could be found at the Chilik rockslide in Kazakhstan triggered by the 1911 Kemin earthquake (see Fig. 8.67).

One more interesting case study (Afghan 035) was found in the lower reaches of the Vardudj River valley upstream of the Ardar village at 36.964° N, 70.943° E. This Ardar rockslide originated on the ca. 1250 m high right-bank slope also composed of the Archean metamorphic rocks and is characterized by 1700 m long, but narrow—less than 480 wide—spoon-shaped headscarp (Fig. 9.117). It is interesting that on its way down, a rockslide body about 17 Mm^3 in volume jumped from a ca. 300 m high bench at ~1800 m a.s.l. (likely from the trough valley shoulder), but afterward moved as a primary rock avalanche (see Section 5.3.1 in Chapter 5), not as a jumping one (see Section 5.3.2 in Chapter 5). Rockslide debris form a compact

FIGURE 9.117 The Ardar rockslide. 3D Google Earth view.

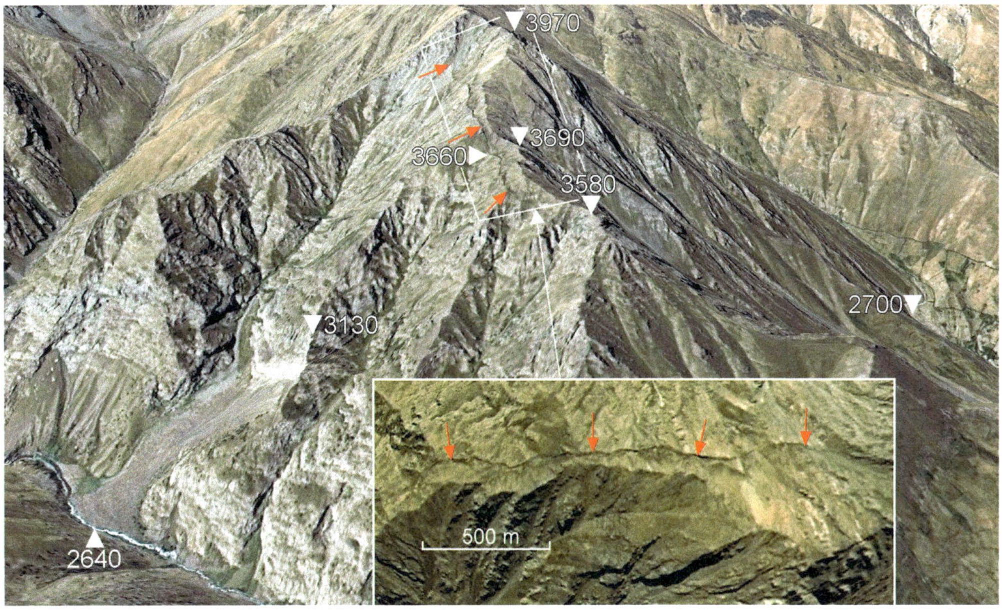

FIGURE 9.118 The 1.4 km long Sackungen on the left watershed of the Khushdara River valley (marked by red arrows) and small Randa-like rock fall at the base of his slope. 3D Google Earth view and plan view in the inset.

oval-shape body with concentric alternating belts of coarser and finer clasts and significant, up to 100 m high frontal run-up (see inset in Fig. 9.117).

Further to the east (Afghan 046, 37.01° N, 71.243° E), the 1.4 km long Sackungen passes along the crest of the local ridge that forms the left side of the Khushdara River valley. It detaches a block of the Archean metamorphic rocks 50–300 m thick that moved towards the north-west, off the Khushdara River (Fig. 9.118). Besides this potentially hazardous site, relatively small Randa-like rock fall (see Gotz and Zimmermann, 1993; Sartori et al., 2003) about 1 Mm^3 in volume (Afghan 042) caused partial damming of the Khushdara River valley at 37.006° N, 71.262° E. Its headscarp crown is at 3130 m a.s.l on Fig. 9.118.

Moving upstream the Vardudj River valley, near the Aqshireh village, at 36.88° N, 71.066° E, one can see the Aqshireh rocks avalanche (Afghan 037, Fig. 9.119). Collapse of about 40 Mm^3 of the Archean metamorphic rocks intruded by granites occurred on the ca. 1500 m high left-bank slope of the valley and spread for more than 2 km^2 over the valley bottom as a sheet of coarse debris about 15–30 m thick. Such spreading of debris over the rather wide valley floor with, likely, shallow ground water table recalls the rock avalanche in Kash River valley in Xinjiang (see Fig. 5.33), the Atdjailau rock avalanche in the Inylchek River valley in Kyrgyzstan (see Fig. 6.4) and the Upper Toguzbulak rockslide in the Tajik part of Pamir (see Fig. 9.71).

Failure affected the upper half of the slope approximately (above ca. 2500 m a.s.l.) and rock debris rushed downslope following the erosional gullies dissecting its lower part (they are marked by the transparent arrows on Fig. 9.119). Portion of debris that passed along one of such gullies forms the upstream part of rockslide and rose for about 145 m on the opposite slope of the valley (see left inset in Fig. 9.119). Distinct varicolored belts of debris visible in more detail in the smaller right inset in Fig. 9.119 demonstrate the

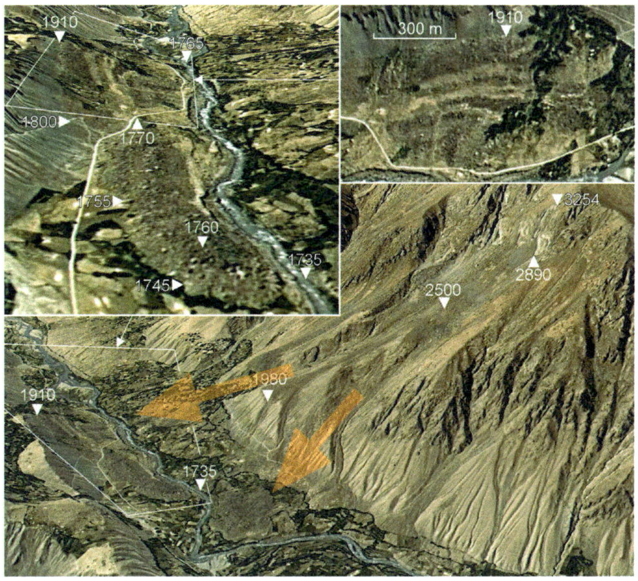

FIGURE 9.119 The Aqshireh rockslide in the Vardudj River valley. The transparent brown arrows show the direction of debris motion through the erosional gullies at the lower part of the slope. 3D Google Earth view.

retention of the original lithological succession in the extremely mobile rock avalanche that can be ascribed to the primary type with frontal confinement (see Section 5.3.1 in Chapter 5). Similar retention was observed at numerous case studies described above all over the Central Asia region.

About 10 km upstream, near the Bashanabad village (36.787° N, 71.078° E), the Vardudj River valley had been dammed at a distance of ~1000 m by the fan-shaped debris accumulation up to 70–80 m thick that came out from the right-side tributary valley (Fig. 9.120). The dammed lake was partially filled and breached but still forms a distinct nick point in the valley profile.

These hazardous phenomena, however, were just the last links of the chain of effects that had started with failure of about 75 Mm^3 of the Archean metamorphic rocks from the ca. 1500 m high rocky spur resembling case studies shown on Figs. 6.12 or 8.257. Rockslide (Afghan 024) blocked the valley forming the small lake upstream that had not been silted yet and collided with opposite slope at an angle of about 64 degrees. This collision produced the deflected Secondary rock avalanche that traveled 2.7 km along the valley leaving not very distinct secondary scar (see Section 5.3.3.1 in Chapter 5) and finally formed the body that caused temporary damming of the Vardudj River. Such chain of effects is similar to that in the Muzart River valley in Chinese Tien Shan where very large rockslide shown on Fig. 4.30 originated on the slope of the tributary valley and produced the deflected secondary rock avalanche that, in turn, blocked the main river valley.

Another interesting peculiarity of the Bashanabad rockslide is the small portion of debris that "jumped over" the headscarp side wall about 600 m below the crown and moved downslope for 500 m up to 3110 m a.s.l. (see inset in Fig. 9.120). The same phenomenon was observed at the Dead Lakes rockslide (see Fig. 8.205), at the Djamantau rock avalanche (see Fig. 8.133), at the Iskanderkul Lake rockslide (see Fig. 8.287) and at the Chagev rockslide (see Fig. 9.84).

Water from the small lake dammed by the main body filters through the uppermost part of the blockage that can be derived from the level of first springs that appear on the downstream face of this natural dam about 5 m below the lake level only (at elevation mark 2570 m a.s.l. shown on Fig. 9.120).

Very high concentration of rockslides was found at the southern part of the ridge between the Vardudj River valley at the west and the Dakhanzar River valley at the east, and also further southward on the left side of the Sanglech River valley and further south in the Munjan and the Yakhshindar River valleys—sources of the Kokcha River (see Fig. 9.115). Some of them are very large with the affected area up to 22 km^2. Sometimes it looks as if the entire valley slopes are unstable being affected by slope failures of different types—block slides and flow slides, gradually moving and catastrophic. The cumulative volume of the largest of them indicated on Fig. 9.115 exceeds 1.5–2.0 km^3. Most of these sites are composed of the Archean metamorphic rocks.

One of such giant complex rockslides affects about 7.2 km^2 on the right side of the Vardudj River valley above the Rabati-Chikhiltan village at 36.63° N, 71.245° E

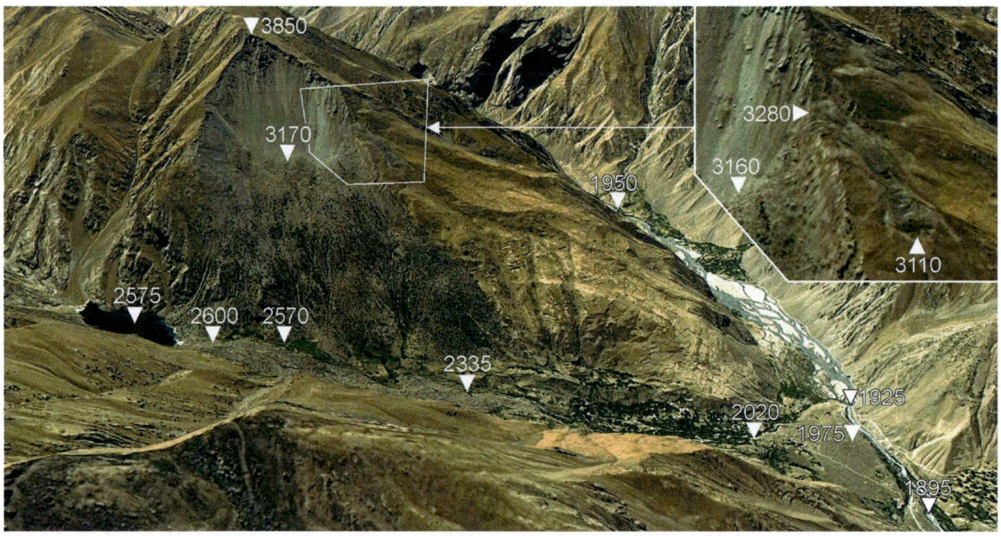

FIGURE 9.120 The Bashanabad rockslide. Portion of rockslide debris that "jumped over" the headscarp side wall is shown in the inset. 3D Google Earth view.

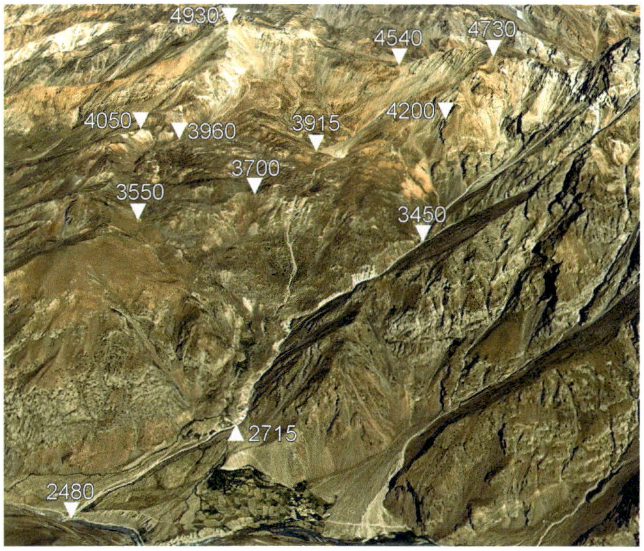

FIGURE 9.121 The Rabati-Chikhiltan rockslide on the right side of the Vardudj River valley. Distance between elevation marks 3450 and 3550 m a.s.l is 2.35 km; between 4930 and 4730—3.2 km. 3D Google Earth view.

FIGURE 9.122 The Razarak multistage rockslide and long runout rock avalanche south of it (A: see also Fig. 9.123). 3D Google Earth view. Evolving DSGSD is shown in the inset.

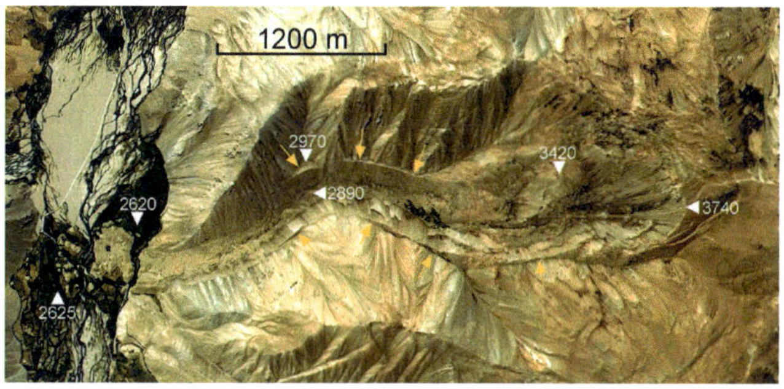

FIGURE 9.123 The 4.3 km long secondary rock avalanche south of the Razarak rockslide. Google Earth image of the area "A" outlined on Fig. 9.122. Trimlines are marked by orange arrows.

(Afghan 021). It did not reach the valley and stopped at 2715 m a.s.l. being about 4.5 km long and up to 3.2 km wide (Fig. 9.121). Rockslide destroyed about 3 km of the ridge watershed and maximal depth of its headscarp is about 350 m. Total volume of the displaced rock mass can be estimated as ~1200 Mm^3.

One more giant rockslide (Afghan 030) affected other side of the same ridge at 36.62° N, 71.35° E, above the Razarak village. Unlike the previous case study that did not reach the valley bottom, the Razarak rockslide had blocked the river valley passing along the northern feet of this ridge. Part of the surface of this complex rockslide is smooth that indicates its relatively large age, while other parts, both close to the watershed and near the modern stream, demonstrate evidence of more resent slope movements (Fig. 9.122). Besides, minor scarps outside the true right side of the headscarp reflect DSGSD that can be activated in future (see small inset in Fig. 9.122). Rough estimate of the total volume of rocks affected by this complex multistage rockslide is about 1500 Mm^3.

South of the gigantic Razarak rockslide, an impressive rock avalanche of the secondary type (see Section 5.3.3.1 in Chapter 5) with 4.3 km long runout and impressive trimlines (Afghan 045) caused partial damming of the Dakhanzar River at 36.589° N, 71.39° E (see Figs. 9.122 and 9.123). Distal part of this rock avalanche was eroded and the debris washed out from its body occupies the 800 m wide floodplain of the river up to 1500 m downstream the damming site (see Fig. 9.123). Judging from the almost absolute lack of vegetation, these deposits accumulated rather recently.

Rock avalanche originated in the upper reaches of the small and, likely, dry creek at the altitude of about 3740 m a.s.l. Volume of the collapsed rock mass likely did not exceed 10 Mm^3 and half of it if not more was involved in flow-like secondary rock avalanche motion. Besides the secondary scar nearly 300 m below the main headscarp, this case study is characterized by the impressive trimlines on both sides of the valley at a distance of ~2.5 km, especially at its true left side (see Fig. 9.123). They are visible about 80 m above the bottom of the ravine along which rock avalanche had passed. It resembles such case studies like the Seit (see Fig. 5.38) and the Mingteke rock avalanches (see Fig. 8.141) in Central Tien Shan and the Karakystiak rock avalanche in Northern Tien Shan (see Fig. 6.7).

About 2 km north from the partial blockage described above and further upstream, at a distance of about 7 km, up to the mouth of the Shulvadar River,

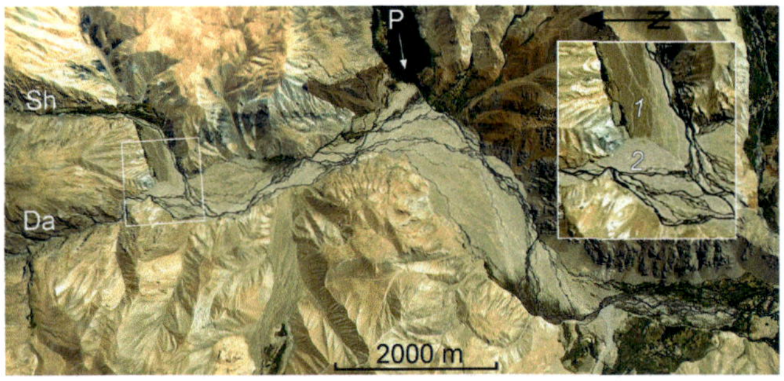

FIGURE 9.124 Google Earth image of the floodplain downstream from the confluence of the Dakhanzar (Da) and the Shulvadar (Sh) Rivers. P: dammed river flowing from the pass between the Pianj and Kokcha River basins. Zoomed image of the outlined area shown in the inset demonstrates temporal relationships between outburst floods from the Shulvadar River (1) and the Dakhanzar River (2).

the floodplain of the Dakhanzar River is covered by the fresh, nonvegetated deposits (Fig. 9.124) similar to those visible on Fig. 9.123. From our point of view, they were left by outburst floods first from the Shulvadar River valley and shortly later from the Dakhanzar River valley. Such relationships can be derived from the fact that the deposits of the second outburst flood cross minor landforms left by the first one (see the inset in Fig. 9.124). These deposits dam the small river that drains the south-western side of the pass between the Pianj and Kokcha River basins and falls into the Dakhanzar River valley from the left, indicating catastrophic character of the outburst floods. It can be also hypothesized that the nonvegetated deposits at the floodplain visible on Fig. 9.123 could be formed not by breach of the low dam just upstream, but by much more powerful flood from upstream that overtopped this low barrier and eroded it significantly.

Both the Dakhanzar and the Shulvadar River valleys upstream this wide section of the floodplain covered by recent deposits are deep, narrow, totally uninhabited canyons that had been dammed by several rockslides 4–8 km upstream from their confluence (Fig. 9.125). All of them originated on the west-facing slopes, except rockslide at 36.698° N, 71.393° E (Afghan 026) that destroyed the rocky prominence above the confluence of the Dakhanzar River and its large right tributary. All slopes are composed of the Archean gneiss and migmatites.

The latter rockslide has nearly a planar not very high headscarp with its crown at 3680 m a.s.l. rising 200–220 m above the compact body about 25 Mm^3 in volume (Fig. 9.126) and slightly recalls the Ananievo rockslide in Northern Tien Shan triggered by the 1911 Kemin Earthquake (see Fig. 8.70), allowing assumption that this slope failure could be seismically triggered as well.

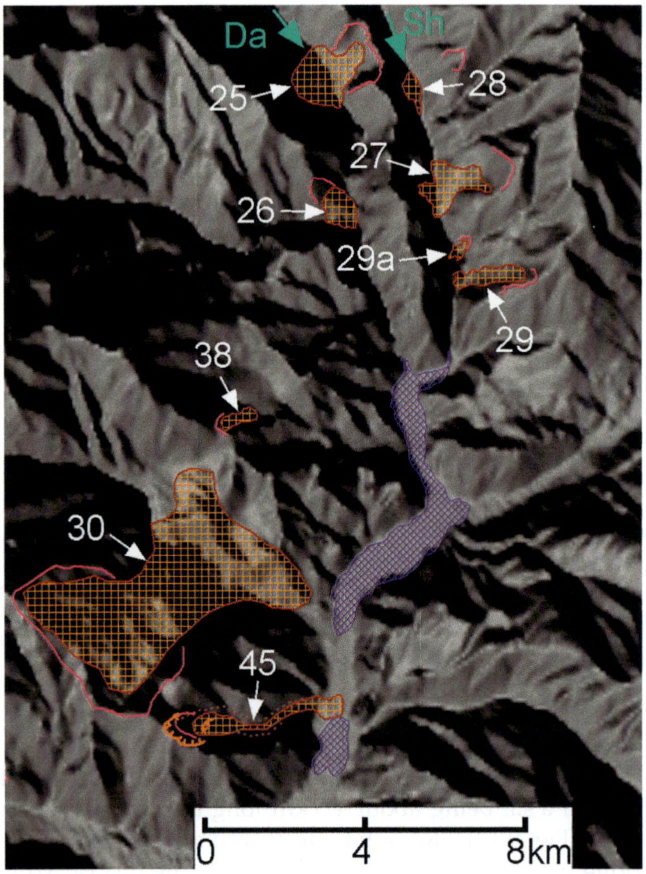

FIGURE 9.125 Numerous river-damming rockslides in the Dakhanzar (Da) and the Shulvadar (Sh) River valleys upstream from floodplain covered by the outburst flood deposits. Cross-hatched areas with brown boundaries: rockslide bodies; pink lines: headscarps; violet cross-hatched areas: sections of floodplain covered by nonvegetated deposits left by outburst flood. Numbers correspond to those in the database without country name. 25: see Fig. 9.128; 26: see Fig. 9.126; 27: see Fig. 9.127; 30: the Razarak rockslide (see Fig. 9.122); 45: see Fig. 9.123. Background: the 3″ SRTM DEM visualized by the Global Mapper software.

Largest rockslide in the Shulvadar River valley (Afghan 027, 36.705° N, 71.422° E) was about 50 Mm^3 in volume and had blocked the valley at a distance of about 1200 m by the dam up to 80 m high approximately (Fig. 9.127). We must notice that 90-m spatial resolution of the SRTM DEM is not enough to reflect geometrical parameters of the bottom of this very narrow gorge adequately. Wider part of the valley bottom upstream of this dam was filled by sediments.

Rockslide that had dammed the Dakhanzar River valley at 36.726° N, 71.389° E (Afghan 025) was much larger, about 180 Mm^3 approximately. It destroyed 1500 of the watershed between the Dakhanzar and the Shulvadar River valleys so that the upper part of the western slope of the Shulvadar River valley 1600 m long and 300–400 m wide in plan view and up to 150–200 m high was affected by the rockslide

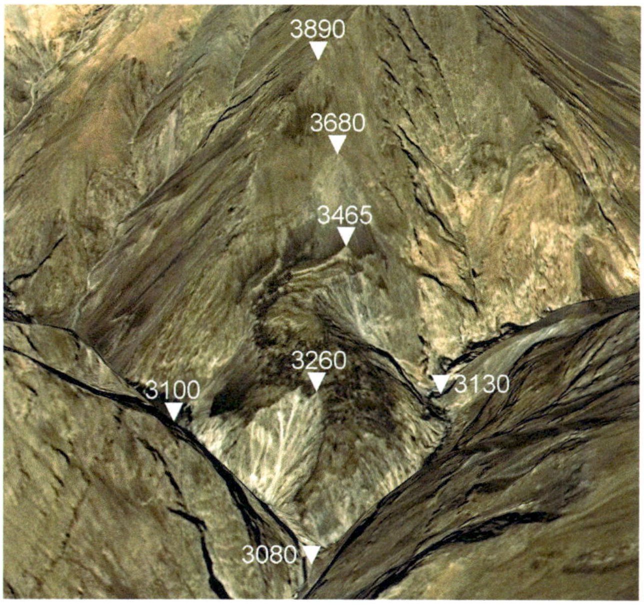

FIGURE 9.126 Rockslide that destroyed rocky spur above the confluence of the Dakhanzar River and its right tributary. 3D Google Earth view.

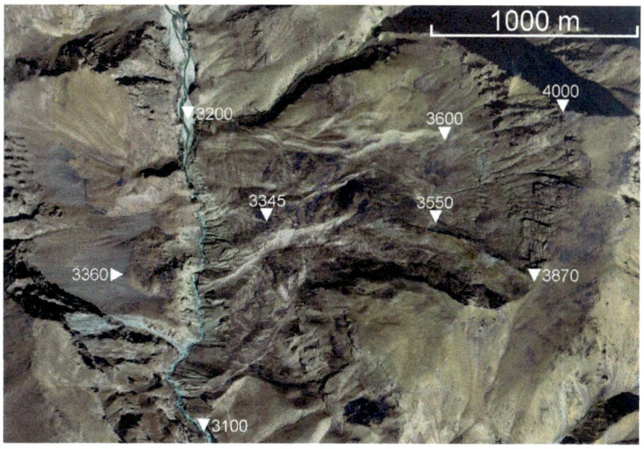

FIGURE 9.127 Largest rockslide that had blocked the Shulvadar River valley upstream from its confluence with the Dakhanzar River. ArcGIS Earth image.

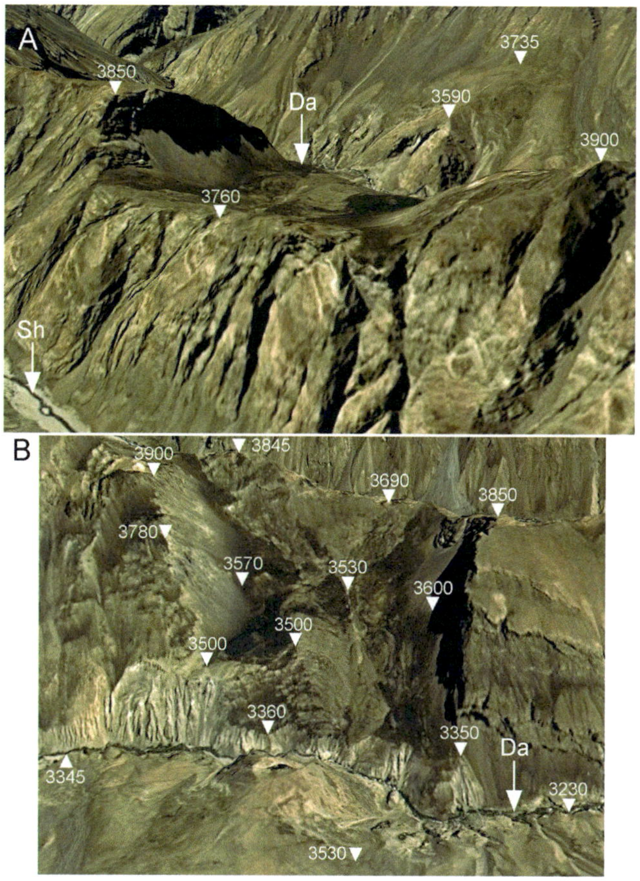

FIGURE 9.128 3D Google Earth views of the largest rockslide that had blocked the Dakhanzar River (Da) valley upstream from its confluence with the Shulvadar River (Sh). (A) Back view on the ridge destroyed by the headscarp; (B) frontal view.

headscarp and collapsed (Fig. 9.128A). This ridge is composed of the Archean gneiss with foliation dipping towards the Dakhanzar River valley (see Fig. 9.128B). Remnants of this rockslide dam on the opposite-right bank of the Dakhanzar River is not very well defined but it can be assumed that its front rose up to 3600 m a.s.l. at least (see Fig. 9.128A). The dam was breached, likely, after its first overtopping.

We mentioned above that outburst floods from both canyons occurred rather recently, but that the first one from the Shulvadar River valley preceded the second one from the Dakhanzar River valley (see inset in Fig. 9.124). It can be hypothesized that all these slope failures occurred simultaneously, likely being triggered by large earthquake, but smaller dams in the Shulvadar River valley had been overtopped and breached in a shorter time than much larger blockage in the Dakhanzar River valley so that the resultant outburst floods had followed each other in a rather quick succession.

Further south, the left-bank slopes of the Sanglech River valley and mouths of its left tributaries are affected by numerous large complex rockslides at a distance of about 20 km at least (from 36.41° to 36.24° N, Fig. 9.129). Unlike most of other bedrock landslides described above, these slope failures (Afghan 002, 002a, 002b, 031, 044) most likely develop gradually, though some sliding episodes could occur catastrophically resulting in temporal river damming, that can be seen on the right inset in Fig. 9.129 downstream from the elevation mark 2960 m a.s.l. Largest of these complex multistage landslides occupy slopes up to

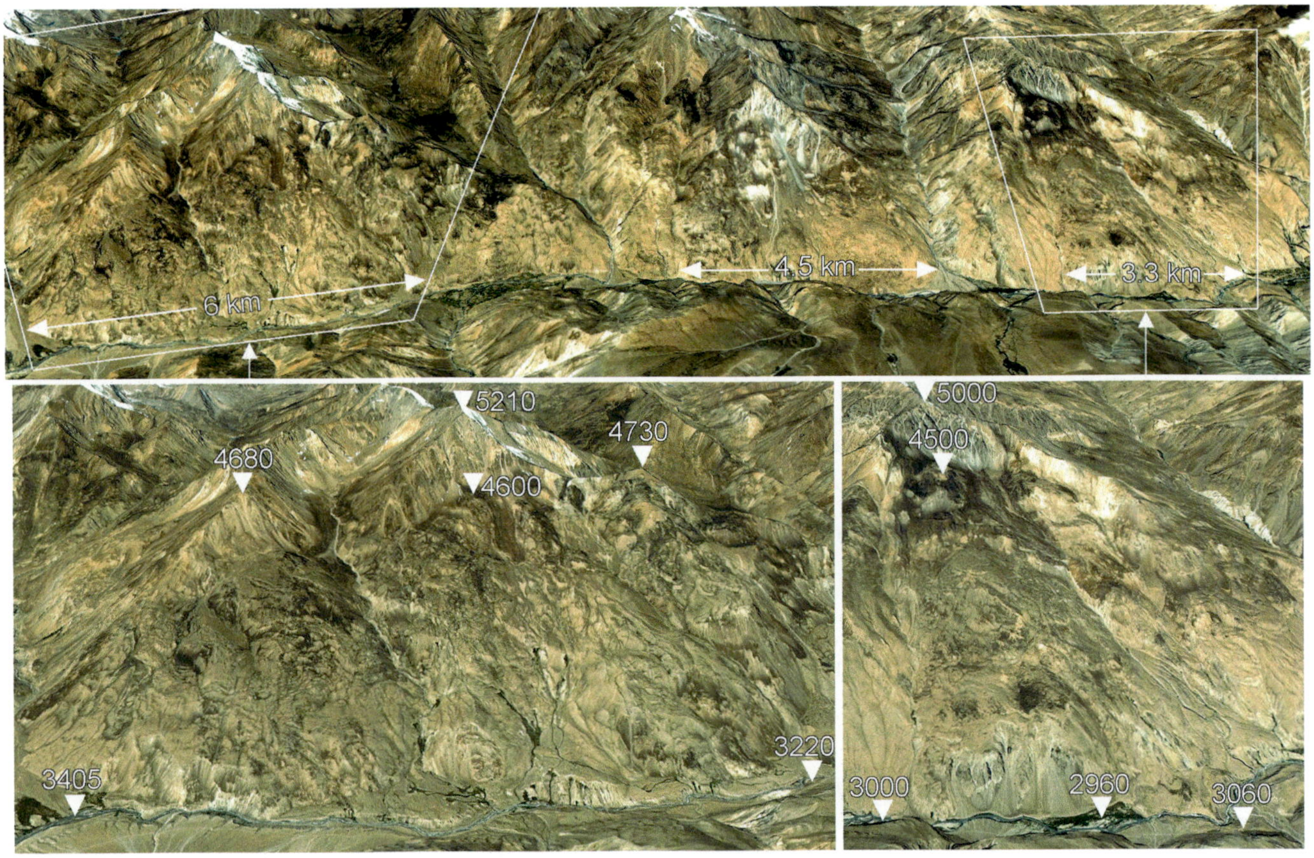

FIGURE 9.129 Large multistage landslides at the left side of the Sanglech River valley. Left inset: Afghan 031; right inset: Afghan 002. 3D Google Earth views.

ca. 2000 m high with an area up to 11–22 km^2 and their volumes could exceed 1 and even 2 km^3.

One rockslide from this cluster, at 36.319° N, 71.134° E (Afghan 003), demonstrates evidence of the catastrophic emplacement. It is a typical flow-like rock avalanche that collapsed from the ca. 800 m high slope and left triangular headscarp about 200 m deep if not more (Fig. 9.130). Large part of its body, at least till 3970 m a.s.l. is formed by the concentric belts of debris with numerous conical hills up to 20–30 m in diameter, and differ significantly from the deposits that fill the valley bottom downstream, up to 3570 m a.s.l. We cannot prove with confidence that the latter is the distal part of the 5.13 km long rock avalanche and cannot exclude glacial origin of the deposits that fill the valley bottom downstream from the elevation mark 3970 m a.s.l. More grounded interpretation requires field inspection of the site. However, we hypothesize that deposits between 3970 and 3570 m a.s.l. were left by rock avalanche and, probably, are the glacial or ice glacier deposits mobilized by the gravitational collapse of about 70 Mm^3 of the Archean metamorphic rocks.

One more similarly oriented rockslide (Afghan 043) destroyed rocky prominence about 1100 m high composed of the Archean metamorphic rocks 8 km south from the abovementioned site at 36.25° N, 71.104° E (Fig. 9.131). Main part of the collapsed rock mass had stopped at the foot of the prominence at ca. 3730 m a.s.l., while part of it, starting from the secondary scar at 4480 m a.s.l., reached the valley and formed a low temporary dam between elevation marks 3720 and 3690 m a.s.l.

The Sanglech River valley inherits fault zone that divides the Archean metamorphic rocks forming western side of the valley from the Proterozoic granite and granodiorite of its eastern side. We must point out that, unlike the left side almost totally affected by numerous very large rockslides, right side of the Sanglech River valley is free of such features except a group of several rockslides and DSGSDs in the lower reaches of the Rastdara River and of the adjacent right tributaries of the Sanglech River. Three of them (Afghan 059, 061, and 062) blocked the Rastdara River mouth at 36.214° N, 71.166° E and their bodies merge forming 20–40 m high blockage up to 3.5 km wide (Fig. 9.132). Two failures occurred on the 500–600 m high left-bank slope of the valley and one on its ca. 300 m high right-bank slope. It is difficult to conclude,

FIGURE 9.130 3D Google Earth view of rock avalanche in the left tributary valley of the Sanglech River. Plan view of the deposits with evidence of the flow-like motion is shown in the inset.

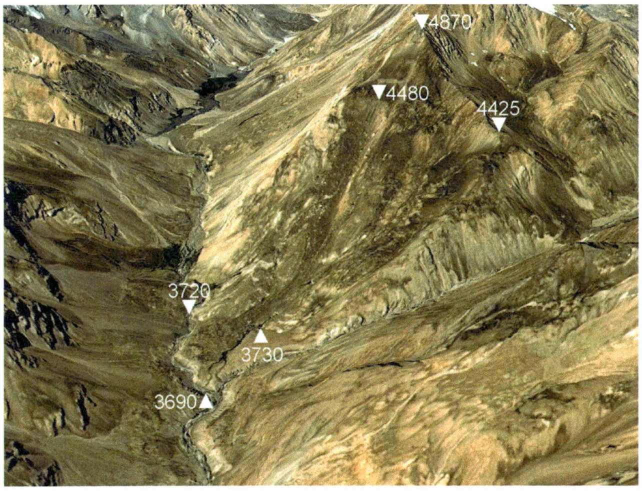

FIGURE 9.131 Rockslide at the rocky prominence accompanied by the secondary rock avalanche that caused temporal river damming. 3D Google Earth view.

FIGURE 9.132 The Rastdara River mouth blocked by merged bodies of three rockslides that caved from both sides of the valley. Headscarp crowns are at 3800, 4110, and 4160 m a.s.l. 3D Google Earth view.

based on the remote sensing data only, if their deposits just merge resting aside, as those of the Bielogorka rockslides (see Fig. 8.99) or of the rockslides that form the Ikhnachkul Lakes (see Figs. 8.226 and 8.227), that would indicate their simultaneous origin, or overlap each other as the Inylchek and Atdjailau rock avalanches (see Fig. 8.45) being formed successively. We should notice, however, that even such overlapping could be caused not by the asynchronical origin of rockslides, but by difference in their runout as in the case of the Khushtarita Western and Eastern rock avalanches (see Fig. 8.268). If all these rockslides really had occurred simultaneously, it would be a sound argument in favor of their seismic triggering.

Besides these adjacent rock slope failures, one rockslide (Afghan 060) was identified at 36.219° N, 71.187° E, about 1.5 km north-east of them on the right-bank slope of the valley that falls in the Rastdara River from the right hand; one more (Afghan 063) on the left-bank slope of the adjacent valley at 36.193° N, 71.143° E. and the last one (Afghan 064) blocks the river 5 km upstream at 36.16° N, 71.196° E forming a small dammed lake.

Apart from well-developed rockslides, about 2.2 km^2 of the 550 m high right-bank slope of the Sanglech River valley downstream from the Rastdara River mouth (36.24° N, 71.174° E) is affected by multiple upslope-facing scarps reflecting the deep-seated gravitational slope deformation (Afghan 065, Fig. 9.133) and, thus, possibility of the future large-scale rock slope failure. Small rockslides at the southern side of this DSGSD are marked by RS on Fig. 9.133. Evidence of the ongoing slope deformations can be also seen near the headscarp of the rockslide with the crown at 3800 m a.s.l. (see Figs. 9.132 and 9.133).

Further south along the same fault zone, 14 km south from the Munjan pass, at 36.046° N, 70.968° E, near the Maghnsavol village, the Dura River valley, right tributary of the Kokcha River, has been blocked by the fan-shape Maghnsavol rockslide dam up to 180 m high (Afghan 039). This slope failure originated from the funnel-shape headscarp at the upper half of the 1 km high right-bank slope of the valley (Fig. 9.134) that resembles at some extent the headscarp of the Karakystiak rock avalanche in Southern Kazakhstan (see Fig. 6.7). Headscarp is located within the tectonic zone where the Archean metamorphic rocks interchange with granites, Permian, and Triassic metasediments. Judging from the thalweg profile, this 2.4 km wide rockslide dam was incised almost completely practically up to the original river level. We must notice that some watersheds in the vicinity of the Maghnsavol rockslide are affected by DSGSDs, one of which, about 2 km east of the headscarp, is shown on the inset in Fig. 9.134.

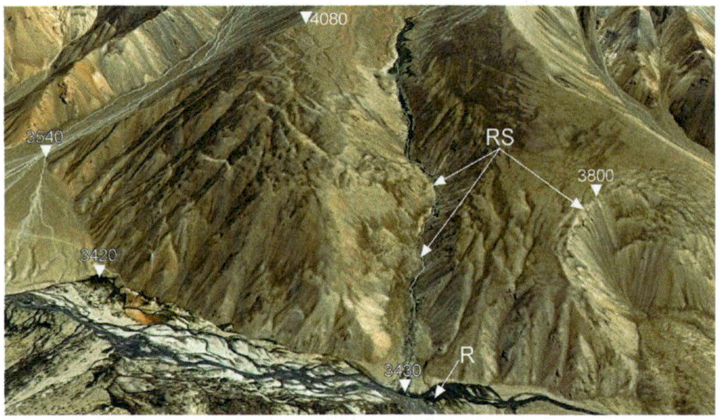

FIGURE 9.133 DSGSD downstream from the Rastdara River mouth (R). Arrows with RS indicate well-developed rockslides. 3D Google Earth view.

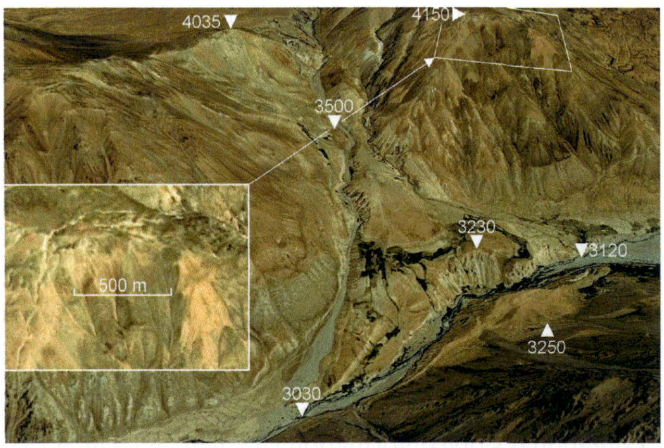

FIGURE 9.134 3D Google Earth view of the breached Maghnsavol rockslide dam in the Dura River valley. Plan view of the DSGSD east of it is shown in the inset.

The tremendous complex Pazhuk rockslide (Afghan 034) about 3 km^3 in volume had blocked the Munjan River valley at 36.02° N, 70.866° E, just upstream from its confluence with the Sakhi River valley (Fig. 9.135). It originated on a more than 1500 m high slope with complex geological structure composed of granites and metamorphic rocks and left 6.2 km wide and more than 500 m deep headscarp triangular in plan view with evidence of the subsequent failures from its 500–700 m high back wall (see, e.g., left part of Fig. 9.135 between elevation marks 3300 and 2720 m a.s.l.). Besides, up to kilometer-long blocks slumped down for several tens of meters (near the elevation mark 3405 m a.s.l.), thus retaining possibility of future failure and posing threat to local infrastructure. Horizontal "depth" of this "bite" exceeds 2.5 km; thus, this feature is larger than the headscarp of the 1911 Usoi rockslide (see Fig. 5.7). Failure affected the watershed so that gullies on the slope facing towards the Sakhi River have been left beheaded (see inset in

FIGURE 9.135 The Pazhuk rockslide. (A) 3D Google Earth view; (B) plan view. Beheaded gullies at the opposite side of the ridge between 4200 and 3760 m a.s.l. can be seen in the inset. Pa: the Pazhuk pass; Sa: the Sakhi River; Du: the Dura River; Mu: the Munjan River.

Fig. 9.135). Rockslide can be classified as the gigantic wedge failure. It recalls similar-shape Damamatsay rockslide in the Tanymas River valley (see Fig. 9.36) even more than the Usoi rockslide, whose source zone is much narrower.

Surface of rockslide body gradually decreased towards the opposite side of the blocked valley, down to 2950 m a.s.l. so that it was incised along its distal edge. Rockslide overlays the ~80 m high river terrace (at 2700 m a.s.l.) though it is not clear either it is the main body or the additional rockslide that collapsed from the headscarp with its crown at 3300 m a.s.l. Besides, there is one more step above the elevation mark 2740 m a.s.l. that probably bounds the main rockslide body from the western side. Such assumption is supported by the presence of whitish carapace blanketing rockslide body at the upper side of this step and absent in its footwall (see Fig. 9.135A near the elevation mark 2860 m a.s.l.).

The upstream side of rockslide spread for about 1.0–1.5 km into the Dura and the Munjan River valleys where it was smoothed and partially reworked by the impounded lake (see Fig. 9.135B). Judging from the thalweg profiles and shape of the formerly dammed valleys upstream, it can be assumed that about 100 m of rockslide body has not been incised yet. It is evident that this outstanding feature requires a detailed field study.

Couple of rockslides on both sides of the narrow (2.2–2.3 km wide only) ridge composed of the Archean metamorphic rocks with well pronounced bedding bounded by parallel glacial valleys up to 1100–1500 m deep with steep walls sloping up to 50 degrees were found about 15 km north of the gigantic Pazhuk rockslide (Afghan 032 at 36.165° N, 70.929° E and Afghan 033 at 36.176° N, 70.898° E). Both originated on the steepest parts of the slopes practically at the same altitude, about 5000 ± 100 m a.s.l., and formed the dams. The western one has not been breached and 3 km long lake is still impounded upstream, while the eastern one was partially filled by outwash from the nearby glaciers that terminates at 4900 m a.s.l. and breached at its frontal edge (Fig. 9.136), still forming a nick point in the river's profile. The latter's body about 10 Mm^3 in volume can be confused with rock glacier but it could not originate on such a steep slope with a nearly vertical, well-defined headscarp visible in the upper inset in Fig. 9.136. We cannot exclude, however, gradual glacier-like motion of debris at the valley bottom after its collapse.

Western rockslide about 50 Mm^3 in volume collapsed from the central part of the ca. 1500 m high slope with bedding dipping towards the valley at a more gentle angle than slopes' inclination and left a headscarp with up to 500 high back wall (in the shadow on Fig. 9.136). It forms a ~210 m high dam that fills the valley bottom at a distance of about 2 km and is characterized by the distinct run-up of debris on the opposite valley slope (see lower inset in Fig. 9.136) that allows classifying it as a primary rock avalanche with frontal confinement (see Section 5.3.1 in Chapter 5). We must notice that DEM resolution does not allow accurate quantitative assessment of all geometric parameters of this blockage. Seeping water appears first 100–120 m below the lake level as bogs at the downstream side of the blockage and visible stream starts about 175 m below lake level at ca. 3830 m a.s.l.

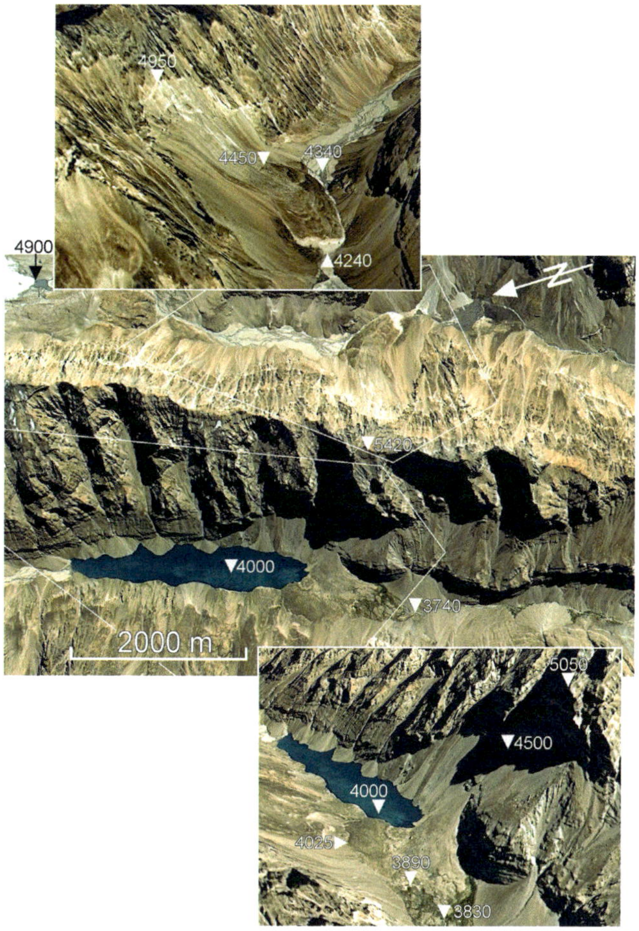

FIGURE 9.136 Google Earth image of rockslides on both sides of the narrow ridge and their 3D views shown in the insets.

If it would be found that both rock slope failures occurred simultaneously (which requires their dating, likely by cosmogenic isotopes method; Walker, 2005), it would be a sound argument in favor of their seismic triggering.

One more rockslide (Afghan 056), whose shape resembles that of the non-breached rockslide dam shown on Fig. 9.136, had blocked the Rakhoi River at

FIGURE 9.137 The Rakhoi Mouth rockslide. 3D Google Earth view.

FIGURE 9.138 Large-scale rockslide in the Rakhoi River basin, whose headscarp affected a large part of the ridge watershed (its assumed position is marked by dotted green line). 3D Google Earth view.

36.044° N, 70.641° E, about 5 km upstream of its confluence with the Yakhshindar River. Failure of about 20 Mm^3 of the Archean metamorphic rocks occurred from the ca. 1000 m high slope and left headscarp with up to 400 m high back wall with its crown at about 3730 m a.s.l. (Fig. 9.137). The Rakhoi Mouth rockslide body that fills the valley bottom is characterized by varicolored belts of debris similar to numerous other rockslides described above (see, e.g., Figs. 9.38, 9.66, and 9.92). Unlike the previous case study, this dam was breached and river forms 1.5 km long and ca. 220 m high rapids close to the base of the collapsed slope. The dam was eroded through its proximal part indicating primary style of debris motion (see Section 5.3.1 in Chapter 5). Small dammed lake in this steeply (more than 5 degrees) inclined valley was filled by sediments carried by numerous debris flows, likely shortly after its impoundment.

About 5 km west from the Rakhoi Mouth rockslide at 36.034° N, 70.6° E, ridge dividing the Rakhoi River basin from the river west of it, was affected by very large rockslide (Afghan 055) with an area of about 6.5 km^2 that destroyed not only part of the eastern slope between two ravines but also the watershed extending about 550 m over it so that the headscarp crown beheaded gullies on the western side of the ridge (Fig. 9.138). Rough estimate of its volume gives a value of ca. 1.0 km^3. The rockslide occurred at the site with complex geological structure: according to the 1:500 000 Geological map of Tajik SSR and adjacent territories, the Archean, Proterozoic, and Triassic metamorphic rocks intruded by the Paleogene diorites are crossed by several tectonic faults with small blocks of Neogene sedimentary rocks between them (Geological map, 1984).

Following one of the large faults that cross the site described above, the Panjshir fault (Wheeler et al., 2005) being inherited by the Yakhshindar River valley,

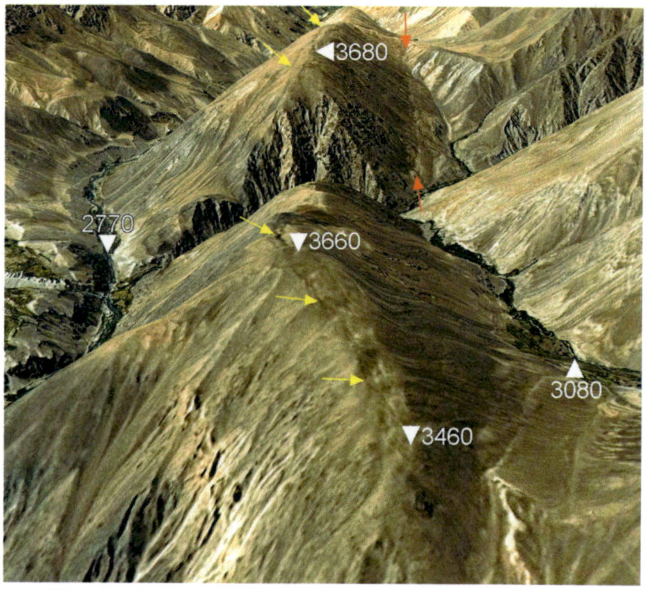

FIGURE 9.139 DSGSDs (marked by yellow arrows) on top of the ridges on the left bank of the Yakhshindar River. Visible trace of the tectonic fault is marked by red arrows. 3D Google Earth view towards south-west.

one can find evidence of the DSGSD on top of the local ridges at 35.99° N, 70.541° E and at 35.963° N, 70.506° E (Afghan 053 and 054) separated from the main mountain range by local valleys 400–500 m deep, and passing just along this fault zone (Fig. 9.139). Locally, there are visible traces of, likely, recent surface rupture along this fault, which are marked on Fig. 9.139. Scarps forming these DSGSD follow ridges' crests for 2.0–2.5 km. Their closeness to the large fault zone with evidence of recent tectonic activity allows assumption that deep-seated slope deformations could be triggered by strong earthquake(s).

The last group of bedrock landslides described in this chapter is located in the upper reaches of the Yakhshindar River valley. Its left-bank slope at a

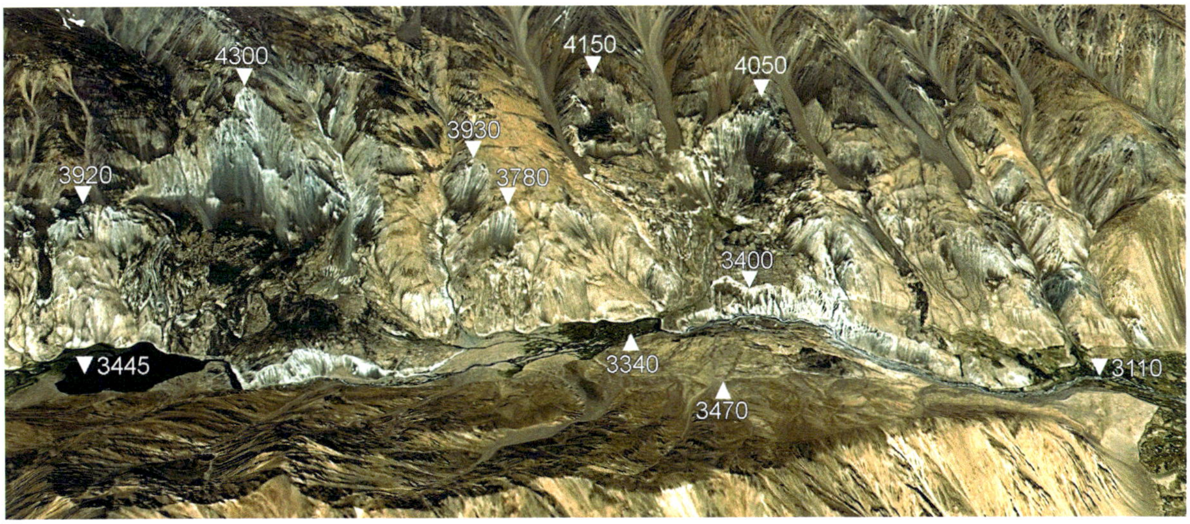

FIGURE 9.140 Rockslides in the upper reaches of the Yakhshindar River valley. 3D Google Earth view.

distance of about 9 km, from ca. 35.863° N, 70.345° E to 35.915° N, 70.415° E, is affected by multiple rockslides which source zones are located at the north-western side of the Panjshir fault zone in the Upper Proterozoic—Lower Proterozoic metasediments. Failures affected the lower part of the 1200-1500 m high valley slope and most of headscarp crowns are at altitudes from 3780 to 4300 m a.s.l. (Fig. 9.140). If simultaneity of these rockslides will be proved by dating methods it will be the sound argument in favor of their seismic origin.

Two largest of these rockslides, about 70 Mm^3 in volume each, blocked the river. The lake dammed by the upstream one (Afghan 049) still exist, despite partial breach of the blockage, while the downstream rockslide dam (Afghan 047) breached completely leaving the 160 m high nick point in the thalweg profile. The upstream river-damming rockslide formed the body with its surface gradually sloping towards the opposite bank, where the erosional canyon was formed. In contrast, the downstream one is characterized by 70–80 m high run-up, allowing its ascription to the Primary rock avalanches (see Section 5.3.1 in Chapter 5). Here breach occurred through the central part of the blockage.

Assuming simultaneous origin of both dams we can hypothesize that partial breach of the upper one led to overfilling of the lower lake and its complete breach. Similar chain effects likely took place in the Unkursai River valley in Central Tien Shan with partially breached Ak-Kiol rockslide dam (see Fig. 5.12) and completely breached Lower Ak-Kiol rockslide dam few kilometers downstream (see Fig. 8.166).

PART III

ANALYSIS

Numerous case studies described earlier that range in volume from slightly less than $1 \times 10^6\ m^3$ to ca. $10 \times 10^9\ m^3$ and in the affected areas from several hectares up to more than 100 km^2 provide excellent opportunity to analyze various aspects of rockslides origination and motion, using both statistical (correlation analysis) and empirical case-by-case approaches. Results of such analyses of the quantitative relationships between parameters characterizing bedrock landslides, of their structure and grain-size composition, emplacement mechanism, of seismically induced landslides and landslide dams, and of the evidence of future large-scale slope failures will be presented in the following chapters.

CHAPTER

10

Quantitative Relationships of Central Asian Bedrock Landslide Parameters

OUTLINE

Hereafter quantitative relationships between various parameters such as rockslide volume, area of the deposits, affected area, height drop, and runout characterizing bedrock landslides of the Central Asian region are presented. Though not all rockslides identified have been quantified at present, large number (more than 400–500) of case studies for which pairs of these parameters are available that allows rather representative assessment of such correlations both for the entire database and for different data sets sampled considering additional criteria such as confinement or rock avalanche type (see Section 5.3 in Chapter 5: Bedrock Landslide Types and Classification Systems). Similar analysis of the database that included about 200 terrestrial, lunar and martial rockslides was performed by P. Shaller in his PhD Thesis (1991). Larger Central Asian database allows compilation of more grounded and statistically representative relationships (Li et al., 2016).

We must point out that selection of case studies whose quantitative parameters have been measured up to now was random—we did not intend to do it for larger features or for rockslides of any special type, etc. It can be demonstrated by practically similar size–frequency distributions compiled for the entire data base (more than 950 cases) about 62% of which were quantified and for 251 case studies from Chinese and Kazakhstan parts of the study region that embrace ~90% of rockslides identified and mapped within this part of Central Asia (Fig. 10.1). Principles of compilation are described in the next section.

10.1 SIZE–FREQUENCY DISTRIBUTION OF BEDROCK LANDSLIDES

Size-frequency distribution is an important characteristics of any natural phenomena—earthquakes, floods, rainstorms, debris flows, rock falls, landslides, etc. (Hadjigeorgiou et al., 1996; Hovious et al., 1997; Pelletier et al., 1997; Stark and Hovious, 2001; Malamud, et al., 2004a, b; Turcotte and Malamud, 2004; Hovious and Stark, 2006; Guzzetti et al., 2008, 2012; Corominas and Moya, 2008; Chen et al., 2011; Korup et al., 2012; Li et al., 2012, 2014, 2016; Mavrouli and Corominas, 2015; Corominas et al., 2017). Rockslide size can be characterized either by area of the deposits, which could be measured with relatively high accuracy (except case studies significantly

Rockslides and Rock Avalanches of Central Asia.
DOI: https://doi.org/10.1016/B978-0-12-803204-6.00010-7

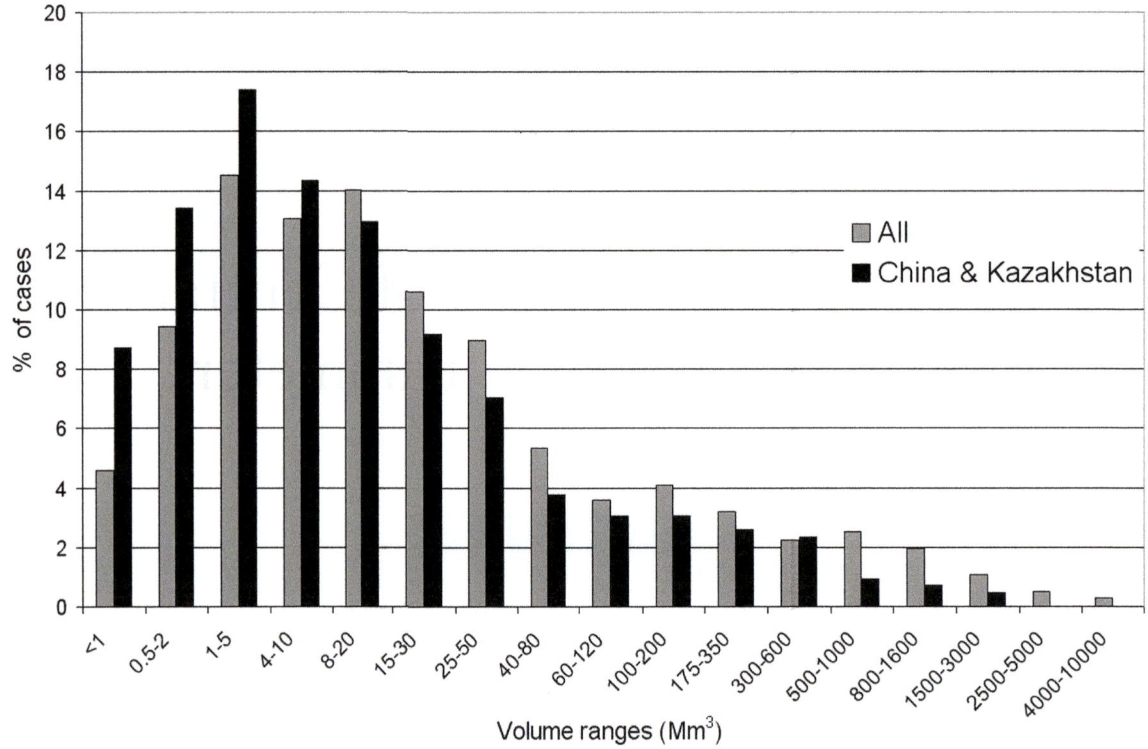

FIGURE 10.1 Histograms demonstrating size (volume)—frequency distributions for the entire Central Asian region (All—595 cases) and for its Chinese and Kazakh parts (251 cases).

reworked by subsequent erosion) or by the deposits volume whose assessment accuracy is much lower and possible errors of these estimates can be up to ±50% (see Section 6.3 in Chapter 6: The Bedrock Landslides Database), though relationships between these parameters are characterized by quite high correlation coefficient $R^2 = 0.8933$ (see Fig. 6.1).

As it was noticed in Corominas and Moya (2008), size (magnitude) versus cumulative frequency of the number of landslides generally has a power-law distribution with a rollover (the cumulative curve flattening) at small landslides (usually less than 10^4 m^2). Analysis of the database of large-scale rockslides, whose deposits' areas range from $\sim 10^4$ to $\sim 10^8$ m^2 shows that they can be characterized by similar distribution, with maximum within the area range of 10^5–3.5×10^6 m^2 (Fig. 10.2A) and with rollover starting at ca. 10^6 m^2 (Fig. 10.2B).

Considering higher accuracy of area measurement, we divided all data sets into nine complementary area ranges with similar intervals in the logarithmic scale (see Fig. 10.2), while for volumes, whose assessment accuracy is much lower (could be about ±50%; see Section 6.3), we used overlapping ranges (see Fig. 10.1). It allowed to obtain more smooth and, as we assume, more realistic distributions.

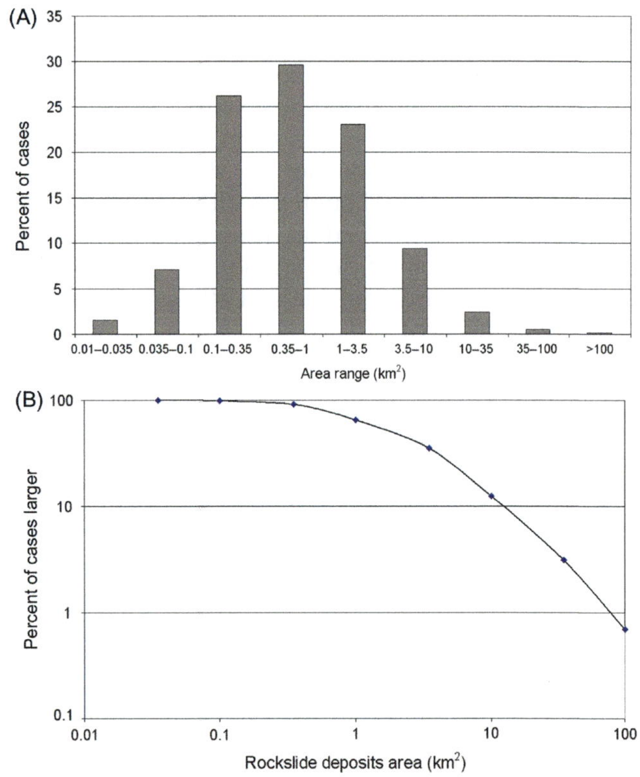

FIGURE 10.2 Size (area) – frequency distribution of rockslides (577 case studies); (A) histogram; (B) cumulative plot.

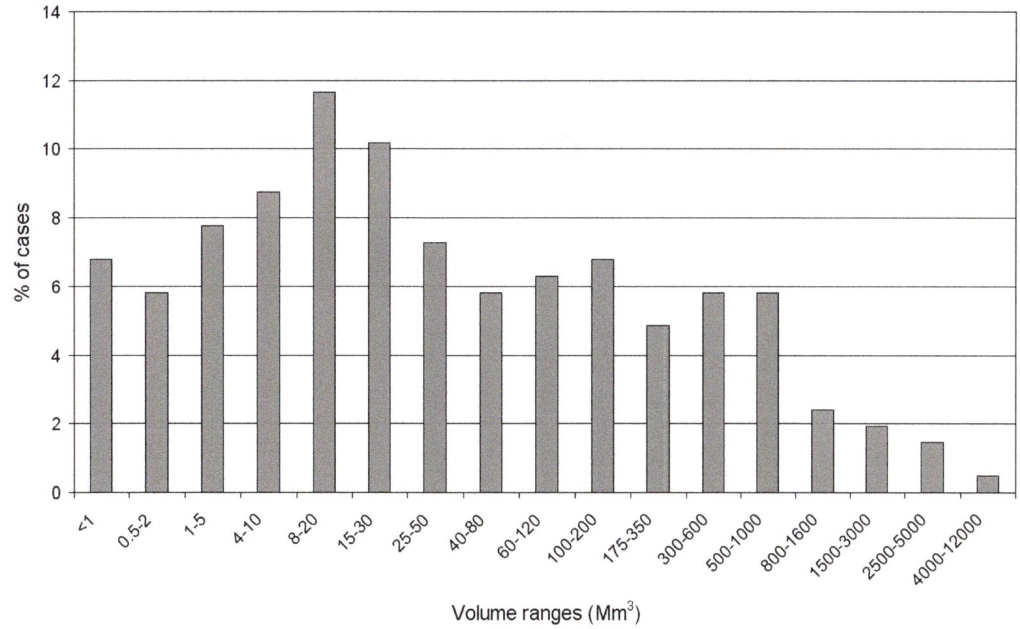

FIGURE 10.3 Size (volume)–frequency distribution of the large-scale Alpine bedrock landslides.

Comparison of size–frequency distributions of large-scale bedrock landslides of the Central Asian region with distribution of the 125 same-size features from the Alps based on the inventories from Heim (1932) and Abele (1974) shows that the latter has much less pronounced maximum and larger proportion of rockslides exceeding ca. 50 Mm^3 in volume (Fig. 10.3) (analyzing Alpine database, we used the same overlapping ranges as for the Central Asian database, see Fig. 10.1).

We expect that such difference in size–frequency distributions could depend on the difference between factors triggering rockslides' formation in both regions. In general, gravity force supporting slope failure (weight of the unstable rock mass) is proportional to the volume of the collapsing rock mass, in other words, to the third power of its linear dimensions, while force preventing landslide formation, i.e., friction force, is proportional to its area—to the second power of linear dimensions. It means that if all other conditions are equal, larger rockslides must occur more frequent than smaller ones. The Alpine rockslides' distribution reflects this regularity better than the Central Asian one.

It is obvious, however, that size–frequency distribution of the features in question is affected also by other factors whose influence can surpass the ratio of forces supporting and preventing rock slope failure. First of all, these are the relief ruggedness and depth of valleys incision—the higher and longer the slopes are, the larger rockslides can be formed on them. It can be demonstrated by the relationship between slope height (vertical distance from the headscarp crown and the lowermost deposits' level) and volume of rockslides that occurred in frontally confined conditions (298 case studies) for which effect of the down valley debris spreading that affects the former parameter (see Section 6.3) should be minimal and can be neglected (Fig. 10.4). This plot demonstrate distinct positive correlation, but, at the same time, significant scatter—up to three orders of volume for the slopes of given height. Special analysis of the distribution of slopes with different height and steepness is needed to quantify effects of these factors on rockslide size-frequency distribution.

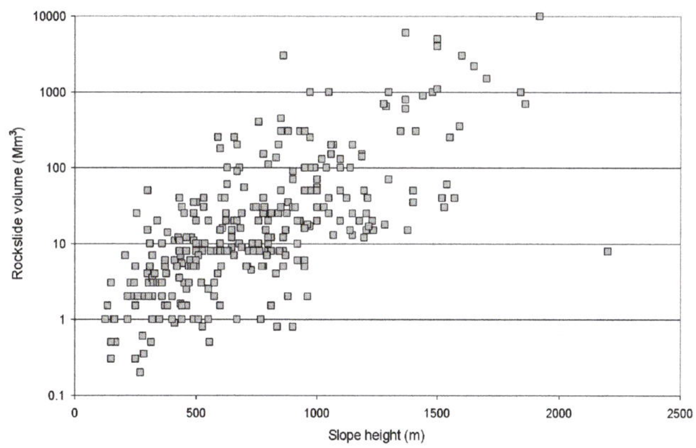

FIGURE 10.4 Relationship between slope height and rockslide volume for the Central Asian region.

Much more complex and hardly quantifiable are various geological factors—lithology, depth of weathering and unloading, tectonic structure, etc. One of the factors that might result in larger proportion of the largest rockslides in the Alps in comparison with Central Asia is much higher neotectonic and seismic activity of the latter region. Considering that most of large Central Asian bedrock landslides occurred in Late Pleistocene and Holocene, many of them, even in those parts of the region where no large historical earthquakes are known, could be triggered by large paleoearthquakes (see Section 3.4 in Chapter 3: Overview of the Study Region, and Chapter 13: Seismically Induced Rockslides: Their Discrimination). In the Alps, in contrast, most of such features are treated as being associated with deglaciation (Brückl and Heuberger, 2001; Poschinger et al., 2006; Poschinger, 2011; Agliardi et al., 2013). We can speculate that well-known effect of the topographic amplification (Puchkov, 1965, 1974; Bugaev et al., 1975; Geli et al., 1988; Pederesen et al., 1994; Assimaki and Gazetas, 2004; Bourdeau and Havenith, 2008; Meunier et al., 2008; Buech et al., 2010; Havenith et al., 2015a) can support destabilizing of the topmost parts of high slopes before their purely gravitational long-term evolution would result in the formation of larger failures that would affect the entire slopes.

One more factor possibly influencing the difference in size–frequency rockslide distributions between Alps and Central Asia is, generally, more intense (relative to the entire areas of these mountain systems) past glaciation of the Alps. It affected practically all mountain valleys up to their mouths and resulted in larger proportion of very steep slopes of trough glacial valleys than in Central Asia where many valleys have been formed either by water erosion or by neotectonic processes (see Chapter 3). However, quantitative assessment of the role of these factors requires more studies.

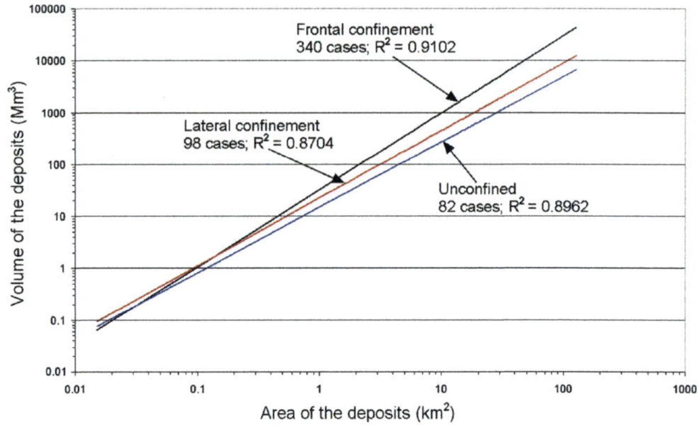

FIGURE 10.5 Power-law charts of the ratio between deposits' volume and mapped area compiled separately for rockslides with frontal and lateral confinements and in unconfined conditions and their correlation coefficients.

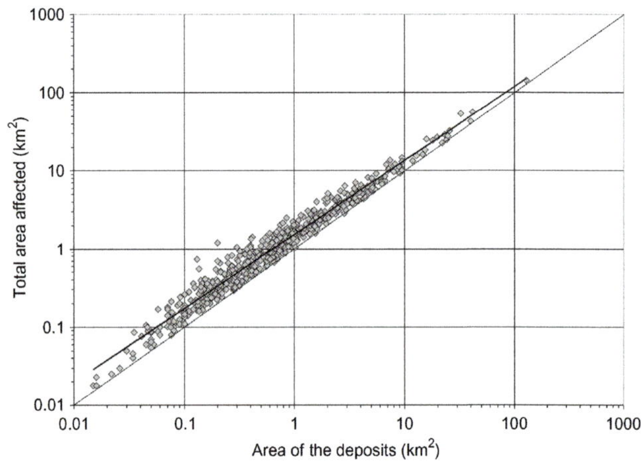

FIGURE 10.6 Relationship between areas of debris accumulation and the entire rockslide affected area.

10.2 AREA–VOLUME RELATIONSHIPS

As it was mentioned already in Chapter 6, where principles of the bedrock landslide database compilation are described, relationship between rockslide deposits areas and their volumes is characterized by good correlation ($R^2 = 0.8933$ for the sample of 551 cases), but, at the same time, by the significant—about one order of magnitude—scatter of the estimated deposits' volume for almost entire area range (see Fig. 6.1). Distributions compiled separately for rockslides emplaced with frontal and with lateral confinements and in unconfined conditions have nearly similar scatters too and correlation coefficients ranging from ca. 0.87 to 0.91. Power-law charts compiled for these samples (Fig. 10.5) show similarity for smallest failures with areas up to ca. 10^5 m^2 and volumes up to 10^6 m^3 and significant (up to five times on average) difference of the latter values for largest rockslides.

We also analyzed the relationship between the area of rockslide deposits (A_{dep}) and the entire area affected by rockslide (A_{total}) that includes its source zone, the travel path often marked by trimline, and the deposition zone. Such analysis performed for the data set of 545 case studies shows that the total area affected is generally less than 2.5 times larger than the area of the deposits within the entire deposits area range (both values are measured in plan view) though in few cases they differ up to three to four and even up to six times (Fig. 10.6). Usually such a significant difference characterizes compact blockages at the feet of high slopes with distinct frontal confinement preventing debris spreading. Lower envelope of this distribution corresponds to the slow-moving bedrock landslides "hanging" on the valley slopes, whose deposits' area is almost equal to the totally affected one. Relationship of these parameters can be described by the power-law chart $A_{total} = 1.5261 \times A_{dep}^{0.9462}$ with high correlation coefficient ($R^2 = 0.9636$).

10.3 PARAMETERS CHARACTERIZING ROCKSLIDE AND ROCK AVALANCHE MOBILITY

Following the pioneering work of Sheidegger (1973), most of the researchers used horizontal projection of runout distance measured from the headscarp crown to the deposits tip (L) and ratio of vertical difference between these points (height drop—H) to L (angle of reach or fahrboschung) as a measure of rockslide mobility (Hsü, 1975; Davies, 1982; Li, 1983; Shaller, 1991; Corominas, 1996; Kilburn and Sorrencen, 1998; Legros, 2002). All these studies demonstrated increase of the runout (L) and decrease of the H/L ratio with the increase of rockslide volume. Similar relationships were found for the Central Asian rockslides, both for the entire database and for highly mobile rockslides that have been classified as secondary rock avalanches (see Section 5.3.3 in Chapter 5).

10.3.1 Overall Mobility

Dependence of runout from rockslide volume (Fig. 10.7) can be characterized by power-law function ($L = 0.803 \times V^{0.2929}$) with the significant—up to half order of magnitude— scatter and correlation coefficient $R^2 = 0.7126$. Interesting results were obtained when we compared such relationships for rockslides with frontal confinement (301 cases), lateral confinement (135 cases),

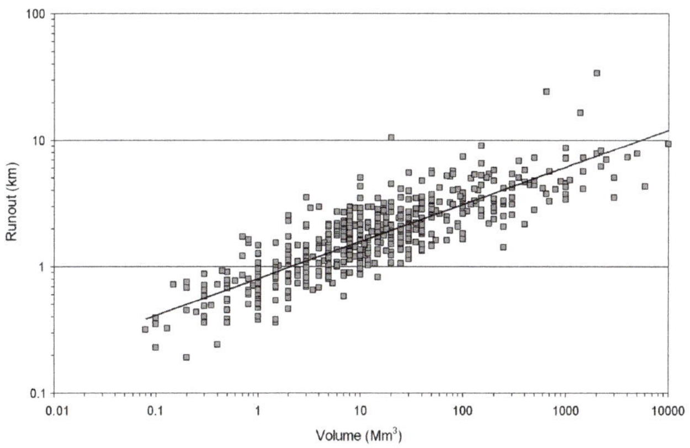

FIGURE 10.7 Relationship between runout and rockslide volume for the entire database available (536 cases).

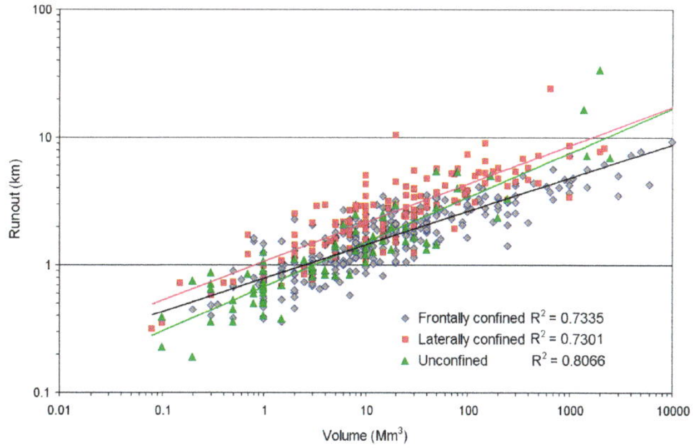

FIGURE 10.8 Relationships between runout and rockslide volume for rockslides with different confinement. Corresponding R^2 values are added to the legend.

and without any pronounced confinement (78 cases) separately (Fig. 10.8). Rockslides with distinct lateral confinement demonstrate significantly higher, about 1.5 times on average, mobility than rockslides with frontal confinement (which is obvious) and even than rockslides that moved over unconfined surfaces. We hypothesize that such an excessive mobility of rockslides and rock avalanches that move along relatively narrow valleys could be associated with larger proportion of secondary rock avalanches (see Section 5.3.3). It will be discussed further in Section 10.3.2 and in Chapter 12, Morphological and Structural Evidence of Rockslides' (Rock Avalanches') Motion Mechanism(s).

Similar analysis was performed to compare rockslide volume with angle of reach. Here we always use elevation difference between headscarp crown and the deposits' tip (see Fig. 5.34). For laterally confined and unconfined features, this value is equal to the elevation difference between the crown and the lowermost point of the deposits (see Section 6.3), but for frontally confined cases they can vary significantly—up to several hundred meters.

These relationships are characterized by much larger scatter—H/L ratio varies from 0.091 to 0.98 that corresponds to the angle of reach between 5.2 and 44 degrees (Fig. 10.9). Scatter decreases with volume growth. Power-law correlation coefficient for the entire database (515 cases) is 0.3524 only. Surprisingly, division of the database considering the confinement type (289 cases with frontal confinement, 134 cases with lateral, and 74 unconfined cases) does not lead to significant improvement of correlations (Fig. 10.10) though power-law correlation coefficient for laterally confined sample increases up to 0.4497 and up to 0.3962 for unconfined, which is, nevertheless, much lower than

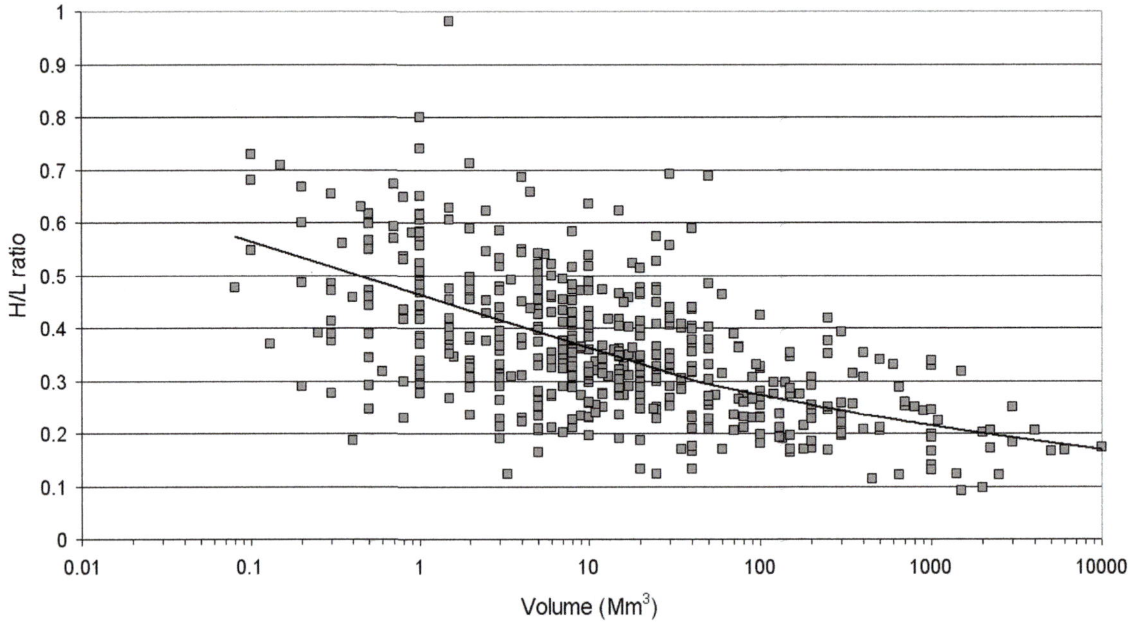

FIGURE 10.9 Relationship between H/L ratio and rockslide volume for the entire database available (536 cases).

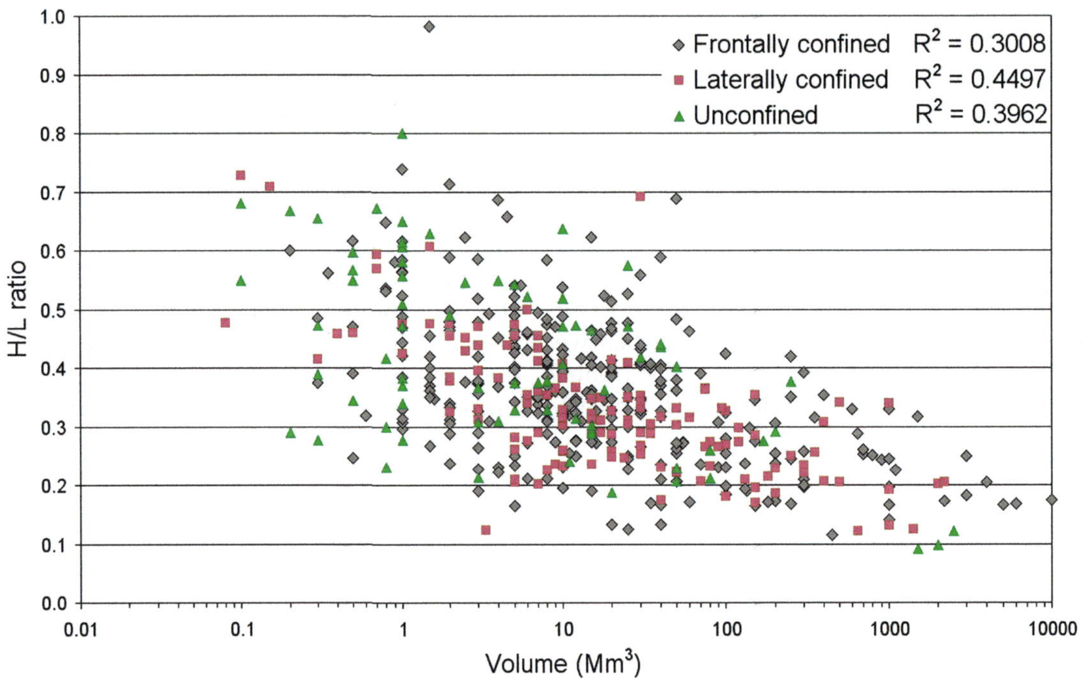

FIGURE 10.10 Relationships between H/L ratio and rockslide volume for rockslides with different confinement. Corresponding R^2 values are added to the legend.

for volume–runout correlations. It supports the conclusion made by Legros (2002) that H/L ratio has no physical meaning and that its decrease with rockslide volume growth is just qualitative.

We also analyzed the relationship between maximal height drop (H_{max}; see Section 6.3) and runout. It is interesting that such correlations appeared to be almost similar for frontally confined features (297 cases) and for unconfined features (74 cases) with correlation coefficient about 0.75 (Fig. 10.11), while rockslides with lateral confinement (68 cases) have much larger scatter with very weak positive correlation.

Similar distributions with slightly higher correlation coefficients were found for the relationships between runout and product of rockslide volume and maximal height drop for the same sample (Fig. 10.12).

It can be explained as follows. Both parameters (height drop and its product with volume) are somehow proportional to the potential energy, though, of course, reflect it with unpredictable error, because the position of the center of gravity and real distribution of the rock mass within the source zone in each particular case are unknown. Runout of the unconfined and frontally confined cases is governed mainly by the kinetic energy gained after the initial descend, by basal and the internal friction and by the rise on the opposite slope (in the frontally confined conditions; see Fig. 5.34).

However, when rockslide moves along the valley with pronounced lateral confinement, its motion is affected by more factors (momentum transfer in case of the secondary rock avalanche formation; see Section 5.3.3), shape, steepness, and asperity of the valley walls, linearity or sinuosity of the travel path, etc., that results in much wider runout scatter typical of this group of rockslides.

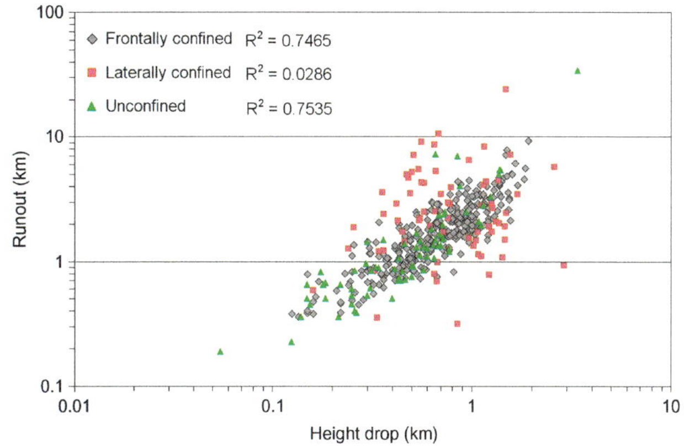

FIGURE 10.11 Relationships between runout and maximal height drop for rockslides with different confinement. Corresponding R^2 values are added to the legend.

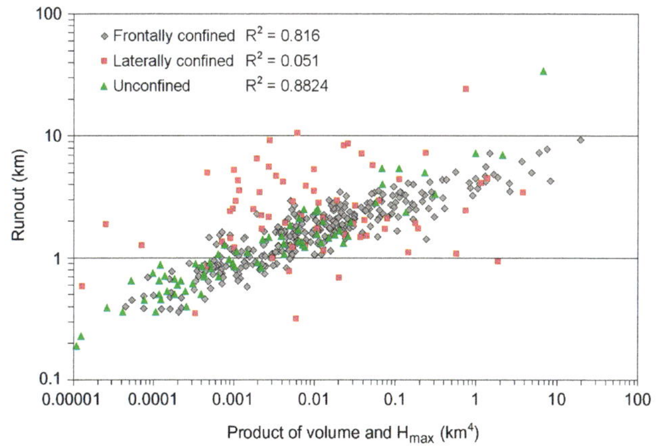

FIGURE 10.12 Relationships between runout and product of rockslide volume and maximal height drop for rockslides with different confinement. Corresponding R^2 values are added to the legend.

10.3.2 Secondary Rock Avalanches' Mobility

As we described in Section 5.3, excessive mobility is typical not only of the primary rock avalanches when it results either in long runout or in the significant run-up (see Fig. 5.34 and further discussion in Section 12.2.1 of Chapter 12) but also of the secondary and jumping rock avalanches with bipartite along-way distribution of the material (Strom, 1996, 2006, 2015b). Number of the distinctly identified jumping case studies is limited, but rock avalanches ascribed to the secondary type are rather widespread (about 60 distinctly identified cases) allowing some quantitative estimates of the effect of such style of motion on rockslide debris mobility.

To analyze it, we compiled plots relating runout of the avalanche-like parts of such features (see Section 6.4) to their volume (Fig. 10.13), height drop (Fig. 10.14), and to the product of these parameters (Fig. 10.15), and compared them with similar relationships revealed for the entire database (runout–volume

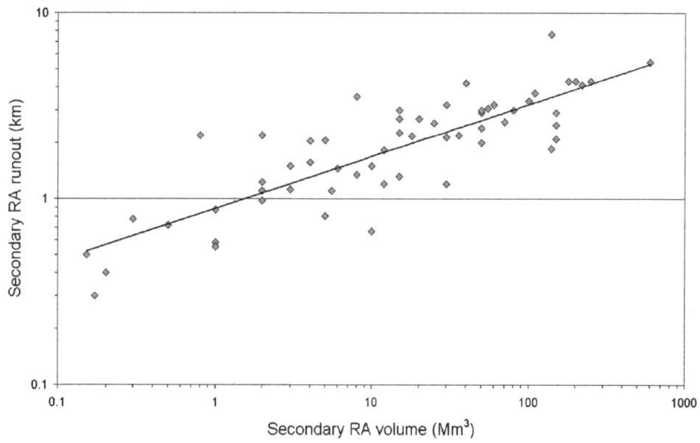

FIGURE 10.13 Relationship between runout and rockslide volume for secondary rock avalanches. $R^2 = 0.6995$ (60 cases).

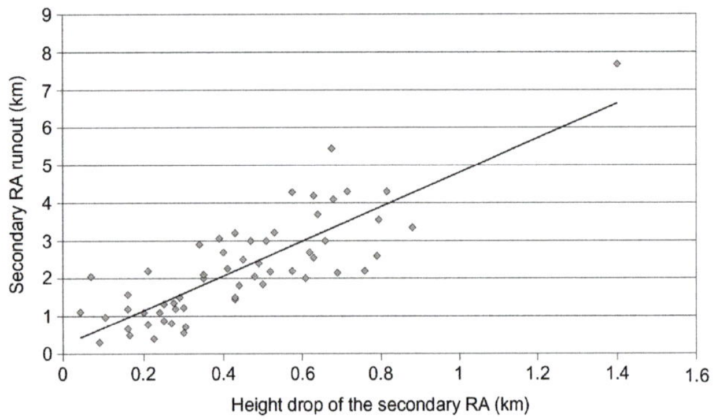

FIGURE 10.14 Relationship between runout and maximal height drop for secondary rock avalanches. $R^2 = 0.6825$ (58 cases).

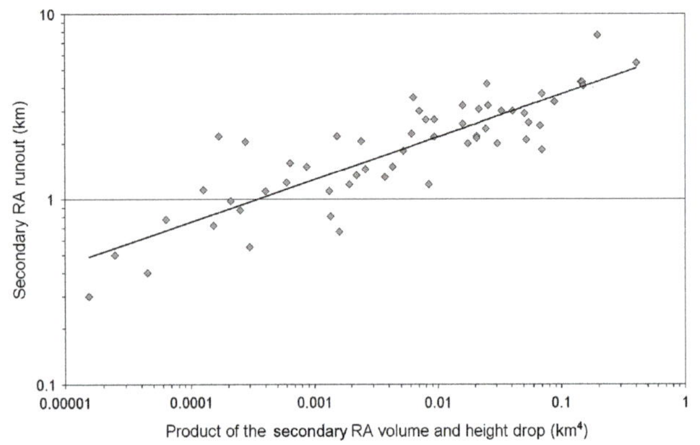

FIGURE 10.15 Relationship between runout and product of rockslide volume and maximal height drop for secondary rock avalanches. $R^2 = 0.7438$ (58 cases).

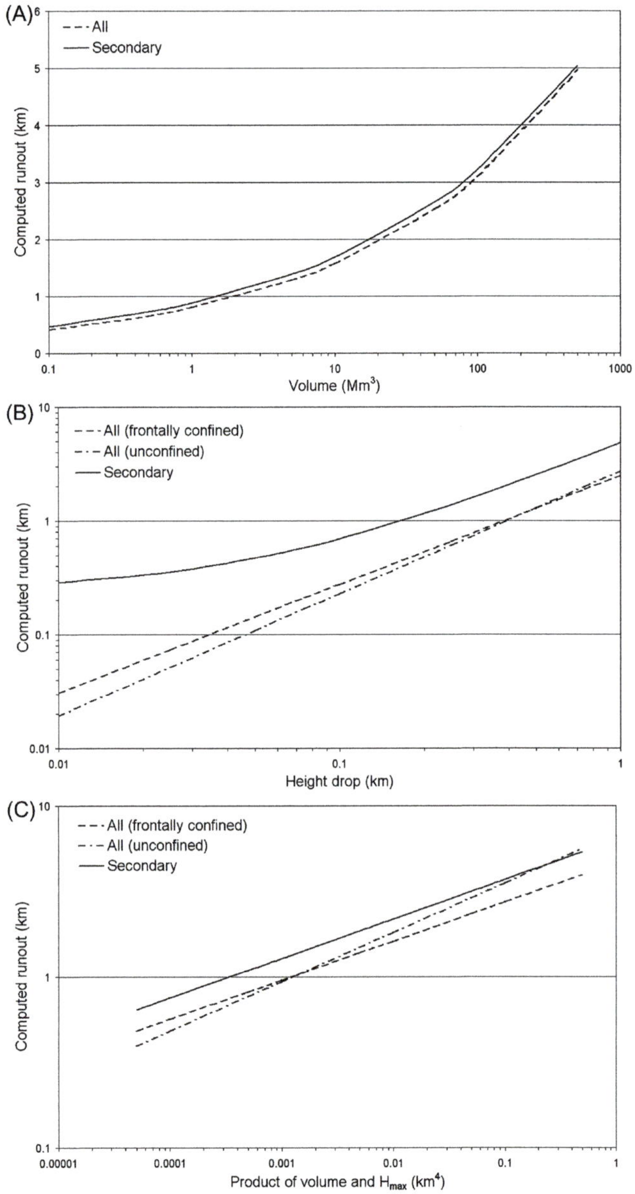

FIGURE 10.16 Relationships of the runout values (L) calculated according to regression equations for the entire database and for the secondary rock avalanches. (A) For the specified volume (V); (B) for the specified height drop (H); (C) for the specified V × H value.

dependency; see Fig. 10.7), for the frontally confined and unconfined case studies (runout – height drop dependency; see Fig. 10.11), and for the dependency of runout from the product of these parameters (see Fig. 10.12).

Most of the plots are in log–log form with power-law trend lines, except the relationship between secondary rock avalanche runout and height drop from the secondary scar crown to rock avalanche tip for which higher correlation coefficient was found for linear regression (see Fig. 10.14).

In all three cases and practically for the entire ranges of the specified volumes, height drops, and V × H values, secondary rock avalanches demonstrate higher mobility (Fig. 10.16 and Table 10.1), whose possible cause will be discussed hereafter in Section 12.2.3.

10.3.3 Area Affected: Additional Parameter Characterizing Rockslide Mobility

Runout (L) as well as its inverse ratio with height drop (H/L) are not the only parameters characterizing rockslide mobility. Debris moves not only forward as, for example, at the Chaartash-3 (see Figs. 5.61 and 8.130) or the Taheman (see Fig. 9.17) rock avalanches but also sidewise forming fan-shaped, as the Chukurchak (see Figs. 8.221 and 8.222) or the Khait

TABLE 10.1 Length Difference (in Percentage and in Kilometers) Between Runout Calculated for the Entire Database (L_{all}) or Samples Selected According to the Confinement and for the Secondary Rock Avalanches (L_{sec})

	Calculated runout difference			
Specified parameters	**Percentage ($L_{sec}/L_{all} \times 100$)**		**Kilometers ($L_{sec} - L_{all}$)**	
Volume (V, Mm^3). Entire database: $L = 0.803 \times V^{0.2929}$; Secondary RA: $L = 0.8877 \times V^{0.2793}$				
0.1	114		0.058	
0.5	112		0.076	
1	111		0.085	
5	108		0.105	
10	107		0.113	
50	105		0.122	
100	104		0.119	
500	102		0.079	
Height drop (H, km). Frontally confined (Entire database): $L = 2.4832 \times H^{0.9558}$; Unconfined (Entire database): $L = 2.7079 \times H^{1.0745}$; Secondary RA: $L = 4.5706 \times H + 0.2401$				
H, km	Frontally confined	Unconfined	Frontally confined	Unconfined
0.01	939	1487	0.256	0.267
0.025	485	689	0.282	0.303
0.05	331	433	0.327	0.360
0.075	279	348	0.374	0.415
0.1	254	306	0.422	0.469
0.25	209	226	0.723	0.772
0.5	197	196	1.245	1.240
0.75	194	185	1.782	1.680
1	194	178	2.328	2.103
$V \times H$ (km^4). Frontally confined (Entire database): $L = 4.6078 \times (V \times H)^{0.2282}$; Unconfined (Entire database): $L = 6.8404 \times (V \times H)^{0.2885}$ Secondary RA: $L = 6.2637 \times (V \times H)^{0.23}$				
$V \times H$ (km^4)	Frontally confined	Unconfined	Frontally confined	Unconfined
0.00005	134	163	0.161	0.249
0.0001	134	157	0.190	0.273
0.0005	134	143	0.277	0.327
0.001	134	137	0.326	0.347
0.005	135	125	0.476	0.368
0.01	135	120	0.561	0.360
0.05	135	109	0.819	0.263
0.1	135	105	0.964	0.168
0.5	136	95	1.407	-0.260

(see Figs. 8.271 and 8.272) rock avalanches, or the pancake-shape bodies, such as rock avalanche in the Kash River valley (see Fig. 5.33). In the latter cases, even if rockslide debris would not spread as far as when it moves straightly forward, it could affect larger area, thus increasing the exposure—one of the basic parameters used for landslide risk assessment (Corominas et al., 2015). That is why correct assessment of the area that could be affected by rockslide seems to be even more important than such a unidimensional parameter as runout. However, statistical relationships of the area affected with rockslide volume and height drop were analyzed quite rarely (Li, 1983).

We must notice that two area estimates are included in the database (see Section 6.3)—area of the deposits (A_{dep}) and total area affected by rockslide (A_{total}) that includes the source, the transition, and the deposition zones. Earlier, we analyzed the relationship between the deposits volume and the deposits area (see Fig. 10.5). However, here it seems more logical to use the total area affected because runout length can be treated as its axial dimension and, thus, comparison of these values and of their relationships with other parameters should have more reliable physical meaning. In particular, the relationships between A_{total} and runout (Fig. 10.17) are surprisingly consistent for rockslides that moved in frontally confined (339 cases) and in unconfined (79 cases) conditions (coefficients in the corresponding power-law equations differ insignificantly: $A_{total} = 0.4134 \times L^{1.7764}$ and $A_{total} = 0.4222 \times L^{1.7644}$). Similar equation for 92 laterally confined rockslides ($A_{total} = 0.2238 \times L^{1.7594}$) returns two times smaller A_{total} estimates practically within the entire runout range.

Areas of the source zones must be independent of the confinement style, being governed by slopes' height, steepness, geological structure of the host massif and, probably, by the triggering factors. Thus, such consistence means that frontally confined rockslides with prevailing up- and down-valley spreading of debris are as mobile as unconfined rock avalanches freely moving forward and, in some cases, sidewise (if considering A_{total} as the mobility measure), while impossibility of the transverse spreading typical of the laterally confined features decreases the deposits area 2.4 times on average, considering that A_{total} is ~1.2 larger than A_{dep} according to the chart shown on Fig. 10.6.

Relationship between the area affected and rockslide volume for the entire database (Fig. 10.18A) can be approximated by the power-law distribution ($A_{total} = 0.1965 \times V^{0.6115}$) with scatter of about half order of area magnitude and correlation coefficient ($R^2 = 0.8993$). Similar analysis demonstrates higher mobility of unconfined rockslides whose volume exceeds ca. 10 Mm^3 (Fig. 10.18B). It differs from the relationship between runout and volume of the collapsed rock mass (compare with Figs. 10.7 and 10.8) that clearly shows larger runout of the laterally confined features.

Similar to the H/L ratio and its relation with rockslide volume (see Figs. 10.9 and 10.10), we analyzed relationship of the H/A_{total} ratio with volume. This chart looks more informative, being plotted in log–log scale (Fig. 10.19), and correlation coefficients of the corresponding relationships are 1.93–2.66 times larger than coefficients of the H/L vs volume relationships (compare Figs. 10.19 and 10.10).

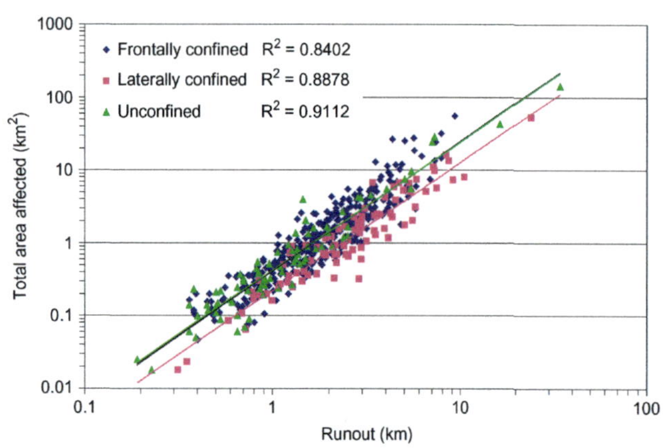

FIGURE 10.17 Relationships between total area affected and runout for rockslide samples with different confinement.

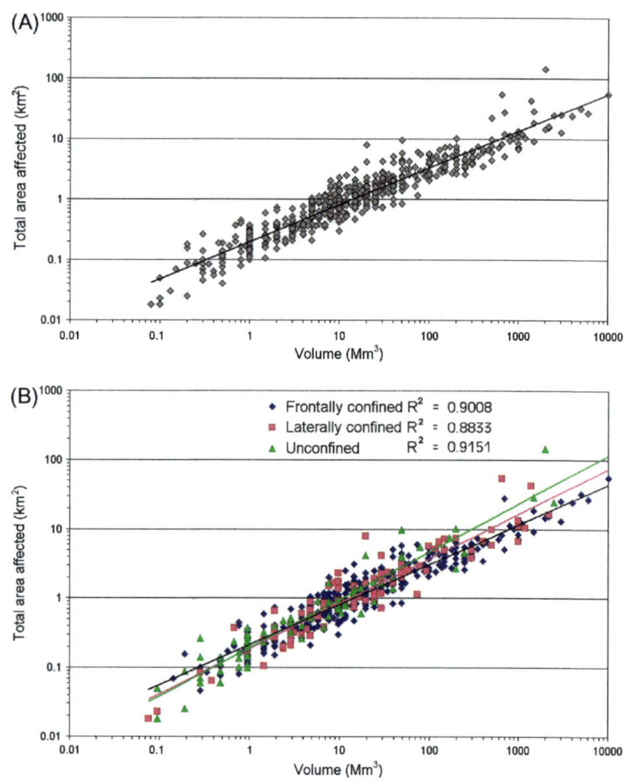

FIGURE 10.18 Relationships between total area affected and rockslide volume. (A) For the entire database (536 cases); (B) For rockslides with frontal confinement (337 cases), lateral confinement (94 cases), and in unconfined conditions (79 cases).

Following the analysis performed for the runout, we compared the total area affected and maximal height drop for rockslides with different confinement (Fig. 10.20). Correlation coefficients for frontally confined (334 cases) and unconfined (76 cases) rockslides are slightly less than for the relationship between maximal height drop and runout, whereas for the laterally confined rockslides (76 cases) it is much higher (compare Figs. 10.20 and 10.11).

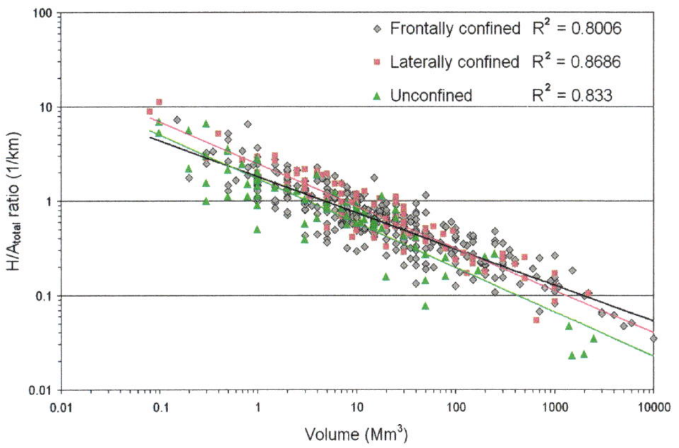

FIGURE 10.19 Relationships between H/A_{total} ratio and rockslide volume for rockslides with frontal confinement (332 cases), lateral confinement (92 cases), and in unconfined conditions (76 cases). Corresponding R^2 values are added to the legend.

Best correlations were found for the relationships between A_{total} and product of rockslide volume and maximal height drop closely related to the potential energy (Table 10.2 and Fig. 10.21). It is interesting that it has high correlation coefficients ($R^2 > 0.925$) for all confinement types unlike similar ratio with runout (see Fig. 10.12) that have good correlation for rockslides in frontally confined ($R^2 = 0.816$) and unconfined ($R^2 = 0.8824$) conditions only.

Summarizing the data presented above, we compared correlation coefficients for various relationships (Table 10.3). Higher correlation coefficients of the relationships between total area affected and H/A_{total} ratio and parameters such as rockslide volume, height drop, and their product support the assumption that area affected reflects rockslide debris mobility better than just its runout.

It was found that some rock avalanches that moved over unconfined planar surfaces of intermontane depressions or wide valley bottoms form tongues of debris whose maximal width is almost the same as the

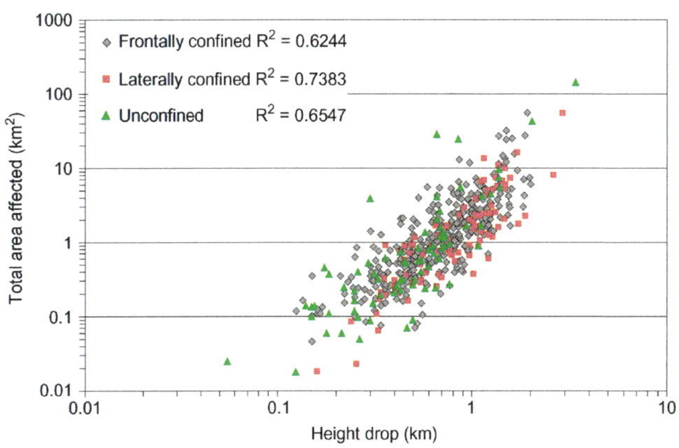

FIGURE 10.20 Relationships between total area affected and maximal height drop for rockslides with different confinement. Corresponding R^2 values are added to the legend.

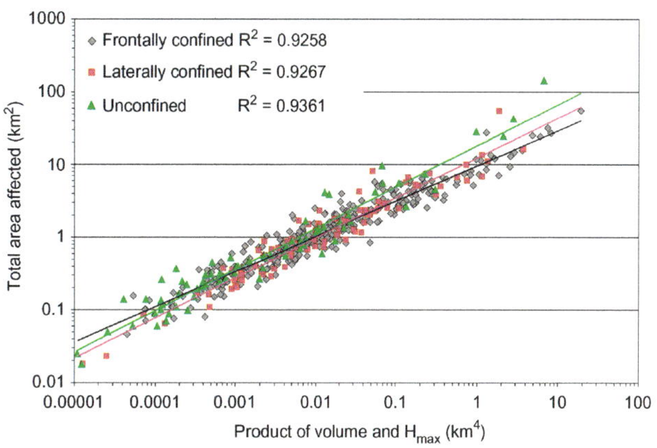

FIGURE 10.21 Relationships between total area affected and product of rockslide volume and maximal height drop for rockslides with different confinement and power-law regression curves. Corresponding R^2 values are added to the legend.

TABLE 10.2 Power-Law Equations Linking Totally Affected Area (A_{total}) and the Product of Rockslide Volume (V) and Maximal Height Drop (H)

Confinement conditions	No. of cases	Equations
Frontally confined rockslides	332	$A_{total} = 9.5295 \times (V \times H)^{0.4849}$
Laterally confined rockslides	92	$A_{total} = 12.257 \times (V \times H)^{0.5497}$
Unconfined rockslides	76	$A_{total} = 17.933 \times (V \times H)^{0.5668}$

TABLE 10.3 Correlation Coefficients for Various Relationships Characterizing Rockslides Mobility (Dependent Parameter First)

	Sample			
Relationship	**Entire database**	**Frontally confined**	**Laterally confined**	**Unconfined**
L/V	0.7126	0.7335	0.7301	0.8066
$H_{max}/L/V$	0.3524	0.3008	0.4497	0.3962
L/H_{max}		0.7465	0.0286	0.7535
$L/V \times H_{max}$		0.8160	0.0510	0.8824
A_{total}/L		0.8402	0.8878	0.9112
A_{total}/V	0.8993	0.9008	0.8833	0.9151
$H_{max}/A_{total}/V$		0.8006	0.8686	0.8330
A_{total}/H_{max}		0.6244	0.7383	0.6547
$A_{total}/V \times H_{max}$		0.9258	0.9267	0.9361

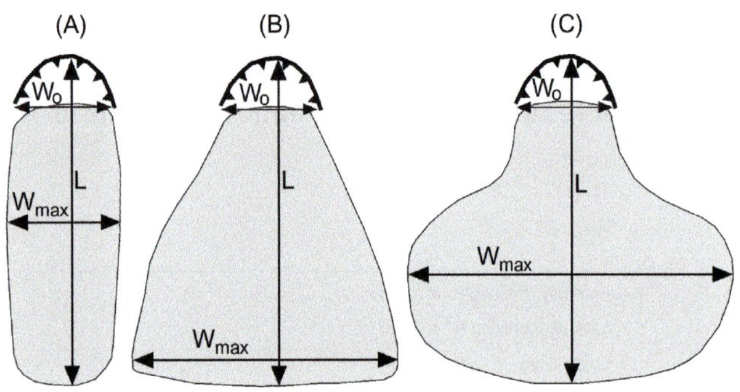

FIGURE 10.22 Main types of rock avalanche deposit shapes in unconfined conditions. (A) monodirectional, (B) fan-shaped, (C) pancake-shaped. L, runout, W_0, headscarp base width, W_{max}, maximal width of the deposits.

headscarp base width (Fig. 12.6A) such as the rock avalanche in Xinjiang west of the Toksun town (see Fig. 5.32), the Chaartash-3 rock avalanche (see Figs. 5.61, 8.130), the Taheman rock avalanche (see Fig. 9.17), and the extraordinary Koman rock avalanche (see Fig. 9.20; here we compare width of the rock avalanche at the Alai intermontane depression bottom with it width at the foot of the Trans Alai Range where it leaves rugged terrain).

In contrast, some rock avalanches most of which had moved over well-developed flood planes or low terraces of wide river valleys spread sidewise significantly forming either fan-shape bodies (Fig. 12.6B) as the Chukurchak rock avalanche (see Figs. 8.221 and 8.222), the Khait rock avalanche (see Figs. 8.271 and 8.272), the Yimake rock avalanche (see Figs. 6.8 or 9.6), rock avalanche in Xinjiang shown on Fig. 8.7, rock avalanche on the Valter glacier (see Figs. 4.36 and 4.38), or pancake-shape bodies (Fig. 12.6-B), such as rock avalanche in the Kash River valley (see Fig. 5.33), and the Atdjailau rock avalanche (see Fig. 6.4, see also (Strom, 2006)) (Fig. 10.22). Features classified as "anomalous fans" were described also in (Jarman et al., 2011).

We believe that the ratio of maximal width of the deposits in the direction transverse to that of the rockslide motion and width of the headscarp base (the latter somehow reflects linear dimension of the zone where the initial accelerated downslope motion of rockslide body converts into its motion under gained momentum) can be used to obtain some quantitative characteristics of these processes (Strom, 2006). To analyses this, we calculated W_{dep}/W_{scar} ratio for 531 case studies for which both values have been measured up to now and analyzed their distribution that appeared to be nearly normal with long tail formed by less than 1.7 % of all cases (six rockslides) whose deposits' width is more than four times larger than the headscarp base (Fig. 10.23).

It is, however, evident that debris spreading depends on a large extent on the confinement. For laterally confined features, transverse spreading is limited by valley walls, though in some cases, when such rock avalanche comes out from the tributary valley onto much wider space of the main valley, it can spread over it much widely such as the 1949 Khait (see Figs. 8.271 and 8.272) or the Chukurchak (see Figs. 8.221 and 8.222) rock avalanches. The latter feature originated from the about 2.5 km wide headscarp, but its distal part entered the Chatkal River valley through the 0.45 km wide gulley and then spread along its bottom for ca. 2.0 km (see Fig. 8.222). Just

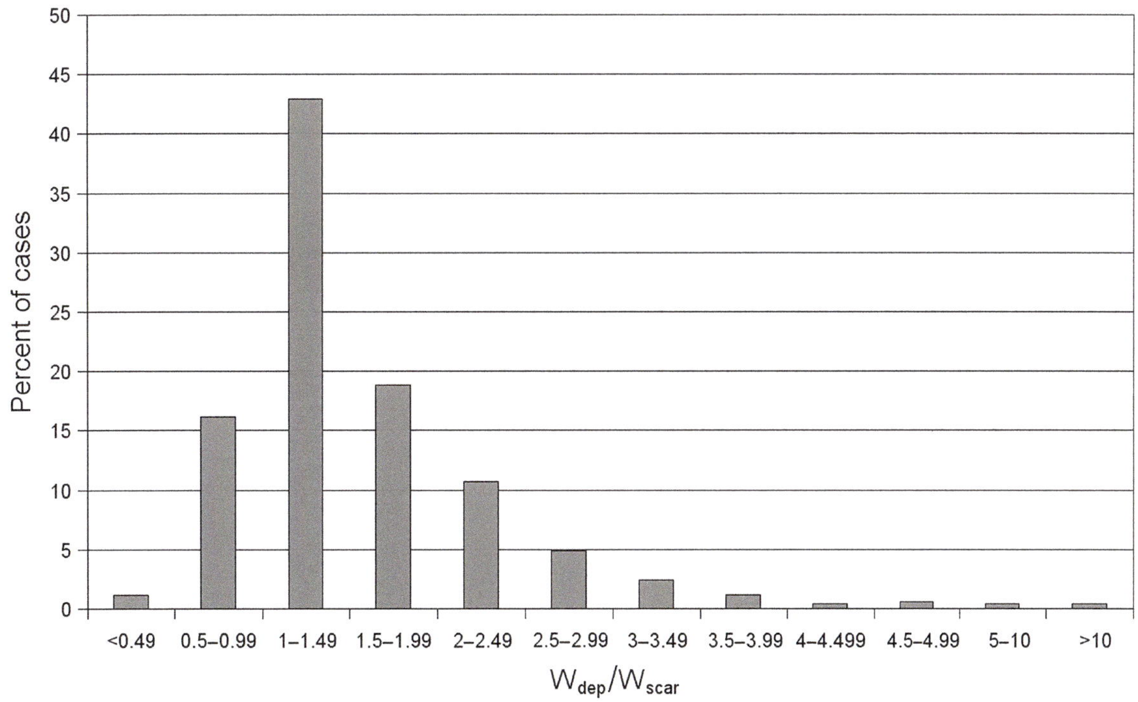

FIGURE 10.23 Nearly normal distribution of case studies characterized by different W_{dep}/W_{scar} ratio.

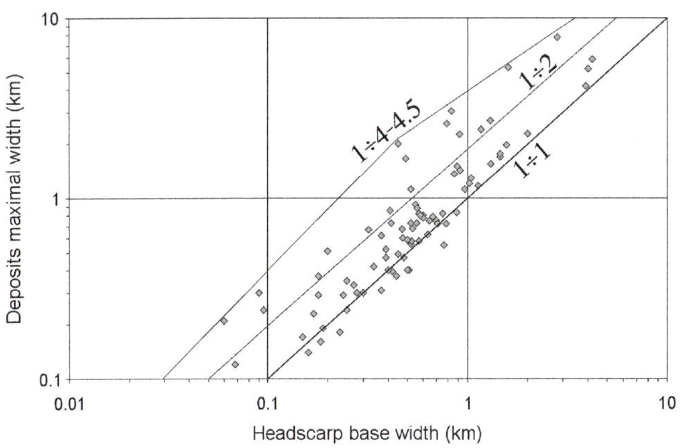

FIGURE 10.24 Comparison of the deposits' maximal width with the headscarp base width for unconfined rockslides and rock avalanches.

these values were used for our analysis. Comparison of the deposits' maximal width with the headscarp base width for 83 rockslides and rock avalanches classified as unconfined features shows that while most of them have the W_{dep}/W_{scar} ratio less than 2, there is rather large proportion of case studies with much larger—up to 4–4.5 times—transverse debris spreading (Fig. 10.24). Possible cause of such bipartition will be discussed in Chapter 12.

CHAPTER

11

Structure and Grain-Size Composition of Large-Scale Bedrock Landslide Deposits

OUTLINE

Despite significant differences in rockslides' size, runout, slope failure mechanisms, confinement, and deposits morphology, the vast majority of them all over the world are characterized by two basic peculiar properties of the deposits: (1) inverse grading, often with sharp boundary between the lower (internal) deeply comminuted part and upper (external) coarse carapace and (2) retention of mutual position of rock types involved in slope failure (Heim, 1932; Erismann, 1979; Hewitt, 1988, 1999, 2006; Strom, 1994a, 2004, 2006a,b; Davies and McSaveney, 1999; Wassmer et al., 2002; Towhata, 2003; Abdrakhmatov and Strom, 2006; Scarascia Mugnozza et al., 2006; McSaveney and Davies, 2006; Dufresne et al., 2009, 2016; Dufresne and Dunning, 2017).

Such peculiarities are typical of numerous case studies described above in Chapters 7–9, Dzungaria, Tien Shan, and Pamir, respectively. Hereafter we will summarize these observations along with description of some physical experiments aimed to reproduce in the lab the phenomena observed in nature.

11.1 RETENTION OF THE ROCK MASS MACROSTRUCTURE

Preservation of host rock mass structure in the large-scale bedrock landslides deposits was described for numerous case studies (Heim, 1932; Johnson, 1978; Erismann, 1979; Yarnold and Lombard, 1989; Capra et al., 2002; Geertsema et al., 2006; Hewitt et al., 2008; Dufresne et al., 2009; Weidinger et al., 2014; Roverato et al., 2015). Numerous excellent examples of such preservation could be found in the study region (Abdrakhmatov and Strom, 2006; Hartvich et al., 2008). Their analysis revealed two basic types of the mutual position of the lithologies involved in slope failure—the "train-like" and the "pseudo-stratified" structures (see Fig. 4.17) (Strom, 1994a,b, 2006).

The first one can be exemplified by the Imom (see Fig. 4.14), the Liangar (see Fig. 4.16), and the Shoripdara (see Fig. 9.78) rockslides in Pamir, a few tens of millions cubic meters in volume each that originated on the slopes composed of metamorphic rocks

Rockslides and Rock Avalanches of Central Asia.
DOI: https://doi.org/10.1016/B978-0-12-803204-6.00011-9

and covered at their bases by moraine material (rounded and semirounded boulders in sandy–silty matrix). The latter forms distal parts of these rockslide bodies being pushed ahead by angular clasts and boulders of rocks from the upper, bedrock-composed parts of the source zones. Similar structure has the body of the Andravdj rockslide, also in Pamir, with distinct outer belt differing from the main part of the deposit (see Fig. 9.66) and the Karachauli rockslide in Central Tien Shan whose frontal part composed of crushed Neogene conglomerates was pushed ahead by the massive rockslide body composed of Paleozoic rocks that form the main part of the source zone (see Figs. 8.172–8.176 and 12.2).

More complex internal structure that, nevertheless, should be ascribed to the same "train-like" type, characterizes the impressive fan-shaped rock avalanche about 18 Mm^3 in volume in Eastern (Chinese) Tien Shan (see Fig. 8.7). Its body is composed of varicolored concentric belts of crushed debris corresponding to the succession of the layers in the host massif. Similar structure with distinct concentric belts of debris of different lithology was observed at the deposits of gigantic, about 300 Mm^3 Kuturgansu-mouth rockslide in Southern Tien Shan (see Fig. 8.246) and of even larger, about 1 km^3 Kudara-Pasor rock avalanche in Pamir (see Fig. 4.20). In the latter case one can observe five successive wide concentric belts composed of more or less homogeneous debris one of which (likely No 1 on Fig. 9.38) corresponds to the glacial deposits that originated from the remnant of old trough valley affected by the headscarp crown, two (No 2 and 3)—to the bedrock debris from the valley slope and one (No 4)—to younger glacial deposits from the slope base. Belt No 5 is, most likely, fluvial material of the valley floor mobilized and bulldozed by rockslide. Besides, thin layers of crushed debris, whose succession corresponds to the bedrock layers, can be identified within belt No 3 on Fig. 9.38 in the central part of this body (see Fig. 9.39).

Retention of the initial order of rock types involved in rockslide motion can be observed quite well in the road cut made through the deposits of the Lower Kokomeren rockslide in Central Tien Shan that affected (from the slope top to its base) varicolored Paleozoic metasediments, fault zone several tens of meters thick composed of cataclasite, and talus displaced from the base of the source zone. All these unmixed lithological units could be identified in the rockslide body (see Figs. 8.181 and 8.183).

Some elements of similar relationships were observed in the trenches excavated at the frontal part of the 1911 Ananievo rockslide in Northern Tien Shan (see Figs. 8.70 and 8.72). However, in this case, it was the abrading of the soil layer from the rockslide travel pass and it's bulldozing, rather than pushing ahead the material from the lower part of the slope initially involved into failure.

In most of the case studies listed herein, one can observe some thrusting of "active" (pushing) debris over the "passive" (pushed) unit. Such successive overthrusting of wide belts composed of debris with different lithology is illustrated by the cross-section of the Kudara-Pasor rock avalanche (see Fig. 9.38). Further development of this phenomena leads to formation of rockslide and rock avalanche bodies with "pseudo-stratified" structure, where debris of lithological units originated from the base of the source zone form the basal "layer" successively overlain by debris from the upper parts of the slope (see Figs. 4.17B, 8.153). Excellent examples of such successions could be found at the Inylchek rockslide in the Highest Tien Shan (see Figs. 8.45 and 8.49), and at the Kokomeren (see Figs. 8.146, 8.147, 8.150–8.152) and the Ornok rockslides (see Figs. 8.168–8.170) in the Central Tien Shan. The latter demonstrates features typical of both "train-like" and "pseudo-stratified" types. As one can see on the schematic cross-section of this rockslide on Fig. 8.171, generally it can be classified as a rotational slide that pushed rocks of the lower part of the slope ahead for about 600 m if not more. At the same time, the frontal avalanche-like part of the Ornok rockslide demonstrates classical "pseudo-stratification" of the deposits that is well expressed on Fig. 8.168 (unit "3") and on Fig. 8.169. It is likely that the tectonic contact between varicolored Paleozoic siltstone and scree shown in the lower right inset in Fig. 8.170 belongs to the frontal, avalanche-like part of this rockslide body (see Fig. 8.171). We should notice that additional studies must be performed to explain the structure and emplacement mechanism of this outstanding feature.

Similar "pseudo-stratification" characterizes the Yashilkul rockslide dam (Strom, 1994), composed of large angular blocks of gneiss underline by moraine material that originated from the headscarp base and was pushed ahead in front of the main collapsing body. Frontally pushed part of this displaced moraine was thrown into the river channel and filled it forming the dam that had been overlaid by coarse blocky carapace, while other part spread from under blocky carapace and moved ahead and aside it (see Figs. 4.18 and 4.19).

Two rockslides in Central Tien Shan, in the Kokomeren River basin, studied by authors in the field, demonstrate retention of different rock types in the longitudinal direction. It is well expressed in the deposits of the Mini-Köfels rockslide, whose left part is composed of reddish-colored gneiss debris (see Figs. 8.159 and 8.160) and the right part — of dark-gray gneiss debris (see Fig. 5.41) that corresponds to the petrographic composition of its source zone.

FIGURE 11.1 (A) Block of augen gneiss forming most of the Kashkasu rockslide body; (B) block of gabbro that form stripes stretching parallel to the direction of debris motion.

Similar regularity has been observed in the upper part of the Kashkasu rockslide body (see Fig. 5.45) composed of augen gneiss mainly (Fig. 11.1A); however, when crossing the upper part of its body along the path visible on Fig. 5.45, at some localities one can see several ten meters wide stripes composed of angular boulders of mafic rocks (Fig. 11.1B) stretching in the direction of debris motion. Most likely these are crushed gabbro dykes existing in the host massif.

Summing up this section we must conclude that, despite kilometers-long travel pass and motion over complex rugged terrain, different lithological units both in coarse carapace (see, e.g., Figs. 9.38 and 11.1) and in the intensively shattered internal part of the deposits (see Figs. 8.49, 8.150, 8.152B) do not mix with each other, indicating laminar style of motion during rockslide emplacement. Even those rockslides that converted into long runout rock avalanches and can be classified as flow-like bedrock landslides (Hungr et al., 2014) represent laminar granular flows. It differs drastically from debris flows—another highly mobile flow-like landslides characterized by much more homogenous internal structure of the deposits due to predominantly turbulent style of flow during their motion (Vinogradov, 1977; Stepanov and Stepanova, 1991; Stepanov and Yafyazova, 2014).

11.2 GRAIN-SIZE COMPOSITION AND MINOR LANDFORMS OF ROCKSLIDE DEPOSITS

As we mentioned above, the analysis of deeply incised rockslides both worldwide and from the study region demonstrates that most of them have coarse carapace and giant internal core or lower zone composed of the intensively comminuted debris. In several cases the additions facies of rockslide deposits could be identified—either basal zone where moving rockslide debris interacted with the substrate (Hewitt, 2002, 2006; Dufresne et al., 2016) or shear zones within the rockslide body along which parts of it slid against each other (Pollet and Schneider, 2004; Pollet et al., 2005; Dufrense and Dunning, 2017). An excellent overview of these peculiarities has been presented recently in Dufresne et al. (2016).

Here we want to summarize data from the study region on grain-size composition typical of these three main facies of rockslide deposits. We want to point out that sharp difference in the extent of grinding between coarse carapace and comminuted interiors cannot be explained by floating of coarser particles typical of mixtures with wide range of grain sizes affected by vibration as it was proposed, for example, to explain formation of molards widely distributed on the surface of the Frank rockslide (Cassie et al., 1988). If it would be like this, then coarse carapace of the "pseudo-stratified" rock avalanche bodies would consist of larger fragments of all lithologies presenting in the deposits, while in fact it consists exclusively of boulders from the uppermost lithological unit (see Fig. 4.17B; the same was mentioned in (Dufresne et al., 2016, page 246)). It can be exemplified by the Inylchek rockslide whose bouldery carapace (see Fig. 8.50) consists of angular boulders of black limestone only that form the uppermost "layer" of this body with comminuted lower part (see Fig. 8.49B). Same lithological and grain-size distribution can be seen at the Kokomeren rockslide where the top of easily attainable right-bank part of the deposits is composed of angular boulder of the red granites overlain by a rather small patch of same size blocks of metasediments that corresponds to their succession on top of the source area (Fig. 11.2; see also Fig. 8.150).

It is noteworthy that we do not find case studies with opposite grading where lower part of rockslide body is composed of more or less intact block overlain

FIGURE 11.2 Composition of the upper section of the Kokomeren rockslide deposits on the right bank of the Kokomeren River. 1 and 2—the carapace facies: 1—blocks and angular boulders of metasediments, 2—same of granites; and 3—intensively comminuted debris of the internal part of rockslide body.

by debris crushed at much higher extent as it was reported for the Tangjiashan rockslide dam triggered by the 2008 Wenchuan earthquake (Subtype III of (Fan et al., 2017), see also (Fan et al. 2012c)). Quite rarely we found dissected rockslide deposits with uniform grain-size distribution (Subtype I of Fan et al., 2017). The rare exception is the left-bank outcrop of the pancake-shape part of the Atdjailau rock avalanche body (see Fig. 8.46), but we must notice that, first, we cannot see its basal part buried by the alluvium, and, second, it demonstrates some evidence of the jigsaw structure, typical of the carapace facies. Practically in all other cases where we could observe rockslide bodies in the cross-section, they have comminuted lower part overlain by much coarser carapace. Such style of grading was observed even at the rather small Lower Toguz-Bulak rockslide about 1 Mm^3 in volume only (see Fig. 9.69), which can be considered as an approximate lower limit of the flow-like bedrock rockslides and at the distal edge of the Bashi-Djaya rockslide body less than 2 m thick (see Fig. 8.77).

Inverse grading can be considered as an important diagnostic indicator that allows identifying even those rockslides that cannot be distinguished morphologically, as the Kiol-Kogur-Northern (see Fig. 8.86) and the Displaced Peneplain (see Fig. 4.6) rockslides.

11.2.1 Blocky Carapace and its Typical Features

11.2.1.1 Grain-Size Composition

At most of the cases studies described in Part 2 of this book, the daylight surface of rockslide and rock avalanche deposits that survived erosion is composed of coarse material, as it can be seen at the Yashilkul rockslide dam in Pamir (see Fig. 4.19) or at the Seit rock avalanche in Central Tien Shan (see Fig. 8.142). Quite often largest blocks exceed 20–30 m and, sometimes, even 50 m in size (see, e.g., Figs. 8.28, 8.36, 8.51, 8.119, 8.251, and 9.52). Sometimes we can see blocks with the jigsaw puzzle structure such as at the surface of the Upper Muzart rock avalanche body (see Fig. 8.36) or on top of the remaining part of the Kokburunsu rockslide dam (see Fig. 8.51) in the "Highest" Tien Shan. Similar jig-saw puzzle structure is very well expressed at the western part of the Lower Kokomeren rockslide body crossed by the road cut (see Fig. 8.182).

Moreover, at several cases, the topmost parts of high compact rockslide bodies are composed of giant blocks up to ca. 1 km^2 in size either of sedimentary rocks with well-preserved bedding, as at the Usoi rockslide (see Fig. 9.42), or of massive blocks of igneous rocks as at the upper part of the Kokomeren rockslide (see Fig. 8.152A). These blocks have been intensively fractured but, nevertheless, retain their integrity.

Overall integrity of fractured jig-saw boulders and giant blocks indicate that they were transported on top of rapidly moving granular flows (see also discussion on "active" or "passive" carapace in Dufresne et al., 2016).

Obviously, grain-size distribution of the carapace facies strongly depends on the block size (fracturing) of the bedrock in the source zone. Rockslides in gneiss, igneous rocks, or coarse-bedded sedimentary rocks with large size of nonfractured blocks usually produce very coarse carapace as, for example, at the Seit rock avalanche (see Figs. 5.37 and 8.142) or at much thinner fan-shaped body of the Chongsu rock avalanche (see Fig. 5.50 and Fig. 5.51), while deposits of the same-size slope failures in thin-bedded or foliated gneiss, schist, or mudstone are covered by much finer debris, which can be exemplified by the Bielogorka rock avalanches (see Fig. 8.99) or by rockslide in Eastern Tien Shan shown on Fig. 8.25.

At several sites, size of the carapace boulders differs significantly at different parts of the deposits. It can be observed at the Djashilkul rockslide in Northern Tien Shan where belt of large boulders can be seen on top of its body surrounded by zones with much more smooth surface composed of finer debris (see Figs. 8.79 and 8.80). We hypothesize that this belt originated from the harder layer of metamorphosed sedimentary rocks that retain its overall integrity despite fragmenting into meter-size angular boulders. One more example is the rockslide in the Kainar River valley, which frontal and, to a lesser extent, proximal (close to the erosional gorge) parts are composed of very coarse material (blocks up to 10 m in size) divided by the 120–150 m wide belt of much finer debris (see Fig. 8.116). Concentration of largest boulders at the most distal part of the deposit could be observed at the Seit rock avalanche (see Figs. 5.37 and 8.142).

Inverse plan-view grain-size distribution of the carapace material characterizes rockslide in the upper reaches of the Dzhergalanty River basin (see Fig. 8.23), whose proximal part is covered by coarse boulders, while distal one that rose more than 200 m on the opposite bank of the river is composed of much finer material. Sharp difference of the grain-size distribution observed on the Korgon rock avalanche deposits surface in Southern Tien Shan (see Figs. 8.251 and 8.252) that collapsed from the up to 1500 m high steep slope could result from more complex emplacement history: judging from the morphology of its deposits, we assume that coarser parts could be the second generation of rockslide deposits, while surface of the first generation is much smoother being composed of finer clasts.

Despite some rockslide are composed of very coarse material with large voids between angular blocks that can be seen, for example on Figs. 4.19, 8.119, or 9.96A, they dam river valleys forming permanent lakes, which means that such bouldery material just blankets much less permeable inner parts of rockslide bodies that will be described hereafter.

What seems to be very important is the sharp change of the grain-size composition of the external (carapace) and internal parts of rockslide deposits, often without distinguishable transition zone. Such difference can be observed both in thick compact rockslide bodies such as those of the Kokomeren rockslide in Central Tien Shan (see Figs. 8.150 and 8.152), of the Djashilkul rockslide (compare belt of large angular boulders on top of its body visible on Figs. 8.79 and 8.80 with the material in the deep erosional gorge; see Fig. 8.82) and the Arashan rockslides in Northern Tien Shan (see Figs. 8.95 and 8.96), in the deposits of more mobile rockslides (rock avalanches)—of the Terek-Suu rockslide in Southern Tien Shan (see Fig. 4.7B), the Lower Aral rockslide in Central Tien Shan (see Fig. 8.162A), the Dead Lakes rockslide in North-Western Tien Shan (see inset in Fig. 8.206). Same phenomenon can be observed in the cross-section of the deposits of the relatively small Lower Toguz-Bulak rockslide in Pamir (see Fig. 9.69) and at the distal edge of the Bashi-Djaya rockslide (see Fig. 8.77). Rather often same difference between the carapace and internal fragmented facies could be seen even on Google Earth images of rockslides such as at the case studies in the Karatag River valley in Southern Tien Shan (see inset in Fig. 8.275) or in the Bartang River valley in Pamir (see Fig. 9.52).

Such abrupt change of the extent of comminution indicates sharp difference of the stress–strain state and fragmentation process in the external and internal parts of rockslide bodies.

11.2.1.2 Carapace Landforms

Besides characteristic (generally quite coarse) grain-size composition, superficial parts of rockslide deposits are often characterized by the specific landforms—longitudinal or transverse linear ridges, furrows, and flow bands (Dufresne and Davies, 2009)—and by much more isomeric hummocks, often classified as molards (Cassie et al., 1988).

Most common are the longitudinal flow bands and small linear ridges indicating direction of debris motion. They are very well expressed at the Seit rock avalanche where such landforms correspond to stripes of coarser and finer material (Fig. 11.3; see also Fig. 5.37). The latter with larger content of fines are more vegetated. Similar striation characterizes surfaces of the Urmochdara rock avalanche in South-Western Pamir (see Fig. 4.41), of the Upper Kashkasu rock avalanche (see Fig. 4.43), and of the secondary rock avalanche of the Karasu Lake rockslide (see Fig. 5.64) in Central Tien Shan. Well-expressed fan-shaped system of longitudinal levees and furrows can be seen at the distal part of the Chaartash-3 rock avalanche (see Fig. 8.130), while most of this more than 3 km long tongue of debris is covered by the transverse, rather than by longitudinally directed minor landforms. It is interesting that much longer Chaartash-1 and Chaartash-2 rock avalanches have less pronounced longitudinal landforms (see Figs. 5.61, 8.128, 8.129). Similar just distal longitudinal striation characterizes the Bayan-Nur rock avalanche in Mongolian Altai with ca. 12 km long runout (see Fig. 11 in Strom, 2006).

At the Djashilkul rockslide longitudinal levees and furrows are less expressed but, nevertheless, can be distinguished on its generally wavy surface (see Fig. 8.80).

Besides rock avalanche deposits with well-pronounced longitudinal directivity of flow bands and other minor landforms, some of them are characterized by the concentric belts of coarser and finer material. It can be exemplified by the secondary rock avalanche of the Southern Karakungey rockslide, which concentric coarser outer and finer inner zones (Fig. 11.4, see also Fig. 5.54) are superimposed on the longitudinal striation that could be seen on Fig. 13.8. This rockslide originated on the slope composed of Paleozoic granite with, generally, uniform blocky structure, thus such grain-size distribution cannot be explained by the inherited difference of the bedrock block sizes that can

FIGURE 11.3 Alternating flow bands of coarser (C) and finer (F) debris at the surface of the Seit rock avalanche.

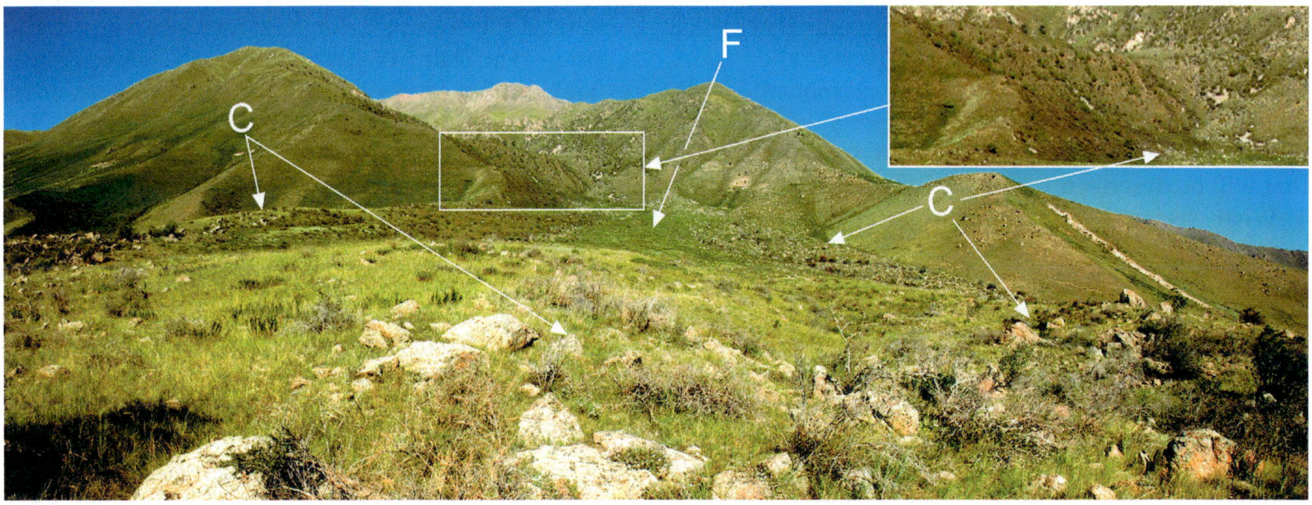

FIGURE 11.4 Upward view on the Secondary rock avalanche of the Southern Karakungey rockslide. Compact initial accumulation and secondary scar with large boulders at its foot are shown in the inset. C, coarse debris; F, fine debris.

be observed at some rockslides in sedimentary or foliated metamorphic rocks (such as the abovementioned Djashilkul rockslide—see Fig. 8.79 or rockslide in Afghan Badakhshan shown on Fig. 9.130). It can be explained quite well, however, if we take in mind that this 1.5 km long tongue of debris is the Secondary rock avalanche (see Section 5.3.3.1 in Chapter 5: Bedrock Landslide Types and Classification Systems), which source zone (the "secondary scar," considering just this, final stage of rockslide emplacement) was at the downstream side of the initial compact rockslide accumulation with grain-size distribution typical of compact rockslide bodies, i.e., with coarse bouldery carapace and comminuted internal part. In such case outer belt of rock avalanche composed of meter-size blocks originated from coarse carapace of the compact body, while its central much finer part originated from its comminuted interior. Presence of large blocks at the foot of the secondary scar (see Fig. 5.53) is in line with this assumption, because these blocks fell from the secondary scar crown (top of the compact body composed of the carapace facies).

Some rock avalanches, however, have quite different morphology with topographic irregularities oriented transverse to the direction of debris motion. One of the most impressive sites where such landforms can be observed is the distal blade of the Snake-Head rock avalanche (see Figs. 5.60 and 11.5A). Landforms developed here are very similar to those of the distal blade of the 43 km long Cerrillos Negros Tomasiri rock avalanche in Peru (Crosta et al., 2012, 2015) (putting to the side quite different size of these features). Well-expressed transverse landforms can be also seen at the central part of rock avalanche in the Kash River valley in Eastern Tien Shan (see Fig. 5.33, south of the elevation mark 2470 m a.s.l.), at a large part of the Yimake rock avalanche in Chinese Pamir where its surface survived the erosion (Fig. 11.5B, see also Figs. 6.8 and 9.6), at the proximal part of the Kumbelsai rock avalanche in Djungaria (see Fig. 7.10), at the Arashansai rock avalanche body in North-Western Tien Shan (see Fig. 8.225), and at the suspicious feature at the foot of the headscarp on the Kungey-Alatau Range northern slope shown on Fig. 8.87. They resembles quite well hummocks that were modeled and described in (Paguican et al., 2012) and, likely, reflect mechanism similar to that proposed by these researchers for large-scale volcanic rock avalanches.

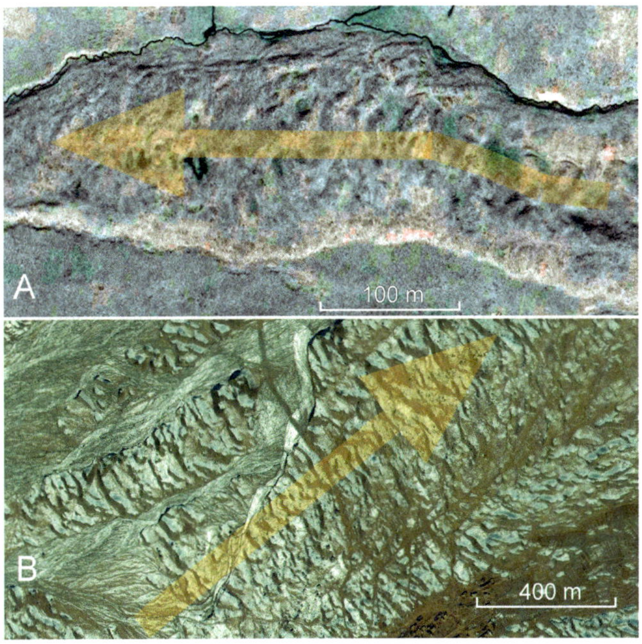

FIGURE 11.5 Minor landforms oriented transverse to the direction of debris motion. (A) At the distal blade of the Snake-Head rock avalanche; (B) on the original surface of the Yimake rock avalanche. Transparent brown arrows show direction of rock avalanches motion. Google Earth images.

Several large-scale rockslide bodies can be characterized by the irregular landforms without visible linearity either longitudinal or transverse and with large jig-saw blocks on their surfaces. They can be exemplified by the Upper Muzart rock avalanche (see Fig. 8.36) and by the Kokburunsu rockslide (see Fig. 8.51) in the Chinese part of the Tien Shan, by some parts of the Korgon rock avalanche in Southern Tien Shan (see Figs. 8.251 and 8.252), and by rockslides in the northern part of the Chinese Pamir (see Fig. 9.2). Revealing of causes of such a structureless carapace formation requires additional investigations.

One of the most common types of minor landforms of rockslide carapace is molards—conical hills from 3–5 m up to several tens of meters in diameter and up to several meters high. They were found on the surface of compact rockslide deposits hundred meters thick such as the Usoi (see Fig. 9.47) or the Shiva Lake (see Figs. 4.22 and 9.110) rockslide dams in Pamir and of long runout rock avalanches forming relatively thin, few tens of meters thick aprons on the valley bottoms. The latter can be exemplified by the Spur rock avalanche in Central Tien Shan covered by hundreds of conical molards from few up to 20 m in diameter (see Fig. 4.23), by the Urmochdara rock avalanche in Pamir where numerous molards could be seen along its more than 4 km long path. Largest of them, up to 50 m in diameter, are at the outer parts of the deposit with well-pronounced longitudinal striation (see Fig. 4.41). One more rock avalanche with widespread molards is in the Arhotyn River valley in Eastern Tien Shan (see Fig. 8.18). Similar features can be observed on the distal blade of the 1949 Khait rock avalanche, especially at its frontal part on the right bank of the Yarhych River (see Fig. 8.272).

Similar landforms can be seen at some parts of the Djamantau rock avalanche (see Fig. 8.134) and the Chaartash-2 rock avalanche (Fig. 8.129) in Central Tien Shan, of the Kudara-Pasor rock avalanche (at its part composed of "nonstratified" debris; see Fig. 9.39), of the Imom rockslide (see Fig. 4.15), and of some other rockslides (see, e.g., Fig. 9.72) in Pamir.

Hummocky relief resembling that of rock avalanches with widespread univocal molards, though with much larger conical hills, is well preserved at the extremely mobile Koman (see Fig. 6.9) and Lenin (see Fig. 9.25) rock avalanches in the Alai valley. Largest of these hummocks are up to 40–50 m in diameter and often group in elongated chains as it can be seen on Fig. 11.6.

Our observations show that, despite morphological similarity of these conical landforms, two types of hummocks or molards could be identified being

FIGURE 11.6 Evidence of linearity in hummocks' distribution on top of the Koman (A) and Lenin (B) rock avalanche deposits. Both rock avalanches had moved from the lower right to the upper left sides of the Google Earth images.

FIGURE 11.7 Molards on top of the Seit rock avalanche with jig-saw puzzle structure reflecting in-situ decomposition of large boulders.

characterized by different structure and likely by different mechanism of their formation. One of them can be exemplified by molards at Spur rock avalanche shown on the inset in Fig. 4.23 and by some molards on top of the Seit rock avalanche body, one of which is shown on Fig. 8.144. In both cases conical hills up to ~2 m high are formed by chaotic accumulation of clasts of different size and, in the latter case, of different lithology. Similar structure characterizes molards on top of the Usoi rockslide (see Fig. 9.47) and, probably, much larger hummocks of the Koman (see Fig. 6.9) and Lenin (see Fig. 9.25) rock avalanches.

We assume that molards of this type could originate, as it was proposed in Cassie et al. (1988), "by vibration of the ground under rockslide avalanches" (we can add—by vibration of the moving rockslide body, considering significant—up to ca. 100 m—thickness of the Seit rock avalanche body and, moreover, 500–600 m thickness of the Usoi blockage). Such gryphon-like model proposed in (Cassie et al., 1988) somehow recalls sand blows caused by liquefaction (see, e.g., Figs. 7.2 and 7.16 of Paleoseismology (McCalpin, 2009)). Material forming such molards seems to be ejected from under rock avalanche blocky carapace. Outside the study region such phenomenon was described, besides the Frank rockslide (Cassie et al., 1988), at the Blackhawk landslide in California by B. Johnson (1978) who pointed out that molards on top of its body and clastic dikes dissecting it were composed by debris of the lithological units that can be found at the base of the deposits and was ejected through the topmost "layers" of debris of this pseudo-stratified body.

Contrary to molard shown on Fig. 8.144, many of numerous molards widely spread over the Seit rock avalanche deposit have completely different jig-saw structure, resulting from the in-situ decomposition of large, meters-size boulders (see Figs. 8.143 and 11.7). We hypothesize that these boulders that originally were much larger than the surrounding blocks of rockslide carapace (see upper photograph on Fig. 11.7) were also ejected on top from the deposits interiors where they underwent significant triaxial compression, by far exceeding lithostatic pressure. Such overloading could be provided by force chains forming within moving rock avalanche and resulting in dynamic fragmentation of rock avalanche debris (Davies and McSaveney, 1999, 2002, 2011; Davies et al., 2006, 2017; McSaveney and Davies, 2006). If such overstressed boulder is ejected upward, on top of the "passively" moving blocky carapace, it undergoes rapid unloading and decomposes along the preexisting fractures forming minor hill with well-observed jig-saw puzzle structure.

Further discussion on the dynamic processes resulting in formation of landforms typical of the rockslide carapace will be presented in Chapter 12, Morphological and structural evidence of rockslides' (rock avalanches') motion mechanism(s).

FIGURE 11.8 One of the outcrops of the distal part of the Koman rock avalanche on the right bank of the Kyzylsu River. Unlike outcrop shown on Fig. 9.22, this one is characterized by uniform lithology (dark-gray metasediments).

11.2.2 Comminuted Interior (Fragmented Facies)

Interiors of almost all deeply incised Central Asian rockslide bodies that have been observed in the field at numerous outcrops are composed of the intensively crushed material. Comminution associated with rock avalanches' formation extends up to nanoscales, which is not reported for the deposits formed by other exogenous processes such as water, glacier, or wind erosion (Reznichenko et al., 2012a). As it was mentioned above, in Section 11.2.1.1, intensity of rock crushing often increases abruptly without transition zone between coarse carapace and comminuted interiors.

Composition of the internal parts of rock avalanches can be seen quite well in the outcrops of the Terek-Suu rockslide in Southern Tien Shan (see Fig. 4.7B), the Inylchek rock avalanche in Highest Tien Shan (see Figs. 8.45 and 8.49), the Mini-Köfels (see Fig. 8.160), the Lower Aral (see Fig. 8.162), the Lower Kokomeren (see Fig. 8.183), the Akterek (see Fig. 8.188), and the younger Djuzumdybulak rockslide (see Figs. 8.192 and 8.193) in Central Tien Shan, the Dead Lakes (see Fig. 8.206) and the Karakul (Fig. 8.209) rockslides in North-Western Tien Shan, the Koman rock avalanche (see Figs. 9.22, 9.23 and 11.8) and of the Lower Toguz-Bulak rockslide (see Fig. 9.69) in Pamir. No doubts that nearly the same could be observed at other case studies all over the Central Asian region, but we listed here those rockslides, studied in the field by authors.

Even at those sites that have been illustrated by space images only, we can often see that material under coarse blocky carapace is much finer (see Figs. 2.1, 4.11, 4.28, 4.31, 5.63, 8.4, 8.20, 8.145, 8.250, 8.255, 8.285, 9.52, 9.76, 9.92, 9.104).

Grain-size composition of the fragmented facies of several Tien Shan rockslides was studied by combined sieving and aerometric analysis. Such analyses were performed for the Djashilkul (see Fig. 8.82 and Table 8.4) and the Arashan (see Fig. 8.96B and Table 8.5) rockslides in Northern Tien Shan and for the attainable right-bank part of the Kokomeren rockslide (see Figs. 8.150, 8.151, 11.9 and Table 8.8). Corresponding grain-size curves shown on Fig. 11.10

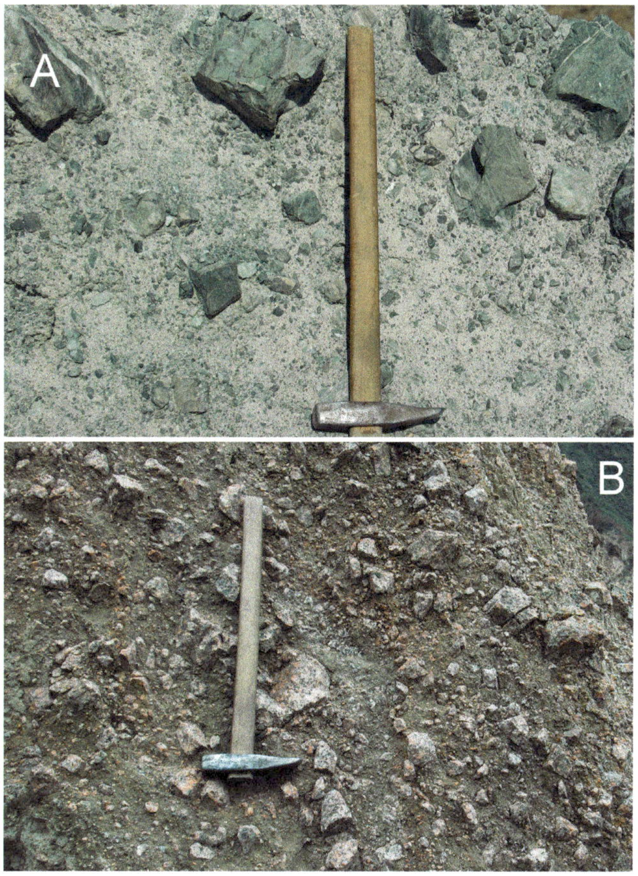

FIGURE 11.9 Typical grain-size composition of the fragmented facies of the Kokomeren rockslide. (A) Crushed metasediments; (B) crushed granite. Data presented in Table 8.8 reflect such material.

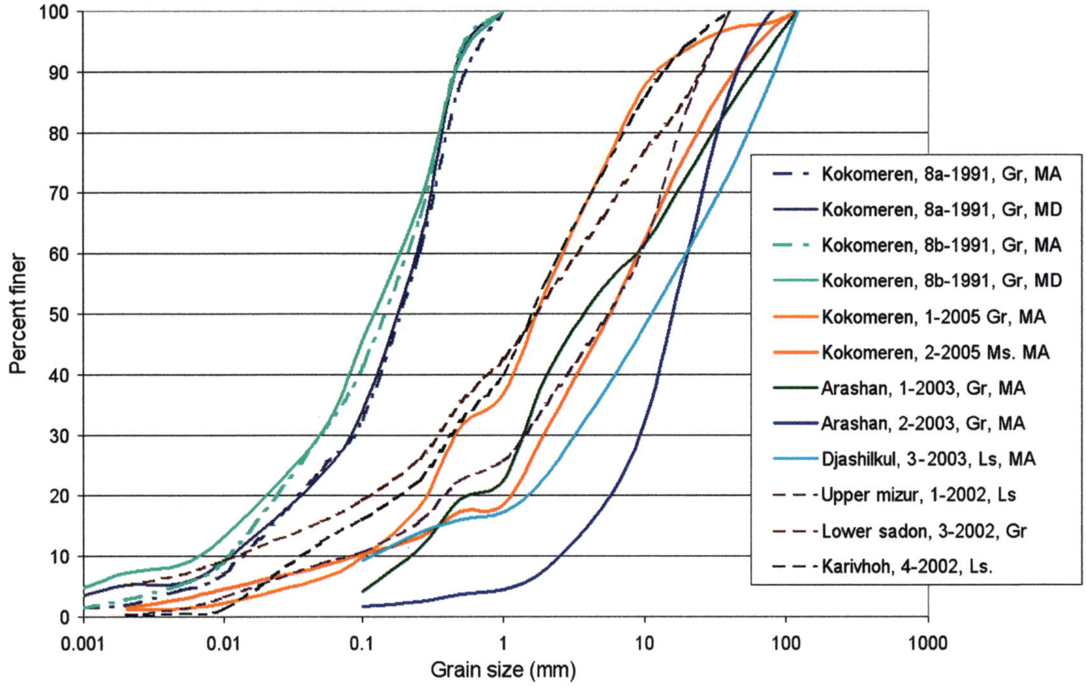

FIGURE 11.10 Grain-size curves of the fragmented facies of the Kokomeren, Djashilkul, and Arashan rockslides from Tien Shan according to data from Tables 8.7, 8.8, 8.4, and 8.5 correspondingly. We added also curves obtained for the Karivhoh, Upper Mizur, and Lower Sadon rockslides in Caucasus (Strom, 2004). The following parameters are indicated in the legend captions that include rockslide name; sample No and year of sampling; lithology (Gr, granite, Ls, limestone); sample preparation method, if available (MA, microaggregation content; MD, maximum dispersion; see explanation in Table 8.7).

recall curves obtained for numerous Alpine and North American rockslides (see, e.g., Crosta et al., 2007, 2011; Shugar and Clague, 2011; Dunning and Armitage, 2011; Dufresne et al., 2016; Dufresne and Dunning, 2017). For the Kokomeren rockslide, such analysis was performed also for the granite samples from the basal facies of this (see Table 8.7 and Fig. 11.10), and its results will be discussed in the following subsection.

We must notice significant difference in grain-size composition of samples collected in 2002–2005 from the fragmented facies forming main parts of rockslide bodies below the external carapace (see figures listed above) and of samples collected in 1991 from the lowermost (basal) unit of the Kokomeren rockslide body just above the alluvium in the old buried channel shown on Fig. 8.148.

As one can see from Fig. 11.10, fragmented facies contain from 1% to 10% of particles less than 10 μm; in other words, tens of kilograms of fine dust per ton of debris. Considering overall quantity of these deposits that could reach hundred millions tons per event and even more, it means that formation of each large-scale catastrophic rockslide or rock avalanche produces millions of tons of dust. It is in line with eyewitnesses' description of the catastrophic rockslide that destroyed Malpa village in India on August 18, 1998, who mentioned whopping emissions of dust (Bhandary and Kumar, 2000). Similar phenomenon was reported after the June 11, 2012, rock/ice avalanche at Lituya Mountain, Alaska (Geertsema, 2012) and during the 2005 Zengziya Rockslide in Chungching region in China (visible on video kindly provided by Prof. Yueping Yin).

Description of rockslide bodies' interiors will be incomplete if we will not mention the quite specific case of the Ak-Kiol rockslide dam in Central Tien Shan described in Section 5.3.1 in Chapter 5, as an example of the Primary rock avalanche with frontal confinement (see Figs. 5.8–5.12). Its unique peculiarity is that the syncline core that slid as a wedge from the left-bank host massif, unfolded in the river valley when it lost lateral confinement and the internal structure of this rockslide dam visible in deep erosional canyon represents the well-preserved stratigraphic succession of almost horizontal layers (see Fig. 5.12). Closer view on these deposits (Fig. 11.11) shows that sandstone, siltstone, and gravelly conglomerate layers forming this body are intensively fractured and fragmented but retain the original bedding. Much stronger granite or gneiss of the internal parts of the Kokomeren (see Figs. 8.150, 8.151, and 11.9B), Mini-Köfels (see Fig. 8.160), or Lower Aral (see Fig. 8.162A) rockslides are crushed and fragmented at a much higher extent.

FIGURE 11.11 Composition of the Ak-Kiol rockslide body visible in the erosional canyon.

FIGURE 11.12 Extent of comminution at the frontal contact of the Kokomeren rockslide body with the opposite valley wall. (A) Crushed metasediments resting directly against the alluvium of high river terrace; (B) crushed granite with metasediment xenoliths; comminuted debris originated from both types of rock did not mix.

Comprehensive explanation of this phenomenon requires additional investigations.

Summing up the description of the rockslide bodies intensively fragmented interiors, we just want to argue with the hypothesis that it could be the effect of collision with an opposite wall of the river valley proposed in (Erismann, 1979). First, such comminution was observed in the lower parts of some rockslides and rock avalanches that stopped before collision with an obstacle such as the Inylchek rock avalanche in Tien Shan (see Figs. 8.45 and 8.49) or the Lower Toguz-Bulak rockslide in Pamir (see Fig. 9.69). Second, in some cases intensive comminution was observed at the proximal (the Djashilkul; see Fig. 8.82), or sidewise (the Mini-Köfels; see Fig. 8.160) parts of rockslide bodies, quite far from their assumed collision zones. Even in those cases where we can observe inner parts of rockslide bodies resting against opposite slope of the blocked valley, such as at the Kokomeren rockslide, it does not look that this material had experienced more intensive comminution than debris of similar lithologies more distant from frontal confinement (compare Figs. 11.9 and 11.12).

Evidence of gradual destruction of the original structure of rock mass can be observed at the frontal (avalanche-like) part of the Ornok rockslide in Central Tien Shan. Here it is visible that thick layers of weak Neogene conglomerates alternating with sand- and siltstone layers disintegrated gradually, converting, first, into meter-size blocks of conglomerates in the sandy−silty matrix and, farther, into more uniform mixture of pebbles, sand, and silt (see Fig. 8.169).

We must also point out that rockslide bodies rarely interact with opposite valley walls in the form of direct impact. Most often they slid along curved surface formed by lower parts of valleys' slopes and their bottoms. It can be exemplified well by the Mini-Köfels rockslide with a significant, 170 m high run-up (see Fig. 5.41 and Table 8.9). Thus usually no "collision" in a strict sense of this word occurs. It supports the assumption that intensive comminution of the internal parts of rockslide bodies reflects "internal" processes evolving in rapidly moving debris during its motion that will be discussed in Chapter 12.

Intensive comminution of the inner parts of most of rockslide dams makes them very sensitive to overtopping and erosion (Dunning et al., 2005; Dunning and Armitage, 2011; Strom, 2010b, 2013a). Removal of coarse blocky carapace leads to intensive erosion of crushed internal "core" of a dam by turbulent flow through the erosional canyon and downslope side of the breaching blockage, rapid incision, and resultant catastrophic breach as it occurred, for example, in 1963 at the Issyk Lake (see brief description of this event in Chapter 2, Brief History of the Bedrock Landslide Studies in the Central Asia, and discussion in Chapter 14, Rockslide Dams—Origin, Evolution, and Failure).

11.2.3 Basal Facies and Rockslide Bodies Interaction with the Substrate

Despite good outcropping and the presence of numerous deeply incised rockslide deposits, we can observe their sliding surfaces and state of the material just above and below them rather rarely—less than at 3% of all studied cases. Many of these outcrops are hardly attainable being either at the inaccessible river banks (see Figs. 8.152B and 8.187B), or in the middle of high unstable cliffs (see Figs. 8.45 and 8.192). Nevertheless, these sites are of utmost interest because their study allows reconstructing interactions of rockslides' basal facies with the substrate and better understanding of rockslides' motion mechanism(s). Hereafter we summarize such observations described in Part 2.

Grain-size composition of the material from the base of a large-scale Central Asian rockslide body was studied at the Kokomeren rockslide. Two samples about 1 kg each of the fragmented debris overlying alluvial boulders of the old river valley (marked as RS on Fig. 8.148) that look in the outcrop as real granite (Fig. 11.13) were collected, and their grain-size composition was measured by combination of sieving and aerometric methods (see Table 8.7). As we mentioned above in Section 8.4 in Chapter 8, Tien Shan, grain-size composition of these samples fits to that of fine sand (!) with a specific gravity of 2.67–2.68 g/cm^3 (see Table 8.7 and curves corresponding to samples Kokomeren 8a, 8b on Fig. 11.10). Qualitatively similar distributions for basal facies of the Tschirgant rockslide deposit in the Austrian Alps were described in Dufresne et al. (2016). We must notice that grain-size composition of samples was, practically, independent of the method of sample preparation. It differs significantly from grain-size composition of clay gauge samples collected from shear zones of many active faults, including those that cross Paleozoic granites (Sample 1 on Fig. 11.14) and sandstones (Sample 2 on Fig. 11.14), with high content of clay minerals whose microaggregation grain-size composition (MA) measured after shaking of the sample in water differs drastically from maximum dispersion composition (MD) that was revealed after boiling of sample in water with natrium pyrophosphate (see Fig. 11.14). Similarity of MA and MD grain-size compositions indicates the absence of clay minerals in the heavily comminuted granite from the base of the Kokomeren rockslide and supports the conclusion that large proportion of fines in its basal facies results from the mechanical fragmentation rather than from the subsequent chemical weathering that could produce clay minerals such as kaolin, montmorillonite or illite.

FIGURE 11.13 Cleaned outcrop of heavily comminuted granite at the base of the Kokomeren rockslide deposits that fill the buried river channel.

Microscopic study of these samples showed that separate particles are angular quartz and feldspar grain fragments with conchoidal surface texture. Most of the smallest particles are quartz scales. Similar surface textures were reported from basal facies of rock avalanches triggered by the 2008 Wenchuan earthquake (Wang et al., 2015). These data also indicate that rock comminution was the result of the dynamic fragmentation (Davies and McSaveney, 1999, 2002, 2011; Davies et al., 2006, 2017; McSaveney and Davies, 2006).

It is interesting to notice that proportion of finest particles (less than 1–2 μm) in the basal facies is just slightly higher than that in the fragmented facies or even equal while a sharp difference was found for sandy and coarser fractions (see Fig. 11.10). As it was reported in Reznichenko et al. (2012a), comminution extends up to nanometer-scale level.

Thickness of the heavily comminuted unit at the base of the Kokomeren rockslide about 1.5 km^3 in volume, whose body was up to ca. 400 m thick is more than 10 m. Basal zone with similarly heavy comminuted granite 10–20 cm thick only, if not less, was found at the base of the much smaller Arashan rockslide near Bishkek City about 6 Mm^3 in volume (Fig. 11.15). These observations allow assumption on the dependence of the basal facies thickness from the overall thickness of rockslide deposits, but more rockslides must be studied to justify this hypothesis and to quantify such relationships if they will be proved. Additional factor affecting thickness and extent of comminution of the most intensively crushed debris at the base of rockslide bodies is the state of host rocks. For example, the Akterek and the Djuzumdybulak rockslides in the Naryn River valley originated on the slopes crossed by large fault zones (see Figs. 8.187 and 8.191) and their deposits (see Figs. 8.188B, 8.192, and 8.193), at least visually, are composed of more

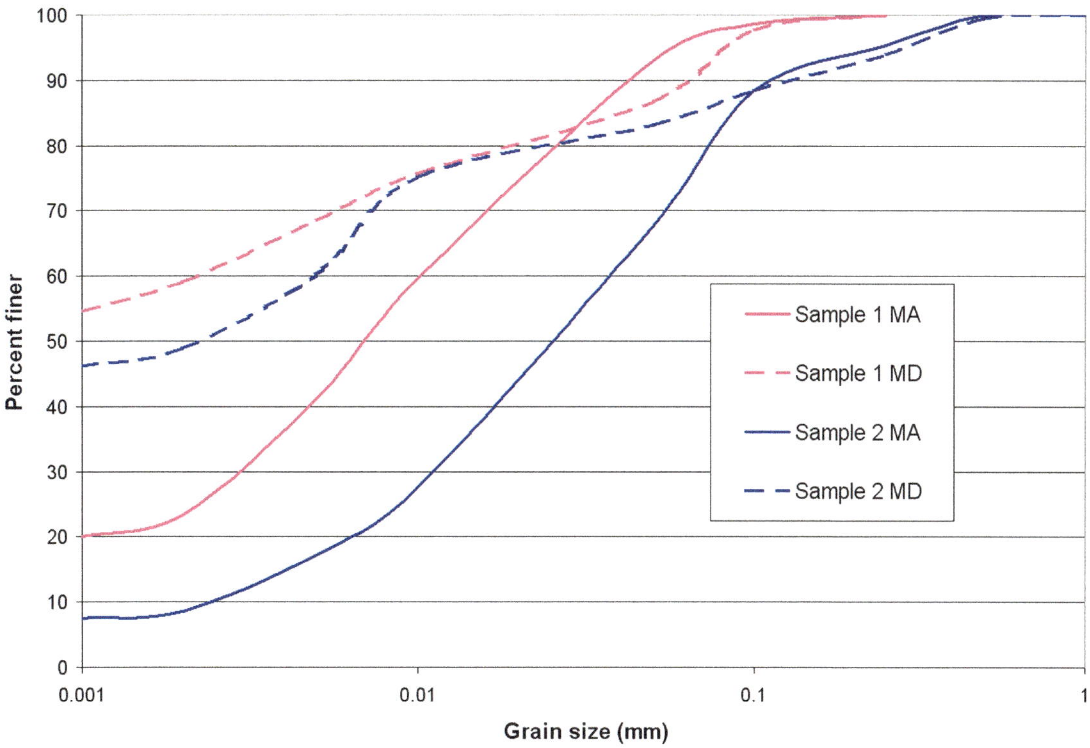

FIGURE 11.14 Grain-size curves of the fault clay gauge demonstrating significant increase of the proportion of finest particles for samples boiled in water with natrium pyrophosphate as against to samples just shook in water.

FIGURE 11.15 Basal facies of the Arashan rockslide deposit resting over glacial till.

FIGURE 11.16 Well-rounded alluvial cobble entrained in the heavily fragmented gneiss of the Mini-Köfels rockslide (zoomed fragment of the outcrop shown on Fig. 8.160).

intensively crushed material than deposits of the Arashan (see Figs. 8.95 and 8.96B) or Lower Aral (see Fig. 8.162) rockslides in granite massifs not affected by such tectonic zones.

At some localities where rock avalanches overlay bouldery alluvium, one can observe entrainment of individual pebbles, cobbles and boulders into comminuted rock avalanche debris (Fig. 11.16). Much more intensive entrainment was observed at several deeply incised Karakoram rockslides (Hewitt, 2006b).

Regarding macro- and meso-structure of the basal sliding surfaces themselves at the majority of the observed outcrops, they look as a sharp linear boundary or very thin zone between heavily comminuted material of rockslide basal unit and underlying substrate (fluvial or glacial deposits). It could be seen at the frontal part of the Terek-Suu rockslide in Southern Tien Shan composed of carbonate debris (see Fig. 4.7), at Mini-Köfels rockslide composed of crushed gneiss (see Figs. 8.160 and 11.16) in the Kokomeren River valley, at the abovementioned Akterek and Djuzumdybulak rockslides (see Figs. 8.187 and 8.191) in the Naryn River valley composed of debris from the intensively tectonized granite (all in Central Tien Shan), at the western part of the Koman rock avalanche (see Fig. 9.23) in Northern Pamir.

At some sites, the structure of this contact is more complex. Interesting interaction of rockslide debris with underlying substrate was observed at the distal parts of the Bashi-Djaya rockslide deposits (see Fig. 8.76) in the Chong-Kemin River valley of Northern Tien Shan. In front of the Bashi-Djaya rockslide, just few meters thick crushed debris overlays soil layer developed on the bouldery till of the terminal $Q_{III\text{-}IV1}$ moraine about 120 m above the stream that can be seen in the road cut (see Fig. 8.77). Rockslide debris moving upslope just before its motion came to a halt had pushed (or pulled) the underlying paleosol about 60 cm thick so that blocks of the horizon "B" of the paleosol profile were thrusted over the blocks of brown organic-rich soil (horizon "A") and all the successions were truncated by advancing rockslide debris.

Something similar, though at a much larger scale, likely occurred at the frontal part of the much older Kiol-Kogur-Northern rockslide in the same valley (see Fig. 8.85) where several thrust shears crosscut the alluvial deposits about 10 m thick deformed by the advancing rockslide body (see Fig. 8.86B).

Interaction of rockslide debris with underlying soil was observed also in the trenches excavated at the front of the 1911 Ananievo rockslide at the base and on top of the 7- to 10-m-high bench composed of light brown loam (see Trenches C and B on Fig. 8.72). It is evident that rockslide debris scraped and bulldozed the soil layer that was in frozen state at that time. As we mentioned in Section 8.3 in Chapter 8, such bulldozing could damp motion of this rockslide at some extent. We want to mention that the substrate entrainment, especially of the substrate from the flat valley bottom, should slow down rockslide motion considering the fundamental low of the momentum conservation ($M \times V = \text{constant}$) because it increases the mass of the moving material without additional transformation of the potential energy of the entrained material into the kinetic one (that might occur if the entrainment occurs when rockslide moves downslope). This effect could be, however, minimized or, even, surpassed if the entrainment results in the significant decrease of the sliding resistance.

One more site where one can see the specific type of rock avalanche/substrate interaction is the northwestern part of the 34-km-long runout Koman rock avalanche where its body together with the underlying alluvium is dissected by the Kyzylsu River draining the Alai intermontane depression. Here dense and well-cemented rock avalanche breccia, locally hanging out over the alluvium in the present-day outcrops (see Fig. 9.23), is underlined by silty–sandy material about 15 cm thick with lenticular texture resting over visually undisturbed fluvial gravels (see inset in Fig. 9.23). Sedimentological analysis of grain size of these lenticular silty sands showed that they belong to the Kyzylsu fluvial sediment and do not reflect exceptional basal fragmentation (Reznichenko et al., 2017). If this conclusion is correct, it means that the basal fragmented zone along the base of the Koman rock avalanche is extremely thin being represented by the centimeter-thick fine-graded pink "layer" just above the sand with lenticular structure visible in the inset, at least at this locality. At the same time, we do not see that the Kyzylsu River terraces outside rock avalanche are covered by such "sandy layer." Thus, there should be some interrelation between its formation and rock avalanche propagation. We can speculate that the extraordinary long Koman rock avalanche could occur during the cold season when not only the Trans Alai Range but also bottom of the Alai intermontane depression was covered by snow that could act as a kind of lubricant and, even, melt due to heat produced by friction of rock avalanche on the substrate (and, likely, by internal crushing and shearing in moving debris). May be this combination could form such a lenticular zone? Anyhow, its true origin is still questionable and requires additional study.

Loess-like loam deformed by moving rock avalanche debris can be observed in the right-bank outcrop of the Inylchek rock avalanche frontal part (see Fig. 8.47). It looks as if this material was scraped from the surface of the older Atjailau rock avalanche by the advancing Inylchek rock avalanche and folded.

Significant variability of the observable interrelations between the rockslide (rock avalanche) deposits and underlying substrate demonstrates complexity and, likely, the multivariate nature of their interaction. As it was proposed in Grigorian (1979), the weakest of the interacting material (i.e., either moving rockslide debris or the substrate) undergoes maximal shearing, which minimizes the resistance force. In future, special attention should be paid to those sites where basal zones of large-scale rockslides could be observed and studied in detail.

11.3 EXPERIMENTAL SIMULATION OF ROCKSLIDES' INTERNAL STRUCTURE

The numerical and physical modeling is an important tool to better understand the processes acting during landslide formation and motion. However, modeling of extra-large rockslides and rock avalanches faces several problems. First, the giant size of these features and extreme velocities of motion intercept their physical modeling, taking in mind the necessity to adhere scaling conditions. On the other hand, lack of clear and univocal understanding of the nature of such rockslides' abnormal mobility complicates selection of motion equation, used for the numerical modeling.

In such a situation, the "conceptual modeling," when we just qualitatively reproduce in the laboratory particular phenomena (or their combination) really observed in nature, seems to be a promising way (Zerkal et al., 2017). It can be exemplified by the successive attempt to model intensive fragmentation of the rockslide/rock avalanche bodies' internal parts along with retention of the initial mutual position of various lithologies involved in slope failure (Dubovskoi et al. 2008). Modeled loads affecting rockslide debris are: (1) the quasi-static (lithostatic) compression produced by the overlying part of rockslide which initial thickness often exceeds the final one (see Figs. 5.38, 5.48, and 8.141 and their explanation in the text); (2) shearing that should take place in moving debris and which is most pronounced in the "Playing card model" proposed in Pollet and Schneider (2004) and Pollet et al. (2005) and based on the detailed study of the Films rockslide deposits.

Three series of experiments were carried out and the quasi-static compression was reproduced at each of them by the pressure of 5.68 Mp (that corresponds to ~200 m of the lithostatic load) on the uniformly graded sand-size (0.63–1 mm) particles of crushed marble in a 30-mm die mold. At two series of experiments, punches with flat face ends were used (Fig. 11.17A and B) and the third series was performed using punches with skewed face ends (Fig. 11.17C).

In the first series (A) after first compression sample was extracted from the mold and sieved to measure proportion of particles that was crushed. Then the material was put back in the mold, compressed, extracted, and sieved again, and so on. In this series of the experiments, materials underwent undamped crushing illustrated by Plot A on Fig. 11.18, but it was totally mixed when extracted that contradicts the phenomenon observed in nature and described in Section 11.1.

Further testing (Series B and C) was performed to reproduce particles crushing without mixing. In the second experimental series (B), we also used flat face punches, but samples were not removed from the die mold between the compressions, and sieving and weighting of crushed material were performed after a definite number of compressions. Results are illustrated by Plot B on Fig. 11.18, and one can see that crushing damps quickly so that after five

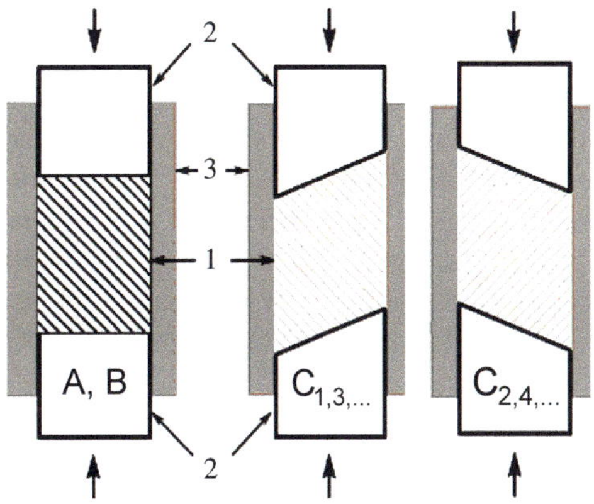

FIGURE 11.17 Scheme of experiments aimed to reproduce rock crushing combined with retention of its internal structure. 1: narrow graded (0.63–1 mm) sand; 2: punches with flat face ends used in Experiments A and B and with skewed face ends used in Experiment C; 3: die mold 30 mm in diameter; $C_{1,3,\ldots}$ and $C_{2,4,\ldots}$ demonstrate positions of the skewed face ends at each odd and even compression, respectively. See explanation in the text. *Modified from Dubovskoi, A.N., Pernik, L.M., and Strom A.L., 2008. Experimental simulation of rockslide fragmentation. Journal of Mining Science 44(2), 123–130, with permission of Springer.*

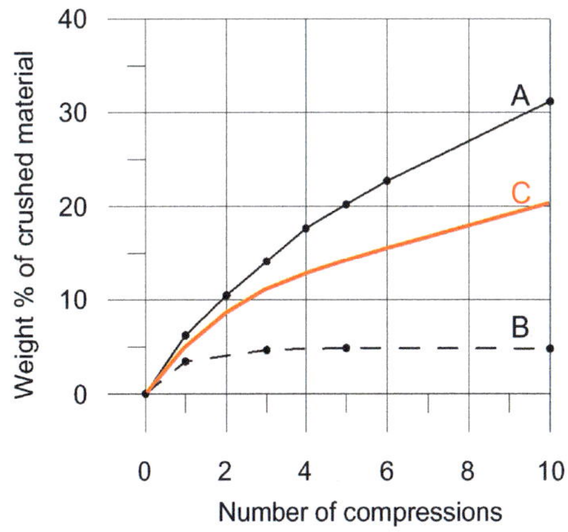

FIGURE 11.18 Plots demonstrating results of Experiments A, B, and C described in the text. *Modified from Dubovskoi, A.N., Pernik, L.M., and Strom A.L., 2008. Experimental simulation of rockslide fragmentation. Journal of Mining Science 44(2), 123–130, with permission of Springer.*

compressions, when about 10% of particles were crushed, fragmentation halts. It remains the same when the die mold with the tested sample was placed on a vibrating table (vibration amplitude 0.7 mm, frequency 20 Hz). In this series of experiments, mixing of the material did not occur, but fragmentation was quite limited, which does not fit the state of the material in the internal parts of rockslide bodies.

The third series of tests (C) was carried out using plunges with faces cut at an angle of 60 degrees to the driving direction (see Fig. 11.17C). After a compression, the upper plunge was removed, turned 180 degrees around vertical axis and put into the die mold again. Then, the assembly was turned up, and the same was done with the lower plunge. Thus, we produced some shearing within the tested sample. These test results are shown on Fig. 11.18 by Plot C. It demonstrates that the proportion of broken particles grows with every next compression, though not so quickly as in the first experimental series but not damping as in the second series.

To reveal whether mixing takes place or not, the tested sand samples were colored and, after some compressions, moistened with water and frozen. Thereby, we could remove it from the mold with its structure intact. It was determined visually that mixing was negligible.

Thus, in the latter experimental series, the progressive crushing of the material with its macrostructure retained unchanged, i.e., conditions observed in nature were reproduced successfully. Based on that, it is possible to explain intense fragmentation of a material in the lower (internal) parts of rockslide bodies by combined action of the lithostatic load and of shearing arising during rockslide motion.

CHAPTER

12

Morphological and Structural Evidence of Rockslides' (Rock Avalanches') Motion Mechanism(s)

OUTLINE

As one can see from the regional description of large-scale bedrock landslides (see Chapters 7–9: Dzungaria, Tien Shan, and Pamir, respectively), threat provided by such phenomena in mountainous regions, where most of settlements and infrastructure are concentrated on river terraces and in intermontane depressions, is predetermined by mechanism of its motion mainly, rather than by mechanism of the initial slope failure, which is the main criterion of most of the landslide classification systems (Varnes, 1954, 1978; Zolotariov, 1964, 1974, 1983; Emelianova, 1972; Cruden and Varnes, 1996; Cruden and VanDine, 2013; Hungr et al., 2012, 2014).

Under specific conditions, almost any type of rock slope failure—translational and rotational sliding, toppling, etc.—can convert into rapid flow-type motion of debris classified as "rock avalanche" (Hungr et al., 2014). Various mechanisms have been proposed to explain their extreme mobility such as air lubrication (Kent, 1966), lubrication by evaporation (Habib, 1975), by rock melting (Erismann, 1979), self-lubrication (Campbell, 1989), mechanical (Davies, 1982; Davies and McSaveney 1999, 2002, 2011; Davies et al., 2006, 2017; McSaveney and Davies, 2006), and acoustic fluidization (Melosh, 1979, 1983, 196, 1990), shearing of the weakest unit at the contact between rock avalanche and substrate (Grigorian, 1979; Ostroumov, 1986), floating of rock avalanche debris in dust (Hsü, 1975, 1978), effect of the basal guided wave (Kobayashi, 1993, 1997), playing cards model (Schneider et al.,

Rockslides and Rock Avalanches of Central Asia.
DOI: https://doi.org/10.1016/B978-0-12-803204-6.00012-0

1979, 1999; Wassmer et al., 2002; Pollet and Schneider, 2004; Pollet et al., 2005), and momentum transfer (Van Gassen and Cruden, 1989). Some of these hypotheses were found to be invalid on their physical grounds, as demonstrated for example by Erismann (1986), and by Erismann and Abele (1999). Peculiarities of rock–ice avalanche motion mechanisms were discussed in Schneider et al. (2011).

In Section 11.3 of Chapter 11, Structure and Grain-Size Composition of Large-Scale Bedrock Landslides, we described the experiments showing that intense fragmentation of a material in the lower (internal) parts of rockslide and retention of the host rock mass macrostructure, characteristic features typical of most of rock avalanches, can be explained by combined action of the lithostatic load and of shearing during rockslide motion.

Here we want to highlight some morphological and structural peculiarities typical of large-scale rockslide deposits and exemplified by numerous case studies from the Central Asian region that can help elaborating consistent and coherent model(s) of their motion mechanism(s) that could be considered as the basics for rockslide hazard assessment.

Unfortunately the eyewitness observations that could shed light on these phenomena are rare and incomplete. That is why most valuable data allowing better understanding of the abnormally high mobility of large-scale bedrock landslides could be derived from careful and rigorous comparative analysis of the morphology and internal structure of numerous historical and past rockslides both from the Central Asian region and worldwide. An attempt of reconstruction of the rock avalanche rheological properties and kinematics of emplacement could be found, for example, in Eppler et al. (1987).

12.1 EYEWITNESS OBSERVATIONS

Just few catastrophic rockslides worldwide have been described based on eyewitness observations. Among them are the 1881 Elm rock avalanche (Heim, 1882; Hsü, 1978), sectoral collapse of the 1980 St. Helens volcano (Voight et al., 1983; Glicken, 1996), the 1962 and 1970 Huascaran ice–rock avalanches (Plafker and Eriksen, 1978; Evans et al., 2009a), the 1998 Malpa rock avalanche (Bhandari and Kumar, 2000), and the 2005 Zengziya Rockslide in Chungching region in Central China (Yueping Yin, personal communication). In the Central Asian region interesting information could be derived from the accounts of eyewitnesses of the Khait rock avalanche that were summarized in Semionov and Semionova (1958) and, likely, can be applied to other similar phenomena (our comments to the following description are in parentheses).

According to these data on July 10, 1949, in the morning, the Khait town and the surrounding area were affected by severe vertical shock accompanied by noisy boom resembling the roar of guns and followed by strong horizontal shocks (those were vibrations of the nearby M7.4 earthquake). People could not remain standing, and some of them were even thrown up (indicating acceleration exceeding 1 g). All buildings in the town collapsed. Along with the continuous shaking and endless boom, some additional noise resembling, as one of the eyewitness said "sound of the stone mill," started far away and increased rapidly. Suddenly dark "wave" of debris up to 100–150 m high appeared from the Obi-Dara-Khauz River valley (such height was indicated according to the emotional description of eyewitnesses, and, judging from the traces of rock avalanche in the upper narrow part of the valley could be less, in authors' opinion). It moved as fast as an express train and buried the town. This surge was accompanied by very strong wind (air blast) that rooted trees out and threw them at a large distance. The same was observed even in the gardens that have not been reached by rock avalanche directly. It was also reported that although frontal parts of this avalanche stopped, the following surges of debris continued moving above and their deposits overlaid each other (these observations are in line with data on the internal structure of numerous rock avalanches described in Part 2 of this book and summarized in Chapter 11).

Additional information about composition of this rock avalanche was provided by G. Panasenko (1952) who found that axial part of the Khait rock avalanche is composed of gneiss and schist debris with blocks up to 300 m^3 and even 700 m^3, whereas its border zones, closer to the valley slopes, are composed of loess-like material that was partially stripped from the valley banks by moving rock debris and partially slid down from the slopes above.

Rapidly moving debris blocked the Yarhych River and spread for about 1.5 km over its right bank (see Figs. 8.271 and 8.272). It was also reported (Semionov and Semionova, 1958) that debris that buried the Khait town was muddy and well saturated, so that at the beginning it was impossible to walk on it. Such state of the material can be explained by a large percentage of loess-like loam saturated by the preceding rains. Besides, on the rock avalanche path, there were several small lakes, whose water was mixed with debris.

This rock avalanche is the only one phenomenon of this type for which we could find the description made by eyewitnesses that allows some assumptions on its motion mechanism. The eyewitnesses' description of the earthflows triggered by the 1887 Vernyi earthquake (see Chapter 2: Brief History of the Bedrock Landslide Studies in the Central Asia) was reported in Mushketov (1890), but these phenomena are beyond the scope of our book. No such description is available for the 1911 Usoi rockslide because three inhabitants of this village who survived were far away from the site, for the Ananievo, Chilik, and Kaindy rockslides triggered by the 1911 Kemin earthquake and for the Belaldy rockslide triggered by the 1992 Suusamyr earthquake.

12.2 GEOMORPHIC INDICATORS OF THE EMPLACEMENT MECHANISM(S)

Well-grounded conclusions on rockslides motion mechanism(s) can be made based on the comparative analysis of the otherwise similar rock slope failures that form rockslides and rock avalanches with the significantly different shape of the deposits. Such analysis of rockslides that originated on the slopes composed of the same type of rocks (granites) with more or less similar volume of the displaced material, slopes height, and steepness allowed selection of the primary, secondary, and jumping types of rock avalanches demonstrating significantly different mechanical processes governing their mobility (see Section 5.3 in Chapter 5: Bedrock Landslide Types and Classification Systems) (Strom, 1996, 2006, 2010a).

Geomorphic indicators allowing some conclusions on the emplacement mechanism(s) of rock avalanches are the along-way distribution of debris, the position of the trimlines (patches of rock avalanche debris that remain on slopes well above the resultant deposits surface) and significant run-up typical of many rock avalanches with frontal confinement. Accumulation of most of the debris involved in the slope failure at the distal part of the transition–deposition zone of the primary rock avalanches indicates flow-like style of motion—crushed rock moves downslope or downvalley like a liquid. Similar debris distribution often characterizes mobile parts of the secondary rock avalanches (see Section 5.3.3 in Chapter 5). Evidence of fluidization was described for Jumping rock avalanches as well (see Section 5.3.2 in Chapter 5).

12.2.1 Indicators of the Extremely High Velocity of Debris Motion

High velocity of rockslide (rock avalanche) debris motion can be derived, besides eyewitness observations ("it moved as fast as an express train"), from the numerous evidence of significant superelevation (run-up) of debris allowing calculating velocity of rock avalanche using simple equation $V = (2 \times g \times h)^{0.5}$, where V—velocity of moving debris, g—gravity acceleration, h—superelevation. Superelevation either left trimlines on the slopes at the travel path bends or result in significant frontal run-up. Trimlines on the slopes at the travel path bends, sometimes up to 100 m above the main debris surface were described at the prehistoric Seit (see Fig. 5.39) or Chongsu (see Fig. 5.48) rock avalanches and resemble effects observed at the 1959 Pandemonium Creek rock avalanche (Evans et al., 1989) or the 2002 Karmadon ice–rock avalanche (Nikitin et al., 2006). In the Chongsu case some boulders ejected from moving rock avalanche were found on the slope above the continuous trimline (see Fig. 5.49). Similar ejection of boulders was described for the 1970 Huascaran ice–rock avalanche (Plafker and Eriksen, 1978).

Velocities calculated from the trimlines superelevation reach ca. 175 km/hour. In fact real velocity should be even higher because the equation $V = (2 \times g \times h)^{0.5}$ ignores kinetic energy consumption needed to overcome the basal friction of debris rising on the slope. Similar and, sometimes, much higher velocities were calculated from the run-up of debris on the opposite valley slopes widely observed at rockslides with pronounced frontal confinement. Such run-up in the study region sometimes exceeds 200 m and reach 370 m at the Big Dragon Lake rockslide (see Fig. 5.65), which correspond to the rockslide debris velocity up to 307 km/hour. Even higher values—up to 360–390 km/hour could be calculated for the Karivhoh rock avalanche in Northern Caucasus with ca. 500 m run-up (Strom, 2004) and for the Avalanche Lake rock avalanche in the Mackenzie Mountains where run-up was about 600 m (Evans et al., 1994).

Forty-six Central Asian rockslides and rock avalanches with frontal run-up equal or more than 100 m described in the previous chapters are listed in Table 12.1. Statistical analysis of this parameter and of its relationship with rockslide volume, with height drop and with their product—parameters proportional to the energy released during rockslide emplacement, was performed for 120 case studies with pronounced run-up (Fig. 12.1), most of which can be ascribed to the primary type (see Section 5.3.1 in Chapter 5). Run-up was always measured from the lowermost part of the dam's crest, not from the valley bottom; height drop—as a vertical distance from the headscarp crown to the deposits tip (H_2 on Fig. 5.34).

At all plots shown on Fig. 12.1, points corresponding to the Big Dragon Lake rockslide ~2000 Mm^3 in volume and the Padjvar rockslide ~ 6000 Mm^3 in

TABLE 12.1 Rockslides and Rock Avalanches with Frontal Run-up Equal or More than 100 m Described in this Book

ID	Name of the feature or of the river valley where it is located	Figure illustrating the phenomenon	Volume (10^6 m^3)	Height drop to the tip (km)	Run-up (km)	Velocity (km/h)
China 015	Big Dragon Lake[a]	Fig. 5.65	2000	0.67	0.37	307
Afghan 001	Padjvar	Fig. 9.104	6000	0.72	0.35	298
China 060	Dzhergalanty River	Fig. 8.23	150	0.87	0.21	231
Kazakh 068	Upper Zhasylkol	Fig. 6.18	140	1.19	0.19	220
China 002	Muzart River	Fig. 5.40	90	0.50	0.17	208
Kyrgyz 016	Akterek	Fig. 8.187	10	0.46	0.17	208
Kyrgyz 085	Mini-Köfels	Fig. 5.41	12	0.55	0.17	208
Kyrgyz 109	Kainar River	Fig. 8.116	130	0.79	0.17	208
Tajik 049	Kuturgansu-mouth	Fig. 8.246	300	0.62	0.17	208
Tajik 198	Chandindara River	Fig. 9.93	12	0.65	0.16	202
China 146	Keqi Kekule	Fig. 8.29	150	0.77	0.15	195
Kyrgyz 071	Chong-Kongur-Debe River	Fig. 8.197	20	0.70	0.15	195
Kyrgyz 086	Kokomeren	Fig. 8.154	1500	1.30	0.15	195
Kyrgyz 357	Djilusuu River	Fig. 8.118A	150	0.57	0.15	195
Tajik 014	Iskanderkul	Fig. 8.287	700	1.42	0.15	195
Tajik 090	Usoi	Figs. 5.7, 9.41, and 13.10	2200	1.06	0.15	195
Tajik 091	Yashilkul	Figs. 4.18 and 9.65	50	0.88	0.15	195
Tajik 099	Ruzhak	Fig. 9.76	13	1.01	0.15	195
Afghan 037	Aqshireh	Fig. 9.119	40	1.35	0.14	189
Tajik 040	Khushtarita-western	Fig. 8.268	150	1.85	0.14	189
Tajik 040a	Khushtarita-eastern	Fig. 8.268	40	1.00	0.14	189
Tajik 061	Big Allo	Fig. 8.269	100	0.70	0.14	189
Tajik 164	Zirgazdara River	Fig. 9.77	15	1.23	0.14	189
Uzbek 014	Badak	Fig. 8.228	13	0.78	0.14	189
Kyrgyz 011	Lower Kokomeren	Fig. 14.3	20	0.73	0.135	185
China 151	Aerpa Aigezi	Fig. 9.8	8	0.65	0.13	182
Kyrgyz 145	Akterek	Fig. 8.253	40	1.26	0.13	182
Tajik 123	Murgab River	Fig. 9.28A	25	0.26	0.13	182
Kazakh 044	Kiol-Sai Lower	Fig. 8.65	135	0.60	0.125	178
Kyrgyz 105	Ukok	Figs. 8.124 and 8.125	200	0.88	0.125	178
Kyrgyz 055	Kutmankul	Fig. 8.203	55	0.44	0.12	175
Kyrgyz 100	Chong-Tash	Fig. 8.135	11	0.31	0.12	175
Tajik 103	Urmochdara	Fig. 4.41	50	1.26	0.12	175
Afghan 005	Yorkhdara	Fig. 9.106	35	1.16	0.11	163
Afghan 035	Ardar[b]	Fig. 9.117	17	1.09	0.11	163
China 005	Urtentai River	Fig. 8.40	300	0.50	0.1	159
China 124	Kash River	Fig. 5.33	20	0.55	0.1	159

(Continued)

TABLE 12.1 (Continued)

ID	Name of the feature or of the river valley where it is located	Figure illustrating the phenomenon	Volume (10^6 m^3)	Height drop to the tip (km)	Run-up (km)	Velocity (km/h)
China 133	Upper Muzart	Fig. 8.35	60	1.50	0.1	159
Kazakh 002	Sarydjayak	Fig. 8.231	30	1.88	0.1	159
Kazakh 023	Great Almaty Lake	Fig. 8.89	450	0.50	0.1	159
Kazakh 075	Northern Dzungaria	Fig. 7.8	16	0.64	0.1	159
Kyrgyz 073	Ai-Kul	Fig. 8.257	75	1.17	0.1	159
Kyrgyz 110	Bashkol	Fig. 6.20	50	0.95	0.1	159
Tajik 052	Kafirnihon	Fig. 8.274	800	0.97	0.1	159
Tajik 092	Kudara-Pasor	Figs. 4.20 and 9.38	1000	0.63	0.1	159
Tajik 107	Gunt River	Fig. 9.67	9	0.58	0.1	159

[a]*Only compact part with primary type of motion is considered.*
[b]*Rockslide likely of the jumping type.*

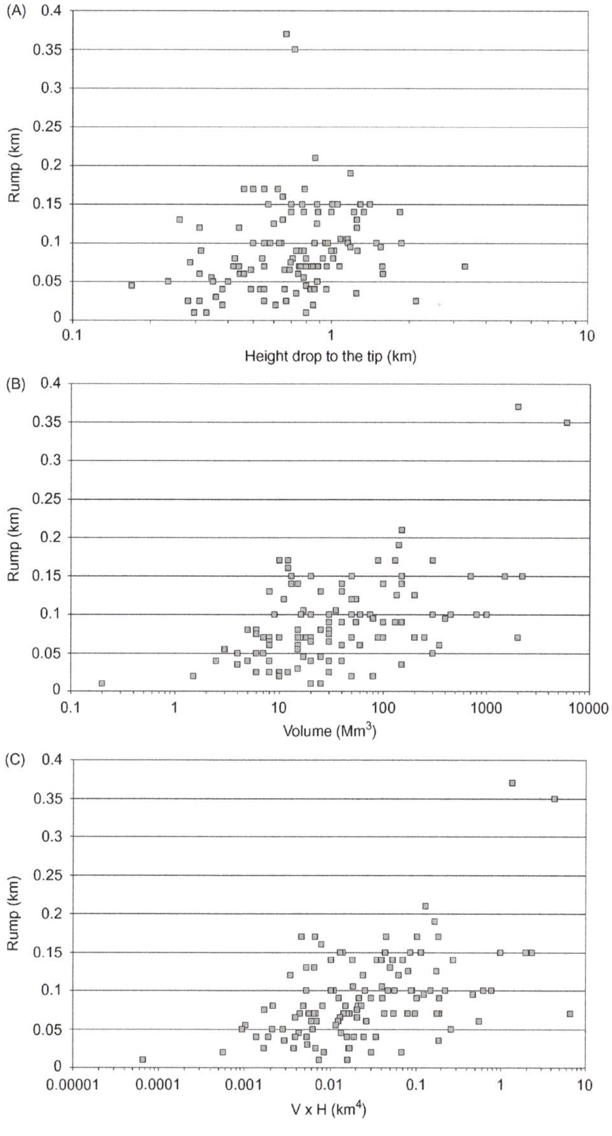

FIGURE 12.1 Relationships between (A) run-up and height drop; (B) run-up and rockslide volume; (C) run-up and product of rockslide volume and height drop.

volume differ significantly from the entire sample. It can be assumed that the cause of their abnormal mobility could be the same as for the Avalanche Lake rock avalanche (Evans et al., 1994)—sliding of the huge rock mass along bedding planes that resulted in the abnormally low basal friction at the initial stage of rock slope failure and, thus, in higher velocity and unit momentum gained after the initial descend.

Analysis of the relationships between these parameters shows that there is no correlation between run-up and height drop (Fig. 12.1A) and quite weak ($R^2 = 0.2781$) positive correlation between run-up and rockslide volume (Fig. 12.1B). Similarly weak positive correlation ($R^2 = 0.2552$) was found for the relationship between run-up and the product of height drop and rockslide volume (Fig. 12.1C) that characterizes the potential energy of the collapsed rock mass better. Even if we ignore the extreme events such as the Koman rock avalanche ~2000 Mm^3 in volume with 3.3 km height drop that had traveled ca. 34 km before rising about 70 m on the northern slope of the Alai intermontane depression or the above-mentioned Big Dragon Lake and Padjvar rockslides with rather low height drop (670 and 720 m only) and extremely high, more than 300 m run-up (see Table 12.1), correlations remain quite poor, indicating that the velocity of rockslide debris motion should be influenced significantly by some additional factors.

We hypothesize that rockslide debris motion velocity can be governed at a large extent by the curvature of the slope base (as it was proposed in Strom, 2006) and of the entire valley bottom along which rockslide moves before coming to the halt. Any impact with an obstacle (terrace cliff, other irregularity on the slope) should consume some momentum, while rockslide body moving along smoothly curved surface should move farther (in the absence of frontal confinement) or rise higher on the opposite slope. Quantitative analysis of the effect of the travel path smoothness and curvature requires additional detailed investigations based on digital elevation models (DEMs) with much higher resolution than available at present. We have to notice also that the absence of runout does not mean low velocity of debris motion before the collision with opposite slope—momentum can be consumed by impact itself or by transverse (along-valley) debris spreading.

12.2.2 Indicators of Rockslide Debris Mechanical Properties Alteration During its Motion

Besides trimlines at the rock avalanche travel path bends described above that allow estimation of debris velocity, one can often find high trimlines along more or less straight sections of rock avalanches indicating significant thinning of debris during its motion. Such trimlines that could be traced for hundreds of meters and kilometers along the travel path of the laterally confined rock avalanches up to 200 m above the resultant surface of the deposits, demonstrate that at the initial stage of flow-like motion rock avalanche body had moved downvalley as a rather compact high "surge" that later on converts in less viscous granular flow. Most impressive examples are listed in Table 12.2. As it was mentioned above, in Section 12.1, rockslide debris can move as a series of surges each of which overlays the deposits left by the previous one.

Relationships between the trimlines' altitude on both sides of the travel pass of the Mingteke rock avalanche and of the resultant surface of debris below them (see Fig. 8.141) allows assumption that debris mobility is correlated directly with the initial thickness of rock avalanche body when it passes any

TABLE 12.2 Rockslides and Rock Avalanches With Trimlines Indicating Significant Decrease of the Initial Debris Thickness

ID	Name of feature or of the area where it is located	Figure illustrating the phenomenon	Level difference between the trimlines and final surface of the deposits (m)
Tajik 093	Imom	Fig. 4.14	80–90
Kyrgyz 240	Seit	Figs. 5.14 and 5.38	100–140
Kazakh 001	Karakystiak	Fig. 6.7	~200
China 037	Sarmin-Ula Range	Fig. 6.15	70–100
Kyrgyz 092	Mingteke	Fig. 8.141	50–100
Afghan 045	Dakhanzar River	Fig. 9.123	80
Kyrgyz 009	Ak-Kiol	Figs. 5.10 and 5.11	115

section of the travel path. At this particular case, we can see that, due to some bending of the travel path, the trimline at the true left side of rock avalanche is 50 m higher than that on its true right part (2950 m a.s.l. vs. 2900 m a.s.l.), while the resultant surface of debris at the left side of this section is much lower. Thus, material at its left side was more mobile and "flowed" downvalley more intensively. We hypothesize that it could be due to more intensive fragmentation of the lower part of rapidly moving rockslide body where lithostatic loading was higher. Such fragmentation could decrease the viscosity of the granular flow that should be considered in numerical modeling of rockslides motion.

We have to notice that such behavior characterizes not only primary rock avalanches but also avalanche-like mobile parts of the secondary rock avalanches. The Mingteke rock avalanche, in particular, is of the secondary type (see its description in Section 8.4 in Chapter 8). Tracing of the trimlines above the mobile part of the Chongsu secondary rock avalanche indicate constancy of moving debris thickness up to the middle part of the transition zone (see Fig. 5.48). After that trimlines sharply come down merging the resultant rock avalanche surface level. Sharp decrease of the moving debris thickness means that debris spreads more intensively either just forward or also sidewise, i.e., its mobility increases. It could occur when fragmentation described in Chapter 11 (in Section 11.2.2) exceeds some critical level resulting in debris rapid fluidization.

12.2.3 Indicators of the Momentum Transfer During Rockslide Emplacement

One of the most characteristic landforms typical of the secondary rock avalanches, at least of the classical subtype (see Section 5.3.3.1 in Chapter 5) are the secondary scars—the concave slopes of the compact bodies above the avalanche-like highly mobile parts. Most impressive examples of such landforms can be seen on Figs. 4.30, 5.47, 5.53, 5.62–5.65, 6.13, 6.15, 7.10, 8.20, 8.88 (at 1550 m a.s.l.), 8.100, 8.250, 8.255, 9.10, 9.52, 9.108, 9.121, 9.123. Such landforms are less pronounced at secondary rock avalanches of the bottleneck subtype (see Section 5.3.3.2 in Chapter 5). As we mentioned above describing these highly mobile rock avalanches, it was hypothesized (Strom, 2010a) that such features originate due to momentum transfer from the rapidly decelerating portion of debris that accumulates at the proximal part of the deposition zone to the portion retaining possibility of further motion. The latter accelerates resulting in additional mobility that is proved by the comparison of the relationships of runout with volume, height drop, and their product for the mobile rock avalanches in their entirety and just for mobile tongues of the secondary rock avalanches (see Table 10.1 and Fig. 10.16). In the proposed model of secondary rock avalanche formation, the secondary scar just marks sharp difference between two successive stages of rockslide debris motion that results in the bipartite along-way debris distribution. Role of momentum transfer (in other words, delivering the kinetic energy of the rear part to the front) in the mobility of rock avalanches was mentioned also in Wang et al. (2016).

However, evidence of the momentum transfer in the course of rockslide motion could be found not only at long runout secondary rock avalanches. Indications of such phenomenon could be derived also from the morphological analysis of compact blockages, in particular, of those that block river valleys at their confluences with tributary valleys. Most informative are those case studies where the collision zone is dissected by the tributary valley mouth at an almost right angle. Comparison of such rockslide dams' shapes shows clear relationships between position of the tributary mouth and of the along-valley blockage profile (Strom, 2010a).

Rockslide dams formed at valleys' sections where the collision zone is dissected by a tributary mouth or by the distinct lowering of the valley slope just at its axial part are most compact with rather steep up- and downstream slopes. They can be exemplified by the Aksu rockslide in the Kyrgyz Range (see Figs. 8.103 and 12.3A), by the Kulun rockslide in the Ferghana Range (see Fig. 8.237), by rockslide in the Khutubi River valley (see Fig. 8.9). Likely similar effect could be observed at the Köfels rockslide in the Tyrolean Alps (Brückl et al. 2001; Dufresne et al., in press). One of the best examples is the Karachauli rockslide in the Kokomeren River valley, whose frontal part composed of Neogene conglomerates had intruded in the tributary valley for more than half kilometer transporting large amount of alluvium (see Figs. 8.172, 8.173, 8.176, and 12.2).

Such morphology of the collision zone plays a role of a trap that, if being large enough, consumes significant part of the overall rockslide momentum due to friction with it slopes and bottom and by the inevitable rise over the inclined bottom of a tributary valley. Such consumption prevents debris spreading along the main blocked valley, so that up- and downstream slopes of the blockage dip at a natural angle of repose typical of bouldery accumulations (see Fig. 12.3A).

An exclusion is the rock avalanche in the central part of the Rivak River valley in Pamir (see Fig. 9.77) that protruded for about 1.0 km upstream the

FIGURE 12.2 View on the frontal part of the Karachauli rockslide from the Kensu River valley (Ke), whose mouth had been filled by the advancing rockslide at a distance of about 500 m. The Kensu River had to erode the new bypass gorge through the bedrock Neogene (N). Rockslide deposits are composed of crushed Neogene conglomerates (N-RS) with inclusions of bouldery Quaternary alluvium (see insets). Ko, the Kokomeren River valley from where rockslide had moved towards the observer.

tributary valley and also spread up and down the main river valley forming relatively thin (ca. 30 m thick only) and 2.3 km wide apron. It can be assumed that such spreading occurred here before the effect of the momentum consumption started playing a significant role because there was not enough material for such effect to be fully displayed. Numerically it can be characterized by ratio between rockslide volume and valley width. In this case this ratio is $\sim 15 \times 10^3$ m^2 that is much less than such ratio in other cases described above ($\sim 67 \times 10^4$ m^2 for the Aksu, $\sim 13 \times 10^4$ m^2 for the Kulun, and $\sim 80 \times 10^4$ m^2 for the Karachauli cases). It is higher even at smaller rockslide in the Khutubi River valley shown on Fig. 8.9 ($\sim 35 \times 10^3$ m^2).

In all the cases mentioned above the "tributary traps" coincide with the axial parts of the collision zones (see Fig. 12.3A). If, however, such trap is located asymmetrically to the collision zone axis, lateral spreading along the main valley takes place at that part of a dam only that is farthest from the "trap," irrespective of whether such slope lowering is at the up- or at the downstream side of the blockage. Another side where part of the momentum was consumed by the "tributary trap" remains much steeper. It can be illustrated by the Usoi rockslide where the tributary Shadau River mouth was upstream of the dam axis (see Figs. 9.41 and 12.3B), and by the Djashilkul rockslide in the Tien Shan where the "tributary trap" was shifted towards the downstream limit of the impact zone (see Figs. 8.79, 8.83, and 12.3C). In the Usoi case such transverse spreading resulted in formation of the impressive secondary scarps (see Fig. 9.43) that are visible well not only on the modern space images (see Fig. 9.41) but also on the topographic map of the blockage made several years after dam's formation (see Fig. 9.44). This case study can be treated as an intermediate type between the primary rock avalanche with frontal confinement and the secondary rock avalanche (see Section 5.3 in Chapter 5).

It can be hypothesized that similar mechanism can be applied to explain the brandung phenomenon (Heim, 1882, 1932; Hewitt, 2002). When rockslide collides with an opposing valley slope, debris spreading due to elastic behavior of rockslide mass that is compressed in the longitudinal direction and expands transversely towards a "free space" can occur not only laterally, resulting in the along-valley debris spreading, but vertically (upslope) as well. Under favorable conditions, the frontal part of the material, being

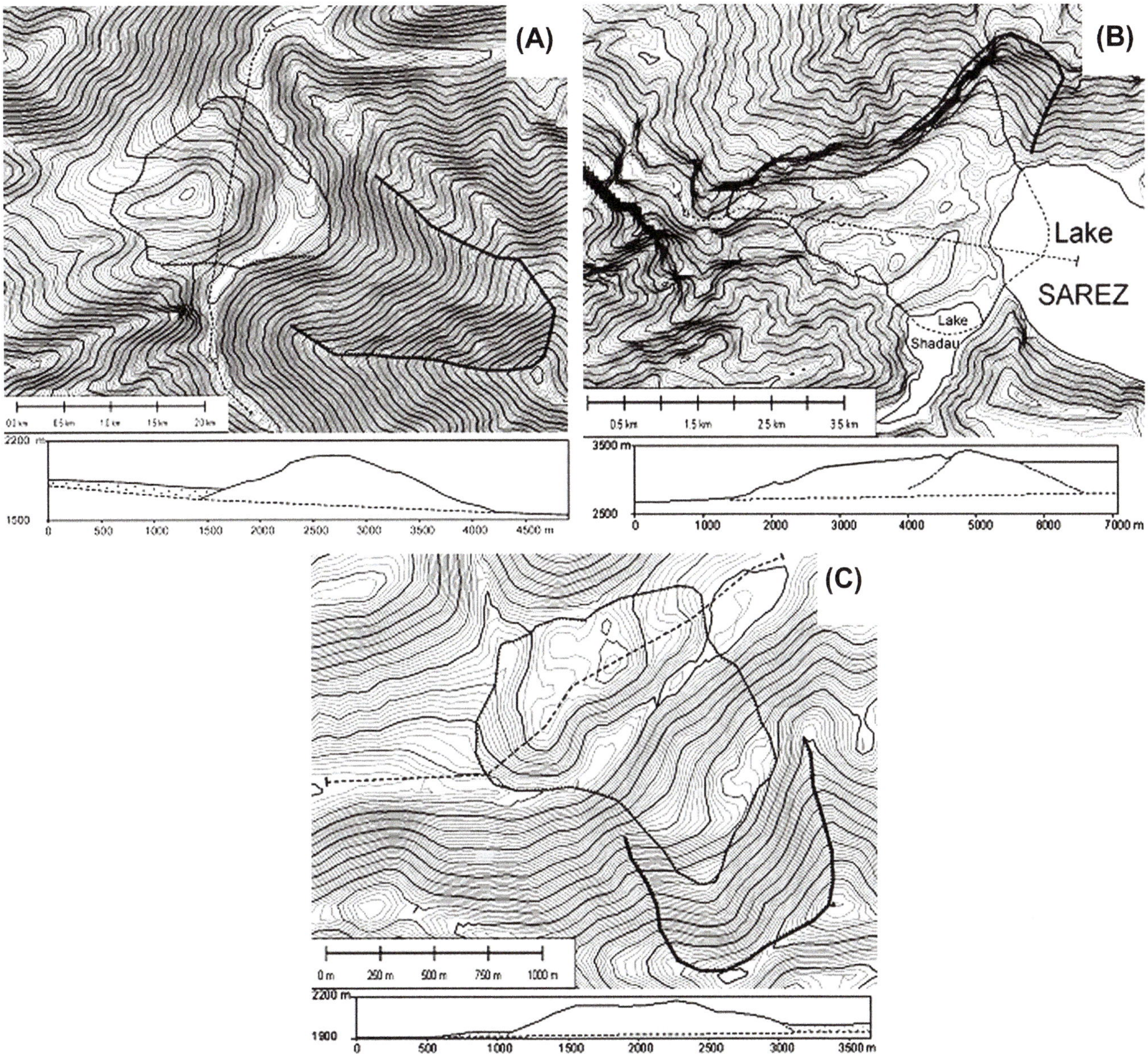

FIGURE 12.3 Plan views and along-valley cross-sections of rockslide dams, whose distal parts entered tributary valleys. (A) The symmetric Aksu rockslide dam (see Fig. 8.103); (B) the Usoi rockslide with distinct downstream spreading (see Figs. 9.41 and 9.43); (C) the Djashilkul rockslide with more pronounced upstream spreading (see Figs. 8.79 and 8.81). *Modified from Strom, A.L., 2010a. Evidence of momentum transfer during large-scale rockslides' motion. In: Williams A.L., Pinches G.M., Chin C.Y., McMorran T.G., Massei C.I. (Eds.), Geologically Active. Proceedings of the 11th IAEG Congress, Auckland, New Zealand, 5–10 September 2010, Tailor & Francis Group, London, 73–86, with permission of Taylor & Francis Group.*

compressed by the advancing rockslide, can gain upward-directed momentum and form a debris "splash" that rises above the main rockslide dam surface. Such mechanism recalls somehow mechanism proposed to explain excessive mobility of jumping rock avalanches (see right sketch on Fig. 5.42, if we rotate it 90 degrees clockwise). In the study region the landform that can be classified as brandung can be seen best at the front of the Big Dragon Lake rockslide (see Fig. 5.65).

12.2.4 Shape of the Unconfined Rock Avalanche Bodies as Indicator of the Rock Avalanche Base Shear Strength

As discussed above in Section 12.2.2, rock avalanche bodies often undergo significant thinning at some stage of their motion. This phenomenon must increase runout distance, especially for rock avalanches in laterally confined conditions when decrease of debris thickness can be compensated by frontward increase of its linear dimensions only. It is confirmed by the statistical relationships between rockslide runout and volume (see Fig. 10.8). Similar increase of debris mobility takes place if debris spreads over flat unconfined area as well (Shemiakin, 1993), and its effect is also well expressed if we consider total area affected (A_{total}) as the measure of rockslide mobility (see Section 10.3.3 in Chapter 10: Quantitative Relationships of Central Asian Bedrock Landslide Parameters). Because in such cases debris can spread in all directions, this parameter increases with volume growth faster for unconfined cases than for rockslides with other confinement conditions (see Fig. 10.18). It also results in lower values of H/A_{total} ratio (that was used as an analogy with the angle of reach) for unconfined rockslides (see Fig. 10.19) and by faster growth of A_{total} when compared with the product of maximal height drop and volume better reflecting the potential energy available (see Fig. 10.21).

Another parameter reflecting debris behavior is the ratio of the deposits maximal width (W_{dep}) and the headscarp base width (W_{scar})—values characterizing linear dimensions transverse to debris motion direction (see Sections 6.2 and 6.3). We mentioned earlier, in Section 10.3.3 that some rock avalanches that moved over unconfined planar surfaces of intermontane depressions or wide valley bottoms form narrow tongues of debris (Fig. 10.22A, see also Figs. 5.32, 5.61, 8.130, 9.17, and 9.20), while other, most of which had moved over well-developed floodplains or low terraces of wide river valleys, spread sidewise significantly forming either fan-shape bodies (see Figs. 4.36, 4.38, 6.8, 8.7, 8.221, 8.222, 8.271, 8.272, 9.6, and 10.22B) or pancake-shape bodies (see Figs. 10.22C, 5.33, and 6.4).

What factor(s) could govern either type of rock avalanche deposits shape—without sidewise spreading (monodirectional rock avalanches) or with significant transverse component of debris motion? It was hypothesized (Strom, 2006) that shape of the debris apron in unconfined conditions depends on the shear strength of the basement over which rockslide (rock avalanche) moves. Most of the monodirectional rock avalanches listed above and similar features from other regions such as the Blackhawk rock avalanche in California (Shreve, 1968; Johnson, 1978), the Sierra Laguna Blanca, and Sierra Aconquija rock avalanches in Argentina (Hermanns and Strecker, 1999; Hermanns et al., 2001) moved over dry bottoms of intermontane depressions in the regions with arid climate. It is interesting that similar morphology have some Martial rockslides (Shaller, 1991), which, most likely, also occurred in dry, extremely arid conditions. It is not surprising that in the Central Asian climatic conditions, monodirectional unconfined rock avalanches with W_{dep}/W_{scar} ratio <2 prevail (see Figs. 10.23 and 10.24).

In contrast, above-mentioned fan-shaped or pancake-shaped rock avalanches formed when rockslides descend onto well-saturated wide floodplains or low terraces of river valleys. Similar shape have the Kazarah, the Satpura Lake-Skardu, the Ghoro Ghoh and some other rock avalanches in the wide formerly glaciated trough valleys of Indus, Gilgit and Hunza Rivers in Karakoram (Hewitt, 1988, 1999), and the 2006 Leyte (Guinsaugon) rock avalanche in Philippines (Catane et al., 2007, 2008; Evans et al., 2007; Sassa et al., 2010). Fan-shaped debris aprons were formed when large-scale rockslides collapsed onto glaciers as in Alaska in 1964 (McSaveney, 1975, 1978) and in 2002 (Jibson et al., 2006). There are of course some exceptions both in the study region (see Figs. 4.32 or 8.7) and in other parts of the world (some rock avalanches that have been found on the glaciers have a stripe-like elongated shape (Evans and Delaney, 2014; Strom, 2014a)), but in the majority of cases such regularity seems to be correct.

Rock avalanche material descending from the high slope gains momentum (which is a vector value) that determines its motion in the same direction as the horizontal projection of the initial collapse. Besides, as we mentioned above, thickness of rapidly moving debris "layer" decreases gradually that also can be considered as the conversion of the potential energy (this time of moving debris) into the kinetic one. Considering that the initial thickness of moving rock avalanche body could reach 100 m, if not more, such conversion can produce additional velocity $V_{add} = (2 \times g \times \Delta h)^{0.5}$, where Δh is the difference between the initial and final thickness of debris. If such difference is ca. 50 m, which is quite reliable estimate, calculated V_{add} value would be of the order of 100 km/hour. Of course in fact it should be less because during debris thinning, some energy is consumed to overcome internal friction. What is critically important, in unconfined conditions such additional velocity and momentum produced by debris thinning should be equidirectional and, thus, should be directed both forward (increasing the velocity of debris front), backward (decreasing that of the proximal part of rock avalanche body), and sidewise, transverse to the overall direction of rock avalanche motion. Such directed addendum should be proportional to the linear dimension of thinning body in corresponding direction.

When rock avalanche moves over dry unconfined surface at the ridge foot formed by merged alluvial fans with numerous, though small-size, longitudinal levees and channels (see Figs. 5.32, 8.130, 9.17, and 9.20) composed of dry, mainly clast-supported deposits with rather high shear strength, this additional momentum is not enough to overcome basal friction. Thus, in such cases rock avalanches' debris spreads forward in the same directions as that of the initial descend and forms the monodirectional bodies. Impossibility of lateral spreading should result in a larger proportion of the additional momentum produced by debris thinning directed straight forward, which should increase the velocity of rock avalanche front and its overal runout significantly.

We have to notice that such model differs somehow from those proposed in Van Gassen and Cruden (1989) and Wang et al. (2016). We hypothesize that additional momentum resulting in acceleration of rock avalanche front is provided by conversion of vertical thinning of debris into increase of its plan dimensions (considering constancy of volume) rather than by the "delivering of the kinetic energy of the rear part to the front" (Wang et al., 2016). The latter mechanism works in case of the secondary rock avalanche formation only (see Section 12.2.3).

However, when rock avalanche moves over the saturated gravels on wide, flat flood plains of postglacial valleys with shallow water table, or over ice and snow on glacier surfaces, situation could be quite different. Shear strength of such basal materials is much less than in the previous case. Moreover, dynamic loading produced by rapidly moving rockslide on its base can cause liquefaction of saturated basal material similar to that caused by earthquakes (Sassa et al., 1994; Fukuoka et al., 2007) and, even, ejection of liquefied material at the base of moving debris. We can speculate that such effect could occur, at least locally, at the base of the Koman rock avalanche ("layer" of fine lenticular sand visible on Fig. 9.23). In case of ice–snow base they can even melt due to heat produced by friction. Such phenomena should decrease friction resistance of these surfaces significantly that can facilitate sidewise debris spreading. It should result in the formation of wide and thin fan- or pancake-shaped aprons. Discussion on the role of the mechanical properties of the substrate material in rock avalanche propagation could be found also in Dufresne (2009, 2014).

We can also hypothesize that some irregularities of the basal shear strength could be the main cause of the "finger deposition" described at the Marquartstein and Altenau rockslides in Bavarian Alps by A. von Poschinger (1994). In the study region "finger-like" deposition characterizes the Lenin rock avalanche in the Alai valley (see Figs. 9.24 and 9.25), but in this case rock avalanche debris followed the erosional depressions in older glacial deposits (Reznichenko et al., 2017).

Possible evidence of mobilization of the substrate material and its involvement in rockslide debris motion were found at the relatively compact Usoi rockslide and at long runout Archabel rock avalanche. Splash of much finer material can be seen at least at one side in front of the downstream slope of the Usoi rockslide about 250 m below dams' crest (see Fig. 9.45). Similar splashes are in front of both branches of the Archabel rock avalanche (see Fig. 9.4). We hypothesize that in both cases it is the material from the valleys' bottom, liquefied and mobilized by advancing rockslide debris.

Analysis of several rock avalanches in the study region that moved over flat areas and stopped without any frontal confinement shows that their distal rims are, usually, smooth in plan view, without significant protrusions (see Figs. 4.14, 4.15, 5.32, 5.51, 5.54, 6.8, 8.7, 8.16, 8.122, 8.130, 8.232, 8.252 (No 2), 8.272, 9.17, 9.18, 9.29, and 9.97). It indicates simultaneous and abrupt halt of motion along the entire debris front, probably when remaining momentum cannot produce driving force large enough to overcome basal and internal friction. We have to notice, however, that there are case studies when some clasts had been thrown out in front of the main body. It was described, in particular, at the Avalanche Lake rock avalanche (Evans et al., 1994). Such distribution was also reproduced in the experiments presented in Wang et al. (2016).

12.2.5 Debris Apron Outside Headscarp Sidewalls—Indicator of Buckling

Several highly mobile rockslides demonstrate quite specific spreading of debris over the ridge slope, indicating that after rock massif destruction debris moved not only downslope, according to the gravity force, but also sidewise, along the slope, outside the headscarp boundaries, and formed thin and rather wide apron blanketing otherwise undisturbed slope. Such phenomenon was observed at the Spur rockslide in Karakudjur River valley (see Fig. 5.4), and at the Beshtash rockslide close to the Talas intermontane depression (see Fig. 8.198). It is assumed that in these cases either rapid or gradual slope deformation starts from the headscarp crown, which result in bulging of the lower part of the slope (the so-called "buckling") and its excessive stressing. When the ultimate strength of this rock mass is exceeded, it spalls in all directions, so that some portions of the resultant debris gains momentum directed parallel to the slope and moves not only forward (downslope) but sidewise as well.

12.2.6 Molards as Indicators of Rockslide Motion Mechanism(s)

Molards described in Section 11.2.1.2 in Chapter 11, are impressive minor landforms that can be interpreted as indicators of localized extension within the blocky carapace through which material from the lower parts of debris is ejected upward (Johnson, 1978). Such mechanism somehow recalls sand blows associated with seismically induced liquefaction of saturated sand (see, e.g., chapter 7 in McCalpin (2009)). It corresponds well with the phenomenon of debris frontward and sidewise spreading due to thinning of moving rock avalanche discussed above in Sections 12.2.2 and 12.2.4. Because intensive fragmentation that can cause change of the mechanical properties of moving debris affecting rockslide/rock avalanche mobility takes place at its inner and basal parts mainly (see Chapter 11), coarse blocky carapace moves more passively and, thus, increase of plan view linear dimensions of the entire feature must cause some extension in this passively moving unit. Another evidence of such extension are the clastic dikes filled by material from the lower units of rock avalanche body described at the Blackhawk rock avalanche in California (Johnson, 1978). We expect that additional information about this process can be provided by the precise mapping of molards and by study of their spatial distribution on the rockslide deposits surface (Yoshida, 2016).

Peculiarity of some molards that is quite important in the context of forces acting within moving rockslide bodies is their jig-saw structure due to disintegration of large single blocks "pushed" upward from rockslide interiors. Such molards were observed at the Seit rock avalanche (see Figs. 8.143 and 11.7). We anticipate that such disintegration was caused by rapid distress of the blocks that had been overstressed inside moving rockslide. When such boulder was ejected upward, high elastic stress likely caused by temporary force chains developing in rockslide interiors releases and boulders disintegrate following their structural weakenings (fractures). These observations support the dynamic fragmentation model proposed in Davies and McSaveney (1999, 2002, 2011), Davies et al. (2006, 2017), McSaveney and Davies (2006), and Bowman et al. (2012).

12.3 SEDIMENTOLOGICAL AND STRUCTURAL INDICATORS OF ROCKSLIDE MOTION MECHANISM(S)

Unlike morphological peculiarities of rockslide and rock avalanche deposits that might be studied and interpreted as indicators of the mechanical processes acting during their emplacement bases on the analysis of the space and aerial images and DEMs mainly, study of rockslide deposits sedimentology and internal structure requires detailed field work and is possible where rockslide bodies have been incised and dissected by erosion. Such case studied in the Central Asian region were described in Abdrakhmatov and Strom (2006), Hartvich et al. (2008), Robinson et al. (2015), and Reznichenko et al. (2017). Results of these studies summarized in Chapter 11 revealed the following main peculiarities:

- preservation of the host rock mass structure in large-scale bedrock landslides' deposits;
- extremely intensive crushing of the body and basal facies of rockslide deposits (up to nanoparticles); and
- distinct difference between grain-size composition of the coarse blocky carapace and intensively comminuted internal (lower) parts of the deposits classified as body and basal facies (Dufresne et al., 2016).

12.3.1 Host Rock Mass Structure Preservation—Indication of the Laminar Granular Flow

Analysis of highly mobile rockslides' and rock avalanches' morphology discussed above clearly demonstrates that most of them can be characterized as granular flows. That is why in the latest classification system such phenomena are designated as flows (Hungr et al., 2012, 2014). Mechanical models of such flow were analyzed in Hungr (1981, 1989, 1995, 2006), Potapov (1991), Crosta et al. (2003), McDougall and Hungr (2004), Iverson (2006), and Sassa et al. (2010, 2014).

An important implication of such preservation is that it indicates the absence of turbulence during rockslide and rock avalanche motion that distinguish these natural phenomena from another highly mobile phenomenon—debris flows most of which represent turbulent flows. Rock avalanche, in contrast, represents laminar flow even if it moves over complex quite irregular surface that can be illustrated well by the Kokomeren rockslide, whose body, after collapse from the headscarp, had passed over the terrace, caved into the river valley, and climbed hundreds of meters on its opposite slope (see Fig. 8.153). Nevertheless, different lithological units after traveling 2–3 km from their initial position did not mix and retain their mutual position both in the carapace (see Fig. 11.2), and in the body facies (see Fig. 8.151). The absence of turbulence during rock avalanche motion must be considered as an important parameter in the numerical modeling of this phenomenon.

Preservation of the succession of lithological units involved in rockslide motion is, however, not always as simple as at the scheme shown on Fig. 4.17A. Many deeply dissected rockslides that had originated on the slopes of variable lithology, such as the Kokomeren (see Figs. 8.146 and 8.154), the Inylchek (see Fig. 8.45), the Ornok (see Figs. 8.168–8.170), demonstrate much more complex internal structure (see Fig. 4.17B). Its formation is shown on the sketch on Fig. 8.153, according to which lithological units forming upper part of the source zone overrun and overlay those units that originate from its lower part. Formation of rock avalanche by the successive surges was observed by eyewitnesses of the 1949 Khait rock avalanche (see Section 12.1). However, because its source zone was much more homogenous, it makes identification of the deposits left by each of these surges quite problematic. Additional studies are necessary to explain why such inner shearing within moving rockslide body occurs just along the boundaries between lithological units.

We want to point out that doubling of the lithological succession typical of the source zone was found also in the deposits of some loess landslides in Xian region, China. Here, in the frontal part of long runout "sheet-like" body, the imbricate structure of the gravels-loess sequence was observed in the outcrops (Gonghui Wang, personal communication). Blocks of loess from the upper part of the slope overrun loess blocks from its lower part, and resultant deposits recall those described at some rockslides mentioned above, of course with due regards to significant difference of the material affected.

12.3.2 Grain-Size Composition and Fragmentation Energy

Enormous amount of rocks in the rockslide deposits' interiors crushed at a very large extent—up to nanoparticles (Adushkin et al., 2006; Dubovskoi et al., 2008; Reznichenko et al., 2012a) requires analysis of energy balance. In other words, what portion of energy available should be consumed by debris crushing? Analysis presented in Davies et al. (2017) demonstrated that grain-size composition of rockslides' body facies contradicts Griffiths' concept of the free surface energy (Griffith, 1921) and that progressive fragmentation process observed in the experiments can be ensured by the initial overstress of debris clasts.

Mechanical model able to reproduce both preservation of the initial structure of the displaced rock mass and intensive crushing is described in Section 11.3 in Chapter 11 (Dubovskoi et al., 2008). Besides, in this paper, the "efficiency" of debris comminution in the course of rockslide emplacement was compared with that by blasting and by the mechanical crushing (ore grinding at preparation plants)—basic technologies used in mining industry.

The potential energy of rock mass before its collapse (E_{pot}) is considered to be the only source of energy released during rockslide emplacement. To compare energy that can be consumed to crush material by moving rockslide and by the above-mentioned technologies, calculations were performed for rockslide about 100 Mm^3 in volume, one half of which ($V = 50\ \mathrm{Mm}^3$) is considered as being intensively comminuted. According to the relationship between maximal height drop (vertical distance from the headscarp crown and the lowermost part of the deposits) and rockslide volume (see Fig. 10.4), such rockslides often occur on slopes of ca. 1000 m high. Because for such calculations the descend of the center of mass (h) is valuable, we used half of slope height—500 m. Mean rock density (ρ) was taken as 2500 $\mathrm{kg/m^3}$. Thus, for such rockslide:

$$E_{pot} = \rho \times 2V \times g \times h = 2500 \times 100 \times 10^6 \times 9.8 \times 500$$
$$= 1.225 \times 10^{15}\,\mathrm{J}\ (\mathrm{m}^2 \times \mathrm{kg/s}^2).$$

Locat et al. (2006) studied grain-size composition of nine rock avalanches from the Canadian Rocky Mountains and the European Alps and found that fragmentation energy averages 20% of the potential energy available (2.45×10^{14} J in our case). We must notice that they use overall grain-size distributions averaging those typical of the basal, body, and carapace facies, while abrupt change of the extent of comminution indicates sharp difference of the stress–strain state and fragmentation process in the external and internal parts of rockslide bodies. Morphology of the micron-size particles from the basal facies of the Kokomeren rockslide (see its grain-size composition in Table 8.7), most of which are quartz scales, is similar to that described for basal facies of rockslides triggered by the 2008 Wenchuan earthquake (Wang et al., 2015). Carapace facies lack such fines and their grain-size distributions are much closer to those of the fractured rocks of the host massifs. Thus, if we consider grain-size distribution of the basal and body facies only, crushed much more intensively than the carapace facies, the degree of rock crushing in these parts of rockslide bodies should be much higher than if it would be averaged for the entire body.

In case of blasting, the potential energy of a rockslide could be equalized to the heat energy of the blast (E_{heat}) that is spent for rock crushing, generation of seismic waves, sound, etc. According to the Blasting Safety Manual (State Unitary Enterprise, 2002), blasting at open cast mines with explosives consumption of $\approx 500\ \mathrm{kg}/1000\ \mathrm{m}^3$ of rock should crush rock mass in the macadam-size clasts with prevailing fraction of 50–150 mm. The share of fine fractions (less than 10 mm) is usually about 10% of the total blasted rock mass, thus its grain-size composition is much coarser than that of debris in the inner/lower parts of large-scale rockslide bodies where fine fractions often comprise more than 50% of the rock mass weight (see Tables 8.4, 8.5, 8.7, 8.8 and Fig. 11.10). To calculate the amount of the explosives required to crush 50 Mm^3 of rock at the same extent as in the inner part of rockslide body, we followed Rodionov et al. (1971), according to which radius of the intensively crushed zone where rock fragments became much smaller than initial block size determined by fracturing can be estimated as the triple radius of the explosive charge. To achieve such conditions, it is necessary to arrange charges in such a way that any point in the blasted massif will be within the bounds of this "near-field" zone. In this case the consumption of explosives should be of the order of 50 kg/1 m^3 of rock—100 times more than in the normal industrial practice! Considering blast heat energy of the ammonite-type industrial explosive as ca. 4×10^6 J/kg, total energy of the explosion that could crush 50 Mm^3 of rock as intensively as by rockslide crushing is about eight times more than the rockslide potential energy calculated above:

$$E_{\text{heat}} \approx 50 \times 10^6 \times 50 \times 4 \times 10^6 \approx 10^{16}\text{J}$$

When comparing the efficiency of rockslide comminution with that of the mechanical grinding at preparation plants, the potential energy can be equalized to the electric energy used by such plants. According to Revnivtsev et al. (1988), fine crushing of rocks (ores) up to the fraction 0.074 mm consumes nearly 10 kWh per ton that corresponds to ca. 25 kWh or 9×10^7 J per cubic meter of rock. Thus, fragmentation of 50 Mm^3 of rock will require 4.5×10^{15}, 3.6 times larger than the potential energy of 100 Mm^3 rockslide with 500 m descend. Though only small portion of rockslide body facies (usually 5%–15%; see Fig. 11.10) is crushed up to such fine fraction, the initial grain-size composition of ore subjected to mechanical grinding is generally finer than blocky composition of the host rock mass involved in rockslide. In any case the efficiency of rockslide comminution appears to be much higher than that of rock crushing both by short-term blasting and by much more prolonged industrial grinding. We expect that reproduction of the processes acting during rockslide motion that result in so intensive debris comminution in mining industry could increase efficiency of some of energy-consuming technological processes.

To conclude, we want to point out that the most powerful (in physical meaning) process—blasting, when energy is released in a fraction of a second—appears to be least efficient in terms of rock crushing in comparison both with industrial grinding and rockslide comminution. It seems that intensive crushing requires multiple cycles of loading–unloading and shearing such as those in the experiments described in Dubovskoi et al. (2008) (see Section 11.3 in Chapter 11) that reproduce phenomena observed in nature.

12.3.3 Some Examples of Rockslide Deposits With Abnormal Internal Structure

Almost each large-scale rockslide described above is, at some extent, unique due to the peculiar combination of its characteristic features. Some case studies, however, reflect quite rare phenomena indicating specific emplacement mechanisms, and it will be pity not to mention here such abnormal peculiarities of rockslides' internal structure.

One of the such unique features is the Ak-Kiol rockslide in the Kokomeren River basin that represents failure of the syncline core composed of the Carboniferous conglomerate and sandstone with thin gypsum interbeds, whose hinge line was inclined towards the river valley about 6–10 degrees only (see Section 5.1 in Chapter 5 and Figs. 5.9 and 5.10). Failure of the syncline core was reported also at Ming-Djilki rockslide in Southern Tien Shan (see Fig. 8.243), but the uniqueness of the Ak-Kiol case is in its deposits' structure, representing succession of nearly horizontal continuous "layers" of sandstones and conglomerates (see Fig. 5.12). It could be formed if the syncline core in the host massif had unfolded within moving rockslide body as soon as lateral confinement had disappeared. Because similar folds in the same rocks remain intact despite deep erosion, and preserve in such state for tens and hundreds of millennia, it is obvious that such an abnormal behavior was caused by highly dynamic processes acting during rapid motion of this rock mass. According to visual impression, "layers" in the rockslide body are fractured more intensively than in the host massif (compare left side of Fig. 5.8 and Fig. 5.12), but retain their integrity, thus indicating quite a specific case of laminar flow.

Another unusual case of rockslide body internal structure is the Karachauli rockslide, whose frontal part is composed of crushed Neogene conglomerates infilling the Pre-Kensu River mouth at a distance of about 500 m (see Figs. 8.172, 8.176, and 12.2). The homogenous mass of crushed Neogene conglomerates have several inclusions up to several tens of meters each of well-rounded (see Figs. 8.173 and 12.2) and angular (see Fig. 8.174) boulders of granite and Paleozoic

FIGURE 12.4 Displaced (bulldozed) alluvial cobbles (alQ) at the frontal part of the Usoi rockslide dam (at the isthmus between the Sarez and Shadau Lakes).

metasediments. Rounded boulders represent alluvium, and angular are, most likely, material of the minor rock falls or scree.

Bulldozing of large amount of alluvium in front of rockslide body took place at the Usoi rockslide (Fig. 12.4; see also Fig. 9.45). Scraping of loose material and its bulldozing was described for the 1911 Ananievo rockslide too (see Fig. 8.72). Involvement of the valley fill that was displaced in front of rapidly moving crushed bedrock debris was found at the bicorn Archabel rock avalanche (see Fig. 9.4) and at the Kudara-Pasor rockslide (see Fig. 9.38, Unit 5). But in the Karachauli rockslide case, it seems that sliding surface had passed below the valley bottom so that giant block of the Paleozoic rocks descending from more than kilometer-high slope had pushed Neogene from the slope base together with river bed and terraces eroded in Neogene and filled by alluvium. Coarse-bedded weak Neogene conglomerates disintegrated in the more or less homogenous mass, while bouldery inclusions, despite significant run out distance, have not been mixed with crushed Neogene but remain as well-delineated inclusions well visible on Fig. 12.2.

One more unique feature in the lower part of the Kokomeren River valley is the Ornok rockslide (see Fig. 8.167) whose internal structure allows considering it as a combination of the rotational block slide at its proximal part and of the flow-like rock avalanche with extremely impressive "pseudo-stratified" internal structure at its front (see Figs. 8.168–8.171).

All case studies mentioned in this section deserve special attention and should be studied in more details within the frames of special research projects.

CHAPTER

13

Seismically Induced Rockslides: Their Discrimination

OUTLINE

Earthquakes are the evident and one of the most common triggers of large-scale bedrock landslides. It has been proved by numerous case studies worldwide, some of which are listed: the 1348 Dobratsch rockslide, Alps, Switzerland (Bonnard, 2011); the 1786 Dadu rockslides, Longmenshan, China (Dai et al., 2005; Lee and Dai, 2011); the Akjar rockslide (see Fig. 8.90) triggered by the 1887 Vernyi earthquake, Northern Tien Shan, Kazakhstan (Mushketov, 1890); the Ananievo (see Figs. 8.70 and 8.71), the Chilik (see Figs. 8.66 and 8.67), and the Kaindy (see Fig. 8.73) rockslides, Tien Shan, Kyrgyzstan and Kazakhstan, triggered by the 1911 Kemin earthquake (Bogdanovich et al., 1914; Delvaux et al., 2001; Havenith et al., 2003); the Usoi rockslide, Pamir, Tajikistan (see Figs. 5.7, 9.40, and 9.47) triggered by the 1911 Sarez earthquake (Bukinich, 1913; Shpilko, 1914; Preobrajensky, 1920; Lange, 1926, 1929; Gaziev, 1984; Fedorenko, 1988; Alford and Schuster, 2000; A. Ischuk, 2011; Ambraseys and Bilham, 2012; Kulikova et al., 2016); the 1929 Falling Mountain rock avalanche, Southern Alps, New Zealand (McSaveney and Davies, 2006); the Khait rock avalanche (see Figs. 5.15, 8.271, and 8.272), Tien Shan, Tajikistan, triggered by the 1949 Khait earthquake, (Panasenko, 1952; Gubin, 1960; Leonov 1960; Evans et al., 2009b); the 1959 Madison rockslide, MT, USA (Witkind et al.,1962; Meyers and Hamilton, 1964; Hadley, 1964); the Sherman Glacier rock avalanche, USA, triggered by the 1964 Alaska earthquake (McSaveney, 1975, 1978); the 1970 Huascaran ice–rock avalanche, Andes, Peru (Plafker and Eriksen, 1978; Evans et al., 2009a); the 1985 Bairaman landslide, Papua New Guinea (King et al., 1989); the Belaldy rockslide (see Fig. 2.4), Tien Shan, Kyrgyzstan, triggered by the 1992 Suusamyr earthquake (Bogachkin et al., 1997; Ghose et al., 1997; Korjenkov et al., 2004); the Tsaoling landslide, Taiwan, triggered by the 1999 Chi-Chi earthquake (Chen et al., 2006); rock avalanches triggered by the 2002 Denali Fault earthquake, AK, USA (Jibson et al., 2006); the 2005 Hattian Bala rockslide, Kashmir, Pakistan (Dunning et al., 2007; Owen et al., 2008; Delaney and Evans, 2011; Schneider et al., 2013); and numerous large rockslides triggered by the 2008 Wenchuan earthquake, Longmenshan, China (Yin et al., 2008; Wang et al., 2009; Cui et al., 2009, Huang and Li, 2009; Chigira et al., 2010; Dai et al., 2011).

Rockslides and Rock Avalanches of Central Asia.
DOI: https://doi.org/10.1016/B978-0-12-803204-6.00013-2

Various factors governing behavior and stability of rock massifs during earthquakes were analyzed, and formation of rockslides listed herein and of other similar case studies was explained by combining methods of engineering geology and engineering seismology (e.g., Keefer, 1984, 1993, 1999, 2000, 2002; Geli et al., 1988; Pederesen et al., 1994; Rodriguez et al., 1999; Havenith, 2002; Havenith et al., 2003, 2013, 2015a, 2016; Assimaki and Gazetas, 2004; Meunier et al., 2008; Huang et al., 2009, 2012; Delgado et al., 2011; Garevski et al., 2013; Yuan et al., 2015; Roy et al., 2016; Tian et al., 2017). However, most of these studies dealt with rockslides whose seismic origin has been known, because they occurred during the historical earthquakes, and were aimed to find qualitative and quantitative interrelation of these phenomena and to select areas with increased likelihood or probability of future landslide occurrence in seismically prone areas.

Much more ambiguous task is the confident and well-grounded discrimination of prehistoric seismically triggered rockslides from large-scale rock slope failures of different, nonseismic origin. Such discrimination is one of the most controversial paleoseimological issues (McCalpin, 2009). It is, however, critically important in the mountainous regions with short historical record of natural phenomena (see, e.g., Penna et al., 2016). Central Asia, whose historical record of seismic events rarely exceeds 200 years that is much shorter than the typical recurrence interval of large earthquakes in the intracontinental seismic zones, and where the vast majority of the identifiable rockslides including those described in Chapters 7–9, Dzungaria, Tien Shan, and Pamir, respectively, are prehistoric, is one of the such regions.

We want to point out that in most of the publications on large prehistoric Alpine rockslides such as Flims, Tamins, or Köfels (e.g., Abele, 1974; Brückl and Heuberger, 2001; Poschinger et al., 2006; Poschinger, 2011) or on deep seated gravitational slope deformations (DSGSDs) in this region (e.g., Agliardi et al., 2013), they are considered to be associated directly or indirectly with Holocene deglaciation. An alternative - seismic triggering has been discussed rarely (Pedrazzini et al., 2016). In contrast, the majority of papers about quite similar features in the presently more seismically active Tien Shan (Korjenkov, 2006), Baikal Rift (Solonenko, 1977), Altai Mountains (Agatova and Nepop, 2011), and the New Zealand (Bull, 1996; Barth, 2013) treat earthquakes as the top-priority triggering factor of large rockslides or DSGSDs.

Additional complexity is introduced by the duality of this problem, influencing both landslide hazard and seismic hazard assessment (Strom, 2015a) (we mentioned it already in the introductory part of Chapter 4: Principles and Criteria of Landslides Identification and Discrimination From the Features Formed by Other Mass Wasting Phenomena). On the one hand, correct identification of seismically induced prehistoric landslides provides input data for more justified landslide susceptibility and hazard assessment in seismically active regions; on the other hand, besides finding just the genetic relationships of the studied rockslide(s) with past earthquake(s), we must estimate basic parameters of seismic event(s) that triggered slope failure (s)—their intensity or magnitude.

Criteria, allowing discrimination of seismically induced rockslides were proposed in Solonenko (1970, 1977), Fedorenko (1968, 1988), Adams (1981a, b), Nikonov (1988), Crozier (1992), Bull (1996), Nepop and Agatova (2008), and Sepúlveda et al. (2016). Good overview can be found in chapter 8 of the second edition of Paleoseismology (McCalpin, 2009). We do not intend to provide the comprehensive analysis of this problem. It requires much more space, than is available in this volume, and much more data that should be collected through the extensive field work in the study region. We just want to highlight its complexity and to propose some characteristic features that might indicate seismic origin of large-scale prehistoric slope failures based on the analysis of case studies from the Central Asian region that could be used worldwide.

13.1 THE HISTORICAL EARTHQUAKE-INDUCED ROCKSLIDES OF THE CENTRAL ASIAN REGION

Most of the large crustal earthquakes in the Central Asian region (though not all of them—see, e.g., Ischuk and Strom, 2016) had triggered large-scale rock slope failures described in previous chapters. Their main parameters are summarized in Table 13.1. Data on rockslides that could be triggered by other seismic catastrophes such as the 1812 Nilki, the 1885 Belovodsk, the 1889 Chilik, the 1902 Kashgar, the 1906 Manas, the 1907 Karatag, and the 1946 Chatkal earthquakes are either scarce or controversial. Discussion on some contradictory information related to the origin and timing of the Bielogorka (see Figs. 8.97–8.100), the Kaindy (see Figs. 8.63 and 8.64), and the Lower Kiol-Sai (see Fig. 8.65) rockslides (all in Northern Tien Shan), and of the Yashinkul rockslide in Southern Tien Shan (see Fig. 2.3) could be found in Chapters 2 and 8, Brief History of the Bedrock Landslide Studies in the Central Asia and Tien Shan, respectively, and will be summarized hereafter. Small number of unquestionable case studies from the study region does not allow any statistical analysis of factors favorable for the formation of seismically induced large-scale rockslides.

TABLE 13.1 Main Parameters of the Unquestionable Historical Seismically Induced Large-Scale Bedrock Landslides in the Central Asian Region

Year, earthquake name, magnitude	Rockslide	Types of rocks	Approximate volume (Mm^3)	Height drop/runout (km) and confinement	Peculiar features
1887, Vernyi, 7.3	Akjar (see Fig. 8.90)	Granite, schist	7–10	0.15–0.4/1.73. Lateral confinement	Very irregular and complex headscarp
1911, Sarez, 7.7	Usoi (see Figs. 5.7, 9.40–9.47)	Terrigenous, carbonate, gypsum, anhydrite	2200	1.67/6.14. Frontal confinement	Very deep wedge failure with its crown at a prominent topographic bench level; upper part of the rockslide body is a block ~1 × 1 km in size; rockslide forms the 500 m deep Sarez Lake
1911, Kemin, 8.2	Ananievo (see Figs. 8.70 and 8.71)	Granite	30	0.42/1.24. Partial lateral confinement	Failure destroyed rocky spur several hundred meters long and wide just at the hanging wall of the surface rupture. Nearly planar back wall of the headscarp
	Kaindy (see Fig. 8.73)	Metasediments	5	0.46/1.4. Partial lateral confinement	Failure occurred just at the hanging wall of the surface rupture
	Chilik (see Figs. 8.66 and 8.67)	Metasediments and granites	8	0.81/2.3. Frontal confinement	Rockslide had blocked the river, but later the dam was breached. Part of displaced material forms the block still hanging on the slope
1949, Khait, 7.4	Khait (see Figs. 5.15, 8.271, and 8.272)	Gneiss, schist, granites	100	1.375/7.41. Lateral confinement	Very deep irregular headscarp that affected the entire slope. Rockslide converted into long runout rock avalanche
1992, Suusamyr, 7.3	Belaldy (see Figs. 2.4, and 13.1).	Granite	20	0.665/1.61. Lateral confinement	Rockslide dammed the valley and ca. 1 year later produced the devastating debris flow

We pointed out above that data on the earthquake-induced landslides of the period before the end of the 19th century are quite controversial. It can be exemplified by the Yashinkul rockslide dam in Southern Kyrgyzstan that breached catastrophically in 1966 (Krukovskiy, 1969, Rezvoi and Rezvoi, 1969; Rezvoi et al., 1971; Reizvikh et al., 1971; Pushkarenko and Nikitin, 1984, Strom, 2010b, 2013a). Several researchers who studied this event mentioned that formation of this rockslide dam had been triggered by strong earthquake. However, according to Pushkarenko and Nikitin (1988), it had occurred in 1835, while Rezvoi and Rezvoi (1969) noticed that rockslide had been triggered by an earthquake about 300 years before dams' breach, thus, in the middle of the 17th century. We have to notice that no data about earthquake in either period in this region have been presented in earthquake catalogues available (Mushketov and Orlov, 1893; Kondorskaya and Shebalin, 1982). Large earthquake in southern part of the Ferghana valley occurred in 1822 or 1823 (Semionov, 1958; Djanuzakov et al., 2003), but its epicenter, according to coordinates provided in Djanuzakov et al. (2003) (40.3° N, 71.5° E) are about 80 km northwest from the Yashinkul rockslide site, and it is unlikely that rock slope failure could be triggered by such a distant shock, which reported macroseismic effects were localized around the assumed epicentral area.

Similarly controversial is the relationship of the well-known Bielogorka rock avalanches (see Figs. 8.97–8.100)

and the 1885 Belovodsk earthquake. As we mentioned above (see Section 8.3 in Chapter 8), these impressive rockslides were not mentioned in the description of this seismic catastrophe made by Mr. Ignatiev (1886), though they could be seen from the populated planes at the foot of the Kyrgyz Range where he inspected settlements affected by this earthquake. Similarly, no indication of these really outstanding phenomena was provided by I. Mushketov in the Annex to his report about the 1887 Vernyi earthquake describing the 1885 event (Mushketov, 1890, pp. 112–115), though several other much smaller rock falls, earth flows, and debris avalanches were mentioned. We hypothesize that these rockslides could be triggered by some older earthquake, though can neither prove nor disprove this assumption at present.

Most of case studies listed in Table 13.1 occurred decades ago, and no data about the conditions that preceded these slope failures are known, i.e., if there were any indications of instability allowing prediction of rockslides formation or, at least, identification of these sites as being most hazardous. Some data are available about the July 10, 1949, M7.4 ± 0.1 Khait earthquake that triggered the devastating 7.4 km long Khait rock avalanche (see Fig. 8.271). It was noticed, in particular, that raining took place sometime before this earthquake (N. Leonov, personal communication). It saturated loess deposits widely developed in the affected area and supported their seismically induced liquefaction resulting in numerous flow-like loess landslides. It is unlikely, however, that those rains could change significantly the mechanical properties of rock massif at a depth of several hundred meters, and ca. 100 Mm^3 in volume that collapsed being triggered by the main shock of this earthquake and transformed into rock avalanche. We have to mention that this shock was just the culmination of the sequence of earthquakes that started on July 8, 1949, by two foreshocks with magnitudes 5.1 ± 0.7 and 5.6 ± 0.3, which intensity near the Khait town was estimated as V–VI points of the Modified Mercalli (MM) or MSK-64 scale. Their most affected areas, where intensity could reach VI-VIII points, were located about 15–20 km west of Khait in the Yasman River valley. Besides, 1949 earthquake was preceded by the 1941 M6.4 ± 0.2 Garm earthquake whose intensity at Khait could reach VII–VIII points (Kondorskaya and Shebalin, 1982). All these shocks could somehow destabilize the slope that collapsed on July 10, though no data on such effects are available now.

Study of rockslides triggered by the historical earthquakes highlights the complexity and ambiguity of the prediction of future failures in seismically prone regions, which is the main goal of landslide susceptibility and hazard mapping. It can be exemplified by the Belaldy rockslide triggered by the 1992 M7.3 Suusamyr earthquake that was the only one large-scale rock slope failure that occurred about 35 km from the epicenter and about 13 km from the surface rupture of this earthquake (Bogachkin et al., 1997; Ghose et al., 1997; Korjenkov et al., 2004). Analysis of the KFA-1000 space images of this area obtained in 70th–80th, and of aerial photographs made in 1961 did not show any abnormal evidence of instability that would allow identifying this site as most hazardous (Fig. 13.1). Even if it could be assumed beforehand that rockslide can originate here, then dozens if not hundreds of the somehow suspicious sites should be selected within the epicentral zone of the 1992 earthquake, while in fact large rockslide originated just here, not anywhere else.

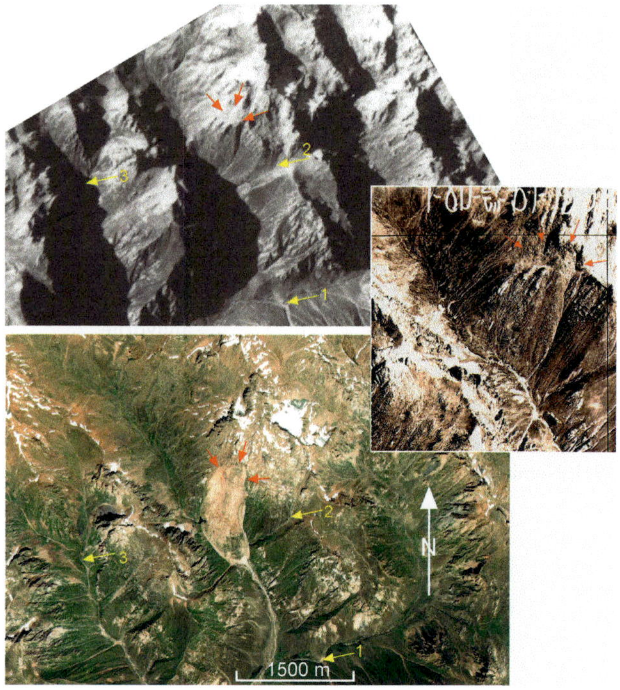

FIGURE 13.1 Space images of the Belaldy rockslide area before (KFA-1000—above) and after (Google Earth—below) the 1992 Suusamyr earthquake. Yellow arrows indicate identifiable points on both images. 1961 aerial photograph of the Belaldy rockslide site is shown in the inset. Red arrows mark the position of the Belaldy rockslide headscarp.

Despite quite limited number of the well-studied historical rockslides that have been triggered by large Central Asian earthquakes, some of them are characterized by quite specific position and shape of their source zones (see last column of Table 13.1), i.e., their coincidence with the prominent pre-earthquake landforms and abnormally large depth of wedge-like or compound failures. Similar characteristics can be considered as possible indicators of seismic triggering of past rock slope failures as well, which will be discussed in Section 13.2.5.

The specific peculiarity of the Pamir section of the study region is the Hindu Kush zone of the deep foci earthquakes with an abnormally high frequency of large events (see Fig. 3.10). It was reported that many landslides in loose soils have been triggered by these earthquakes some times 400–500 km far from their epicenters (see, e.g., Niazov, 2009). Being quite deep—70 to more than 200 km—they affect vast areas exceeding those affected by crustal earthquakes with the comparable magnitudes. Strong motions caused by these earthquakes are characterized by long duration of shaking and lower frequencies prevailing in earthquake spectra, while peak ground acceleration values could be lower than those of shallow earthquakes. Such combination of vibration parameters seems to be favorable for triggering landslides. Though we could not find direct indications that these earthquakes had triggered large-scale rockslides in Afghanistan in the available publications, it can be hypothesized that frequently occurring shaking should decrease slopes' stability in this part of the region that can explain high density of large-scale rock slope failures in the upper reaches of the Kokcha River basin (see Fig. 9.115) and, perhaps, in some other parts of the Pamir mountains in Tajikistan and Afghanistan.

13.2 JUSTIFICATION OF SEISMIC TRIGGERING OF THE PREHISTORIC ROCKSLIDES

The presence of large and/or long runout prehistoric rockslides is often treated as an unequivocal evidence of strong past earthquakes without providing any additional reasons supporting such assumption—just due to their enormous size and expressiveness. As an example we want to notice that Prof. Zolotariov—one of the leading Soviet experts in landslide science—mentioned in his lecture in 70th, attended by one of the authors of this book, that if he would not know that the 1964 Aini rockslide occurred without an earthquake, he would consider it as a seismically induced feature.

V.P. Solonenko—one of the founders of paleoseismological method—considered the extreme mobility typical of many large-scale rockslides as a sound indicator of their seismic triggering (Solonenko, 1977). However, the empirical data demonstrate well that large and highly mobile rock slope failures often occur without seismic triggering. It can be exemplified by the 1806 Goldau and 1881 Elm rock avalanches in Switzerland (Heim, 1982, 1932; Abele, 1974), the 1893 Gohna Tal rockslide in India (Weidinger, 1998), the 1903 Frank rock avalanche in Canada (McConnell and Brock, 1904; Krahn and Morgenstern, 1976; Cruden and Hungr, 1986), the 1962 Huascaran ice–rock avalanche in Peru (Plafker and Eriksen, 1978; Evans et al., 2009a), the above-mentioned 1964 Aini rockslide in Tajikistan (Fedorenko, 1988), the 1974 Mayunmarka landslide about 1 km^3 in volume in Peru (Hutchinson and Kojan, 1975; Kojan and Hutchinson, 1978), the 1987 Val Pola rock avalanche in Italy (Crosta et al., 2011), the 1993 La Josefina landslide in Ecuador (Plaza et al., 2011), the 2002 Yigong rock avalanche with ca. 10 km long runout in China (Shang et al., 2003; Xu et al., 2008; Evans and Delaney, 2011; Delaney and Evans, 2015), and the 2010 Attabad rockslide in Pakistan (Delaney and Evans, 2011; Schneider et al., 2013).

Incorrectness of considering rockslides long runout as an indicator of seismic triggering can also be substantiated by the following reasons. Rockslide debris after its initial descend moves by inertia, releasing gained momentum, and its runout must depend on the kinetic energy available. The latter is converted from the potential energy of rock mass resting on a slope, regardless of the cause of its failure, and can also be supplied by seismic shaking to the rock mass of seismically triggered rockslide.

The potential energy of rock mass (E_{pot}) with volume V ranging from 10^6 to 10^{10} m^3 resting on slope of different height so that its center of gravity descends 10, 100, 500, and 1000 m that can be converted into the kinetic energy is calculated as:

$$E_{pot} = m \times g \times H = V \times \rho \times g \times H \qquad (13.1)$$

The kinetic energy that can be supplied to the same rock mass by seismic vibrations with peak ($v_{peak} \sim 1$ m/s) and spectral ($v_{spectr} \sim 3$ m/s) velocities typical of seismic strong motion of the MM (MSK-64) 9-point earthquakes (Aptikaev, 2012) is calculated as:

$$E_{kin} = m \times v^2/2 = V \times \rho \times (v_{peak})^2/2 \qquad (13.2a)$$

$$E_{kin} = m \times v^2/2 = V \times \rho \times (v_{spectr})^2/2 \qquad (13.2b)$$

Their comparison (Fig. 13.2) shows that if during the rockslide emplacement center of gravity descends for more than 100 m, which is typical of most large-scale rockslides in mountainous regions, the kinetic energy that could be converted from the potential one would be 2–3 orders (!) larger than the kinetic energy that could be supplied to this rock mass by a very strong earthquake. Thus, seismic shaking provides less than a percent of the total kinetic energy and cannot affect motion of rock mass detached from the host massif significantly.

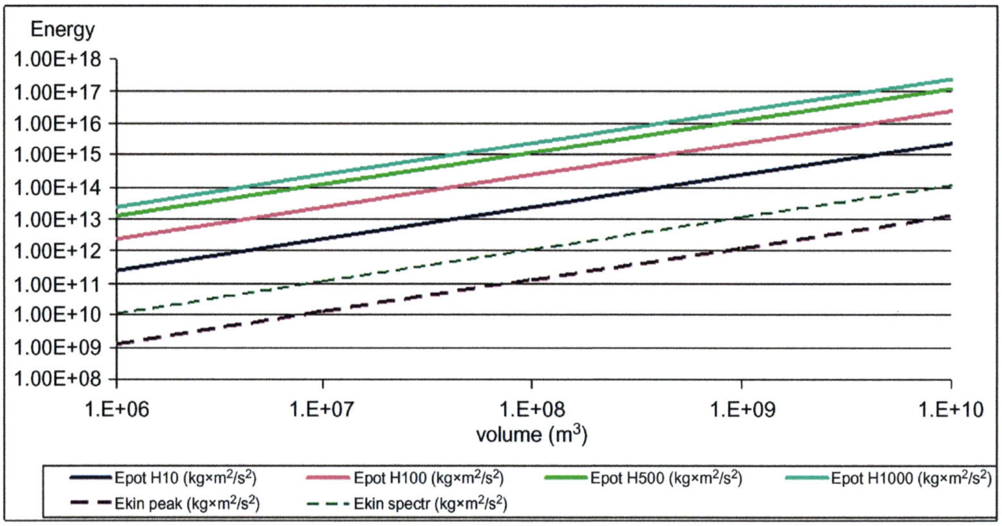

FIGURE 13.2 Comparison of the potential energy of rock mass and the kinetic energy that can be supplied to it by seismic shaking. Solid lines —the potential energy of rock mass, whose center of gravity descends 10, 100, 500, and 1000 m; dashed lines—the kinetic energy corresponding to peak and spectral velocities of 1 m/s and 3 m/s, typical of the earthquakes with MM intensity ~9; gravity acceleration $g = 9.8\ m/sec^2$; mean density of rocks $\rho = 2500\ kg/m^3$ (this value was just used to simplify calculations. In fact rocks' density is usually higher, about 2.65–2.90 (Clark, 1966)).

Because neither volume of a rockslide nor its mobility (runout) can prove its seismic origin, interpretation of any large-scale prehistoric slope failure as a seismically induced feature must be justified by some additional evidence, otherwise it would remain more emotional rather than scientific. Thus, we should analyze other peculiarities allowing discrimination of seismically induced rockslides.

Six criteria indicating seismic origin of landslides were proposed in Crozier (1992) based on the studies carried out in the New Zealand mainly:

1. **Modern seismic activity in the study region that resulted in triggering landslides.** Though the use of this criterion looks logical, rockslides not being triggered by earthquakes are known in many seismically active regions, including the New Zealand (McSaveney, 1993, 2002; McSaveney et al., 1992; Cox et al., 2015) and Central Asia where such phenomena can be exemplified by the 1964 Aini rockslide (see Figs. 2.2 and 8.266) or the 2008(?) Djamansu rockslide described in Section 8.4 in Chapter 8 (see Figs. 8.110–8.112). Besides, it is unclear if such seismic activity should be known just for the nearest vicinity of the site where the feature in question is located or for the broader region. For example, no earthquakes with M > 5.5 have been recorded in the historical times in the Central Tien Shan east of the Talas-Ferghana fault zone (see Fig. 3.10). At the same time both number and size of rockslides identified in this part of the Tien Shan exceed those in other, much more seismically active (in present) parts of this mountainous system (compare Figs. 8.107 and 8.58). Thus, this criterion itself can neither prove nor disprove anything.
2. **Coincidence of landslide distribution with an active fault or seismic zone.** This criterion seems to be one of the most convincing. Spatial coincidence of rockslides and active faults in the study region will be discussed hereafter in Section 13.2.1. Additional argument supporting assumption of seismic triggering of a landslide is its not only spatial but also temporal coincidence with active faulting, as it was demonstrated in (Bull, 1996) for the Alpine fault in the New Zealand and in (Belousov et al., 1994) for the younger Djuzumdybulak rockslide (see Fig. 8.190) within the Naryn-Sonkul fault zone and for the Karasu Lake (see Fig. 5.64), and the Kapkatash (see Fig. 8.196) rockslides associated with the Talas-Ferghana active fault. Contemporaneity is especially important because presently active faults often inherit long-lived fault zones, with wide belts of intensive fracturing and weakening of rocks (see, e.g., Fig. 8.191) that form geotechnical conditions favorable for slope failures regardless of the exact trigger.
3. **High safety factor of the slope indicating impossibility of its failure without the additional (i.e., seismic) loading.** From our point of view, this criterion can be used in cases with rather simple and well-known engineering–geological conditions where back analysis of the failed slope stability can

be performed with reasonable accuracy (see, e.g., section 8.5 of the McCalpin's Paleoseismology (McCalpin, 2009)). However, in most of the cases described in Chapters 7–9, reliable assessment of safety factor of the slopes prior to their failure requires input data on host massifs structure and mechanical properties that can be provided with very poor confidence, taking into account size of the features, height of the mountain slopes, complexity of their geological structure, and ambiguity of the pre-slide topography reconstruction. Some qualitative considerations on impossibility of rock slope failure without applying the external force (seismic triggering) based on peculiarities of several historical seismically triggered rockslides will be discussed in Section 13.2.5.

4. **Large size of the feature in question.** From our point of view, based on the empirical data listed in the beginning of this section, this criterion does not work. Besides the empirical observations, such conclusion has some physical grounds. It is known that on a first approximation, forces supporting slope failure are proportional to landslide weight (volume), in other words to the third degree of the landslide linear dimensions, whereas forces preventing sliding are proportional to its area (area of the sliding surface), i.e., to the second degree of linear dimensions. It means that if other factors would be equal, the larger the rock massif is, the more susceptible for sliding it should be, regardless of the particular trigger.
5. **The presence of liquefaction features associated with landslides.** This criterion could work well in the planes with widespread fine sediments prone to liquefaction or in the volcanic regions where liquefaction can develop in weak layers of ash that forms sliding surfaces (Sassa et al., 1994). In the Central Asian region where Quaternary deposits are, usually, much coarser and are developed much more locally, evidence of seismically induced liquefaction associated in any way with large-scale landslides could be found just by chance. One of such cases was described in Section 8.4 in Chapter 8 (see Fig. 8.178) and will be discussed hereafter in Section 13.2.3.
6. **Landslides distribution that cannot be described solely on the basis of geological and geomorphic conditions.** As it can be seen on Figs. 7.1, 7.5, 8.2, 8.10, 8.32, 8.58, 8.106, Fig. 8.107, 8.138, 8.199, 8.236, 8.259, 9.1, 9.64, and 9.115, rockslides distribution within the study region is quite uneven. There are, however, so many factors governing landslides formation that it is very difficult to formalize this criterion for practical use. Moreover, large crustal earthquakes in Central Asia are associated with active fault zones, and it would be difficult to associate landslide distribution with seismicity, not associating them with such fault zones.

Nevertheless, despite inevitable complexity and ambiguity of seismically induced rockslides discrimination, some geological and geomorphic evidence that can be used to substantiate rockslides' association with seismic shaking will be discussed in the following sections being exemplified by case studies described in the previous chapters.

13.2.1 Spatial and Temporal Coincidence of Rockslides With Active Faults

Active faults with traces of the recurrent and, rarely, single-event Late Quaternary offsets are the main and the most studied geological and geomorphic indicators of large (high magnitude) past earthquakes (McCalpin, 2009). It is evident that large-scale bedrock landslides that have occurred close to active faults and, especially, those, whose headscarps coincide with such faults, can be considered as seismically induced features most confidently. Despite shortage of age determinations of both surface faulting episodes and of rockslides, allowing more unequivocal proof of their genetic association, there are many case studies where such association could be substantiated directly or indirectly.

In some cases, we were able to reveal sequencing of fault displacement and of rock slope failure events as, for example, in Kyrgyzstan, at the southern slope of the Nuratau Range, just west of the Lesser Naryn River valley, where the Kalmak-Ashu surface fault displaces deposits of the small rock fall (see Fig. 3.12). In this case surface rupture is definitely younger; otherwise rock fall would fill the lowering at the downthrown side of this fault scarp before expanding downslope. Thus, rather small rock fall, whose deposits were ruptured, could be triggered either by the penultimate earthquake that produced ca. 4 m high upslope-facing fault scarp or it could occur immediately before the last surface fault movement being triggered by P-waves that could reach the daylight surface faster than rupture appeared at the slope foot producing additional 4 m offset.

Opposite relationships are observed in some cases too. They can be exemplified by the Djaya rock avalanche in the Chong-Kemin River valley, whose body, not being affected by the latest rupturing event, overlies impressive trace of recent surface rupture (see Fig. 8.78), by the Karasu Lake rockslide (see Fig. 5.64) resting over the Talas-Ferghana active fault (both sites are in Kyrgyzstan), and by several rockslides on both sides of the small ridge in Xinjiang (Fig. 8.38), one of which, at its northern side, overlays the trace of surface rupture (see the inset on Fig. 8.39). Such rockslides are younger than the latest faulting episode at each particular site. Nevertheless, they can be attributed to the latest rupturing events too, assuming that large-scale rock slope failures most likely were triggered by the strongest shaking usually caused by the S-waves and surface waves. Besides, large rockslide bodies millions of cubic meters in volume need more time to be detached from the host massif than minor rock fall of several thousands of cubic meters. Thus surface faulting could just precede emplacement of voluminous rockslides.

We cannot also exclude that slope failure could occur later than the main shock accompanied by surface faulting, being triggered, for instance, by aftershock. Such delayed effect was observed, for example, after the M6.4 Tsambagarav earthquake of July 23, 1988, in Mongolia. Seventeen days later, on August 9, eyewitness observed formation of about 12 Mm^3 ice–rock avalanche that traveled 5 km along the Zuslan Creek valley (Avdeev et al., 1989), resembling those that collapsed in 1962 and 1972 from the Huascaran peak in Peru (Plafker and Eriksen, 1978; Evans et al., 2009a) or the 2002 Karmadon disaster in Caucasus (Nikitin et al., 2006). Within the context of the problem discussed herein, it is important that such large-scale failure occurred in the epicentral zone of a significant earthquake more than 2 weeks later than its main shock.

Active faults that cross the headscarp areas directly have been found at the Kokomeren rockslide in Kyrgyzstan (see Figs. 8.146 and 8.155) that will be described in more details in Section 13.2.6, and in Gunt River valley in Tajikistan where the impressive surface rupture passes close to the Chartym rockslide (see Fig. 9.73) and reaches the headscarp of the very large Vir-Tangiv rockslide (see Fig. 9.75). The latter active fault stretches in the north-eastern direction for about 60 km with distinct evidence of recent movements at some sections (see Fig. 3.13). About 19 km east-northeast from the Vir-Tangiv site, this fault zone is overlain by the suspicious accumulation shown on Fig. 9.70.

Most often large-scale failures in the study region took place on the slopes rising above active fault zones that divide neotectonic uplifts and depressions. It can be exemplified by the Ananievo (see Fig. 8.70) and the Kaindy (Fig. 8.73) rockslides triggered by the 1911 Kemin earthquake, whose rupture passes just at the feet of the collapsed slopes. Such relationships are typical of the Tien Shan where most of ridges and valleys are neotectonic structures divided by active faults. Good examples can be seen at the northern sides of the Chong-Kemin (see Fig. 8.78), the Karakudjur (see Figs. 8.120 and 8.123), and the Minkush–Kokomeren depressions (see Fig. 8.186). One of such sites is at the northern side of the Western Akshiyriak Range affected by the outstanding Beshkiol rockslide about 10 km^3 in volume (see Fig. 4.1) at the east, and by the long runout Chaartash rock avalanches (see Figs. 5.61, 8.128, and 8.130) at the west. An impressive system of surface ruptures passes along its northern feet just below the large-scale DSGSD (see Fig. 8.132). Peculiarity of this site that needs additional study to be explained, is that these fault scarps look more as normal rather than as reverse faults typical of the Tien Shan neotectonic structures (Shultz, 1948; Makarov, 1977; Chedia, 1984).

Same spatial relationships could be found in Eastern and Northern Pamir ranges rising for 3–4 km above neotectonic depressions divided by active faults (see Figs. 3.6 and 3.7A). The most exciting case studies are the 34-km long Koman and 24-km long Lenin rock avalanches in the Alai intermontane depression (see Figs. 9.20 and 9.24), and the Taheman rock avalanche that originated on the slope directly undercut by surface rupture (see Fig. 9.17).

Interesting case study was found in the Tsagaan Giunge Mountains (see Fig. 8.8), where two rockslides 18 and 10 Mm^3 in volume originated on the slopes composed of granite at the downthrown side of the assumed 2.5- to 3.0-km long surface rupture with presumably normal kinematics, which, likely, is a segment of a longer system of active faults. It can be hypothesized that rapid acceleration–deceleration of the subsiding tectonic block during rupturing event could "dislodge" some weathered rock from the free face of the slope.

Several large bedrock and soil slides were identified just nearby active faults with well-defined surface ruptures like the Urmochdara rock avalanche in Southwestern Pamir (see Fig. 4.40) or the Chaek rotational landside in Central Tien Shan (see Figs. 5.5 and 13.3). In such cases their genetic association is quite possible but its justification needs dating of both rupturing event and slope failure.

Describing rockslides distribution in more regional scale, one can see that they often group within some localized areas, many of which, though not all, coincide with active faults. Such clusters composed of multiple rockslides and rock avalanches have been identified all over the study region being separated from each other by vast areas almost free of such features. It can be seen on the regional maps of rockslides distribution (see Figs. 7.1, 7.2, 7.5, 8.2, 8.10, 8.32, 8.58, 8.107, 8.138, 8.199, 8.236, 8.259, 9.1, and 9.64).

Some active fault zones such as those bounding the Minkush–Kokomeren intermontane depression in the

FIGURE 13.3 Above—river terraces displaced by active fault that passes along the northern boundary of the uplift (marked by red arrows) north of the Chaek town and 2–3 km from the Chaek landslide shown on Fig. 5.5. The increasing difference between terrace levels (1, 2, and 3) indicates recurrent recent displacements along this surface rupture. Fault scarps displacing Terraces 1 and 2 are partially buried by sediments accumulated at its footwall. Below—thrust fault visible in the quarry wall about 600 m east from the site shown above.

Central Tien Shan (see Fig. 8.138), passing along the Zeravshan River in Southern Tien Shan (see Fig. 8.259) or along the Sanglech and the Yakhshindar River valleys in Afghan Badakhshan (see Fig. 9.115) are affected by large-scale rock slope failures almost over their entire length. We must mention that no large historical earthquakes are known within the first two areas, while the third one passes over the deep-foci Hindu Kush seismic zone (see Fig. 3.10).

Some other fault zones, also featuring clear evidence of recent tectonic activity, demonstrate localized clustering of landslides with 10–20 km long gaps between these clusters (Strom and Abdrakhmatov, 2004). Such spatial distribution characterizes, in particular, system of active faults stretching along the Chong Kemin and Chilik River valleys north of the Issyk-Kul Lake (see Fig. 8.58). It is interesting that fault passing along these rivers was ruptured at a distance of about 100 km by the 1911 M8.2 Kemin earthquake (Bogdanovich et al., 1914; Delvaux et al., 2001), and, likely, by previous prehistoric earthquakes, while rockslides identified in this area form three distinct clusters (Numbers 6, 10, and 9 on Fig. 8.58) divided by rockslide-free sections. Similar irregular rockslide distribution characterizes fault zone that forms the northern boundary of the Suusamyr intermontane depression (see Fig. 8.106).

Some fault zones, however, are almost free of large-scale slope failures. It can be exemplified by the Dzungarian and Lepse faults (see Figs. 3.8, 3.9, and 7.1). Some other are accompanied by very few rockslides, such as, for example, the fault zone passing along the northern feet of the Trans Alai Range (see Fig. 3.7A) or bounding intermontane depressions of the Eastern (Chinese) Pamir (see Fig. 3.6) where quite limited number of large-scale rockslides was identified despite extreme topographic gradient (see Fig. 9.1).

Summing up the description of the relationships between active faults and large-scale slope failures, we must mention that sometimes impressive scarps at the feet of mountain ranges resembling those of active reverse or thrust faults are in fact the frontal boundaries of large rotational landslides originating on the adjacent slopes, whose sliding surfaces plunge below the piedmont and daylight several hundred meters ahead. It can be observed at the northern feet of the Peter the First Range—the northern boundary of the outer Pamir mountain chain. One of the most impressive sites was found near the Hufak village at 38.935° N, 70.145° E (Tajik 001) where arcuate frontal scarp 20–30 m high passes up to 670 m off the foot of the slope (Fig. 13.4). This scarp can be treated as the front of the DSGSD that affects more than 5 km wide lower part of the Peter the First Range northern slope, likely being associated with the neotectonic thrust fault passing along its feet (see Fig. 5.21).

In such situations it could be difficult to determine real nature of such landforms—are they tectonic or landslide-related features (compare Fig. 13.4 with Fig. 3.7A—real surface rupture at the eastern continuation of the same fault zone in the Alai valley)? Such possibility should be considered and analyzed carefully when interpreting data on slip rate measured by the GPS (Ischuk et al., 2013) or by INSAR, otherwise sliding-induced deformations can be treated erroneously as the effect of modern tectonic movements.

Similar modern phenomenon was observed by one of the authors of this book in 1998 at the Himalayan front in India not far from the Chandigarh city where formation of the horseshoe-shaped scarp up to ca. 1.0 m high in front of the cliff foot (Fig. 13.5) was interpreted first as manifestation of modern nonseismic tectonic faulting while in fact it was just the front of the rotational landslide.

13.2.2 Position of the Rockslide Source Zone

One of the commonly cited criterion proposed to identify seismically induced rockslides—failure of just the upper part of the slope (see, e.g. Fedorenko, 1988; Meunier et al., 2008)—is based on the well-known effect of seismic strong motion topographic amplification resulting in significant increase of the oscillations

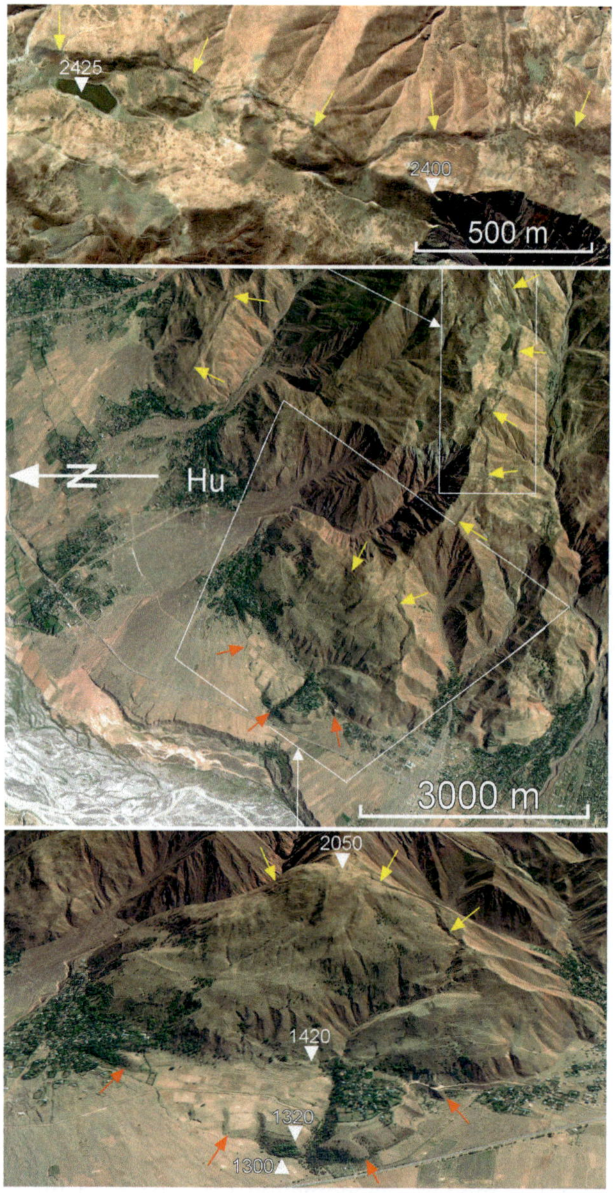

FIGURE 13.4 DSGSD landforms at the northern slope of the Peter the First Range near the Hufak village (Hu). Red arrows mark the frontal fault-looking scarp; yellow arrows indicate landslide headscarps and scarp bounding DSGSD.

FIGURE 13.5 Frontal scarp of the rotational landslide that appeared on the daylight surface in front of the subsiding hilly area near the Chandigarh City in India. Above—general view (arrows mark the fault-like scarp); below—closer view.

amplitudes at the near-watershed parts of mountain ridges (Puchkov, 1965, 1974; Bugaev et al., 1975; Geli et al., 1988; Pederesen et al., 1994; Assimaki and Gazetas, 2004; Bourdeau and Havenith, 2008; Meunier et al., 2008; Buech et al., 2010; Havenith et al., 2015a). Significant topographic amplification was proved to be the main factor that influenced formation of the Ananievo rockslide triggered by the 1911 Kemin earthquake (Havenith, 2002; Havenith et al., 2003). Besides, if failure affects only the near-watershed zone, slope undercutting by the stream erosion as an alternative cause of slope failure can be excluded.

At the same time, however, numerous aseismic rockslides occurred in the recent years in Asia, Europe, and New Zealand being sourced at the upper parts of high slopes (Shang et al., 2003; Xu et al., 2008; Evans and Delaney, 2011; Delaney and Evans, 2015; McSaveney, 1993, 2002; McSaveney et al., 1992; Barla et al., 2000; Cox et al., 2015). In the study region, such phenomenon can be exemplified by the 2008(?) Djamansu rock avalanche in Central Tien Shan (see Figs. 8.110–8.112) where failure of about 8 Mm^3 of bedrock occurred from the top of the 1100-m-high ridge without seismic triggering and, moreover, without any visible indicators of the preceding instability (see Fig. 8.112). Though such case studies indicate ambiguity of this criterion, good physical grounds force us to take it into consideration.

Rockslides that caved from the ridges crests have been identified all over the study region (see Figs. 6.17, 6.20, 7.3, 7.9, 8.5, 8.6, 8.9, 8.13, 8.14, 8.15, 8.34, 8.64, 8.74, 8.186, 8.187, 8.205, 8.231, 8.243, 8.260, 8.268, 8.276, 8.278, 9.4, 9.53, 9.57, 9.76, 9.92, 9.116, and 9.117). One of the most striking examples is the Lower Aral rockslide (see Fig. 8.161) that originated on top of the 1-km-high mountain composed of granite. Dip angle of its kilometer-long sliding surface was 30° on average. The collapsed rock mass moved initially over the ca. 800 m long nearly horizontal bench before falling down in

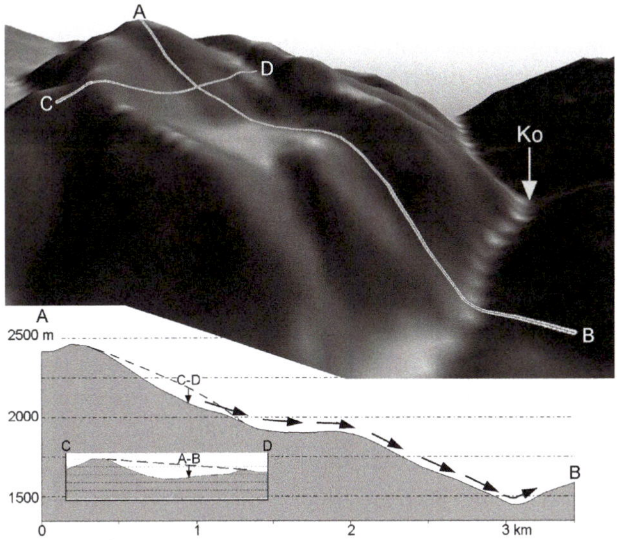

FIGURE 13.6 3D relief and schematic cross-sections A-B and C-D of the Lower Aral rockslide (Central Tien Shan). Dashed line – assumed pre-slide slope surface. Ko – the Kokomeren River. Black arrows on the cross-section A–B show the travel path of rockslide. 3″ SRTM DEM visualized by Global Mapper software.

the Kokomeren River valley (Fig. 13.6). It is really difficult to propose any cause of such slope failure other than strong earthquake; moreover, this site is located not far—about 4–5 km only from the active reverse fault bounding the Minkush-Kokomeren intermontane depression from the north.

Failure of the topmost part of the ridge could be predetermined by its internal structure, which can be exemplified by the Mingdjilki rockslide—collapse of the syncline core from the watershed part of the ~1200-m-high ridge in the Isfairamsay River basin in Southern Tien Shan (see Fig. 8.243). In such situation, the necessity of seismic shaking for its triggering can be substantiated or rejected by back analysis of slope stability, but it would require detailed data on the rock massif structure, mechanical properties of the bedding surface, and reconstruction of the pre-slide topography.

Sometimes a seismically triggered failure affects not only the topmost summit part but also the top of the convex bench on the ridge slope as in the case of the 1911 Usoi rockslide (see Fig. 5.7). Such localization of seismically induced failure can be explained by resonance of natural vibrations of the wider part of the mountain range with seismic waves produced by large earthquake (H.-B. Havenith, personal communication). Similar position of the headscarp crown has the Kudara-Pasor rockslide in the adjacent Kudara River valley (see Fig. 4.20).

Judging from the presence of both pro- and contra-case studies and considering that rock slope failure can be predetermined by combination of many factors, both internal and external, the position of the source zone of past rockslides at the upper part of a slope should be treated just as an indirect indicator of its possible seismic origin.

13.2.3 Rockslides and Seismically Induced Liquefaction

Seismically induced liquefaction within the sliding zone is one of the most probable mechanisms of landslides formation in loose soils and in some types of weak or deeply weathered rocks (Sassa et al., 1994). Its applicability to bedrock landslides that originate in hard rock (granite, gneiss, sandstone, and other types of strong igneous, metamorphic, and sedimentary rocks) is quite questionable. This mechanism, however, could support debris motion when a rockslide (rock avalanche) moves over valley bottom composed of loose well-saturated sediments or with low water table (see Chapter 12: Morphological and Structural Evidence of Rockslides' (Rock Avalanches') Motion Mechanism(s)), but at this stage, liquefaction can be induced by intensive vibrations produced by debris motion itself (Melosh, 1979, 1983, 1986, 1990) and may not be related to the process that really triggered rock slope failure.

We could find only one case study—the Karachauli rockslide (see Fig. 8.172) in the lower part of the Kokomeren River valley, where some logic chain could be developed, linking the formation of rockslide dam and liquefaction that occurred in lacustrine sediments accumulated in the dammed lake (see Section 8.4 in Chapter 8). Outcrop of the lacustrine silt with intensive liquefaction that affected its basal layers (see Fig. 8.178) was found about 30 km upstream of the dam—near the Aral village at the confluence of the Kokomeren and Djumgal Rivers (Fig. 13.7). A lake with assumed water level of about 1580 m a.s.l. had stored nearly 7 km^3 of water. Judging from the present-day mean annual discharge of the Kokomeren River (2.5×10^9 m^3/year; State Water Inventory, 1987c), it could be impounded in ca. 3 years and could reach the level of the liquefied strata several months earlier. Lacustrine sedimentation had started immediately after the river banks were submerged by rising water, and initially it was just the reworking of the colluvial deposits on the river banks that formed the basal layer of coarse sand up to 1 m thick (Unit 1 on Fig. 8.178). When the water level raised further, the sedimentation style had changed abruptly and silt laminae started accumulating at the lake bottom. Soon after that, when thickness of the silt deposits had reached 10 cm approximately, this site had been

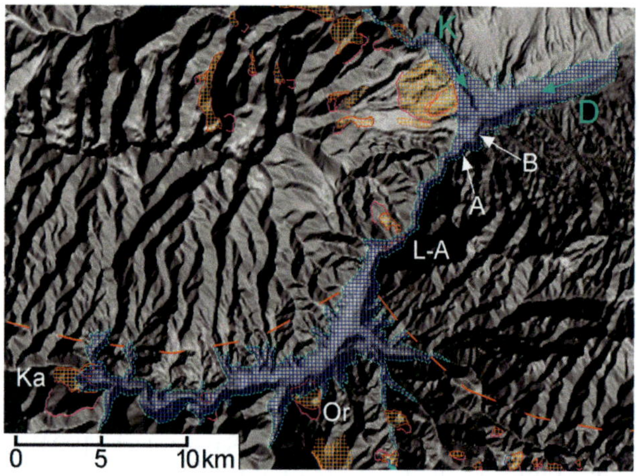

FIGURE 13.7 Possible extent of the lake dammed by the Karachauli rockslide (Ka). The transparent area hatched by blue lines with bright blue dashed boundary—past lake at 1580 m a.s.l.; pink lines—rockslide headscarps; hatched areas with brown boundaries—rockslide bodies; comb orange line and area hatched by brown lines without boundary—DSGSD; dashed red line—approximate position of the fault zone bounding the Minkush–Kokomeren intermontane depression from the north. A—outcrop shown on Fig. 8.178; B—outcrop shown on Fig. 8.177; K—the Kokomeren River; D—the Djumgal River; Or—the Ornok rockslide; L-A—Lower Aral rockslide.

shaken intensively so that the basal coarse sand had been liquefied, ejected through the cylindrical vents, and spread over the lake bottom (Unit 1′ on Fig. 8.178) forming the so-called event horizon (McCalpin, 2009). The lowermost silt laminae that had been deposited at this time (Unit 2 on Fig. 8.178) sank into the liquefied sand. After this event, lake sedimentation had continued without any interruptions and up to ca. 10 m of silt had been accumulated above the event horizon (Unit 3 on Fig. 8.178; see also Fig. 8.177). As we mentioned already in Section 8.4 in Chapter 8, judging from the small thickness of the affected silt, and considering time required for the impoundment, liquefaction should take place, likely, within the first years after the lake formation. This succession of events will be illustrated hereafter in a graphical form in Fig. 13.17.

The most reliable cause of shaking that resulted in this liquefaction was strong earthquake (Strom, 2013b), and we hypothesized that it could be an aftershock of an earthquake that had triggered the Karachauli rockslide itself. The significant—about 30 km—distance between the described outcrop and the Karachauli rockslide dam (see Fig. 13.7) does not contradict this assumption because it is well known that the strongest aftershocks of large (M ~ 7.0) earthquakes often occur several tens kilometers from the epicenters of their main shocks. It was reported for the 1992 M7.3 Suusamyr earthquake in Kyrgyzstan (Bogachkin et al., 1997) and for the 1991 M7.0 Racha earthquake in Georgia (Arefiev et al., 2006). Moreover, east-west trending faults bounding the Minkush–Kokomeren intermontane depression, with which such earthquake could be associated, passes only 10–12 km south from the site where liquefaction was observed (see Figs. 3.1, 3.3 and 13.7). All these peculiarities do not contradict to the hypothesis that liquefaction within the lacustrine sediments could be caused by the strong aftershock of an earthquake that had triggered the river-damming Karachauli rockslide.

It would be interesting to study the basal parts of the sedimentary fill of other breached lakes in the Central Asian region that had been dammed by rockslides, where they are observable due to rivers' incision, to find similar liquefied strata.

13.2.4 Geological Evidence of Rockslides Simultaneous Formation

Since Central Asia is characterized by arid climate and rarity of the extreme rainstorms that also can trigger the widespread slope failures, the synchronous collapse of rocky slopes within some localized area or zone is a sound argument in favor of their triggering by a strong (with high intensity) or large (with high magnitude) earthquake. If the synchronous landslides are distributed within some area as it occurred during the 1911 M8.2 Kemin earthquake (Bogdanovich et al., 1914; Delvaux et al., 2001) or the 2008 M8 Wenchuan earthquake (Yin et al., 2008; Huang and Li, 2009; Chigira et al., 2010; Fan et al., 2012b), magnitude of a past earthquake can be estimated using the approach proposed by D. Keefer (1983, 1994, 1999, 2000, 2002) and his followers (Rodriguez et al., 1999). We should notice, however, that such relationships have been derived from spatial distribution of landslides of all sizes mapped after the historical earthquakes, while dealing with past earthquakes that occurred centuries or millennia ago, we can base on largest slope failures only, likely associated with these events, that remain identifiable for such a long time. Special analysis of the empirical data on spatial distribution of just large slope failures should be performed in order to derive relationships valid for paleoseismological application of this method in mountainous regions such as Dzungaria, Tien Shan, or Pamir. Judging from the spatial distribution of large-scale rockslides triggered by high-magnitude earthquakes, the 2008 Wenchuan earthquake at first, we should notice that distance from the causative faults (surface ruptures) of such earthquakes should be considered instead of the epicentral distance (Havenith et al., 2016).

Additional complexity is provided by the fact that catastrophic natural phenomena capable for triggering landslides (not only large earthquakes but also tropical storms and hurricanes) often result in formation of only one really large slope failure, that would remain identifiable for a very long time, accompanied with numerous much smaller features (Malamud et al., 2004a, b; Turcotte and Malamud, 2004). Such distribution can be exemplified by the 1911 M7.4–7.7 Sarez earthquake or the 1992 M7.3 Suusamyr earthquake—each triggered only one really large rockslide—the Usoi (see Figs. 5.7, 9.40–9.43) and the Belaldy (see Fig. 2.4) correspondingly. Analysis of the prehistoric case study with likely one large rockslide will be presented hereafter in Section 13.2.6.

Lack of age determinations forces us to analyze the alternative approaches to substantiate the simultaneity of the adjacent large rock slope failures. The first step is to look for nearby rockslides similarly expressed in the landscape—with the same level of minor landforms preservation, same extent of vegetation cover, etc. Numerous rockslide clusters with such characteristics were identified in Dzungaria (see Figs. 7.3, 7.4, 7.6, and 7.8), in different parts of the Tien Shan (see Figs. 5.52, 5.61 and 8.128, 6.5, 6.15, 8.12–8.15, 8.8, 8.38, 8.97, 8.226 and 8.227, 8.268, and 8.269), and of the Pamir (see Figs. 4.21, 9.8, 9.11, 9.51, 9.79, 9.83, 9.107, 9.109, and 9.140).

Additional characteristics supporting assumption on the simultaneous formation of several bedrock landslides due to strong motion of the same seismic event is the similarity of headscarps' orientation and of the direction of the initial motion of the descending rock massifs. It can be exemplified by the Karakungey rock avalanches in Central Tien Shan (see Fig. 5.52). Both rockslides whose deposits look similarly "fresh" (see Figs. 5.43 and 5.54), originated on the north-facing slopes of the adjacent ridges and moved initially almost in same direction (Fig. 13.8). We can speculate, taking into consideration distance between the Belaldy rockslide triggered by the 1992 M7.3 Suusamyr earthquake and its causative fault zone (see Section 13.1) that both Karakungey rock avalanches could be triggered by an earthquake associated with active faults bounding the Minkush–Kokomeren intermontane depression that passes 10–15 km south of them.

Same similarity of orientation of the nearby rockslides with similar landscape characteristics of the deposits was found in Dzungaria (see Fig. 7.8), at the Sarmin-Ula rockslide cluster in Eastern Tien Shan where most of the recent rockslides originated on west- and south-west-facing slopes (see Fig. 8.11) and often started moving in the same direction (see Fig. 8.15), in the Terskey Alatau Range where they form the Akkiol and the Karakiol rockslide dammed lakes in the adjacent valleys (see Fig. 8.42), in Pyandj River valley south

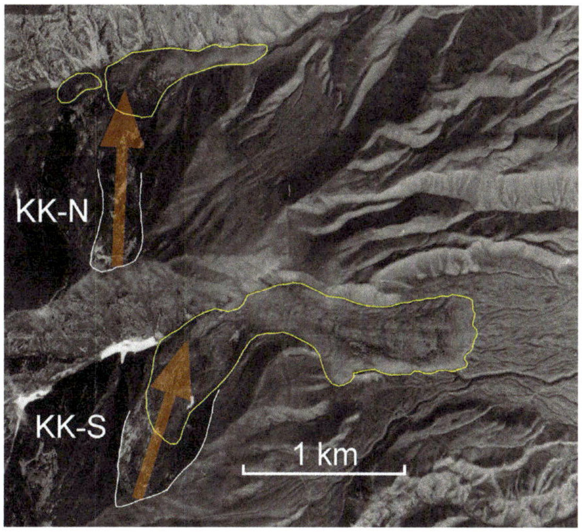

FIGURE 13.8 Nearly similar directivity of slope failure of the Karakungey rock avalanches at the western termination of the Djumgal intermontane depression. Aerial photograph with outlined headscarps (white lines) and rock avalanche deposits (yellow lines). Transparent brown arrows show the directions of the initial slope failures.

of the Shiva Lake (see Fig. 9.113) and in the Kokcha River basin (see Figs. 9.130 and 9.131).

Sometimes failures took place on both sides of the same ridge, which also could indicate their seismic triggering (Fedorenko, 1988). It can be exemplified by several case studies such as those in the Eastern Tien Shan (see Figs. 8.26 and 8.38), in the Northwestern Tien Shan (see Fig. 8.202), in the Southern Tien Shan not far from the Aini town (see Fig. 8.265), or by the Kokomeren rockslide (see Fig. 8.149) and the Sarysu rock avalanche in the Central Tien Shan (see Figs. 8.156 and 8.157) that originated on the opposite sides of the same watershed. Of course, to make an ultimate decision on the simultaneity of such similarly or contrarily oriented rockslides, both features should be dated by any "absolute" dating method—radiocarbon, optically stimulated luminescence (OSL), cosmogenic isotopes, etc. (Walker, 2005).

There are, however, situations where a sound conclusion on the simultaneity of the nearby large rockslides can be drawn even if absolute ages of both features are unknown. An ultimate decision can be made if bodies of two (or more) rockslides that originated on different slopes merge without overlapping each other. Such relationships characterize the Bielogorka rockslides in Northern Tien Shan (see Figs. 8.97 and 8.99), the Ikhnachkul rockslides in Northwestern Tien Shan (see Figs. 8.226 and 8.227) or the Ukok rockslide in the northern part of Central Tien Shan that collapsed from the twofold headscarp and formed the body with a distinct linear boundary between the two parts originated nearby (see Fig. 8.125). Similar relationships of the deposits were found at some rockslides in Dzungaria (see Fig. 7.6) and in Afghan Badakhshan (see Fig. 9.132).

It could occur only if rockslides from different slopes collapsed simultaneously, so that their deposits reached the confluence site at the same time. Opposite relationships can be exemplified by the Atdjailau and the Inylchek rock avalanches in the Highest Tien Shan (see Section 8.2), where the latter body overlays that of the former one (see Fig. 8.45) with the significant time span between their formation. Such interval is required to form the small erosional feature filled by the alluvial gravels in the older body (alQ on Fig. 8.45) that is overlain by the younger rock avalanche, and to accumulate the loess-like loam that covers the Atdjailau rock avalanche body being abraded and deformed by the Inylchek rock avalanche front (see Fig. 8.47). One more example of definitely multi-aged rockslides was found in the Naryn River valley in Central Tien Shan where two Djuzumdybulak rockslides occurred successively so that the older one rests on the ca. 45 m to 100 m high terraces and its body is overlaid by the younger one which lowermost part rests on the terrace not more that 15–20 m high (see Figs. 8.190 and 8.192). It was assumed that the second failure could occur between 111 BC and 240 AD (with 95.4% confidence) being triggered by a large earthquake (Belousov et al., 1994) (age was determined from the surface rupture study).

Sometimes, however, overlapping of rockslide bodies could result not from the significantly different age of their formation as in the above cases, but also from the different (real) time required for their emplacement due to the different runout distance. It can be exemplified by two rockslides from the Sarmin-Ula cluster in Eastern Tien Shan shown in Fig. 8.14. It is clearly seen that frontal part of rock avalanche 042 overlays body of blockslide 050. Both failures occurred in granite and had nearly similar inclination of their sliding surfaces—about 33 to 35 degrees, allowing assumption of similar velocity of their motion after initial descent. However, the distance that rockslide body 050 had traveled before coming to the halt was about 800 m (measured from the base of the collapsed slope to the deposits front), while such distance of rockslide 042 was about 1100 m, almost 300 m longer. Thus, the latter could collide with the former one that was already "in place" and slightly overlap its body.

Similar explanation could be applied for the Khushtarita rock avalanches in the Southern Tien Shan (see Fig. 8.268). The distal part of the Khushtarita western rock avalanche deposit that traveled up to 5.23 km (runout measured from the headscarp crown) overlaps the Khushtarita eastern rock avalanche body about 1.2 km from the base of the latter source zone only. Such difference in the travel distance would result in several minutes' delay of the western one to reach the area where otherwise morphologically similar deposits merged.

Sometimes same conclusion could be drawn even if deposits of the adjacent rockslides do not merge as those of the Djashilkul and Kiol-Kogur rockslides in Northern Tien Shan that destroyed northern and eastern slopes of the rocky prominence above the confluence of the Chong-Kemin River and is a small left tributary (see Fig. 8.83). In this case the assumption of the simultaneous formation of both slope failures is based on the unified system of fractures visible above both headscarp crowns.

Argumentation on the simultaneity of rockslides that dams the same river at a rather short distance can be provided also by the analysis of valley sedimentation due to dammed lakes siltation and outburst floods based on the examination of the thalweg profiles (see Figs. 4.27 and 9.99). Such analysis for the Shidz and Voznavd rockslide dams in the Pianj River valley (see Figs. 9.98 and 9.102) and for the Vir-Tangiv and Zuvor rockslide dams in the Gunt River valley (see Fig. 4.42) was presented in the beginning of Section 9.3 in Chapter 9, Pamir.

Concluding this section, we want to mention one more interesting case study from the Afghan Badakhshan—evidence of the successive outburst floods likely caused by breach of rockslide dammed lakes formed in the adjacent Shulvadar and the Dakhanzar River valleys that had occurred within a relatively short time (see Figs. 9.124 and 9.125). Though detailed field work should be performed to prove (or disprove) the hypothesis of the simultaneous (seismic) origin of rockslide dams in both valleys, the proposed explanation seems to be self-consistent, especially considering the location of this area just above the Hindu Kush focal zone (see Figs. 3.10 and 9.115).

13.2.5 Specific Features of the Source Zones of Some Seismically Triggered Rockslides

As it was mentioned already in Section 13.1, the headscarps of some historical seismically induced rockslides are characterized by the specific morphological peculiarities such as extreme depth and confinedness to the prominent landforms (see Table 13.1). We assume that such peculiarities could be used as indirect indicators of seismic triggering of the prehistoric rockslides with similar headscarp shapes.

The characteristic feature of the source zone of the Ananievo rockslide triggered by the 1911 M8.2 Kemin earthquake is that the entire rocky prominence about 400 m long if not more was "chopped off" during the

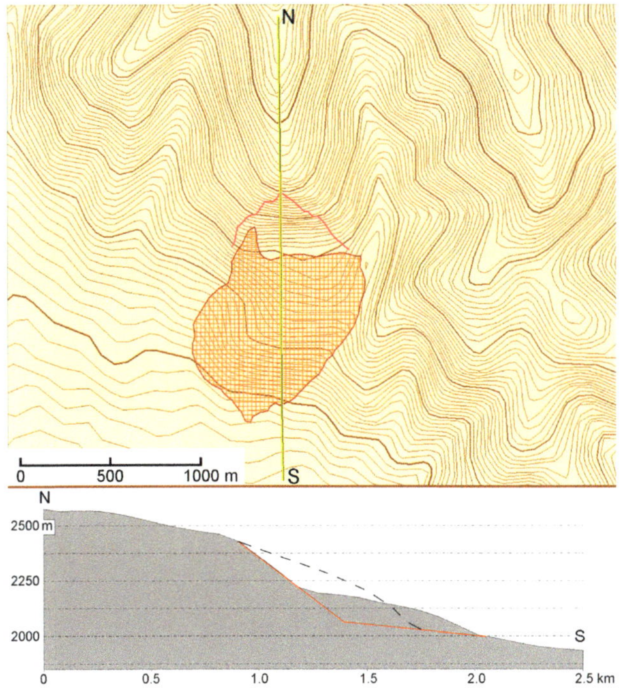

FIGURE 13.9 Schematic map with 10-m contour line spacing and cross-section of the Ananievo rockslide. Dashed black line—the schematically reconstructed pre-slide daylight surface; red lines on cross-section—planar segments of the assumed sliding surface; same on the map—headscarp crown.

earthquake leaving nearly the planar back wall (see Fig. 8.71). It looks as its sliding surface consists of two more or less straight segments—steep back wall up to ca. 370 m high and very gentle basal plane (Fig. 13.9). Short runout distance of the Ananievo rockslide could be caused by the bulldozing of frozen ground in front of it (see Fig. 8.72), which had damped its mobility (the Kemin earthquake occurred in January, when upper part of the soil up to 1.0–1.5 m thick was frozen). Besides, such bulldozing argues for planar shape of the basal segment of the sliding surface rather than for its cylindrical shape typical of the rotational landslides (compare with Figs. 13.4 and 13.5).

Several prehistoric rockslides in the study region have similar source zones at "chopped off" rocky spurs with nearly planar back walls—the Aidyn-Ula rockslide in Eastern Tien Shan (see Fig. 4.4) where the assumed length of the destroyed rocky prominence could reach ca. 1000 m (see Fig. 6.3), rock avalanche in the upper reaches of the Korumdy River valley in Northern Tien Shan (see Fig. 8.60), the Chaartash-3 (see Fig. 8.130), and the Mingteke (see Fig. 8.140) rock avalanches in Central Tien Shan, the Ai-Kul rock avalanche in Southern Tien Shan (see Fig. 8.257), and, likely, one rockslide in the Khidardjakdara River valley in Afghan part of Pamir (see Fig. 9.112A), whose body recalls that of the Ananievo rockslide. Following the proposition of R. Huang that seismically induced rockslides should have deeper headscarps than those of nonseismic origin (Huang, 2014), such shape of rockslide source zones could be considered as an argument in favor of their seismic triggering.

The source zone of another well-known seismically induced slope failure—the 1911 Usoi rockslide—is also very deep, but quite different in shape, even without mentioning nearly two-order differences in size ($\sim 30\ Mm^3$ vs $\sim 2200\ Mm^3$). We want to point out that the earthquake that triggered this outstanding slope failure was by far not the largest one in the study region. Its magnitude estimates range from 7.4 to 7.7 (Kondorskaya and Shebalin, 1982; Ambraseis and Bilham, 2012; Kulikova et al., 2016), while many other earthquakes, including the 1911 Kemin, had magnitudes up to 8.0 and even more (see Section 3.4.1 in Chapter 3: Overview of the Study Region). Nevertheless, volume of the Usoi rockslide is about 1.5 order of magnitude larger than that of any other historical seismically induced rockslide in the entire Central Asian region.

The Usoi rockslide represents a wedge-like type of failure (see Section 5.1 in Chapter 5: Bedrock Landslide Types and Classification Systems) with the extremely deep headscarp—more than 1 km in the vertical direction (see Fig. 5.7) and up to 2 km in plan view, measured from the headscarp crown to the foot of the collapsed slope (see Figs. 9.41 and 13.10). Taking in mind that in such case the detaching rock mass has to overcome friction force not only along the basal sliding surface (as in the Ananievo rockslide case) but also along side walls it can be assumed that formation of rock slope failures should be supported by the action of the external seismic force. Same morphology of the wedge-like source zones with high, long, and steep side walls was found at several other case studies.

One of them—the Ak-Kiol rockslide in Central Tien Shan (see Figs. 5.9 and 5.10)—represents rare case of the dropped out syncline core. Another impressive rockslide with wedge-shape headscarp in the Afghan Badakhshan destroyed ridge composed of the Achaean metamorphic rocks with a well-pronounced foliation dipping gentler than valley slope (see Fig. 9.128). Foliation formed the basal sliding surface, while 250–330 m high side walls most likely inherited steep fracture systems.

Many other rock slope failures with very deep wedge-like or compound source zones occurred on slopes composed of igneous, volcanic, or metamorphic rocks with less regular internal structure. They can be exemplified by the 1949 Khait rock avalanche in Southern Tien Shan (see Fig. 5.15), the Tianchi rockslide in Eastern Tien Shan (see Fig. 8.3), the Aksai-Kiolsu rock avalanche (see Fig. 8.119), and two other

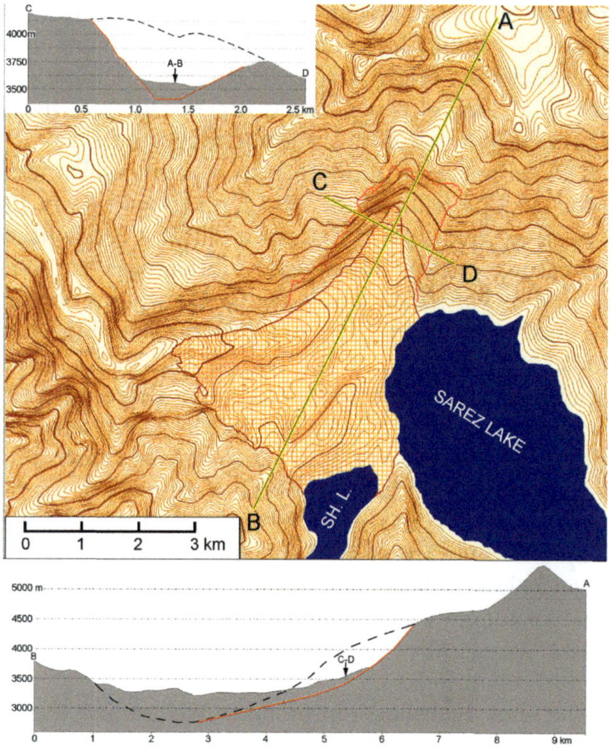

FIGURE 13.10 Schematic map with 20-m contour line spacing and cross-sections A-B and C-D of the Usoi rockslide. Dashed black lines—the schematically reconstructed pre-slide daylight surface; red lines on cross-sections—assumed sliding surface; same on the map—headscarp crown; SH. L.—the Shadau Lake.

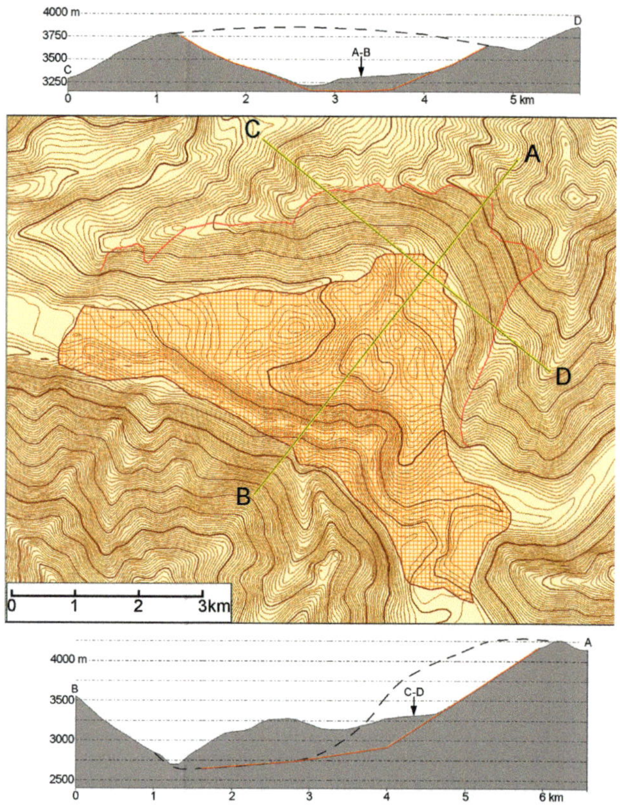

FIGURE 13.11 Schematic map with 20-m contour line spacing and cross-sections A-B and C-D of the Pazhuk rockslide. Dashed black lines—the schematically reconstructed pre-slide daylight surface; red lines on cross-sections—assumed sliding surface; same on the map—headscarp crown.

assumed rockslides in Central Tien Shan (see Fig. 8.118), by the Damamatsay rockslide (Fig. 9.36) in the Tanymas River valley and the Tusen rockslide in Shakhdara River valley (Fig. 9.83) in Tajik part of the Pamir, and by the tremendous Pazhuk rockslide (Fig. 9.135) in Afghan Badakhshan. The latter, being the same order of size as the Usoi dam and even larger—about 3 km^3 in volume—has a wider headscarp, whose longitudinal profile recalls that of the Ananievo rockslide with steep linear back wall and a long and very gentle basal segment (compare Figs. 13.11 and 13.9), while the Usoi rockslide had, likely, more smoothed sliding surface (see Fig. 13.10).

We agree with the assumption of R. Huang (2014) that the extreme depth of wedge or compound rock slope failures can be considered as a possible indicator of their seismic triggering because the detaching rock massifs have to overcome very high basal (and in some cases, sidewise) friction. Of course, a more sound conclusion on the necessity of seismic loading to produce such rockslides could be done based on the backward calculations of the slope stability. Such analysis, however, requires detailed field mapping and laboratory tests of samples to obtain mechanical properties of rock mass and of the slip surface.

13.2.6 Substantiation of Seismic Triggering of the Kokomeren Rockslide

Finishing this chapter, we want to present an example of the Late Pleistocene Kokomeren rockslide in Central Tien Shan about 1.5 km^3 in volume (see Fig. 8.146), whose seismic triggering justification was based on the combination of various criteria (Strom and Stepanchikova, 2008; Strom, 2013b,c). Rockslide originated on the 1600-m high left bank slope of the Kokomeren River valley and formed a dam with a maximal height up to 400 m (the effective height had been much less) that was breached after some time and, finally, deeply eroded (Fig. 13.12).

First, as it was indicated in Section 8.4 in Chapter 8, the Kokomeren rockslide collapsed onto wide river terrace about 100 m high above the riverbed (see Figs. 8.146 and 8.147) and filled the river channel (see Fig. 8.148). At the time of failure, this terrace was about 50 m high that excludes slope undercutting as a possible cause of its instability. Position of these outcrops is marked by 1 in Fig. 13.12.

Second, about 1.5 km downstream of the rockslide body, the series of strath terraces of the Kokomeren

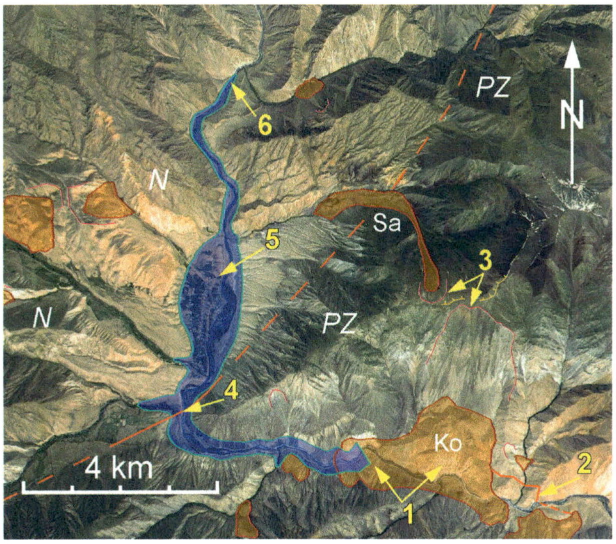

FIGURE 13.12 Schematic map of the Kokomeren rockslide dam and dammed lake. Location of features and outcrops explained in the text to substantiate rockslide seismic triggering are marked by numbered yellow arrows. The transparent blue area—dammed lake at minimal possible water level of ~1800 m a.s.l.; the transparent brown areas—landslide deposits (Ko, Kokomeren rockslide, Sa, Sarysu rock avalanche); pink lines—headscarps; orange comb line—sackungen; solid red lines—active faults identified in the field; dashed red lines—their assumed extensions; PZ, Paleozoic metasediments on the left side of the river valley; N, Neogene red beds on the right side of the valley; arrows with numbers 1-6—see explanation in the text. Google Earth image as a background. *Reprinted with some modification from Strom, A., 2013b. Use of indirect evidence for the prehistoric earthquake-induced landslide identification. In: Ugai K., Yagi H., Wakai A. (Eds.), Earthquake-induced landslides. Springer, Heidelberg, New York, Dordrecht, London, 21–30 (ISBN 978-3-642-32237-2), with the permission of Springer.*

River is displaced by recent reverse fault evidencing the recurrent offsets (see Fig. 8.155). This fault stretches towards the base of the headscarp area (2 in Fig. 13.12) providing a sound argument that the slope failure could be triggered by one of rupturing episodes, most likely associated with a strong earthquake.

Third, both sides of the ridge had been affected by large-scale slope failures—by the Kokomeren rockslide from the south and by the Sarysu rock avalanche from the north (3 in Fig. 13.12). Besides, there is a scarp along this watershed (sackungen). Unfortunately, none of these landforms have been dated.

And the last but not the least indicator of seismic triggering of the Kokomeren rockslide was derived from a thorough analysis of the extremely informative outcrop in the road cut about 5 km upstream from the rockslide dam on the right bank of the Kokomeren River (41.93° N, 74.15° E, 4 in Fig. 13.12); see Fig. 13.13.

The stratigraphic succession in this outcrop starts with crushed and mylonitized Paleozoic metasediments of the large northeastward trending fault zone (1 in Fig. 13.13) that bounds the Kyzyl-Oi intermontane depression from the southeast (see Fig. 13.12). Just its presence here, at the confluence of the Kokomeren River and its powerful right tributary—the Koviuksu River where the significant erosion can be expected, looks strange, considering the weakness of this material. We assume that outcropping of mylonite at this locality reflects long-term recurrent neotectonic movements.

Bedrock mylonite is overlain by well-rounded boulders representing stream facies of the alluvium, locally mixed with angular boulders of Paleozoic metasediments (2 in Fig. 13.13), some of which were smoothed later by water (Fig. 13.14). These angular clasts can be nothing else than rock fall deposits that fell directly in the stream channel from the left-bank slope of the Kokomeren River valley composed of such rocks. They were mixed with rounded alluvial boulders but were not brought away from this site, most likely due to sudden stop of water flow in the dammed lake. We have to notice that this outcrop is located rather far from the left-bank slope of the valley and it is unlikely that "normal" scree could reach such a distance from the slope base. More likely, it was a real rock fall episode that occurred practically at the same time as the master rockslide had dammed the river, stopping the water flow (Strom and Stepanchikova 2008; Strom 2013b).

This mixture is overlaid, in turn, by the layer of basal matrix-supported or clast-supported poorly bedded breccia up 1.0–1.5 m thick (3 in Figs. 13.13 and 13.15) composed of the up to 5 cm clasts of the left-bank dark gray Paleozoic metasediments mixed with fragments and matrix of the red claystone and siltstone eroded from the Neogene red beds outcropping now on the right side of the valley mainly. This basal breccia layer is exposed at a distance of about 3 km (up to Point 5 in Fig. 13.12 approximately). At some locations, the breccia layer is replaced by lenses of well-rounded alluvial cobbles and boulders that could be brought by debris flows from the adjacent higher river terraces.

At a distance of several tens of meters, closer to the fault zone, the breccia layer becomes much thinner and is overlain by the alternation of fine gravel, sand, and silt up to ~1.5 m thick (4 in Fig. 13.13), which is overlain, in turn, by classical laminated lacustrine silt (5 in Fig. 13.13). Further upstream, this silt overlays basal breccia layer directly. The age of these deposits determined from the single OSL sample is about 60 Ka (S. Rosenwinkel personal communication). The entire succession was tilted and ruptured (Fig. 13.16). 20- to 30-degree tilting could result from the predominantly vertical displacement along large active fault bounding the neotectonic intermontane depression. We must notice, however, that vertical separation of layers along the active fault visible in the outcrop resulted from the significant, about 7 m, left-lateral offset of the layers

FIGURE 13.13 Above—the outcrop near the confluence of the Kokomeren and the Koviuksu Rivers with succession of lacustrine sediments affected by a fault zone with evidence of the recurrent recent tectonic movements. Below—upstream extension of the same outcrop several dozen meters aside the fault zone. 1—crushed and mylonitized Paleozoic bedrock—fault zone material; 2—angular boulders of Paleozoic metasediments mixed with rounded alluvial boulders (see Fig. 13.14); 3—basal horizon of lake sediments—mixture of the detritus of Paleozoic metasediments and of Neogene red beds (see Fig. 13.15); 4—tilted and ruptured succession of the lacustrine and fluvial sediments accumulated in the dammed lake (see Fig. 13.16); 5—laminated whitish lacustrine silt; 6—debris flow deposits blanketing the entire succession. A—approximate position of the outcrop cleaned and shown in Fig. 13.14.

FIGURE 13.14 Rockfall deposits (angular metasediment boulders) mixed with well-rounded alluvial cobbles (al). Angular boulder smoothed from one side by water subsequently is marked by "S."

FIGURE 13.15 Composition of the lacustrine sedimentary succession basal layer. Dark gray fragments—detritus of the left-bank Paleozoic metasediments; red fragments and fine matrix—material eroded from the right-bank Neogene red beds.

FIGURE 13.16 Surface rupture displacing the tilted succession of lacustrine and fluvial sediments.

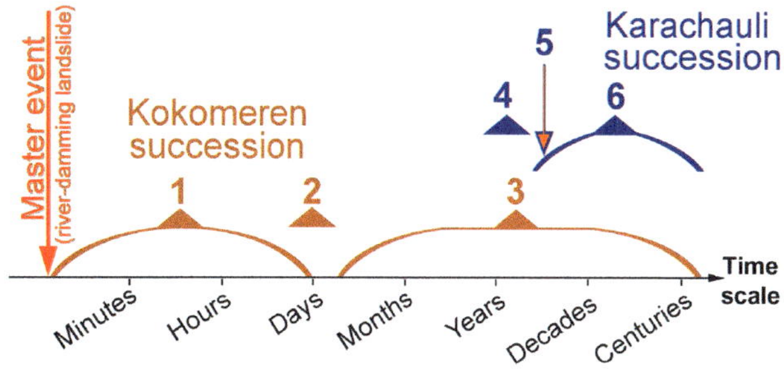

FIGURE 13.17 Succession of events that occurred after the formation of the Kokomeren (brown) and the Karachauli rockslides (blue). Large red arrow—master event likely triggered by a strong earthquake. 1—rock falls and talus formation; 2 and 4—deposition of the basal unit at the base of the dammed lake sediments (mixed breccia of Paleozoic and Neogene debris in the Kokomeren dammed lake, coarse sand in the Karachauli dammed lake); 3 and 6—laminated lacustrine silt accumulation; 5—formation of the liquefaction-induced features likely triggered by strong aftershock. *Reprinted with some modification from Strom, A., 2013b. Use of indirect evidence for the prehistoric earthquake-induced landslide identification. In: Ugai K., Yagi H., Wakai A. (Eds.), Earthquake-induced landslides. Springer, Heidelberg, New York, Dordrecht, London, 21–30 (ISBN 978-3-642-32237-2), with permission from Springer.*

that have been tilted already. Such kinematics was proved by the presence of nearly horizontal striae found on the rupture surface. It is likely that these successive deformations were associated with large past earthquakes. Considering that these tectonic movements deformed sediments accumulated in the rockslide dammed lake, both vertical and lateral displacements should occur later than formation of the river damming Kokomeren rockslide. Afterward, all these deformed units were eroded and overlain by typical debris flow deposits (6 on Fig. 13.13).

Judging from the uppermost level of lacustrine sediments that have been identified (6 in Fig. 13.12), water rose at least up to ca. 1800 m a.s.l., indicating the dams' effective height not less than 130 m. Because the river had cut a new bypass gorge through the bedrock and there are indications that the lake had been silted significantly, it was hypothesized that the water body had existed for about 1000 years (Hartwich et al., 2008).

The basal breccia traced at the base of the lacustrine sediments for a large distance can be considered as the event horizon whose accumulation had started immediately after appearance of the water rising in the dammed lake at this level. Because there was the narrow gorge between the damming site and the site shown on Fig. 13.13, and, considering the present-day mean discharge of the Kokomeren River of about 80 m^3/s (State Water Inventory, 1987c), water should appear at this site within 5 days after river damming or even faster. Thus, rock falls that brought angular boulders directly into the river channel should occur within this time span, which is much less than the precision of any numerical dating method (Walker, 2005).

Debris flows whose deposits locally interrupt basal breccia could originate due to the breach of local dams formed in the vicinity of the study site. We want to point out that the formation of debris flows is typical of the epicentral zones of large earthquakes that could be exemplified by the events following the 2008 Wenchuan earthquake (Tang et al., 2011).

All these spatially distributed depositional phenomena should occur either simultaneously with the Kokomeren rockslide formation or within a very short period after it (Fig. 13.17). Such a close timing, in turn, along with sound reasoning that spatially distributed rock fall and debris flow deposits could be triggered by seismic shaking, provides a strong, though indirect, argument of seismic triggering of the master event—the Kokomeren rockslide.

Consistent, noncontradictory combination of various evidence—failure on high terrace excluding slope undercutting by stream erosion as possible cause of its destabilizing, presence of active fault that stretches into the headscarp area, failure on both sides of the watershed, and simultaneity of main collapse and of minor slope failures in its vicinity allows reasonable assumption on seismic triggering of the Kokomeren rockslide.

CHAPTER

14

Rockslide Dams—Origin, Evolution, and Failure

OUTLINE

Rockslides that block river valleys and landslide-dammed lakes pose major threat to population in mountainous areas and to various infrastructure objects such as dams, roads, lifelines, etc., due to valleys' inundation and, first of all, due to possibility of a catastrophic breach of natural blockages. Most recent overview of various aspects of this problem could be found in the book "Natural and Artificial Rockslide Dams" published by Springer in 2011 (see Evans et al., 2011). Several examples of the disasters caused by such outburst floods were mentioned in Chapters 1 and 2, Introduction and Brief History of the Bedrock Landslide Studies in the Central Asia, respectively. Basic characteristics of the Central Asian river-damming landslides and associated lakes were listed in Section 6.5 in Chapter 6, The Bedrock Landslides Database, and are included in the rockslide database (see the Appendix). Here we summarize data on characteristic features governing formation, evolution, and failure of rockslide-dammed lakes that can be derived based on the case studies described in Chapters 7–9, Dzungaria, Tien Shan, and Pamir, respectively, and in Part 1, General Information.

Inventories of landslide dams were compiled for many regions (Swanson et al., 1986; Casagli and Ermini, 1999; Ermini and Casagli, 2002, 2003; Korup, 2004; Hermanns et al., 2011a; Fan et al., 2012b, 2017; Tacconi Stefanelli et al., 2016). Different indices allowing prediction of stability (or instability) of blockages were proposed based on their analysis (see an overview in Dufresne et al., in press). Applicability of one of them—the dimensionless blockage index (DBI) (Ermini and Casagli, 2003) for Central Asian region was tested in Strom (2010b) based on a limited number of case studies, and it was shown that several rockslide dams demonstrate an abnormal behavior. Hereafter, possible causes of such deviations will be proposed.

The important step of the river-damming rockslides analysis is their classification. The first attempt to classify these features was presented in the well-known papers (Schuster and Costa, 1986; Costa and Shuster, 1988) where six types of rockslide dams were selected. More complex classification proposed in Hermanns et al. (2011b) was described in brief in Section 5.2 of Chapter 5, Bedrock Landslide Types and Classification Systems (see Figs. 5.29 and 5.30). We must notice,

Rockslides and Rock Avalanches of Central Asia.
DOI: https://doi.org/10.1016/B978-0-12-803204-6.00014-4

however, that, despite the complexity of this system, the variability of the relationships of landslide bodies with the preexisting topography is so high that we can find case studies that do not fit exactly to any of the types or subtypes of this classification. It can be exemplified by the Twin-lakes rock avalanche in Xinjiang (see Fig. 8.20) that can be ascribed to subtypes IVb, IVc, or IVd shown on Fig. 5.29 (Hermanns et al., 2011b) but does not correspond to each of them exactly.

Rather widespread group of water bodies associated with rockslides that was not included in the proposed classification are lakes formed not in the dammed valley upstream of the blockage, but within the rockslide body, either in the proximal depression at the foot of the headscarp, which is typical of the Primary rock avalanches mainly (see Section 5.3.1), or in small closed depressions on the irregular surface of rockslide deposits. It was suggested (Dufresne et al., in press) to classify them as "landslide lakes" differing from the "backwater lakes" (the latter are real "landslide-dammed lakes"). Landslide lakes in the proximal depressions can be exemplified by water bodies at the Arashansai rock avalanche in the North-Western Tien Shan (see Fig. 8.225) or the Kolduk rockslide in the Southern Tien Shan (lake at 2540 m a.s.l. on Fig. 8.241). Lakes on top of the rockslide deposits somehow (but not fully, because they are not fed by the pre-slide streams) correspond to Subtypes I and IIIa on Fig. 5.29 and can be found on top of the Beshkiol landslide (see Fig. 4.1), the Dead Lakes rockslide (see Fig. 8.205), or the Sarychelek rockslide (see Fig. 8.215). Similar proximal or surficial depressions at other rockslides are dry as, for example, at the Bielogorka-left rock avalanche (see Fig. 8.98B).

Traces of past landslide lakes can be found at the Ornok rockslide in the Central Tien Shan were laminated silt partially mixed with sorted debris flow deposits rest both in the deepest part of the proximal depression formed by the rotational blockslide at the foot of the main headscarp (marked by L on Fig. 8.167) and on top of the avalanche-like part of its body (marked by L on Fig. 8.169, see also Fig. 14.1).

Generally landslide lakes do not pose a significant threat due to their limited volume and the absence of the feeding streams with large catchments, whose discharge could increase significantly due to rainstorm or snow melt resulting in catastrophic outburst flood.

As it was mentioned above, in Section 5.2, classifications of landslide dams proposed in Hermanns et al. (2011b) reflects the relationships between the shape and size of the rockslide deposit and those of the river valley (see Figs. 5.29 and 5.30) and consider neither type of slope failure described in Section 5.1, nor rockslide internal structure (see Chapter 11: Structure and Grain-Size Composition of Large-Scale Bedrock Landslides), nor motion mechanism(s) classified in Section 5.3 and discussed in Chapter 12, Morphological and Structural Evidence of Rockslides' (Rock Avalanches') Motion Mechanism(s). More comprehensive characteristics of natural blockages that open the way to predict their evolution, i.e., the longevity of the stream damming and possible style of dams' breach requires joint analysis of rockslide dams morphology and of peculiarities of their internal structure and grain-size composition (Dunning et al., 2005; Dunning and Armitage, 2011; Strom, 2010b, 2013a; Fan et al., 2017). In fact, both characteristics are predetermined by the rockslide dam emplacement mechanism(s).

14.1 FORMATION OF ROCKSLIDE-DAMMED LAKES

Analysis of case studies described in the previous chapters shows that rockslide dams could be formed

FIGURE 14.1 Lacustrine sediments (L) on top of the Ornok rockslide (rock avalanche) deposits (RS). Left—laminated lacustrine silt; right—lacustrine silt alternating with coarser layers left by debris flows.

by rock slope failures practically of all types—planar, rotational, wedge, or compound, moving either as blockslides (rarely) or, more often, transforming in the flowslides (rock avalanches) of the primary, jumping, or secondary types. Natural dam formed by the large planar rock slope failure can be exemplified by the Lower Ak-Kiol rockslide (see Figs. 5.6 and 8.166) in the Central Tien Shan. The most common type of large-scale rock slope failures resulting in the formation of rockslide dams are wedge failures characterized by deep sliding surfaces. They can be exemplified by the 1911 Usoi rockslide that created the Sarez Lake in Pamir (see Fig. 5.7), by the Aik-Kiol (see Fig. 4.3), and the Ak-Kiol rockslides (see Figs. 5.9 and 5.10) in the Central Tien Shan. The more irregular compound river-damming rockslides can be exemplified by the gigantic Beshkiol landslide (see Fig. 4.1) and the Bashkol Lake rockslide (see Fig. 6.20) in the Central Tien Shan, by the Twin-lakes rockslide (see Fig. 8.20) in the Eastern Tien Shan, or by the Lower Kiol-Sai rockslide (see Fig. 8.65) in the Northern Tien Shan. Landslide lakes associated with the rotational slides can be exemplified by the above-mentioned Ornok rockslide (see Figs. 8.167 and 8.171). This rockslide, however, did not block wide Kokomeren River valley completely and, thus, did not form a real dammed lake.

Most of the rockslides identified in the study region occurred in the narrow and deep erosional river valleys and form or had formed permanent or temporary water-retaining blockages if their volumes were large enough to block the valley. Rockslides with the pronounced frontal confinement form rather compact blockages. Shape and main morphological characteristics of such dams—the effective height (minimal level of the dam's crest over blocked valley bottom), position of the dams' lowermost part where incision caused by overtopping starts (Fig. 14.2), dams' compactness (ratio of dams' effective height and its width along the stream—H_d/W_d), and steepness of the slopes, especially of the downstream one, depend on the style of rockslide debris motion (see Section 5.3) and on the conditions of its collision with an opposite valley slope described in Section 12.2.3 (see also (Strom, 2010a)). These parameters, along with the dam's size, internal structure, material properties, and hydrological characteristics of the stream, predetermine dammed lake evolution (see also (Korchevskiy et al., 2011)).

Sometimes dammed lakes originate in river valleys passing along wide intermontane neotectonic depressions being dammed by long runout rock avalanches. It can be exemplified by the 9-km long Chaartash-2 rock avalanche in the Central Tien Shan that had crossed the Kiok-Art River valley just before its motion

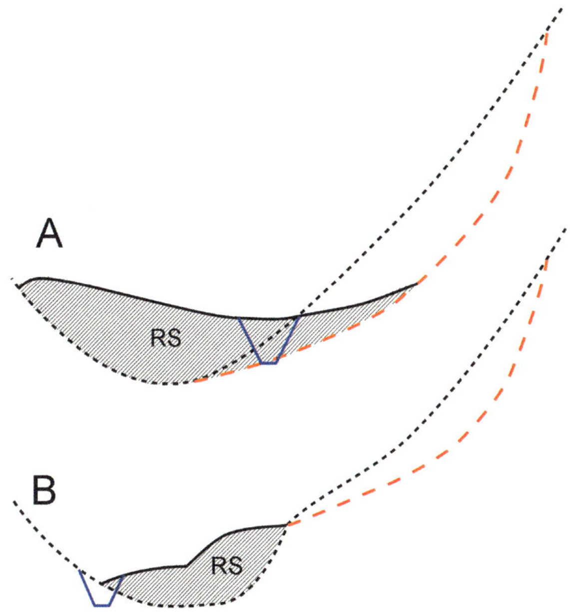

FIGURE 14.2 Rockslide dams with different across-valley debris distribution. (A) Formed by Primary rock avalanche with distal run-up and pronounced proximal lowering; (B) formed by Jumping rock avalanche with distal lowering (as the Kashkasu rockslide shown on Fig. 5.45). Dashed back line—the pre-slide valley profile; dashed red line—sliding surface; bold black line—dam's crest; hatched area marked by RS—rockslide deposits; blue line indicates the position of the erosional canyon that would be formed in case of overtopping.

halted (see Figs. 5.61 and 8.129), by the 34-km long Koman rock avalanche in the Northern Pamir that had blocked the Kyzylsuu River (see Fig. 9.20) or by the ca. 4-km long bicorn Archabel rock avalanche in the Northeastern Pamir (see Fig. 9.4). Most of such natural dams were rather low and spread over valleys' bottoms. Blockages with similar morphological characteristics have been formed by rock avalanches that entered wide valleys of the major rivers after passing through their tributaries such as the 7.5-km long 1949 Khait rock avalanche (see Figs. 8.271 and 8.272) or the prehistoric, also 7.5-km long Chukurchak rock avalanche (see Figs. 8.221 and 8.222). To the same type of river-damming, long runout rock avalanches we can ascribe the 2000 Yigong rock avalanche in Tibet (Shang et al., 2003; Wang, 2008; Xu, 2008; Yin and Xing, 2011; Evans and Delaney, 2011).

Long runout rock avalanches with pronounced lateral confinement do not form large dammed lakes. Most of them passed either along valleys such as the Urmochdara rock avalanche in Pamir (see Fig. 4.41) or the Upper Kashkasu rock avalanche in the Central Tien Shan (see Fig. 4.43) and only small ponds form sometimes in the proximal depressions at the feet of their headscarps. One more good example is the 6.5-km long Karakystiak rock avalanche in Northern Tien Shan that originated in a tributary valley and then entered larger but similarly narrow valley and followed it up to the halt. Rock avalanche had dammed the latter one and a small gulley on the way, but these water bodies are rather small too (see Fig. 6.7).

Practically all extremely mobile rock avalanches can be ascribed either to primary or to secondary types (see Sections 5.3.1 and 5.3.3 in Chapter 5). If a river dammed by such a long runout rock avalanche is powerful enough, it, usually, erode the blockage rather shortly as it occurred after the 1949 Khait earthquake, because such dams are relatively low and are composed of the material that underwent intense fragmentation during its motion. At the same time, large width of such blockages and their gentle slopes resulting in low-gradient overflow might support slow gradual incision of the stream and even almost complete siltation of the dammed lakes as at the ca. 4-km long Archabel rock avalanche shown on Fig. 9.4.

Type of rockslide/rock avalanche (see Section 5.3) that dams narrow valley with frontal confinement predetermines its further evolution at a large extent. Primary rock avalanches in such valleys often produce blockages with prominent proximal lowerings whose effective height is much less than it could be if the dams' crests would be more even. Such crest profiles with proximal lowering and well-pronounced distal run-up (see Fig. 14.2) are most typical of the Central Asian blockages and can be exemplified by the Mini-Köfels (see Fig. 5.41), the Lower Kokomeren (Fig. 14.3), the Akterek (see Fig. 8.187 and Table 8.9), the Chong-Tash (see Fig. 8.135), the Yorkhdara (see Fig. 9.106), the Aqshireh (see Fig. 9.119), rockslides, and by some other no-name features in different parts of the study region (see Fig. 5.40 and 8.253). Same style of motion characterizes the Usoi (see Fig. 9.44), the Aksu (see Figs. 8.102 and 8.103), the Kudara-Pasor (see Fig. 4.20), the Kafirnihon (see Fig. 8.274) rockslide dams, rockslide dam in the Kainar River valley (see Fig. 8.116), and other case studies with significant, but, relative to these dams height, less pronounced run-up (see Table 12.1). Similar across-valley profiles characterize also compact river-damming parts of several large-scale secondary rock avalanches (see Section 5.3.3) with primary style of the initial motion such as the Big Dragon (see Fig. 5.65) or the Karasu Lake (see Fig. 5.64) cases.

FIGURE 14.3 View on the Lower Kokomeren rockslide from upstream. Distal part of its body rose on the right bank of the river up to ca. 1435 m a.s.l., while the lowest part of the dam's crest was at ca. 1300 m a.s.l. only.

Besides decreased effective height of rockslide dams formed by primary rock avalanches with pronounced run-up, such blockages being overtopped are dissected by the erosional canyons cut through the loose material at the proximal parts of their bodies (see Fig. 14.2A), while in those cases where incision starts at dams' fronts, there is higher possibility that it will evolve through the bedrock (see Fig. 14.2B) overlain by relatively thin distal parts of the deposits as it occurred at the largest rockslide of the entire Central Asian region—the Beshkiol rockslide (see Figs. 4.1 and 5.13). Because incision through the bedrock is the gradual long-term process, such dammed lakes can exist for a long time; the Beshkiol rockslide, in particular, had blocked the powerful Naryn River for about 2200–3000 years (see Section 8.4 in Chapter 8; Korup et al., 2006). Long-term (~1500 years) river damming due to incision through the bedrock took place at the Kokomeren rockslide (see Figs. 8.146 and 13.12) described in Section 8.4 and Section 13.2.6 in Chapter 13, Seismically Induced Rockslides: Their Discrimination, though in this case the Kokomeren River had eroded new channel through the bedrock at a small section at upstream part of the blockage (see Fig. 8.154). The bypass channel was incised in the bedrock at the front of the Displaced Peneplain rockslide (see Fig. 4.5), few kilometers downstream from the Kokomeren one. However, no lake sediments that could be attributed to this damming event have been identified.

We must notice that frontal lowering of the dams' crest, regardless of its cause, does not mean that incision will go through the bedrock. If the opposite (confining) valley slope is steep enough, bottom erosion will affect dams' body and, thus, could be as rapid as in case of the proximal incision. It occurred during the 1963 Issyk (see Fig. 2.1) and the 1966 Yashinkul (see Fig. 2.3) disasters. Frontal incision took place at the Arashan (see Fig. 8.95), the Kokomeren (see Fig. 8.146), the Tianchi (see Fig. 8.4), and Pazhuk (Fig. 9.135) rockslide dams, at the rockslide dam shown on Fig. 8.113. Similar frontal incision will develop with time at the Kiol-say (Fig. 5.63) or the Pairon (Fig. 8.277) rockslide dams.

Case studies where dams' crest is nearly horizontal are rare. They can be exemplified by the Aik-Kiol (see Fig. 4.3) or the Shiva Lake (see Fig. 4.22) rockslide dams.

The shape of rockslide dams depends also on the along-valley debris distribution, in other words on dam compactness. It is evident that for the given rockslide volume and valley width, more compact dam would be higher and with steeper slopes. These parameters depend at a large extent on the morphology of the opposite valley wall with which rapidly moving rockslide body collides (Strom, 2010a). As it was described in Section 12.2.3 of Chapter 12, rockslides entering the tributary valley that falls in the main dammed valley close to the blockage axis are more compact, i.e., higher and with steeper slopes than those that just collided with an opposite slope of the valley, producing up- and/or downstream spreading of debris. Compact blockages trapped in the tributary mouths can be exemplified by the intact Kulun rockslide dam (see Fig. 8.237), by the breached Aksu (see Fig. 8.103), the Karachauli (see Figs. 8.172 and 12.2) rockslide dams, and by breached rockslide dam in the Khutubi River valley shown on Fig. 8.9.

Dams, whose bodies spread along the dammed valley, besides well-pronounced Secondary rock avalanches such as the Karasu Lake rockslide dam (see Fig. 5.64) or the Big Dragon rockslide dam (see Fig. 5.65), can be exemplified by the rockslide dam in Xinjiang shown on Fig. 6.5 or by the Upper Zhasylkol Lake in Dzungaria (see Fig. 6.18) that can be ascribed to the Primary type (see Section 5.3.1).

14.2 EVOLUTION OF ROCKSLIDE-DAMMED LAKES

As it was mentioned in Schuster and Costa (1986) and Costa and Shuster (1988), 85% of landslide dams have lasted less than 1 year. However, their statistics included blockages of all sizes created by various slope processes and, thus, cannot be applied directly for the river-damming barriers formed by large-scale bedrock landslides as pointed out in Evans et al. (2011).

About 10% of rockslides identified in the Central Asian region form dams with backwater lakes; 80 of them can be seen on Figures cited in the previous chapters. The most well-known are the 1911 Usoi rockslide dam (see Figs. 9.41 and 9.43) and 500-m deep Sarez Lake impounded upstream that contains ca. 17 km^3 of water (see Fig. 6.21) (Preobrajensky, 1920, Alford and Schuster, 2000; A. Ischuk, 2011). Numerous large, though much smaller than the Sarez, dammed lakes impounding hundred millions cubic meters of water are in Tien Shan—the Sarychelek (see Fig. 8.215), the Karasu (see Fig. 5.64) the Tianchi (see Fig. 8.3) the Big Dragon (see Fig. 5.65), the Iskanderkul (see Figs. 8.288 and 8.290), and the Aksai-Kiolsu (see Fig. 8.119) Lakes; in Dzungaria—the Upper Zhasylkol Lake (see Fig. 6.18); and in Pamir—the Yashilkul (see Figs. 4.18, 4.19 and 9.65) and the Shiva (Fig. 4.22) Lakes.

Much more rockslide-dammed lakes had been silted completely or partially, or emptied after dam's breach. More than 130 Figures cited above demonstrate such case studies. Some of the past long-living landslide dammed lakes were comparable with the Sarez Lake such as the lake in the Kokomeren River valley dammed by the Karachauli rockslide that stored about 7 km^3 of water (see Figs. 8.172 and 13.7). Moreover, several prehistoric dammed lakes were even larger, such as the silted and then breached lake in the Naryn River valley dammed by the gigantic Beshkiol landslide (see Fig. 4.1) whose volume could reach 25 km^3 (see Figs. 8.126 and 8.127). It survived for 2000–3000 years according to the amount of lacustrine sediments accumulated (Korup et al., 2006). Another example of an extremely large dammed lake impounded by Shidz rockslide that stored up to 31 km^3 of water is shown on Fig. 9.101. Its possible evolution is described in Section 9.3 in Chapter 9.

The key question related to any rockslide dam is: can it be considered as stable and, thus, safe feature, or as unstable hazardous one? However, identification of a rockslide dam and landslide-dammed lake as "Stable" or "Unstable" is complex and, sometimes, controversial. There are many case studies when rockslide-dammed lake that had existed for centuries if not millennia breached one day producing disastrous outburst flood. It can be exemplified by the 1914 breach of the Rio Barrankos Lake in Argentina (Groeber, 1916; Gonzalez Diaz et al., 2001; Hermanns et al., 2004). In the study region similar delayed breaches occurred in 1963 with the Issyk Lake in Kazakhstan (see Fig. 2.1) and in 1966 with the Yashinkul Lake in Kyrgyzstan (see Fig. 2.3) (see the description in Chapter 2). Both lakes had existed for a very long time and their catastrophic breaches have not been preceded by any medium-term phenomena allowing prediction of such disaster. On the other hand, the well-known 2010 Attabad rockslide dam in Pakistan (Delaney and Evans, 2011) that blocks the powerful Hunza River, should be breached soon after its overtopping, according to expectations based on its high (4.62 – 4.85) DBI value. However, it still exist, though, of course, its lifetime is still rather short to make sound conclusion on its long-term evolution.

Some lakes, both existing and past, demonstrate evidence of partial breach such as the outwash accumulations just downstream from the erosional canyon (see Fig. 8.135), series of terraces in the lacustrine deposits upstream (see Figs. 6.6, 8.57, 8.263), or distinct trace of the original lake level above the present-day level (see Fig. 8.290). Considering all these circumstances, it is better to use the collocation "Long-living lake" instead of such terms as "Stable" or "Permanent" lake. It can be used to characterize both the existing and the past dammed lakes. Two main types of the long-living lakes could be selected: those with the existing water bodies that originated centuries or millennia ago and those that had been silted completely or partially regardless if they have been breached or not, because such siltation requires long-term stream damming. Similarly, the collocation "Short-living lake" is more preferable than the "Unstable lake."

Structural/sedimentological and morphometric characteristics of Central Asian rockslide dams' predetermining their evolution are summarized in the following sections.

14.2.1 Influence of Dam's Internal Structure and Grain-Size Composition on Their Evolution

It is evident that stability of any natural blockage depends at a large extent on the mechanical properties of the material that forms the barrier, first of all on its grain-size composition and bulk density that predetermine dams' permeability and erosion resistance. Effects of internal structure on dams' stability were analyzed in Dunning et al. (2005) and Dunning and Armitage (2011). As it was mentioned above repeatedly, vast majority of Central Asian rockslide dams have coarse blocky carapace with large voids, sometimes with gigantic integral, though, fractured blocks that are hundreds of meters in size (see, e.g., Fig. 9.42). These carapace facies (see Section 11.2.1 in Chapter 11) characterized by high permeability overlay much less permeable internal core formed by the intensively fragmented and compacted material. It results in a rapid rise of water in the dammed lake immediately after its formation until it reaches the permeable carapace zone through which filtration starts. After that, depending on the blockage dimensions and relationship between the inflow and seepage discharge, lake level either continue increasing or stabilizes. It can be exemplified well by the Sarez Lake dammed on February, 18, 1911. For about 3.5 years the Usoi rockslide had blocked the Murgab River completely and only in spring of 1914, when the lake was about 200 m deep, first springs with ca. 2 m^3/s discharge appeared on its downstream slope (Preobrajensky, 1920; A. Ischuk, 2011). After that this natural reservoir continued growing, and now its level reaches ca. 3263 m a.s.l. with annual variation between summer and winter levels up to 12 m and continues increasing though quite slowly (A. Ischuk, 2011). We must mention that despite gradual rise of lake level, the recorded mean inflow rate (47.1 m^3/s) is lower than that of the outflow (47.7 m^3/s) [ibid]. This controversy appeared, likely, because discharges of several minor tributaries that bring their water to the lake have not been measured. According to hydrological investigations and the position of the most powerful springs (see Figs. 9.43 and 9.46) that are about 140–150 m below the lake level, most of the seepage takes place through the upper carapace facies of this blockage. Same can be observed at the gigantic Shiva Lake rockslide dam where springs appear about 100 m below the lake level and ca. 170 m below the crest of the nearly 500-m high blockage (see Fig. 4.22).

Sometimes seeping water appears either at the base of the downstream slope of the blockage as at the Kutmankul rockslide dam in the Northwestern Tien Shan (see Fig. 8.203) or at the contact between rockslide body and the valley wall as at the Karasu Lake case (see Fig. 5.64). It can be assumed that in such cases water found its way through the coarse pre-slide talus accumulations at the valley floor and its slopes' base overlaid by rockslide deposits.

An important peculiarity of rockslide dams' internal structure that can have very negative effect on their stability is sharp transition from the coarse carapace facies to intensively fragmented body facies (see Section 11.2). Unlike bouldery carapace that can resist erosion caused by overflow, fine comminuted debris is much more erodible being affected by rapid turbulent flow. That is why as soon as carapace is incised or removed as it occurred in 1963 at the Issyk rockslide dam (see Fig. 2.1) where it was washed out by waves caused by debris flow surges that entered the dammed lake, erosion through the body facies evolves incrementally resulting in catastrophic lake outburst.

14.2.2 Morphometric Parameters Determining Rockslide Dams' Evolution

General similarity of the internal structure and grain-size composition of most of the Central Asian river-damming rockslide bodies should level out effect of their lithology, which is quite variable, on the stability of water-retaining blockages. Their evolution depends at a major extent on the hydrology of a dammed river and on rockslide dam morphometric parameters.

Analysis of some Central Asian rockslide dams, both intact or breached, allows identifying several morphological factors affecting evolution of blockages and rockslide-dammed lakes that result in their abnormally long-term or short-term lifetime that contradicts the expectations derived according to their DBI values (Ermini and Casagli, 2002, 2003) (see also Strom, 2010b).

Besides, we have to mention that while the entire study region is characterized by arid climate, main hydrological characteristics (mean annual and peak discharge, yearly runoff, and sediment yield) of rivers draining these mountain systems vary drastically influencing the fate of past, existing, and inevitable future blockages. In particular, long-term existence of the Sarez Lake is stipulated by low mean discharge of the Murgab River that is balanced by seepage through the extremely long (across the valley) permeable zone of the Usoi rockslide. Several other high blockages in large river valleys such as the Shidz rockslide dam in the Pianj River valley (see Figs. 9.98 and 9.101) or the Karachauli rockslide dam in the Kokomeren River valley (see Figs. 8.172 and 13.7) were not as large as the Usoi rockslide dam, and these rivers had much larger mean discharges (about $100\,m^3/s$ for Kokomeren and about $845\,m^3/s$ for Pianj) than the contemporary Murgab that drains the Pamir Plateau with a mean precipitation less than 400 mm/year only and without large glaciers. Such discharges could not be released by filtration and, thus, these dams could not survive. Anomalously long-term survival of the Beshkiol dammed lake (DBI = 2.97; Strom, 2010b) in the Naryn River valley with the mean discharge of about $200\,m^3/s$ was possible just due to the lucky combination of the landslide dam size and the dammed valley shape resulting in the formation of the bypass gorge incised in bedrock in front of the blockage (see Fig. 4.1 and the description in Section 8.4).

One more very large rockslide about $4\,km^3$ in volume that covers the area of ca. $18\,km^2$ forms the Sarychelek Lake in the Northwestern Tien Shan (see Fig. 8.215) with an extremely low DBI value (0.57 (Strom, 2010b)). The lake with an area of $5.4\,km^2$ only is much smaller than the rockslide dam that ensures its long-term stability. Large dimensions and shape of the blockage (see Figs. 8.215 and 8.216) result in low gradients of the stream flowing over the dam and of its very low erosion rate. We can anticipate that even in the worst case, if next large seismic event comparable with the 1946 M7.5 Chatkal earthquake would cause some large-scale slope failure on the lake bank (we must notice that there are no indications of such instability), amount of water that can be splashed out from the lake would not be able to erode the canyon deep enough to produce the catastrophic outburst. Small catchment and the absence of present-day glaciers result in quite limited sediments yield and low rate of lake siltation.

Long-living blockage that still forms dammed lake despite high DBI value (3.71; Strom, 2010b), typical of the unstable features, can be exemplified by the Yashilkul rockslide in Pamir (see Figs. 4.18 and 9.65). Its stability is guaranteed, besides coarse blocky carapace (see Fig. 4.19) by so large voids between them that water flows almost free, and by the morphology of this blockage that is very wide (about 4 km along the stream) and relatively low ($\sim$50 m high). Such dams' profile results in a low gradient of the flow passing through the carapace that could not erode it. Shape of the wide glacial valley, whose lower part is filled by the lake, almost excludes the possibility of any large-scale scale slope failure in the lake that could produce surge wave. Large area occupied by this water body prevents significant rise of water level during flood period. Besides, small sediment yield of the Alichur River and other streams falling into the lake prevents its siltation. All these factors guarantee very long-term lifetime of this water body that is used now to regulate the Gunt River flow (see Fig. 4.18).

On the other hand, some rockslide dams, which, according to very low DBI values should be "stable," had been breached completely and, judging from the absence of lacustrine sediments upstream, their breach occurred soon after dams' formation—likely due to first overtopping. Such evolution can be exemplified by the ca. $1\,km^3$ and 400-m high Aksu rockslide in Northern Kyrgyzstan (see Fig. 8.103) that had blocked the stream with the catchment area of $329\,km^2$ only. Its DBI is 2.12 that explicitly places this blockage in the stability domain (Ermini and Casagli, 2003; Korup, 2004; Tacconi Stefanelli et al., 2016) (previously its DBI was calculated as 1.94 due to larger—up to $1.5\,km^3$ volume estimate; Strom, 2010b). However, the dam failed. Considering the size and shape of the valley and the present-day annual runoff of $140\,Mm^3$ (State Water Inventory, 1987c), it had been overtopped in less than 1 year.

We assume that such abnormal behavior could be governed by combination of two main factors. First, because rockslide body had entered the wide lowering on the opposite side of the valley (see Figs. 8.103 and 12.3A, and discussion in Section 12.2.3 in Chapter 12), it had formed a very compact blockage with relatively high H_d/W_d ratio and with inevitably steep slopes.

Second, the upstream lake was deep but rather small due to the valley shape and high channel grade and, thus, during flood period, its level should increase significantly so that peak discharge passing through (over) this dam should be much bigger than in the Yashilkul case, despite similarity of mean annual runoff of both rivers (State Water Inventory, 1987c, 1990). Such overflow eroded steep downstream slope of the dam rapidly resulting in complete breach of the blockage.

We anticipate that dam compactness (H_d/W_d) and capability of the natural reservoir to reduce peak discharge that can be quantified, as a first approximation, by the runoff during flood period (R_f) divided by the lake area (A_L) can be used as additional parameters of the blockage stability assessment. More compact dams with steeper downstream slopes as well as dammed lakes with smaller areas should be considered, if all other factors held equal, as more susceptible for breach by overtopping.

Compactness, however, might have dual, not only negative, but also positive effect on dams' stability. Indeed, the higher the dam is, the larger the carapace infiltration zone is and, if the latter is large enough, filtration through the permeable blocky carapace (see subsection 11.2.1) can balance the inflow as it takes place at the Sarez Lake (Alford and Schuster, 2000; A. Ischuk, 2011).

Rockslides that, contrariwise, underwent significant along-valley spreading after collision with an opposite valley walls and formed dams with long and gentle downstream slope can resist the incision caused by the low-gradient overflow and, thus, represent stable and safe features. Such case studies can be exemplified by the Upper Zhasylkol rockslide dam and lake in Dzungaria (see Fig. 6.18), as well as by several completely silted lakes with almost intact dams as in the upper reaches of the Dzhergalanty River basin in the Eastern Tien Shan (see Fig. 8.23) and at the confluence of the Keskentas and the Alaigyr Rivers in the Highest Tien Shan (see Fig. 8.41).

14.2.3 Cascades of Dammed Lakes

An important factor predetermining rockslide dams' evolution is the presence of several dammed lakes along the same river. We rarely can prove their simultaneous formation though sometimes this assumption is quite logical (see the description of the Shidz and Voznavd rockslides in Section 9.3 in Chapter 9 and of the Vir-Tangiv and Zuvor rockslides in Sections 4.2 and 9.2 in Chapters 4 and 9: Principles and Criteria of Landslides Identification and Discrimination From the Features Formed by Other Mass-Wasting Phenomena, and Pamir, respectively). Anyhow, if river is blocked at several sites forming the cascade of dammed lakes, the fate of the downstream ones depends at a large extent on the stability of the upstream blockage. If the upper one is large enough, it can transform flood runoff, accumulating summer water flow and releasing it gradually that decreases the erosional capacity supporting long-term stability of the dam(s) located downstream.

It can be exemplified by rockslide-dammed lakes in the Kulun River in Southern Tien Shan. Upper part of its basin with catchment area of 140 km^2 is blocked by the 260-m high Kulun rockslide dam about 130 Mm^3 in volume (see Fig. 8.237) with DBI = 2.45. Nearly 3.5-km downstream, another small lake exists being dammed by rockslide about 2 Mm^3 in volume only (ibid); 22.5 km downstream, near river mouth, one more small rockslide, also about 2 Mm^3 in volume, had blocked it by the ca. 70-m high intact dam, which lake was silted completely (see Figs. 6.23 and 8.238), despite its high DBI value of 4.09 that put it in the instability domain (Ermini and Casagli, 2003; Strom, 2010b). While the small lake just downstream of the Kulun rockslide dam evaded siltation being so close that almost no material can be transported in it, the Lower Kulun Lake have accumulated sediments from the 213 km^2 catchment but peak discharge of the inflow has been reduced significantly, and this blockage remains almost intact.

Similar protection of the downstream blockage by the much larger upstream rockslide dam can be observed in the upper reaches of the Sarkat River basin, also in the Southern Tien Shan. Here the Ai-Kul rock avalanche dam about 75 Mm^3 in volume (see Fig. 8.257) had protected the much smaller rockslide dam about 3 Mm^3 in volume only located 7 km downstream, which dammed lake had been silted completely before the incision started (see Fig. 8.258).

One more example of such cascade can be found in the Eastern Karasu River basin in the North-Western Tien Shan (see Section 8.5), whose mouth had been blocked by the Karakul rockslide dam about 200-m high, and this dammed lake had been silted completely (see Figs. 6.22 and 8.208), despite rather high (2.85) DBI value that puts this feature in the uncertainty domain (Ermini and Casagli, 2003; Strom, 2010b). Dammed lake had been fed by two rivers—the Eastern Karasu and by its tributary—the Karakol River, both of which had been also dammed by large-scale rockslides—the Karasu Lake rockslide that forms intact dam (see Fig. 5.64) and by the partially breached Dead Lakes rockslide dam (see Figs. 8.205 and 8.206). Their assumed temporal relationships were discussed in Section 8.5. Long-term survivance of the Karakul rockslide dam was predetermined by its size and by the dammed valley profile—the blockage was so high that overflow had stated at the opposite—the left bank of the Eastern Karasu River composed of bedrock (see Fig. 6.22), which is much more resistant to the erosion. It recalls situation at the Beshkiol rockslide dam (see Fig. 4.1) bypassed by the post-damming erosional gorge.

We must point out, however, that in case of the upstream dam breach, lake(s) located downstream have little or even no chance to survive. It can be exemplified by the 1963 Issyk Lake breach caused by debris flow generated by the breach of the small Jarsay glacial lake upstream (see Chapter 2).

Likely similar succession of events occurred in the Unkursai River valley in Central Tien Shan dammed by partially breached Ak-Kiol rockslide dam (see Figs. 5.9 and 5.12) and by almost completely breached Lower Ak-Kiol rockslide dam ~2 km downstream (see Figs. 5.6, 5.62, and 8.166). The latter's breach could be caused by partial breach of the upstream Ak-Kiol dam. Water released from the upper lake overfilled the lower one, likely overtopped its dam composed of weak crushed Neogene conglomerates and eroded it.

Cascades of rockslide-dammed lakes are rather common in the South-Western Pamir where numerous rockslides had blocked the Rivak, the Dirumdara, and the Chandindara River valleys (see Fig. 9.64). Here the influence of the state of the upstream blockage on the lower lakes is not as distinct as in the Kulun or Unkursay River valleys described above. Several cascades of small dammed lakes were found in Dzungaria (see Chapter 7, Fig. 7.2). Some of them still exist (see Fig. 7.3), and others have been breached (see Fig. 7.4).

The largest cascade of seven existing and one filled lakes is located in the Southern Tien Shan in Shing River valley—the so-called Marguzor Lakes (see Fig. 8.281). Description of these river-damming bodies, whose origin is debatable (see N. Ischuk, 2016) can be found at the end of Section 8.6 in Chapter 8. We want to mention that sound indication of their rockslide or glacial origin could be derived from the relative age of these features—if chain of these dams was really formed by stepped glacier retreat, each upper dam should be younger than the lower one, otherwise they could not survive being affected by advancing glacier, while visual impression of these blockages (see Figs. 8.282–8.286) counts in favor of their simultaneous origin. Solution of this debate requires dating of all Marguzor dams, most likely by cosmogenic nuclides method.

14.3 BREACHED LAKES AND EVIDENCE OF THE PAST CATASTROPHIC OUTBURST FLOODS

We pointed out above (see Chapter 1) that outburst floods caused by catastrophic breach of large rockslide dams can be much more disastrous than rockslides and rock avalanches themselves because they can affect the entire river valleys (Costa and Schuster, 1988, 1991; Evans et al., 2011). Even if people living downstream will be evacuated timely, short-term release of millions and billions cubic meters of water can devastate valley floor destroying settlements, roads, lifelines, and arable land.

Few historical case studies of such disastrous rockslide dam failures in the Central Asian mountains—the 1963 Issyk (see Fig. 2.1), the 1964 Aini (see Fig. 2.2), the 1966 Yashinkul (see Fig. 2.3), the 1977 and 1998 disasters in the Shahimardan River basin, the 1993 Belaldy and Torkent debris flow were described in brief in Chapter 2. Floods caused by breach of the 1911 Chilik (see Figs. 8.66 and 8.67) and of the 1949 Khait (see Fig. 8.271) rockslide dams that occurred soon after their formation were less disastrous due to the small height of these blockages, scarce population of the Chilik River valley in the former case, and short distance to the much larger Surkhob River valley that accommodated the released water in the latter case.

Here we summarize observations on the past (prehistoric) outburst floods described in Chapters 7–9, some of which were really outstanding, with estimated peak discharges by far exceeding those recorded during the historical events mentioned above. Their study demonstrates what could happen if some of the existing or newly formed large rockslide dams would fail.

Despite large number of the existing rockslide-dammed lakes (more than 80 features over the entire study region), most of the rockslide dams had been breached either shortly (months–years) or centuries–millennia after their formation, leaving erosional canyons, emptied, or partially silted lake basins upstream, and, rarely, well-recognizable deposits and minor landforms left by the outburst floods downstream. In favorable cases, careful analysis of both remote sensing data and of the outcrops downstream from past blockages allows estimating peak discharge of these outburst floods. Such analysis was done for several case studies—the ~80-m high Lower Aral rockslide dam in the Kokomeren River in the Central Tien Shan (see Section 8.4, Figs. 8.161–8.164) that produced outburst flood with a peak discharge of about 30,000 m^3/s (Strom and Zhirkevich, 2013) and for the ~300-m high Sarydjaz-Aksu rockslide dam on the same name river in the southern part of the Highest Tien Shan (see Section 8.2, Figs. 8.53–8.56) with enormous estimated peak discharge up to 400,000–500,000 m^3/s (!).

One more similar case study is in the Sokh River valley just downstream from the confluence of the Ak-Terek and the Khodja-Achkan Rivers in the Southern Tien Shan. Here, peak discharge of the outburst flood can be estimated from the presence of the dendritic amygdaliform ravines left by the surge wave on the 30- to 40-m high river terrace (see Fig. 8.252). It was hypothesized that these impressive landforms about 40 m above the present-day floodplain were formed by an outburst flood caused by breach of the ~150-m high rock avalanche dam about 40 Mm^3 in volume in the Ak-Terek River valley ~4.5 km upstream. This case study is, however, controversial. Unlike the Lower Aral and the Sarydjaz-Aksu outburst floods that had emptied dammed lakes almost completely, the Ak-Terek dammed lake had been silted practically up to the dam's crest level (see Fig. 8.253; its longitudinal profile is similar to that shown on Fig. 8.24) that excludes simultaneous release of large amount of water required to produce so high surge wave. At the same time, area of the valley cross-section below the eroded terrace up to the terrace level is about 8000 m^2, and, considering high velocity of the outburst flood wave, corresponds to the peak discharge of several tens of thousands cubic meters per second for which there was no water that could be released. Possible explanation could be that flood occurred when the Sokh River valley was blocked by the second generation of the Korgon rockslide (marked by 2 on Fig. 8.252). This dam could fill the river canyon almost up to the terrace level, so that when outburst flood occurred the valley was inundated already. Thus, the discharge of the surge wave that curved the amygdaliform ravines could be much lower than if it would pass through the original "empty" valley. Besides, this wave could trigger the breach of the blockage formed by the second Korgon rockslide.

Interesting case study of two recent powerful outburst floods, one of which, likely, occurred slightly later than another one, was found in the upper reaches of the Dakhanzar River at its confluence with the Shulvadar River (Kokcha River basin) in Afghan Badakhshan (see Fig. 9.124). Outburst floods were quite powerful because their deposits dammed the small river that falls into the Dakhanzar River valley downstream. We anticipated that both outburst floods could be caused by closely timed breach of rockslide dams in the upper reaches of two rivers shown on Figs. 9.125–9.128. The absence of landforms identifiable on space images upstream from these breached dams that could be associated with lacustrine sedimentation indicates short-term river damming. Of course, more grounded reconstruction of the entire succession of events in this part of the Afghan Badakhshan requires detail field study.

Most of the prehistoric breached dams had stored millions, rarely several tens or a few hundred millions of cubic meters of water. Study of the traces of such prehistoric outburst floods could provide data that can be used to estimate potential effects of similar existing or future rockslide dammed lakes breach.

However, several past dammed lakes were much larger and could be compared with the largest Central Asian rockslide-dammed water body—the 500-m deep Sarez Lake. Though at present this lake is considered as more or less safe feature (Alford and Schuster, 2000; A. Ischuk, 2011; N. Ischuk, 2013b, 2014), its long-term (decades—centuries) fate is questionable and could be hardly predicted (Strom, 2014b). Estimates of the flood wave parameters in case of the Usoi dam overtopping due to collapse of the anticipated (though debatable) Right-bank landslide in the lake or in case of the dam breach caused by any other process could be found in Risley et al. (2006a,b). There is no doubt that study of the prehistoric outburst floods produced by emptying of the kilometers cube-size dammed lakes can provide empirical data that could be used as an analogue for the assessment of the potential hazards of similar events that might occur in future. One of such case studies is the above-mentioned outburst flood caused by the Sarydjaz-Aksu rockslide dam breach and release of ca. 2 km^3 of water (see the description in Section 8.2; Figs. 8.53–8.56).

An abnormally large recent alluvial fan was identified in the Muzart River valley where it leaves the Tien Shan mountains at 41.78° N, 81.03° E (Fig. 14.4). The 27-km long and up to 7.5-km wide area is free of vegetation contrasting with the surrounding arable land and has slightly convex profile typical of the debris flow fans. Its surface is curved by the dendritic amygdaliform ravines similar to those left by the Sarydjaz-Aksu flood (see Figs. 8.55 and 8.56). Upstream from the head of this alluvial fan, the Muzart River flows in the ca. 50 m deep valley eroded in the fluvial–glacial deposits (fl-glQ on Fig. 14.4) framing the distinct end moraine (gl Q) (we have to notice that it is a very rare case in the Central Asian region where the valley glacier came out from the mountain range and formed the lobe at its foot like in the Alps). Though we have no data either on the age of this alluvial fan or on the direct cause of the outburst flood

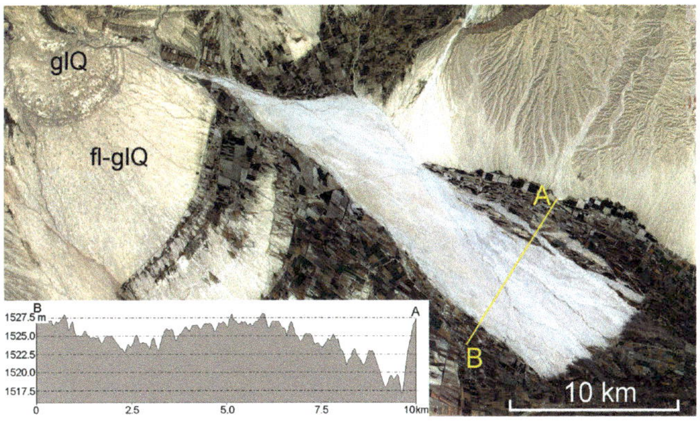

FIGURE 14.4 Abnormally large fresh alluvial fan of the Muzart River. glQ—end moraine of the Muzart valley glacier; fl-glQ—fluvial–glacial deposits. Google Earth image. Profile A–B across the alluvial fan is shown in the inset.

that could form it, we can hypothesize that it could be associated with the breach of the ca. 150-m high fan-shaped rockslide dam in the Muzart River valley shown on Fig. 4.30. The impounded lake could store up to 1.1 km^3 of water.

The largest outburst flood in the entire region had to occur when the lake that had inundated the Pianj River valley for more than 100 km being dammed by the Shidz rockslide (see Fig. 9.98) had been emptied catastrophically. Volume of water accumulated in this relatively short-lived natural reservoir was estimated as 31 km^3 (see Fig. 9.100). Possible reconstruction of the formation and breach of this lake was presented in Section 9.3. It is obvious that outburst flood during rapid release of such enormous amount of water should result in the outstanding catastrophe exceeding that in the Sarydjaz-Aksu River valley described in Section 8.2 (see Figs. 8.53–8.56) that resulted from the outburst of the dammed lake with volume of about 2 km^3 only.

Possible evidence of this prehistoric natural catastrophe could be found about 350 km downstream from the dam site, near the Moskovskyi town where Pianj River leaves the Darvaz Range—the frontal mountain chain of the Pamir. We hypothesize that the dendritic amygdaliform microrelief of the ca. 15-km wide Pianj River floodplain (Fig. 14.5) recalling those in the Sarydjaz-Aksu, the Muzart, and the Sokh River valleys described above could be formed by this outburst flood. Comparison of the topographic profiles across the floodplain at the head of this giant alluvial fan (A–B on Fig. 14.6) and ~25 km downstream (ibid, C–D) shows that the second one is flat, with multiple 5- to 10-m deep channels, which is typical of the floodplains of large Central Asian rivers where they leave mountains. However, the first profile (A–B) has the distinct convex shape. Its central part just opposite Pianj gorge outlet is 18–29 m higher than its side

FIGURE 14.5 Google Earth space image of the Pianj River floodplain near Moskovskyi town (M) and its dendritic amygdaliform microrelief shown in the zoomed inset. Lines A-B and C-D mark position of the profiles shown on Fig. 14.6.

parts. Such shape is typical of the debris flow fans rather than of the floodplain of a large river with the mean discharge of 845 m^3/s (State Water Inventory, 1990).

From our point of view, such landform could reflect, at least partially, aggradation caused by deposition of the material transported by the outburst flood from the breached Shidz Lake. This material was eroded both from the Shidz rockslide dam, from the Voznavd rockslide dam 11 km downstream (see Fig. 9.102) and, likely, from the river banks when this extraordinary flood (debris flow?) had passed along the Pianj River valley. Rough estimate of the alluvial fan volume gives value of ca. 3.2 km^3, which is several times larger than the amount of debris that could be eroded from both blockages (up to 500–600 Mm^3 approximately). Besides, large part of the material eroded from the Shidz rockslide dam was deposited just downstream from it (see Fig. 9.99). On the other hand, we must consider the possibility of the significant entrainment of the material that, in case of debris flows caused by dams' breach, could exceed 10-fold the initial water discharge (Stepanov and Stepanova, 1991; Stepanov and Yafyazova, 2014). Detailed analysis of the microrelief and topography of this fan reveals some evidence of its two-stage formation (see dashed envelopes on Profile A–B on Fig. 14.6). Southern

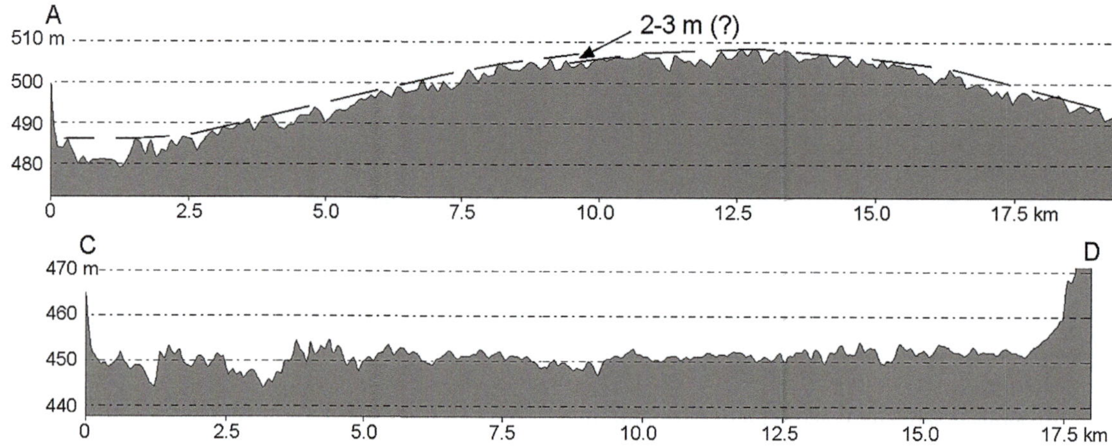

FIGURE 14.6 Topographic profiles of the Pianj River floodplain. See their position on Fig. 14.5.

alluvial fan, which generalized surface is 2–3 m lower, is incised in the older northern fan. We can speculate that the older fan could be created by some older outburst flood whose source is not recognizable now. Of course additional studies are necessary to prove or reject all these assumptions and to reconstruct real history of the Pianj River evolution.

As we described earlier in Section 9.3, thalweg profiles of the Pianj River downstream of the Bartang River mouth (see Fig. 9.99) and of the Gunt River downstream of the Chartym rock avalanche dam (see Fig. 4.27) allows assumption of the simultaneous formation of the Shidz and Voznavd rockslide dams at the first site and of the Vir-Tangiv and Zuvor rockslide dams at the second site. In both cases, water impounded upstream from the upper dam could not be released free before the lower one had been breached too. During the first stage of outburst, water flow had been confined and the material eroded from the upper dam had been redeposited first in the valley section between two blockages. When the lower dam failed, this redeposited material that formed some counterforce prevailing complete breach of the upper dam could be eroded partially. Now its remaining part fills valley sections between two breached dams that has steeper river slope.

Similar phenomenon, though of much smaller scale, was observed in the Unkursai River valley dammed by the Ak-Kiol and the Lower Ak-Kiol rockslides (see Fig. 8.166) and described above as an example of the domino effect triggered by failure, this time partial, of the upper dam. One more case study with same succession of the nearby damming bodies is the Lower Aral rockslide dam that had split into two branches, upper of which had blocked the valley while the lower one likely caused just partial blockage (see Fig. 8.161). We assume that preservation of the terrace-like landform composed of debris eroded from the breached upper dam (see Fig. 8.164) was facilitated by partial confinement provided by the lower blockage.

We must notice, however, that deposition of debris eroded from the rockslide body could occur downstream of a single blockage as well, especially in those cases when the erosion that have started on the downstream side of the dam did not reach its crest. It can be observed immediately downstream from the Chong-Tash (see Fig. 8.135) and the Bash-Kugandy (see Fig. 8.136) rockslide dams in Central Tien Shan. Similar accumulation took place below the partially silted and breached Kuturgansu-mouth rockslide dam in the Southern Tien Shan (see Fig. 8.246). One more example could be found just downstream from the completely breached Djashilkul rockslide dam in the Northern Tien Shan (see Figs. 8.81 and 14.7). Based on the presence of such accumulation, we can speculate that this dam could be breached in several episodes, first of which had redeposited the material eroded from the blockage, while more powerful subsequent outburst had eroded both the remaining part of the dams' body and the material accumulated at its foot.

Evidence of such multistage breach has been found at several sites where series of the erosional terraces have been eroded in thick lacustrine sediments (see Figs. 6.6, 8.57, and 8.263). Each terrace level was formed by a rather long stand of water at this altitude when bank erosion prevailed. Sometimes we can see traces of lake level much above the present-day level as at the Iskanderkul Lake (see Fig. 8.290).

Some of the existing rockslide-dammed lakes with deep canyons incised in the downstream slopes should be considered as potentially hazardous despite their long-term existence in such state (see Figs. 8.135, 8.136, or 8.287). One day gradual back erosion would reach the lake, and drastically increasing discharge could cause progressive growth of this outlet and catastrophic breach.

FIGURE 14.7 View on the breached Djashilkul rockslide dam from the downstream. OD?—terrace-like landform presumably composed of the outwah deposits eroded from the dam during its breach.

14.4 ROLE OF ROCKSLIDE DAMMING IN RIVER VALLEYS' EVOLUTION

River valleys' evolution in mountainous regions, i.e., bottom and lateral erosion, formation of river terraces, etc., depends on the combination of the tectonic and climatic factors mainly. Neotectonic uplifts and subsidences along with water flow characteristics (discharge and its annual variations, annual runoff, and sediment yield) determine the prevalence of erosion or accumulation along the river valley. Tectonic and climatic factors are somehow interrelated, e.g., intensive uplift results in the formation of high mountains subjected to glaciation that change the hydraulic regime of the rivers; increased erosion caused by climate change may intensify isostatic uplift, etc. These interacting processes form a typical shape of river valleys with narrow gorges and steeper thalwegs where they cross actively growing mountain ranges, and widened valleys with well-developed fluvial deposits where they pass through neotectonic depressions.

However, large-scale rockslides, damming river valleys, can disrupt this topographic regularity significantly. It was demonstrated by Kenneth Hewitt (2006a) who proposed the "Disturbance regime" concept to describe evolution of river valleys of the Indus basin in Pakistan affected by numerous large-scale rockslides. Somehow similar effects can be provided by glacial (ice and moraine) river damming (Korup and Tweed, 2007), but in Central Asia with arid climate such phenomena are not as widespread as in mountainous systems located in more humid zones.

Rivers' sections characterized by the disturbed regime have the specific long profiles with distinct nick points (see Figs. 4.27, 6.19, 8.24, 8.127, 8.289, 9.89, and 9.99). One of the streams affected by multiple damming episodes is the Shing River in Central Tien Shan that was dammed by eight blockages at a distance of about 15 km, seven of which still form lakes while the lowermost one was sited completely (see Fig. 8.281).

A good example of the river with disturbed regime, where the influence of rockslide damming on river's thalweg profile could overreach effect of neotectonic movements, is the Kokomeren River in Central Tien Shan (Fig. 14.8). In its upper reaches, it passes through the western part of the Suusamyr intermontane depression where it is named the Suusamyr River, and after its confluence with the Western Karakol River continues as Kokomeren. On its way towards the confluence with the Naryn River, it passes along the boundary between complex neotectonic uplifts of the Suusamyr Range at the west and the Djumgal Range at the east with minor Kyzyl-Oi intermontane depression inside and enters western termination of the much larger Djumgal intermontane depression where Kokomeren merges with its large left tributary—the Djumgal River. After this confluence, it follows the boundary between the southern branch of the Suusamyr neotectonic anticline and the Kavaktau neotectonic anticline and, after merging with the Minkush River, turns to the west and passes along the Minkush–Kokomeren neotectonic depression. Neotectonic uplifts (anticlines) are composed of Paleozoic granites (mainly) and metasediments, while neotectonic depressions are composed of Neogene terrigenous deposits and Quaternary fluvial sediments (Fig. 14.9).

The thalweg profile of the Kokomeren River (Fig. 14.10) has several sections that differ significantly from each other. The upper one, within the Suusamyr intermontane depression, is gentle, with a stream gradient of about 0.4%. River valley here is wide with thick Quaternary fill. Valley shape and profile change

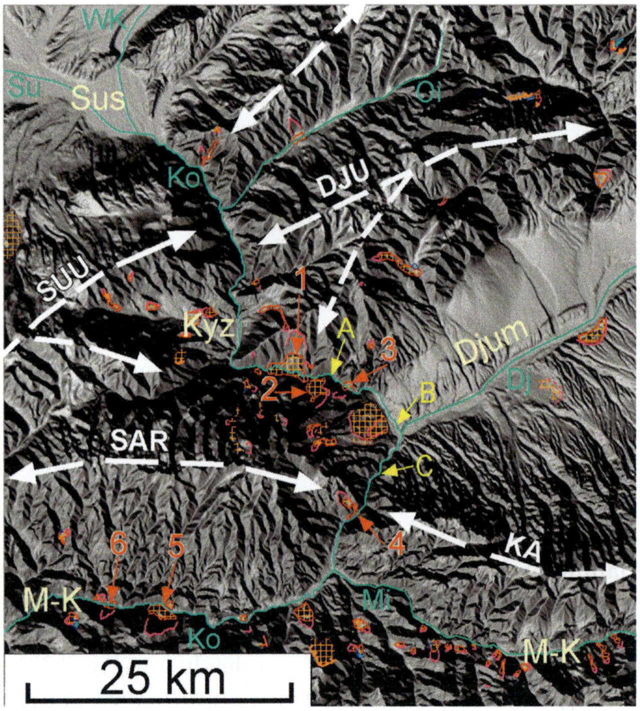

FIGURE 14.8 Rockslides and landslides in the Kokomeren River valley (Ko). Rivers: Su, Suusamyr; WK, Western Karakol; Oi, Oikaing; Dj, Djumgal; Mi, Minkush. Profile shown on Fig. 14.10 was compiled along the Suusamyr and Kokomeren Rivers. Thick white double-headed arrows, axes of neotectonic uplifts: SUU, Suusamyr Range; SAR, Sarykamysh Range; KA, Kavaktau Range; DJU, Djumgal Range. Intermontane depressions: Sus, Suusamyr; Kyz, Kyzyl-Oi; Djum, Djumgal; M-K, Minkush-Kokomeren. A, B, and C, location of photographs shown on Fig. 14.11; red arrows mark river-damming rockslides: 1, the Kokomeren, 2, the Displaced Peneplain, 3, the Mini-Köfels, 4, the Lower Aral, 5, the Karachauli, 6, the Lower Kokomeren.

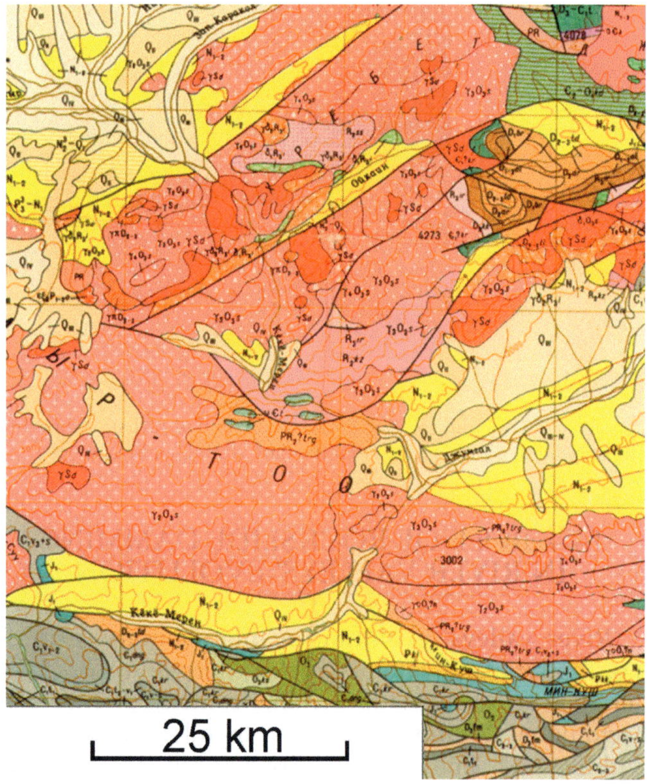

FIGURE 14.9 1:200 000 State geological map of the USSR of the area shown on Fig. 14.8.

abruptly when river enters mountain range. Gradient sharply increases up to ~2% and flattens gradually up to 0.7% towards the Kyzyl-Oi intermontane depression. Stream is full of rapids and is incised in the bedrock, except short section in the Kyzyl-Oi depression where floodplain becomes wider and is filled with alluvium. Downstream from this depression gradient increases again up to 1.5%. All these changes reflect the style of neotectonic deformations quite well—steep sections correspond to neotectonic uplifts and flattened sections correspond to intermontane neotectonic depressions. No river-damming rockslides have been identified at these sections of the Kokomeren River.

At the section between the Kyzyl-Oi and the Djumgal intermontane depressions, the Kokomeren River had been dammed at least three times by the Kokomeren (see Fig. 8.149), the Displaced Peneplain (see Figs. 4.5 and 4.6), and the Mini-Köfels (see Fig. 5.41) rockslides. The Displaced Peneplain one even forced the Kokomeren River to erode a bypass channel (see Fig. 8.158). However, all three blockages are rather old (likely Pleistocene) and do not form nick points in the modern thalweg profile that is rather steep (~1.5%) upstream from the Kokomeren rockslide and flattens gradually up to ~0.5% at the western termination of the Djumgal intermontane depression. Stream at the upper part of this section is fast, with rapids (Fig. 14.11A) and became more tranquil near the confluence with the Djumgal River (Fig. 14.11B).

However, downstream from this confluence, where the Kokomeren River enters the V-shape gorge between the Sarykamysh and Kavaktau neotectonic anticlines, its behavior becomes totally abnormal. Instead of steepening, which is typical of river sections where they cross neotectonic uplifts, stream remains tranquil (with the gradient of ~0.4%), with small alluvial islands (Fig. 14.11C). Such a profile extends for about 30 km, up to the Karachauli rockslide dam (see Fig. 8.172). Minor irregularity exist at stream section where the river had been blocked by the Lower Aral rockslide (see Fig. 8.161), but resolution of the DEMs available is not enough to reflect it adequately. Downstream from this site and near the confluence with the Minkush River, floodplain of the Kokomeren River becomes much wider, and here and further downstream river flows through the alluvial deposits. This section coincides with the Minkush–Kokomeren

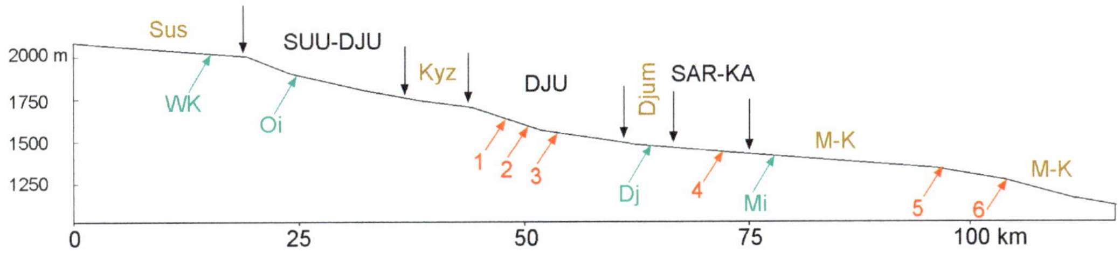

FIGURE 14.10 Thalweg profile of the Kokomeren River based on the 3″ SRTM DEM. Indications are the same as on Fig. 14.8.

FIGURE 14.11 Different styles of the Kokomeren River stream. Locations are marked on Fig. 14.8.

neotectonic depression composed of the intensively folded Neogene conglomerates overlying Jurassic coal-bearing sediments. We anticipate that such an abnormal behavior of the stream could be caused by river damming at the end of this section by the Karachauli rockslide that had displaced the entire valley with its alluvial fill (see Figs. 8.173–8.176 and 12.2). Giant dammed lake had extended for more than 30 km upstream (see Fig. 13.7). If this assumption is correct, base of the alluvium close to the Karachauli breached dam could be several tens of meters below the present-day riverbed.

One more nick point where stream gradient changes from ca. 0.8% to 1.4% coincides with the breached Lower Kokomeren rockslide dam (see Fig. 14.3). Stream here becomes faster with numerous rapids. We should notice that complete reconstruction of the Kokomeren River valley Late Quaternary evolution requires special research project and, in particular, dating of numerous river-damming bodies and of various fluvial landforms.

Besides direct effects of river damming resulting in formation of the distinct nick points in the thalweg profile, powerful outburst floods caused by dams' breach also produce specific landforms, same as described in Procter et al. (2009), whose impressive examples in the study region were summarized in Section 14.3.

River damming or stream diversion could be also caused either by surface faulting associated with large earthquakes or due to rapid, though gradual, neotectonic uplift or subsidence. Indeed, temporary diversion of the Suusamyr River stream occurred during the 1992 Suusamyr earthquake (see the upper part of Fig. 3.5; the stream at the foreground was diverted from the main water course of the river by surface rupture that crossed it). The recurrent temporary damming was observed north of the Chaek town where the small valley is crossed by an active fault (see Figs. 5.5 and 13.3). However, because the largest terrestrial single-event vertical offsets associated with earthquakes do not exceed ca. 10 m (Wells and Coppersmith, 1994; Strom and Nikonov, 1997), such phenomena provide quite limited hazard though it might affect the evolution of rather small streams significantly.

Though the tectonic damming is out of the scope of this book, we have to mention it because sometimes lacustrine accumulation in large river valleys associated with landslide damming have been interpreted erroneously as the effect of neotectonic movements. Such misinterpretation was made, in particular, for thick lacustrine sediments in the Naryn River valley near the Alabuga River mouth. This abnormal sedimentation was treated as the result of the continuous, long-term, but at the same time, relatively rapid uplift of the Western Akshiyriak Range (Makarov, 1977), while in fact it was the siltation of the lake dammed by a gigantic Beshkiol landslide about 10 km^3 in volume located about 15 km downstream from the site where active fault crosses the Naryn River (Korup et al., 2006; Strom and Korup, 2006) (see Figs. 8.126 and 8.127).

CHAPTER

15

Evidence of Slope Instabilities—Potential of Future Large-Scale Slope Failures

Most of the case studies described above are bedrock landslides that had occurred in the past, in the prehistoric times mainly, and, thus, do not pose now a threat to local communities and infrastructure unless they still block river valleys and have the potential to be breached producing devastating outburst floods or could be the source of loose material for debris flows. The latter cases were discussed in Chapter 14, Rockslide Dams—Origin, Evolution, and Failure. However, from the practical point of view, the most challenging topic is the identification of sites where large-scale rock slope failures can be anticipated in future.

Complexity and ambiguity of the ongoing or future rockslide danger assessment (we consciously do not use here term "hazard" that implies some quantitative characteristics of the probability of landslide occurrence within some given period of time (Corominas et al., 2015)) can be exemplified by the Belaldy rockslide triggered by the 1992 Suusamyr earthquake and described in Section 13.1 in Chapter 13, Seismically Induced Rockslides: Their Discrimination (see Figs. 2.4 and 13.1) and by the aseismic 2008 (?) Djamansu rockslide described in Section 8.4 in Chapter 8, Tien Shan (see Figs. 8.110–8.112). As it was pointed out in Sections 13.1 and 8.4, both rock slope failures could be predicted neither in space nor in time.

Methods of landslide susceptibility assessment based on the statistical correlation between spatial distribution of known landslides and of factors that might have influence on slope instability and are widely used for identification of the soil slide–prone areas (Havenith et al., 2006a, 2015b; Günther et al., 2013; Kavzoglu et al., 2014; Aristizábal et al., 2016) can be hardly applied to select sites where large-scale bedrock landslides can occur—the spatial distribution of the known case studies is too irregular (see Figs. 7.1, 7.5, 8.2, 8.10, 8.32, 8.58, 8.106, 8.107, 8.138, 8.199, 8.236, 8.259, 9.1, 9.64, and 9.115).

Unlike landslides in soils, whose formation is predetermined at a larger extent by the overall mechanical properties of the soil masses, which, in turn, depend on their generic type and composition, rock types themselves seem to have minor influence on rockslides distribution because formation of such slope failures depends to a greater extent on the structure of the host rock massif (bedding, foliation, fracturing, etc.) rather than on lithology. Such structural peculiarities can be characterized adequately based on detailed large-scale mapping and site investigations only. Rockslides identified in the study region affected slopes composed practically of all types of rock—carbonate (see, e.g., Figs. 5.64, 6.4, 8.45, 6.22, and 8.115), terrigenous (see Figs. 5.12, 8.27, 8.53, and 8.98), metamorphic (see Fig. 4.19, 4.28, 5.15, 9.107, 9.121, and 9.122), and igneous, in granite mainly (see Figs. 4.43, 5.35, 5.37, 5.45, 5.47, 5.50, 5.54, 5.63, 8.8, and 8.70; we must take into account that granite massifs dominate among all other types of the plutons), and of variable ages—from Precambrian (see Fig. 9.135) to Neogene (see Fig. 5.6).

Even such parameters as slope height and steepness can be hardly used for the overall susceptibility assessment. For example, density of large-scale rockslides in the Eastern and Northern Pamir ranges that rise 3–4 km above the adjacent intermontane depressions (see Fig. 9.1) is significantly lower than in many other parts of the study region with less-pronounced relief.

Distance of rockslides from active faults is quite variable too—some active fault zones are almost totally affected by rockslides, which can be exemplified by the frontal thrust of the Pamir at the Peter the First Range (see Fig. 5.20) or by faults bounding the Minkush–Kokomeren intermontane depression in the Central Tien Shan (see Fig. 8.138), while other, such as the univocally active Talas-Ferghana fault (see Fig. 8.107), the fault zone passing along the Kemin and

Rockslides and Rock Avalanches of Central Asia.
DOI: https://doi.org/10.1016/B978-0-12-803204-6.00015-6

Chilik Rivers in the Northern Tien Shan (see Fig. 8.58), or faults bounding intermontane depressions of Eastern Pamir (see Fig. 9.1)—just locally, with long gaps between rare rockslides or rockslide clusters. This irregularity was mentioned in Section 13.2.1 of Chapter 13, when we described the relationships between seismically induced rockslides and active faults.

From our point of view, most reasonable identification of sites where large-scale catastrophic rock slope failures could be anticipated is based on site-specific evidence of the ongoing slope instabilities. Some examples of such expectations can be found in Hutchinson (2006, Section 3.3). Evidence of the potential instability from the study region has been described case by case in Chapters 7–9, Dzungaria, Tien Shan, and Pamir, respectively, and here we just summarize these data.

Gradually evolving landslides of planar or rotational types mainly (Hungr et al., 2014) on high bedrock slopes composed predominantly of well-bedded sedimentary or metamorphic rocks were identified at several localities in the Eastern Tien Shan (see Figs. 8.30 and 8.31), in the Southern Tien Shan (see Figs. 8.239, 8.240, and 8.256), in Tajik (see Figs. 9.68, 9.85, and 9.86), and Afghan parts of Pamir (see Figs. 9.114 and 9.129). All these slopes are undercut by rivers providing conditions favorable for slide activation. However, considering that in the past these rockslides moved progressively, it is most likely that activation would affect these slopes partially, rather than they would collapse catastrophically "in one gulp" resulting in most adverse effects such as large-scale river damming.

The increased capability of large rockslide formation characterizes slopes with well-pronounced systems of fractures above or aside the headscarps of relatively recent catastrophic failures. It can be exemplified by the slope adjacent to the Khait rockslide headscarp (see Fig. 5.15) that was disrupted by the 1949 earthquake but did not collapse at that time. Similar disrupted rock massifs were identified above or just aside headscarps of rockslide in the Kaindy River valley in Kazakhstan (see Fig. 8.64), of the Bielogorka Right (see Fig. 8.98) and the Djamno (see Fig. 8.186) rockslides in Kyrgyzstan, of the Vir-Tangiv rockslide in Tajik part of Western Pamir (see Fig. 9.75), and of the rockslide at the Afghan side of the Pianj River valley shown in the lower inset in Fig. 9.113. Similar disrupted rock mass, the right part of which had collapsed already can be seen on the right bank of the Sarez Lake (see Fig. 9.48). We could not find information if this slope failure originated simultaneously with the Usoi rockslide or had existed already in 1911. According to the data available, no significant failures have occurred on the lake banks after the 1911 earthquake.

In the most unfavorable cases volume of such anticipated rockslides can exceed volume of the slope failures that had occurred at these sites earlier. It is especially threatening if such potential rockslide could block river valley producing landslide dam and lake, whose impoundment and subsequent breach could be much more disastrous than slope failure itself. Such chain effects can be anticipated at the Ragnow-mouth rockslide in the Obi-Khingou River basin (see Figs. 9.59 and 9.60) and at the Satsu rockslide in the Muksu River valley (see Fig. 9.54) where belts of well-developed arcuate fractures hundreds of meters wide can be seen on the slopes above the recent rockslide headscarps.

Evolution of the Satsu rockslide from 1962 till mid-1980s due to slope undercutting by the Muksu River was reported in Pyzhyanov and Presnuhin (1987) (see discussion in Section 9.2 in Chapter 9). As for the Ragnow-mouth rockslide, we could trace its development for more than 50-year-long period by comparison of the modern space image (see Fig. 9.59) with a very detailed aerial photograph of 1956 and with KFA-1000 space image with 5–8 m spatial resolution obtained in the late 1970s (Fig. 15.1). It is visible that while system of well-developed fractures accompanied by small landslides has existed before 1956, rather large frontal slope failure that resulted in temporal river damming visible on Fig. 9.59 had occurred between 1956 and late 1970s. Further monitoring of this site could be performed by the use of the INSAR technology (Burgmann et al., 2000; Colesanti and Wasowski, 2006; Cascini et al., 2009; Cigna et al., 2010; Lauknes et al., 2010; Bozzano et al., 2011).

Both large evolving rockslides—the Ragnow-mouth and the Satsu, are located in the Northern part of the Pamir in the catchment area of the Vakhsh River where the world highest earth fill dams—the existing 300-m high Nurek dam and the 335-m high Rogun dam, which is under construction now, are located. Considering sensitivity of such type of water retaining structures to the overflow, both sites require monitoring or, at least, regular inspection aimed not to miss large-scale slope failure occurrence and to undertake timely measures required to guarantee dams' safety. Taking in mind that accumulation of significant amount of water (hundreds of millions cubic meters if not more) would last months before the blockage could be overtopped and breached, the threatened reservoirs could be partially emptied that will allow accumulation of the outburst flood without dams' overtopping. We must notice that such adverse scenarios of these rockslides evolution are hypothetical. Nevertheless, considering high risks for hydraulic

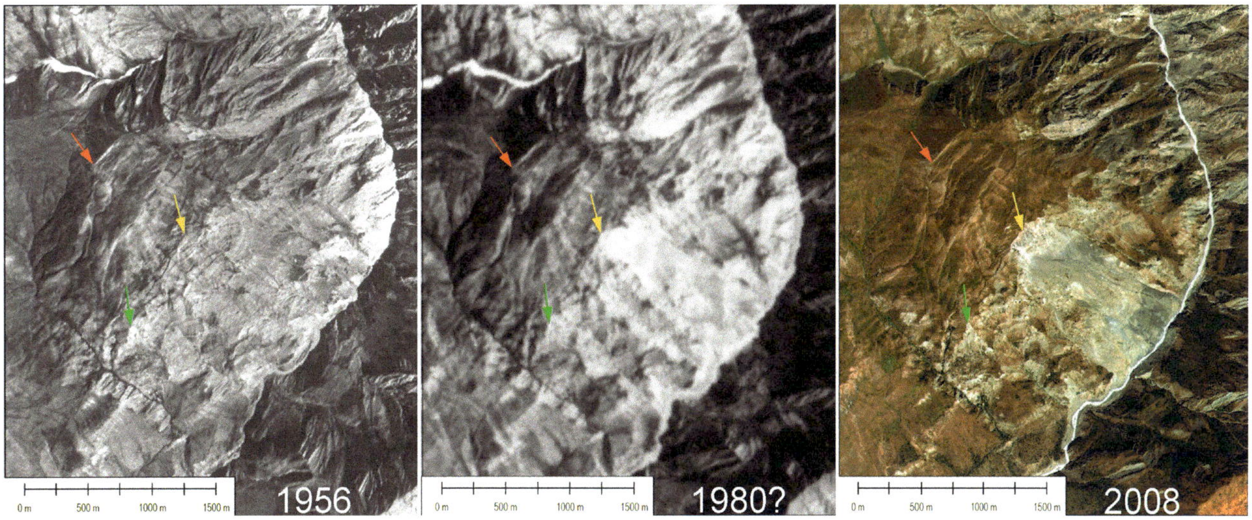

FIGURE 15.1 The Ragnow-mouth rockslide. (A) 1956 aerial photograph; (B) fragment of the KFA-1000 space image of late 1970s; (C) 2008 Google Earth image. Red arrows mark the most distant fracture visible on all images; green arrows mark the crown of the small landslide also visible on all images; orange arrows mark the position of the rockslide crown, which was absent in 1956.

schemes downstream, they should not be excluded and these sites must be investigated and monitored to provide more substantiated assessment of the potential hazards.

The specific subtype of voluminous rock mass "ready to collapse" can be exemplified by the Djamno rock avalanche in the Central Tien Shan (see Fig. 8.186) and the Bakharak rock avalanche in Afghan part of the Pamir (see Fig. 9.116). Here large bedrock blocks millions of cubic meters in volume slid downslope for several hundreds of meters and, most likely, stopped abruptly, dislodging surficial weathered material and forming highly mobile rock/debris avalanches in front of them. However, the subsided blocks themselves remain more or less intact and still rest high above the valley floors. Considering significant, about 300 m, descend of the displaced blocks, they should be underlined by well-developed sliding surfaces, and the next triggering phenomenon, such as large earthquake, can provoke their collapse. Similar "hanging block" remains in the headscarp of the Bielogorka Left rockslide in the Northern Tien Shan (see Fig. 8.98B), but its possible collapse will be captured, at least partially, by the ca. 20-m deep proximal depression at the foot of the present-day headscarp.

One more site where very large rock massif (likely about 2 km^3 in volume) subsided significantly—up to 140 m, being accompanied by the system of fractures stretching along the ridge crest—was found at the central part of the Western Akshiyriak Range in the Central Tien Shan (see Figs. 8.131 and 8.132), between the outstanding Beshkiol landslide—the largest slope failure of the entire Central Asian region, and the Chaartash rock avalanches with the extremely long runout. Unlike case studies described above there are no distinct indications of slope deformation below this topographic step. Nevertheless, we consider this site as one of those where very large slope failure might occur with potentially quite adverse consequences. We should notice that similar step can be seen above the Beshkiol landslide headscarp (see Figs. 4.1 and 5.13).

Another type of the ongoing slope instability is the so-called toppling that could transform, under specific conditions, into brittle catastrophic rockslide. According to Nichol et al. (2002), two distinct types of these phenomena can be distinguished: (1) the ductile, self-stabilizing flexural toppling in a weak rock with a single dominant joint set and (2) the brittle, catastrophic block toppling in a strong rock containing persistent, downslope oriented or horizontal cross joints". The characteristic geomorphic expression of such type of slope deformation is the presence of the numerous regular upslope-facing scarps. Impressive examples could be seen in the Kokomeren River basin in the Central Tien Shan (see Fig. 5.3), in the Zeravshan River valley in the Southern Tien Shan (see Fig. 8.261), and in the Sanglech River valley in Afghan Badakhshan (see Fig. 9.133).

We should point out that intensive seismic shaking can cause rapid slope failure even at sites featuring the first of the above two mechanisms. It occurred at the small Chet-Korumdu Ridge composed of weak Neogene deposits where part of the slope affected by distinct toppling (indicated by series of upslope-facing scarps) failed in 1992 during the M7.3 Suusamyr earthquake (Fig. 15.2).

The most common type of landforms indicating the ongoing slope disturbances are the so-called

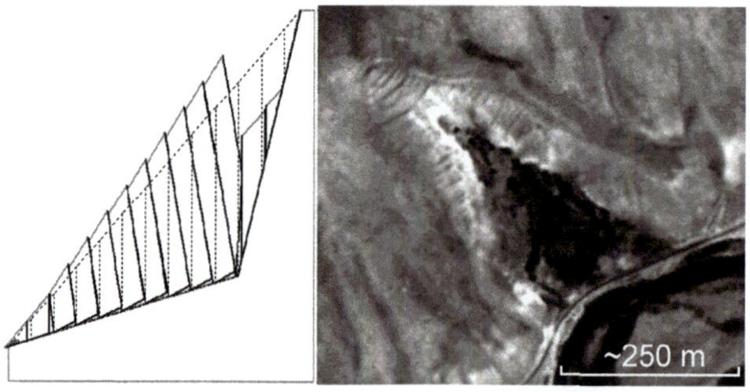

FIGURE 15.2 General scheme of flexural toppling and example of such phenomenon at the southern slope of the Chet-Korumdu Ridge. The KFA-1000 space image was obtained several years prior to the 1992 Suusamyr earthquake.

deep-seated gravitational slope deformations (DSGSD) (Agliardi et al., 2001, 2013; Crosta et al., 2013; Jomard et al., 2014; Pánek and Klimeš, 2016) described briefly in Section 5.1 in Chapter 5, Bedrock Landslide Types and Classification Systems.

At several sites, evidence of the potential instability is rather simple being represented by a single elongated upslope facing scarp or by a system of such scarps (sackungen), sometimes even with small ponds in the lowerings dammed by the these scarps. Such case studies were found on the crest of the Shet-Merke River valley in the Northern Tien Shan (see Fig. 8.59), in the Naryn River valley near the Kyzylkiol cavity in the Central Tien Shan (see Fig. 8.184), on the ridge crest above the Aini town in the Southern Tien Shan (see Fig. 8.266), in Eastern (Chinese) (see Fig. 9.3), and Western (Tajik) parts of Pamir (see Figs. 9.51, 9.62, and 9.81), and in the Afghan Badakhshan (see Fig. 9.139).

The specific example of such DSGSD was found in the Central Tien Shan south of the Koshbulak Creek where thick stratum of the Neogene red beds slid along the inclined planation surface of the Paleozoic basement and forms the sackungen with several ponds in it. This DSGSD is accompanied by numerous shallow landslides that affect the uppermost layers of the Neogene unit (see Fig. 8.137). Further sliding on this slope, both local, in the form of shallow landslides, and overall, as a large blockslide, can be anticipated.

Many of the DSGSDs described above demonstrate features typical of both gradually sagging blocks—bow-shaped downslope-facing scarps and of large-scale toppling—sets of upslope-facing scarps. Some of them are well pronounced on ridge's tops but gradually disappear downslope, such as landforms on the left bank of the Betdjin Creek in the Highest Tien Shan (see Fig. 5.17), in the Sokuluk River basin in the Northern Tien Shan (see Fig. 8.104), in the Naryn River basin in the Central Tien Shan (see Fig. 8.185), or in the Zeravshan River basin in the Southern Tien Shan (see Fig. 5.18). Many of the large DSGSDs not only affect one side of the ridge but also extend hundreds of meters on the opposite side of the ridge crest. Sometimes deformations affect the entire slope as in the Ridjistdara River valley in the South-Western part of the Pamir (see Fig. 9.82) or downstream from the Rastdara River mouth in Afghan Badakhshan (see Fig. 9.133).

Unlike the Alps where slopes' debuttressing due to deglaciation is considered as the main cause of the DSGSDs (see, e.g., Agliardi et al., 2013; Pánek and Klimeš, 2016), in the Central Asian region, it is not so obvious. Sometimes slopes are destabilized by channel shift caused by its partial damming by landslides or debris flows as it occurred in the Aksu River valley in the Northern Tien Shan resulting in the formation of the Aksu rockslide about 1 km^3 in volume (see Figs. 8.102 and 8.103 and explanation in Section 8.3). In this case, failure affected the entire slope and it is unknown whether it was preceded by DSGSD. Another scenario evolves, likely, in the Kokomeren River valley upstream of the Djumgal intermontane depression at 41.924° N, 74.267° E. Here the old valley had been blocked by the Displaced Peneplain rockslide (see Figs. 4.5 and 4.6), so that the Kokomeren River had to erode a new bypass valley (see Fig. 8.158). Such a rapid incision decreases the stability of the left-bank rock massif weakened by neotectonic faulting and forms the prerequisites for future failure. Its manifestation can be represented by the suspicious landforms on top of the left-bank slope of the modern valley (see inset in Fig. 8.158).

Very large DSGSD associated with large fault zone was identified in the Central Tien Shan above the Ornok rockslide that can be treated as the catastrophic collapse at its frontal part. Here deep-seated deformations of the hanging wall of large neotectonic fault can be localized quite well between two river valleys dissecting the southern side of the Minkush–Kokomeren intermontane depression (see Fig. 8.167).

One more quite special case study of the potential slope instability is the gigantic Ters rockslide 5 × 6 km in size in the Northwestern Tien Shan, in Kyrgyzstan (see Figs. 4.2 and 5.16). Despite gentle inclination of the daylight surface, it has a very distinct horseshoe-shape scarp up to 50- to 70-m high and several kilometers long and numerous secondary deformational features. If activation of this landslide will take place, it could result in Ters River blocking. This site requires special study and monitoring.

Sometimes large landslides originate in older rockslide deposits, whose thickness reaches hundreds of meters. It can be exemplified by the Karakudjur rockslides in the Central Tien Shan with numerous

FIGURE 15.3 Recent landslides in the deposits of the western Karakudjur rockslide. K-K - the Karakudjur River. SAS Planet image visualized in 3D by Global Mapper software.

fractures indicating ongoing instability (see Fig. 8.121) on the steep river-facing slope worked out by the stream erosion in the deposits of the Eastern rockslide (KK-E on Fig. 8.120) and with a rather recent large landslide up to ca. 20 Mm^3 in volume (Fig. 15.3) in the deposits of the western twin (KK-W on Fig. 8.120). Besides, large landslides can originate in this river valley on its ca. 150-m high right bank at the hanging wall of the active fault passing along the Karakudjur River (see Fig. 8.123).

Similar subsequent failures originated in the up to 400-m thick deposits of the deeply eroded Aksu rockslide dam in the Northern Tien Shan about 50 km southwest of Bishkek City (see Fig. 8.102). One of such failure most likely was triggered by the 1885 Belovodsk earthquake (see Section 8.3).

Future landslides can be anticipated in the headscarp of the Djuzumdybulak rockslide in the Central Tien Shan within loose material accumulated after rockslide formation and displaced by up to 200-m long fresh fissure (Fig. 8.192). We hypothesize that this sliding could be triggered by the 1992 Suusamyr earthquake, whose intensity in this part of the Naryn River valley could reach VII points of the MSK-64/Modified Mercalli (MM) scale (Korjenkov, 2006), because no evidence of such sliding was observed prior to this earthquake. Judging from the length of this fissure, it could be related to the sliding along the base of the loose material. In the worst case, such sliding could cause temporal separation of the Kambarata-2 HPP reservoir into two parts. To avoid it, an artificial bypass channel was designed and its excavation is on the way now.

The largest ongoing large-scale slope deformations within the entire Central Asian region, whose total volume could reach hundreds of cubic kilometers (!) affect the entire northern slope of the Peter the First Range more than 100-km long and up to 10-to 11-km wide, being associated with neotectonic thrust faulting (see Figs. 5.20 and 5.21).

Similar phenomena, though at a much smaller scale, likely took place at the northern slope of the Surhku Range that stretches in the southwestern direction for about 40 km between the Obi-Garm and Faizabad towns east of the Dushanbe City. Thick, nearly horizontal nappe zone, locally with lenticular structure along which the Upper Jurassic and Cretaceous sedimentary rocks overlay mixed Quaternary deposits can be seen at a distance of about 2 km in the Obi-Chushon River valley just north of the Rogun Town—from 38.704° N, 69.76° E, at ~1200 m a.s.l. (Fig. 15.4) to 38.697° N, 69.74° E, at ~1350 m a.s.l. (Fig. 15.5). Such relationships could result either from large-scale overthrusting or from the same-scale sliding. It was proposed that landslides had blocked the ancient Vakhsh River valley that flowed towards the Gissar intermontane depression and forced it to change its river course towards its present position (Kolichko, 1975). From our point of view, blocking of the pre-Vakhsh valley was caused by the combined effect of the neotectonic thrusting and of the large-scale sliding at its front. Both the Peter the First and the Surhku gigantic sliding zones are rich of local landslides and DSGSDs shown on Figs. 9.26, 9.27, and 13.4 that can be activated in future.

At several sites, future rockslides might affect settlements and critical facilities directly. One of these sites is on the right-bank slope of the Gunt River valley just above the Khorog town in Tajik Pamir (see Fig. 9.80). Here large displaced block bounded from the top by distinct headscarp about 100-m high "hangs" ca. 500 m above the built-up area of the valley floor. This hazardous site requires monitoring and elaboration of the emergency plan for local community in case of its activation.

Extremely hazardous effects have landslides affecting mining waste disposals, especially those of the abandoned uranium nines in the Mailisu and Minkush River basins (Torgoev and Alioshin, 2001, 2005, 2009; Havenith, 2006b). Adverse consequences follow landslides that occur either directly on the slopes above such waste disposals as in the Tuyuksu River valley (see Fig. 8.165) or up- or downstream from them as in Mailisu River valley that could cause erosion or inundation of waste storage facilities and pollution of the vast area downstream. These sites require not only monitoring but also urgent protection measures.

There are no doubts that brief listing of sites where evidence of the ongoing slope deformations that could evolve into catastrophic large-scale failures has been identified during regional reconnaissance does not

FIGURE 15.4 Tectonized Upper Jurassic siltstone and claystone with lenticular structure—T(J_3) overlying compacted Quaternary (Q) sediments (talus or debris flow deposits). γPZ at the background—Paleozoic granites.

FIGURE 15.5 Nearly horizontal Upper Jurassic layers overlying Quaternary debris flow accumulations.

exhaust areas where such phenomena could originate in future. We want to accentuate once more that location, magnitude, or timing of some of them cannot be predicted, especially considering large earthquakes as one of the main triggering factor, because their prediction is an even more equivocal task. Further progress in this field requires elaboration of national or international programs of regular mapping and identification of potentially hazardous sites and of their detailed site-specific investigations.

Conclusive Remarks

Numerous case studies described in this book show that Central Asia can be considered not only as the "landslide hotspot" (Nadim et al., 2006) but also as the "rockslide sanctuary." A combination of rugged topography, active tectonics, seismicity, arid climate, and relatively scarce population makes this region extremely favorable for studying peculiarities of these natural phenomena and their causes, triggers, and consequences. Data presented herein are just the first, brief overview of large-scale Central Asian bedrock landslides, only a few of which have been studied in the field. Many if not most of them deserve much more comprehensive study than it have been performed up to now, including detailed large-scale mapping, dating, analysis of the deposits' grain-size composition, of the traces of upstream inundation and of outburst floods like the dendritic amygdaliform landforms on the terraces downstream (see Figs. 8.55 and 8.252). Such investigations can be performed within the frames of numerous Master's or PhD projects.

The first priority task is to finish compilation of the regional database, whose preliminary version is presented in the Appendix. As we mentioned in Chapter 10, Quantitative Relationships of Central Asian Bedrock Landslide Parameters, basic quantitative parameters have been derived for only 62% of case studies identified, though in some parts of the region this proportion increases up to 90%. Larger number of case studies quantified will allow more representative correlations between various parameters characterizing rockslides' mobility. It could be a part of the global rockslides database, whose analysis will help deriving both general relationships and their regional variations, if any.

We should notice that all known historical rockslides occurred during the last 150 years approximately. Most of them are seismically triggered—the 1887 Akjar (see Fig. 8.90), the 1911 Usoi (see Figs. 5.7, 9.40–9.47), the same year Chilik (see Figs. 8.66 and 8.67), the Ananievo (see Figs. 8.70 and 8.71), the Kaindy (see Fig. 8.73), and several more rockslides likely associated with the M8.2 Kemin earthquake (see Figs. 8.68, 8.69, and 8.74), the 1949 Khait (see Figs. 8.271 and 8.272), and the 1992 Belaldy (see Fig. 2.4) rockslides. Several large slope failures occurred without earthquakes—the 1964 Aini (see Figs. 2.2 and 8.266) and the 2008(?) Djamansu (see Figs. 8.110–8.112) rockslides, as well as some rockslides identified on modern glaciers in presently glaciated regions (see Figs. 4.32–4.38). Number of the historical rockslides on modern glaciers could be larger, but those older than few decades were reworked by moving ice and cannot be identified at present.

Keeping in mind that oldest dated rockslides occurred ca. 50–60 Ka ago (Sanhueza-Pino et al., 2011) the modern rate of rockslides occurrence—about 15 known cases in ca. 150 years—seems to be quite high—1 event in 10 years on average, while mean rate during Late Quaternary time appears to be about 1 event in 55 years, if not considering missed features (rockslides that collapsed on the glaciers during Late Quaternary time; some of the Late Pleistocene rockslides out of the glaciated areas, whose size was close to the minimal threshold of 1 Mm^3). At the same time, spatial distribution of the historical rockslides is very irregular—most of them occurred in Northern and Southern Tien Shan, while in Dzungaria, Pamir, and in other parts of Tien Shan they are much rarer. Dating of as many of such features as possible will allow better understanding of temporal distribution of large-scale rock slope failures within the entire Central Asia and in its' parts, which, along with data on their spatial distribution, will serve as input data for rockslide hazard assessment.

Another important task of future studies is to increase number of Central Asian rockslides for which data on their internal structure and grain-size composition are available (see Chapter 11: Structure and Grain-Size Composition of Large-Scale Bedrock Landslide Deposits). It will be very interesting to collect such detailed data on rockslides of different types (Hungr et al., 2014), which originated on slopes composed of different types of rocks and formed bodies of different morphological types (see Section 5.3 in Chapter 5: Bedrock Landslide Types and Classification Systems). Such data, along with well-grounded relationships between rockslides' volume, area, runout, etc. (see Chapter 10), and results of extensive physical and numerical modeling will help better understanding of rockslide motion mechanism(s) and of their mobility that, as we mentioned in Chapter 12, Morphological and Structural Evidence of Rockslides' (Rock Avalanches') Motion Mechanism(s), predetermine hazard provided by these natural phenomena. We want to point out that numerical modeling should consider basic peculiarities of rockslide deposits' internal structure reflecting mechanics of debris motion (see Chapter 11), otherwise even good reproduction of the known rock slope failures dimensions and morphology will not guarantee the correctness of the prediction of future rockslides' behavior.

Further studies will be necessary to find unequivocal interpretation of several natural blockages for which some researchers propose glacial origin (N. Ischuk, 2008, 2011, 2013a, 2016), while other, including authors of this book—rockslide origin. As we mentioned in Sections 4.2 and 13.2 in Chapters 4 and 13, Principles and Criteria of Landslides Identification and Discrimination From the Features Formed by Other Mass Wasting Phenomena, and Seismically Induced Rockslides: Their Discrimination, respectively, its correct solution is important both for landslide and for seismic hazard assessment of the study region.

Well-grounded identification of seismically triggered rockslides is one of the most controversial topics of landslide studies. In Chapter 13 we discuss several criteria that can be used for this purpose, but a lot of work has to be done for ultimate discrimination of rock slope failures triggered by earthquakes from those of different, non-seismic origin. We have to admit that at present in most of cases such an unequivocal solution is impossible. No doubts that significant progress in this field will be based on the precise dating of the nearby rock slope failures (see, e.g., Bull, 1996). We anticipate, however, that careful examination of landforms and outcrops in the areas that had been inundated by rockslide-dammed lakes could also provide valuable information allowing grounded solution of this really controversial issue as it was demonstrated in Section 13.2.6 in Chapter 13. Increased number of features, whose simultaneity will be proved, will help adaptation of method proposed by D. Keefer (1993, 1999, 2000) for paleoseismological studies in Central Asian region.

Special emphasis (and also a lot of additional field and office work) should be placed on determining various morphological parameters of rockslide dams and dammed lakes as well as of the hydrological characteristics of blocked rivers and their catchments. It will allow quantitative evaluation of landslide dams according to various indices proposed to assess their stability (Korup, 2004; Tacconi Stefanelli et al., 2016; Dufresne et al., in press). Besides previously proposed indices, it will be interesting to calculate dam's compactness and ratio between flood runoff and lake area and to analyze how these parameters affect dam's stability as it is mentioned in Chapter 14, Rockslide Dams—Origin, Evolution, and Failure.

The critically important topic is identification of sites where large-scale catastrophic rock slope failures can occur in future. At a regional scale, such sites can be hardly selected unmistakably, without false alarms or missed targets, as it was exemplified by case studies summarized in Chapter 15, Evidence of Slope Instabilities—Potential of Future Large-Scale Slope Failures. Data on spatial distribution of past rockslides and of the deep-seated gravitational slope deformations where large-scale slope failures can be anticipated, provided in Chapters 7–9, Dzungaria, Tien Shan, and Pamir, respectively, can be used as inputs for large rockslides' susceptibility assessment, though, as we mentioned in Chapter 15, the approaches traditionally used to assess landslide susceptibility (Havenith et al., 2006a, 2015b; Günther et al., 2013; Kavzoglu et al., 2014; Aristizábal et al., 2016) are less efficient for large-scale rock slope failures.

Selection of sites where such failures can be anticipated in near future can be based on the combined analysis of high-resolution optical space images and radar images. While optical images allow selecting sites with geomorphic evidence of both past and recent slope deformations that can evolve into catastrophic failure, the INSAR method provides data on the present-day slope movement. Sites, identified by such regional scale monitoring can be monitored farther by more precise local control systems. Because it is practically impossible to prevent failure of high mountainous slopes that affect rock massifs millions and billions cubic meters in volume, monitoring and early warning will help to reduce risks associated with such natural hazards.

Authors will be grateful for any comments and corrections regarding both particular case studies and general issues discussed in our book.

References

Abdrakhmatov, K.Y., 1995. Tectonic movements of the territory of Kyrgyzstan in Late Pleistocene – Holocene. Doctor of Science Thesis. Bishkek, 250 pp. (in Russian).

Abdrakhmatov, K.Y., Aldazhanov, S.A., Hager, B.H., Hamburger, M.W., Herring, T.A., Kalabaev, K.B., et al., 1996. Relatively recent construction of the Tien Shan inferred from GPS measurements of present-day crustal deformation rates. Nature 384, 450–453.

Abdrakhmatov, K., Lemsin, I., Strom, A., 1994. Neotectonics of the Ketmen-Tiube depression. The Tien Shan in the Recent Orogeny Period. Ilim, Bishkek, pp. 86–96 (in Russian).

Abdrakhmatov, K., Strom, A., 2006. Dissected rockslide and rock avalanche deposits; Tien Shan Kyrgyzstan. In: Evans, S.G., Scarascia Mugnozza, G., Strom, A., Hermanns, R.L. (Eds.), Landslides from Massive Rock Slope Failure. NATO Science Series: IV: Earth and Environmental Sciences, vol. 49. Springer, New York, pp. 551–572.

Abdrakhmatov, K.E., Thompson, S., Weldon, R., 2007. Active Tectonics of the Tien Shan. Ilim Publishers, Bishkek, 71 pp. (in Russian).

Abdrakhmatov, K.E., Walker, R.T., Campbell, G.E., Carr, A.S., Elliott, A2, Hillemann, C., et al., 2016. Multi-segment rupture in the July 11th 1889 Chilik earthquake (Mw 8.0-8.3), Kazakh Tien Shan, interpreted from remote-sensing, field survey, and palaeoseismic trenching. J. Geophys. Res. Solid Earth. Available from: https://doi.org/10.1002/2015JB012763.

Abele, G., 1974. Bergstürze in den Alpen, ihre Verbeitung, Morphologie und Folgeerscheinungen, Wissenschaftliche Alpenverein 25, 230 pp.

Adams, J., 1981a. Earthquake-dammed lakes in New Zealand. Geology 9, 215–219.

Adams, J., 1981b. Earthquakes, landslides, and large dams in New Zealand. Bull. N. Z. Nat. Soc. Earthquake Eng. 14, 93–95.

Adushkin, V.V., 2000. About initiation of natural creative processes by explosions. Combust. Explo. Shock Waves 36 (6), 21–30.

Adushkin, V.V., 2006. Mobility of rock avalanches triggered by underground nuclear explosions. In: Evans, S.G., Scarascia Mugnozza, G., Strom, A., Hermanns, R.L. (Eds.), Landslides from Massive Rock Slope Failure. NATO Science Series: IV: Earth and Environmental Sciences, vol. 49, 267–284.

Adushkin, V.V., 2011. Russian experience with blast-fill dam construction. In: Evans, S.G., Hermanns, R., Strom, A.L., Scarascia-Mugnozza, G. (Eds.), Natural and Artificial Rockslide Dams. Lecture Notes in Earth Sciences, 133. Springer, Heidelberg, pp. 595–616.

Adushkin, V.V., Pernik, L.M., Popel', S.I., Strom, A.L., Shishaeva, A. S., 2006. Formation of nano- and micro-range particles in rocky slopes collapse. Nano- and Micro-Scale Particles in Geophysical Processes. Institute of Geospheres Dynamics, RAS, Moscow (in Russian).

Agakhanjanz, O.E., 1989. Sarez. Leningrad Press, Leningrad, 112 pp. (in Russian).

Agatova, A.R., Nepop, R.K., 2011. Assessing the rate of seismogravitational denudation of the relief of Southeastern Altai: the Chagan-Uzun R. Basin. J. Volcanol. Seismol. 5 (6), 53–62.

Agliardi, F., Crosta, G.B., Frattini, P., Malus, M.G., 2013. Giant non-catastrophic landslides and the long-term exhumation of the European Alps. Earth. Planet. Sci. Lett. 365, 263–274.

Ahmed, M.F., Rogers, J.D., Abu Bakar, M.Z., 2016. Hunza River watershed landslide and related features inventory mapping. Environ. Earth Sci. 10 75 (6). Available from: https://doi.org/10.1007/s12665-015-5172-2.

Ahmed, M.F., Rogers, J.D., Ismail, E.H., 2014. Historic landslide dams along the upper indus river, Northern Pakistan. Nat. Hazards Rev. ASCE. Available from: https://doi.org/10.1061/(ASCE)NH.1527-6996.0000165.

Aizen, V.B., 2011. Pamir glaciers. In: Sigh, V.P., Singh, P., Haritashya, U.K. (Eds.), Encyclopedia of Snow, Ice and Glaciers. Springer, Dordrecht, pp. 813–815.

Aizen, V.B., Aizen, E.M., Melack, J.M., Dozier, J., 1997. Climate and hydrologic changes in the Tien Shan, central Asia. J. Clim. 10, 1393–1404.

Alford, D., Schuster, R. (Eds.), 2000. Usoi Landslide Dam and Lake Sarez. An assessment of Hazard and Risk in the Pamir Mountains, Tajikistan. UN Publication, Sales No. E.00.III.M.1.

Allen, C.R., 1975. Geological criteria for evaluating seismicity. Geol. Soc. Am. Bull. 86, 1041–1057.

Ambraseys, N., Bilham, R., 2012. The Sarez-Pamir earthquake and landslide of 18 February 1911. Seismol. Res. Lett. 83, 294–314.

Aptikaev, F.F., 2012. Instrumental Scale of Seismic Intensity. Nauka & Obrazovanie, Ltd, Moscow, 175 pp. (in Russian).

Arefiev, S.S., Rogozhin, E.A., Bykova, V.V., Dorbath, C., 2006. Deep structure of the Racha earthquake source zone from seismic tomography data. Izvestiya Phys. Solid Earth 42 (1), 27–40.

Aristizábal, E., Vélez, J.I., Martines, H.E., Jaboyedoff, M., 2016. SHIA_Landslide: a distributed conceptual and physically based model to forecast the temporal and spatial occurrence of shallow landslides triggered by rainfall in tropical and mountainous basins. Landslides 13, 497–517. Available from: https://doi.org/10.1007/s10346-015-0580-7.

Arrowsmith, J.R., Strecker, M.R., 1999. Seismotectonic range-front segmentation and mountain-belt growth in the Pamir-Alai region, Kyrgyzstan, India-Eurasia collision zone). Geol. Society of Amer. Bull. 111, 1665–1683.

Arrowsmith, J.R., Crosby, C.J., Korjenkov, A.M., Mamyrov, E., Povolotskaya, I., 2005. Surface rupture of the 1911 Kebin (Chon-Kemin) earthquake, Northern Tien Shan, Kyrgyzstan. EosTrans. AGU, Fall Meet. Suppl.: Abstr. 86 (52), San-Francisco, USA, T51F-05.

Assimaki, D., Gazetas, G., 2004. Soil and topographic amplification on canyon banks and the 1999 Athens earthquake. J. Earthquake Eng. 8 (1), 1–43.

Atlas of Geological maps of Northern-Central-Eastern Asia and Adjsacent Areas, 2012. Geological Map of Northern-Central-Eastern Asia and Adjsacent Areas 1:2 500 000. www.vsegei.ru/ru/info/inter-proj/geo-asia2500.

Avdeev, V.A., Nartov, S.V., Baljinniam, I., Monhoo, D., Erdenbileg, B., 1989. Tsambagarav earthquake of July 23, 1988. Geol, Geophys, (11), 118–124 (in Russian).

Avouac, J.P., Tapponnier, P., Bai, M., You, H., Wang, G., 1993. Active thrusting and folding along the northern Tien Shan and late Cenozoic rotation of the Tarim relative to Dzungaria and Kazakhstan. J. Geophys. Res. Solid Earth 98 (B4), 6755–6804. Available from: https://doi.org/10.1029/92JB01963.

Barla, G., Dutto, F., Mortara, G., 2000. Breva Glacier rock avalanche of 18 January 1997 on the Mount Blsanc Range, Northern Italy. Landslide News (No 13), 2–5.

Barsch, D., 1977. Nature and importance of mass-wasting by rock glaciers in alpine permafrost environments. Earth Surface Proc, 2, 231–245.

Barsch, D., 1988. Rock glaciers. In: Clark, M.J. (Ed.), Advances in Periglacial Geomorphology. Wiley, Chichester, pp. 69–90.

Barth, N.C., 2013. The Cascade Rock Avalanche: a very large Alpine Fault-triggered failure, South Westland, New Zealand. Landslides. Available from: https://doi.org/10.1007/s10346-013-0389-1.

Beetham, R.D., McSaveney, M.J., Read, S.A.L., 2002. Four extremely large landslides in New Zealand. In: Rybar, J., Stemberk, J., Wagner, P. (Eds.), Landslides. A.A. Balkema, Rotterdam, pp. 97–102.

Belousov, T.P., Skobelev, S.F., Strom, A.L., 1994. On estimation of the recurrence period of strong earthquakes of the central Tien Shan (according to the data of absolute geochronology). J. Earthquake Pred. Res. 3, 226–236.

Břežný, M., Pánek, T. Deep-seated landslides affecting monoclinal flysch morphostructure: evaluation of LIDAR-derived topography of the highest range of the Czech Carpatians. Paper, submitted to Geomorphology.

Bhandari, R.H., Kumar, K., 2000. Malpa rock avalanche of 18 August 1998. Landslide News (No 13), 18–21.

Bochkariov, V.P., 2006. Natural and artificial rockslide dams in mountain geosystems of Kazakhstan orogenic belt. Italian J. Eng. Geol. Environ. (Special Issue), 203–205.

Bochkariov, V.P., Medeuov, A., 1990. Slope processes in mountainous regions of the South-eastern Kazakhstan. Eng. Geol. (No 6), 50–58 (in Russian).

Bogachkin, B.M., Korjenkov, A.M., Mamyrov, E., Nechaev, Yu.V., Omuraliev, M., Petrosian, A.E., et al., 1997. Structure of the 1992 Suusamyr earthquake source on the basis of its geological and seismological effects analysis. Phys. Earth. (No 11), 3–18 (in Russian, English translation).

Bogdanovich, M.C., Kark, J., Korolkov, B., Muchketov, D., 1914. Earthquake of the 4th January 1911 in the northern districts of the Tien Shan. Trans. Geol. Com. New Series 89, 270 pp. (in Russian with extended French abstract).

Bonnard, C., 2011. Technical and human aspects of historic rockslide-dammed lakes and landslide dam breaches. In: Evans, S.G., Hermanns, R., Scarascia-Mugnozza, G., Strom, A.L. (Eds.), Natural and Artificial Rockslide Dams. Lecture Notes in Earth Sciences 133, 101–122.

Bourdeau, C., Havenith, H.B., 2008. Site effects modelling applied to the slope affected by the Suusamyr earthquake (Kyrgyzstan, 1992). Eng. Geol. 97, 126–145.

Bowman, E.T., Take, W.A., Rait, K.L., Hann, C., 2012. Physical models of rock avalanche spreading behaviour with dynamic fragmentation. Can. Geotech. J. 49, 460–476.

Bozzano, F., Cipriani, I., Mazzanti, P., Prestininzi, A., 2011. Displacement patterns of a landslide affected by human activities: insights from ground-based InSAR monitoring. Nat. Hazards. Available from: https://doi.org/10.007/s11069-011-9840-6.

Bronnikov, M.M., 1908. The Karatag earthquake. Proc. Geol. Committee 27 (No 147), 475–515 (in Russian).

Brückl, J., Heuberger, H., 2001. Present structure and prefailure topography of the giant rockslide of Köfels. Zeitschrift für Gletscherkunde und Glazialgeologie 37, 49–79.

Buech, F., Davies, T., Pettinga, J.R., 2010. The Little Red Hill seismic experimental study: topographic effects on ground motion at a bedrock-dominated mountain edifice. Bull. Seismol. Soc. Am. 100 (5A), 2219–2229.

Bugaev, E.G., Lavrov, I.M., Kharlov, E.M., 1975. Oscillations of the Chirkey canyon during aftershocks. Bull. Eng. Seismol. (No 9), 99–110 (in Russian).

Bukinich, D.D., 1913. Usoi earthquake and its consequences. Russian records, No. 187, Wednesday, August 14, 1913 (In Russian).

Bull, W.B., 1996. Prehistorical earthquakes on the Alpine fault, New Zealand. J. Geophys. Res. 101 (No B3), 6037–6050.

Burbank, D.W., 1983. Multiple episodes of catastrophic flooding in the Peshawar basin during the past 700,000 tears. Geol. Bull. Uniu. Peshawar 16, 43–49.

Burgmann, R., Rosen, P.A., Fielding, E.J., 2000. Synthetic aperture radar interferometry to measure Earth's surface topography and its deformation. Annu. Rev. Earth. Planet. Sci. 28, 169–209.

Burtman, V.S., 1964. The Talas-Ferghana strike-slip fault. Proc. Geol. Inst. Acad. Sci. USSR 104, 143 pp. (in Russian).

Burtman, V.S., 1975. Structural geology of variscan Tien Shan, USSR. Am. J. Sci. 275, 157–186.

Burtman, V.S., 2012. Tien Shan and High Asia: Geodynamics in the Cenozoic. Transactions of the Geological Institute of RAS, vol. 603 Moscow, GEOS, 187 pp. (in Russian).

Burtman, V.S., Skobelev, S.F., Molnar, P., 1996. Late Cenozoic slip on the Talas-Fergana fault, the Tien Shan, central Asia. Geol. Soc. Am. Bull. 108, 1004–1021.

Calhoun, N., Poschinger, A.V., Clague, J.J., Giardino, M., Masera, D., Perotti L., 2015. New Pieces to the Flims-Tamins Rockslide Puzzle. In: Lollino, G., Giordan, D., Crosta, G., Corominas, J., Azzam, R., Wasowski, J., Sciarra, N. (Eds.), Engineering Geology for Society and Territory - Volume 2 Landslide Processes, 899-903. doi: 10.1007/978-3-319-09057-3_155.

Campbell, C.S., 1989. Self-lubrication for long runout landslides. J. Geol. 97, 635–665.

Campbell, G.E., Walker, R.T., Abdrakhmatov, K., Jackson, J., Elliott, J.R., Mackenzie, D., et al., 2015. Great earthquakes in low strain rate continental interiors: an example from SE Kazakhstan. J. Geophys. Res. Solid Earth 120, 5507–5534. Available from: https://doi.org/10.1002/2015JB011925.

Campbell, G.E., Walker, R.T., Abdrakhmatov, K., Schwenninger, J.L., Jackson, J., Elliott, J.R., et al., 2013. The Dzhungarian fault: late quaternary tectonics and slip rate of a major right-lateral strike-slip fault in the northern Tien Shan region. J. Geophys. Res. Solid Earth 118, 1–18. Available from: https://doi.org/10.1002/jgrb.50367.

Canuti, P., Casagli, N., Garzonio, C.A., Vannocci, P., 1990. Lateral spreads and landslide hazards in the Northern Apennine: the example of the Mt. Fumaiolo (Emilia-Romagna) and Chiusi della Verna (Tuscany). In: Proceedings 6th Congr. IAEG, Amsterdam, 3, 1525–1533.

Cao, M.S., 1998. Detection of abrupt changes in glacier mass balance in the Tien Shan Mountains. J. Glaciol. 44 (147), 352–358.

Capra, L., Macías, J.L., Scott, K.M., Abrams, M., Garduño-Monroy, V. H., 2002. Debris avalanches and debris flows transformed from collapses in the Trans-Mexican Volcanic Belt, Mexico - behavior, and implications for hazard assessment. J. Volcanol. Geotherm. Res. 113 (1–2), 81–110.

Cascini, L., Peduto, D., Fornaro, G., Lanari, R., Zeni, G., Guzzetti, F., 2009. Spaceborne radar interferometry for landslide monitoring. Int. J. Remote. Sens. 27, 1709–1716.

Cassie, J.W., Van Gassen, W., Cruden, D.M., 1988. Laboratory analogue of the formation of molards, cones on rock-avalanche debris. Geology 16, 735–738.

Catane, S.G., Cabria, H.B., Tomarong, C.P., Saturay, R.M., Zarco, M.A. H., Pioquinto, W.C., 2007. Catastrophic rockslide-debris avalanche at St. Bernard, Southern Leyte, Philippines. Landslides 4, 85–90.

Catane, S.G., Cabria, H.B., Zarco, M.A.H., Saturay, R.M., Mirasol-Robert, A.A., 2008. The 17 February 2006 Guinsaugon rock slide-debris avalanche, Southern Leyte, Philippines: deposit characteristics and failure mechanism. Bull. Eng. Geol. Environ. 67 (3), 305. Available from: https://doi.org/10.1007/s10064-008-0120-y.

Chedia, O.K., 1971. South of the Central Asia in the newest orogeny epoch, Book 1. Continental Cenozoic accumulations and geomorphology. Frunze, Ilim Publishing house. 332 pp. (in Russian).

Chedia, O.K., 1972. South of the Central Asia in the newest orogeny epoch, Book 2. Newest tectonics and paleogeography. Frunze, Ilim Publishing house. 225 pp. (in Russian).

Chedia, O.K., 1986. Morphostructures and Neotectonic of the Tien Shan. Frunze, Ilim Publishing House, 314 pp. (in Russian).

Chedia, O.K., Lemzin, I.N., 1980. Seismogenerating faults of the Chatkal Depression. In: Seismotectonics and seismicity of the Tien Shan. Frunze, Ilim, 18–28. (in Russian).

Chen, R.F., Chang, K.J., Angelier, J., Chan, Y.C., Deffontaines, B., Lee, C.T., et al., 2006. Topographical changes revealed by high-resolution airborne LiDAR data: the 1999 Tsaoling landslide induced by the Chi–Chi earthquake. Eng. Geol. 88 (3), 160–172.

Chen, Q., Cheng, H., Yang, Y., Liu, G., Liu, L., 2014. Quantification of mass wasting volume associated with the giant landslide Daguangbao induced by the 2008 Wenchuan earthquake from persistent scatterer InSAR. Remote Sens. Environ. 152, 125–135.

Chen, X., Cui, P., You, Y., Cheng, Z., Khan, A., Ye, C., et al., 2017. Dam-break risk analysis of the Attabad landslide dam in Pakistan and emergency countermeasures. Landslides 14, 675–683.

Chen, C.C., Telesca, L., Lee, C.T., Su, Y.S., 2011. Statistical physics of landslides: new paradigm. EPL 95 (4), 49001. Available from: https://doi.org/10.1209/0295-5075/95/49001.

Chigira, M., Wu, X.Y., Inokuchi, T., Wang, G.H., 2010. Landslides induced by the 2008 Wenchuan earthquake, Sichuan, China. Geomorphology 118, 225–238.

Chipizubbov, A.V., 1998. Identification of single-event and even-aged past surface ruptures and assessment of past earthquakes magnitudes on the base of their dimensions. Russian Geol. Geophys. 39, 386–398.

Chongshao, S., Hongwei, C., 1997. Developing and distributing characteristics of collapses and landslides during strong historic earthquakes in China. J. Nat. Dis. 6 (No 1), 25–30 (in Chinese).

Cigna, F., Del Ventisette, C., Liguori, V., Casagli, N., 2010. Advanced radar interpretation of InSAR time series for mapping and characterization of geological processes. Nat. Hazards Earth Syst. Sci. 11, 865–881.

Clark, S.P. (Ed.), 1966. Handbook of Physical Constants. Revised Edition, 97. Yale University, New Haven, Connecticut, The Geol. Soc. Am. Inc. Memoir.

Colesanti, C., Wasowski, J., 2006. Investigating landslides with space-borne Synthetic Aperture Radar (SAR) interferometry. Eng. Geol. 88 (3–4), 173–199.

Cook, S.J., Porter, P.R., Bendall, C.A., 2013. Geomorphological consequences of a glacier advance across a paraglacial rock avalanche deposit. Geomorphology 189, 109–120.

Corominas, J., 1996. The angle of reach as a mobility index for small and large landslides. Can. Geotech. J. 33, 260–271.

Corominas, J., Einstein, H., Davis, T., Strom, A., Zuccaro, G., Nadim, F., Verdel, T., 2015. Glossary of terms on landslide hazard and risk. In: Lollino, G., et al., (Eds.), Engineering Geology for Society and Territory, Volume 2. Springer International Publishing, Switzerland, pp. 1775–1779. Available from: https://doi.org/10.1007/978-3-319-09057-3_314.

Corominas, J., Mavrouli, O., Ruiz-Carulla R., 2017. Rockfall Occurrence and Fragmentation. In: K. Sassa, et al. (Eds.), Advancing Culture of Living with Landslides, 75–97. DOI: 10.1007/978-3-319-59469-9_4.

Corominas, J., Moya, J., 2008. A review of assessing landslide frequency for hazard zoning purposes. Eng. Geol. 102, 193–213. Available from: https://doi.org/10.1016/j.enggeo.2008.03.018.

Costa, J.E., Schuster, R.L., 1988. The formation and failure of natural dams. Geol. Soc. Am. Bull. 100, 1054–1068.

Costa, J.E., Schuster, R.L., 1991. Documented historical landslide dams from around the world. United States Geological Survey Open-File Report 91–239, 486 pp.

Cox, S.C., McSaveney, M.J., Spencer, J., Allen, S.,K., Ashraf, S., Hancox, G.,T., et al., 2015. Rock avalanche on 14 July, 2014 from Hillary Ridge, Aoraki/Mount Cook, New Zealand. Landslides 12, 395–402.

Crosby, C., Arrowsmith, J., Korjenkov, A., Guralnik, B., Mamyrov, E., Povolotskaya, I., 2007. The hunt for surface rupture from the 1889 Ms8.3 Chilik earthquake, Northern Tien-Shan, Kyrgyzstan and Kazakhstan. In: AGU Fall Meeting Abstracts, F5, poster presentation.

Crosta, G.B., Frattini, P., Fusi, N., 2007. Fragmentation in the Val Pola rock avalanche, Italian Alps. J. Geophys. Res. 112, F01006. Available from: https://doi.org/10.1029/2005JF000455.

Crosta, G.B., Frattini, P., Fusi, N., Sosio, R., 2011. Formation, characterisation and modeling of the Val Pola rock-avalanche dam (Italy). In: Evans, S.G., Hermanns, R.L., Strom, A.L., Scarascia Mugnozza, G. (Eds.), Natural and Artificial Rockslide Dams. Springer, Berlin, Heidelberg, pp. 347–368.

Crosta, G.B., Frattini, P., Valbuzzi, E., Hermanns, R.L., 2015. E–Cerrillos Negros Giant Rock Avalanches (Tacna, Peru). In: Lollino, G., Giordan, D., Crosta, G.B., Corominas, J., Azzam, R., Wasowski, J., Sciarra, N. (Eds.), Engineering Geology for Society and Territory, Volume 2. Springer International Publishing, Switzerland, pp. 921–924.

Crosta, G., Hermanns, R.L., Murillo, P.V., 2012. Large rock avalanches in southern Perù: the Cerro Caquilluco – Cerrillos Negros rock slide - avalanche (Tacna, Tomasiri, Perù). Geophys. Res. Abs. 14, EGU 2012-11469-1.

Crosta, G.B., Imposimato, S., Roddeman, D.G., 2003. Numerical modelling of large landslides stability and runout. Nat. Hazards Earth Syst. Sci. 3, 523–538.

Crozier, M.J., 1992. Determination of paleoseismicity from landslides. In Landslides (Glissements de terrain) (D. H. Bell, Ed.), Proceedings of the 6th International Symposium, Christchurch, New Zealand, 1992, vol. 2, pp. 1173–1180. A. A. Balkema, Rotterdam.

Cruden, D.M., Hungr, O., 1986. The debris of the Frank Slide and theories of rockslide-avalanche mobility. Can. J. Earth. Sci. 23, 425–432.

Cruden, D., VanDine, D.F., 2013. Classification, description, causes and indirect effects. In: Canadian Technical Guidelines and Best Practices related to Landslides: a national initiative for loss reduction. Geological Survey of Canada, Open File 7359. 22 pp.

Cruden, D.M., Varnes, D.J., 1996. Landslide types and processes. In: Turner, A.K., Schuster, R.L. (Eds.), Landslides: Investigation and Mitigation, U.S. Transportation Research Board, Special Report, no. 247, 36–75.

Cui, P., Chen, X., Chen, Z., 2009. The Barrier Lakes created during the Wenchuan Earthquake and its disaster mitigation works. Rock Characterisation, Modelling and Engineering Design Methods. Proc. of the ISRM-Sponsored International Symposium on Rock Mechanics, Hong Kong, 19–22 May 2009, CD-ROM.

Cui, Y., Deng, J., Dai, F., Fu, W., Wei, J., Zhang, Z., Xu C., In press. Mahu Landslide: a Huge Landslide in Sichuan Province of China. Paper submitted for Geoenvironmental Disasters.

Dai, F.C., Lee, C.F., Deng, J.H., Tham, L.G., 2005. The 1786 earthquake-triggered landslide dam and subsequent dam-break flood on the Dadu River, southwestern China. Geomorphology 65, 205–221.

Dai, F.C., Xu, Q., Yao, X., Xu, L., Tu, X.B., Gong, Q.M., 2011. Spatial distribution of landslides triggered by the 2008 Ms 8.0 Wenchuan earthquake, China. J. Asian Earth Sci. 40, 833–895.

Davies, T.R., 1982. Spreading of rock avalanche debris by mechanical fluidisation. Rock Mech. 15, 9–24.

Davies, T.R., McSaveney, M.J., 1999. Runout of dry granular avalanches Can. Geotech. J. 36, 313–320.

Davies, T.R., McSaveney, M.J., 2002. Dynamic simulation of the motion of fragmenting rock avalanches. Can. Geotech. J. 39, 789–798.

Davies, T.R., McSaveney, M.J., 2011. Rock-avalanche size and runout—implications for landslide dams. In: Evans, S.G., Hermanns, R., Strom, A.L., Scarascia-Mugnozza, G. (Eds.), Natural and Artificial Rockslide Dams. Lecture Notes in Earth Sciences, 133. Springer, Heidelberg, pp. 441–462.

Davies, T.R., McSaveney, M.J., Beetham, R.D., 2006. Rapid block glides: slide-surface fragmentation in New Zealand's Waikaremoana landslide. Quat. J. Eng. Geol. Hydrogeol. 39, 115–129.

Davies, T.R., McSaveney, M.J., Reznichenko, N., 2017. The fate of elastic strain energy in brittle fracture. In: Leith, K., Ziegler, M., Perras, M., Loew, S. (Eds.), Progressive rock failure. An ISRM Specialized Conference, Monte Verità, 5–9 June 2017. ETH, Extended abstracts, 107–108.

Delaney, K.B., Evans, S.G., 2011. Rockslide dams in the northwest Himalayas (Pakistan, India) and the adjacent Pamir mountains (Afghanistan, Tajikistan), Central Asia. In: Evans, S.G., Hermanns, R., Strom, A.L., Scarascia-Mugnozza, G. (Eds.), Natural and Artificial Rockslide Dams. Lecture Notes in Earth Sciences, vol. 133. Springer, Heidelberg, pp. 205–242.

Delaney, K.B., Evans, S.G., 2015. The 2000 Yigong landslide (Tibetan Plateau), rockslide-dammed lake and outburst flood: review, remote sensing analysis, and process modelling. Geomorphology 246, 377–393.

Delgado, J., Garrido, J., López-Casado, C., Martino, S., Peláez, J.A., 2011. On far field occurrence of seismically induced landslides. Eng. Geol. 123, 204–213.

Delvaux, D., Abdrakhmatov, K.E., Lemzin, I.N., Strom, A.L., 2001. Landslides and surface breaks of the 1911, Ms 8.2 Kemin earthquake, Kyrgyzstan. Russ. Geol. Geophys. 42, 1167–1177.

Deng, X.F., Ding, Y.J., 1990. Landslide dam and Tianchi Lake near Urumchi, Sinkiang Province, China. Landslide News (No 4), 23–25.

Denikaev, Sh.Sh, 1970. About ancient blockage in the Sarez Lake region. Proc. Acad. Sci. Tajik Soviet Socialist Republick 13, 32–40 (in Russian).

Djanuzakov, K.D., Omuraliev, M., Omuralieva, A., Iliasov, B.I., Grebennikova, V.V., 2003. Strong Earthquakes in Tien Shan within the territory of Kyrgyzstan and the adjacent regions of Central Asia. Bishkek, Ilim, 201 pp. (in Russian).

Djumabaeva, A., 2012. Seimogenic ruptures of the Kochkor depression. Trans. Nat. Acad. Sci. Kyrgyz Republic (No 3), 41–44 (in Russian).

Drew, F., 1875. The Jummoo and Kashmir territories; a geographical account. Edward Stanford, London. 568 pp.

Dubovskoi, A.N., Pernik, L.M., Strom, A.L., 2008. Experimental simulation of rockslide fragmentation. J. Mining Sci. 44 (2), 123–130.

Dufresne, A., 2009. Influence of Runout Path Material on Rock and Debris Avalanche Mobility: Field Evidence and Analogue Modeling. PhD Thesis. University of Canterbury, Christchurch, New Zealand, 271 pp.

Dufresne, A., 2014. An overview of rock avalanche – substrate interaction. In: Sassa, K., Canuti, P., Yin, Y. (Eds.), Landslide Science for a Safer Geoenvironment, 1. Springer, Beijing, pp. 345–349.

Dufresne, A., Bösmeier, A., Prager, C., 2016. Sedimentology of rock avalanche deposits – case study and review. Earth-Sci. Rev. 163, 234–259.

Dufresne, A., Davies, T.R., 2009. Longitudinal ridges in mass movement deposits. Geomorphology 105 (3–4), 171–181.

Dufresne, A., Davies, T.R., McSaveney, M.J., 2009. Influence of runout-path material on emplacement of the Round Top rock avalanche, New Zealand. Earth Surf. Proc. Landf. 35, 190–201.

Dufresne, A., Dunning, S.A., 2017. Process dependence of grain size distributions in rock avalanche deposits. Landslides. Available from: https://doi.org/10.1007/s10346-017-0806-y.

Dufresne, A., Ostermann, M.A., Preusser, F., In press. River-damming, late-1 Quaternary rockslides and rock avalanches in the Ötz Valley region (Tyrol, Austria).

Dunning, S.A., Armitage, P.J., 2011. The grain-size distribution of rock-avalanche deposits: implications for natural dam stability. In: Evans, S.G., Hermanns, R., Scarascia-Mugnozza, G., Strom, A. L. (Eds.), Natural and Artificial Rockslide Dams, 133. Lecture Notes in Earth Sciences, pp. 479–498.

Dunning, S.A., Mitchell, W.A., Rosser, N.J., Petley, D.N., 2007. The Hattian Bala rock avalanche and associated landslides triggered by the Kashmir Earthquake of 8 October 2005. Eng. Geol. 93, 130–144.

Dunning, S.A., Petley, D.N., Rosser, N.J., Strom, A.L., 2005. The morphology and sedimentology of valley confined rock-avalanche deposits and their effect on potential dam hazard. In: Hungr, O., Fell, R., Couture, R., Eberhardt, E. (Eds.), Landslide Risk Management. A.T. Balkema, Amsterdam, pp. 691–704.

Ecological State of the Ferghana Valley Regions. Popular Report, 2006. http://enrin.grida.no/htmls/ferghana_valley/ferghana_valley_soe/

Eisbacher, G.H., 1979. Cliff collapse and rock avalanches (sturzstroms) in the Mackenzie Mountains, northwestern Canada. Can. Geotech. J. 16, 309–334.

Emelianova, E.P., 1972. Main regularities of landslide processes. Moscow, Nedra Publishing House, 310 pp. (in Russian).

Engdahl, E.R., Villaseñor, A., 2002. Globas seismicity: 1900–1999. International handbook of earthquakes and engineering seismology, 81A. International Association Seismol. & Phys. Earth's Interior. Committee on Education, 665–690.

Eppler, D.B., Fink, J., Fletcher, R., 1987. Rheologic properties and kinematics of emplacement of the Chaos Jumbles rockfall avalanche, Lassen Volcaic National Park, California. J. Geophys. Res. B95, 3623–3633.

Ercanoglu, M., Gokceoglu, C., 2004. Use of fuzzy relations to produce landslide susceptibili y map of a landslide prone area (West Black Sea Region, Turkey). Eng. Geol. 75, 229–250.

Erismann, T.H., 1979. Mechanisms of large landslides. Rock Mech. 12, 15–46.

Erismann, T.H., 1986. Flowing, rolling, bouncing, sliding: synopsis of basic mechanisms. Acta Mech. 64, 101–110.

Erismann, T.H., Abele, G., 1999. Dynamics of Rockfalls and Rockslides. Springer, Heidelberg, 316 pp.

Ermini, L., Casagli, N. 2002. Criteria for a preliminary assessment of landslide dams evolution. In: Stemberk, J., Wagner, P. (Eds.), Landslides, Proc. of the First European Conf. on Landslides, Prague, June 24–24, 2002. Swets & Zeitlinger, Lisse, 157–162.

Ermini, L., Casagli, N., 2003. Prediction of the behavior of landslide dams using a geomorphological dimensionless index. Earth Surface Proc. Landforms 28, 31–47.

Evans, S.G., 2006. Single-event landslides resulting from massive rock slope failure: characterizing their frequency and impact on society. In: Evans, S.G., Scarascia Mugnozza, G., Strom, A., Hermanns, R.L. (Eds.), Landslides from Massive Rock Slope Failure. NATO Science Series: IV: Earth and Environmental Sciences, vol. 49. Springer, Dordrecht, pp. 53–73.

Evans, S.G., Bishop, N.F., Smoll, L.F., Murillo, P.V., Delaney, K.B., Oliver-Smith, A., 2009a. A reexamination of the mechanism and human impact of catastrophic mass flows originating on Nevado Huascarán, Cordillera Blanca, Peru in 1962 and 1970. Eng. Geol. 108, 96–118.

Evans, S.G., Clague, J.J., 1988. Catastrophic rock avalanches in glacial environments. In: Bonnard, C. (Ed.), Proceedings of the 5th International Symposium on Landslides, vol. 2. Lausanne, Switzerland, pp. 1153–1158.

Evans, S.G., Clague, J.J., 1999. Rock avalanches on glaciers in the Coast and St. Elias Mountains, British Columbia. In: Slope Stability and Landslides, Proceedings of the 13th Annual Vancouver Geotechnical Conference, Vancouver, BC.

Evans, S.G., Clague, J.J., Woodsworth, G.J., Hungr, O., 1989. The Pandemonium Creek rock avalanche, British Columbia. Can. Geotech. J. 26, 427–446.

Evans, S.G., Delaney, K.B., 2011. Characterization of the 2000 Yigong Zangbo River (Tibet) Landslide Dam and Impoundment by Remote Sensing. In: Evans, S.G., Hermanns, R., Scarascia-Mugnozza, G., Strom, A.L. (Eds.), Natural and Artificial Rockslide Dams, 133. Lecture Notes in Earth Sciences, pp. 543–559.

Evans, S.G., Delaney, K.B., 2014. The 1997 Mount Munday landslide (British Columbia) and the behaviour of rock avalanches on glacier surfaces. Landslides. Available from: https://doi.org/10.1007/s10346-013-0456-7.

Evans, S.G., Delaney, K.B., Hermanns, R.L., Strom, A., Scarascia-Mugnozza, G., 2011. The formation and behaviour of natural and artificial rockslide dams; implications for engineering performance and hazard management. In: Evans, S.G., Hermanns, R., Scarascia-Mugnozza, G., Strom, A.L. (Eds.), Natural and Artificial Rockslide Dams, 133. Lecture Notes in Earth Sciences, pp. 1–75.

Evans, S.G., Guthrie, R.H., Roberts, N.J., Bishop, N.F., 2007. The disastrous 17 February 2006 rockslide-debris avalanche on Leyte Island, Philippines: a catastrophic landslide in tropical mountain terrain. Nat. Hazards Earth Syst. Sci. 7, 89–101.

Evans, S.G., Hungr, O., Enegren, E.G., 1994. The Avalanche Lake rock avalanche, Mackenzie Mountains, Northwest Territories, Canada: description, dating and dynamics. Can. Geotech. J. 31, 749–768.

Evans, S.G., Roberts, N.J., Ischuck, A., Delaney, K.B., Morozova, G.S., Tutubalina, O., 2009b. Landslides triggered by the 1949 Khait earthquake, Tajikistan, and associated loss of life. Eng. Geol. 109 (3–4), 195–212.

Evans, S.G., Scarascia Mugnozza, G., Strom, A.L., Hermanns, R.L., Ischuk, A., Vinnichenko, S., 2006. Landslides from massive rock slope failure and associated phenomena. In: Evans, S.G., Scarascia Mugnozza, G., Strom, A., Hermanns, R.L. (Eds.), Landslides from Massive Rock Slope Failure. NATO Science Series: IV: Earth and Environmental Sciences, vol. 49. Springer, Dordrecht, pp. 3–52.

Fan, X., van Westen, C.J., Korup, O., Gorum, T., Xu, Q., Dai, F., et al., 2012a. Transient water and sediment storage of the decaying landslide dams induced by the 2008 Wenchuan earthquake, China. Geomorphology v. 171–172, 58–68.

Fan, X., van Westen, C.J., Xu, Q., Gorum, T., Dai, F., 2012b. Analysis of landslide dams induced by the 2008 Wenchuan earthquake. J. Asian Earth Sci. 57 (5), 25–37.

Fan, X., Tang, C.X., van Westen, C.J., Alkema, D., 2012c. Simulating dam-breach scenarios of the Tangjiashan landslide dam induced by the Wenchuan earthquake. Nat. Hazards Earth System Sci. 12, 3031–3044.

Fan, X., Xu, Q., van Westen, C.J., Huang, R., Tang, R., 2017. Characteristics and classification of landslide dams associated with the 2008 Wenchuan earthquake. Geoenviron. Disasters 4, 12. Available from: https://doi.org/10.1186/s40677-017-0079-8.

Federico, F., Cesali, C., 2015. Sliding of rapid granular masses: analytical modeling and empirical criteria. In: Lollino, G., et al., (Eds.), Engineering Geology for Society and Territory –, Volume 2. © Springer International Publishing Switzerland, pp. 1641–1650. Available from: https://doi.org/10.1007/978-3-319-09057-3_292.

Fedorenko, V.S., 1968. Tectonic and seismic phenomena and their significance in formation of catastrophic rock falls and landslides (in the regions of the Chatkal and Alai Mountain ranges). In: Problems of Engineering Geology and soil science. Issue 2. Moscow, 229–244 (in Russian).

Fedorenko, V.S., 1976. Main regularities of formation and types of landslides and rock falls in the valleys of the Zeravshan River basin. In: Geological regularities of landslides, rock falls and debris flows evolution. Issue 2. Moscow, 26–38 (in Russian).

Fedorenko, V.S., 1988. Rockslides and Rock Falls and Their Prediction. Moscow State University Publishing House, Moscow, 214 pp. (in Russian).

Fedorenko, V.S., Kozhevnikov, A.V., 1974. Landslides, rockfalls and debris flows of the middle part of the Zeravshan and Shing Rivers (according to engineering-geological investigsations of 1971–1974). Unpublished report available in the archives of the Geological Survey of Tajikistan (in Russian).

Fedorenko, V.S., Denikaev, Sh., Lim, V.V., 1981. Main engineering-geological aspects of the Sarez Lake problem. Eng. Geol. (No 3), 70–80 (in Russian).

Fedorenko, V.S., Nikulin, F.V., Kalinin, E.V., Lipilin, V.I., 1979. Mechanism of large rockslides displacement. Eng. Geol. (No 6), 30–46 (in Russian).

Fen, C., 1987. Traces of past earthquakes in the Tien Shan region. Sibei Dichjen Suebao. 9 (No 3), 113–115.

Feng, X., 1990. The Nilke earthquake fault zone in Xinjiang. Inland Earthquake. No 3 (in Chinese).

Florensov, N.A., Solonenko, V.P. (Eds.), The Gobi Altai earthquake, 1963. Publishing House of the Academy of Sciences of the USSR, Moscow, 391 pp. (in Russian). English edition is available: Israel Program for Scientific Translations. U.S. Department of Commerce, Washington, DC, 424 pp.

Fort, M., Peulvast, J., 1995. Catastrophicmass-movements and morphogenesis in the Peri-Tibetan Ranges: examples for west Kunlun, east Pamir and Ladakh. In: Slaymaker, O. (Ed.), Steepland Geomorphology. Wiley, New York, pp. 171–198.

Frattini, P., Riva, F., Crosta, G.B., Scotti, R., Greggio, L., Brardinoni, F., et al., 2016. Rock-avalanche geomorphological and hydrological impact on an alpine watershed. Geomorphology 262. Available from: https://doi.org/10.1016/j.geomorph.2016.03.013.

Fukuoka, H., Sassa, K., Wang, G., 2007. Shear behaviour and shear zone structure of granular materials in naturally drained ring shear tests. In: Sassa, K., Fukuoka, H., Wang, F., Wang, G. (Eds.), Progress in Landslide Science. Springer-Verlag, Berlin, Heidelberg, pp. 99–111.

Galitzin, P.B., 1915. Sur le tremblement de terre du 18 de février 1911. Comptes Rendus de l'Academie des Sciences de Paris 160, 810–813.

Garevski, M., Zugic, Z., Sesov, V., 2013. Advanced seismic slope stability analysis. Landslides 10, 729–736. Available from: https://doi.org/10.1007/s10346-012-0360-6.

Gaziev, E., 1984. Study of the Usoi Landslide in Pamir. In: Proc. 4th Int. Symp. on Landslides, Toronto, 1, 511–514.

Geertsema, M., 2012. Initial observations of the 11 June 2012 rock/ice avalanche, Lituya Mountain, Alaska. In: TheFirst Meeting of Cold Region Landslides Network, Harbin, China. 2012. https://www.researchgate.net/publication/237589579.

Geertsema, M., Hungr, O., Schwab, J.W., Evans, S.G., 2006. A large rockslide-debris avalanche in cohesive soil at Pink Mountain, Northeastern British Columbia, Canada. Eng. Geol. 83, 64–75.

Geli, L., Bard, P.Y., Jullien, B., 1988. The effect of topography on earthquake ground motion: a review and new results. Bull. Seismol. Soc. Am. 78, 42–63.

Geological map of China. 1 :250 000 scale. http://en.ngac.org.cn/Map/List

Geological map of Kazakh SSR, 1:500 000 scale, 1979. All-Union research and production association «Aerogeology). Leningrad (in Russian).

Geological map of Kyrgyz SSR, 1:500 000 scale, 1978. All-Union Geological Institute. Leningrad (in Russian).

Geological map of Tajik SSR and adjacent territories, 1:500 000 scale, 1984. All-Union Geological Institute, Leningrad (in Russian).

Geological map of Uzbek SSR. 1 :1500 000 scale, 1978. Ministry of Geology of the USSR. (in Russian).

Geological map of Xinjiang Uigur Autonomous Region, China, 1:2000 000, 1985. Beijing, Geological Publishing House (in Chinese).

Gerasimov, V., 1965. The Issyk catastrophe in 1963 and its effects on geomorphology of the Issyk River Valley. Trans. All-union Geogr. Soc. 97 (6), 541–547 (in Russian).

Gerasimov, I.P., Rantsman, E.Ya, 1964. Neotectonics of seismic regions of the Tien Shan and the Pamir-Alay according to the data of the geomorphic analysis. Activated Zones of Earth Crust, Newest Tectonic Movements and Seismicity. Nauka Publishers, Moscow (in Russian).

Ghose, S., Mellors, R.J., Korjenkov, A.M., Hamburger, M.W., Pavlis, T.L., Pavlis, G.L., et al., 1997. The Ms = 7.3 1992 Suusamyr, Kyrgyzstan earthquake: 2. Aftershock Focal Mechanisms and Surface Deformation. Bull. Seismol. Soc. Am. 87, 23–38.

Glazirin, G.E., 2010. A century of investigations on outbursts of the icedammed lake Merzbacher (central Tien Shan). Aust. J. Earth Sci. 103 (2), 171–179.

Glicken, H., 1996. Rockslide-debris avalanche of May 18, 1980, Mount St. Helens Volcano, Washington, U.S. Geol. Surv. Open-File Rep. 96-0677.

Golovko, D., Roessner, S., Behling, Kleinschmit, B., 2017. Automated derivation and spatio-temporal analysis of landslide properties in southern Kyrgyzstan. Nat. Hazards 85 (3), 1461–1488. Available from: https://doi.org/10.1007/s11069-016-2636-y.

Golovko, D., Roessner, S., Behling, R.,Wetzel, H.U., Kaufmann, H., 2014. GIS-based integration of heterogeneous data for a multi-temporal landslide inventory. In: Sassa, K., Canuti, P., Yin, Y. (Eds.), Landslide Science for a Safer Geoenvironment, 2, 799–804.

González Diaz, E.F., Giaccardi, A.D., Costa, C.H., 2001. La avalancha de rocas del rio Barrancas (Cerro Pelan), norte del Neuquen; su relacion con la catastrofe del Rio Colorado (29/12/1914). Rev. Asoc. Geol. Argentina 56, 466–480.

Gorbunov, A.P., Titkov, S.N., 1989. Rock glaciers of the Central Asian mountains. Jakutsk, IMZ of the Siberian branch of the Academy of Sciences of the USSR, 164 pp. (in Russian).

Gorbunov, A.P., Titkov, S.N., Polyakov, V.G., 1992. Dynamics of rock glaciers of the northern Tien Shan and Dzungar Ala Tau, Kazakhstan. Permafrost Periglacial Proc, 3, 29–39.

Gori, S., Falcucci, E., Dramis, F., Galadini, F., Galli, P., Giacciod, B., et al., 2014. Deep-seated gravitational slope deformation, large-scale rock failure, and active normal faulting along Mt. Morrone (Sulmona basin, Central Italy): geomorphological and paleoseismological analyses. Geomorphology 208, 88–101.

Gotz, A., Zimmermann, M., 1993. The 1991 rock slides in Randa: causes and consequences. Landslide News 7, 22–25.

Griffith, A.A., 1921. The phenomena of rupture and flow in solids. Philos. Trans. Royal Soc. London, A 221, 163–198.

Grigorian, S.S., 1979. New friction law and mechanism of large-scale rockfalls and landslides, Proceedings of Academy of Sciences of USSR 244, 846-849 (in Russian).

Groeber, P., 1916. Informe sobre las causas que han producido las crecientes del Rio Colorado (Territorios del Neuquen y La Pampa) en 1914, Direccion General de Minas, Geologia e Hidrologia, 11 (Serie B, Geologia), Buenos Aires, 29 pp.

Grützner, C., Carson, E., Walker, R.T., Rhodes, E.J., Mukambayev, A., Mackenzie, D., et al., 2017. Assessing the activity of faults in continental interiors: palaeoseismic insights from SE Kazakhstan. Earth. Planet. Sci. Lett. 459, 93–104.

Guang-hua, Y.N., Jun, L.I., Zhang, Y., Yang, Z., 2001. Statistical analysis and research of landslides caused by Nilki (Nilke) earthquake. Inland Earthquake 15 (1), 56–63 (in Chinese).

Gubin, I.E., 1943. The Garm Earthquake in 1941. Published by the Tajik branch of the Academy of Sciences of the USSR, 132 pp. (in Russian).

Gubin, I.E., 1960. Patterns of Seismic Manifestations of the Tajikistan Territory (Geology and Seismicity). Academy of Sciences of USSR Publishing House, Moscow, 464 pp. (in Russian).

Günther, A., Reichenbach, P., Malet, J.-P., Eeckhaut, M.V.D., Hervás, J., Dashwood, C., et al., 2013. Tier-base approaches for landslide susceptibility assessment in Europe. Landslides 10, 529–546. Available from: https://doi.org/10.1007/s10346-012-0349-1.

Guschenko, O.I., Leonov, N.N., 1972. Paleoseisodislocations of the Chatkal River basin and their interrelation with the newest stress field. In: Schukin, Y.K. (ed.), Seismogenic structures and seismic dislocations, Moscow, 49–51 (in Russian).

Guzzetti, F., Ardizzone, F., Cardinali, M., Galli, M., Reichenbach, P., Rossi, M., 2008. Distribution of landslides in the Upper Tiber River basin, central Italy. Geomorphology 96 (1–2), 105–122. Available from: https://doi.org/10.1016/j.geomorph.2007.07.015.

Guzzetti, F., Ardizzone, F., Cardinali, M., Rossi, M., Valigi, D., 2009. Landslide volumes and landslide mobilization rates in Umbria, Central Italy. Earth. Planet. Sci. Lett. 279, 222–229. Available from: https://doi.org/10.1016/j.epsl.2009.01.005.

Guzzetti, F., Mondini, A.C., Cardinali, M., Fiorucci, F., Santangelo, M., Chang, K.-T., 2012. Landslide inventory maps: new tools for an old problem. Earth-Sci. Rev. 112, 42–66.

Habib, P., 1975. Production of gaseous pore pressure during rock slides. Rock Mech. 7, 193–197.

Hadian-Amri, M., Solaimani, K., Kavian, A., Afzal, P., Glade, T., 2014. Curve estimation modeling between area and volume of landslides in Tajan River basin, north of Iran. Ecopersia 2 (3), 651–665.

Hadjigeorgiou, J., Couture, R., Locat, J., 1996. In-situ block size distributions as tools for the study of rock avalanche mechanisms,". Proceedings of the 2nd North American Rock Mechanics Symposium 1, Montreal.

Hadley, J.B., 1964. Landslides and related phenomena accompanying the Hebgen Lake earthquake of August 17, 1959. USGS Professional Paper 345-K, 107–138.

Haeberli, W., 1985. Creep of mountain permafrost: internal structure and flow of alpine rock glaciers. Mitteilungen der Versuchsantalt für Wasserbau, Hydrologie und Glaziologie 77, 142.

Hancox, G.T., Perrin, N.D., 2009. Green Lake Landslide and other giant and very large postglacial landslides in Fiordland, New Zealand. Quat. Sci. Rev. 28 (11–12), 1020–1036.

Harrison, J.V., Falcon, N.L., 1937. The Saidmarreh landslip, southern Iran. J. Geogr. 89, 42–47.

Hartvich, F., Mugnai, F., Proietti, C., Smolkova, V., Strom, A., 2008. A reconstruction of a former rockslide-dammed lake: the case of the Kokomeren River valley (Tien Shan, Kyrgyzstan). Poster presentation at the EGU conference, Vien.

Havenith, H.-B., 2002. Landslides Triggered by Earthquakes. Experimental Studies in the Tien Shan Mountains (Central Asia) and Dynamic Modelling. Ph. D. Thesis. Université de Liège, Faculté des Sciences Appliquées, 239 pp.

Havenith, H.-B., Strom, A., Caceres, F., Pirard, E., 2006a. Analysis of landslide susceptibility in the Suusamyr region, Tien Shan: statistical and geotechnical approach. Landslides 3, 39–50.

Havenith, H.-B., Strom, A., Jongmans, D., Abdrakhmatov, K., Delvaux, D., Trefois, P., 2003. Seismic triggering of landslides,

Part A: field evidence from the Northern Tien Shan. Nat. Hazards Earth System Sci. 3, 135–149.

Havenith, H.-B., Torgoev, I., Meleshko, A., Alioshin, Y., Torgoev, A., Danneels, G., 2006b. Landslides in the Mailuu-Suu valley, Kyrgyzstan: hazards and impacts. Landslides 3, 137–147.

Havenith, H.-B., Fan, X., Torgoev, A., 2013. What we can learn from a blind-test for predicting earthquake-triggered landslides applied to the Wenchuan area, China – also for paleoseismological studies. In: Grützner, C., Rudersdorf, A., Pérez-López, R., Reicherter, K. (Eds.), Proc. 4th International INQUA Meeting on Paleoseismology, 99. Active Tectonics and Archeoseismology (PATA), Aachen, Germany, 9-14 October 2013.

Havenith, H.-B., Strom, A., Torgoev, I., Torgoev, A., Lamair, L., Ischuk, A., et al., 2015a. Tien Shan geohazards database: earthquakes and landslides. Geomorphology 249, 16–31.

Havenith, H.-B., Torgoev, A., Schlögel, R., Braun, A., Torgoev, I., Ischuk, A., 2015b. Tien Shan Geohazards Database: landslide susceptibility analysis. Geomorphology 249, 32–43.

Havenith, H.-B., Torgoev, A., Braun, A., Schlögel, R., Micu, M., 2016. A new classification of earthquake-induced landslide event sizes based on seismotectonic, topographic, climatic and geologic factors. Geoenviron. Dis. 3 (6). Available from: https://doi.org/10.1186/s40677-016-0041-1.

Heim, A., 1878 Untersuchungen über den Mechanismus der Gebirgsbildung im Anschluß an die Geologische Monographie der Tödi-Windgällen-Gruppe Basel, Schwabe, 346 pp.

Heim, A., 1882. Der Bergsturz von Elm. Deutsch. Geol. Gesell. Zeitschr. 34, 74–115.

Heim, A., 1883 Der alte Bergsturz von Flims, Jb. d. Schweizer Alpenclubs, 18. Jg., 295–309.

Heim, A., 1932. Bergsturz und Menschenleben. Fretz and Wasmuth, Zürich. 218 pp.

Hermanns, R.L., Folguera, A., Penna, I., Fauqué, L., Niedermann, S., 2011a. Landslide dams in the Central Andes of Argentina (Northern Patagonia and the Argentine Northwest). In: Evans, S. G., Hermanns, R., Strom, A.L., Scarascia-Mugnozza, G. (Eds.), Natural and Artificial Rockslide Dams. Lecture Notes in Earth Sciences, vol. 133. Springer, Heidelberg, pp. 147–176.

Hermanns, R.L., Hewitt, K., Strom, A., Evans, S.G., Dunning, S.A., Scarascia-Mugnozza, G., 2011b. The classification of rockslide dams. In: Evans, S.G., Hermanns, R., Strom, A.L., Scarascia-Mugnozza, G. (Eds.), Natural and Artificial Rockslide Dams. Lecture Notes in Earth Sciences, vol. 133. Springer, Heidelberg, pp. 581–593.

Hermanns, R.L., Niedermann, S., Garsia, A.V., Gomes, J.S., Strecker, M.R., 2001. Neotectonics and catasrophic failure of mountain fronts in the southern intra-Andean Plateau, Argentina. Geology 29, 619–623.

Hermanns, R.L., Niedermann, S., Ivy-Ochs, S., Kubik, P.W., 2004. Rock avalanching into a landslide-dammed lake causing multiple dam failure in Las Conchas valley (NW Argentina) – evidence from surface exposure dating and stratigraphic analyses. Landslides 1, 113–122.

Hermanns, R.L., Strecker, M.R., 1999. Structural and lithological controls on large Quaternary rock avalanches (sturzstroms) in arid northwestern Argentina. Geol. Soc. Am. Bull. 111, 934–948.

Hermanns, R.L., Trauth, M.H., Niedermann, S., McWilliams, M., Strecker, M.R., 2000. Tephrochronologic constraints on temporal distribution of large landslides in northwest Argentina. J. Geol. 108, 35–52.

Hewitt, K., 1988. Catastrophic landslide deposits in the Karakoram Himalaya. Science 242, 64–67.

Hewitt, K., 1998. Catastrophic landslides and their effects on the Upper Indus streams, Karakoram Himalaya, Northern Pakistan. Geomorphology 26, 47–80.

Hewitt, K., 1999. Quaternary moraines vs catastrophic rock avalanches in the Karakoram Himalaya, Northern Pakistan. Quat. Res. 51, 220–237.

Hewitt, K., 2001. Catastrophic rockslides and the geomorphology of the Hunza and Gilgit river valleys, Karakoram Himalaya. Erdkunde 55, 72–93.

Hewitt, K., 2002. Styles of rock avalanche depositional complex in very rugged terrain, Karakoram Himalaya, Pakistan. In: Evans S. G. (Ed.), Catastrophic Landslides: effects, occurrence and mechanisms, Reviews in Engineering Geology, 15, 345–378.

Hewitt, K., 2006a. Disturbance regime landscapes: mountain drainage systems interrupted by large rockslides. Prog. Phys. Geogr. 30, 365–393.

Hewitt, K., 2006b. Rock avalanches with complex run out and emplacement, Karakoram Himalaya, Inner Asia. In: Evans, S.G., Scarascia Mugnozza, G., Strom, A., Hermanns, R.L. (Eds.), Landslides from Massive Rock Slope Failure. NATO Science Series: IV: Earth and Environmental Sciences, vol. 49. Springer, pp. 521–550.

Hewitt, K., 2009. Rock avalanches that travel onto glaciers and related developments, Karakoram Himalaya, Inner Asia. Geomorphology 103, 66–79.

Hewitt, K., 2011. Rock avalanche dams on the Trans Himalayan Upper Indus streams: a survey of Late Quaternary events and hazard-related characteristics. In: Evans, S.G., Hermanns, R., Strom, A.L., Scarascia-Mugnozza, G. (Eds.), Natural and Artificial Rockslide Dams. Lecture Notes in Earth Sciences, 133. Springer, Heidelberg, pp. 177–204.

Hewitt, K., Clague, J.J., Orwin, J.F., 2008. Legacies of catastrophic rock slope failures in mountain landscapes. Earth-Sci. Rev. 87, 1–38.

Hodjaev, A., 1974. About influence of the neotectonic movements on the development of the geodynamic processes and phenomena in the Chatkal Range south-western offshoots. In: Problems of the engineering geodynamics. Tashkent, 47–55 (in Russian).

Hodjaev, A., 1985. Paleoseismology of the Chatkal-Kuraminskiy Region. Fan Publihing House, Tashkent, 140 pp. (in Russian).

Hovius, N., Stark, C.P., 2006. Landslide-driven erosion and topographic evolution of active mountain belts. In: Evans, S.G., Scarascia Mugnozza, G., Strom, A., Hermanns, R.L. (Eds.), Landslides from Massive Rock Slope Failure. NATO Science Series: IV: Earth and Environmental Sciences, vol. 49. Springer, pp. 573–590.

Hovius, N., Stark, C.P., Allen, P.A., 1997. Sediment flux from a mountain belt derived by landslide mapping. Geology 25, 801–804.

Hsü, K.J., 1975. Catastrophic debris streams (sturzstroms) generated by rock falls. Geol. Soc. Am. Bull. 86, 129–140.

Hsü, K.J., 1978. Albert Heim: observations on landslides and relevance to modern interpretations. In: Voight, B. (Ed.), Rockslides and Avalanches, 1. Elsevier, Amsterdam, pp. 71–93. Natural Phenomena.

Huan, W.-L., Shi, Z.-L., 1987. Great earthquakes of M > 8 in the mainland of China and their evolution. Tectonophysics 138 (1), 55–68.

Huang, R.Q., 2014. Understanding the Mechanism of Large-scale Landslides. Engineering Geology for Society and Territory, Proc. of the 11th IAEG Congress, Volume 2. Springer, Torino, Italy, pp. 13–32.

Huang, R.Q., Li, W.L., 2009. Analysis of the geo-hazards triggered by the 12 May 2008 Wenchuan earthquake, China. Bull. Eng. Geol. Environ. 68, 363–371.

Huang, R., Pei, X., Fan, X., Zhang, W., Li, S., Li, B., 2012. The characteristics and failure mechanism of the largest landslide triggered by the Wenchuan earthquake, May 12, 2008. China. Landslides 9, 131–142.

Huang, R., Pei, X., Zhang, W., 2009. Further examination on characteristics and formation mechanism of Daguangbao Landslide. J. Eng. Geol. 6, 725–736.

Huggel, C., 2008. Recent extreme slope failures in glacial environments: effects of thermal perturbation. Quat. Sci. Rev. Available from: https://doi.org/10.1016/j.quascirev.2008.06.007.

Huggel, C., Salzmann, N., Allen, S., Caplan-Auerbach, J., Fischer, L., Haeberli, W., et al., 2010. Recent and future warm extreme events and high-mountain slope stability. Phil. Trans. R. Soc. A 368, 2435–2459. Available from: https://doi.org/10.1098/rsta.2010.0078.

Hungr, O., 1981. Dynamics of rock avalanches and other types of mass movements. Ph.D. thesis, Edmonton, University of Alberta, 500 pp.

Hungr, O., 1989. Mobility of rock avalanches. Reports of the National Research Institute for Earth Science and Disaster Prevention, Tsukuba, Japan, 46, 11–20.

Hungr, O., 1995. A model for the runout analysis of rapid flow slides, debris flows, and avalanches. Can. Geotech. J. 32, 610–623.

Hungr, O., 2006. Rock avalanche occurrence, process and modeling. In: Evans, S.G., Scarascia, Mugnozza, G., Strom, A., Hermanns, R. L. (Eds.), Landslides from Massive Rock Slope Failure. NATO Science Series: IV: Earth and Environmental Sciences, vol. 49. Springer, pp. 243–266.

Hungr, O., Leroueil, S., Picarelli, L., 2012. Varnes classification of landslide types, an update. In: Eberhardt, E., Froese, C., Turner, A.K., Leroueil, S. (Eds.), Landslides and Engineered Slopes: Protecting Society through Improved Understanding. Balkema, Rotterdam, pp. 47–58.

Hungr, O., Leroueil, S., Picarelli, L., 2014. Varnes classification of landslide types, an update. Landslides 11, 167–194.

Hutchinson, J.N., 2006. Massive rock slope failure: perspective and retrospectives on state-of-the-art. In: Evans, S.G., Scarascia-Mugnozza, G., Strom, A.L., Hermanns, R.L. (Eds.), Landslides from Massive Rock Slope Failure. Nato Science Series. Series IV, Earth and Environmental Sciences 49, 619–662.

Hutchinson, J.N., Kojan, E., 1975. The Mayunmarca landslide of 25th April, 1974, Peru, UNESCO Report Serial No. 3124, UNESCO.

Ignatiev, I.V., 1886. The earthquake in the Tokmak district in the 1885. Proc. Russian Imperial Geogr. Soc. 22 (2), (in Russian).

Ischuk, N.R., 2008. Blockages of river valleys in Tajikistan and their role in assessment of the regional seismicity. In: Proceedings of the 4th international symposium "geodynamics of the intracontinental orogens and geoecological problems", Bishkek, June 15–20, 2008, 244–245.

Ischuk, A.R., 2011. Usoi Rockslide Dam and Lake Sarez, Pamir Mountains, Tajikistan. In: Evans, S.G., Hermanns, R., Strom, A.L., Scarascia-Mugnozza, G. (Eds.), Natural and Artificial Rockslide Dams. Lecture Notes in Earth Sciences, 133. Springer, Heidelberg, pp. 423–440.

Ischuk, N.R., 2011. The role of glacial deposits in forming modern mountain lake dams in the Pamirs. Georisk 1, 16–29 (in Russian).

Ischuk, N.R., 2013a. The origin of the mountain river dams in Tajikistan. In: Margottini, C., Canuti, P., Sassa, K. (Eds.), Landslide Science and Practice, Volume 6 - Risk Assessment, Management and Mitigation. Springer, Heidelberg, pp. 13–17.

Ischuk, N.R., 2013b. "Landslides" on the Sarez Lake banks (Tajikistan). Georisk. 2, 48–60 (in Russian).

Ischuk, N.R., 2014. The Usoy dam at the Sarez Lake. Georisk. 1, 46–52 (in Russian).

Ischuk, N.R., 2016. Origin of lake dams at the northern slope of the Zeravshan Range. GeoRisk 2, 44–55 (in Russian).

Ischuk, A., Bendick, R., Rybin, A., Molnar, P., Khan, S.F., Kusikov, S., et al., 2013. Kinematics of the Pamir and Hindu Kush regions from GPS geodesy. J. Geophys. Res. Solid Earth 118. Available from: https://doi.org/10.1002/jgrb.50185.

Ischuk, N.R., Strom, A.L., 2016. Geological effects of the earthquake at the Sarez Lake on the 7th of December, 2015. GeoRisk 3, 56–74 (in Russian).

Iturrizaga, L., 2012. Hummocky debris landforms in the Chapursan Valley (Karakoram range, Pakistan): a glaciogeomorphological investigation. Geomorphology 169–170, 1–16.

Iverson, R.M., 2006. Forecasting runout of rock and debris avalanches. In: Evans, S.G., Scarascia Mugnozza, G., Strom, A., Hermanns, R.L. (Eds.), Landslides from Massive Rock Slope Failure. NATO Science Series: IV: Earth and Environmental Sciences, vol. 49. Springer, pp. 197–209.

Jansky, B., Sobr, M., Engel, Z., 2010. Outburst flood hazard: case studies from the Tien-Shan Mountains, Kyrgyzstan. Limnol. Ecol. Manag. Inland Waters 40 (4), 358–364.

Jarman, D., Agliardi, F., Crosta, G.B., 2011. Megafans and outsize fans from catastrophic slope failures in Alpine glacial troughs: the Malser Haide and the Val Venosta cluster, Italy. In: Jaboyedoff, M. (Ed.), Slope Tectonics., 351. Geological Society, London, pp. 253–277. , Special Publications.

Jibson, R.W., Harp, E.L., Schulz, W., Keefer, D.K., 2006. Large rock avalanches triggered by the M 7.9 Denali Fault, Alaska, earthquake of 3 November 2002. Eng. Geol. 83, 144–160.

Johnson, B., 1978. Blackhawk landslide, California, U.S.A. In: Voight, B. (Ed.), Rockslides and Avalanches, 1, Natural Phenomena. Elsevier, Amsterdam, pp. 481–504.

Jomard, H., Lebourg, Th, Guglielmi, Y., 2014. Morphological analysis of deep-seated gravitational slope deformation (DSGSD) in the western part of the Argentera massif. A morpho-tectonic control? Landslides 11, 107–117. Available from: https://doi.org/10.1007/s10346-013-0434-0.

Kariya, Y., Sato, G., Komori, J., 2011. Landslide-induced terminal morainelike landforms on the east side of Mount Shiroumadake, Northern Japanese Alps. Geomorphology 127, 156–165.

Kavzoglu, T., Kutlun Sahin, E., Colkesen, I., 2014. Landslide susceptibility mapping using GIS-based multi-criteria decision analysis, support vector machines, and logistic regression. Landslides 11, 425–439. Available from: https://doi.org/10.1007/s10346-013-0391-7.

Keefer, D.K., 1984. Landslides caused by earthquakes. Geol. Soc. Am. Bull. 95, 406–421.

Keefer, D.K., 1993. The susceptibility of rock slopes to earthquake-induced failure. Assoc. Eng. Geol. Bull. 30, 353–361.

Keefer, D.K., 1999. Earthquake-induced landslides and their effects on alluvial fans. J. Sed. Res. 69, 84–104.

Keefer, D.K., 2000. Statistical analysis of an earthquake-induced landslide distribution – The 1989 Loma Prieta, California event. Eng. Geol. 58, 231–249.

Keefer, D.K., 2002. Investigating landslides caused by earthquakes – a historical review. Surv. Geophys. 23 (6), 473–510.

Kellerer-Pirklbauer, A., Kaufmann, V., 2012. About the relationship between rock glacier velocity and climate parameters in central Austria. Aust. J. Earth Sci. 105/2, 94–112.

Kent, P.E., 1966. The transport mechanism in catastrophic rockfalls. J. Geol. 74, 79–83.

Khain, V.E., 1979. Regional Geotectonics. Exoalpine Asia and Australia. Nedra Publishing House, Moscow, p. 356 (in Russian).

Khain, V.E., 1984. Regional Geotectonics. The Alpine Mediterranian Belt. Nedra Publishing House, Moscow, p. 3344 (in Russian).

Khromovskikh, V.S., 1984. Rocky Dragon. Mysl Publishers, Moscow, 156 pp. (in Russian).

Kilburn, C.R.J., Sørensen, S.-A., 1998. Runout length of sturzstroms: the control of initial conditions and of fragment dynamics. J. Geophys. Res. 103 (No B8), 17877–17884.

King, J., Loveday, I., Schuster, R.L., 1989. The 1985 Bairaman landslide dam and resulting debris flow Papua New Guinea. Q. J. Eng. Geol. 22, 257–270.

Kobayashi, Y., 1993. A hypothesis for reduced resistance in large landslides. In: Safety and Environmental Issues in Rock Engineering, Proc. of the ISRM Int. Symp. Lisboa, June 21–24, 1993 1, Balkema Rotterdam, 835–839.

Kobayashi, Y., 1997. Long runout landslides riding on basal guided wave. In: Marinos, K., Tsiambaos, Stoumaras (Eds.), Engineering Geology and the environment. Balkema, Rotterdam, pp. 761–766.

Kojan, E., Hutchinson, J.N., 1978. Mayunmarca rockslide and debris flow, Peru. In: Voight, B. (Ed.), Rockslides and Avalanches 1, Developments in Geotechnical Engineering 14A. Elsevier, Amsterdam, pp. 315–361.

Kolesnikov, V.S., 1929. Brief description of the trip to the Sarez Lake in 1925. Proceedings of the Central Asian Geographical Society, XIX, Tashkent (in Russian).

Kolichko, A.V., 1975. History of evolution and formation mechanism of bedrock landslides in the north-eastern part of the Southern-Tadjik depression. Trans. Acad. Sci. USSR 222, 1417–1419 (in Russian).

Kolichko, A.V., Fil, V.N., 1980. Engineering-geological conditions of the Rogun HPP on Vakhsh River. Unpublished report of the Central Asian brunch of Hydroproject Institute. Tashkent 286. http://www.rfgf.ru/catalog/docview.php?did = cba33f2e8fd621fd2fa7ecd1628c6fe9 (in Russian).

Kondorskaya, N.V., Shebalin, N.V., 1982. New catalog of strong earthquakes in the U.S.S.R. from ancient times through 1977. NOAA, Boulder.

Koppes, M., Gillespie, A.R., Burke, R.M., Thompson, S.C., Stone, J., 2008. Late Quaternary glaciation in the Kyrgyz Tien Shan. Quat. Sci. Rev. 27, 846–866.

Korchevskiy, V.F., Kolichko, A.V., Strom, A.L., Pernik, L.M., Abdrakhmatov, K.E., 2011. Utilisation of data derived from large-scale experiments and study of natural blockages for blast-fill dam design. In: Evans, S.G., Hermanns, R., Strom, A.L., Scarascia-Mugnozza, G. (Eds.), Natural and Artificial Rockslide Dams. Lecture Notes in Earth Sciences, 133. Springer, Heidelberg, pp. 617–637.

Korchevskiy, V.F., Muratova, M.H., 1991. Construction of dams by collapse, Hydrotechnical Construction, No 3, 6–11 (in Russian).

Korchevskiy, V.F., Petrov, G.N., 1989. Designing and Investigation of Blastfill Dams. Energoatomisdat, Moscow, 170 pp. (in Russian).

Korjenevskiy, N.L., 1930. Catalogue of the Central Asian glaciers. Central Asian Meteorological Institute, Tashkent, 200 pp. (in Russian).

Korjenevskiy, N.L., 1979. Geomorphology and Glaciation of the Pamir-Alai. Fan Publishing House, Tashkent, 175 pp. (in Russian, reprinted).

Korjenkov, A.M., 2000. Seismogenic convolutionsin soft lacustrine sediments of the Issyk-Kul Lake, Tien Shan, Kyrgyzstan – initial report. J. Earthquake Pred. Res. 8, 514–519.

Korjenkov, A.M., 2006. Seismogeology of the Tien Shan (within the limits of the Kyrgyztan and ajacent regions. Bishkek, Ilim, 289 pp. (in Russian).

Korjenkov, A.M., Chedia, O.K., 1986. Neotectonic structure and seismic dislocations of the Tiulek and Karakudjur depressions (Tien Shan). Trans. Acad. Sci. Kyrgyz Republic (No 5), 26–33 (in Russian).

Korjenkov, A.M., Mamyrov, E., Omuraliev, M., Kovalenko V.A., Usmanov S.F., 2004. Rock avalanches and landslides formed in result of strong Suusamyr (1992, M = 7,4) earthquake in the Northern Tien Shan - test structures for mapping of paleoseismic deformations by satellite images. In: Buchroithner, M.F. (ed.), High Mountain Remote Sensing Cartography VII (HMRSC VII). Kartographische Bausteine, Band 23, Dresden, 117–136.

Korjenkov, A.M., Rust, D., Tibaldi, A., Abdieva, S.V., 2012. Parameters of the strong paleoearthquakes along the Talas-Fergana Fault, the Kyrgyz Tien Shan. In: D'Amico, S. (Ed.), Earthquake Research and Analysis – Seismology, Seismotectonics and Earthquake Geology. InTech Publishers, Rijeka, Croatia, pp. 33–84. , ISBN: 978-953-307-656-0.

Korolkov, B.Ya, 1912. The Karadag earthquake (with map and 2 tables). Trans. Turkestan Brunch Russian Geogr. Soc. 7, 35–47 (in Russian).

Korup, O., 2002. Recent research on landslide dams – a literature review with special attention to New Zealand. Prog. Phys. Geog. 26, 206–235.

Korup., O., 2004. Geomorphometric characteristics of New Zealand landslide dams. Eng. Geol. 73, 3–35.

Korup, O., 2005a. Geomorphic imprint of landslides on alpine river systems, southwest New Zealand. Earth Surface Proc. Landforms 30, 783–800.

Korup, O., 2005b. Distribution of landslides in southwest New Zealand. Landslides 2, 43–51.

Korup, O., Görüm, T., Hayakawa, Y., 2012. Without power? Landslide inventories in the face of climate change. Earth Surf. Proc. Landforms 37, 92–99.

Korup, O., Strom, A.L., Weidinger, J.T., 2006. Fluvial response to large rock-slope failures – examples from the Himalayas, the Tien Shan and the New Zealand Southern Alps. Geomorphology 78, 3–21.

Korup, O., Tweed, F., 2007. Ice, moraine, and landslide dams in mountainous terrain. Quat. Sci. Rev. 26, 3406–3422.

Kotlyakov, V.M. (ed.) 2006. Glaciation in Nortern and Central Eurasia at present time. Nauka Publishers, Moscow, p. (in Russian).

Kotlyakov, V.M., Desinov, L.V., Osipova, G.B., Hauser, M., Tsvetkov, D.G., Schneider, J.F., 2003. The 2002 events at the Geographical Society (RGO) glacier, Pamir. Mater. Glaciol. Investigat. 95, 221–230 (in Russian).

Krahn, J., Morgenstern, N.R., 1976. Mechanics of the Frank slide, Rock Engineering for Foundations and Slopes, 1. American Society of Civil Engineers, pp. 309–331.

Krukovskiy, G.L., 1969. On the role of slope failure phenomena in formation of high mountain Yashinkul and Kokkul lakes located in the southern part of Ferghana valley in the Tegermach River Basin. Hydrogeology and engineering geology of the arid zones of the USSR. Tashkent, Issue 9, part II, 25–31 (in Russian).

Kuchai, V.K., 1969. The results of repeated inspection of residual deformations in the pleistoseismal area of the Kebin earthquake. Geol. Geophys. 116 (N°8), 101–108 (in Russian).

Kuchai, V.K., 1971. Use of paleoseismic dislocations for studying seismic regime (exemplified by the pleistoseismal zone of the 1946 Chatkal earthquake). Geol. Geophys. (No 4), 124–129 (in Russian).

Kulikova, G., Schurr, B., Krüger, F., Brzoska, E., Heimann, S., 2016. Source parameters of the Sarez-Pamir earthquake of 1911 February 18. Geophys. J. Int. 205, 1086–1098.

Kurdiukov, K.V., 1950. Ancient rockfalls in the Alay Range valleys. Prob. Geogr. 21, 191–204 (in Russian).

Kurdiukov, K.V., 1956. Lepse reverse fault – recent displacement of the earths crust in East Balkhash area. In: Proc. Acad. Sci. USSR, Geol. Series No 5, 22–28 (in Russian).

Kurdiukov, K.V., 1964. Newest tectonic movements and traces of the extreme seismic shocks on the northern slope of the Zaalai Range. In: Activated zone of Earth crust, newest tectonic movements and seismicity. Moscow, 153–159. (in Russian).

Landgraf, A., Dzhumabaeva, A., Abdrakhmatov, K., Macaulay, E.A., Strecker, M.R., Arrowsmith, J.R., et al., 2016. Repeated large-magnitude earthquakes in a tectonically active, low-strain continental interior: the northern Tien Shan, Kyrgyzstan. J. Geophys. Res., Solid Earth 121 (5), 3888–3910. Available from: http://dx.doi.org/10.1002/2015JB012714.

Landslides and mudflows, 1984. UNESCO/UNEP. 2 volumes. Moscow (in Russian).

Landslides and slopes' stability. Bibliogaraphical Reference, 2010. Geoinformmark, 525 p., ISBN 5988770421X, 9785988770411 (in Russian).

Lange, O.K., 1926. Expedition on the survey of the Usoi blockage and the Sarez Lake. Nat. Econ. Central Asia (No 10), 76–79 (in Russian).

Lange, O.K., 1929. Modern State of the Usoi Blockage, 19. Proceedings of the Central Asian Geographical Society, Tashkent (in Russian).

Larsen, S.H., Davies, T.R.H., McSaveney, M.J., 2005. A possible coseismic landslide origin of late Holocene moraines of the Southern Alps, New Zealand. New Zealand J. Geol. Geophys. 48, 311–314.

Lauknes, T.R., Piyush Shanker, A., Dehls, J.F., Zebker, H.A., Henderson, Larsen, Y., 2010. Detailed rockslide mapping in northern Norway with small baseline and persistent scatterer interferometric SAR time series methods. Remote Sens. Environ. 114, 2097–2109.

Lee, C.F., Dai, F.C., 2011. The1786 Dadu River Landslide Dam, Sichuan, China. In: Evans, S.G., Hermanns, R., Scarascia-Mugnozza, G., Strom, A.L. (Eds.), Natural and Artificial Rockslide Dams. Lecture Notes in Earth Sciences 133, 369–388.

Legros, F., 2002. The mobility of long-runout landslides. Eng. Geol. 63, 301–331.

Lemzin, I.N., 1988. Cenozoic geomorphology and paleogeography of the Chatkal depression. Frunze, Ilim 91, (in Russian).

Leonov, N.N., 1960. The Khait 1949 earthquake and geological conditions of its origin. Proc. Acad. Sci. USSR Geophys. Series 3, 409–424 (in Russian).

Leonov, N.N., 1965. Role of the geogical structure in formation of the residual deformations of soils during earthquakes. Seismic Mycrozoning. Proceedings of the Institute of Physics of the Earth 203 (No 36), 132–136 (in Russian).

Leonov, N.N., 1970. The 1946 Chatkal earthquake, Prob. Eng. Seismol., 13. Nauka, Moscow, 64–77 (in Russian).

Leonov, N.N., Strom, A.L., 1979. Use of Paleoseismological Method for Seismic Hazard Assessment (Exemplified by the Rogun HPP). Proceedings of Hydroproject Institute, Moscow, pp. 64–70 (in Russian).

Li, T., 1983. A mathematical model for predicting the extent of a major rockfall. Z. für Geomorphologie N.F. 27, 473–482.

Li, L.P., Lan, H.X., Wu, Y.M., 2012. Comment on "Statistical physics of landslides: New paradigm" by Chen C.-c. et al. EPL 100 (2), 29001.

Li, L.P., Lan, H.X., Wu, Y.M., 2014. The volume-to-surface-area ratio constrains the rollover of the power law distribution for landslide size. Eur. Phys. J. Plus 129 (5), 89. Available from: https://doi.org/10.1140/epjp/i2014-14089-y.

Li, L.P., Lan, H.X., Wu, Y.M., 2016. Effect of sample size on exploiting landslide size distribution. Geoenviron. Dis. 3, 18. Available from: https://doi.org/10.1186/s40677-016-0052-y.

Litovchenko, A.F., 1964. Catastrophic debris flow along the Issyk River. Meteorol. Hydrol. 4, 39–42 (in Russian).

Locat, P.l., Couture, R., Leroueil, S., Locat, J., Jaboyedoff, M., 2006. Fragmentation energy in rock avalanches. Can. Geotech. J. 43, 830–851.

Luknitskiy, P.N., 1955. Travel over the Pamirs. Molodaya Gvardia Publishers, Moscow, 502 pp. (in Russian).

Makarov, V.I., 1977. Newest tectonic structure of Central Tien Shan. Trans. Geol. Inst. RAS 307, 171 pp. (in Russian).

Makarov, V.I., Abdrakhmatov, K.E., et al., 2005. Recent Geodynamics of Intracontinental Areas of Collision Mountaing Building (Central Asia). Scientific World, Moscow, 399 pp. (in Russian).

Malamud, B.D., Turcotte, D.L., Guzzetti, F., Reichenbach, P., 2004a. Landslides inventories and their statistical properties. Earth Surface Proc. Landforms 29, 687–711.

Malamud, B.D., Turcotte, D.L., Guzzetti, F., Reichenbach, P., 2004b. Landslides, earthquakes and erosion. Earth. Planet. Sci. Lett. 229, 45–59.

Mallitskiy, N.G., 1929. Information about breach of some temporary mountain lakes in Central Asia and adjacent countries. Trans. Central Asia Brunch Russian Geogr. Soc. 19, 29 (in Russian).

Mason, K., 1929. Indus floods and shyok glaciers. Himalayan J. 1, 10–29.

Matveev, Y.D., 1968. Regulationg of the colossal rock falls and landslides in the Toktogul dam construction region. In: Problems of engineering geology and soil science. Issue 2. Moscow, 245–258 (in Russian).

Mavlianov, G.A., 1968. Landslides of Central Asia, state of their knowledge and research tasks. In: Hydrogeology and engineering geology of the arid zone of the USSR. Proceedings of Central Asian meeting on landslides' study and protection measures. Donish Publishers, Dushanbe, issue 12. (in Russian).

Mavrouli, O., Corominas J., 2015. Comparing kinematically detachable rock masses and rockfall scar volumes. International Symposium on Geohazards and Geomechanics (ISGG2015). IOP Conf. Series: Earth and Environmental Science 26 DOI:10.1088/1755-1315/26/1/012020.

McCalpin, J.P., 2009. second ed. Paleoseismology, vol. 95. Elsevier, p. 613, International Geophysics Series.

McColl, S.T., 2012. Paraglacial rock-slope stability. Geomorphology 153–154, 1–16.

McColl, S., Cody, E., Draebing, D., 2017. Development of an ice-buttressed slope failure: mueller rockslide, New Zealand, Progressive rock failure. Extended abstracts, 24–26. ETH Zürich. Available from: www.prf2017.ethz.ch.

McConnell, R.G., Brock, R.W., 1904. Report on the Great Landslide at Frank, Alberta, Annual Report of the Department of Interior for the year 1902–1903, Part VIII, 17 pp.

McDougall, S., Hungr, O., 2004. A model for the analysis of rapid landslide motion across three-dimensional terrain. Can. Geotech. J. 4, 1084–1097.

McSaveney, M.J., 1975. Sherman Glacier Rock Avalanche of 1964: The Emplacement and Subsequent Effects on the Glacier Beneath It. PhD thesis. Ohio State University, 426 pp.

McSaveney, M.J., 1978. Sherman glacier rock avalanche. In: Voight, B. (Ed.), Rockslides and Avalanches 1, Natural Phenomena. Elsevier, Amsterdam, pp. 197–258.

McSaveney, M.J., 1993. Rock avalanches of 2 May and 16 September 1992, Mount Fletcher, New Zealand. Landslide News, No 7, 2–4.

McSaveney, M.J., 2002. Recent rockfalls and rock avalanches in Mount Cook National Park, New Zealand. In: Evans, S.G., DeGraff, J.V. (Eds.), Catastrophic Landslides: Effects, Occurrence, and Mechanisms, v. XV. Geol. Soc. Am. Rev. Eng. Geol., pp. 35–70.

McSaveney, M.J., Chinn, T.J., Hancox, G.T., 1992. Rock Avalanche of 14 December 1991, New Zealand. Landslide News, No 6, 32–34.

McSaveney, M.J., Davies, T.R.H., 2006. Rapid rock-mass flow with dynamic fragmentation: inferences from the morphology and internal structure of rockslides and rock avalanches. In: Evans, S. G., Scarascia Mugnozza, G., Strom, A., Hermanns, R.L. (Eds.), Landslides From Massive Rock Slope Failure. NATO Science Series: IV: Earth and Environmental Sciences, vol. 49. Springer, Heidelberg, pp. 285–304.

Melosh, H.J., 1979. Acoustic fluidization: a new geologic process? J. Geophys. Res. 84, 7513–7520.

Melosh, H.J., 1983. Acoustic fluidisation. Am. Sci. 71, 158–168.

Melosh, H.J., 1986. The physics of very large landslides. Acta Mech. 64, 89–99.

Melosh, H.J., 1990. Giant rock avalanches. Nature 348, 483–484.

Meunier, P., Hovious, N., Haines, J.A., 2008. Topographic site effects and the location of earthquake induced landslides. Earth Planet Sci. Letters 275, 221–232.

Meyers W.B., Hamilton W., 1964. Deformation accompanying the Hebgen Lake earthquake of August 17, 1959. U.S. Geological Survey Professional Paper, No 435, 55–98.

Ming, Z.Q., Wu, G., Jiang, Z.X., Liu, C.S., Yang, Y.L., 1995. The Catalogue of Chinese Historical Strong Earthquakes (B.C. 23-A.D. 1911). Seismological Publishing House, Beijing, 514 pp. (in Chinese).

Muller, L., 1964. The rock slide in the Vaiont valley. Rock Mech. and Eng. Geol. 6, 148–212.

Murge, M.R., 1965. Rockfall-avalanche and rockslide-avalanche deposits at Sawtooth Ridge, Montana. Geol. Soc. Am. Bull. 76, 1003–1014.

Mushketov, I.V., 1890. The Vernyi earthquake of 28/05 (9/06) 1887. Proceedings of Geological Committee, 10(1), St. Petersburg, 154 pp (in Russian).

Mushketov, I.V. 1891. Materials for study of Russian earthquakes. St. Petersburg, Issue 1, 11–46. (in Russian).

Mushketov, I.V., Orlov, A.P., 1893. Catalogue of Earthquakes of Russian Empire. Proc. Russian Geogr. Soc. 26, 582 pp. (in Russian).

Nadim, F., Kjekstad, O., Peduzzi, P., Herold, C., Jaedicke, C., 2006. Global landslide and avalanche hotspots. Landslides 3, 159–173.

Nemčok, A., 1982. Zosuvy v Slovenskych Karpatov (Landslides in the Slovak Carpathians). Slovak Academy of Sciences, Bratislava.

Nepop, R.K., Agatova, A.R., 2008. Estimating magnitudes of prehistoric earthquakes from landslide data: first experience in south-eastern Altai. Russian Geol. Geophys. 49 (2), 144–151.

Neshikhovskiy, R.A., 1988. Floods on rivers and lakes. Hydrometeoizdat, Leningrad, 184 pp. (in Russian).

Niayazov, R.A., 2015. Landslides triggered by the Pamir-Hindukush earthquakes. Tashkent, ISBN 978-9943-4519-2-6. 223 pp. (in Russian).

Nichol, S., Hungr, O., Evans, S.G., 2002. Large-scale brittle and ductile toppling of rock slopes. Can. Geotech. J. 39, 773–780.

Nicoletti, P.G., Sorriso-Valvo, M., 1991. Geomorphic controls of the shape and mobility of rock avalanches. Geol. Soc. Am. Bull. 103, 1365–1373.

Nikitin, M., Huggel, C., Schwarz, M., Goncharenko, O., Galushkin, I. V., 2006. The analysis of the remotely sensed materials for the reconstruction of the Kolka Glacier collapse. In: Proc. Of the International Conference on High Mountain Hazard Prevention, Vladikavkaz – Moscow, June 23–26, 2004, 160–171.

Nikonov, A.A., 1974. Modern and Holocene seismotectonic dislocation in the Southern Tien Shan seismically active zone (Tajikistan). Izvestia Acad. Sci. USSR, Phys. Earth 12, 71–76 (in Russian).

Nikonov, A.A., 1977. Holocene and Modern Motions of Earth crust. "Nauka" (Science) Publishing House, Moscow, 240 pp. (in Russian).

Nikonov, A.A., 1988. Reconstruction of the main parameters of old large earthquakes in Soviet Central Asia using the paleoseismogeological method. Tectonophysics 147, 297–312.

Nikonov, A.A., 1995. Active faults definition and problem of identification. Geoecol. Eng. Geol. Hydrogeol. Geocryol. (4), 16–27 (in Russian).

Nikonov, A.A., Shebalina, T.Yu, 1979. Lichenometry and earthquake age determination in Central Asia. Nature 280, 675–677.

Nikonov, A.A., Shebalina, T.Yu, 1986. Lichenometric Method of Seismic Dislocations Dating. Institute of Physics of Earth, Moscow, 185 pp. (in Russian).

Nikonov, A.A., Vakov, A.V., Veselov, I.A., 1983. Seismotectonics and Earthquakes of the Zone of the Pamirs and Tien Shan Junction. Nauka Publishing House, Moscow, 240 pp. (in Russian).

Niyazov, R.A., 1974. Landslides in loess. Tashkent, 148 pp. (in Russian).

Niyazov, R.A., 1982. Formation of large landslides in Central Asia. Tashkent, 156 pp. (in Russian).

Niyazov, R.A., 2009. Landslides in Uzbekistan (tendencies of evolution at the turn of the XXI Century). Tashkent, Gidroingeo, 206 pp. (in Russian).

Okeke, A.C., Wang, F., 2016a. Hydromechanical constraints on piping failure of landslide dams: an experimental investigation. Geoenviron. Dis. 3, 4. Available from: https://doi.org/10.1186/s40677-016-0038-9.

Okeke, A.C., Wang, F., 2016b. Critical hydraulic gradients for seepage-induced failure of landslide dams. Geoenviron. Dis. 3, 9. Available from: https://doi.org/10.1186/s40677-016-0043-z.

Omori, F., 1907. Note on the Kashgar (Turkestan) Earthquake of Aug. 22, 1902. Bull. Imp. Earthquake Investigat. Committee 1 (4), 161–166.

Omuraliev, M., Omuralieva, A., 2004. Late Cenozoic Tectonics of the Tien Shan. Bishkek, 166 pp. (in Russian).

Orlov, L.N., 1980. On the Kinematics and Dynamics of the Overthrusts at the Boundary of the Northern and Middle Tien Shan. Seismotectonics and Seismicity of Tien Shan. Ilim Publishing House, Frunze, pp. 50–59 (in Russian).

Ostroumov, A.V., 1986. Mechanism of friction in rockfall flows. Problems of Soil Thermomechanics. Moscow State University, pp. 37–48 (in Russian).

Owen, L.A., England, J., 1998. Observations on rock glaciers in the Himalayas and Karakoram mountains of northern Pakistan and India. Geomorphology 26, 199–213.

Owen, L.A., Kamp, U., Khattak, G.A., Harp, E.L., Keefer, D.K., Bauer, M.A., 2008. Landslides triggered by the 8 October 2005 Kashmir earthquake. Geomorphology 94, 1–9.

Pacheco, J.F., Sykes, L.R., 1992. Seismic moment catalogue of large shallow earthquakes, 1900 to 1989. Bull. Seism. Soc. Am. 82, 1306–1344.

Paguican, E.M.R., van Wyk de Vries, B., Lagmay, A.M.F., 2012. Hummocks: how they form and how they evolve in rockslide-debris avalanches. Landslides 11, 67–80.

Panasenko, G.D., 1952. Collapse of the Chokurak Mountain. Trans. Acad. Sci. USSR (No 3), 35 (in Russian).

Pederesen, H., Le Brun, B., Hatzfeld, D., Campillo, M., Bard, P.-Y., 1994. Ground-motion amplitude across ridges. Bull. Seismol. Soc. Am. 84, 1786–1800.

Pedrazzini, A., Humair, F., Jaboyedoff, M., Tonini, M., 2016. Characterisation and spatial distribution of gravitational slope deformation in the Upper Rhone catchment (Western Swiss Alps). Landslides 13, 259–277. Available from: https://doi.org/10.1007/s10346-015-0562-9.

Pelletier, J.D., Malamud, B.D., Blodgett, T., Turcotte, D.L., 1997. Scale-invariance of soil moisture variability and its implications for the frequency–size distribution of landslides. Eng. Geol. 48 (3–4), 255–268.

Penna, I., Hermanns, R.L., Jaboyedoff, M., Fauqué, L., 2016. Large scale rockslides in the Argentinean Andes. Distribution and forcing factors. In: Aversa, et al., (Eds.), Landslides and Engineered Slopes. Experience, Theory and Practice. Associazione Geotecnica Italiana, Rome, Italy, ISBN 978-1-138-02988-0.

Pettley, D., The Landslide Blog. http://blogs.agu.org/landslideblog/.

Philip, H., Meghraoui, M., 1983. Structural analysis and interpretation of the surface deformations of the El Asnam earthquake of October 10, 1980. Tectonics 2, 17–49.

Philip, H., Ritz, J.F., 1999. Gigantic paleolandslide associated with active faulting along the Bogd fault (Gobi-Altay, Mongolia). Geology 27, 211–214.

Picarelli, L., Russo, C., 2004. Mechanics of slow active landslides and interaction with manmade works. In: Lacerda., W.A., Ehrlich, M., Fontoura, S.A.B., Sayao, A.S.F. (Eds.), Landslides. Evaluation &

Stabilization, 9th International Symposium on Landslides. A.A. Balkema, Rotterdam, pp. 1141–1176. , Rio de Janeiro (28 June -2 July, 2004).

Plafker, G., Eriksen, G.E., 1978. Nevados Huascaran avalanches, Peru. In: Voight, B. (Ed.), Rockslides and Avalanches, vol. 1. Elsevier, Amsterdam, pp. 277–314.

Plaza, G., Zevallos, O., Cadier, É., 2011. La Josefina landslide dam and its catastrophic breaching in the Andean region of Ecuador. In: Evans, S.G., Hermanns, R., Strom, A.L., Scarascia-Mugnozza, G. (Eds.), Natural and Artificial Rockslide Dams. Lecture Notes in Earth Sciences, 133. Springer, Heidelberg, pp. 389–406.

Pánek, T., Klimeš, J., 2016. Temporal behavior of deepseated gravitational slope deformations: a review. Earth Sci. Rev. 156, 14–38. Available from: https://doi.org/10.1016/j.earscirev.2016.02.007.

Pollet, N., Cojean, R., Couture, R., Schneider, J.-L., Strom, A.L., Voirin, C., et al., 2005. A slab-on-slab model for the Flims rockslide (Swiss Alps). Can. Geotech. J. 42, 587–600.

Pollet, N., Schneider, J.-L.M., 2004. Dynamic disintegration processes accompanying transport of the Holocene Flims Sturzstrom (Swiss Alps). Earth Planet Sci. Lett. 221, 433–448.

Poschinger, A. von., 1994. Some special aspects of the "impact" of a landslide on the valley floor. Landslide News (No 8), 26–28.

Poschinger, A. von., 2011. The Flims rockslide dam. In: Evans, S.G., Hermanns, R., Strom, A.L., Scarascia-Mugnozza, G. (Eds.), Natural and Artificial Rockslide Dams. Lecture Notes in Earth Sciences, 133. Springer, Heidelberg, pp. 407–421.

Poschinger, A., von, Wassmer, P., Maisch, M., 2006. The Flims rockslide: history of interpretation and new insights. In: Evans, S.G., Scarascia Mugnozza, G., Strom, A., Hermanns, R.L. (Eds.), Landslides from Massive Rock Slope Failure. NATO Science Series: IV: Earth and Environmental Sciences, vol. 49. Springer, Heidelberg, pp. 329–356.

Potapov, A.V., 1991. Numerical Modelling of the Nonstationary Geomechanical Processes with Low Internal Friction, Ph.D. Thesis. Institute of Geospheres Dynamics, RAS, 118 pp. (in Russian).

Preobrajensky, I.A., 1920. The Usoi Landslide, Geol. Comm., Papers on Applied Geol., 14, 21 p. (in Russian).

Presnuhin, V.I., 1971. Regional pattern of distribution and evolution of the landslide phenomena in Tajikistan. PhD abstract. Tashkent, 25 p. (in Russian).

Prochaska, A.B., Santi, P.M., Higgins, J.D., Cannon, S.H., 2008. A study of methods to estimate debris flow velocity. Landslides . Available from: https://doi.org/10.1007/s10346-008-0137-0.

Procter, J.N., Cronin, S.J., Zernack, A.V., 2009. Landscape and sedimentary response to catastrophic debris avalanches, western Taranaki, New Zealand. Sediment. Geol. 220 (3-4), 271–287.

Puchkov, S.V., 1965. Role of the local topography for seismic microzoning. Proceedings of Institute of Physics of Earth, Academy of Sciences of the USSR, No 36, 105–114 (in Russian).

Puchkov, S.V., 1974. Regulations of ground vibrations during earthquakes. Moscow, 119 pp. (in Russian).

Pushkarenko, V.P., Nikitin, A.M., 1984. Experience in the Regional Investigation of the State of Mountain Lake Dams in Central Asia and the Character of Breach Mudflow Formation. Landslides and Mudflows (in two volumes). UNESCO/UNEP, Moscow, pp. 17–32 (in Russian).

Pyzhyanov, I.V., Presnuhin, V.I., 1987. Landslide phenomena in the Muksu valley (Northern Pamir). Proc. Acad. Sci. Tajik SSR, Sect. Phys. Math. Chem. Geol. Sci. (3), 64–69 (in Russian).

Rahman, A., Hashimoto, T., Konagai, K., 2014. An attempt for velocity estimation of Nebukawa debris flow triggered by the Great Kanto earthquake, 1923. In: Proc. 34th Earthquake Engineering Symposium, Nagaoka, Japan. DOI: 10.2208/jscejseee.71.I_387.

Rastvorova, V.A., Enman, S.V., 1997. Garm Geodynamic Field Test Site. Joint Institute of Physics of the Earth of Russian Academy of Sciences, Moscow, 134 pp. (in Russian).

Reizvikh, V.N., Krukovskiy, G.L., Nikitin, A.M., 1971. Causes and conditions of the Yashinkul Lake outburst. Collect. Papers Tashkent Hydro-Meteorol. Obs. 4, 49–95 (in Russian).

Revnivtsev, V.I., Gaponov, G.V., Zarogadskiy, L.P., 1988. Selective destruction of minerals. Nedra, Moscow, 285 pp. ISBN 5-247-00088-9 (in Russian).

Reznichenko, N.V., Andrews, G.R., Geater, R.E., Strom, A., 2017. Origin of large hummock deposits "chukuryi" in Alai Valley, Northern Pamir: geomorphological and sedimentological investigation. Geomorphology 285, 347–362.

Reznichenko, N., Davies, T., 2015. The Gigantic Komansu rock avalanche deposit in the glaciated Alai Valley, Northern Pamir of Central Asia. In: Lollino, G., et al., (Eds.), Engineering Geology for Society and Territory, Volume 2. Springer International Publishing Switzerland, pp. 895–898. Available from: https://doi.org/10.1007/978-3-319-09057-3_154.

Reznichenko, N.V., Davies, T.R.H., Shulmeister, J., Larsen, S.H., 2012a. A new technique for identifying rock-avalanche-sourced sediment in moraines and some palaeoclimatic implications. Geology 40 (4), 319–322.

Reznichenko, N.V., Davies, T.R.H., Shulmeister, J., Winkler, S., 2012b. Influence of rock avalanches upon the formation of moraines and their subsequent palaeoclimatic interpretation: a critical appraisal. Zeitschrift für Geomorphologie (Supplementary Issue), Available from: https://doi.org/10.1127/0372-8854/2012/S-00079 (fast track article published online).

Reznichenko, N.V., Davies, T.R.H., Winkler, S., 2015. Revised Palaeoclimatic Significance of Mueller Glacier moraines, Southern Alps, New Zealand. Earth Surf. Process. Landforms. Published online in Wiley Online Library (wileyonlinelibrary.com) https://doi.org/10.1002/esp.3848.

Rezvoi, D.P., Rezvoi, P.D., 1969. Lake that vanished .... Priroda (Nature) (No 7), 81–83 (in Russian).

Rezvoi, D.P., Alexeenko, A.V., Bolgar, B.D., Ermolenko, Yu.A., Marushkin, I.A., Rezvoi, P.D., et al., 1971. Rockfall-Dammed Lakes of the Alai Range—Important Factor of the High Mountain Relief Formation. Reports and News of the L'vov Branch of the Ukraine Geographical Society (1968–1969). Lvov University Publishing House, pp. 71–73 (in Russian).

Risley, J.C., Walder, J.S., Denlinger, R.P., 2006a. Usoi Dam wave overtopping and flood routing in the Bartang and Panj Rivers, Tajikistan: USGS Water-Resources Investigations Report 03-4004, 37 p. http://pubs.usgs.gov/wri/wri03-4004/pdf/wri034004.pdf.

Risley, J.C., Walder, J.S., Denlinger, R.P., 2006b. Usoi Dam wave overtopping and flood routing in the Bartang and Panj Rivers, Tajikistan. Natural Hazards 38 (3), 375–390.

Rivers and Lakes of the World, 2012. Enciclopedia. "Enciclopedia Publishers" Ltd., Moscow, 928 pp. (in Russian).

Roberts, N.J., Evans, S.G., 2013. The gigantic Seymareh (Saidmarreh) rock avalanche, Zagros Fold-Thrust Belt, Iran. J. Geol. Soc. Available from: https://doi.org/10.1144/jgs2012-090.

Robinson, T.R., Davies, T.R.H., Reznichenko, N., De Pascale, G.P., 2015. The extremely long runout Komansu rock avalanche in the Trans Alay range, Pamir Mountains. Southern Kyrgyzstan. Landslides 12, 523–535.

Robinson, A.C., Yin, A., Manning, C.E., Harrison, M.T., Zhang, S.-H., Wang, X.-F., 2004. Tectonic evolution of the northeastern Pamir: constraints from the northern portion of the Cenozoic Kongur Shan extensional system, western China. Geol. Soc. Am. Bull. 116, 953–973.

Robinson, A.C., Yin, A., Manning, C.E., Harrison, T.M., Zhang, S.H., Wang, X.F., 2007. Cenozoic evolution of the eastern Pamir:

implications for strain-accommodation mechanisms at the western end of the Himalayan–Tibetan orogen. Geol. Soc. Am. Bull. 119, 882–896.

Rodionov, N.E., 1931. Hydrology of the Usoi Blockage, LXIII. Transactions of the State Geographical Society, Moscow (in Russian). Available from: http://www.nikzdaru.com/rodionov/text_rodionov.htm.

Rodionov, V.N., Adushkin, V.V., Kostyuchenko, V.N., Nikolaevskii, V.N., Romashov, A.N., Tsvetkov, V.M., 1971. The Mechanical Effect of Underground Explosion. Nedra Publishers, Moscow, 221 pp. (in Russian).

Rodriguez, C., Bommer, J., Chandler, R., 1999. Earthquake-induced landslides: 1980-1997. Soil Dynam. Earthquake Eng. 18, 325–346.

Roessner, S., Wetzel, H.-U., Kaufmann, H., Samagoev, A., 2005. Potential of satellite remote sensing and GIS for landslide hazard assessment in Southern Kyrgyzstan (Central Asia). Nat. Hazards 35, 395–416.

Roverato, M., Cronin, S., Procter, J., Capra, L., 2015. Textural features as indicators of debris avalanche transport and emplacement, Taranaki volcano. GSA Bull. 127 (1–2), 3–18.

Roy, R., Ghosh, D., Bhattacharya, G., 2016. Influence of strong motion characteristics on permanent displacement on slopes. Landslides 13, 279–292. Available from: https://doi.org/10.1007/s10346-015-0568-3.

Ruzhentsev, S.V., 1968. Tectonic Evolution of Eastern Pamir and Role of Horizontal Movements in the Formation of its Alpine Structure. Nauka Publishing House, Voscow, 204 pp. (in Russian).

Sadybakasov, I., 1972. Neotectonics of the Central part of the Tien Shan. Frunze, Ilim 116 pp. (in Russian).

Sadybakasov, I., 1990. Neotectonics of High Asia. Nauka, Moscow, 180 pp. (in Russian).

Sanders, D., Ostermann, M., Brandner, R., Prager, C., 2010. Metheoric lithification of Catastrophic rockslide deposits: diagenesis and significance. Sediment. Geol. 223, 150–161.

Sanhueza-Pino, K., Korup, O., Hetzel, R., Munack, H., Weidinger, J. T., Dunning, S., et al., 2011. Glacial advances constrained by 10Be exposure dating of bedrock landslides, Kyrgyz Tien Shan. Quat. Res. 76, 295–304.

Sapov, O.M., 1970. The Pamir Landslides and rockfalls. Proc. Of the scientific-technical meeting on the problems of debris flows, rockfalls and landslides study and presiction. Dushanbe. (in Russian).

Sartori, M., Baillifard, F., Jaboyedoff, M., Rouiller, J.-D., 2003. Kinematics of the 1991 Randa rockslides (Valais, Switzerland). Nat. Hazards Earth Syst. Sci. 3, 423–433.

Sassa, K., Fukuoka, H., Lee, J.-H., Shoaei, Z., Zhang, D., Xie, Z., et al., 1994. Prediction of landslide motion based on the measurement of geotechnical parameters. In: Development of a new Cyclic Loading Ring Shear Apparatus to study earthquake-induced-landslides. Report for Grant-in-Aid for Developmental Scientific Research by the Ministry of Education, Science and Culture, Japan (Project No 03556021), DPRI, Kyoto, 72–106.

Sassa, K., He, B., Khang, D., Nagai, O., Takara, K., 2014. Progress in Landside Dynamics. Proceedings of World Landslide Forum 3, Beijing, 1, 37–70.

Sassa, K., Nagai, O., Solidum, R., Yamazaki, Y., Ohta, H., 2010. An integrated model simulating the initiation and motion of earthquake and rain induced rapid landslides and its application to the 2006 Leyte landslide. Landslides 7, 219–236.

Scarascia Mugnozza, G., Bianchi Fasani, G., Esposito, C., Martino, S., Saroli, M., Di Luzio, E., et al. 2006. Rock Avalanche and Mountain Slope Deformation in a Convex Dip-Slope: The Case of the Maiella Massif, Central Italy. In: Evans, S.G., Scarascia Mugnozza, G., Strom, A., Hermanns, R.L. (Eds.), Landslides from Massive Rock Slope Failure. NATO Science Series: IV: Earth and Environmental Sciences 49, 357–376.

Schlögel, R., 2009. Detection of Recent Landslides in Maily-Say Valley, Kyrgyz Tien Shan, Based on Field Observations and Remote Sensing Data. (Master's Thesis in Geological Sciences). University of Liege, Belgium, 133 pp.

Schlögel, R., Torgoev, I., de Marneffe, C., Havenith, H.-B., 2011. Evidence of a changing distribution of landslides in the Kyrgyz Tien Shan, Central Asia. Earth Surf. Proc. Landf. 36, 1658–1669.

Schneider, J.F., Gruber, F.E., Mergili, M., 2013. Recent cases and geomorphic evidence of landslide-dammed lakes and related hazards in the mountains of Central Asia. In: Margottini, C., Canuti, P., Sassa, K. (Eds.), Landslide Science and Practice, Volume 6 - Risk Assessment, Management and Mitigation. Springer, Heidelberg, pp. 57–64.

Schneider, D., Huggel, C., Haeberli, W., Kaitna, R., 2011. Unraveling driving factors for large rock–ice avalanche mobility. Earth Surface Proc. Landf. 36, 1948–1966.

Schneider, J.-L., Wassmer, P., Ledésert, B., 1999. La fabrique interne des dépôts du sturzstrom de Flims (Alpes suises): caractéristiques et implications sur les méchanismes de transport. C.P. Acad. Sci. Paris. Earth & Plan. Sci 328, 607–613.

Schröder, H., 1992. Active Blockgletscher im zentralen Teil des nördlichen Tienschan. Petermanns. Geogr. Mitt. 136, 109–119.

Schröder, H., Kokarev, A., Harrison, S., 2005. Rock glaciers in the northern Tien Shan, Kazakhstan: new data on movement rates and distribution. *Glacial Geology and Geomorphology* rp01/2005, pp. 10.

Schuster, R.L., Costa, J.E., 1986. A perspective on landslide dams. In: Schuster, R.L. (Ed.), Landslide Dams: Processes, Risk and Mitigation. American Society of Civil Engineers, New York, NY, pp. 1–20. , Geotechnical Special Publication No. 3.

Schuster, R.L., Evans, S.G., 2011. Engineering measures for the hazard reduction of landslide dams. In: Evans, S.G., Hermanns, R., Strom, A.L., Scarascia-Mugnozza, G. (Eds.), Natural and Artificial Rockslide Dams. Lecture Notes in Earth Sciences, vol. 133. Springer, Heidelberg, pp. 77–100.

Selivanov, P.I., Andreev, V.I., 1959. About water level behaviour and possible evolution of the Sarez Lake. Meteorol. Hydrol. (No 3), (in Russian). Available from: http://www.nikzdaru.com/selivanov/text_selivanov.htm.

Semenza, E., 2010. The story of Vaiont told by the geologist who discovered the landslide. K-flash, ISBN 978-88-89288-02-3.

Semionov, A.A., 1958. List of eartquakes in Central Asia and adjacent countries from ancient times till 1830. Proc. Inst. Seismol. Acad. Sci. Tajik SSR 94, 37–52 (in Russian).

Semionov, P.G., Semionova, V.A., 1958. Catalogue of earthquakes felt on the territory of Tajikistan during the periods (of time) 1865-1940 and 1941-1952. Proc. Inst. Seismology, Acad. Sci. Tajik SSR 86 (3), 143 pp. (in Russian).

Seong, Y.B., Owen, L.A., Yi, C., Finkel, R.C., Schoenbohm, L., 2009. Geomorphology of anomalously high glaciated mountains at the northwestern end of Tibet: Muztag Ata and Kongur Shan. Geomorphology 103, 227–250.

Sepúlveda, S.A., Pastén, C., Moya, S., García, M., Lara, M., Montalva, G., et al. 2016. Landslides and engineered slopes. Exp. Theory Pract. 3, 1823–1828.

Shaller, P.J., 1991. Analysis and Implications of Large Martian and Terrestrial Landslides, Ph.D. Thesis. California Institute of Technology, 569 pp.

Shang, Y., Yang, Z., Li, L., Liu, D., Liao, Q., Wang, Y., 2003. A super-large landslide in Tibet in 2000: background, occurrence, disaster, and origin. Geomorphology 54, 225–243.

Sheidegger, A.E., 1973. On the prediction of the reach and velocity of catastrophic landslides. Rock Mech. 5, 231–236.

Sheko, A.I., 1968. Assessment of the Usoi blockage stability and possibility of the Sarez Lake breach. Bulletin of Moscow Society of Natural Scientists. Sect. Geol. 43 (4), 151–159 (in Russian).

Sheko, A.I., Lekhatinov, A.M., 1970. Present date state of the Usoy blockadge and tasks of future investigations, in Materials of Scientific-Technical Meeting on the Methodical Problems of Mud Flows, Rockfalls and Landslides Investigation and Forecast, Dushanbe, Donish, 219–223. (in Russian).

Shemiakin, E.I., 1993. On the mobility of large landslides. Proc. Russian Acad. Sci. 331, 742–744 (in Russian).

Sherard, J.L., Woodward, R.J., Gizienski, S.F., Clevenger, W.A., 1963. Earth and Earth-Rock Dams. John Wiley and Sons, New York, 722 pp.

Shoaei, Z., Ghayoumian, J., 2000. Seimareh landslide, Western Iran, one of the World's largest complex landslides. Landslide News 13, 23–27.

Shpilko, G.A., 1914. The 1911 earthquake in Pamirs and its effects. Proc. Russian Geogr. Soc. 3, 69–94 (in Russian).

Shreve, R.L., 1968. The Blackhawk landslide, Geological Society of America Special Paper. N 108.

Shugar, D.H., Clague, J.J., 2011. The sedimentology and geomorphology of rock avalanche deposits on glaciers. Sedimetology. Available from: https://doi.org/10.1111/j.1365-3091.2011.01238.x.

Shulmeister, J., Davies, T.R., Evans, D.J.A., Hyatt, O.M., Tovar, D.S., 2009. Catastrophic landslides, glacier behaviour and moraine formation – a view from an active plate margin. Quat. Sci. Rev. 28, 1085–1096.

Shultz, S.S., 1948. Analysis of newest tectonics and relief of the Tien Shan. Proc. All-Union Geogr. Soc. New Series 3, 222 pp.

Simonett, D.S., 1967. Landslide distribution and earthquakes in the Bewani and Torricelli Mountains, New Guinea. In: Jennings, J.N., Mabutt, J.A. (Eds.), Landform studies from Australia and New Guinea. Cambridge University Press, Cambridge, pp. 64–84.

Sippl, C., Schurr, B., Yuan, X., Mechie, J., Schneider, F., Gadoev, M., et al., 2013. Geometry of the Pamir-Hindu Kush intermediate-depth earthquake zone from local seismic data. J. Geophys. Res. Solid Earth 118, 1438–1457. Available from: https://doi.org/10.1002/jgrb.50128.

Soeters, R., Van Westen, C.J., 1996. Slope stability recognition analysis and zonation. In: Turner, A.K., Schuster, R.L. (Eds.), Landslides: Investigation and Mitigation. Transportation Research Board Special Report., 247. National Academy Press, Washington, DC, pp. 129–177.

Solomina, O.N., 1999. Mountain glaciation of Northern Eurasia in the Holocene. Moscow, Scientific World Publishers, 272 pp. (in Russian).

Solomina, O.N., Barry, R.G., Bodnya, M., 2004. The retreat Of Tien Shan glaciers (Kyrgyzstan) since the Little Ice Age estimated from aerial photographs, lichenometric and historical data. Geografiska Annaler. Series A 86 (2), 205–216.

Solonenko, V.P., 1970. Scars on the Earth's face. Priroda (Nature) (No 9), 17–25 (in Russian).

Solonenko, V.P., 1977. Landslides and collapses in seismic zones and their prediction. Bull. Int. Assoc. Eng. Geol. 15, 4–8.

Sørensen, S.-A., Bauer, B., 2003. On the dynamics of the Köfels Sturzstrom. Geomorphology 54, 11–19.

Stark, C.P., Hovius, N., 2001. The characterization of landslide size distributions. Geophys. Res. Lett. 28, 1091–1094.

Stark, C.P., Wolovick, M., Ekstrom, G., 2012. Glacier Surge Triggered by Massive Rock Avalanche: Teleseismic and Satellite Image Study of Long-Runout Landslide onto RGO Glacier, Pamirs, Abstract C32A-07 Presented at 2012 Fall Meeting. AGU, San Francisco, Calif.

State Unitary Enterprise Scientific and Technical Center for Industrial Safety of the Gosgortekhnadzor of Russia, 2002. Blasting safety. Series 13, Issue 1, Moscow (in Russian).

State Water Inventory, 1987a. Perennial Data on the Mainland Superficial Water Regime and Resources. IV, Uzbek SSR. Gidrometeoizdat Publishing House, Leningrad (in Russian).

State Water Inventory, 1987b. Perennial Data on the Mainland Superficial Water Regime and Resources. V, Issue 3, Kazakh SSR. Gidrometeoizdat Publishing House, Leningrad (in Russian).

State Water Inventory, 1987c. Perennial Data on the Mainland Superficial Water Regime and Resources. XI, Kyrgyz SSR. Gidrometeoizdat Publishing House, Leningrad (in Russian).

State Water Inventory, 1988. Perennial Data on the Mainland Superficial Water Regime and Resources. V, Issue 4, Kazakh SSR. Gidrometeoizdat Publishing House, Leningrad (in Russian).

State Water Inventory, 1990. Part 1. Surface Water. Series 2. Annual Data on the Mainland Surface Water Regime and Resources. XII, Tajik SSR. Gidrometeoizdat Publishing House, Leningrad (in Russian).

Stepanov, B.S., Stepanova, T.S., 1991. Mechanics of debris flows. Moscow, Gidrometeoizdat, 379 pp. (in Russian).

Stepanov, B.S., Yafyazova, R.K., 2014. Debris flows of the Southeastern Kazakhstan, Debris Flow Processes and Counter Measures From Debris Flows Hazards, vol. 3. Almaty, 433 pp. (in Russian). ISBN 978-601-7150-72-3.

Strecker, M.R., Hilley, G.E., Arrowsmith, J.R., Coutand, I., 2003. Differential structural and geomorphic mountain-front evolution in an active continental collision zone: the northwest Pamir, southern Kyrgyzstan. Geol. Soc. Am. Bull. 115 (2), 166–181.

Strom, A.L., 1982. About New Zone of Paleoseismic Dislocations in the Northern part of the Central Tien Shan. In: Main Problems of Seismotectonics. Moscow, 4–13, deposited in VINITI, No 3290–83 (in Russian).

Strom, A.L., 1994a. Mechanism of stratification and abnormal crushing of rockslide deposits. Proc. 7th International IAEG Congress 3. Balkema, Rotterdam, pp. 1287–1295.

Strom, A.L., 1994b. Formation of structure of large rockslide deposits, Geoecology. Eng. Geol. Hydrogeol. Geocryol. (No 5), 64–77 (in Russian).

Strom, A.L., 1996. Some morphological types of long-runout rockslides: effect of the relief on their mechanism and on the rockslide deposits distribution. In: Senneset, K. (Ed.), Landslides. Proc. of the Seventh International Symposium on Landslides. Balkema, Trondheim, Norway, Rotterdam, pp. 1977–1982. In: Senneset, K. (Ed.), Landslides. Proc. of the Seventh International Symposium on Landslides. Balkema, Trondheim, Norway, Rotterdam, pp. 1977–1982.

Strom, A.L., 1998. Giant ancient rockslides and rock avalanches in the Tien Shan Mountains, Kyrgyzstan. Landslide News 11, 20–23.

Strom, A.L., 2000. Caldera-like collapses at the watersheds in the Central Tien Shan. Landslides in Research, Theory and Practice. Balkema.

Strom, A.L., 2004. Rock avalanches of the Ardon River valley at the southern foot of the Rocky Range, Northern Caucasus, North Osetia. Landslides 1, 237–241.

Strom, A.L., 2006. Morphology and internal structure of rockslides and rock avalanches: grounds and constraints for their modelling. In: Evans, S.G., Scarascia Mugnozza, G., Strom, A., Hermanns, R.L. (Eds.), Landslides from Massive Rock Slope Failure. NATO Science Series: IV: Earth and Environmental Sciences, 49, 305–328.

Strom, A.L., 2010a. Evidence of momentum transfer during large-scale rockslides' motion. In: Williams, A.L., Pinches, G.M., Chin, C.Y., McMorran, T.G., Massei, C.I. (Eds.), Geologically Active.Proc. of the 11th IAEG Congress, Auckland, New Zealand, 5–10 September 2010, Tailor & Frensis Group, London, 73–86.

Strom, A.L., 2010b. Landslide dams in Central Asia region. Landslides. J. Japan Landslide Soc. 47 (6), 309–324.

Strom, A.L., 2012. Effects of rockslides and rock avalanches on hydropower schemes in the Naryn River valley. In: Eberhardt, et al., (Eds.), Landslides and Engineered Slopes: Protecting Society Through Improved Understanding. Proc. of the 11th ISL and 2nd NASL. Taylor & Francis Group, pp. 1867–1872.

Strom, A., 2013a. Geological prerequisites for Landslide Dams' Disaster Assessment and Mitigation in Central Asia. In: Wang, F., Miyajima, M., Li, T., Fathani, T.F. (Eds.), Progress of Geo-Disaster Mitigation Technology in Asia. Springer-Verlag, Berlin, Heidelberg, pp. 17–53 (ISBN 978-3-642-29106-7).

Strom, A., 2013b. Use of indirect evidence for the prehistoric earthquake-induced landslide identifi-cation. In: Ugai, K., Yagi, H., Wakai, A. (Eds.), Earthquake-Induced Landslides. Springer, Heidelberg, New York, Dordrecht, London, pp. 21–30. (ISBN 978-3-642-32237-2).

Strom, A.L., 2013c. Constraints and promises of earthquake-triggered landslides discrimination. In: Grützner, C., Rudersdorf, A., Pérez-López, R., Reicherter, K. (Eds.), Proc. 4th International INQUA Meeting on Paleoseismology, Active Tectonics and Archeoseismology (PATA), 9–14 October 2013, Aachen, Germany, 267–270.

Strom, A.L., 2013d. Rockslides and rock avalanches in the Kokomeren River valley (Kyrgyz Tien Shan). In: Arbanas, S.M., Arbanas, Ž. (Eds.), Landslide and Flood Hazard Assessment. Proceedings of the 1st Regional Symposium on Landslides in the Adriatic-Balkan Region. Zagreb, pp. 245–250.

Strom, A., 2014a. Catastrophic Slope Processes in Glaciated Zones of Mountainous Regions. In: Shan, W., et al., (Eds.), Landslides in Cold Regions in the Context of Climate Change, Environmental Science and Engineering. Springer International Publishing, Switzerland, pp. 3–10. Available from: https://doi.org: 10.1007/978-3-319-00867-7_1.

Strom, A., 2014b. Sarez Lake problem: ensuring long-term safety. In: Sassa, K., Canuti, P., Yin, Y. (Eds.), Proceedings of World Landslide Forum 3, 2–6 June 2014, Beijing, v. 3, 633–639.

Strom, A., 2015a. Natural river damming: climate-driven or seismically induced phenomena: basics for landslide and seismic hazard assessment. In: Lollino, G., Giordan, D., Crosta, G.B., Corominas, J., Azzam, R., Wasowski, J., et al., Engineering Geology for Society and Territory, Volume 2. Springer International Publishing Switzerland, pp. 33–41. Available from: https://doi.org/10.1007/978-3-319-09057-3_3.

Strom, A.L., 2015b. Possible causes of accelerating landslide motion in confined environment. In: Lollino, G., Giordan, D., Crosta, G. B., Corominas, J., Azzam, R., Wasowski, J., et al., Engineering Geology for Society and Territory, Volume 2. Springer International Publishing, Switzerland, pp. 883–885. Available from: https://doi.org/10.1007/978-3-319-09057-3_151.

Strom, A.L., 2015c. Clustering of large bedrock landslides and recurrent slope failure: implications for landslide and seismic hazard assessment. Proc. of the 4th International Symposium on Mega Earthquake Induced Geo-disasters and Long Term Effects. Chengdu University of Technology, China, May 9-13, 2015.

Strom, A.L., 2017. Active faults at structure's foundations: definition and its influence on hazard assessment. Proc. 16th World Conference on Earthquake, 16WCEE, Santiago Chile, January 9th to 13th, 2017.

Strom, A.L., Abdrakhmatov, K.E., 2004. Clustering of large rockslides: the phenomenon and its possible causes. In: Lacerda, W. A., Ehrlich, M., Fontoura, A.B., Sayao, A. (Eds.), Landslides: Evaluation and Stabilization. Taylor and Francis Group, London, pp. 317–320.

Strom, A.L., Abdrakhmatov, K.E., 2009. International summer school on rockslides and related phenomena in the Kokomeren River Valley, Tien Shan, Kyrgyzstan. In: Sassa, K., Canuti, P. (Eds.), Landslides. Disaster Risk Reduction. Springer, Berlin-Heidelberg, pp. 223–227.

Strom, A.L., Abdrakhmatov, K.E., 2017. Large-scale rockslide inventories - from Kokomeren River basin to the entire Central Asia region - IPL106-2 and WCoE. Proc. 4WLF.

Strom, A.L., Groshev, M.E., 2009. Mysteries of rock massifs destruction. In: Abbie, M., Bedford, J.S. (Eds.), Rock Mechanics: New Research. Nova Science Publishers, New York, pp. 211–231.

Strom, A.L., Korup, O., 2006. Extremely large rockslides and rock avalanches in the Tien Shan, Kyrgyzstan. Landslides 3, 125–136.

Strom, A.L., Nikonov, A.A., 1997. Relations between the seismogenic fault parameters and earthquake magnitude. Izvestia Phys. Solid Earth 33 (No 12), 1011–1022.

Strom, A.L., Nikonov, A.A., 2000. Displacement distribution along seismic surface ruptures and application of slip non-uniformity for paleoseismogeological studies. Volcanol. Seismol. 21, 705–722.

Strom, A., Pernik, L., 2013. Modeling of debris crushing during rock avalanche motion. Geophys. Res. Abst. 15, EGU2013-1373.

Strom, A.L., Stepanchikova, P., 2008. Seismic triggering of large prehistoric rockslides: Pro and Con case studies. In: Proceedings of the International Conference on Management of Landslide Hazard in the Asia-Pacific Region (Satellite symposium of the First World Landslide Forum, Tokyo), Sendai, 11th – 12th November 2008, pp. 202–211.

Strom, A., Zhirkevich, A., 2013. "Remote" landslide-related hazards and their consideration for the hydraulic schemes design. In: Genevois, R., Prestinnizi, A. (Eds.), Proc. Int. conference on Vajont – 1963–2013. Italian Journal of Engineering Geology and Environment, Book series 6. 295–303.

Suvorov, A.I., 1963. Main faults of Kazakhstan and Central Asia. In Faults and Horizontal Movements of the Earth's Crust. Trans. Geol. Inst. 80, 173–237 (in Russian).

Swanson, F.J., Oyagi, N., Tominaga, M., 1986. Landslide dams in Japan. Landslide Dams: Processes, Risk, and Mitigation. ASCE, pp. 131–145.

Tacconi Stefanelli, C., Segoni, S., Casagli, N., Catani, F., 2016. Geomorphic indexing of landslide dams evolution. Eng. Geol. 208, 1–10.

Tadjibekov, M., 1991. Origin of the Shing Lakes in Central Tajikistan. Proc. Acad. Sci. Tajik SSR, Sect. Phisics-Math. Geol. Chem. 119 (2), 48–53 (in Rissian).

Tadjibekov, M., Tokarev, V.V., 1990. New data on the origin of the Iskanderkul blockage (Seravshan-Gissar region). Proc. Acad. Sci. Tajik SSR, Sect. Phisics-Math. Geol. Chem. 118 (4), 47–51 (in Rissian).

Tang, C., Zhu, J., Ding, J., Cui, X., Chen, L., Zhang, J., 2011. Catastrophic debris flows triggered by a 14 August 2010 rainfall at the epicenter of the Wenchuan earthquake. Landslides 8, 485–497. Available from: https://doi.org/10.1007/s10346-011-0269-5.

ten Brink, U.S., Geist, E.L., Andrews, B.D., 2006. Size distribution of submarine landslides and its implication to tsunami hazard in Puerto Rico. Geophys. Res. Lett. 33, L11307.

Tian, Y., Xu, C., Chen, J., Zhou, Q., Shen, L., 2017. Geometrical characteristics of earthquake-induced landslides and correlations with control factors: a case study of the 2013 Minxian, Gansu, China, Mw 5.9 event. Landslides. Available from: https://doi.org/10.1007/s10346-017-0835-6.

Tibaldi, A., Corazzato, C., Rust, D., Bonali, F.L., Pasquarè Mariottoc, F.A., Korzhenkov, A.M., et al., 2015. Tectonic and gravity-induced deformation along the active Talas–Fergana Fault, Tien Shan, Kyrgyzstan. Tectonophysics 657, 38–62.

Tibaldi, A., Graziotto, E., Forcella, F., Gapich, V., 1997. Morphotectonic indicators of Holocene faulting in central Tien-Shan, Kazakstan, and geodynamic implications. J. Geodynamics 23 (1), 23–45.

Titkov, S., 1979. Variations of the rock glaciers in the Sailijskij Alatau. Cryogenic phenomena of the Kazakstan and Central Asia. Jakutsk, pp. 34–42 (in Russian).

Torgoev, I.A., Alioshin, Yu.G., 2001. Ecology of mining complex of Kyrgyzstan. Bishkek, Ilim, 182 pp. (in Russian).

Torgoev, I.A., Alioshin, Yu.G., 2005. Geodynamic and geoecological monitoring of the geological environment for landslide hazard reduction. In: Makarov, V.I., et al., (Eds.), Recent Geodynamics of Intracontinental Areas of Collision Mountaing Building (Central Asia). Scientific World, Moscow, pp. 299–317 (in Russian).

Torgoev, I.A., Alioshin, Yu.G., 2009. Geoecology and waste products of mining complex of Kyrgyzstan. Bishkek, Ilim, 239 pp. (in Russian).

Torgoev, A., Lamair, L., Torgoev, I., Havenith, H.-B., 2013. A review of recent case studies of landslides investigated in the Tien Shan using microseismic and other geophysical methods. In: Ugai, K., Yagi, H., Wakai, A. (Eds.), Earthquake-Induced Landslides, pp. 285–294.

Towhata, I., 2003. Instability of submarine slope deposits; case studies and discussion on mass movement over long distance. Proceedings of the Workshop on Seismic Design for Sakhalin to Japan Gas Pipeline. Japan Sakhalin Pipeline Co., Ltd.

Trifonov, V.G., Machette, M.N., 1993. The world map of major active faults project. Ann. Geophys. (Annali di Geofisica) XXXVI (3–4), 225–236.

Trifonov, V.G., Makarov, V.I., Skobelev, S.F., 1992. The Talas-Ferghana active right-lateral fault. Ann. Tectonicae 6, 224–237.

Trifonov, V.G., Soboleva, R.V., Trifonov, R.V., Vostrikov, G.A., 2002. Recent geodynamics of the Alpine-Himalayan collision belt, Transactions of the Geological Institute of RAS, 541. GEOS, Moscow, 224 pp. (in Russian).

Turcotte, D.L., Malamud, B.D., 2004. Landslides, forest fires, and earthquakes: examples of self-organized critical behaviour. Physica. A. 340, 580–589.

Utirov, Ch.U., 1975. Seismodislocations and residual deformations of rocks. Experience of Coplex Seismic Zoning by Example of the Chudepression. Ilim Publishers, Frunze, pp. 109–111 (in Russian).

Utirov, Ch.U., 1991. Seismic dislocations. Recent Geodynamics of the Tien Shan Lithosphere. Nauka Publishers, Moscow, pp. 158–164 (in Russian).

Van Gassen, W., Cruden, D.M., 1989. Momentum transfer and friction in the debris of rock avalanches. Can. Geotech. J. 26, 623–628.

Van Westen, C.J., Castellanos, E., Kuriakose, S.L., 2008. Spatial data for landslide susceptibility, hazard, and vulnerability assessment: an overview. Eng. Geol. 102, 112–131.

Varnes, D.J., 1954. Landslide types and processes. In: Eckel, E.B. (Ed.), Landslides and Engineering Practice, Special Report 176. Transportation Research board, National Academy of Sciences, Washington, DC, pp. 11–33.

Varnes, D.J., 1978. Slope Movement types and processes. In: Schuster, R.L., Krizek, R.J. (Ed.), Landslides – Analysis and Control: National Academy of Sciences Transportation Research Board Special Report 176, 12–33.

Vinnichenko, S.M., 1977. Paleoseismodislocations – evidences of past strong earthquakes in the Upper Vakhsh region. Proc. Tajik Acad. Sci. 20 (7) (in Russian).

Vinnichenko, S.M., 1989. Seismogenic Landslides and Rockfalls of the Southern Tien Shan – Pamir Convergence Zone and Their Engineering-Geological Implication. PhD thesis. Moscow State University, 321 pp. (in Russian).

Vinnichenko, S.M., 2006. Landslide blockages in Tadjikistan Mountains (Gissar-Alai & Pamirs): their origin and development. Italian J. Eng. Geol. Environ. (Special Issue 1), 81–85.

Vinnichenko, S.M., Kuhtikov, M.M., Yakubov, D.H., Yarmukhamedov, A.R., 1990. Bedrock landslides and rockfalls of the South-Western Pamir. Proc. Acad. Sci. Tajik SSR 33 (No 6), 396–399 (in Russian).

Vinogradov, Yu,B., 1977. Glacial outburst floods and debris flows. Leningrad, Gidrometeoizdat 1977, 143 pp. (in Russian).

Voight, B., Janda, R.J., Glicken, H., Douglass, P.M., 1983. Nature and mechanics of the Mount St. Helens rockslide-avalanche of 18 May 1980. Geotechnique 33, 243–273.

Voitovich, V.S., 1969. Nature of the Dzungarian deep fault. Trans. Geol. Inst. 183, 189 pp. (in Russian).

Walker, M., 2005. Quaternary Dating Methods. John Willey & Sons, Ltd, Chichester, 286 pp.

Wang, Z.H., 2008. A thunder at the beginning of the 21st century – the giant Yigong landslide. In: Chen, Z., Zhang, J., Li, Z., Wu, F., Ho, K. (Eds.), Landslides and Engineered Slopes, 2111–2118.

Wang, F., Cheng, Q., Highland, L., Miyajima, M., Wanmg, H., Yan, C., 2009. Preliminary investigation of some large landslides triggered by the 2008 Wenchuan earthquake, Sichuan Province, China. Landslides 6, 47–54.

Wang., Y.-F., Cheng, Q.-G., Zhu, Q., 2015. Surface microscopic examination of quartz grains from rock avalanche basal facies. Can. Geotech. J. 52 (2), 167–181. Available from: https://doi.org/10.1139/cgj-2013-0284.

Wang, Y.-F., Xu, Q., Cheng, Q.-G., Li, Y., Luo, Z.-X., 2016. Spreading and deposit characteristics of a rapid dry granular avalanche across 3D topography: experimental study. Rock Mech. Rock Eng. Available from: https://doi.org/10.1007/s00603-016-1052-7 July 2016.

Wassmer, P., Schneider, J.-L., Pollet, N., 2002. Internal structure of huge Mass Movements: a key for a better understanding of long runout. The multy-slab theoretical model. Proc. Int. Symp. Landslide Risk Mitigation and Protection of Cultural and Natural Heritage. Kyoto University, Kyoto, pp. 97–107, 21-25 Jan.

Watson. R.A., Wright, H.E. 1969. The Saidmarreh Landslide, Iran. In: Schumm, S.A., Bradley, W.C. (Eds.) United States contributions to quaternary research, Geol. Soc. Am. Special Paper 123, pp. 115–139.

Weidinger, J.T., 1998. Case history and hazard analysis of two lake-damming landslides in the Himalayas. J. Asian Earth Sci. 16, 323–331.

Weidinger, J.T., Korup, O., Munack, H., Altenberger, U., Dunning, S., Tippelt, G., et al., 2014. Giant rockslide from the inside. Earth. Planet. Sci. Lett. 389, 62–73.

Wells, D.L., Coppersmith, K.J., 1994. New empirical relationships among magnitude, rupture length, rupture width, rupture area, and surface displacement. BSSA 84, 974–1002.

Whalley, W.B., 1974. The origin of rock glaciers. J. Glaciol. 13, 323–324.

Wheeler, R.L., Bufe, C.G., Johnson, M.L., Dart R.L., 2005. Seismotectonic Map of Afghanistan, with Annotated Bibliography. USGS Open-File report 2005–1264. U.S. Geological Survey, Reston, Virginia, 34 pp.

Witkind, I.J., Meyers, W.B., Hadley, J.B., Hamilton, W., Fraser, G.D., 1962. The earthquake at Hebgen Lake, Montana, on August 18, 1959 (GCT): geological features. Seismol. Soc. Am. Bull. 52, 163–180.

Wolter, A., Stead, D., Ward, B.C., Clague, J.J., Ghirotti, M., 2016. Engineering Geomorphological Characterisation of Vajont Slide, Italy. Landslides 13, 1067–1081.

WP/WLI (International Geotechnical Societies' UNESCO Working Party on World Landslide Inventory), 1993a. Multilingual Landslide Glossary. Bitech Publishers, Richmond, British Columbia, 59 pp.

WP/WLI (International Geotechnical Societies' UNESCO Working Party on World Landslide Inventory), 1993b. A suggested method for describing the activity of a landslide. Bull. Int. Assoc. Eng. Geol. 47, 53–57.

WP/WLI (International Geotechnical Societies' UNESCO Working Party on World Landslide Inventory), 1994. A suggested method for describing the causes of a landslide. Bull. Int. Assoc. Eng. Geol. 50, 71–74.

Wu, S., 2006. Inland earthquake and earthquake prevention and disaster reduction. Rec. Dev. World Seismol. No 5 (Serial No329), 25–28 (in Chinese).

Wu, S., Wang, T., Shi, L., 2010. Study on catastrophic landslides triggered by Wenchuan Earthquake, Sichuan, China in 2008. J. Eng. Geol. (No. 2), 145–159.

Xu, Q., Fan, X.-M., Jiang, L.-W., Liu, P., 2008. Yigong rock avalanche – flow landslide event, Tibet, China. In: Chen, Z., Zhang, J., Li, Z., Wu, F., Ho, K. (Eds.), Landslides and Engineered Slopes, 1675–1680.

Yan, Ch, Cheng, H., Yuing, G., Lee, D., Zang, Y., 1985. The big earthquake at Nilki, Xinjiang, on March, 8, 1812 (M = 8.0). Sibey Dichzen Ciuebao 7, 59–64 (in Chinese).

Yang, F., Zhari, M., Feng, L., 2005. Formation of geological hazards in Yili area and measures for controls and prevention. J. Xinjiang Normal Univ. (Nat. Sci. ed.) 24 (No 3), 117–120 (in Chinese).

Yarnold, J.C., Lombard, J.P., 1989. Faciesmodel for large rock avalanche deposits formed in dry climates. In: Colburn, I.P., Abbott, P.L., Minch, J. (Eds.), Field Trip Guidebook – Pacific Section, 62. Society of Economic Palaeontologists and Mineralogists, pp. 9–31.

Yi, C., Zhu, I., Seong, Y.B., Owen, L.A., Finkel, R.C., 2006. A late glacial rock avalanche event, Tianchi Lake, Tien Shan, Xinjiang. Quat. Int. 154–155, 26–31.

Yin, G.H., Jiang, J.X., Zhang, Y., Li, J., 2002. The character of seismic fault of 1812 Nilke earthquake. Northwestern Seismol. J. 02, (in Chinese).

Yin, G.H., Li, J., Zhang, Y., Yang, Z., 2001. Statistical analysis and research of landslides caused by Nilke earthquake. Inland Earthquake. 2001-01.

Yin, Y., Wang F., Sun, P., 2008. Landslide hazards triggered by the 12 May 2008 Wenchuan earthquake, Sichuan, China. In: Proc. of the First World Landslide Forum., 18–21 November 2008, UNU, Tokyo, Japan. Parallel Session Volume, 1–17.

Yin, Y., Xing, A., 2011. Aerodynamic modeling of the Yigong gigantic rock slide-debris avalanche, Tibet, China. Bull. Eng. Geol. Environ. Available from: https://doi.org/10.1007/s10064-011-0348-9.

Yoshida, H., 2016. Magnitude-frequency distribution of hummocks on rockslide-debris avalanche deposits and its geomorphological significance. Geosciences 6, 5. Available from: https://doi.org/10.3390/geosciences6010005.

Yuan, Z., Chen, J., Li, W., Owen, L.A., Schoenbohm, L.M., 2012. ^{10}Be dating of Taheman large-scale landslide in eastern Pamir and paleoseismic implications. Quat. Sci. 32, 409–416 (in Chinese with English abstract).

Yuan, Z., Chen, J., Owen, L.A., Hedrick, K.A., Caffee, M.W., Li, W., et al., 2013. Nature and timing of large landslides within an active orogen, eastern Pamir, China. Geomorphology 182, 49–65.

Yuan, R.-M., Tang, C.-L., Deng, Q.-H., 2015. Effect of the acceleration component normal to the sliding surface on earthquake-induced landslide triggering. Landslides 12, 335–344. Available from: https://doi.org/10.1007/s10346-0141-0486-9.

Zerkal, O., Frolova, Y., Strom, A., 2017. The conceptual modeling of the style of rock massif destruction and of its influence on rockslides and rock avalanche formation. In: Leith, K., Ziegler, M., Perras, M., Loew, S. (Eds.), Progressive rock failure. An ISRM Specialized Conference, Monte Verità, 5-9 June 2017. ETH, Extended abstracts, 113–114.

Zolotariev, G.S., Kalinin, E.V., Fedorenko, V.S., Sheshenia, N.L., 1969. Engineering-Geological Study of Rockfalls and Other Gravitational Phenomena on Mountainous Slopes. Moscow State University, 188 pp. (in Russian).

Zolotariov, G.S., 1964. Genetic Types of Landslides, Their Evolution and Study. Proceedings of the Conference on Landslides Study and Protection Measures. Kiev State University Publishing House, Kiev, pp. 165–170 (in Russian).

Zolotariov, G.S., 1974. Geological regulations of landslides and rock falls evolutoion – basics of their study and prediction theory. Bull. Moscow State Univ. Geol. Series (No 4), 3–15 (in Russian).

Zolotariov, G.S., 1983. Engineering Geodynamics. Moscow State University Publishing House, Moscow, 328 pp. (in Russian).

Zubovich, A.V., Wang, X., Scherba, Y.G., Schelochkov, G.G., Reilinger, R., Reigber, C., et al., 2010. GPS velocity field for the Tien Shan and surrounding regions. Tectonics 29. Available from: https://doi.org/10.1029/2010TC002772.

Index

Note: Page numbers followed by "*f*" refer to figures.

Chon-Kemin intermontane depression, 150, 388Elm rock avalanche, 1, 75, 366, 385

L

M

T